T0276186

INSECT CONTROL

BIOLOGICAL AND SYNTHETIC AGENTS

INSECT CONTROL

BIOLOGICAL AND SYNTHETIC AGENTS

EDITED BY

LAWRENCE I. GILBERT

SARJEET S. GILL

ELSEVIER

Amsterdam • Boston • Heidelberg • London • New York • Oxford
Paris • San Diego • San Francisco • Singapore • Sydney • Tokyo
Academic Press is an imprint of Elsevier

ACADEMIC
PRESS

Academic Press, 32 Jamestown Road, London, NW1 7BU, UK
30 Corporate Drive, Suite 400, Burlington, MA 01803, USA
525 B Street, Suite 1800, San Diego, CA 92101-4495, USA

© 2010 Elsevier B.V. All rights reserved

The chapters first appeared in *Comprehensive Molecular Insect Science*, edited by
Lawrence I. Gilbert, Kostas Iatrou, and Sarjeet S. Gill (Elsevier, B.V. 2005).

*All rights reserved. No part of this publication may be reproduced or transmitted in any form or by any
means, electronic or mechanical, including photocopy, recording, or any information storage and
retrieval system, without permission in writing from the publishers.*

*Permissions may be sought directly from Elsevier's Rights Department in Oxford, UK:
phone (+44) 1865 843830, fax (+44) 1865 853333, e-mail permissions@elsevier.com.*

Requests may also be completed on-line via the homepage (http://www.elsevier.com/locate/permissions).

Library of Congress Cataloging-in-Publication Data
Insect control : biological and synthetic agents / editors-in-chief: Lawrence I. Gilbert, Sarjeet S. Gill. – 1st ed.
p. cm.
Includes bibliographical references and index.
ISBN 978-0-444-63827-4 (alk. paper)
1. Insect pests–Control. 2. Insecticides. I. Gilbert, Lawrence I. (Lawrence Irwin), 1929- II. Gill, Sarjeet S.
SB931.I42 2010
632'.7–dc22
2010010547

A catalogue record for this book is available from the British Library

ISBN 978-0-444-63827-4

Cover Images: (Top Left) Important pest insect targeted by neonicotinoid
insecticides: Sweet-potato whitefly, Bemisia tabaci; (Top Right) Control (bottom) and
tebufenozide intoxicated by ingestion (top) larvae of the white tussock moth,
from Chapter 4; (Bottom) Mode of action of Cry1A toxins, from Addendum A7.

This book is printed on acid-free paper
Printed and bound in China

CONTENTS

PREFACE

When Elsevier published the seven-volume series *Comprehensive Molecular Insect Science* in 2005, the original series was targeted mainly at libraries and larger institutions. While this gave access to researchers and students of those institutions, it has left open the opportunity for an individual volume on a popular area in entomology: insect control. Such a volume is of considerable value to an additional audience in the insect research community — individuals who either had not had access to the larger work or desired a more focused treatment of these topics.

As two of the three editors of the *Comprehensive* series, we felt that it was time to update the field of insect control by providing an updated volume targeted specifically at professional researchers and students. In editing the original series, we expended a great deal of effort in finding the best available authors for each of those chapters. In most instances, authors who contributed to *Comprehensive Molecular Insect Science* also provided updates on *Insect Control*. The authors carefully reviewed recent data and cited the most relevant literature. We as coeditors reviewed and edited the final addenda.

The chapters incorporate all major insecticide classes in use for insect control, with the exception of organophosphates and carbamates. Pyrethroids continue to contribute to insect control in agriculture, and more recently have seen increased use in bed nets for the control of mosquito-borne diseases, such as malaria, and an update is provided in the addendum to "Pyrethroid Insecticides and Resistance Mechanisms." The chapters "Neonicotinoid Insecticides," "Indoxycarb – Chemistry, Mode of Action, Use and Resistance," "Spinosyns," and "Insect Growth Regulators and Development Disrupting Insecticides" cover newer classes of insecticides in use. "Azadirachtin, a Natural Product in Insect Control" deals with a plant extract that has also seen increased use. A number of chapters deal with viral, bacterial, and fungal control of insects, and these topics are discussed in the chapters titled "Genetically Modified Baculoviruses for Pest Insect Control," "*Bacillus thuringiensis*, with Resistance Mechanisms," "*Bacillus sphaericus* Taxonomy and Genetics," and "Entomopathogenic Fungi and Their Role in Regulation of Insect Populations." Finally, "Insect Transformation" outlines critical issues on using transformed insects for control programs. Additional chapters of interest are also available in a companion volume titled *Insect Pharmacology*.

Several years of effort was expended by both of us and our colleagues in choosing topics for the seven-volume series, in the selection of authors, and in the editing of the original manuscripts and galley proofs. Each and every chapter in those volumes was important, and even essential, to make it a "Comprehensive" series. Nevertheless, we feel strongly that having this volume with the updated material and many references on these important aspects of insect control will be of great help to professional insect biologists, to graduate students conducting research for advanced degrees, and even to undergraduate research students contemplating an advanced degree in insect science.

– LAWRENCE I. GILBERT,
Department of Biology,
University of North Carolina,
Chapel Hill

– SARJEET S. GILL,
Cell Biology and Neuroscience,
University of California,
Riverside

CONTRIBUTORS

J T Andaloro
E. I. Du Pont de Nemours and Co., Newark, DE, USA

P W Atkinson
University of California, Riverside, CA, USA

C Berry
Cardiff University Park Place, Wales, UK

A Bravo
Universidad Nacional Autónoma de México,
Cuernavaca Morelos, Mexico

J-F Charles
Institut Pasteur, Paris, France

P Daborn
University of Bath, Bath, UK

I Darboux
INRA, Sophia Antipolis, France

T S Dhadialla
Dow AgroSciences LLC, Indianapolis,
IN, USA

A J Dowling
University of Exeter in Cornwall, Tremough
Campus, Penryn, UK

J E Dripps
Dow AgroSciences, Indianapolis, IN, USA

J Eilenberg
University of Copenhagen, Frederiksberg,
Denmark

R H ffrench-Constant
University of Exeter in Cornwall, Tremough
Campus, Penryn, UK

M H N L Silva Filha
Centro de Pesquisas Aggeu Magalhães-Fundação
Oswaldo Cruz, Recife-PE, Brazil

S S Gill
University of California, Riverside, USA

T R Glare
Lincoln University, Lincoln, Christchurch,
New Zealand

M S Goettel
Agriculture & Agri-Food Canada, Lethbridge,
Canada

B D Hammock
University of California, Davis, USA

A B Inceoglu
University of California, Davis, USA

P Jeschke
Bayer CropScience AG, Monheim am Rhein,
Germany

P J Jewess
Rothamsted Research, Harpenden,
Hertfordshire, UK

S G Kamita
University of California, Davis, USA

B P S Khambay
Rothamsted Research, Harpenden,
Hertfordshire, UK

K-D Kang
University of California, Davis, USA

S F McCann
E. I. Du Pont de Nemours and Co., Newark, DE,
USA

A J Mordue (Luntz)
University of Aberdeen, Aberdeen, UK

E D Morgan
Keele University, Keele, UK

R Nauen
Bayer CropScience AG, Monheim am Rhein, Germany

C Nielsen-Leroux
Institut Pasteur, Paris, France

A J Nisbet
University of Melbourne, Werribee, Australia

D A O'Brochta
University of Maryland, College Park, MD, USA

D Pauron
INRA, Sophia Antipolis, France

H Ranson
Liverpool School of Tropical Medicine, Liverpool, UK

A Retnakaran
Canadian Forest Service, Sault Ste. Marie, ON, Canada

A S Robinson
FAO/IAEA Agriculture and Biotechnology Laboratory, Seibersdorf, Austria

V L Salgado
BASF Corporation, Research Triangle Park, NC, USA

G Smagghe
Ghent University, Ghent, Belgium

M Soberón
Universidad Nacional Autónoma de México. Cuernavaca Morelos, Mexico

T C Sparks
Dow AgroSciences, Indianapolis, IN, USA

J Vontas
University of Crete, Greece

N Waterfield
University of Bath, Bath, UK

G B Watson
Dow AgroSciences, Indianapolis, IN, USA

P A Wilkinson
University of Exeter in Cornwall, Tremough Campus, Penryn, UK

M S Williamson
Rothamsted Research, Harpenden, UK

K D Wing
E. I. Du Pont de Nemours and Co., Newark, DE, USA

1 Pyrethroids

B P S Khambay and P J Jewess, Rothamsted
Research, Harpenden, Hertfordshire, UK

© 2010, 2005 Elsevier B.V. All Rights Reserved

1.1. Introduction

The introduction of synthetic pyrethroids in the 1970s signalled a new era for environmentally friendly, highly effective and selective insecticides. For example, deltamethrin when introduced was the most effective insecticide on the market, being over 100 times more active than DDT and with the benefit of not accumulating in the environment. More than 30 new pyrethroids, including several new structural types, have been commercialized in the past 20 years (Bryant, 1999; Tomlin, 2000). Despite the evolution of resistance to pyrethroids in many insect pests, pyrethroids continue to be an important class. Part of their success has been due to the implementation of resistance-management strategies and the exploitation of new applications.

Since the previous review in this series (Ruigt, 1985), research activity in the discovery of novel pyrethroids peaked in the early 1990s and has now largely ceased. Major advances have been made in understanding the mode of action, especially by the application of molecular biology techniques, and in the management of resistance in the field. This chapter will review the advances made in these areas with the emphasis on commercial compounds.

1.1.1. Pyrethrum Extract

Despite the introduction of a number of other natural insecticides, pyrethrum extract from the flowers of the pyrethrum daisy (*Tanacetum cinerariaefolium* Schulz –Bip.; syns *Chrysanthemum cinerariaefolium* Vis., *Pyrethrum cinerariaefolium* Trevir.) remains commercially the most important, with demand continually outstripping supply. Supplies from Kenya, the major producer, have been irregular for many years. A new pyrethrum industry has been established in Tasmania, Australia, which is currently the second biggest producer and is set to increase further. Their industry was founded on the basis of extensive research leading to the development of plant lines that flower simultaneously, thus allowing mechanical harvesting and thereby a significant saving in labor costs. In addition they produce a very high-quality product by further purification of the hexane extract with a new procedure based on partitioning with liquid carbon dioxide. Major applications of pyrethrum extract continue to center on control of flying insects (e.g., mosquitoes), pre-harvest clean-up, and human- and animal-health pests. With an increasing demand for organically grown produce, use of pyrethrum extract has also increased.

1.1.2. Synthetic Pyrethroids

The market share of synthetic pyrethroids remained for many years at around 20–25% of the global insecticide market but in the past few years has declined to around 17%. Cotton remains the biggest market for pyrethroid applications; however, due to resistance build-up, this market has steadily declined from nearly 50% to less than 40% of the total pyrethroid tonnage. With the exception of some Japanese and Chinese companies, most major agrochemical companies ceased research effort in the discovery of novel pyrethroids by the late 1990s.

Since 1985, major advances have been made in identifying novel pyrethroids that have substantially increased the spectrum of activity to include control of mites, soil pests, rice pests (previously limited by the high fish toxicity of pyrethroids), and newer problem pests such as the cotton (also called the tobacco or sweet-potato) whitefly (*Bemisia tabaci*). In the past decade, commercial effort has been directed at introducing enriched isomer mixtures and the exploitation of niche markets including head-lice, termites, and bednets for the control of mosquitoes. Rather remarkably, sea-lice on salmonid fish are controlled by established pyrethroids (e.g., deltamethrin) which, despite their high fish toxicity, are being used at carefully selected doses. Another area of progress has been the development of novel formulations to extend the utility of established pyrethroids. For example, lambda-cyhalothrin formulated as micro-capsules allows slow release for public-health uses requiring long-term residual effects and when impregnated in plastic films controls termites with minimal operator exposure.

At the commercial level, global sales of pyrethroids in 2002 have been estimated at $1280M. Over 95% of the pyrethroid sales are accounted for by 12 active ingredients, of which deltamethrin continues to be the market leader with >16% share and sales exceeding $200M. Lamda-cyhalothrin is a close second with nearly 16% share; permethrin and cypermethrin account for 14% and 11% respectively, with all other pyrethroids contributing less than 7% each. Notably, although tefluthrin accounts for only 6% of the pyrethroid market, it is one of the most widely used soil insecticides in maize in the USA (Cheung and Sirur, 2002). The biggest concern in the sustained use of pyrethroids is the development of resistance. Though successful integrated pest-management strategies have been utilized, overall levels of resistance have continued to increase steadily. However, a greater understanding of the resistance mechanisms and the development of simple and high-throughput diagnostic procedures have aided the recognition of resistance mechanisms and hence facilitated the management of resistance.

1.2. Development of Commercial Pyrethroids

There are several reviews (Elliott, 1977, 1995, 1996; Davies, 1985; Henrick, 1995; Pap *et al.*, 1996a; Matsuo and Miyamoto, 1997; Pachlatko, 1998; Bryant, 1999; Katsuda, 1999; Khambay, 2002; Soderlund *et al.*, 2002) covering aspects of structure–activity relationships in pyrethroids. To appraise the advances made since 1985, it is useful to recollect the systematic approaches adopted earlier, in particular by Elliott, in the discovery of synthetic pyrethroids (see **Figure 1** for structures). The key to his success was the selection of pyrethrin I (**1**) over pyrethrin II (**2**) as the lead structure and the choice of test species (housefly *Musca domestica* and mustard beetle *Phaedon cochleariae*), which are model insects rather than agricultural pests. Pyrethrin I has superior insecticidal activity to pyrethrin II, though the latter gives better knockdown. This chapter will briefly review the progress in the development of pyrethroids using commercial pyrethroids as examples and for ease of readability these are not cited in chronological order.

Elliott's work culminated in the discovery of bioresmethrin (**3**) in 1967 followed by permethrin (**4**), cypermethrin (**5**), and deltamethrin (**6**) in 1977. These compounds have greater insecticidal activity than pyrethrin I, yet exhibit lower mammalian toxicity (**Table 1**) and initiated a major interest in pyrethroid research in agrochemical companies. These early synthetic compounds still account for a substantial share of the pyrethroid market despite the subsequent introduction of over 30 new pyrethroids. The majority of these compounds are closely related analogs. Replacement of the *E* (but not *Z*) vinyl chloride in cypermethrin with a CF_3 or a 4-chlorophenyl group retains activity and led to the discovery of cyhalothrin (**7**) and flumethrin (**8**), respectively. Both have good acaricidal activity and the latter is especially active against a range of tick species and is extensively used in animal health. Bromination of deltamethrin led to tralomethrin (**9**), a proinsecticide that breaks down to deltamethrin in the insect. Though most of the commercial pyrethroids are based on pyrethrin I, two have been developed using pyrethrin II as the lead structure. Kadethrin (**10**) emulates the high knockdown activity of pyrethrin II but acrinathrin (**11**) is more noted for its high activity against mites, ticks, lepidopteran, and sucking pests.

The other major influence in the development of pyrethroids came with the announcement of fenvalerate (**12**), a compound lacking the cyclopropane moiety. It was announced in 1976 by researchers at Sumitomo, around the same time as those disclosed by Elliott. Subsequently, many analogs were discovered, mainly with different substituents in the acyl moiety, especially at position 4. Examples include flucythrinate (**13**) and its derivative flubrocythrinate (**14**), also referred to as ZXI 8901) (Jin *et al.*, 1996) with Br introduced at the 4″ position. Introduction of NH and optimization of substituents in the acyl moiety led to fluvalinate (**15**), which exhibits broad-spectrum insecticidal and miticidal activity. In contrast to most other pyrethroids, it is essentially nontoxic to bees and is used to control mites in honeybee colonies and orchards.

Advances since 1985 in the development of new alcohol moieties have been limited. Recognition of the need for unsaturation in the side chain of the alcohol moiety of pyrethrin I had led to the development of the first commercial synthetic pyrethroid, allethrin (16). Substitution of the alkene with an alkyne group resulted in the more active analog prallethrin (17). Further systematic development resulted in the discovery of imiprothrin (18) and the more volatile, albeit only moderately insecticidal, empenthrin (19). The latter has low mammalian toxicity and is used in controlling insects that eat fabrics. Pyrethroids containing fluorinated benzyl alcohol moieties have long been recognized for their high vapor pressure but many exhibit high mammalian toxicity. An exception is transfluthrin (20) (acute LD_{50} >5000 mg kg^{-1} to male rats), containing a tetrafluorobenzyl alcohol moiety, which has only recently been commercialised for domestic use. Although many pyrethroids exhibit high activity in contact assays against soil pests (e.g., the corn rootworm *Diabrotica balteata*), the majority are ineffective under field conditions because they do not possess the required physical properties, especially high volatility (Khambay et al., 1999). Tefluthrin (21), having a 4-methyltetrafluorobenzyl alcohol moiety, is sufficiently volatile and, despite its relatively high mammalian toxicity (acute LD_{50} 22 mg kg^{-1} to male rats), is the only pyrethroid to have been commercialized as a soil insecticide. Two analogs of tefluthrin, metofluthrin (22) and profluthrin (23), have recently been developed for non-agricultural use by Sumitomo.

An interesting development has been the replacement of 3-phenoxybenzyl alcohol by 3-phenylbenzyl alcohol, which is sterically more restricted. Activity in this series is enhanced by the introduction of a CH_3 group at position 2 as in bifenthrin (24), which forces noncoplanarity of the phenyl rings presumably so matching the conformation of the target site. It has broad-spectrum activity and is particularly effective against some mite species. Although it is has one of the highest levels of mammalian and fish toxicity, it is commercially very successful.

The development of "hybrid" pyrethroids containing structural features both of DDT and of pyrethroids has been discussed previously (Ruigt, 1985). One compound, cycloprothrin (25), was eventually commercialized in 1987. Though it is much less insecticidal than deltamethrin, it has lower mammalian and fish toxicity and is therefore used to control pests in a range of applications including paddy rice, vegetables, tea, and cotton.

The most noteworthy advance in the past 20 years has been the discovery of the nonester pyrethroid etofenprox (26) in 1986. Further innovations led to silicon (e.g., silafluofen, (27)) and tin (e.g., (28)) analogs, the latter being nearly ten times more active than the corresponding carbon compounds. It had been a major objective of many researchers to replace the ester linkage in pyrethroids so as to achieve increased metabolic stability. A key feature of the nonester pyrethroids is their low toxicity to fish, and indeed etofenprox is used in rice paddy. However, it may be noted that even ester pyrethroids, despite having high intrinsic toxicity to fish, often manifest lower toxicity in the field than might be expected. This lower toxicity is observed because their concentration in water is limited by very low solubility, together with mitigation by strong sorption into organic matter such as pond sediments and river vegetation. However, the nonester pyrethroids have not yet made a significant commercial impact.

Of the many other variations, introduction of fluorine at the 4-position in the alcohol moiety has been the most significant. In general it leads to an appreciable enhancement of activity, albeit only to some species especially ticks (as in flumethrin (8)). Cyfluthrin (cypermethrin with a 4'-F) appears to have been introduced mainly on commercial considerations rather than on the grounds of efficacy. Furthermore, this substitution reduces metabolic attack at the 4'-position, thereby reducing the level of resistance in some insect species.

Commercial effort in the discovery of novel pyrethroids had all but ceased by the late 1990s, although efforts have continued to introduce single- or enriched-isomer mixtures (Table 2). Deltamethrin was the first pyrethroid to be commercialized as a single enantiomer by Rousell-Uclaf. This was followed by esfenvalerate (1S, αS component of fenvalerate) and acrinathrin. A single isomer of cyhalothrin, gamma-cyhalothrin (also called super-cyhalothrin), is currently under development. For cypermethrin, commercial production of a single isomer has not been possible; however, enriched isomer mixtures have been introduced to exploit specific properties and market opportunities. For example, the *trans* isomers of cypermethrin are known to be more readily metabolized in mammals than the *cis* isomers. Thus theta-cypermethrin, containing only the *trans* isomers, has a much higher safety margin than cypermethrin and, in addition to its use in agriculture, also has uses in veterinary and public health. Prallethrin also has potentially

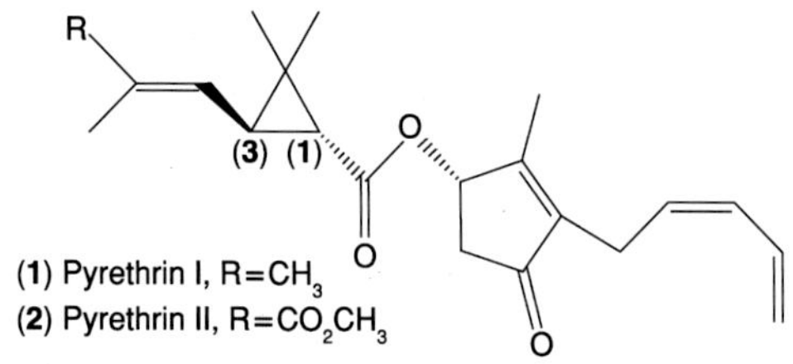

(1) Pyrethrin I, R=CH$_3$
(2) Pyrethrin II, R=CO$_2$CH$_3$

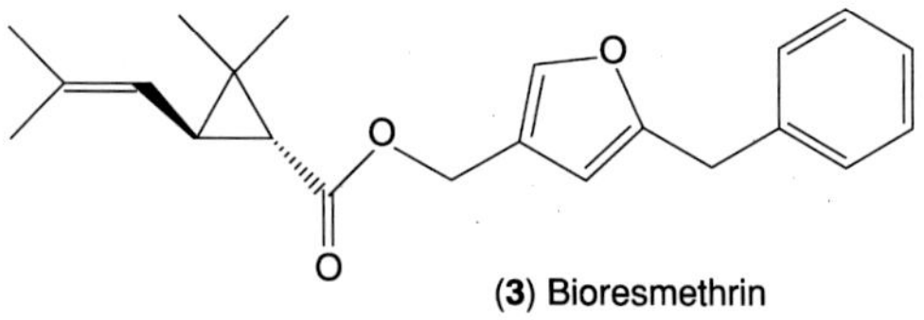

(3) Bioresmethrin

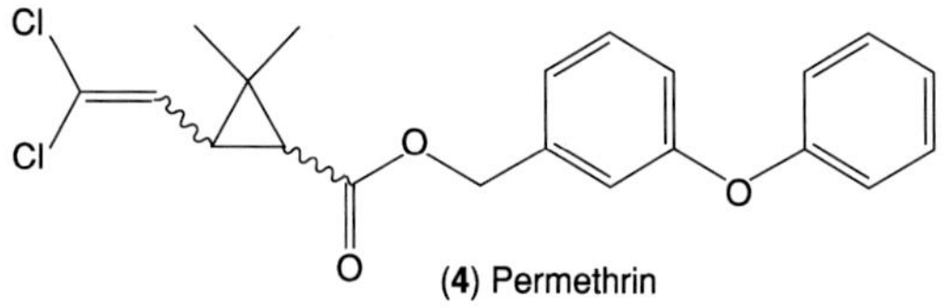

(4) Permethrin

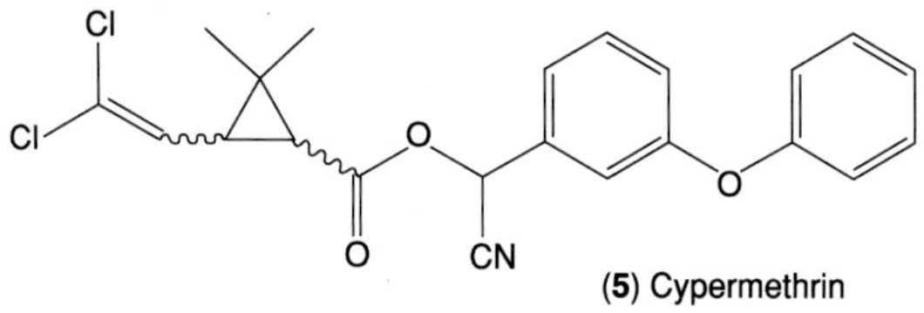

(5) Cypermethrin

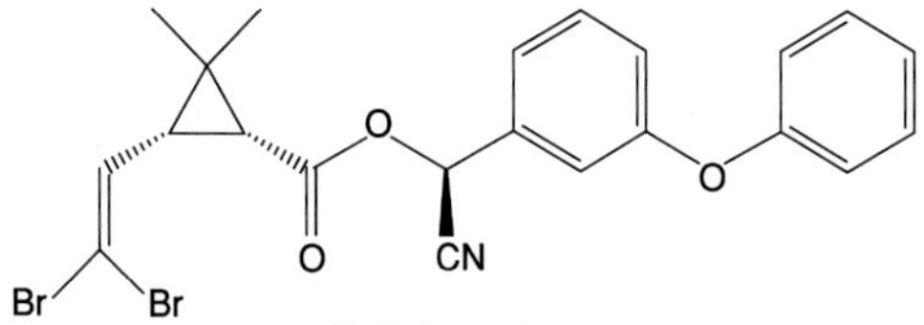

(6) Deltamethrin

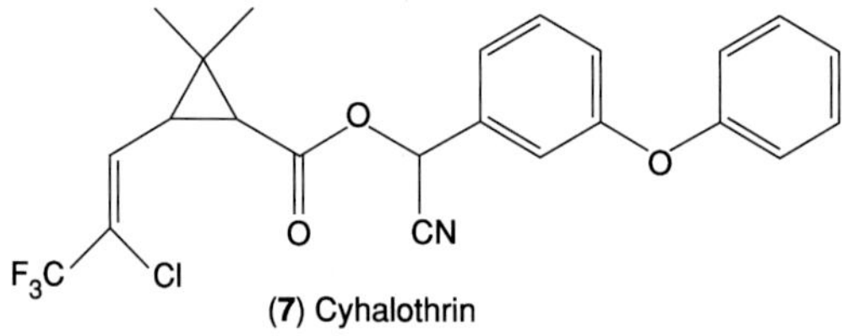

(7) Cyhalothrin

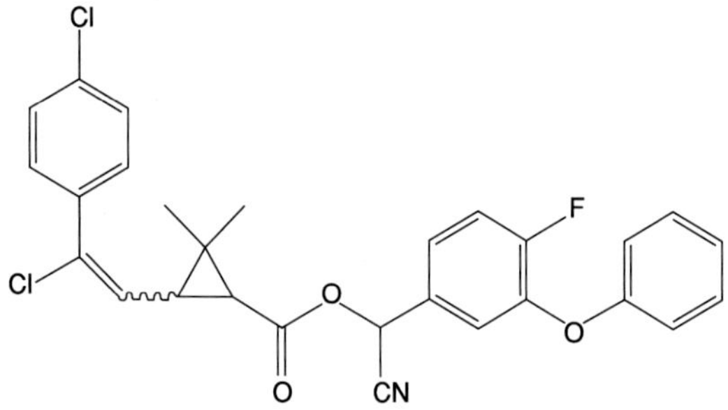

(8) Flumethrin

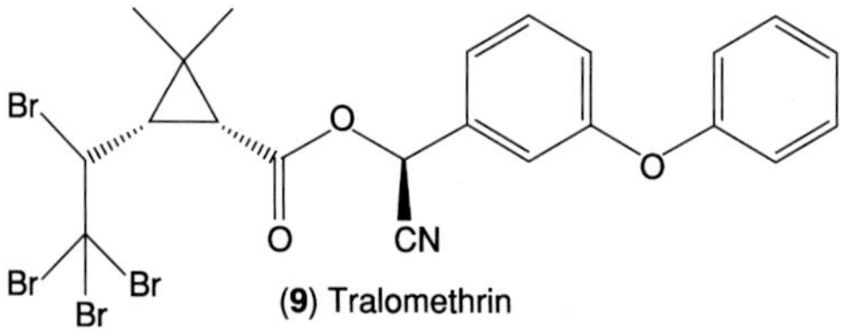

(9) Tralomethrin

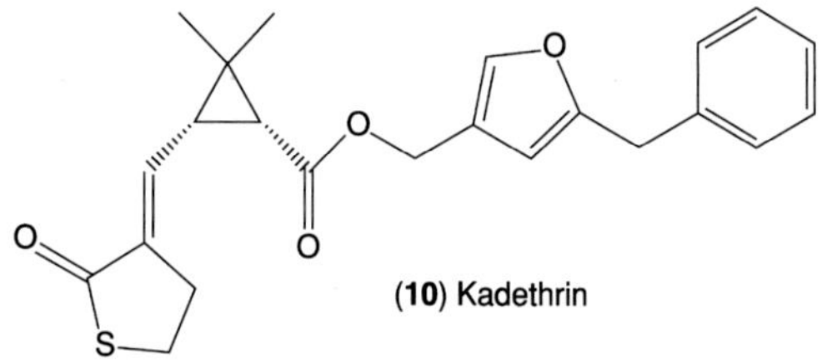

(10) Kadethrin

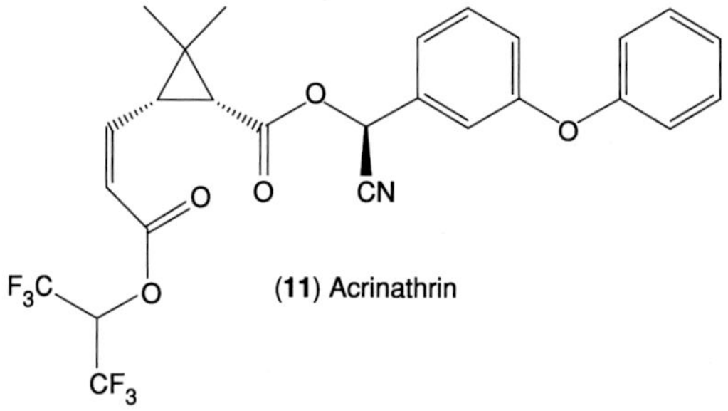

(11) Acrinathrin

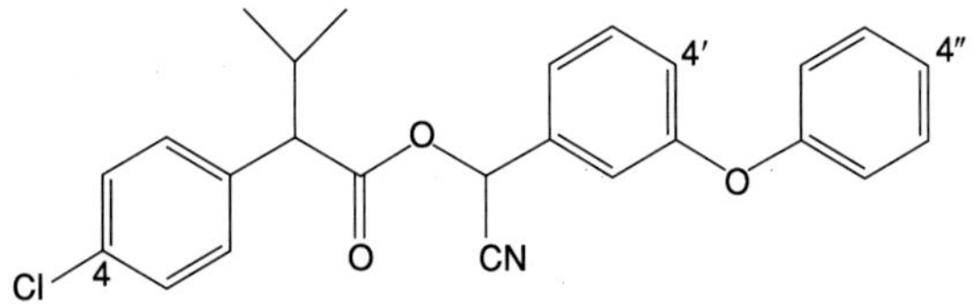

(12) Fenvalerate

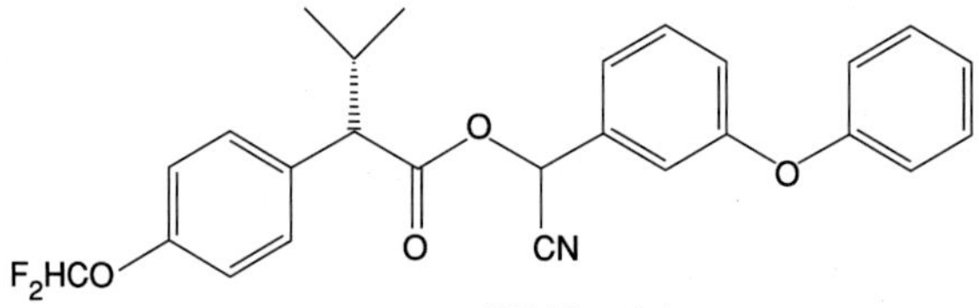

(13) Flucythrinate

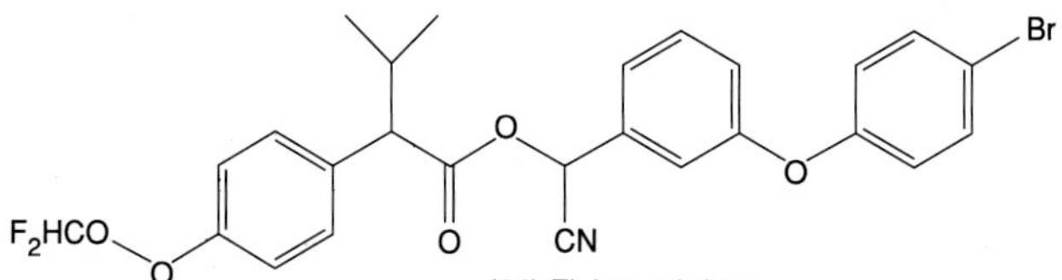
(**14**) Flubrocythrinate

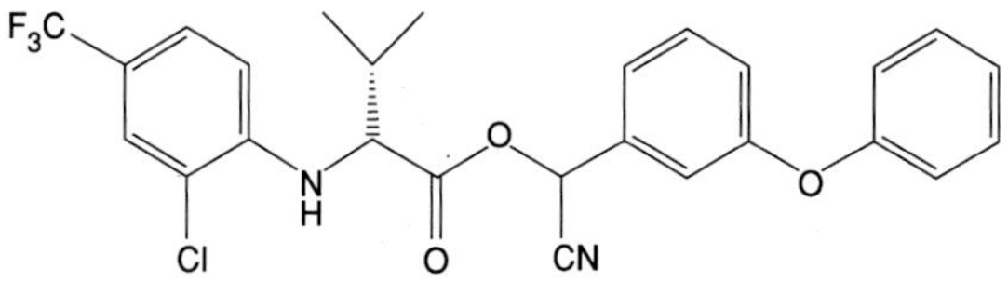
(**15**) Fluvalinate

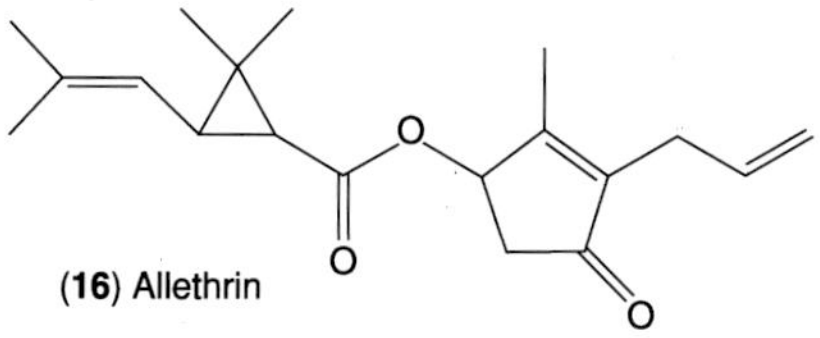
(**16**) Allethrin

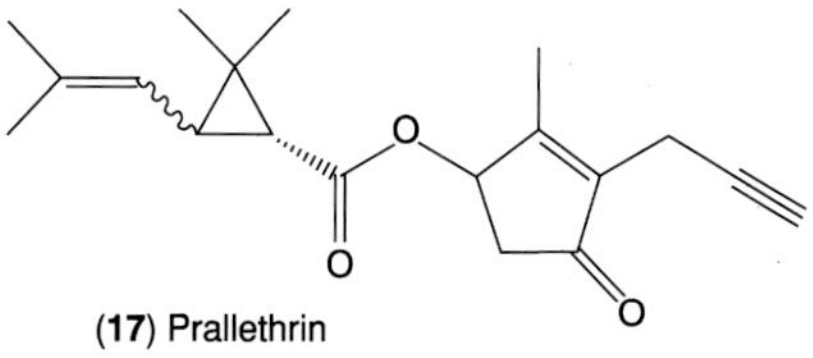
(**17**) Prallethrin

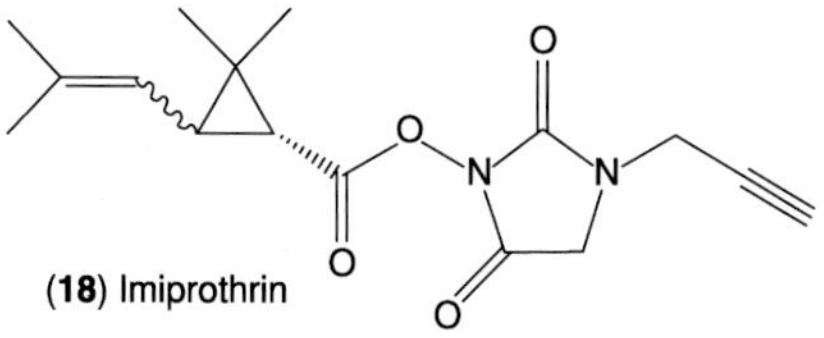
(**18**) Imiprothrin

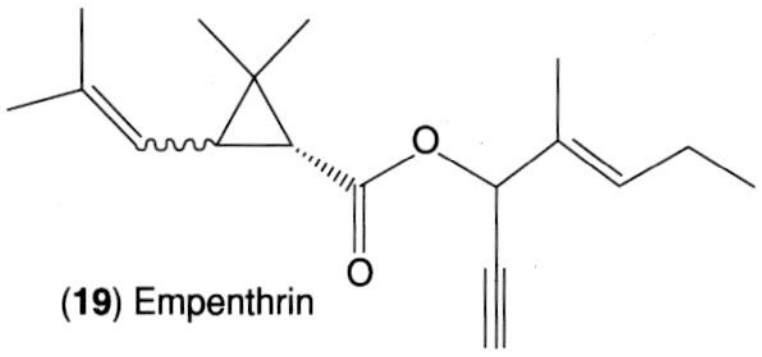
(**19**) Empenthrin

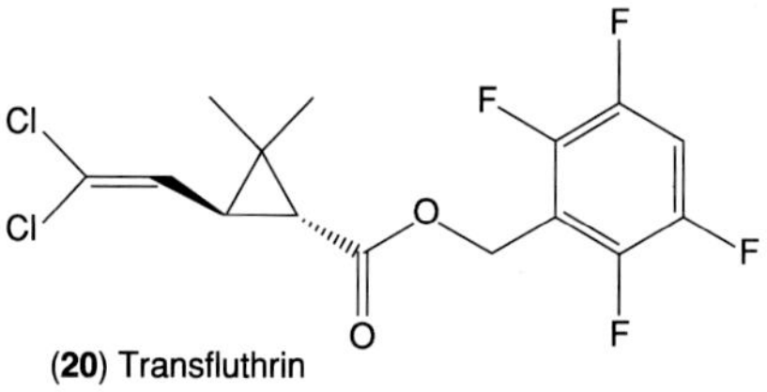
(**20**) Transfluthrin

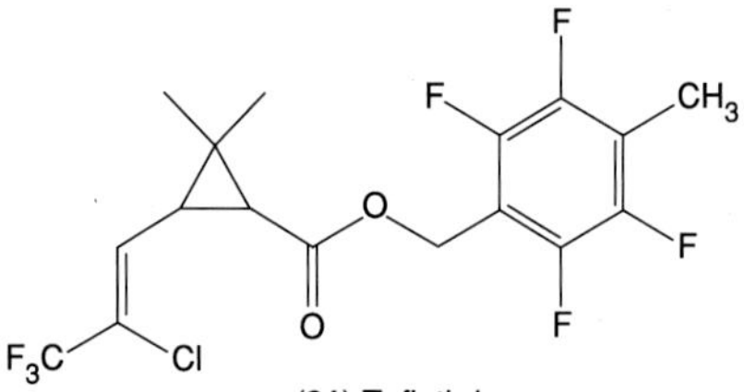
(**21**) Tefluthrin

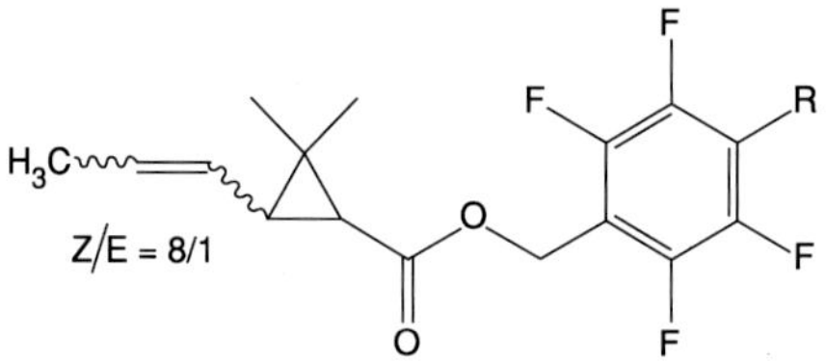
(**22**) Metofluthrin R=CH_2OCH_3
(**23**) Profluthrin R=CH_3

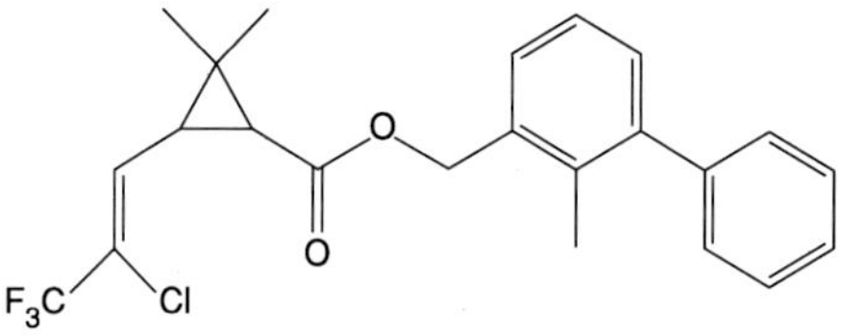
(**24**) Bifenthrin

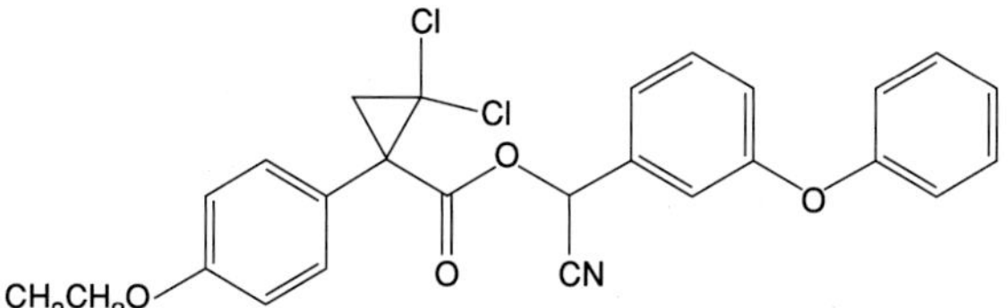
(**25**) Cycloprothrin

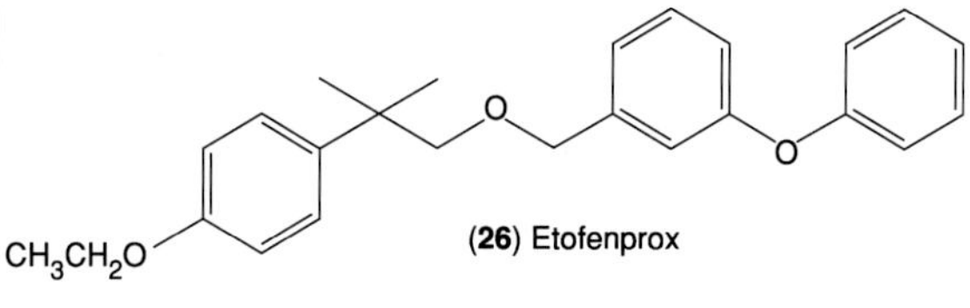
(**26**) Etofenprox

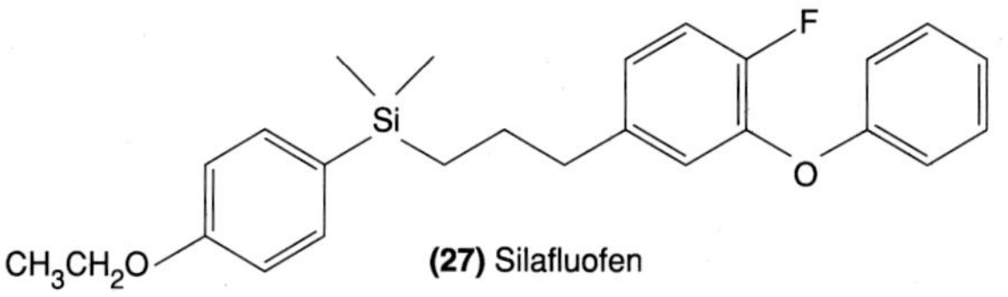
(27) Silafluofen

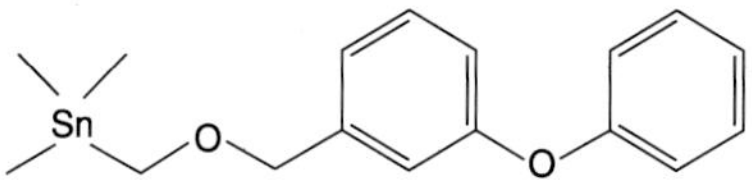
(28)

Figure 1 Commercial and novel pyrethroids (structures **1–28**).

Table 1 Insecticidal and mammalian toxicity of pyrethrin I and selected pyrethroids

Pyrethroid	Representative median lethal dose (mg kg^{-1})	
	Housefly, Musca domestica	Rat, oral
Pyrethrin I (1)	30	420
Bioresmethrin (3)	6	800
Permethrin (4)	0.7	2000
Deltamethrin (6)	0.02	~100

Table 2 Isomer mixtures of pyrethroids

Pyrethroid	Product	Number of isomers	Stereochemistry
Cypermethrin		8	(1R)-cis, αS; (1S)-cis, αR
			(1R)-cis, αR; (1S)-cis, αS
			(1R)-trans, αS; (1S)-trans, αR
			(1R)-trans, αR; (1S)-trans, αS
	Alpha	2	(1R)-cis, αS; (1S)-cis, αR
	Beta	4	(1R)-cis, αS; (1S)-cis, αR
			(1R)-trans, αS; (1S)-trans, αR
	Theta	2	(1R)-trans, αS; (1S)-trans, αR
	Zeta	4	(1R)-cis, αS; (1S)-cis, αS
			(1R)-trans, αS; (1S)-trans, αS
Cyhalothrin		4	(1R)-cis, αS; (1S)-cis, αR
			(1R)-trans, αS; (1S)-trans, αR
	Lambda	2	(1R)-cis, αS; (1S)-cis, αR
	Gamma	1	(1R)-cis, αS
Fluvalinate		4	(2R), αR; (2R), αS
			(2S), αS; (2S), αS
	Tau	2	(2R), αR; (2R), αS

eight isomers but is sold as a mixture of two active *cis/trans* isomers, each with the 1R center at C1 and αS in the alcohol moiety.

1.3. Structure–Activity Relationships in Pyrethroids

As described above, pyrethroids form a diverse class of insecticides highly effective against a broad spectrum of insect and acarine pests. The level of activity is determined by penetration, metabolism within the insect, and the requirements at the target site. As an

illustration (see **Figure 2** for structures), pyrethrin I has good activity against houseflies but poor activity against mites, whereas an experimental compound (**29**) has very high miticidal but poor insecticidal activity. Despite this diversity, it is possible to make some generalizations regarding structural requirements for high activity in pyrethroids. The seven-segment model (**Figure 3**) based on pyrethrin I, the lead structure originally adopted by Elliott (Elliott and Janes, 1978), serves as a good model to illustrate these general requirements and exceptions to them.

Segment A: Unsaturation generally leads to high activity. However, exceptions are known; for example fenpropathrin (**41**), containing the tetramethylcyclopropane acyl moiety (lacking in unsaturation), has high insecticidal and acaricidal activity. In general, large variations (steric and electronic) can be tolerated in this region. For example, (**30**) has an extended side-chain compared with cypermethrin and yet their activity is similar; flumethrin (**8**), possessing a bulky aromatic substituent, is highly active towards ticks.

Segment A + B: In ester pyrethroids, the gem-dimethyl group or its steric equivalent must be beta to the ester group. The required geometrical stereochemistry (*cis/trans*) across the cyclopropane ring (segment B) is influenced by the nature of the alcohol moiety, substituent at C3 and test species. In general, IR-*trans*-chrysanthemates (as in bioresmethrin) are more active then the corresponding *cis* isomers. The vinyl dihalo substituents are generally more effective than the chrysanthemates; however, the relative activity of the *cis* and *trans* isomers varies considerably. For alcohols containing two rings (e.g., 3-phenoxybenzyl alcohol as in deltamethrin), the 1R-*cis* isomers are generally more active than the *trans* isomers, but with alcohol moieties that contain a single ring (e.g., pentafluorobenzyl as in fenfluthrin (**31**) and cyclopentenone as in allethrin) the converse is true, with the *trans* isomers being more active.

In pyrethrin II, the vinyl carbomethoxy group in the acid moiety has the *E* configuration. Its corresponding analogs, *cis* and *trans* esters of α-cyano-3-phenoxybenzyl alcohol, are equally active; the Z isomer of the nor homolog (e.g., (**32**)) of the *cis* (but not the *trans*) isomer was unexpectedly much more active. Acrinathrin incorporates these two features and, although only half as active as deltamethrin against houseflies, it is more active towards some species of mites, ticks, lepidopteran insects, and sucking pests. Furthermore, stereochemistry also influences species selectivity of pyrethroids, a process primarily determined by selective metabolism. For example, contrary to the above generalization for pyrethroids containing 3-phenoxybenzyl

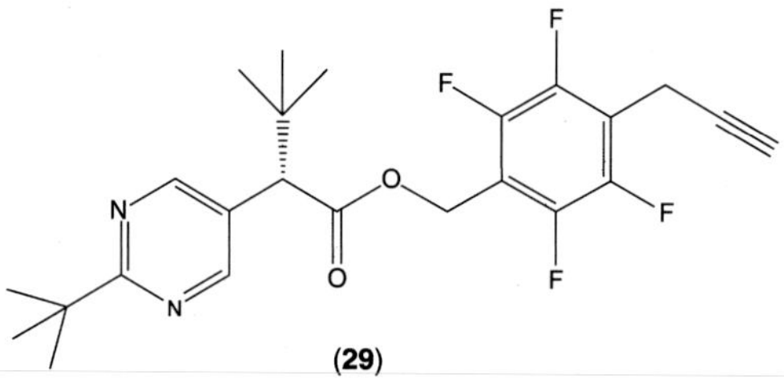

(**29**)

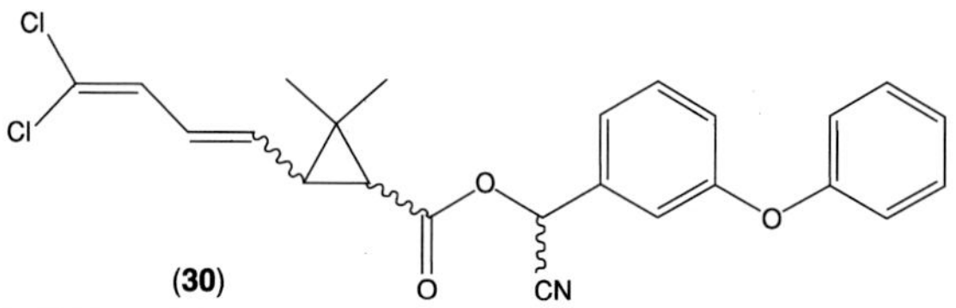

(**30**)

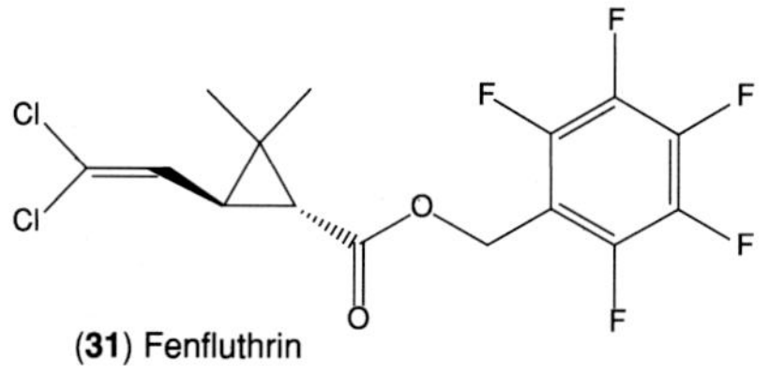

(**31**) Fenfluthrin

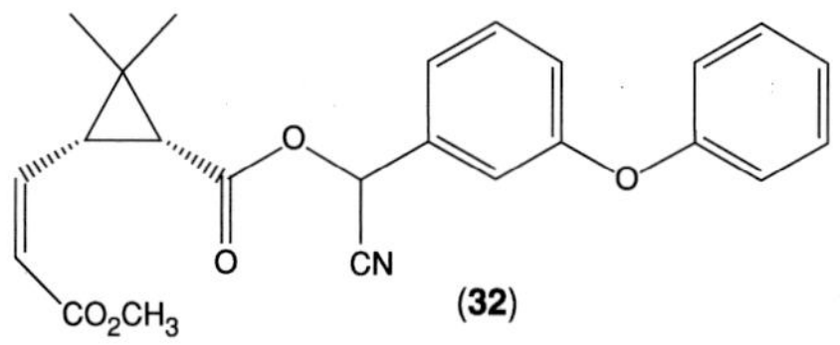

(**32**)

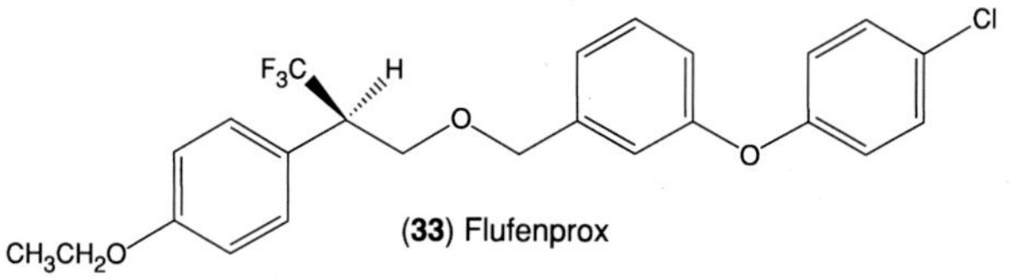

(**33**) Flufenprox

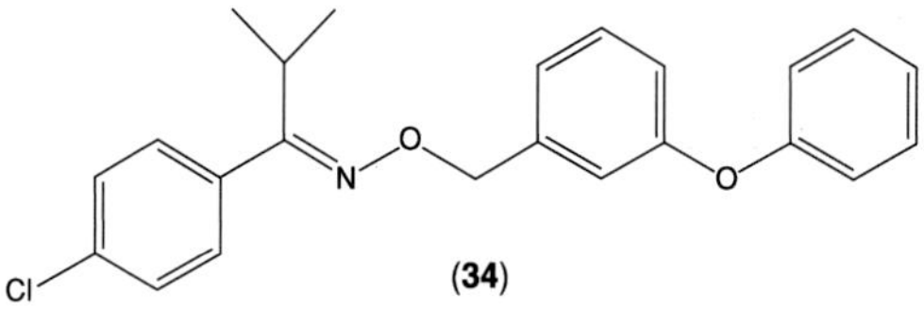

(**34**)

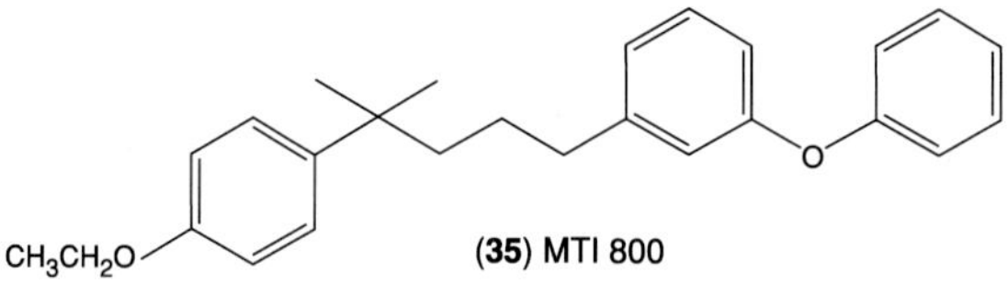

(**35**) MTI 800

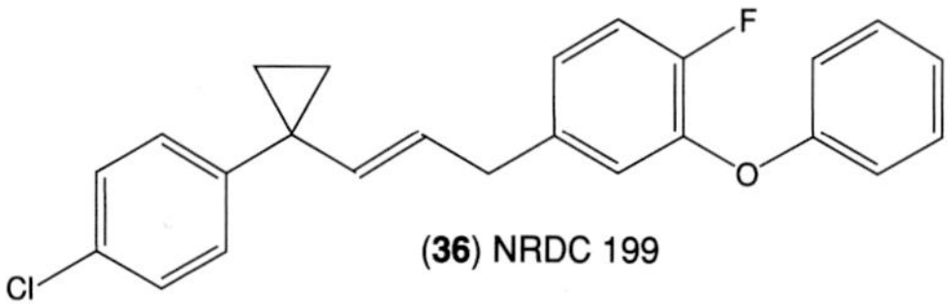

(**36**) NRDC 199

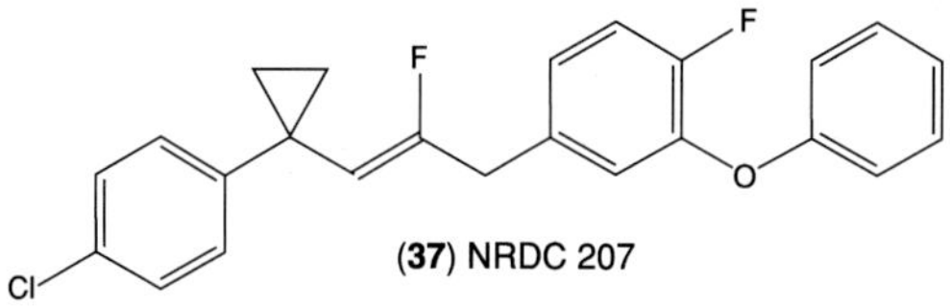

(**37**) NRDC 207

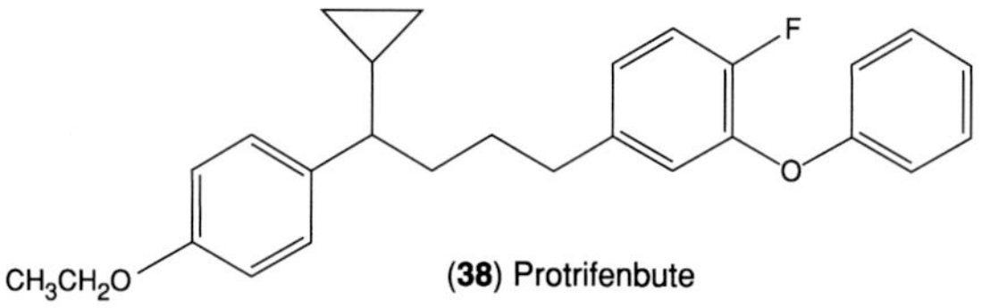

(**38**) Protrifenbute

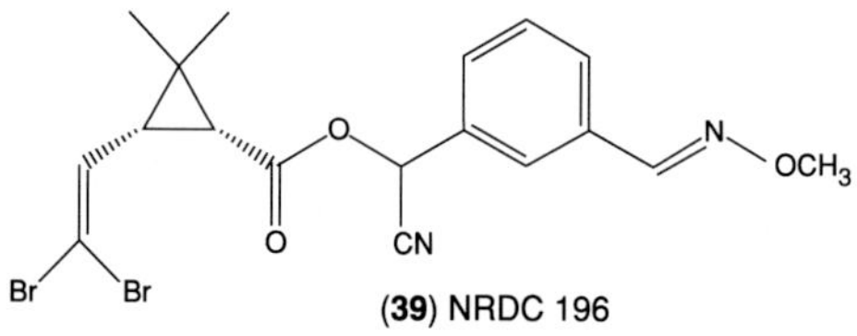

(**39**) NRDC 196

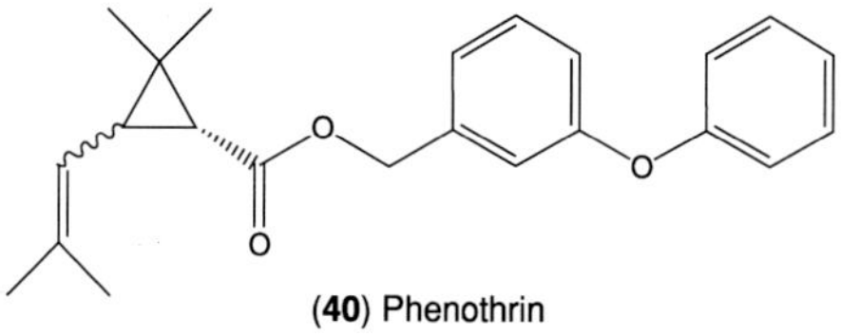

(**40**) Phenothrin

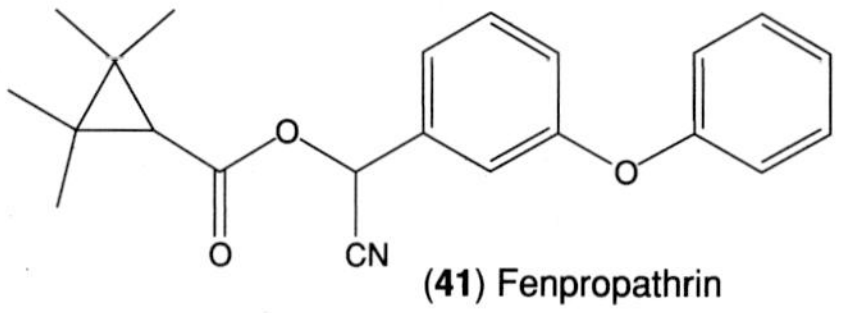

(**41**) Fenpropathrin

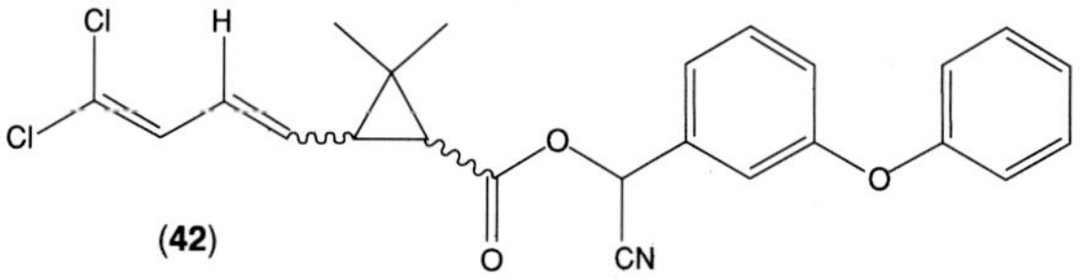

(**42**)

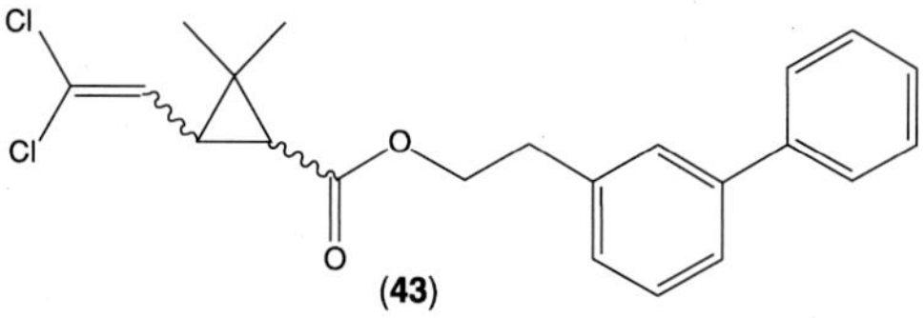

(**43**)

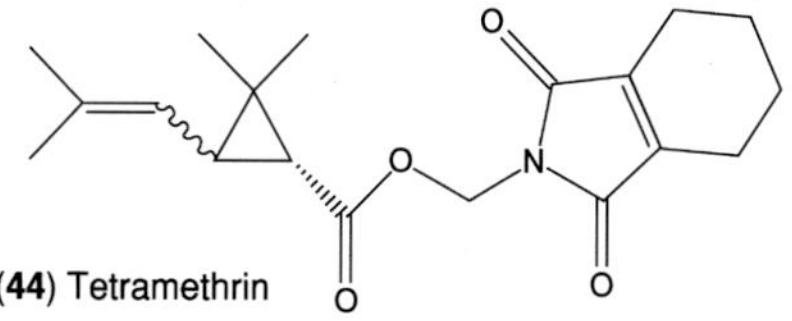

(**44**) Tetramethrin

Figure 2 Pyrethroids referred to in structure–activity relationships section (structures **29–44**).

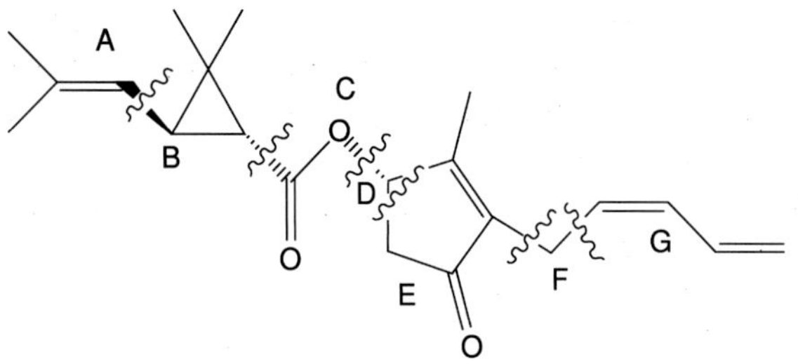

Figure 3 Schematic breakdown of the pyrethrin I molecule into seven segments.

alcohol, theta-cypermethrin (*trans* isomers) controls lepidopteran pests more effectively than α-cypermethrin (*cis* isomers), whereas the converse is true for dipterans.

When present, correct chirality at C1 of the acyl moiety (segment B) is crucial for activity. For both the esters of cyclopropane acids and for the fenvalerate series, activity requires the same spatial configuration at C1 (note that this is designated by *R* in the former and *S* in the latter by the nomenclature rules). Flufenprox (**33**), with an *R* configuration at C1, appears to contradict this generalization but apparently it is more similar to esfenvalerate than its corresponding *S* isomer.

Segment C: The ester linkage can be replaced by an isosteric group. However, changes are directly affected by the nature of segment B. For the cyclopropane esters, all isosteric replacements of the ester linkage substantially reduce activity. However, in the fenvalerate series, it can be replaced with retention of activity, for example (**34**) in which a ketoxime group replaces the ester linkage. The first such non-ester pyrethroid to be commercialized was etofenprox (**26**). Successful replacement of the ester group by an ether linkage was found to be largely dependent on the spatial arrangement of the gem-dimethyl group. In addition, the gem-dimethyl group could be replaced by cyclopropane and the ether link by other isosteric groups such as alkyl (MTI 800, (**35**)) and alkenyl (e.g., NRDC 199 (**36**) and NRDC 201 (**37**)). A subsequent important observation was that, in contrast to fenvalerate, replacement of the dimethyl with a cyclopropyl group in the nonester analogs (as in protrifenbute, (**38**)) resulted in a substantial (almost ×100) increase in activity.

Segments D and E: The carbon joined to the ester oxygen must be sp3 hybridized, with at least one proton attached to it and linked to an unsaturated center (e.g., an alkene). This can be a cyclic moiety (as in pyrethrin I) or substituted aromatic (as in permethrin). In the latter, introduction of an α-cyano group (e.g., deltamethrin) enhances activity

only with benzyl alcohols with a 3-phenoxy or an equivalent cyclic substituent (e.g., 3-benzyl and 3-benzoyl). With other alcohols containing cyclic moieties (e.g., 5-benzyl-3-furylmethyl alcohol as in bioresmethrin and 3-phenyl as in bifenthrin), it reduces activity. With a single cyclic moiety (e.g., NRDC 196, (**39**)) containing an acyclic substituent, introduction of an α-cyano group has a negligible effect on activity. However, in the absence of a side-chain, introduction of a cyano group in such compounds leads to loss of activity (e.g., as with tefluthrin (**21**)). An ethynyl group is generally less effective, though in empenthrin it is more effective than a cyano group.

The chirality at the oxygenated carbon of the alcohol moiety (the α carbon) is also important for activity but, in contrast to the strict requirement at C1 of the acyl moiety, the required chirality appears to depend on the structure of the alcohol. For example, the orientation conferring high activity in the cyclopentanone (as in pyrethrin I) and acyclic (as in empenthrin) alcohol moieties is the opposite to that for the α-cyano-3-phenoxybenzyl alcohols (as in deltamethrin). Furthermore, the difference in the level of activity between the epimers at this center is much larger for the latter (see below). In addition, selectivity in pyrethroids between insect species can be due to selective metabolic processes. A good illustration is an elegant investigation by Pap *et al.* (1996b). They investigated the insecticidal activity of three structurally related series of pyrethroids (phenothrin (**40**), permethrin (**4**), and cypermethrin (**5**)) toward six insect species. The chirality requirement at C1 was as indicated above, and, as expected at the α-cyano center, the *S* isomer was more effective than *R* in all species; however, in the presence of metabolic inhibitors, the *R* isomer had comparable activity to the *S* towards mosquitoes but not the other species. Against houseflies, permethrin and phenothrin were less active than the corresponding α-cyano analogs. However, in the presence of the synergist piperonyl butoxide (PBO), the difference in the levels of activity was much lower. Both examples demonstrate that the α*R* isomer is more sensitive to metabolism and that in mosquitoes, but not the other species, unexpectedly binds to the sodium channel receptor in a comparable manner to the *S* isomer.

Segment F: The CH2 group can be replaced by an oxygen or sulfur atom, or by an isosterically equivalent group such as carbonyl, with retention of variable levels of activity. In the case of benzyl alcohols, both cyclic and acyclic substituents can be attached either at position 3 or 4. However, cyclic substituents such as 3-phenoxybenzyl in the

4-position result in almost complete loss of activity, though a subsequent introduction of an α-cyano group enhances activity only for 3-substituted benzyl alcohols.

Segment G: Unsaturation in this region was initially considered essential for high activity. However, subsequent developments have indicated anomalies. Replacement of the diene in pyrethrin I with an alkene had led to the discovery of allethrin, the first synthetic pyrethroid to be commercialized. Although against houseflies it was as effective as pyrethrin I, it was significantly less effective against a range of other insect species. Indeed, this was the key observation made by Elliott that led to the development of bioresmethrin. However, since then many other highly effective alcohols have been discovered that contain unsaturated acyclic side-chains e.g., prallethrin and imiprothrin, both containing an alkynyl group. Two commercial compounds, trans-fluthrin (**20**) and empenthrin (**19**), both devoid of a side-chain, appear to contradict this generalization.

There are constraints in the overall shape, size, and the relative positioning of the various structural features within the pyrethroid molecule. Thus, despite the general belief, it is clear that the requirements for overall size are not stringent, though the alcohol moiety is limited to an overall size of no more than two benzene rings. The acid moiety can be varied substantially with retention of activity, for example flumethrin (**8**), fenpropathrin (**41**), and compound (**42**). In addition, although some of the regions of pyrethroid molecules can be varied independently of others, many require simultaneous modifications in adjacent regions (e.g., as in nonester pyrethroids). With regard to the relative positioning of the segments, compound (**43**), despite having the correct overall size, is nontoxic; this indicates that the relative positioning of the unsaturated center in the alcohol moiety to the ester linkage is crucial for activity. Thus, overall shape and size are both important for activity in pyrethroids.

Selectivity of pyrethroids is influenced by several factors, including rate of penetration, susceptibility to detoxification, and potency at the site of action. Finally, synergism in pyrethroids has been reported for a combination of active and inactive isomers, and between two structurally different pyrethroids (e.g., bioallethrin and bioresmethrin). Synergy is also known to arise with other classes of insecticides (e.g., organophosphates) that inhibit metabolic degradation (Naumann, 1990; Katsuda, 1999).

Quantitative structure–activity relationships (QSAR), using both *in vivo* and *in vitro* bioassay data with a range of pyrethroids, have been extensively investigated to gain a better understanding of structural/physical requirements for transport and binding of pyrethroids at the site of action (Ford et al., 1989; Ford and Livingstone, 1990; Hudson et al., 1992; Mathew et al., 1996; Chuman et al., 2000b). The most comprehensive QSAR studies with pyrethroids have been undertaken over the past 20 years by researchers at Kyoto University and collaborators (Nishimura et al., 1988, 1994; Matsuda et al., 1995; Nishimura and Okimoto, 1997, 2001; Akamatsu, 2002). For example, early studies (Nishimura et al., 1988 and references therein), using benzyl chrysanthemates and bioassays with the American cockroach *Periplaneta americana*, indicated that knockdown activity was strongly correlated with steric bulkiness (Van der Waals' volume) and both the hydrophobicity and position of substitution in the benzyl alcohol moiety. Thus, for *meta*-substituted derivatives, both hydrophobicity and steric factors were important, but for *para*-substituents activity varied parabolically with substituent volume. Both the rate of development of intoxication and the potency of the symptoms directly influenced knockdown activity and were dependent primarily on the pharmacokinetics of the transport process.

More recently, a direct relationship between changes in membrane potentials and physical properties of pyrethroids has been established (Matsuda et al., 1995). Using a diverse set of 39 ester pyrethroids, incorporating nine types of acyl and five types of alcohol moieties, they found a parabolic relationship between membrane potentials in crayfish (*Procambarus clarkii*) giant axons and the hydrophobicity of the compounds tested. They concluded that nonspecific penetration from outside the nerve membrane to the target site was the major factor influencing the rate of development of knockdown effects. The optimum lipophilicity for knockdown effect was calculated to be log K_{ow} 5.4 (where K_{ow} is the partition coefficient between 1-octanol/water). This was slightly higher than predicted values of about log K_{ow} 4.5 (Briggs et al., 1976; Kobayashi et al., 1988), indicating that other factors may also influence *in vivo* activity. Hence, less lipophilic pyrethroids can still exhibit high knockdown activity. Of particular note is imiprothrin, which gives high knockdown though having an exceptionally low log K_{ow} of 2.9. Interestingly, compounds of such low lipophilicity are expected to exhibit systemic movement in plants, though such activity has not been reported in pyrethroids. Given the complexity of structure–activity relationships in pyrethroids, it is not surprising that such QSAR studies have not led to the new commercial compounds.

It should be noted that commercial pyrethroids can exhibit significant phytotoxicity which can limit their application. For example, deltamethrin is too phytotoxic to be used on Sitka spruce (*Picea sitchensis*) transplants to control the large pine weevil (*Hylobius abietis*), and black pine beetles (*Hylastes* spp). Permethrin exhibits little phytotoxicity and cypermethrin is intermediate between the two (Straw *et al.*, 1996).

1.4. Mode of Action of Pyrethroids

1.4.1. Classification of Pyrethroids

Pyrethroids, which comprise a diverse range of structures, have historically been classified into two broad groups (Type I and Type II) on the basis of their biological responses (**Table 3**). Interpretation of most mode of action studies on insects has been predicated on this classification, though this is now considered to be an overly simplistic approach.

The two types of symptoms have been associated with the absence (Type I) or presence (Type II) of an α-cyano group. For example, permethrin exhibits Type I behavior; introduction of an α-cyano group gives cypermethrin which shows Type II behavior. This modification is accompanied by an increase in activity of an order of magnitude, and indeed all commercial Type II compounds have an α-cyano group in conjunction with a 3-phenoxybenzyl alcohol moiety. However, introduction of an α-cyano group does not always lead to an increase in activity. For example, in bioresmethrin it leads to a loss in activity and in NRDC 196 activity is only marginally increased. Furthermore, the level of activity is influenced by the choice of test species. Thus

classification as Type I or II cannot be based solely on the presence/absence of an α-cyano group. It is most likely that Type I and Type II represent the extremes of a continuum, with many pyrethroids exhibiting intermediate properties.

Despite the uncertainty over such a classification, it has been suggested (Soderlund *et al.*, 2002; Vais *et al.*, 2003) that, on the basis of electrophysiological properties, there are two distinct binding sites for pyrethroids on the sodium channel. However, this would not explain why some pyrethroids (e.g., fluvalinate and bifenthrin, respectively with and without an α-cyano group) appear to manifest intermediate responses in insects, which behavior might indicate the presence of only a single binding site. The most likely explanation for the differential responses of Type I and Type II pyrethroids in *in vitro* assays lies in the occurrence of pharmacologically different voltage-gated sodium channels in insects (see below).

1.4.2. Site of Action

Pyrethroids are recognized for their rapid knockdown action on insects but are generally poor at killing them, a fact not fully appreciated for many years. Measurements of LD_{50} are usually made 24 and 48 h after treatment, at which times paralyzed insects are regarded as dead. However, insects often recover fully, even at higher doses (e.g., $5 \times LD_{50}$) (Bloomquist and Miller, 1985), over the following few days. In practice, such long-term recovery does not limit pyrethroid efficacy in the field, as death occurs by secondary processes including desiccation and predation.

Pyrethroids are nerve poisons that disrupt nerve conduction in both invertebrates and vertebrates

Table 3 Classification of pyrethroids

Response/action	Type I	Type II
Poisoning symptoms	• Rapid onset of symptoms even at sublethal levels	• Slow onset of symptoms
	• Hyperactivity often leading to knockdown	• Convulsion followed by paralysis
	• Low kill with high recovery	• High kill with low recovery
	• Inversely related to changes in temperature	• Little effect of temperature change
Electrophysiological response in nerve tissue	• Repetitive discharges in axons	• Blockage of conduction at synapses
Action on sodium-channel function	• Monophasic and rapid decay of tail currents	• Biphasic and very slow decay of tail currents
	• Bind preferentially to closed channels	• Bind preferentially to open channels
Level of resistance due to resistant houseflies with *super-kdr* mechanism	• Below 100-fold	• Over 200-fold

and their primary site of action is the voltage-gated sodium channels. However, there is controversy regarding the precise mode of action of pyrethroids and their binding site on the sodium channels.

Much of the work on the mode of action was reported during the 1980s but investigations on sodium channels have continued and a great deal of progress made. Many *in vitro* and *in vivo* investigations on mode of action and metabolism have been carried out on mammalian systems because of the interest in toxicology of these pesticides to mammals including humans and the expertise present in many laboratories for electrophysiology on mammalian neurons and muscles. Experiments have also used model systems such as squid and crustacean giant nerve fibers because of their large size and robustness. However, such studies may not be relevant to insects since there is considerable variation, not only in the metabolic capacities but also in the amino-acid sequences of the sodium channel proteins and thus their pharmacological properties. Another complication is that, in mammals, several genes express different sodium-channel isoforms, which exhibit differential responses to the two types of pyrethroids (Tatebayashi and Narahashi, 1994; Soderlund *et al.*, 2002). Furthermore, each gene can express channels (splice variants) with varying pharmacological and biophysical properties (Tan *et al.*, 2002). In insects, the voltage-gated sodium channels are encoded by only a single gene, which can also express splice-variants. Such channels, though having a common binding site, might nevertheless exhibit differences in their responses to pyrethroids. Such differences have been demonstrated in mammals but not yet in insects, though there is ample indirect evidence for this (see below).

These different factors give rise to the complex nature of pyrethroid action and therefore difficulty in defining in detail the mode of action. Early investigations into their mode of action were conducted on insect and mammalian nerve preparations but the interpretation of these measurements is somewhat controversial. Broad agreement has only been possible with measurements from sodium channels (see below) (Soderlund *et al.*, 2002). Thus the summary presented below is a brief personal overview of the area.

1.4.2.1. Investigations based on nerve preparations

Of the various symptoms induced by pyrethroids (hyperactivity, tremors, incoordination, convulsions followed by paralysis and death), only the initial toxicity symptoms can be linked to neurophysiological measurements and then only for Type I pyrethroids. They induce repetitive firing (short bursts of less than 5 s duration) in axonal nerve preparation in the sensory nerves accompanied by occasional large bursts of action potential in the ganglia (Nakagawa *et al.*, 1982; Narahashi, 2002; Soderlund *et al.*, 2002). Repetitive firing has been correlated to hyperactivity and uncoordinated movements leading to rapid knockdown in insects. This is also thought to stimulate the neurosecretory system causing an excessive release of diuretic hormones which eventually results in the disruption of the overall metabolic system in insects (Naumann, 1990 and references therein). The initial symptoms are manifested at very low concentrations ($0.1–0.001 \times LD_{50}$), it being estimated that less than 1% of the sodium channels need to be modified to induce repetitive firing (Song and Narahashi, 1996). In contrast, Type II pyrethroids, even at rather high concentrations, initially cause much less visible activity in insects, convulsions and rapid paralysis being the main symptoms. They cause slow depolarization of nerve membranes leading to block of nerve conduction. The concentration required ultimately to kill insects is much lower for Type II than for Type I compounds.

In general, Type I compounds have been viewed as more active than Type II in producing both visible symptoms and repetitive discharges in the peripheral nerves, though exceptions are known, especially when considering results from noninsect nerve preparations. For example, repetitive firing has been observed for both Type I and Type II pyrethroids in frog-nerve preparations (Vijverberg *et al.*, 1986).

The first claimed correlation between a neurophysiological response and the poisoning action of pyrethroids related to the increased production of miniature excitatory postsynaptic potentials (MEPSPs) in response to depolarization of the nerve terminal caused by both Type I and Type II compounds (Salgado *et al.*, 1983a, 1983b; Miller, 1998). This increase in MEPSPs results in depletion of neurotransmitter at the nerve-muscle synapses leading to nerve depolarization (paralysis) and also blockage of flight reflex responses. At low threshold concentrations, Type I pyrethroids induce repetitive firing and hyperactivity but not neurotransmitter release so that these processes are not linked to the toxic action of pyrethroids. These observations thus accounted for the apparent contradiction that though initial symptoms of Type I pyrethroids are positively correlated with temperature, the toxic action (release of neurotransmitter and conduction block) is negatively correlated with temperature. At high concentrations, Type I pyrethroids behave similarly to Type II pyrethroids and elicit toxic symptoms in insects with neurotransmitter release and neuromuscular block. Thus, Type I pyrethroids affect the overall sensory nervous system but Type IIs affect only a specific part. This was elegantly demonstrated by Salgado *et al.* (1983a, 1983b), who showed that deltamethrin applied to the abdomen of houseflies took over an hour to reach the CNS, so that that the knockdown effect seen within 20 min could only be attributed to its action at the peripheral nervous system. Moreover, flies that had been knocked down by deltamethrin were able to fly when thrown in the air because communication along the CNS between the brain and the flight system had not yet been disrupted. In contrast, flies knocked down by Type I pyrethroids cannot fly when thrown in the air, indicating transmission block at the motor neurons. This effect can be reversed by increasing the temperature with Type I pyrethroids but not for Type II, in line with the observed *in vivo* toxicity symptoms.

1.4.2.2. Investigations based on sodium channels
The most significant advances since 1985 in understanding pyrethroid action have been made by the application of molecular biological approaches to the functioning of voltage-gated channels. The ability to introduce specific mutations in the voltage-sensitive channels and to measure inter- and intra-cellular changes (by voltage- and patch-clamp experiments) has provided conclusive evidence that the primary target sites for pyrethroids are the voltage-gated sodium channels. In summary, pyrethroids interact with the voltage-gated sodium channels to produce tail currents, which can be used to quantify the potency of the insecticides, higher tail-current being associated with higher insecticidal activity (Nishimura and Okimoto, 2001). This action appears to account for symptoms of pyrethroid poisoning and actions on the peripheral and central neurons. The prolongation of these tail-currents is much longer and the effect is less reversible for Type II pyrethroids than Type I. The presumption that Type I and Type II pyrethroids represent the extremes of a continuum is supported by the observation that many pyrethroids exhibit an intermediate response. Thus, deltamethrin (Type II) and permethrin (Type I) exhibit Hill slopes (an indicator of the molar binding ratio) of ~2 and ~1, respectively but bifenthrin exhibits an intermediate slope of ~1.4 (Khambay *et al.*, 2002). Finally, the link between single point mutations in sodium channels and resistance in insects has provided the most conclusive evidence that the voltage-gated sodium channels are the primary target for pyrethroids.

Other measurements on single sodium channels that confirm their being the primary target for pyrethroids include the negative temperature coefficient with toxicity being paralleled by a negative temperature dependence of channel depolarization. Furthermore, hyperactivity induced by Type I pyrethroids can be explained by the tail currents exceeding a certain threshold that then triggers repetitive firing and hyperactivity. Interestingly, it has recently (Motomura and Narahashi, 2001) been demonstrated that tetramethrin (**44**) (Type I) displaces fenvalerate (Type II) from mammalian sodium channels, implying that the former has a higher binding affinity. Also, tetramethrin displaces deltamethrin in single-channel studies. At first sight, this observation appears contradictory in that Type II pyrethroids are significantly more toxic than Type I pyrethroids in insects. However, the explanation may lie in the differences between sodium channels from insects and mammals.

1.4.3. Modeling Studies on Prediction of Putative Pharmacophores
Sodium channels are lipid-bound trans-membrane proteins and consequently it is difficult to obtain their crystal structure, especially with bound ligands. However, a recent publication (Jiang *et al.*, 2003) of the crystal structure of the voltage-gated potassium channel should aid progress. Indirect techniques involving binding of pyrethroids to the channel protein (e.g., with photo-reactive and radio-labelled pyrethroids) have been unsuccessful, primarily due to nonspecific binding to the channel and adjacent protein. Given these limitations, effort has been directed at modeling studies.

The "bioactive" conformation has been defined as that most probable at the binding site, and which is energetically stable and three-dimensionally similar to other compounds acting at the same site. Although ester pyrethroids such as deltamethrin have seven torsional axes, most conformational studies have focused on three axes (T2, T3, and T4 in **Figure 4**) around the ester linkage. Most studies have investigated common conformations of minimum energy and correlated them with toxicity/knockdown, overall electronic effects, geometric bulk, and partitioning coefficients (Naumann, 1990 and references therein; Chuman *et al.*, 2000a, 2000b). A "folded" form (**Figure 4**), also referred to as the "clamp" and "horseshoe", was identified as the active conformation.

As an example of the approaches used, Chuman *et al.* (2000a, 2000b) chose the nonester pyrethroid MTI 800 (representative of the most flexible series of pyrethroids) to identify the possible conformers. By comparison, they then deduced the likely

(a)

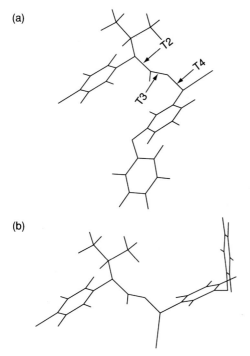

(b)

Figure 4 Folded and extended conformations of fenvalerate.

common conformations (within 10 kcal/mol of the ground energy) for the active isomers of pyrethrin I, deltamethrin, and fenvalerate (representatives of each of the three other main series of pyrethroids). To define the bioactive conformer, they also considered a further seven pyrethroid structures, both active and inactive. Similarity/dissimilarity searches using Cosine and Tanimoto coefficients together with superimposition considerations indicated the folded form as the common bioactive conformation for all four types of pyrethroids.

For deltamethrin (a cyclopropane ester), this bioactive folded conformation is close to minimum energy, and is similar to that both of the crystal structure and as observed in solution by NMR (under certain conditions). However, for fenvalerate (not containing a cyclopropane acid moiety), the extended form (**Figure 4**) has lower minimum energy than the bioactive folded conformation.

However, another recent study appears to contradict the above findings. Using new computational methodologies based on cluster analysis of molecular dynamics trajectories, Ford *et al.* (2002) proposed an "extended" conformation for the lethal action of both Type I and Type II pyrethroids, which would also account for the negative temperature coefficient. For knockdown activity (Type I compounds only), he proposed a different higher-energy conformation which is not accessible to Type II pyrethroids. Not being a minimum-energy conformation, this would be transient in nature, and would occur

with a higher probability at elevated temperatures (i.e., positive temperature correlation with knockdown activity). Once again the key torsion angles, T2 and T4, lie around the ester bond, T3 being invariant throughout the dynamics simulations undertaken in this study.

1.5. Resistance to Pyrethroids

Under selection from repeated sprays of insecticides, individuals possessing biochemical mechanisms that can detoxify the insecticide more rapidly or are less sensitive to it are likely to be favoured. These resistant insects survive doses that would kill normally sensitive individuals. Genes encoding these mechanisms will then be passed on to the succeeding generations, resulting in pest populations that are not controlled effectively. This can lead to farmers increasing the rate or frequency of applications, imposing further selection pressure and ultimately leading to a situation whereby the pests become totally immune. Removal of selection pressure may result in the pest populations regaining some degree of sensitivity, particularly if there is a fitness cost to resistance such as longer development times or reduced over-wintering ability. Usually, however, the population never regains the degree of sensitivity of the naïve population, and often there appears to be little fitness cost so that high numbers of resistant pests remain in the population. Pesticide resistance has occurred with all insecticides; the existence of resistance mechanisms common to pyrethroids and other older insecticides meant that onset of resistance to the new compounds was quite rapid.

Many factors, in particular the type of crop and the history of insecticide use, can affect the selection and dominance of resistance mechanism(s) within an insect species. In addition, the relative importance of these mechanisms can change over time (Gunning *et al.*, 1991). Thus, management of resistance requires knowledge of the mechanism(s) present and ideally also their relative contribution. Though the nature of the resistance mechanisms can be identified relatively straightforwardly, there are, as yet, no established methods for predicting their relative importance in field strains of resistant insects. Such prediction requires a multidisciplinary approach based on both *in vivo* and *in vitro* assays. The concept of using novel selective inhibitors, designed using structure–activity investigations and ideally devoid of insecticidal activity, could prove particularly useful in determining the biochemical mechanism(s) responsible for resistance. The design of such inhibitors will need to take into consideration the variations in specificities of metabolic enzymes amongst insect species. A greater input from chemists would be a key factor in such investigations.

There are three main mechanisms by which insects become less sensitive to pesticides: increased detoxification ("metabolic resistance"), reduced penetration, and target-site resistance. For many pests, the most important resistance mechanism for pyrethroids is considered to be site insensitivity, often referred to as knockdown resistance (*kdr*), caused by the inheritance of point mutations in the *para*-type sodium channel which is the target-site of the pyrethroids. The principal mechanisms of detoxification are esteratic cleavage of the ester bond and oxidative mechanisms catalysed by cytochrome P450. The other common mechanism of xenobiotic detoxification, dealkylation catalysed by glutathione-*S*-transferases (GSTs), is of lesser importance. Also, reduction in cuticular penetration and/or increased excretion has long been recognized as a resistance mechanism, though its contribution is likely to be as a modifier of metabolic or target-site resistance rather than as a major mechanism in its own right.

Advances in molecular techniques have led to the development of simple and efficient diagnostic kits for the detection of *kdr* resistance. Only a limited amount of work has been done to develop tests for resistance caused by metabolic enzyme(s), for which synergists are commonly used to detect their presence. The biggest problem currently is that the established inhibitors (e.g., piperonyl butoxide, PBO) are not specific and thus may give unreliable results. For example, PBO has been shown to inhibit both cytochrome P450 monooxygenases and esterases in Australian *H. armigera* (Gunning *et al.*, 1998a).

When present, the site-insensitivity (*kdr* and especially *super-kdr*) mechanisms often predominate, although populations of pests frequently have multiple resistance mechanisms. The situation regarding the two principal metabolic mechanisms is much less clear and the relative importance of esterase or oxidative mechanisms may vary between species or even between populations of the same species.

Details of the resistance mechanisms and their management in the field are discussed in the following sections.

1.5.1. Resistance Mechanisms

1.5.1.1. Metabolic resistance
1.5.1.1.1. *Metabolism of pyrethroids in insects*
The principal metabolic pathways by which pyrethroids are degraded in insects were mostly evaluated prior to 1985 and are summarized in Roberts and Hutson (1999). Metabolism is conveniently divided into two phases: the initial biotransformation of a pesticide is referred to as Phase I metabolism, comprising mainly oxidative, reductive, and hydrolytic processes; Phase II metabolism is biotransformation, in which the pesticide or Phase I metabolite is conjugated with a naturally occurring compound such as a sugar, sugar acid, glutathione, or an amino acid. In general, the major Phase I degradative routes in insects and other animals (mostly mammals and birds) are similar. It is only in the details of their Phase II reactions whereby polar conjugates with sugars or amino acids are formed that there are major qualitative differences in the nature of the metabolites. As degradation studies on insects are not required for the registration of insecticides, such studies are usually only undertaken in order to understand specific questions concerning structure–activity relationships or to evaluate problems associated with resistance caused by enhanced metabolic breakdown. It is the latter reason that has seen the majority of insect metabolism studies performed.

Since 1985 there has been a vast increase in the knowledge of insect molecular genetics. The publication of the draft genome sequences for the fruit fly *Drosophila melanogaster* in 2000 and the mosquito *Anopheles gambiae* in 2002 has greatly increased the knowledge of the enzymes involved in metabolic degradation, particularly cytochromes P450. Consequently, recent research is beginning to make inroads into understanding which of the isozymes of cytochrome P450 are responsible for metabolizing different substructures of the pyrethroid molecule. The situation with respect to the nature of the carboxyesterases responsible for catalysing the hydrolysis of the pyrethroid ester bond is less clear and the carboxyesterases that catalyze the hydrolysis of pyrethroids in different species of insects may be different and nonhomologous.

Insects detoxify pyrethroids at varying rates and this degradative metabolism is important in understanding the detailed toxicology. Indeed, part of the relatively modest insecticidal activity of the natural pyrethrins is attributable to their rapid metabolic breakdown. Consequently, household sprays are usually formulated with synergists that inhibit the enzymes that catalyze this metabolic degradation and thereby enhance the insecticidal activity. Selection of insect strains possessing elevated levels of catabolizing enzymes is also an important mechanism in the development of decreased sensitivity (resistance) toward pyrethroids. There are two principal routes of detoxification of pyrethroids in insects: de-esterification catalysed by both esterases and cytochromes P450, and the hydroxylation of aromatic rings or methyl groups by cytochromes P450, and these two mechanisms will be considered separately. The points of the chemical structure where a generalized 3-phenoxybenzyl pyrethroid

molecule is detoxified are shown in **Figure 5**. The width of the arrow indicates the approximate extent of metabolic attack. It is in the detailed enzymology and molecular biology of the enzymes responsible for pyrethroid metabolism where there have been the most important advances.

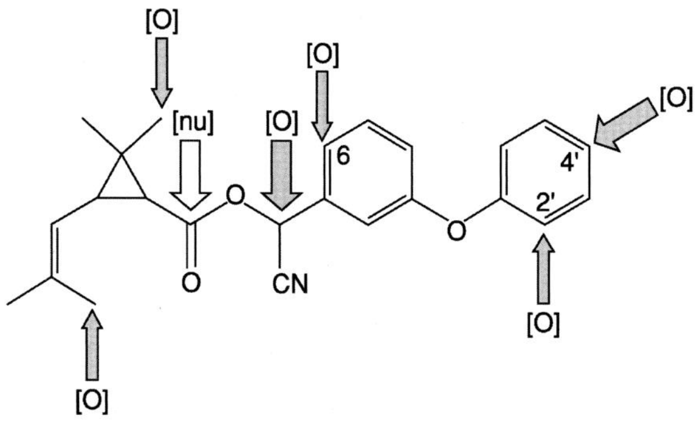

1.5.1.1.2. Metabolic pathways The most complete analysis of the metabolic pathways of pyrethroid degradation in insects has been by Shono (Shono *et al.*, 1978; reviewed by Soderlund *et al.*, 1983), who studied the metabolism of permethrin in American cockroaches (*Periplaneta americana*), houseflies (*Musca domestica*), and caterpillars (*Trichoplusia ni*). Forty-two metabolites (including conjugates) were identified, and the most important Phase 1 metabolites are shown in **Figure 6**. Metabolism occurred by ester cleavage to 3-phenoxybenzyl

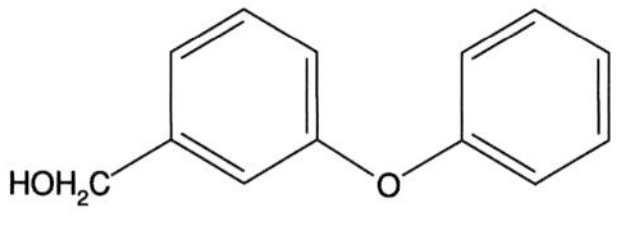

Figure 5 Points of metabolic attack in a generalized pyrethroid structure.

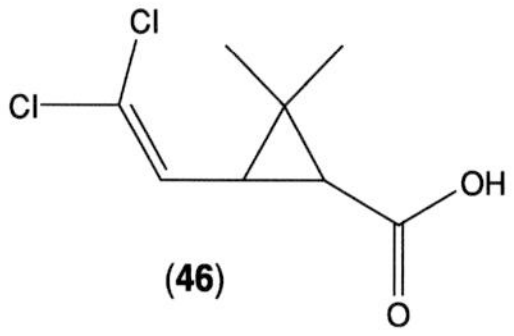

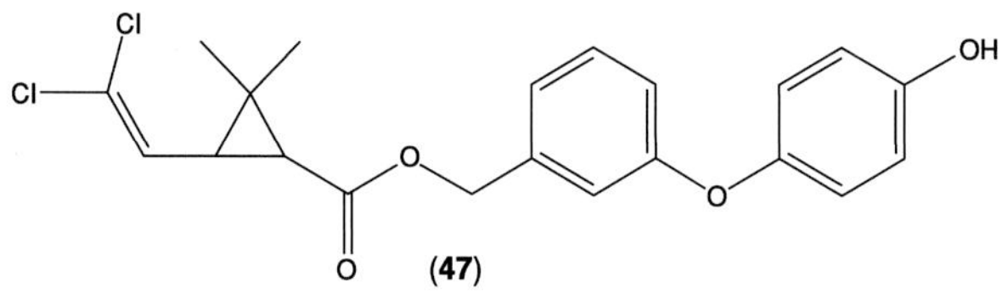

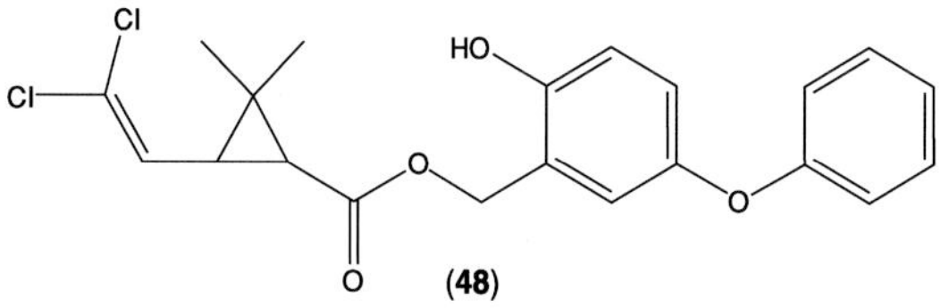

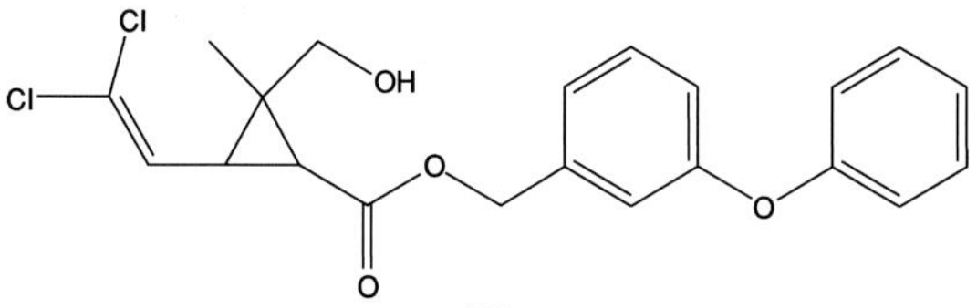

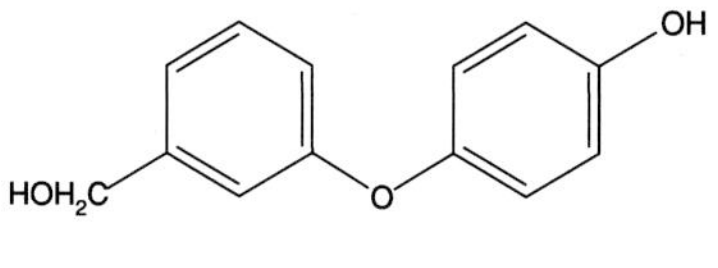

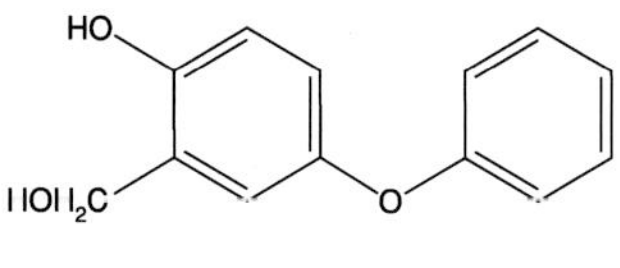

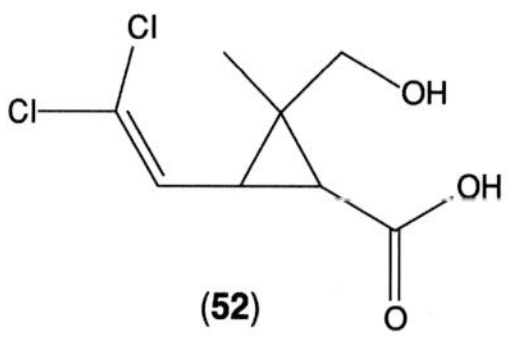

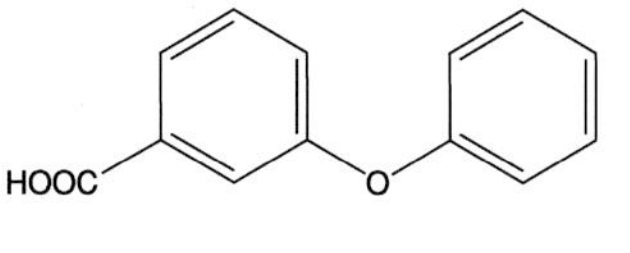

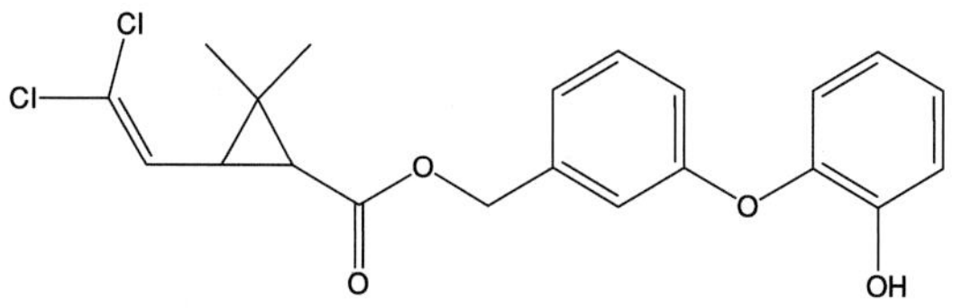

Figure 6 Metabolites from oxidative attack on permethrin (structures **45–54**).

alcohol (3-PBA) (45) and 3-(2,2-dichlorovinyl)-2,2-dimethylcyclopropanecarboxylic acid (DCVA) (46). Hydroxylation also occurred at the 4'- and 6-positions of the phenoxybenzyl moiety and the *cis* or *trans* methyl groups of the DCVA moiety to give (47), (48), and (49) respectively. Hydroxylation of the *trans*-methyl group rather than the *cis*- was preferred. Ester cleavage of these hydroxylated metabolites gave the structures (50), (51), and (52) respectively. 3-PBA (50) was further oxidized to 3-phenoxybenzoic acid (53), as were the hydroxylated analogs of 3-PBA to their analogous benzoic acids.

A similar pattern of Phase I metabolites was observed with both *cis*- and *trans*-permethrin, although the relative proportion of certain of the conjugated structures was influenced by the stereochemistry. All metabolites found in the cockroach were also found in houseflies but the series of metabolites arising from the 6-hydroxylation of the 3-PBA moiety was only found in flies. Metabolites consisting of the whole hydroxylated molecule (e.g., (47), (48), and (49)) were exclusively found as their glucosides, whereas the ester cleavage products were found both free and as their glucoside and amino acid conjugates. All three insects conjugated DCVA with one or more of the amino acids glycine, glutamic acid, glutamine, and serine, in addition to forming glucose esters. This study did not identify the 2'-hydroxylated metabolite (54), a significant metabolite in mammalian studies. However, this metabolite and compounds derived from it were identified in a study of the metabolism of permethrin by the American bollworm *Helicoverpa zea*, the tobacco budworm *Heliothis virescens* (Lepidoptera, Noctuidae) (Bigley and Plapp, 1978) and the Colorado potato beetle, *Leptinotarsa decemlineata* (Soderlund *et al.*, 1987), in which it was the principal metabolite. Other later studies have confirmed this overall pattern in different insects, although some studies have shown more differences in the pattern of metabolites between *cis*- and *trans*-permethrin isomers.

Holden (1979) showed that *trans*-permethrin was ester-cleaved at a higher rate than the *cis*-isomer in *P. americana*, an observation consistent with the findings that the *trans*-isomers of cyclopropane-containing pyrethroids are much better substrates for esterases than the *cis*-isomers (see below). This pattern of permethrin metabolism by insects can be taken as a template for the breakdown of other pyrethroids comprising esters of 2,2-dimethyl-3-(substituted)vinylcarboxylic acid with 3-PBA. Thus, the principal mechanisms of Phase I metabolism involve ester cleavage, both hydrolytic and oxidative, and aromatic substitution of one or

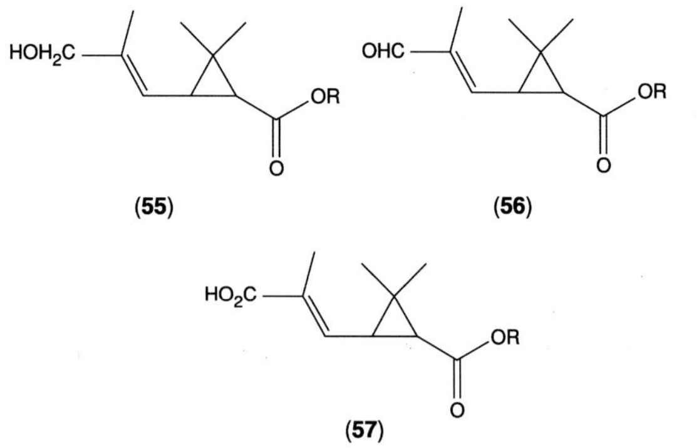

Figure 7 Metabolites from oxidative attack on the chrysanthemic acid moiety of allethrin (structures **55–57**).

other of the 3-PBA rings (the 4'-position is usually the major site) and the 2,2-dimethyl group of the acid moiety. Analogous reactions have been shown to take place with bifenthrin and deltamethrin in the bulb mite *Rhizoglyphus robini* to give the ester-cleaved products and the 4'-hydroxy metabolites (Ruzo *et al.*, 1988). Similarly, *trans*-cypermethrin gave *trans*-DCVA and both 2'- and 4'-hydroxy-cypermethrin in the cotton bollworm *Helicoverpa armigera* and in *H. virescens* (Lepidoptera) (Lee *et al.*, 1989). Fenvalerate, which lacks a cyclopropyl group, was metabolized via ester cleavage and 4'-hydroxylation in houseflies (Funaki *et al.*, 1994). In pyrethroids containing the chrysanthemic acid moiety (rather than a 2,2-dimethyl-3-dihalovinylcarboxylic acid), such as allethrin, phenothrin, and tetramethrin, the methyl groups of the isobutylene group are also subject to oxidative attack. Hydroxylation of these groups in allethrin to the alcohol (55) (**Figure 7**), followed by successive oxidation to the aldehyde (56) and carboxylic acid (57), occurred in houseflies (Yamamoto *et al.*, 1969). Tetramethrin was mainly metabolized via ester cleavage, with the production of the carboxyl derivative (57) of the *trans*-methyl group as a minor product (Yamamoto and Casida, 1968).

1.5.1.1.3. Ester hydrolysis All pyrethroids with the exception of the nonester compounds, such as MTI 800 which has a hydrocarbon linkage between the "acid" and "alcohol" moieties, are subject to esterase-catalyzed breakdown. Even etofenprox, which has an ether linkage in this position, is cleaved via oxidation of the benzylic carbon and hydrolysis of the resultant ester in the rat (Roberts and Hutson, 1999). The stereochemistry of the cyclopropanecarboxylic acid is important in determining the rate of esteratic detoxification, since the *trans*-isomers are very much better substrates for esterases than are the *cis*-isomers (Soderlund and Casida, 1977). Indeed, *cis*-pyrethroids possessing an α-cyano group (Type II pyrethroids) and consequently a secondary ester are generally degraded via a microsomal P450 mechanism through oxidation of the α-carbon. Evidence from work with mammalian liver indicates that the esterases responsible for pyrethroid hydrolysis are also microsomal (Soderlund and Casida, 1977). This study also showed that the Type II pyrethroids cypermethrin and deltamethrin were the least susceptible to both esterase-catalysed hydrolysis and oxidation. Generally, the major route of Type II pyrethroid metabolism in mammals is via de-esterification catalysed by microsomal oxidases (e.g., Crawford and Hutson, 1977). Studies on insects have also concluded that ester cleavage is often mainly by an oxidative mechanism (Casida and Ruzo, 1980; Funaki et al., 1994). The latter study concluded that the de-esterification of fenvalerate by pyrethroid-resistant houseflies was principally due to over-expression of cytochromes P450; only a small portion of the ester bond cleavage was caused by hydrolases.

Whether the ester cleavage is hydrolytic or oxidative is apparently largely dependent on the species. For example, T. ni (Ishaaya and Casida, 1980), S. littoralis, (Ishaaya et al., 1983), S. eridana (Abdelaal and Soderlund, 1980) (Lepidoptera), and B. tabaci (Jao and Casida, 1974; Ishaaya et al., 1987) (Hemiptera) all degraded trans-pyrethroids via a hydrolytic mechanism. Conversely, houseflies (Diptera) and the rust-red flour beetle Tribolium castaneum (Coleoptera) used an oxidative pathway (Ishaaya et al., 1987). A rather special case is that of the green lacewing Chrysoperla carnea agg. (Neuroptera), which is highly resistant to pyrethroids. This insect has been shown to contain high levels of a pyrethroid-hydrolyzing esterase that is able to catalyze the hydrolysis of cis-isomers, including those of Type II ester pyrethroids such as cypermethrin (Ishaaya and Casida, 1981; Ishaaya, 1993). How the over-production of esterases induces resistance to cis-pyrethroids such as deltamethrin, which are poor substrates for the enzyme(s) from other species, has been the cause of some conjecture. Devonshire and Moores (1989) presented evidence that cis-pyrethroids bind tightly to the active site and are thus sequestered by the large amounts of the resistance-associated esterases (E4 and FE4) produced by resistant peach-potato aphids M. persicae. Consequently, deltamethrin acts as a competitive inhibitor of esterase activity but is removed by binding to the protein rather than by hydrolysis. Selection pressure, whether artificially or from frequent commercial use of pyrethroids, frequently leads to multifactorial mechanisms of insecticide resistance, including esterases, cytochromes P450, target-site (kdr and super-kdr) mechanisms, and sometimes reduced penetration (e.g., Anspaugh et al., 1994; Pap and Toth, 1995; Ottea et al., 2000; Liu and Pridgeon, 2002). Such insect strains have extremely high resistance and usually show some degree of cross-resistance to all pyrethroids.

Most insect esterases that catalyze the hydrolysis of pyrethroid esters are soluble nonspecific B-type carboxylesterases. These have a wide substrate-specificity and model substrates, such as 1-naphthyl acetate, have been used to visualize electrophoretically distinct protein bands in B. tabaci (Byrne et al., 2000) and many other insect species. There is little information to date on the occurrence of microsomal esterases in insects analogous to those in mammals (e.g., Prabhakaran and Kamble, 1996). Most evidence for the involvement of esterases in pyrethroid breakdown is indirect and has used model substrates; however, direct evidence for pyrethroid hydrolysis has been obtained for esterases from several species, including the horn fly Haematobia irritans (Diptera) (Pruett et al., 2001). Concern has been raised of the validity of using model substrates to predict the breakdown of pyrethroids, and consequently some researchers have designed model substrates with pyrethroid-like structures. Thus, Shan and Hammock (2001) developed sensitive fluorogenic substrates based on DCVA coupled to the cyanohydrin of 6-methoxynaphthalene-2-carboxaldehyde. Hydrolysis of this ester and decomposition of the cyanohydrin regenerates this fluorescent aldehyde. An enzyme preparation that used this substrate from a cypermethrin-selected resistant strain of H. virescens (Lepidoptera) gave better selectivity (×5) than did 1-naphthyl acetate (×1.4). A similar approach, but using 1-naphthyl esters of the four stereoisomers of DCVA with detection of released 1-naphthol by diazo-coupling to Fast Blue RR, has been reported (Moores et al., 2002). In these experiments, only the (1S)-trans-isomer acted as a substrate for aphid esterases, in agreement with other experiments on the stereospecificity of insect pyrethroid esterases. Activity staining of electrophoretic gels using this novel substrate showed that the pyrethroid-hydrolyzing activity was distinct from the resistance-associated esterases visualized with 1-naphthyl acetate, indicating that the main mechanism for resistance was binding/sequestration rather than hydrolysis.

Work since 1985 has concentrated on understanding the nature of the esterases responsible for pyrethroid hydrolysis. In vivo studies have often yielded equivocal results and, using such tools as specific inhibitors of oxidative or hydrolytic metabolism, it has frequently been difficult to prove which mechanism is primarily responsible for pyrethroid catabolism. These problems have arisen in part because there is no such thing as a specific inhibitor and some compounds thought to be specific inhibitors of cytochromes P450 (e.g., piperonyl butoxide) may also inhibit esterases (Gunning et al., 1998b; Moores et al., 2002).

Studies have included both in vivo work on whole insects or their tissues and in vitro studies with isolated enzymes. In resistant populations of many insect species, the mechanisms are most frequently due to both enhanced esterase and cytochrome P450 levels, so that dissection of the proportions of the different mechanisms is difficult. However, evidence that resistance ratios are reduced or abolished in the presence of reliably specific inhibitors such as organophosphates (Gunning et al., 1999; Corbel et al., 2003) can be taken as a good indication that enhanced esterase levels are responsible for metabolic resistance.

1.5.1.1.4. Cytochrome P450 monooxygenases

These are a class of Phase I detoxification enzymes that catalyse various NADPH- and ATP-dependent oxidations, dealkylations, and dehydrogenations. Both microsomal and mitochondrial forms occur in insects. They are probably responsible for the most frequent type of metabolism-based insecticide resistance (Oppenoorth, 1985; Mullin and Scott, 1992; Scott and Wen, 2001). They are also a major mechanism for pyrethroid catabolism (Tomita and Scott, 1995). Their occurrence and importance in insect xenobiotic metabolism has been reviewed by Scott and Wen (2001). The super-family of cytochrome P450 genes has probably evolved by gene duplication and adaptive diversification, and comprises 86 functional genes in D. melanogaster. The large number of substrates metabolized by P450s is due both to the multiple isoforms and to the fact that each P450 may have several substrates (Rendic and DiCarlo, 1997). Because these enzymes may have overlapping substrate specificities, it is difficult to ascribe the function to individual P450 enzymes. In insects, although the importance of oxygenases in the metabolism of many substrates is known, the particular P450 isoforms involved have rarely been identified.

Several P450 iso-enzymes have been isolated or expressed from insect sources. Regarding pyrethroid metabolism, the best-characterized P450 isoform is CYP6D1. This was originally purified from a strain of highly resistant (ca. ×5000) houseflies designated "Learn pyrethroid resistant" (LPR) selected by the continuous usage of permethrin to control flies in a New York State dairy. A reduced-penetration mechanism and kdr were also present in the strain. CYP6D1 has been purified (Wheelock and Scott,

1989) and sequenced via the use of degenerate primers derived from known protein sequences and PCR amplification (Tomita and Scott, 1995). Overproduction of this P450 isozyme was found to be the major mechanism of deltamethrin detoxification in microsomes derived from the LPR flies (Wheelock and Scott, 1992). The enzyme requires cytochrome b5 as a co-factor and is specific in its action, because only the 4'-hydroxy metabolite was produced from cypermethrin (Zhang and Scott, 1996). CYP6D1 was found to be the major and possibly the only P450 isoform responsible for pyrethroid metabolism in this strain of houseflies; consequently, the resistance ratios are very much less for pyrethroids such as fenfluthrin that do not have the 3-phenoxybenzyl group (Scott and Georghiou, 1986). The same mechanism was found to be responsible for PBO suppressible resistance to permethrin from a Georgia poultry farm in the USA (Kasai and Scott, 2000). In both these housefly strains, the mechanism was due to an increased (ca. ×10) transcription of the gene, leading to increased levels of CYP6D1 mRNA and higher levels of the enzyme. CYP6D1 is expressed in the insect nervous system and has been shown to protect the tissue from the effects of cypermethrin (Korytko and Scott, 1998). Clearly, from the metabolic specificity of CYP6D1, other isoforms of cytochromes P450 must also be implicated in pyrethroid metabolism, although which reactions are catalyzed by which isoform has yet to be determined.

It is characteristic of monooxygenases that they are inducible within an individual animal. The use of phenobarbitone to induce monooxygenase activity in rat liver is well known, and many other agents are capable of transiently up-regulating cytochromes P450. Phytophagous insects are exposed to many plant xenobiotics, for example monoterpenes which also induce P450 production. Such induction of P450s may incidentally induce an isoform also capable of metabolizing pyrethroids. For example, feeding larvae of H. armigera on mint (Mentha piperita) leaves induced a 4× resistance to pyrethroids compared with those fed on a semi-defined diet (Hoque, 1984; Terriere, 1984; Schuler, 1996; Scott et al., 1998). CYP6B2 mRNA, a P450 isoform also implicated in pyrethroid resistance, is inducible by peppermint oil and specifically α-pinene in larvae of H. armigera (Ranasinghe et al., 1997). This induction was rapid (ca. 4h) and disappeared within a similar period of removing the stimulus. Clearly, the mechanism for the transient induction of P450s (Ramana, 1998) is different from the situation with the LPR houseflies, in which CYPD1 is permanently up-regulated (Liu and Scott,

1998), although the precise mechanistic details still remain to be elucidated. In nonphytophagous arthropods for example the cattle tick *Boophilus microplus* which is not subjected to a barrage of allelochemicals in its diet, P450s were found to be of little importance in the induction of resistance to pyrethroids (Crampton *et al.*, 1999), although the converse has been found for adults of the mosquito *Culex quinquefasciatus* (Kasai *et al.*, 1998).

1.5.1.1.5. Model substrates

When managing insecticide resistance, it is important that resistant alleles can be detected at low frequency in populations. Data from bioassays will only detect a quite high proportion of individuals with reduced sensitivity in the population. Consequently, it is useful to design biochemical or DNA (molecular) tests that can identify resistance in individual insects. Thus, model substrates commonly used to measure P450 levels *in vitro* need to be substrates of the relevant P450 isoforms that degrade the insecticide. Amongst such model substrates are the sensitive fluorogenic reagents 7-ethoxycoumarin, ethoxyresorufin, and methoxyresorufin, and the chromogenic substrate 4-nitroanisole. These compounds are dealkylated by monooxygenases to yield a fluorescent or colored product. Unfortunately, it has generally been found that these substrates are also isozyme specific, so that they may not be good indicators of P450-induced pyrethroid resistance. In the case of CYP6D1, methoxyresorufin was found to be a substrate, but ethoxyresorufin and 7-ethoxycoumarin were not (Wheelock and Scott, 1992). A similar variation in the activity of elevated oxygenases to 4-nitroanisole, benzo(a)pyrene, benzphetamine, and methoxyresorufin in the mid-gut of a multi-resistant (cypermethrin and thiodicarb) strain of *H. virescens* larvae was noted (Rose *et al.*, 1995). In this strain, demethylation rates of both 4-nitroanisole and methoxyresorufin were useful as indicators of insecticide resistance; however, on such multi-resistant strains, several P450 isoforms are probably elevated in tandem making comparisons difficult. Consequently, the use of model substrates to estimate the levels of P450 monooxygenases in individual insects may only give equivocal information on levels of P450-derived metabolic resistance to pyrethroids.

1.5.1.1.6. Glutathione-S-transferases (GSTs)

GSTs are important in the detoxification of organophosphorus insecticides and other electrophilic compounds, which are dealkylated and conjugated with glutathione. Pyrethroids are not electrophilic and would not be expected to be detoxified by this mechanism. However, there are several reports that have correlated enhanced levels of GSTs with pyrethroid resistance in a number of species, *S. littoralis* (Lagadic *et al.*, 1993), *T. castaneum* (Reidy *et al.*, 1990), and *Aedes aegypti* (Grant and Matsumura, 1989). Additionally, pyrethroids have been shown to induce production of GSTs in the honeybee *Apis mellifera* (Yu *et al.*, 1984), fall armyworm *Spodoptera frugiperda* (Punzo, 1993), and German cockroach *Blatella germanica* (Hemingway *et al.*, 1993). The role of the glutathione-S-transferase system as a mechanism of defence against pyrethroids is not fully understood, but it is thought that GST proteins sequester the pyrethroids (Kostaropoulos *et al.*, 2001) or possibly protect the tissues from pyrethroid-induced lipid peroxidation (Vontas *et al.*, 2001).

1.5.1.2. Cuticle penetration

As a resistance mechanism, reduced penetration of the insect cuticle has been studied less and has generally been considered of subordinate importance to enhanced detoxification and target-site mutations. Where detected, it is usually found with other mechanisms, e.g., the "LPR strain" of houseflies referred to above. Reduced cuticular penetration of pyrethroids has been detected in resistant strains of a number of other species, including *H. armigera* (Gunning *et al.*, 1991), *H. zea* (AbdElghafar and Knowles, 1996), *H. virescens* (Little *et al.*, 1989), *S. exigua* (Delorme *et al.*, 1988), the diamond-back moth *Plutella xylostella* (Noppun *et al.*, 1989), *B. germanica* (Wu *et al.*, 1998), and *B. microplus* (Schnitzerling *et al.*, 1983). In all these cases, reduced cuticular penetration was found in addition to other resistance mechanisms, enhanced metabolism, and/or target-site resistance (*kdr*). Mechanisms involving reduced uptake appear not to give significant resistance to the lethal effects of insecticides but provide a more than additive effect when combined with other mechanisms. Thus, decreased penetration, although a minor factor on its own, can, when coupled with other mechanisms, increase resistance many-fold (Ahmad and McCaffery, 1999).

1.5.1.3. Target-site resistance

Target-site resistance is the most important mechanism of pyrethroid resistance and its selection and spread have compromised the use of pyrethroid insecticides on many insect pests. It is a particularly important mechanism of pyrethroid resistance, as it confers a degree of loss of sensitivity to all members of the class. It is characterized by a reduction in the sensitivity of the insect nervous system and has been termed "knockdown resistance" or *kdr* (Sawicki, 1985). The *kdr* trait was first identified in the early 1950s in houseflies (Busvine, 1951). It causes a loss of sensitivity to DDT and its analogs, and to pyrethrins and pyrethroids, which all owe their activity to interaction with the *para*-type voltage-gated sodium channel in nerve membranes. This loss of activity is characterized by a reduction in the binding of these insecticides to the sodium channel (Pauron *et al.*, 1989). An enhanced form of this resistance termed *super-kdr* has also been characterized in houseflies (Sawicki, 1978). Both the *kdr* and *super-kdr* traits were mapped to chromosome 3 and found to occupy the same allele, the *para*-type sodium channel. This has been confirmed by molecular cloning studies of these channels in *kdr* and *super-kdr* houseflies (Williamson *et al.*, 1996) and *kdr B. germanica* (Miyazaki *et al.*, 1996; Dong, 1997). The *super-kdr* resistance in houseflies is due to a methionine to threonine (M918T) point mutation in the gene encoding the *para* sodium channel. Both mutations were located with domain II of the ion channel. The L1014F mutation in IIS6 was found in both housefly and cockroach strains, and confers *kdr* resistance. To date, the M918T mutation has not been detected as a single substitution in any housefly strain, but only occurs in conjunction with L1014F. Mammals are intrinsically much less susceptible to pyrethroids and DDT. Significantly, mammalian neuronal sodium channels have an isoleucine rather than methionine in the position (874) that corresponds to the housefly *super-kdr* site (918). Site-directed mutation of this residue to methionine gives rise to a channel with >100x increased sensitivity to deltamethrin, suggesting that differential pyrethroid sensitivity between mammals and insects may be due in part to structural differences between the mammalian and insect sodium channels (Vais *et al.*, 2000). In some insect species, the existence of *kdr*-type target-site resistance has often been masked by efficient metabolic resistance mechanisms; an example is *M. persicae*, in which resistance-associated esterase is responsible for much of the reduced sensitivity towards organophosphates, carbamates, and pyrethroids. Any inference that target-site resistance might also be a factor has usually been based on the sensitivity of the insects in the presence of synergists such as PBO or DEF *S,S,S,*-tributyl phosphorotrithiolate, and the assumption that any residual decreased sensitivity is due to *kdr*-type target-site resistance.

Use of degenerate DNA primers for the *para*-type sodium channel and sequencing of the gene have resulted in the unequivocal identification of resistance-inducing mutations in the trans-membrane domain II (Martinez-Torres *et al.*, 1999b) of the channel in *M. persicae*, in which a leucine to

phenylalanine (L1014F) mutation associated with *kdr* was identified. Insects containing this mutation could also be identified by the use of a discriminating dose of DDT, as DDT is unaffected by the enhanced-esterase mechanisms also present in most of the aphid clones. Indeed, of 58 aphid clones analysed for both *kdr*- and esterase-based mechanisms, only four contained an esterase (E4) mutation and not *kdr*. Estimates for resistance factors in aphids containing both mechanisms are 150–540-fold, in comparison to 3–4-fold (FE4 esterase alone) or 35-fold for *kdr* alone. Consequently, these dual-resistance mechanisms afford a level of decreased sensitivity whereby insects become totally immune to field dosages of pyrethroids. Similar *kdr* mutations have also been detected in cockroaches (Miyazaki *et al.*, 1996; Dong, 1997), *H. irritans* (Guerrero *et al.*, 1997), *P. xylostella* (Schuler *et al.*, 1998), and *An. gambiae* (Martinez-Torres *et al.*, 1998). In a pyrethroid-resistant strain of the tobacco budworm *H. virescens*, the same locus was mutated to histidine rather than phenylalanine (Park and Taylor, 1997)

As with metabolic resistance mechanisms, it is important to establish methods that can identify sodium-channel *kdr*-type mechanisms in single insects so that it is possible to adjust insect-control methods. The *kdr*-mutation of nerve insensitivity was originally identified by electrophysiology, and this method still remains a fundamental way of confirming nerve insensitivity. However, it is a specialized and rather cumbersome technique that is out of the question when attempting to test large numbers of an agricultural pest species. The DDT bioassay using a discriminating dose remains a useful method but may not completely discriminate between homozygous and heterozygous individuals. The most useful technique has been direct diagnosis of the mutation(s) based on PCR amplification and sequencing. The identification of the L1014F mutation in knockdown-resistant housefly strains has led to the development of several diagnostic assays for its occurrence in other species, including *H. irritans* (Guerrero *et al.*, 1997), the mosquitoes *An. gambiae* (Martinez-Torres *et al.*, 1998) and *Culex pipiens* (Martinez-Torres *et al.*, 1999a), as well as *M. persicae*. However, this technique will only identify known mutations and the test designed to detect L1014F in *An. gambiae* (Ranson *et al.*, 2000) did not detect the additional L1014S.

The molecular biology of knockdown resistance to pyrethroids has been reviewed (Soderlund and Knipple, 2003). Sequencing of the *para*-sodium channel gene from several arthropod species has led to the discovery of a number of amino-acid polymorphisms in the protein. Of the 20 amino-acid polymorphisms each uniquely associated with pyrethroid resistance, those occurring at four sites have so far been found as single mutations in resistant populations. These are valine 410 (V410M in *H. virescens*; Park *et al.*, 1997), methionine 918 (M918V in *B. tabaci*; Morin *et al.*, 2002), leucine 1014 (L1014F in several species; L1014H in *H. virescens*; Park and Taylor, 1997; L1014S in *C. pipiens*; Martinez-Torres *et al.*, 1999a), *An. gambiae* (Ranson *et al.*, 2000), and phenylalanine 1538 (F538I, in *B. microplus*; He *et al.*, 1999). All these sites have been located on trans-membrane regions of the channel and it is inferred that they are close to the binding site(s) of pyrethroids and DDT. Additionally, other polymorphisms have been found that only occur in the presence of L1014F and appear to act in a similar way to M918T in houseflies, causing enhanced resistance.

1.5.2. Resistance to Pyrethroids in the Field

The main cause of development of resistance in insects in the field is a persistent and high selection pressure as a consequence of repeated applications of a single class of insecticide (or another with the same mode of action). Therefore, despite effective control in the initial stages, a small number of survivors with innate resistance then rapidly multiply until control fails. In this regard, pyrethroids are no different from other insecticide classes. It is noteworthy that it was the development of resistance to pyrethroids that first prompted companies to take resistance seriously and to take joint action. Pyrethroids suffered an inherent disadvantage at the outset in that *kdr* also confers resistance to DDT, and prior use of DDT had already selected *kdr* alleles to significant levels in the same pests. Presently over 80 species have developed resistance to pyrethroids (Whalon *et al.*, 2003).

Until the late 1970s, the major agrochemical companies had seen the development of resistance to established classes as commercially beneficial and motivation for the introduction of new classes. However, the increasing cost of discovery, together with the realization that pyrethroids could be rendered ineffective in the field within a much shorter period of time than other classes of insecticides, forced them to take collective action. In 1979 they set up the Pyrethroid Efficacy Group (PEG), which in 1984 become a sub-group of a larger international organization called the Insecticide Resistance Action Committee (IRAC). This group communicates its actions through a website (http://www.irac-online.org) and also sponsors the biannual Resistant Pest Management newsletter. Before the establishment of PEG, the investigation of resistance mechanisms and their consequences had remained the domain of academic research, and the development of resistance to pyrethroids was largely overlooked and even denied by some companies. The primary aim of PEG was to prolong the effectiveness of pyrethroids in the field. It encouraged and assisted in the investigation of all aspects of resistance to pyrethroids and particularly the development and implementation of insecticide-resistance management (IRM) strategies.

Persistent selection with a single class of insecticide will invariably lead to resistance, even if synergists are used. Therefore the overall aim is to minimize use of any single insecticide class so as to limit selection pressure and thereby conserve susceptibility in pest insects. This requires an in-depth knowledge of factors ranging from resistance mechanisms to genetic and ecological attributes of both pest and beneficial insects. Any strategy has also to integrate the judicious use of different insecticide types (namely those with different modes of action and synergy where possible) with other pest-management options (e.g., agronomic practices), together with regular monitoring of both the levels of resistance and the nature of the resistance mechanisms. Finally, the key requirement for the success of any strategy is the cooperation of the growers. By way of example, two IRM strategies are considered below which encompass these factors.

The first successful and the most publicized IRM strategy came from Australia. Synthetic pyrethroids were introduced in Australia in 1977 when there was already widespread resistance to virtually all established classes. However, within six years of introduction, resistance to these pyrethroids had also developed in commercially important insect species (Forrester *et al.*, 1993). A resistance-management strategy for *H. armigera* was implemented in the 1983/4 season. A different approach was used for each class of insecticide, depending on the severity of the resistance risk and predicted selection pressure. It integrated the use of chemical and nonchemical control methods, especially biological and cultural. For chemical control, unrelated chemistries were used with a strong emphasis on pyrethroid/ovicide mixtures. In essence, a three-stage strategy was implemented. In Stage I (September to February), only endosulfan (an organochlorine) was permitted. Stage II (January to February) allowed a maximum of three pyrethroid sprays within a 42-day window (later reduced to 38 days), enough to control only one of the five *H. armigera* generations present within a single growing season. In Stages I and II, ovicidal compounds (e.g., methomyl) could also be used. Organophosphates and carbamates were used in Stage III and also if required for additional sprays during Stages I and II.

The use of a "softer" insecticide (endosulfan) in the early season was deliberate to minimize disruption of beneficial parasitoids and predators and to avoid a potential upsurge of secondary pests such as mites, aphids, and whiteflies, which were not controlled by the pyrethroids available at the time. Examples of nonchemical countermeasures incorporated in the strategy to reduce selection pressure included the use of early-maturing crops, avoiding early-growing crops (e.g., maize) in adjacent fields, which may act as early-season nurseries for resistant *H. armigera*, and utilization of host plants in refugia to maintain a large pool of susceptible individuals, which would continually dilute the resistant population in the crop. Resistance was monitored on a weekly basis using a discriminating dose of fenvalerate with and without the synergist PBO. Later monitoring showed a rise in metabolic resistance attributed to MFOs and this resulted in the inclusion of PBO in the last of the maximum of three pyrethroid sprays. This strategy was successful for many years but the underlying trend of upward increase in the proportion of resistant insects continued and finally led to a complete reorganization of the strategy in the mid 1990s with a shift away from reliance on pyrethroids.

Initial resistance to pyrethroids was thought to be due to the presence of the knockdown site-insensivity resistance (*kdr*) mechanism probably as a direct result of cross-resistance to DDT. Over the years of the IRM strategy, metabolic resistance mechanisms appear to have taken over. However, there is still controversy over whether it is primarily due to elevated levels of esterases or mixed-function oxidases (MFOs). The main reason for the uncertainty concerns the role of synergists used in the studies. For example, PBO is now thought to inhibit both these types of enzymes. Furthermore, it has been suggested that such inhibitors may themselves have become resisted in the field over time (McCaffery, 1998), thus obscuring the mechanism of resistance.

The second IRM strategy involved the whitefly, *B. tabaci*, a representative sucking pest on a wide range of crops (Denholm *et al.*, 1998). In 1995, overreliance on a limited range of pyrethroids in Arizona had led to a classic treadmill scenario, with farmers responding to rising levels of resistance by increasing the number of sprays (as many as 8–12 applications per season). In this pest, the haplodiploid breeding system encourages rapid selection and fixation of resistance genes. Males are produced

uniparentally from unfertilized, haploid eggs, and females are produced biparentally from fertilized diploid eggs. In addition, for this (and other) highly polyphagous species, the interaction between pest ecology and resistance is complex and generally not well understood, making formulation of IRM strategies even more difficult. The strategy of Dennehy and Williams (1997) implemented in 1996 had several features in common with that for *H. armigera*. It too relies on the continuous availability of a pool of susceptible whiteflies in refugia (e.g., Brassica crops) throughout the year and alteration of agronomic practices (e.g., timing of planting) in crops to minimize whitefly numbers whilst still maintaining the level of natural enemies. The first of the three-stage IRM strategy involved use of a single spray each of pyriproxyfen and buprofezin (then newly available insect-growth regulators) with a 14- or a 21-day gap. The second stage allowed nonpyrethroid conventional insecticides and the third used pyrethroids synergized with acephate as late as possible in the season. A threshold of infestation was defined to initiate insecticide applications. Mixtures were limited to no more than two compounds and any one active ingredient was restricted to no more than two applications in one season. This strategy has been extremely successful in reducing the number of sprays required and regaining susceptibility both to synergized pyrethroids and key nonpyrethroid insecticides.

As for *H. armigera*, the need for monitoring, establishing threshold levels to trigger spray applications, and cooperation of the growers was key to the strategy. Long-term success of any IRM strategy depends on many factors because, even when an IRM strategy has been successful (McCaffery, 1998), effectiveness of synergized pyrethroids can be lost after just two applications in areas with a history of resistance to pyrethroids.

In conclusion, the key requirement for the development and sustainability of an IRM strategy is diagnosis of the resistance mechanism(s) in field populations. This is especially important when assessing the relative importance of individual mechanisms when several are present, which influences changes to the strategy with time. As alluded to earlier, there is still much uncertainty in the diagnosis of mechanisms. The use of established inhibitors or just *in vitro* bioassay data has been shown to be unreliable in this regard. For example in *M. persicae*, elevated esterase was considered for many years to be the main resistance mechanism. It was only recently shown (Devonshire *et al.*, 1998) that this mechanism made only a minor contribution to the observed resistance to pyrethroids, with a *kdr* mechanism being the main contributor. This case further emphasized the need for not only accurately identifying the resistance mechanisms present but also their relative contribution. One way forward would be to design selective inhibitors on the basis of *in vivo* and *in vitro* structure–activity relationship studies from susceptible and resistant insects.

1.5.2.1. Chemistry-led options for circumventing resistance The established IRM strategies (considered above) successfully exploit a range of options including use of insecticides with different modes of action, agronomic practices, and biological control. However, there are two additional options involving insecticides that have hitherto not been fully exploited.

1. Negative cross-resistance is a phenomenon in which a resistant pest is rendered more susceptible to another class of pesticide. This phenomenon is rare in insects but for pyrethroids, two different classes, the *N*-alkylamides (Elliott *et al.*, 1986) and dihydropyrazoles (and related classes) (Khambay *et al.*, 2001), have been found to cause this effect. They both exhibit similar, albeit small (resistance factors of 0.4), levels of negative cross-resistance to houseflies with nerve insensivity (*super-kdr*) as the main resistance mechanism. The mechanisms involved are not properly understood. However this phenomenon is apparent only in the presence of the *super-kdr* mechanism, with *kdr* alone having little impact. BTG 502 (**Figure 8**; (**58**)) is representative of the *N*-alkylamides, a class with which this phenomenon was first identified. No compounds of this class have been commercialized to date, but see **Chapter 2**.

In insects that have developed resistance to pyrethroids as a consequence of increased levels of metabolic enzymes (MFOs and esterases), negative cross-resistance has been demonstrated towards insecticides that require *in vivo* activation (i.e., pro-insecticides). For example, negative cross-resistance has been observed with indoxacarb in *H. armigera* with increased levels of esterases (Gunning, 2003). However, this compound has little effect in *M. persicae* with elevated levels of esterase (Khambay, unpublished data). This difference may be accounted for by the kinetics of the hydrolysis reaction. In susceptible *H. armigera*, hydrolysis is incomplete by the end of the bioassay time. The increased levels of esterases in resistant *H. armigera* generate more active compound over the same period, resulting in a higher level of activity. In *M. persicae*, the hydrolysis is complete during the test period even in susceptible insects and therefore no additional effects are observed if increased levels of esterases are present. Negative cross-resistance has also been observed for pyrethroid-resistant insects with increased levels of MFOs. Examples of such insecticides include the pyrazole chlofenapyr, the organophorothioates diazinon and chlorpyrifos, and the methylcarbamate propoxur.

2. Pyrethroids constitute a diverse class of chemical structures and levels of resistance to them also vary considerably. For example, in houseflies with the *super-kdr* resistance mechanism, the resistance factor for different pyrethroids can vary from less than 10 to over 500. Structure–activity relationships have been discerned and the levels of resistance factors (RFs, derived by dividing the LD50 of resistant insects by that of the corresponding susceptible insects) have been shown to be directly influenced by the nature of the alcohol group and the α substituent, but independent of the nature of the acid group (Farnham and Khambay, 1995a, 1995b; Beddie *et al.*, 1996). This is reflected in the description of Types I and II. Thus, highest RFs are shown by compounds with a 3-phenoxybenzyl (or

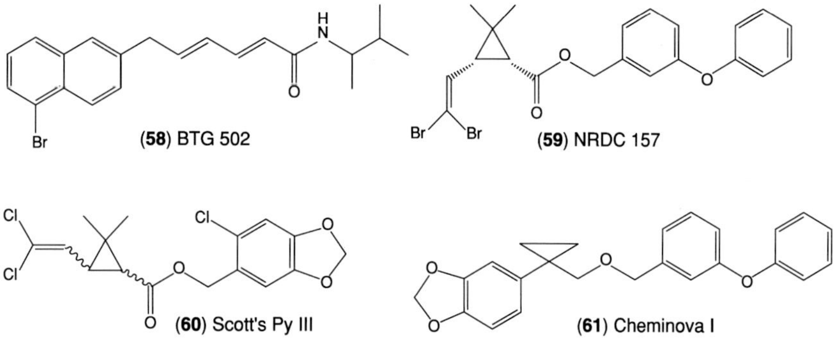

Figure 8 Compounds referred to in the resistance section (structures **58–61**).

equivalent) alcohol ester with an α-cyano substituent (e.g., deltamethrin) with an RF of 560. The RF of 91 for NRDC 157 (**59**) (deltamethrin analog lacking the α-cyano group) is significantly smaller, and indeed surprisingly the LD_{50} values against *super-kdr* flies are similar for the two compounds. This indicates that the additional target-site binding, and hence increased insecticidal activity, in susceptible flies gained by introducing an α-cyano group into NRDC 157 has been lost in *super-kdr* flies. The lowest RFs are observed for pyrethroids with alcohols containing a single ring and no α substituent, an example being tefluthrin with RF < 4.

In *H. armigera*, there has been some controversy regarding the resistance mechanism. Though it is accepted that the major mechanism is due to increased metabolism, it is unclear whether MFOs or esterases prevail. The highest RFs were shown by ester pyrethroids containing a phenoxybenzyl alcohol moiety in combination with an aromatic acid. The lowest RFs were with pyrethroids containing simple alcohol moieties as in bioallethrin. Resistance-breaking pyrethroids (RF < 1) were also identified e.g., those containing a methylenedioxyphenyl moiety as in Scott's Py III (**60**) and Cheminova 1 (**61**).

This approach clearly has potential to prolong the usefulness of pyrethroids as a class of insecticides. However, the agrochemical industries are unlikely to develop new pyrethroids purely for a role in "breaking" resistance to established compounds due to commercial considerations.

Despite the detection of resistance to pyrethroids soon after their introduction over 30 years ago, it has proved possible to manage it in many field situations. Therefore it is likely that pyrethroids will continue to be a useful tool for the foreseeable future.

Acknowledgments

We thank Dr Richard Bromilow for assistance in preparing this manuscript. Rothamsted Research receives grant-aided support from the Biotechnology and Biological Sciences Research Council, UK.

References

Abdelaal, Y.A.I., Soderlund, D.M., **1980**. Pyrethroid-hydrolyzing esterases in southern armyworm larvae – tissue distribution, kinetic properties, and selective inhibition. *Pestici. Biochem. Physiol.* 14, 282–289.

AbdElghafar, S.F., Knowles, C.O., **1996**. Pharmacokinetics of fenvalerate in laboratory and field strains of *Helicoverpa zea* (Lepidoptera: Noctuidae). *J. Econ. Entomol.* 89, 590–593.

Ahmad, M., McCaffery, A.R., **1999**. Penetration and metabolism of *trans*-cypermethrin in a susceptible and a pyrethroid-resistant strain of *Helicoverpa armigera*. *Pestici. Biochem. Physiol.* 65, 6–14.

Akamatsu, M., **2002**. Studies on three-dimensional quantitative structure–activity relationships in pesticides. *J. Pestici. Sci.* 27, 169–176.

Anspaugh, D.D., Rose, R.L., Koehler, P.G., Hodgson, E., Roe, R.M., **1994**. Multiple mechanisms of pyrethroid resistance in the German cockroach, *Blattella germanica* (L). *Pestici. Biochem. Physiol.* 50, 138–148.

Beddie, D.G., Farnham, A.W., Khambay, B.P.S., **1996**. The pyrethrins and related compounds. 41. Structure-activity relationships in non-ester pyrethroids against resistant strains of housefly (*Musca domestica* L). *Pestici. Sci.* 48, 175–178.

Bigley, W.S., Plapp, F.W., **1978**. Metabolism of cis-[C-14]-permethrin and trans-[C-14]-permethrin by tobacco budworm and bollworm. *J. Agric. Food Chem.* 26, 1128–1134.

Bloomquist, J.R., Miller, T.A., **1985**. Carbofuran triggers flight motor output in pyrethroid-blocked reflex pathways of the house-fly. *Pestici. Biochem. Physiol.* 23, 247–255.

Briggs, G.G., Elliott, M., Farnham, A.W., Janes, N., Needham, P.H., *et al.*, **1976**. Insecticidal activity of the pyrethrins and related compounds VIII. Relationship of polarity with activity in pyrethroids. *Pestici. Sci.* 7, 236–240.

Bryant, R., **1999**. Agrochemicals in perspective: analysis of the worldwide demand for agrochemical active ingredients. The Fine Chemicals Conference 99 [synopsis on the Internet] http://agranova.co.uk/agtalk99.asp Accessed 23 April 2004.

Busvine, J.R., **1951**. Mechanism of resistance to insecticide in houseflies. *Nature* 168, 193–195.

Byrne, F.J., Gorman, K.J., Cahill, M., Denholm, I., Devonshire, A.L., **2000**. The role of B-type esterases in conferring insecticide resistance in the tobacco whitefly, *Bemisia tabaci* (Genn). *Pest Manage. Sci.* 56, 867–874.

Casida, J.E., Ruzo, L.O., **1980**. Metabolic chemistry of pyrethroid insecticides. *Pestici. Sci.* 11, 257–269.

Cheung, K.W., Sirur, G.M., **2002**. Agrochemical Products Database, 2002. Cropnosis Limited, Edinburgh, UK.

Chuman, H., Goto, S., Karasawa, M., Sasaki, M., Nagashima, U., *et al.*, **2000a**. Three-dimensional structure–activity relationships of synthetic pyrethroids: 1. Similarity in bioactive conformations and their structure–activity pattern. *Quant. Struct.-Activ. Relationships* 19, 10–21.

Chuman, H., Goto, S., Karasawa, M., Sasaki, M., Nagashima, U., *et al.*, **2000b**. Three-dimensional structure–activity relationships of synthetic pyrethroids: 2. Three-dimensional and classical QSAR studies. *Quant. Struct.-Activ. Relationships* 19, 455–466.

Corbel, V., Chandre, F., Darriet, F., Lardeux, F., Hougard, J.M., **2003**. Synergism between permethrin and propoxur against *Culex quinquefasciatus* mosquito larvae. *Med. Vet. Entomol. 17*, 158–164.

Crampton, A.L., Green, P., Baxter, G.D., Barker, S.C., **1999**. Monooxygenases play only a minor role in resistance to synthetic pyrethroids in the cattle tick, *Boophilus microplus. Exp. Appl. Acarol. 23*, 897–905.

Crawford, M.J., Hutson, D.H., **1977**. Metabolism of pyrethroid insecticide (+/–)-alpha-cyano-3-phenoxybenzyl 2,2,3,3-tetramethylcyclopropanecarboxylate, WL 41706, in rat. *Pestici. Sci. 8*, 579–599.

Davies, J.H., **1985**. The pyrethroids: an historical introduction. In: Leahey, J.P. (Ed.), The Pyrethroid Insecticides. Taylor & Francis, London, pp. 1–41.

Delorme, R., Fournier, D., Chaufaux, J., Cuany, A., Bride, J.M., *et al.*, **1988**. Esterase metabolism and reduced penetration are causes of resistance to deltamethrin in *Spodoptera exigua* Hub (Noctuidae, Lepidoptera). *Pestici. Biochem. Physiol. 32*, 240–246.

Denholm, I., Cahill, M., Dennehy, T.J., Horowitz, A.R., **1998**. Challenges with managing insecticide resistance in agricultural pests, exemplified by the whitefly *Bemisia tabaci. Phil. Trans. R. Soc. Lond. B Biol. Sci. 353*, 1757–1767.

Dennehy, T.J., Williams, L., **1997**. Management of resistance in *Bemisia* in Arizona cotton. *Pestici. Sci. 51*, 398–406.

Devonshire, A.L., Field, L.M., Foster, S.P., Moores, G.D., Williamson, M.S., *et al.*, **1998**. The evolution of insecticide resistance in the peach-potato aphid, *Myzus persicae. Phil. Trans. R. Soc. Lond. B Biol. Sci. 353*, 1677–1684.

Devonshire, A.L., Moores, G.D., **1989**. Detoxication of insecticides by esterases from *Myzus persicae* – is hydrolysis important? In: Reiner, E., Aldridge, W.N. (Eds.), Enzymes Hydrolysing Organophosphorus Compounds. Ellis Horwood, Chichester, UK, pp. 180–192.

Dong, K., **1997**. A single amino acid change in the para sodium channel protein is associated with knockdown-resistance (kdr) to pyrethroid insecticides in German cockroach. *Insect Biochem. Mol. Biol. 27*, 93–100.

Elliott, M., **1977**. Synthetic pyrethroids. In: Gould, R.F. (Ed.), Synthetic Pyrethroids. American Chemical Society, San Francisco, pp. 1–28.

Elliott, M., **1995**. Chemicals in insect control. In: Casida, J.E., Quinstad, G.B. (Eds.), Pyrethrum Flowers: Production, Chemistry, Toxicology, and Uses. Oxford University Press, New York, pp. 3–31.

Elliott, M., **1996**. Synthetic insecticides related to natural pyrethrins. In: Copping, L.G. (Ed.), Crop Protection Agents from Nature: Natural Products and Analogues. Royal Society of Chemistry, Cambridge, UK, pp. 254–300.

Elliott, M., Farnham, A.W., Janes, N.F., Johnson, D.M., Pulman, D.A., *et al.*, **1986**. Insecticidal amides with selective potency against a resistant (Super-Kdr) strain of houseflies (*Musca domestica* L). *Agric. Biol. Chem. 50*, 1347–1349.

Elliott, M., Janes, N., **1978**. Synthetic pyrethroids – a new class of insecticide. *Chem. Soc. Rev. 7*, 473–505.

Farnham, A.W., Khambay, B.P.S., **1995a**. The pyrethrins and related compounds. 39. Structure-activity relationships of pyrethroidal esters with cyclic side-chains in the alcohol component against resistant strains of housefly (*Musca domestica*). *Pestici. Sci. 44*, 269–275.

Farnham, A.W., Khambay, B.P.S., **1995b**. The pyrethrins and related-compounds. 40. Structure-activity relationships of pyrethroidal esters with acyclic side-chains in the alcohol component against resistant strains of housefly (*Musca domestica*). *Pestici. Sci. 44*, 277–281.

Feyereisen, R., **1999**. Insect P450 enzymes. *Ann. Rev. Entomol. 44*, 507–533.

Ford, M.G., Greenwood, R., Turner, C.H., Hudson, B., Livingstone, D.J., **1989**. The structure–activity relationships of pyrethroid insecticides. 1. A novel approach based upon the use of multivariate QSAR and computational chemistry. *Pestici. Sci. 27*, 305–326.

Ford, M.G., Hoare, N.E., Hudson, B.D., Nevell, T.G., Banting, L., **2002**. QSAR studies of the pyrethroid insecticides Part 3. A putative pharmacophore derived using methodology based on molecular dynamics and hierarchical cluster analysis. *J. Mol. Graphics Modelling 21*, 29–36.

Ford, M.G., Livingstone, D.J., **1990**. Multivariate techniques for parameter selection and data analysis exemplified by a study of pyrethroid neurotoxicity. *Quant. Struct.-Activ. Relationships 9*, 107–114.

Forrester, N.W., Cahill, M., Bird, L.J., Layland, J.K., **1993**. Management of pyrethroid and endosulfan resistance in *Helicoverpa armigera. Bull. Ent. Res* (Lepidoptera: Noctuidae) in Australia.

Funaki, E., Dauterman, W.C., Motoyama, N., **1994**. *In-vitro* and *in-vivo* metabolism of fenvalerate in pyrethroid-resistant houseflies. *J. Pest. Sci. 19*, 43–52.

Grant, D.F., Matsumura, F., **1989**. Glutathione S-transferase 1 and 2 in susceptible and insecticide resistant *Aedes aegypti. Pestici. Biochem. Physiol. 33*, 132–143.

Guerrero, F.D., Jamroz, R.C., Kammlah, D., Kunz, S.E., **1997**. Toxicological and molecular characterization of pyrethroid-resistant horn flies, *Haematobia irritans*: Identification of kdr and super-kdr point mutations. *Insect Biochem. Mol. Biol. 27*, 745–755.

Gunning, R.V., Devonshire, A.L., **2003**. Negative cross-resistance between indoxacarb and pyrethroids in the cotton bollworm, *Helicoverpa armigera*, in Australia: a tool for resistance management. The BCPC International Congress. Glasgow, British Crop Protection Council. pp. 789–794.

Gunning, R.V., Easton, C.S., Balfe, M.E., Ferris, I.G., **1991**. Pyrethroid resistance mechanisms in Australian *Helicoverpa armigera. Pestici. Sci. 33*, 473–490.

Gunning, R.V., Moores, G.D., Devonshire, A., **1999**. Esterase inhibitors synergise the toxicity of pyrethroids in Australian *Helicoverpa armigera* (Hubner) (Lepidoptera: Noctuidae). *Pestici. Biochem. Physiol. 63*, 50–62.

Gunning, R.V., Moores, G.D., Devonshire, A.L., **1998a**. Inhibition of pyrethroid resistance related esterases by piperonyl butoxide in Australian *Helicoverpa armigera* and *Aphis gossypii*. In: Jones, G. (Ed.), Piperonyl

Butoxide: the Insecticide Synergist. Academic Press, London, pp. 215–226.

Gunning, R.V., Moores, G.D., Devonshire, A.L., **1998b**. Insensitive acetylcholinesterase and resistance to organophosphates in Australian *Helicoverpa armigera*. *Pestici. Biochem. Physiol.* 62, 147–151.

He, H.Q., Chen, A.C., Davey, R.B., Ivie, G.W., George, J.E., **1999**. Identification of a point mutation in the *para*-type sodium channel gene from a pyrethroid-resistant cattle tick. *Biochem. Biophys. Res. Commun.* 261, 558–561.

Hemingway, J., Dunbar, S.J., Monro, A.G., Small, G.J., **1993**. Pyrethroid resistance in German cockroaches (Dictyoptera, Blattelidae) – resistance levels and underlying mechanisms. *J. Econ. Entomol.* 86, 1631–1638.

Henrick, C.A., **1995**. Pyrethroids. In: Godfrey, C.R.A. (Ed.), Agrochemicals from Natural Products. Marcel Dekker, New York, pp. 63–145.

Holden, J.S., **1979**. Absorption and metabolism of permethrin and cypermethrin in the cockroach and the cotton leafworm larvae. *Pestici. Sci.* 10, 295–307.

Hoque, M.R., **1984**. Effect of different host plants and artificial diet on the tolerance levels of *Heliothis armigera* (Hubner) and *H. punctigera* (Wallengern) to various insecticides. MSc Thesis, University of Queensland, Australia.

Hudson, B.D., George, A.R., Ford, M.G., Livingstone, D.J., **1992**. Structure-activity relationships of pyrethroid insecticides. 2. The use of molecular dynamics for conformation searching and average parameter calculation. *J. Computer-Aided Mol. Design* 6, 191–201.

Ishaaya, I., **1993**. Insect detoxifying enzymes – their importance in pesticide synergism and resistance. *Arch. Insect Biochem. Physiol.* 22, 263–276.

Ishaaya, I., Ascher, K.R.S., Casida, J.E., **1983**. Pyrethroid synergism by esterase inhibition in *Spodoptera littoralis* (Boisduval) larvae. *Crop Protect.* 2, 335–343.

Ishaaya, I., Casida, J.E., **1980**. Properties and toxicological significance of esterases hydrolyzing permethrin and cypermethrin in *Trichoplusia ni* larval gut and integument. *Pestici. Biochem. Physiol.* 14, 178–184.

Ishaaya, I., Casida, J.E., **1981**. Pyrethroid esterase(s) may contribute to natural pyrethroid tolerance of larvae of the common green lacewing (Neuroptera, Chrysopidae). *Environ. Entomol.* 10, 681–684.

Ishaaya, I., Mendelson, Z., Ascher, K.R.S., Casida, J.E., **1987**. Cypermethrin synergism by pyrethroid esterase inhibitors in adults of the whitefly *Bemisia tabaci*. *Pestici. Biochem. Physiol.* 28, 155–162.

Jao, L.T., Casida, J.E., **1974**. Esterase inhibitors as synergists for (+)-trans-chrysanthemate insecticide chemicals. *Pestici. Biochem. Physiol.* 4, 456–464.

Jiang, Y.X., Lee, A., Chen, J.Y., Ruta, V., Cadene, M., et al., **2003**. X-ray structure of a voltage-dependent K⁺ channel. *Nature* 423, 33–41.

Jin, W.-G., Sun, G.-H., Xu, Z.-M., **1996**. A new broad spectrum and highly active pyrethroid-ZX 8901. Brighton Crop Protection Conference – Pests and Diseases, Brighton, British Crop Protection Council. pp. 455–460.

Kasai, S., Scott, J.G., **2000**. Overexpression of cytochrome P450CYP6D1 is associated with monooxygenase-mediated pyrethroid resistance in house flies from Georgia. *Pestici. Biochem. Physiol.* 68, 34–41.

Kasai, S., Weerashinghe, I.S., Shono, T., **1998**. P450 monooxygenases are an important mechanism of permethrin resistance in *Culex quinquefasciatus* Say larvae. *Arch. Insect Biochem. Physiol.* 37, 47–56.

Katsuda, Y., **1999**. Development of and future prospects for pyrethroid chemistry. *Pestici. Sci.* 55, 775–782.

Khambay, B.P.S., **2002**. Pyrethroid insecticides. *Pestici. Outlook* 13, 49–54.

Khambay, B.P.S., Boyes, A., Williamson, M.S., Vias, H., Usherwood, P.N.R., **2002**. Defining Type I and Type II activity in insecticidal pyrethroids, 10th International Congress on the Chemistry of Crop Protection, Basel 4–9 August 2002, Abstract no. 3C. 49.

Khambay, B.P.S., Denholm, I., Carlson, G.R., Jacobson, R.M., Dhadialla, T.S., **2001**. Negative cross-resistance between dihydropyrazole insecticides and pyrethroids in houseflies, *Musca domestica*. *Pest Manage. Sci.* 57, 761–763.

Khambay, B.P.S., Farnham, A.W., Liu, M.G., **1999**. The pyrethrins and related compounds. Part XLII: Structure-activity relationships in fluoro-olefin non-ester pyrethroids. *Pestici. Sci.* 55, 703–710.

Kobayashi, T., Nishimura, K., Fujita, T., **1988**. Quantitative structure–activity studies of pyrethroids. 13. Physicochemical properties and the rate of development of the residual and tail sodium currents in crayfish giant-axon. *Pestici. Biochem. Physiol.* 30, 251–261.

Korytko, P.J., Scott, J.G., **1998**. CYP6D1 protects thoracic ganglia of houseflies from the neurotoxic insecticide cypermethrin. *Arch. Insect Biochem. Physiol.* 37, 57–63.

Kostaropoulos, I., Papadopoulos, A.I., Metaxakis, A., Boukouvala, E., Papadopoulou-Mourkidou, E., **2001**. Glutathione S-transferase in the defence against pyrethroids in insects. *Insect Biochem. Mol. Biol.* 31, 313–319.

Lagadic, L., Cuany, A., Berge, J.B., Echaubard, M., **1993**. Purification and partial characterization of glutathione-S-transferases from insecticide-resistant and lindane-induced susceptible *Spodoptera littoralis* (Boisd) larvae. *Insect Biochem. Mol. Biol.* 23, 467–474.

Lee, K.S., Walker, C.H., McCaffery, A., Ahmad, M., Little, E., **1989**. Metabolism of trans-cypermethrin by *Heliothis armigera* and *Heliothis virescens*. *Pestici. Biochem. Physiol.* 34, 49–57.

Little, E.J., McCaffery, A.R., Walker, C.H., Parker, T., **1989**. Evidence for an enhanced metabolism of cypermethrin by a monooxygenase in a pyrethroid-resistant strain of the tobacco budworm (*Heliothis virescens* F). *Pestici. Biochem. Physiol.* 34, 58–68.

Liu, N., Scott, J.G., **1998**. Increased transcription of CYP6D1 causes cytochrome P450 mediated insecticide resistance in house fly. *Insect Biochem. Mol. Biol. 28*, 531–535.

Liu, N.N., Pridgeon, J.W., 2002. Metabolic detoxication and the kdr mutation in pyrethroid resistant house flies, *Musca domestica* (L.). *Pestici. Biochem. Physiol. 73*, 157–163.

Martinez-Torres, D., Chandre, F., Williamson, M.S., Darriet, F., Berge, J.B., et al., 1998. Molecular characterization of pyrethroid knockdown resistance (kdr) in the major malaria vector *Anopheles gambiae* S.S. *Insect Mol. Biol. 7*, 179–184.

Martinez-Torres, D., Chevillon, C., Brun-Barale, A., Berge, J.B., Pasteur, N., et al., 1999a. Voltage-dependent Na$^+$ channels in pyrethroid-resistant *Culex pipiens* L mosquitoes. *Pestici. Sci. 55*, 1012–1020.

Martinez-Torres, D., Foster, S.P., Field, L.M., Devonshire, A.L., Williamson, M.S., 1999b. A sodium channel point mutation is associated with resistance to DDT and pyrethroid insecticides in the peach-potato aphid, *Myzus persicae* (Sulzer) (Hemiptera: Aphididae). *Insect Mol. Biol. 8*, 339–346.

Mathew, N., Subramanian, S., Kalyanasundaram, M., 1996. Quantitative structure–activity relationship studies in pyrethroid esters derived from substituted 3-methyl-2-phenoxybutanoic acids against *Culex quinquefasciatus*. *Indian J. Chem. Sect. B Org. Chem. Med. Chem. 35*, 40–44.

Matsuda, K., Oimomi, N., Komai, K., Nishimura, K., 1995. Quantitative structure–activity studies of pyrethroids. 32. Rates of change in membrane-potentials of crayfish and cockroach giant-axons induced by kadethric acid-esters and related-compounds. *Pestici. Biochem. Physiol. 52*, 201–211.

Matsuo, N., Miyamoto, J., 1997. Development of synthetic pyrethroids with emphasis on stereochemical aspects. In: Hedin, P.A., Hollingworth, R.M., Masler, E.P., Miyamoto, J., Thompson, D.G. (Eds.), Phytochemicals for Pest Control. American Chemical Society, Washington, pp. 182–194.

McCaffery, A.R., 1998. Resistance to insecticides in heliothine Lepidoptera: a global view. *Phil. Trans. R. Soc. Lond. B Biol. Sci. 353*, 1735–1750.

Miller, T., 1998. Pyrethroid Insecticides. In: Chemistry & Toxicology of Insecticides. Web course at University of California, Riverside, Entomology 128 (Section 001). >http://www.wcb.ucr.edu/wcb/schools/CNAS/entm/tmiller/1/modules/pages28.html> 23 April 2004.

Miyazaki, M., Ohyama, K., Dunlap, D.Y., Matsumura, F., 1996. Cloning and sequencing of the para-type sodium channel gene from susceptible and kdr-resistant German cockroaches (*Blattella germanica*) and house fly (*Musca domestica*). *Mol. Gen. Genet. 252*, 61–68.

Moores, G.D., Jewess, P.J., Boyes, A.L., Javed, N., Gunning, R.V., 2002. Use of novel substrates to characterise esteratic cleavage of pyrethroids. The BCPC Conference: Pests and Diseases, Volumes 1 and 2. Proceedings of an international conference held at the Brighton Hilton Metropole Hotel, Brighton, UK, 18–21 November 2002. British Crop Protection Council, Farnham, UK, pp. 799–804.

Morin, S., Williamson, M.S., Goodson, S.J., Brown, J.K., Tabashnik, B.E., et al., 2002. Mutations in the *Bemisia tabaci* para sodium channel gene associated with resistance to a pyrethroid plus organophosphate mixture. *Insect Biochem. Mol. Biol. 32*, 1781–1791.

Motomura, H., Narahashi, T., 2001. Interaction of tetramethrin and deltamethrin at the single sodium channel in rat hippocampal neurons. *Neurotoxicology 22*, 329–339.

Mullin, C.J., Scott, J.G., 1992. Biomolecular basis for insecticide resistance classification and comparisons. In: Mullin, C.J., Scott, J.G. (Eds.), Molecular Basis for Insecticide Resistance Diversity Among Insects. American Chemical Society, Washington, D.C., pp. 1–13.

Nakagawa, S., Okajima, N., Kitahaba, T., Nishimura, K., Fujita, T., et al., 1982. Quantitative structure–activity studies of substituted benzyl chrysanthemates. 1. Correlations between symptomatic and neurophysiological activities against American cockroaches. *Pestici. Biochem. Physiol. 17*, 243–258.

Narahashi, T., 2002. Nerve membrane ion channels as the target site of insecticides. *Mini Rev. Med. Chem. 2*, 419–432.

Naumann, K., 1990. Synthetic Pyrethroid Insecticides: Structures and Properties. Springer, Berlin.

Nishimura, K., Kato, S., Holan, G., Ueno, T., Fujita, T., 1994. Quantitative structure–activity studies of pyrethroids. 31. Insecticidal activity of *meta*-substituted benzyl-esters of 1-(*para*-ethoxyphenyl)-2,2-dichlorocyclopropane and 2,2,3,3-tetrafluorocyclobutane-1-carboxylic acids. *Pestici. Biochem. Physiol. 50*, 60–71.

Nishimura, K., Kitahaba, T., Ikemoto, Y., Fujita, T., 1988. Quantitative structure–activity studies of pyrethroids. 14. Physicochemical structural effects of tetramethrin and its related-compounds on knockdown activity against house-flies. *Pestici. Biochem. Physiol. 31*, 155–165.

Nishimura, K., Okimoto, H., 1997. Quantitative structure–activity relationships of DDT-type compounds in a sodium tail-current in crayfish giant axons. *Pestici. Sci. 50*, 104–110.

Nishimura, K., Okimoto, H., 2001. Effects of organosilicon pyrethroid-like insecticides on nerve preparations of American cockroaches and crayfish. *Pest Manage. Sci. 57*, 509–513.

Noppun, V., Saito, T., Miyata, T., 1989. Cuticular penetration of s-fenvalerate in fenvalerate-resistant and susceptible strains of the diamondback moth, *Plutella xylostella* (L). *Pestici. Biochem. Physiol. 33*, 83–87.

Oppenoorth, F.J., 1985. Biochemistry and genetics of insecticide resistance. In: Kerkut, G.A., Gilbert, L.I. (Eds.), Comprehensive Insect Physiology, Biochemistry and Pharmacology. Pergamon, Oxford, UK, pp. 731–774.

Ottea, J.A., Ibrahim, S.A., Younis, A.M., Young, R.J., 2000. Mechanisms of pyrethroid resistance in larvae and adults from a cypermethrin-selected strain of *Heliothis virescens* (F.). *Pestici. Biochem. Physiol. 66*, 20–32.

Pachlatko, J.P., **1998**. Natural Products in Crop Protection. 2nd International Electronic conference on Synthetic Organic Chemistry (ECSOC-2) <http://www.mdpi.org/ecsoc> September 1–30, 1998.

Pap, L., Bajomi, D., Szekely, I., **1996a**. The pyrethroids, an overview. *Internat. Pest Control 38*, 15–19.

Pap, L., Kelemen, M., Toth, A., Szekely, I., Bertok, B., **1996b**. The synthetic pyrethroid isomers II. Biological activity. *J. Environ. Sci. Health B Pestic. Food Contam. Agric. Wastes 31*, 527–543.

Pap, L., Toth, A., **1995**. Development and characteristics of resistance in the susceptible WHO/SRS house fly (*Musca domestica*) strain subjected to selection with beta-cypermethrin. *Pestici. Sci. 45*, 335–349.

Park, Y., Taylor, M.F.J., **1997**. A novel mutation L1029H in sodium channel gene hscp associated with pyrethroid resistance for *Heliothis virescens* (Lepidoptera: Noctuidae). *Insect Biochem. Mol. Biol. 27*, 9–13.

Park, Y., Taylor, M.F.J., Feyereisen, R., **1997**. A valine421 to methionine mutation in IS6 of the hscp voltage-gated sodium channel associated with pyrethroid resistance in *Heliothis virescens* F. *Biochem. Biophys. Res. Commun. 239*, 688–691.

Pauron, D., Barhanin, J., Amichot, M., Pralavorio, M., Berge, J.B., *et al.*, **1989**. Pyrethroid receptor in the insect Na$^+$ channel – alteration of its properties in pyrethroid-resistant flies. *Biochemistry 28*, 1673–1677.

Prabhakaran, S.K., Kamble, S.T., **1996**. Biochemical characterization and purification of esterases from three strains of German cockroach, *Blattella germanica* (Dictyoptera: Blattellidae). *Arch. Insect Biochem. Physiol. 31*, 73–86.

Pruett, J.H., Kammlah, D.M., Guerrero, F.D., **2001**. Variation in general esterase activity within a population of *Haematobia irritans* (Diptera: Muscidae). *J. Econ. Entomol. 94*, 714–718.

Punzo, F., **1993**. Detoxification enzymes and the effects of temperature on the toxicity of pyrethroids to the fall armyworm, *Spodoptera frugiperda* (Lepidoptera, Noctuidae). *Comp. Biochem. Physiol. C Pharmacol. Toxicol. Endocrinol. 105*, 155–158.

Ramana, K.V., Kohli, K.K., **1998**. Gene regulation of cytochrome P450 – an overview. *Indian J. Exp. Biol. 36*, 437–446.

Ranasinghe, C., Headlam, M., Hobbs, A.A., **1997**. Induction of the mRNA for CYP6B2, a pyrethroid inducible cytochrome P450, in *Helicoverpa armigera* (Hubner) by dietary monoterpenes. *Arch. Insect Biochem. Physiol. 34*, 99–109.

Ranson, H., Jensen, B., Vulule, J.M., Wang, X., Hemingway, J., *et al.*, **2000**. Identification of a point mutation in the voltage-gated sodium channel gene of Kenyan *Anopheles gambiae* associated with resistance to DDT and pyrethroids. *Insect Mol. Biol. 9*, 491–497.

Reidy, G.F., Rose, H.A., Visetson, S., Murray, M., **1990**. Increased glutathione-S-transferase activity and glutathione content in an insecticide-resistant strain of *Tribolium castaneum* (Herbst). *Pestici. Biochem. Physiol. 36*, 269–276.

Rendic, S., DiCarlo, F.J., **1997**. Human cytochrome P450 enzymes: a status report summarizing their reactions, substrates, inducers, and inhibitors. *Drug Metabol. Rev. 29*, 413–580.

Roberts, T.R., Hutson, D.H. (Eds.), **1999**. Metabolic Pathways of Agrochemicals, Vol. 2. Royal Society of Chemistry, Cambridge, UK, pp. 579–725.

Rose, R.L., Barbhaiya, L., Roe, R.M., Rock, G.C., Hodgson, E., **1995**. Cytochrome P450–associated insecticide resistance and the development of biochemical diagnostic assays in *Heliothis virescens*. *Pestici. Biochem. Physiol. 51*, 178–191.

Ruigt, G.S.F., **1985**. Pyrethroids. In: Kerkut, G.A., Gilbert, L.I. (Eds.), Comprehensive Insect Physiology, Biochemistry, and Pharmacology. Pergamon Press, Oxford, UK, pp. 183–262.

Ruzo, L.O., Cohen, E., Capua, S., **1988**. Comparative metabolism of the pyrethroids bifenthrin and deltamethrin in the bulb mite *Rhizoglyphus robini*. *J. Agric. Food Chem. 36*, 1040–1043.

Salgado, V.L., Irving, S.N., Miller, T.A., **1983a**. Depolarization of motor-nerve terminals by pyrethroids in susceptible and kdr-resistant house-flies. *Pestici. Biochem. Physiol. 20*, 100–114.

Salgado, V.L., Irving, S.N., Miller, T.A., **1983b**. The importance of nerve-terminal depolarization in pyrethroid poisoning of insects. *Pestici. Biochem. Physiol. 20*, 169–182.

Sawicki, R.M., **1978**. Unusual response of DDT-resistant houseflies to carbinol analogues of DDT. *Nature 275*, 443–444.

Sawicki, R.M., **1985**. The relevance of resistance in the study of the insecticidal activity of pyrethroids. *Abst. Pap. Am. Chem. Soc. 190*, 23–AGO.

Schnitzerling, H.J., Nolan, J., Hughes, S., **1983**. Toxicology and metabolism of some synthetic pyrethroids in larvae of susceptible and resistant strains of the cattle tick *Boophilus microplus* (Can). *Pestici. Sci. 14*, 64–72.

Schuler, M.A., **1996**. The role of cytochrome P450 monooxygenases in plant-insect interactions. *Plant Physiol. 112*, 1411–1419.

Schuler, T.H., Martinez-Torres, D., Thompson, A.J., Denholm, I., Devonshire, A.L., *et al.*, **1998**. Toxicological, electrophysiological, and molecular characterisation of knockdown resistance to pyrethroid insecticides in the diamondback moth, *Plutella xylostella* (L.). *Pestici. Biochem. Physiol. 59*, 169–182.

Scott, J.G., Georghiou, G.P., **1986**. Mechanisms responsible for high-levels of permethrin resistance in the housefly. *Pestici. Sci. 17*, 195–206.

Scott, J.G., Liu, N., Wen, Z.M., **1998**. Insect cytochromes P450: diversity, insecticide resistance and tolerance to plant toxins. *Comp. Biochem. Physiol. C Pharmacol. Toxicol. Endocrinol. 121*, 147–155.

Scott, J.G., Wen, Z.M., **2001**. Cytochromes P450 of insects: the tip of the iceberg. *Pest Manage. Sci. 57*, 958–967.

Shan, G.M., Hammock, B.D., 2001. Development of sensitive esterase assays based on alpha-cyano-containing esters. *Analyt. Biochem. 299*, 54–62.

Shono, T., Unai, T., Casida, J.E., 1978. Metabolism of permethrin isomers in American cockroach adults, house-fly adults, and cabbage-looper larvae. *Pestici. Biochem. Physiol. 9*, 96–106.

Soderlund, D.M., Casida, J.E., 1977. Effects of pyrethroid structure on rates of hydrolysis and oxidation by mouse-liver microsomal-enzymes. *Pestici. Biochem. Physiol. 7*, 391–401.

Soderlund, D.M., Clark, J.M., Sheets, L.P., Mullin, L.S., Piccirillo, V.J., et al., 2002. Mechanisms of pyrethroid neurotoxicity: implications for cumulative risk assessment. *Toxicology 171*, 3–59.

Soderlund, D.M., Hessney, C.W., Jiang, M., 1987. Metabolism of fenvalerate by resistant Colorado potato beetles. *J. Agric. Food Chem. 35*, 100–105.

Soderlund, D.M., Knipple, D.C., 2003. The molecular biology of knockdown resistance to pyrethroid insecticides. *Insect Biochem. Mol. Biol. 33*, 563–577.

Soderlund, D.M., Sanborn, J.R., Lee, P.W., 1983. Metabolism of pyrethrins and pyrethroids in insects. In: Hutson, D.H., Roberts, T.R. (Eds.), Progress in Pesticide Biochemistry and Toxicology. John Wiley, Chichester, UK.

Song, J.H., Narahashi, T., 1996. Modulation of sodium channels of rat cerebellar Purkinje neurons by the pyrethroid tetramethrin. *J. Pharmacol. Exp. Ther. 277*, 445–453.

Straw, N.A., Fielding, N.J., Waters, A., 1996. Phytotoxicity of insecticides used to control aphids on Sitka spruce, *Picea sitchensis* (Bong) Carr. *Crop Protect. 15*, 451–459.

Tan, J.G., Liu, Z.Q., Nomura, Y., Goldin, A.L., Dong, K., 2002. Alternative splicing of an insect sodium channel gene generates pharmacologically distinct sodium channels. *J. Neurosci. 22*, 5300–5309.

Tatebayashi, H., Narahashi, T., 1994. Differential mechanism of action of the pyrethroid tetramethrin on tetrodotoxin-sensitive and tetrodotoxin-resistant sodium channels. *J. Pharmacol. Exp. Ther. 270*, 595–603.

Terriere, L.C., 1984. Induction of detoxication enzymes in insects. *Annu. Rev. Entomol. 29*, 71–88.

Tomita, T., Scott, J.G., 1995. cDNA and deduce protein-sequence of Cyp6d1 – the putative gene for a cytochrome-P450 responsible for pyrethroid resistance in the house fly. *Insect Biochem. Mol. Biol. 25*, 275–283.

Tomlin, C.D.S. (Ed.) 2000. The Pesticide Manual: a World Compendium, 12th edn. British Crop Protection Council, Farnham, UK.

Vais, H., Atkinson, S., Eldursi, N., Devonshire, A.L., Williamson, M.S., et al., 2000. A single amino-acid change makes a rat neuronal sodium channel highly sensitive to pyrethroid insecticides. *FEBS Letts 470*, 135–138.

Vais, H., Atkinson, S., Pluteanu, F., Goodson, S.J., Devonshire, A.L., et al., 2003. Mutations of the para sodium channel of *Drosophila melanogaster* identify putative binding sites for pyrethroids. *Mol. Pharmacol. 64*, 914–922.

Vijverberg, H.P.M., de Weille, J.R., Ruigt, G.S.F., 1986. The effect of pyrethroid structure on the interaction with the sodium channel in the nerve membrane. In: Ford, M.G., Lunt, G.G., Reay, R.C., Usherwood, P.N.R. (Eds.), Neuropharmacology and Pesticide Action. Ellis Horwood, Chichester, UK.

Vontas, J.G., Small, G.J., Hemingway, J., 2001. Glutathione S-transferases as antioxidant defence agents confer pyrethroid resistance in *Nilaparvata lugens*. *Biochem. J. 357*, 65–72.

Whalon, M., Mota-Sanchez, D., Hollingworth, R.M., Bills, P., Duynslager, L., 2003. The Database of Arthropods Resistant to Pesticides [database on the Internet] Michigan State University <http://www.pesticideresistance.org/ DB/index.php> Accessed 23 April 2004.

Wheelock, G.D., Scott, J.G., 1989. Simultaneous purification of a cytochrome P450 and cytochrome 65 housefly, *Musca domestica* L. *Insect Biochem. 19*, 481–488.

Wheelock, G.D., Scott, J.G., 1992. The role of cytochrome-P450$_{ipr}$ in deltamethrin metabolism by pyrethroid-resistant and susceptible strains of house flies. *Pestici. Biochem. Physiol. 43*, 67–77.

Williamson, M.S., Martinez-Torres, D., Hick, C.A., Devonshire, A.L., 1996. Identification of mutations in the housefly *para*-type sodium channel gene associated with knockdown resistance (kdr) to pyrethroid insecticides. *Mol. Gen. Genet. 252*, 51–60.

Wu, D.X., Scharf, M.E., Neal, J.J., Suiter, D.R., Bennett, G.W., 1998. Mechanisms of fenvalerate resistance in the German cockroach, *Blattella germanica* (L.). *Pestici. Biochem. Physiol. 61*, 53–62.

Yamamoto, I., Casida, J.E., 1968. Syntheses of 14C-labeled pyrethrin I, allethrin, phthalthrin, and dimethrin on a submillimole scale. *Agric. Biol. Chem. 32*, 1382–1391.

Yamamoto, I., Kimmel, E.C., Casida, J.E., 1969. Oxidative metabolism of pyrethroids in houseflies. *J. Agric. Food Chem. 17*, 1227–1236.

Yu, S.J., Robinson, F.A., Nation, J.L., 1984. Detoxication capacity in the honey bee, *Apis mellifera* L. *Pestici. Biochem. Physiol. 22*, 360–368.

Zhang, M.L., Scott, J.G., 1996. Cytochrome b(5) is essential for cytochrome P450 6D1-mediated cypermethrin resistance in LPR house flies. *Pestici. Biochem. Physiol. 55*, 150–156.

A1 Addendum: Pyrethroid Insecticides and Resistance Mechanisms

J Vontas, University of Crete, Greece
H Ranson, Liverpool School of Tropical Medicine, Liverpool, UK
M S Williamson, Rothamsted Research, Harpenden, UK

© 2010 Elsevier B.V. All Rights Reserved

A1.1. Introduction

Since the previous review in this series (Khambay and Jewess, 2005), research emphasis has shifted from discovery of novel pyrethroids to exploration of new commercial applications. The latter includes development of improved formulations, such as slow-release polymers and resistance-breaking inhibitors, particularly in the relatively small public health market. However, research efforts to understand the nature of resistance have intensified due to increased reports of pyrethroid resistance in insect pests, and also aided by advances in insect genomics and a clearer elucidation of pyrethroid binding at the sodium channel target.

A1.2. Market – Applications in Agricultural and Nonagricultural Fields

A wide range of synthetic pyrethroids are used in crop protection and for the control of veterinary and medically important pests species, with annual sales figures in excess of $1.5 billion (approximately 16% of the world insecticide market in 2007). Their main use is in agriculture, particularly for controlling bollworms (members of the *Helicoverpa/Heliothis* complex), aphids, whiteflies and mites, with deltamethrin, lambda cyhalothrin, permethrin and cypermethrin accounting for the majority of sales. The lead crop for the pyrethroid market remains cotton (45%), followed by rice (28%) and vegetables (14%). Pakistan (34%), Indonesia (28%), Brazil (15%) and China (15%) are the major consumer countries (Ralf Nauen, Bayer Crop Science, personal communication). Pyrethroids are also used extensively in veterinary medicine for agricultural and domestic purposes, mainly to control insect and tick parasites (see recent review by Anadon *et al.*, 2009). Because of its low mammalian toxicity, pyrethroid is the preferred insecticide class used for control of medically important arthropods such as mosquitoes, lice, and cockroaches. Although the public health market constitutes only a small fraction of total insecticide usage (<1% of that spent on agriculture), the role of pyrethoids is nevertheless vital, since they are the only class of insecticides approved for insecticide-treated bednets (ITNs) and the preferred choice for indoor residual spraying (IRS).

A1.3. Research for Novel Commercial Compounds – Products

A novel group of methyl ester pyrethroid acids containing an aromatic ring on the acid moiety with high insecticidal activity against a diverse range of insects has been reported recently (Silverio *et al.*, 2009); but these products are still at an early stage of development. There has been investment in the development of new formulations to overcome insecticide resistance in crop pests (Bingham *et al.*, 2007) and public health products, such as the slow release plastic strip formulation of metofluthrin (Sumitomo Chemical Co.), a highly volatile pyrethroid which, in contrast to coil/liquid vaporizers, evaporates without heating (Kawada *et al.*, 2005). An establishment of a public–private partnership, the Innovative Vector Control Consortium (IVCC, www.ivcc.com), has also provided a welcome stimulus for investment in new pyrethroid chemistry for the public health market with a particular focus on products targeting mosquitoes.

A1.4. Resistance

The increasing number of reports of pyrethroid resistance in the field has fuelled further studies to determine the molecular basis of resistance, and recent progress in this area is summarized in **Table A1**. Several cytochrome P450s that are over-expressed in insecticide-resistant strains and are able to metabolize pyrethroids have been identified. For example, the *Anopheles gambiae* enzyme, CYP6P3, is upregulated in pyrethroid-resistant populations from Ghana and Benin and it metabolizes both α-cyano (type 2) and non-α-cyano (type 1) pyrethroids (Muller *et al.*, 2008). Another CYP6 enzyme from the mosquito *An. minimus* has also been demonstrated to have activity against deltamethrin, supporting its likely role in resistance (Boonsuepsakul *et al.*, 2008). A number of P450s from *Helicoverpa armigera* have been implicated in pyrethroid resistance by metabolism studies (**Table A1**), although genetic mapping has suggested that the *in vitro* ability of certain P450s (such as the CYP6B7 and CYP6B6) to metabolize pyrethroids may not be sufficient evidence to prove their role in resistance (Grubor and Heckel, 2007). The application of RNA silencing techniques has demonstrated the importance of CYP6BG1 from the diamondback moth, *Plutella xylostella*, in conferring permethrin resistance (Bautista *et al.*, 2009). Recently, the gene duplication events in *An. funestus* as a possible contributing mechanism for the increased P450 activity in pyrethroid resistance populations were identified in this enzyme family in insects (Wondji *et al.*, 2009).

The number of insect and mite species documented as containing mutations in the pyrethroid target site, the voltage-gated sodium channel of nerve membranes, continues to grow (see **Table A1**). Many of these are clustered within localized "hotspots," including the domain IIS4-S5 linker, IIS5 and IIS6 helices and the corresponding regions of domains I and III (reviewed in Davies *et al.*, 2007), and have been instrumental to identify a clearly defined binding site for these compounds at the sodium channel pore (see Section 1.5). The functionality of many of these mutations has now been confirmed by analysis of their properties in *Xenopus* oocyte-expressed insect channels, and some of the more commonly identified mutations at residues, such as M918, T929, L1014 and F1538, confer insensitivity across a range of pyrethroid structures. The resistance mechanism is corroborated when the same, or similar, mutations are found to occur in a new species; for example, the recent identification of F1538I in the spider mite *Tetranychus urticae* (Tsagkarakou *et al.*, 2009). Other mutation sites

have been identified that are clearly linked to the resistance phenotype but have not been functionally characterized by *in vitro assay*; for example, V1016I in *Aedes aegypti* (Saavedra-Rodriguez *et al.*, 2007). A note of caution, however, is that, not all reported mutations have been so clearly linked to pyrethroid resistance and some may represent natural polymorphisms within the sodium channel sequence. Also, secondary processing events known to generate developmental and/or tissue-specific diversity of mature sodium channel isoforms may also affect the level of resistance imparted by a particular mutation; for example, the expression of alternatively spliced sodium channel transcripts in resistant diamondback moth populations produced distinct channels with different sensitivities to pyrethroids (Sonoda *et al.*, 2008). Finally, in populations with multiple resistance mechanisms, the relative contribution of each mutation to the resistance phenotype is often poorly defined. This has been explored in *Culex* mosquitoes and revealed a complex interplay between metabolic and target site resistance (Hardstone *et al.*, 2009).

A1.5. Pyrethroid–Sodium Channel Interactions

One area in which there has been significant progress over the past 5 years is in the development of a detailed molecular model showing a putative binding site for pyrethroids at the insect sodium channel. This was made possible by publication of a series of high-resolution crystal structures for closely related potassium channels, including the eukaryotic voltage-gated K^+ channel Kv1.2 (Long *et al.*, 2005), that provided a structural template on which to model the insect (housefly) sodium channel sequence (O'Reilly *et al.*, 2006). Close examination of the pore region of this model revealed a long, narrow cavity between the IIS4-S5 linker, IIS5 helix and IIIS6 helix that could accommodate a range of pyrethroid structures as well as DDT (see **Figure A1**). Several known features of pyrethroid mode of action and resistance are supported by this model (O'Reilly *et al.*, 2006), including (1) a hydrophobic cavity accessible to lipid-soluble pyrethroids, (2) the observation that the binding site is formed by movement of the IIS4-S5 linker towards the other two helices during channel activation, which is consistent with electrophysiological evidence that pyrethroids bind preferentially to the open state of the channel, (3) that this high-affinity binding would stabilize the open state to slow channel inactivation, consistent with the pyrethroid-induced tail currents seen following membrane repolarization, (4) the proposed binding site includes many known mutation sites in the

Table A1 Summary of main reports of pyrethroid resistance and underlying molecular mechanisms in major pests

Species	Pyrethroid[a]	RR[b]	Genes implicated in metabolic resistance	Target site mutations[c]	Reference[d]
Agricultural pests					
Bemisia tabaci	Cypermethrin	>1000		M918V, **L925I, T929V**	Davies et al. (2007)
Cydia pomonella	Deltamethrin	80		**L1014F**	Davies et al. (2007)
Frankliniella occidentalis	Deltamethrin	1300		T929I/V/C, **L1014F**	Davies et al. (2007)
Heliothis virescens	Deltamethrin	>100		**V410M**, L1014H, D1549V, E1553G	Davies et al. (2007)
Helicoverpa zea	Permethrin	>100	**CYP321A1, CYP6B8**		Davies et al. (2007)
Helicoverpa armigera	Fenvalerate, Cypermethrin	>1000	CYP6B7, **CYP9A12, CYP9A14**, CYP4G8	D1549V, E1553G	Grubor and Heckel (2007), Yang et al. (2008), Davies et al. (2007)
Leptinotarsa decemlineata	Permethrin	>200		**L1014F**	Davies et al. (2007)
Myzus persicae	Deltamethrin	540		L1014F, F979S, **M918T**	Davies et al. (2007), Soderlund (2008)
Plutella xylostella	Permethrin	>500	**CYP6BG1**	T929I, **L1014F**	Soderlund (2008), Bautista et al. (2009)
Tetranychus urticae	Bifenthrin	>1000		F1538I, A1215D	Tsagkarakou et al. (2009)
Nonagricultural pests (public health, veterinary medicine)					
Aedes aegypti	Deltamethrin	>100		I1011M, V1016G	Saavedra-Rodriguez et al. (2007)
Anopheles funestus	Permethrin	>200	CYP6P9, CYP6P4		Wondji et al. (2009)
Anopheles gambiae	Permethrin	>50	**CYP6P3**, CYP6M2	L1014F, L1014S	Muller et al. (2008)
Anopheles minumus	Deltamethrin	>50	**CYP6AA3**		Boonsuepsakul et al. (2008)
Blattella germanica	Permethrin	>1000		L1014F, **E435K,C785R**, D59G, P1899L	Davies et al. (2007), Soderlund (2008)
Boophilus microplus	Permethrin	>1000		**F1538I**	Khambay and Jewess (2005)
Cimex lectularius	Deltamethrin	264		V419I, **L925I**	Yoon et al. (2008)
Ctenocephalides felis	Permethrin	>50		**T929V, L1014F**	Davies et al. (2007)
Haematobia irritans	Cypermethrin	>500		**M918T, L1014F**	Khambay and Jewess (2005)
Musca domestica	Permethrin	5000	**CYP6D1**	L1014F, L1014H, **M918T**	Soderlund (2008)
Pediculus capitis	Phenothrin	>100		M827I, **T929I, L932F**	Khambay and Jewess (2005)
Sarcoptes scabiei	Permethrin	>20		G1535D	Pasay et al. (2008)
Varroa destructoor	Fluvalinate	500		F1528L, I1752V, M1823I, **L1770P**	Soderlund (2008)

[a]Representative and/or first report of pyrethroid resistance.
[b]Resistance ratio (Max reported in the species).
[c]Bold indicate functionally validated gene/mutation responsible for resistance.
[d]Review articles have been cited where available.

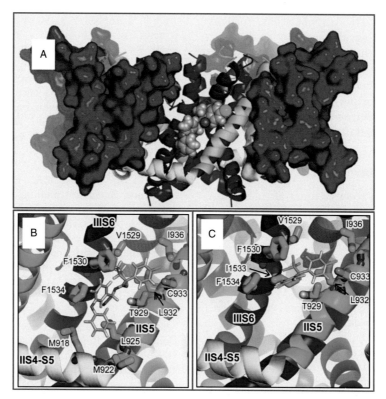

Figure A1   Structural model of the housefly sodium channel showing the putative binding site for pyrethroids and DDT (Reproduced with permission, from O'Reilly, A.O., Khambay, B.P.S., Williamson, M.S., Field, L.M., Wallace, B.A., Davies, T.G.E., 2006. Modelling insecticide-binding sites in the voltage-gated sodium channel. *Biochem. J. 396*, 255–263. © The Biochemical Society). (a) Homology model with the voltage sensor domains coloured red and central pore helices in blue, cyan and yellow. The pyrethroid fenvalerate (grey space fill) is shown docked at the binding site. (b, c) Close-up docking predictions for fenvalerate (b) and DDT (c) within the binding site, highlighting the residues involved in ligand–target interactions.

IIS4-S5 linker, IIS5 helix and IIIS6 helix that cause reduced sensitivity to pyrethroids and DDT in resistant insect and mite strains, and (5) the binding site will also accommodate DDT, albeit with a reduced number of contacts, which in turn is consistent with the lower potency of this compound compared with pyrethroids.

The publication of this model, and further dockings of a range of pyrethroid and DDT structures, identifies a number of additional residues, above and beyond those already highlighted from resistance studies, that are also predicted to form important ligand–target interactions within the binding site. These predictions can be investigated experimentally and two recent mutagenesis/expression studies have confirmed an important role for several new residues in the IIS4-S5 linker/IIS5 helix (Usherwood et al., 2007) and IIIS6 helix (Du et al., 2009), thereby reinforcing the validity of the original model. This research provides an opportunity to re-examine species-specific differences in sodium channel sequences between insects and mammals that are responsible for the exquisite insect (and

mite) selectivity of these compounds. By combining increased processing power of modern computers with improved programs for in silico ligand/protein predictions, it may now be possible to re-model structural features that contribute to insect selectivity at the species level (e.g., in comparing crop pests with beneficial insects), and to design novel "resistance-defeating" pyrethroid derivatives that bind with higher affinity within the binding site of strains that carry preexisting resistance mutations.

A1.6. Conclusions and Future Prospects

Although research activity in pyrethroid chemistry has largely ceased, with the industry focus shifting more towards less neurotoxic modes of action, the pyrethroids, nevertheless, remain major player in the world insecticide market and are particularly important for the more restricted public health market. Significant progress has been achieved in our understanding of pyrethroid sodium channel pharmacology and molecular characterization of pyrethroid resistance mechanisms. Translation of this research into practical applications, such as rapid and sensitive molecular diagnostics for resistance monitoring and the development of synergistic inhibitors for existing insecticide formulations will hopefully improve the sustainability of pyrethroid applications by supporting the implementation of effective resistance management strategies that combat the onset and spread of resistance.

Acknowledgments

Rothamsted Research receives grant-aided support from the Biotechnology and Biological Research Council of the United Kingdom.

References

Anadon, A., Martinez-Larranaga, M.R., Martinez, M.A., 2009. Use and abuse of pyrethrins and synthetic pyrethroids in veterinary medicine. *Vet. J. 182*, 7–20.

Bautista, M.A.M., Miyata, T., Miura, K., Tanaka, T., 2009. RNA interference-mediated knockdown of a cytochrome P450, CYP6BG1, from the diamondback moth, *Plutella xylostella*, reduces larval resistance to permethrin. *Insect. Biochem. Mol. Biol. 39*, 38–46.

Bingham, G., Gunning, R.V., Gorman, K., Field, L.M., Moores, G.D., 2007. Temporal synergism by micro-encapsulation of piperonyl butoxide and alpha-cypermethrin overcomes insecticide resistance in crop pests. *Pest. Manag. Sci. 63*, 276–281.

Boonsuepsakul, S., Luepromchai, E., Rongnoparut, P., 2008. Characterization of *Anopheles minimus* CYP6AA3 expressed in a recombinant baculovirus system. *Arch. Insect Biochem. Physiol. 69*, 13–21.

Davies, T.G.E., Field, L.M., Usherwood, P.N.R., Williamson, M.S., 2007. DDT, pyrethrins, pyrethroids and insect sodium channels. *IUBMB Life 59*, 151–162.

Du, Y., Nomura, Y., Luo, N., Liu, Z., Lee, J.-E., Khambay, B., Dong, K., 2009. Molecular determinants on the insect sodium channel for the specific action of type II pyrethroid insecticides. *Toxicol. Appl. Pharmacol. 234*, 266–272.

Grubor, V.D., Heckel, D.G., 2007. Evaluation of the role of CYP6B cytochrome P450s in pyrethroid resistant Australian *Helicoverpa armigera*. *Insect. Mol. Biol. 16*, 15–23.

Hardstone, M.C., Leichter, C.A., Scott, J.G., 2009. Multiplicative interaction between the two major mechanisms of permethrin resistance, kdr and cytochrome P450-monooxygenase detoxification, in mosquitoes. *J. Evol. Biol. 22*, 416–423.

Kawada, H., Yen, N.T., Hoa, N.T., Sang, T.M., Dan, N.V., Takagi, M., 2005. Field evaluation of spatial repellency of metofluthrin impregnated plastic strips against mosquitoes in Hai Phong city, Vietnam. *Am J. Trop. Med. Hyg. 73*, 350–353.

Khambay, B., Jewess, P., 2005. Pyrethroids. In: Gilbert, L.I., Iatrou, K., Gill, S.S. (Eds.), Comprehensive Molecular Insect Science, Vol. 6. Elsevier, Oxford, UK, pp. 1–29.

Long, S.B., Campbell, E.B., Mackinnon, R., 2005. Crystal structure of a mammalian voltage-dependent Shaker family K+ channel. *Science 309*, 897–903.

Muller, P., Warr, E., Stevenson, B.J., Pignatelli, P.M., Morgan, J.C., Steven, A., Yawson, A.E., Mitchell, S.N., Ranson, H., Hemingway, J., Paine, M.J.I., Donnelly, M.J., 2008. Field-caught permethrin-resistant *Anopheles gambiae* overexpress CYP6P3, a P450 that metabolises pyrethroids. *Plos Genetics* 4, Genet 4(11): e1000286. doi:10.1371/journal.pgen.1000286.

O'Reilly, A.O., Khambay, B.P.S., Williamson, M.S., Field, L.M., Wallace, B.A., Davies, T.G.E., 2006. Modelling insecticide-binding sites in the voltage-gated sodium channel. *Biochem. J. 396*, 255–263.

Pasay, C., Arlian, L., Morgan, M., Vyszenski-Moher, D., Rose, A., Holt, D., Walton, S., McCarthy, J., 2008. High-resolution melt analysis for the detection of a mutation associated with permethrin resistance in a population of scabies mites. *Med. Vet. Entomol. 22*, 82–88.

Saavedra-Rodriguez, K., Urdaneta-Marquez, L., Rajatileka, S., Moulton, M., Flores, A.E., Fernandez-Salas, I., Bisset, J., Rodriguez, M., McCall, P.J., Donnelly, M.J., Ranson, H., Hemingway, J., Black, W.C., 2007. A mutation in the voltage-gated sodium channel gene associated with pyrethroid resistance in Latin American *Aedes aegypti*. *Insect. Mol. Biol. 16*, 785–798.

Silverio, F.O., Alvarenga, E.S.D., Moreno, S.C., Picanco, M.C., 2009. Synthesis and insecticidal activity of new pyrethroids. *Pest. Manag. Sci. 65*, 900–905.

Soderlund, D.M., 2008. Pyrethroids, knockdown resistance and sodium channels. *Pest. Manag. Sci. 64*, 610–616.

Sonoda, S., Igaki, C., Tsumuki, H., 2008. Alternatively spliced sodium channel transcripts expressed in field strains of the diamondback moth. *Insect. Biochem. Mol. Biol. 38*, 883–890.

Tsagkarakou, A., Van Leeuwen, T., Khajehali, J., Ilias, A., Grispou, M., Williamson, M.S., Tirry, L., Vontas, J., 2009. Identification of pyrethroid resistance associated mutations in the para sodium channel of the two-spotted spider mite *Tetranychus urticae* (Acari: Tetranychidae). *Insect. Mol. Biol. 18*, 583–593.

Wondji, C.S., Irving, H., Morgan, J., Lobo, N.F., Collins, F.H., Hunt, R.H., Coetzee, M., Hemingway, J., Ranson, H., 2009. Two duplicated P450 genes are associated with pyrethroid resistance in *Anopheles funestus*, a major malaria vector. *Genome Res. 19*, 452–459.

Usherwood, P.N.R., Davies, T.G.E., Mellor, I.R., O'Reilly, A.O., Peng, F., Vais, H., Khambay, B.P.S., Field, L.M., Williamson, M.S., 2007. Mutations in DIIS5 and the DIIS4-S5 linker of *Drosophila melanogaster* sodium channel define binding domains for pyrethroids and DDT. *FEBS Lett. 581*, 5485–5492.

Yoon, K.S., Kwon, D.H., Strycharz, J.P., Hollingsworth, C.S., Lee, S.H., Clark, A.J.M., 2008. Biochemical and molecular analysis of deltamethrin resistance in the common bed bug (Hemiptera: Cimicidae). *J. Med. Entomol. 45*, 1092–1101.

Yang, Y., Yue, L., Chen, S., Wu, Y., 2008. Functional expression of *Helicoverpa armigera* CYP9A12 and CYP9A14 in *Saccharomyces cerevisiae*. *Pest. Biochem. Physiol. 92*, 101–105.

2 Indoxacarb and the Sodium Channel Blocker Insecticides: Chemistry, Physiology, and Biology in Insects

K D Wing, J T Andaloro, and S F McCann,
E. I. Du Pont de Nemours and Co., StineHaskell
Research Laboratories, Newark, DE, USA
V L Salgado, Bayer CropScience AG, Monheim am
Rhein, Germany

© 2010, 2005 Elsevier B.V. All Rights Reserved

2.1. Introduction

There is an increasing need to discover novel chemical insecticides which act on unique biochemical target sites. This is necessary for insecticide resistance management, to maintain agricultural food and fiber production at reasonable economic levels, and also to retain the effectiveness of older and new classes of chemistry as useful tools.

The modes of action of the major insecticide classes are reviewed in Ishaaya (2001) and in this Encyclopedia. These insecticidal compounds are required not only for economic pest control, but also as critical probes to elucidate their respective biochemical target sites, allowing insights into their fundamental biological function.

While there is an increasing need for new agrochemicals, regulatory requirements for the registration

of new insecticides continue to escalate, and society's requirements for safety to nontarget organisms and the environment and compatibility with ongoing agricultural practice, pose increasing challenges.

The voltage-gated sodium (Na$^+$) channel is a primary insecticide target for the synthetic and natural pyrethroids, DDT, *N*-alkylamides, and a host of natural product and peptide toxins (Zlotkin, 2001). At least ten independent target sites exist for these compounds, based on functional studies, and the compounds can exert allosteric effects on one another via these independent sites. In addition, it has been definitively shown that specific genetic mutations in the Na$^+$ channel in both laboratory and field insects can confer resistance to pyrethroids and DDT (see **Chapter 1**). The discovery of additional new Na$^+$ channel toxins acting at novel sites thus creates the possibility of useful invertebrate control agents as well as the development of additional tools for understanding ion channel function.

The identification of the Na$^+$ channel blocker insecticides (SCBIs) represents such a discovery. These compounds all act at a unique site in insect voltage-gated Na$^+$ channels, which may correspond to the local anesthetic/anticonvulsant site. Our knowledge of the mode of action of indoxacarb derives largely from studies on the pyrazoline (also known as dihydropyrazole) forerunners of this family, carried out in the 1980s at the Rohm and Haas Company (Salgado, 1990, 1992). Pyrazolines were found to paralyze insects by blocking action potential initiation in nerve cells. The mechanism of this block was due to a voltage-dependent blocking action on voltage-gated Na$^+$ channels. This mechanism is also observed with many therapeutically useful local anesthetic, antiarrhythmic, and anticonvulsant drugs. The bioactivated form of indoxacarb, N-decarbomethoxylated DPX-MP062 (DCMP), also works in this manner. Indoxacarb's selective toxicity towards pest insects is due to its rapid bioactivation to the active metabolite DCMP, while higher animals primarily degrade indoxacarb to inactive metabolites via alternate routes.

This article will trace the chemical evolution of these compounds and their physiological activity in invertebrates, which eventually led to the commercial introduction of a member of this class, indoxacarb (Harder et al., 1996; Wing et al., 2000). The unique metabolic, insecticidal, and pest management control properties of indoxacarb will also be discussed.

2.2. Chemistry of the Na$^+$ Channel Blockers

2.2.1. Chemical Evolution and Structure–Activity of the Na$^+$ Channel Blocker Insecticides

The first insecticidal pyrazoline Na$^+$ channel blockers were reported in patents from Philips-Duphar in

1973 and are represented by PH 60–41 (Mulder and Wellinga, 1973) (Figure 1). These compounds were reported to have high levels of efficacy against coleopteran and lepidopteran pests. In 1985, Rohm and Haas reported pyrazolines such as RH-3421 with ester substituents on the pyrazoline 4-position (Jacobson, 1985). These compounds were later reported to have high insecticidal efficacy, low mammalian toxicity, and a rapid rate of dissipation in the environment (Jacobson, 1989). Subsequent work by DuPont on the SCBIs resulted in the discovery of several classes of related structures, all with similarly high levels of insecticidal activity (Figure 2). It was found that transposition of the N1 and C3 atoms in the pyrazoline core gave active compounds (compound B). Conformationally constrained pyrazolines, resulting from bridging the pyrazoline C4 postion with the C3 aryl substituent (forming indazoles and oxaindazoles) (compound C in Figure 2), were also highly active. Semicarbazones D are also insecticidal; these compounds are structurally similar to pyrazolines, but with the C5 carbon removed. Expanding the pyrazoline ring by one carbon gave highly active pyridazine compounds (compound E). Substituting oxygen or nitrogen for the C5 pyridazine atom gives oxadiazines and triazines, also with high insecticidal activity (compound F). Indoxacarb is, in fact, a representative of the oxadiazine subclass of SCBIs.

The insecticidal structure–activity relationships for the oxadiazines are summarized in Figure 3. For substituent R1, 4- or 5-Cl, Br, OCH$_2$CF$_3$, and CF$_3$ groups gave compounds with the highest activity. In the R2 position, 4-halo-phenyl and CO$_2$Me were the most active substituents. For groups at R3, 4-OCF$_3$, 4-CF$_3$, and 4-Br were best for activity. For the substituents on nitrogen (R4), CO$_2$CH$_3$ and COCH$_3$ were the most active, followed by H, Me, and Et.

2.2.2. Chemistry and Properties of Indoxacarb

Indoxacarb is synthesized as described in McCann et al. (2001, 2002) and Shapiro et al. (2002). Its

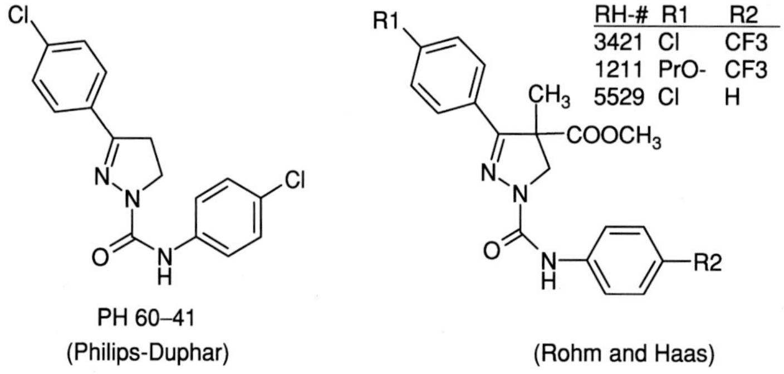

Figure 1 Structures of original pyrazoline Na$^+$ channel blockers.

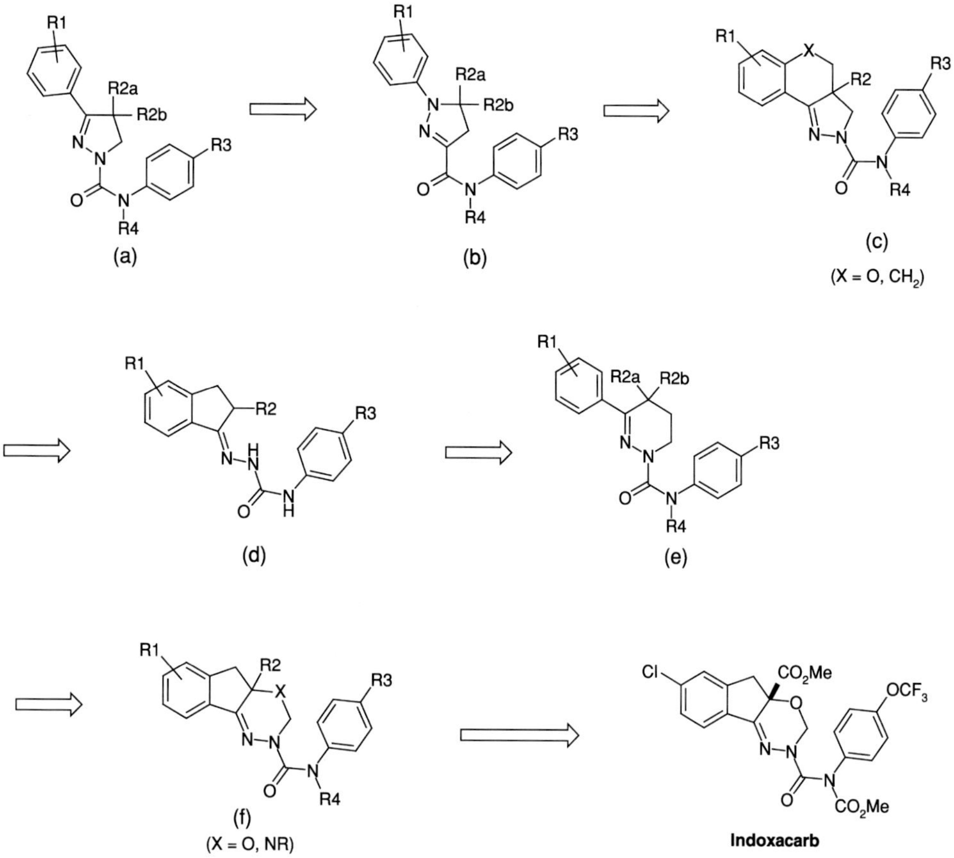

Figure 2 Structures of pyrazoline-like Na⁺ channel blockers.

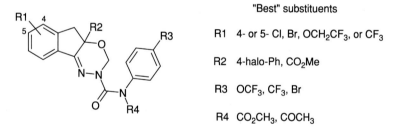

"Best" substituents

R1 4- or 5- Cl, Br, OCH₂CF₃, or CF₃

R2 4-halo-Ph, CO₂Me

R3 OCF₃, CF₃, Br

R4 CO₂CH₃, COCH₃

Figure 3 Structure–activity relationships for oxadiazine insecticides vs. *Spodoptera frugiperda*.

physical, toxicological, and environmental properties are shown in **Table 1**; the profile indicates a safe, environmentally benign compound with relatively low water solubility.

2.3. Metabolism and Bioavailability of Indoxacarb

2.3.1. Bioactivation of Indoxacarb

Indoxacarb requires metabolic activation by insects before it acts as a strong Na⁺ channel blocker (**Figure 4**). An esterase or amidase-type of enzyme(s) cleaves the carbomethoxy group from the urea linkage, liberating the free urea DCMP, which then

acts as the voltage-dependent Na⁺ channel blocker. This was first demonstrated for the racemic compound DPX-JW062 (50 : 50 mixture of active *S* and inactive *R* enantiomers) being converted to the corresponding *N*-decarbomethoxylated DPX-JW062 (DCJW) (Wing *et al.*, 1998). The enzyme involved has been only partially characterized. It is clear that the enzyme is a hydrolase of some type, may be serine dependent. In Lepidoptera the enzyme is found in high concentrations in midgut cells (but not the gut contents) and in fat body, and also found in high concentrations in several subcellular fractions (nuclear, mitochondrial, microsomal, and cytosolic). The enzyme can be inhibited by several

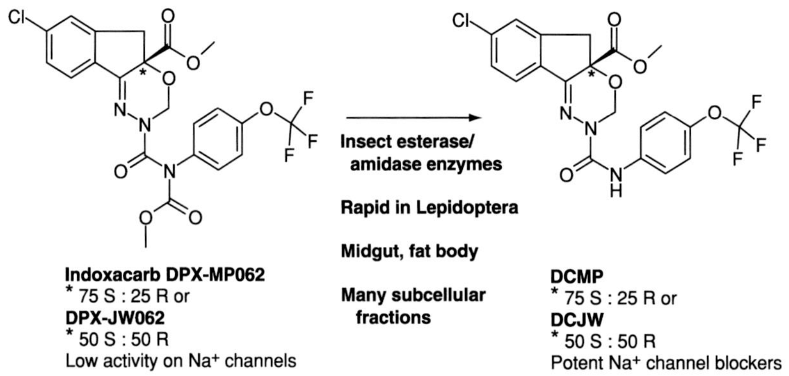

Figure 4 Bioactivation of oxadiazines to *N*-decarbomethoxylated Na$^+$ channel blocker metabolites in insects.

Table 1 Properties of indoxacarb (DPX-MP062)

Physical state (99% active enantiomer DPX-KN128)	Solid, powder
Melting point	88.1 °C
Solubility (for DPX-MP062, 75% DPX-KN128 + 25% R-enantiomer)	Water – 0.20 ppm *n*-Heptane – 1.72 mg ml^{-1} 1-Octanol – 14.5 mg ml^{-1} Methanol – 103 mg ml^{-1} Xylene – 117 mg ml^{-1} Acetonitrile – 139 mg ml^{-1} Ethyl acetate – 160 mg ml^{-1} Dichloromethane – >250 g kg^{-1} Acetone – >250 g kg^{-1} Dimethylformamide – >250 g kg^{-1}
Partition coefficient in octanol/water (KPX-KN128)	log K_{ow} = 4.65
Vapor pressure (DPX-KN128)	9.8 × 10^{-9} Pa at 20 °C 2.5 × 10^{-8} Pa at 25 °C
Acute toxicity of indoxacarb (DPX-MP062)	Oral LC$_{50}$: 1730 mg kg^{-1} (male rats); 268 mg kg^{-1} (female rats) Dermal LD$_{50}$: greater than 5000 mg kg^{-1} in rats Inhalation LC$_{50}$, 4 h: greater than 5.5 mg l^{-1} Dermal irritation: nonirritant Eye irritation: moderate eye irritant Ames test: negative
Environmental fate characteristics	"No major issues in the areas of soil persistence, mobility, and fish bioaccumulation for indoxacarb."

Data from EPA, **2000**. Fact Sheet on Indoxacarb. Issued Oct. 30, 2000, Environmental Protection Agency, Washington, DC.

organophosphate esterase inhibitors but not the cytochrome p450 monooxygenase inhibitors piperonyl butoxide or 1-phenyl imidazole, nor the glutathione transferase inhibitor *N*-ethyl maleimide. Formal structural proof of the indoxacarb active metabolite was obtained by ^{14}C cochromatography and reversed phase high-performance liquid chromatography (HPLC)/electrospray mass spectrometry with authentic standards.

The bioactivation reaction is widespread throughout several orders of insects. Wing *et al.* (1998) first demonstrated that Lepidoptera representing *Trichoplusia*, *Spodoptera*, *Helicoverpa/Heliothis*, and *Manduca* species perform this reaction after oral administration of [^{14}C]DPX-JW062 within 3 h, by which time all larvae were observed to halt feeding, and to suffer a lack of righting, mild convulsions, and a distinct and irreversible pseudoparalysis. A more comprehensive study including Lepidoptera, Homoptera (*Nephotettix*, *Nilaparvata*, *Peregrinus*, *Myzus* species), and Hemiptera (*Lygus* species) indicate that onset of toxic symptoms in all of these insects is well correlated with formation of [^{14}C]DCJW from [^{14}C]DPX-JW062 or unlabeled indoxacarb to DCMP after oral administration (Wing *et al.*, 2000). However, it was noted that the bioactivation rate as well as speed of onset of symptoms was noticeably more rapid with Lepidoptera, and that the Lepidoptera quantitatively converted far more of the parent oxadiazine into the active metabolite than the other orders examined. In addition, there is no apparent stereoselectivity of activation for either the *S* (active Na$^+$ channel blocker) or *R* enantiomers of parent oxadiazine – both enantiomers appeared to be cleaved rapidly to the corresponding free ureas. However, it appears that all of the insecticidal activity resides in the *S*-enantiomer of DCJW or indoxacarb (Wing *et al.*, 1998) as has been observed for dihydropyrazoles (Hasan *et al.*, 1996). Bioactivation and subsequent cessation of feeding and pseudoparalysis in lepidopteran larvae was slower and required higher doses after topical, compared to oral administration.

While only these free ureas demonstrated strong Na$^+$ channel blocking activity, it was frequently observed that both indoxacarb and its activated metabolite were similarly potent in intoxicating Lepidoptera (Wing *et al.*, 1998; Tsurubuchi *et al.*, 2001). This again argues for efficient bioactivation in this order of insects.

Indoxacarb bioactivation was also studied in the hemipteran cotton pest, the tarnished plantbug (*Lygus lineolaris*), compared to its hemipteran predator, big-eyed bug (*Geochoris punctipes*) (Tillman *et al.*, 2001). These studies again demonstrate that the onset of neurotoxic symptoms in both bugs was well correlated with the appearance of the activated metabolite DCMP (albeit at slower rates than observed in Lepidoptera). The importance of oral uptake as a major route of bioactivation/intoxication, compared to tarsal absorption of residues or through dermal uptake, was clearly evident. Interestingly, laboratory studies showed both pest and predator to be similarly sensitive to indoxacarb, although *Geochoris* could tolerate higher levels of indoxacarb before showing signs of intoxication. However, in treated cotton fields, *Lygus* has been found to be more sensitive to indoxacarb treatment while *Geochoris* has been relatively refractory; further studies showed that this was largely due to infield behavior and the preference of *Geochoris* to feed on pest insects and their eggs as opposed to foliage, which significantly slowed oral indoxacarb uptake (Tillman *et al.*, 2001).

In addition, all current evidence points to a SCBI structural requirement for an underivatized urea linkage to be present to exert strong, quasi-irreversible insect Na$^+$ channel blocking effects, as opposed to derivatized ureas.

2.3.2. Catabolism of Indoxacarb and Other Na$^+$ Channel Blocker Insecticides

Relatively little is known about the metabolic breakdown of indoxacarb or other SCBIs in insects. On balance, it is clear that bioactivation is occurring much more rapidly than degradation after field application, when satisfactory insect control is observed. However, like any other novel insecticide, resistance management will be a key to maintaining the agricultural utility of this insecticide class.

Indeed, several field populations of North American oblique-banded leafroller (*Choristoneura rosaceana*) have shown a degree of insensitivity to indoxacarb, which is unique for Lepidoptera (Ahmad *et al.*, 2002). Though this insect has never been exposed to indoxacarb for commercial control in the field, it has been treated with several other insecticide classes, including a number of organophosphates. Current studies indicate that resistant strains of this leafroller may be able to catabolize indoxacarb more rapidly

than susceptible strains; the specific mechanisms and metabolites are currently being characterized (Hollingworth *et al.*, unpublished data).

Interestingly, indoxacarb is apparently degraded primarily by higher animals via routes including cytochrome p450 mediated attack of the indanone and oxadiazine rings, while *N*-decarbomethoxylation is a relatively minor pathway (EPA, 2000; Scott, 2000). The rapid metabolic degradation is a critical factor responsible for the high nontarget animal safety of indoxacarb.

2.4. Physiology and Biochemistry of the Na$^+$ Channel Blockers

2.4.1. Symptoms of SCBI Poisoning in Insects: Pseudoparalysis

Pyrazolines and indoxacarb produce identical acute neurotoxic symptoms in the American cockroach *Periplaneta americana*, progressing through initial incoordination (5–20 min after 1–10 μg injection), then tremors and prostration, and finally a distinctive pseudoparalysis, so named because the apparently paralyzed insects produce violent convulsions when disturbed. This pseudoparalysis was maintained for 3–4 days, after which the ability to move when disturbed waned. By 4–6 days, the insects could be considered dead. Lower doses caused a similar progression of symptoms, but the time before appearance of symptoms was delayed. This unusual pseudoparalysis was key to unraveling the complex mode of action of this family in insects.

Irreversible pseudoparalysis was also observed in lepidopterous larvae, with disturbance during the pseudoparalyzed state leading to squirming, and tremors of legs and mandibles evident upon microscopic examination. Higher doses of pyrazolines, especially in lepidopterous larvae, led within a few hours to flaccid paralysis (Salgado, 1990). Injection of 10 μg g^{-1} indoxacarb into fifth instar *Manduca sexta* larvae produced excitatory neurotoxic symptoms leading to convulsions and then relatively rapidly to flaccid paralysis (Wing *et al.*, 1998). As discussed above, indoxacarb itself is converted *in vivo* to DCMP by esterase-like enzymes. The pseudoparalytic symptoms caused by indoxacarb have also been clearly observed in *Spodoptera frugiperda* larvae treated by injection of 20 μg g^{-1} indoxacarb or DCJW, and also in male *P. americana* injected with 1, 3, or 10 μg g^{-1} DCJW (Salgado, unpublished data).

Under field conditions, after ingesting or being directly sprayed with indoxacarb, insects will irreversibly stop feeding within a few minutes to 4 h; higher doses cause more rapid onset of symptoms.

Higher temperatures increase desiccation and speed of kill. Unlike the pyrethroids, indoxacarb exhibits a positive temperature coefficient. Uncoordinated insects may fall off the plant and desiccate, drown, or become subject to predation. Affected insects can stay alive for 4–96 h, depending on the dose of indoxacarb and susceptibility of the insect. Insects exposed to subparalytic doses eat much less than untreated larvae, develop more slowly, gain less weight, and pupate and emerge later than untreated insects.

2.4.2. Block of Spontaneous Activity in the Nervous System

RH-3421 and RH-1211, two highly active pyrazolines, were used as model compounds in an electrophysiological analysis of poisoned *P. americana*, together with RH-5529, which was useful for some experiments because of its ready reversibility (**Figure 1**). The upper panel in **Figure 5** compares extracellular recordings from three parts of the nervous system before and shortly after paralysis with 5 μg g^{-1} RH-3421. The connectives between the second and third thoracic ganglia are major pathways for interneurons within the central nervous system (CNS); the crural nerve is the major leg nerve, with both sensory and motor traffic; and the cercal nerve is exclusively sensory. In all cases,

the complete absence of neural activity upon dihydropyrazole poisoning is striking. It has likewise been shown that activity in the CNS connectives of *M. sexta* larvae is completely blocked after paralysis with indoxacarb or DCJW via injection (Wing *et al.*, 1998). The lower panel in **Figure 5** shows recordings from *S. frugiperda* larvae before and shortly after paralysis with both DCJW and indoxacarb. In the CNS connectives and also in the ventral nerve roots of the abdominal ganglia, which carry motor and sensory axons, there is no nerve activity in the paralyzed insects. The assessment of the state of the nervous system during poisoning is completed with **Figure 6**, which shows that even in a cockroach paralyzed for 24 h, tactile stimulation of the trochanter could elicit sensory spikes in the crural nerve that in turn initiate reflex motor activity in the same nerve. This is a clear demonstration that axonal conduction and synaptic transmission function more or less normally in paralyzed insects; the evident defect is that the nerves no longer generate action potentials spontaneously.

Normally, even in a quiescent insect, there is background or spontaneous action potential activity in the nervous system arising from pacemaker cells in the CNS and from tonic sensory receptors. Mechanoreceptors exist in many places on the cuticle to detect cuticular deformations or hair

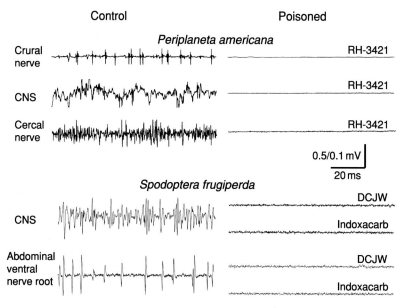

Figure 5 SCBI poisoning inhibits all spike activity in the nervous system. Representative extracellular recordings from the central nervous system (CNS) and peripheral nerves of paralyzed insects and untreated DMSO-injected controls. Adult male cockroaches were treated by injection of 5 μg g^{-1} RH-3421 and fifth instar *Spodoptera* larvae weighing 500 mg were treated with 5 μg g^{-1} DCJW or indoxacarb, 3–4 h before dissection of and recording from the paralyzed insects. (Cockroach data: reproduced with permission from Salgado, V.L., **1990**. Mode of action of insecticidal dihydropyrazoles: selective block of impulse generation in sensory nerves. *Pestic. Sci.* 28, 389–411; © Society of Chemical Industry, permission is granted by John Wiley & Sons Ltd. on behalf of the SCI. *Spodoptera* data previously unpublished.)

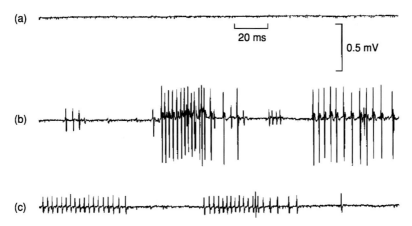

Figure 6 Sensory and motor nerve activity could still be elicited long after the onset of paralysis. These recordings, from nerve 5 (crural nerve) of a cockroach injected 24 h earlier with $5\,\mu g\,g^{-1}$ RH-3421, show that although there was no background activity (trace a), tactile stimulation of the ipsilateral trochanter elicited activity in several axons (trace b). Furthermore, cutting nerve 5 proximal to the recording site abolished the larger spikes (trace c), showing that the remaining smaller spikes were in primary sensory neurons while the larger ones in trace (b) were reflexly evoked motor spikes. (Reproduced with permission from Salgado, V.L., **1990**. Mode of action of insecticidal dihydropyrazoles: selective block of impulse generation in sensory nerves. *Pestic. Sci. 28*, 389–411; © Society of Chemical Industry, permission is granted by John Wiley & Sons Ltd. on behalf of the SCI.)

deflections resulting from stresses due to the body's own movements or to external stimuli, and internally in the form of chordotonal organs or muscle stretch receptor organs to report on joint angles. Mechanoreceptors in general have both phasically and tonically responding sensory cells or units. The phasic units respond with a burst of activity at the start or end of a stimulus, but rapidly adapt in the presence of constant stimuli, because their function is to sensitively detect changes in stimulus level. The tonic units maintain a steady firing rate during stimulus presentation and give rise to slowly adapting sensations. In insect muscle stretch receptor organs, both phasic and tonic functions are subserved by a single neuron. A sudden elongation of this receptor induces firing initially at a high rate, which then declines to a constant rate that is proportional to static elongation, so that both velocity and position are encoded (Finlayson and Lowenstein, 1958).

The complete absence of neural activity in poisoned insects indicates that SCBIs block not only tonic sensory activity, but also pacemaker activity in the CNS. Both of these effects involve action potential generation in regions of neurons that are able to generate action potentials repetitively in response to constant stimuli. The ability of phasic receptors to respond long after paralysis at a high dose (**Figure 6**), suggests that phasic receptors are not as sensitive as the tonic ones, although they may also be affected at higher doses in Lepidoptera, when insects are completely paralyzed, as mentioned earlier. At this point in the electrophysiological analysis, the pseudoparalysis resulting from SCBI poisoning is apparently due to inhibition of spontaneous activity

in tonic sensory receptors and pacemaker neurons. However, excitatory symptoms are also seen during the course of poisoning; a plausible hypothesis is that early in poisoning, before the compound reaches the CNS, sensory receptors in the periphery become blocked. In the absence of adequate sensory feedback to the CNS in response to attempted movements, there would be a tendency to overaccentuate these movements, resulting in the altered posture and gait that is observed. The quiet periods that predominate later in pseudoparalysis may be due to block of pacemaker activity in the CNS. An explanation of the tremors will be proposed below, after the cellular effects of the compounds that cause the observed block are considered. What is noteworthy, however, about the excitatory symptoms arising from a blocking action is the long time period over which apparent excitatory effects can be seen. Compounds with primary excitatory effects on neurons lead to continuous excitation, and within a few hours to complete paralysis due to, among other things, neuromuscular block, as seen for example with pyrethroids (Schouest *et al.*, 1986) or spinosad (Salgado, 1998). Insects poisoned with SCBIs, alternatively, do not suffer this pysiological exhaustion, and can therefore show strong tremors intermittently for several days after poisoning.

2.4.3. Block of Na⁺ Channels in Sensory Neurons

A deeper understanding of the mode of action of the SCBIs required the investigation of their cellular effects on sensory neurons. Abdominal stretch receptors from cockroach adults and *M. sexta* larvae were shown to be blocked by various pyrazolines

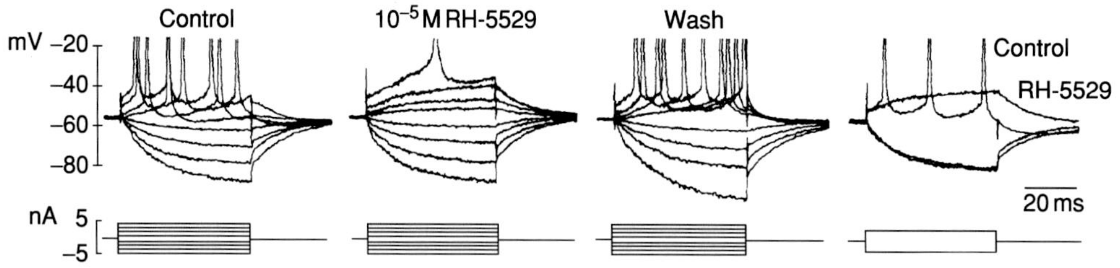

Figure 7 RH-5529 raised the threshold in current-clamped crayfish stretch receptors, without affecting passive membrane resistance. Records show voltage (upper) and current (lower) traces for current steps before, during, and after application of 10 µM RH-5529. Corresponding traces superimposed on the right show that the spike threshold was raised by RH-5529, with no change in resting potential or in the response to hyperpolarizing pulses. (Reproduced with permission from Salgado, V.L., **1990**. Mode of action of insecticidal dihydropyrazoles: selective block of impulse generation in sensory nerves. *Pestic. Sci. 28*, 389–411; © Society of the Chemical Industry, permission is granted by John Wiley & Sons Ltd., on behalf of the SCI.)

(Salgado, 1990), but the slowly adapting stretch receptor (SASR) of the crayfish *Procambarus clarkii* (Wiersma *et al.*, 1953) was selected for studies of the cellular mechanism of action, because the greater size of its neuron enables intracellular recordings at the site of spike initiation.

Pyrazolines blocked the stretch receptor in crayfish with a potency similar to that in insects. In mechanoreceptors, membrane deformation resulting from elongation activates stretch-sensitive ion channels that induce a so-called generator current that depolarizes the membrane of the spike initiation zone to the threshold for action potential generation. In other words, the generator current resulting from mechanosensory transduction is encoded as a spike train in the spike initiation zone. Insertion of a microelectrode into the spike-initiation zone of an SASR neuron allowed the function of this transduction and spike-encoding process to be studied. **Figure 7** shows a SASR neuron studied under current clamp, in which current pulses of varying amplitude were injected to either hyperpolarize or depolarize the membrane. Such currents bypass the sensory transduction process and allow direct assessment of the spike-encoding process. The voltage traces in response to the injected currents are shown in the upper row. For negative or hyperpolarizing pulses, represented as downward deflections of both current and voltage, RH-5529 did not affect the membrane response, indicating that it did not affect passive membrane properties (**Figure 7**). The singular effect of RH-5529 was to raise the threshold for spike generation in response to depolarizing pulses, making it more difficult for injected currents and, by inference, for generator currents, to elicit spikes. From this result, it was immediately clear that voltage-dependent Na^+ channels, whose activation determines the threshold and initiates the action potential, were blocked by the pyrazoline. This was also likely the mechanism of block in the spike initiation zones

of CNS pacemaker neurons, where spikes are also generated by the activation of Na^+ channels in response to depolarization by summation of synaptic inputs and pacemaker currents.

2.4.4. Mechanism of Na^+ Channel Block

In order to better understand the action of SCBIs on Na^+ channels, further studies were carried out on crayfish giant axons treated with pyrazolines, with techniques that allowed the study of Na^+ channels under highly controlled conditions. A first hypothesis to explain the insensitivity of axonal Na^+ channels to pyrazolines was that the compounds block Na^+ channels in a voltage-dependent manner, and are therefore selective for channels in the spike initiation zone. Voltage-gated Na^+ channels are complex proteins, whose function is regulated by membrane potential through voltage-dependent conformational changes occurring on a timescale from less than a millisecond to several seconds (Pichon and Ashcroft, 1985). At the spike initiation zone, neurons operate near the threshold for action potential generation, in the range of −70 to −50 mV, where important conformational changes in Na^+ channels occur prior to opening. In contrast, axons have a resting potential near −90 mV, where Na^+ channels are predominantly in the resting state. However, when axons are depolarized to the range where Na^+ channels begin to undergo conformational changes, they become sensitive to pyrazolines (Salgado, 1990). Likewise, the extracellularly recorded compound action potential from motor nerves of *M. sexta* abdominal ganglia, which was likewise highly insensitive to DCJW, was rendered sensitive by depolarization of the nerves with a high K^+ saline (Wing *et al.*, 1998).

Figure 8 shows membrane currents evoked by test pulses to 0 mV before and after equilibration with 10 µM RH-3421 at various holding potentials. For each trace, after initial rapid downward and upward transients, the current became inward and peaked

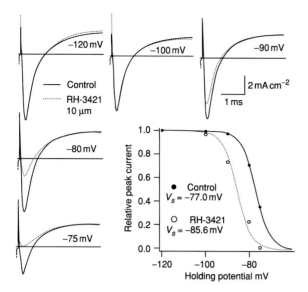

Figure 8 Dihydropyrazole block appears as a parallel shift of the steady state slow inactivation curve in the direction of hyperpolarization. Ionic current traces were scaled by a common factor so that the peak at −120 mV matched the peak before treatment with the dihydropyrazole. Peak I_{Na} was depressed most at depolarized potentials, whereas outward current, I_K, was not affected by the treatment. The graph shows plots of peak current normalized to the value at −120 mV. RH-3421 (10 µM) appears to shift the steady state inactivation relation to the left by 8.6 mV. (Reproduced with permission from Salgado, V.L., 1992. Slow voltage-dependent block of Na⁺ channels in crayfish nerve by dihydropyrazole insecticides. *Mol. Pharmacol. 41*, 120–126; © American Society for Pharmacology and Experimental Therapeutics.)

within 0.5 ms, then reversed and became outward. The inward (downward) peak is the Na⁺ current and the outward (upward) steady state current is the K⁺ current (Pichon and Ashcroft, 1985). Depolarization of the axon to potentials more positive than −90 mV in the control suppressed the Na⁺ current by a process known as inactivation, which in this case had a midpoint potential of −77 mV. Na⁺ channels have two partially independent inactivation processes, known as fast and slow inactivation. Fast inactivation occurs on a millisecond timescale and serves to terminate the action potential, while slow inactivation occurs on a much slower timescale, over hundreds of milliseconds, and performs a slow, modulatory function. Slow inactivation occurs at more negative potentials than fast inactivation, and is responsible for the inactivation seen in **Figure 8**. After equilibration of the axon with RH-3421, there was no effect on either Na⁺ or K⁺ current at −120 or at −100 mV (**Figure 8**). At more depolarized potentials, however, the Na⁺ current was specifically depressed by RH-3421, whereas the K⁺ current was unaffected. When peak Na⁺ current is plotted against membrane potential, it appears that

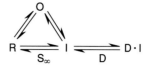

Figure 9 A model showing the resting (R), Open (O), and inactivated (I) states of the Na⁺ channel, and specific interaction of the SCBIs (D) with the inactivated state.

RH-3421 has shifted the slow inactivation curve by 9 mV to the left.

The next step in the analysis is to consider the mechanism of this apparent shift of slow inactivation. Whereas fast inactivation occurs with a time course of hundreds of microseconds to a few milliseconds, and slow inactivation on the order of tens to hundreds of milliseconds, the changes in peak Na⁺ current in the presence of SCBIs occur on a much slower timescale, on the order of 15 min. This slow readjustment of peak current in response to voltage change can therefore be attributed to a new process: the binding and unbinding of SCBIs to Na⁺ channels.

The Na⁺ channel undergoes transitions between many different conformations or states, which can be grouped into resting (R), open (O), and inactivated (I) states, each of which may have several substates. The transitions between these naturally occurring states are shown on the left in **Figure 9**. In this simplified model, $S_\infty = [R]/[R] + [I]$, is the steady-state slow inactivation parameter, which depends on membrane potential according to

$$S_\infty = \frac{1}{1 + \exp[(V - V_S)/k]} \qquad [1]$$

where V is membrane potential, V_S is the potential at which $S_\infty = 0.5$ (half of the channels are in the inactivated state), and k is a constant that includes the Boltzmann constant and describes the voltage sensitivity of the inactivation process (Hodgkin and Huxley, 1952).

This model ignores the fast-inactivated state, but interaction of the SCBIs with the channels is so slow, that it and the open state, O, can be ignored in the analysis. The observed enhancement of slow inactivation can be explained if we assume that the SCBI drug (D) binds selectively to an inactivated state, as also shown on the right-hand side in the model (**Figure 9**). Drug binding would then effectively remove channels from the I pool, which would be compensated for by mass action rearrangements among the other states, leading to net flux of more receptors from the R pool into the I and D•I pools. $K_I = [D] \times [I]/[D•I]$ is defined as the equilibrium dissociation constant of the D•I complex.

In the presence of SCBI drug, two processes lead to the decrease in current upon depolarization: slow inactivation and drug binding. Fortunately, slow inactivation can be removed from the measurements by hyperpolarizing to −120 mV before applying the depolarizing pulse to measure the peak current. **Figure 10a** shows the time course of block after stepping from −120 mV to various depolarized potentials, in an axon equilibrated with 1 μM RH-1211. The pulse protocol, shown in the inset in panel A (**Figure 10**), includes a 500 ms hyperpolarizing

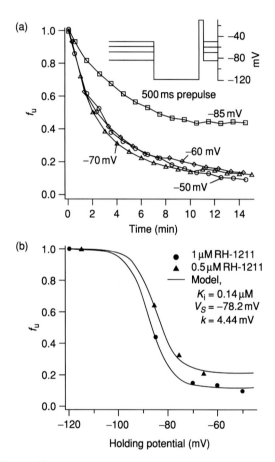

Figure 10 Block is asymptotic at strong depolarizations. The model predicts that block of Na⁺ channels by SCBIs is voltage dependent only over the voltage range over which inactivation occurs. In order to observe block directly at strong depolarizations, where inactivation is complete, the axon was held at various potentials as in **Figure 8**, but a 500 ms hyperpolarizing prepulse to −120 mV was applied just before the 10 ms test pulse, during which I_{Na} was measured (see inset protocol). The prepulse was of sufficient amplitude and duration to remove inactivation completely, without interfering with block. (a) Time course of f_u, the fraction of channels unblocked, following steps from −120 mV to various other potentials, in an axon equilibrated with 1 μM RH-1211. Steady state block from this and another axon treated with 0.5 μM RH-1211 are plotted in (b). The solid curves in (b) were plotted according to eqn [2], with $K_i = 0.14$ μM, $V_h = 78.2$ mV, and $k = 4.44$ mV. (Reproduced with permission from Salgado, V.L., **1992**. Slow voltage-dependent block of Na⁺ channels in crayfish nerve by dihydropyrazole insecticides. *Mol. Pharmacol.* *41*, 120–126; © American Society for Pharmacology and Experimental Therapeutics.)

prepulse to −120 mV before each test pulse, which was long enough to completely remove slow inactivation, but not long enough to permit significant dissociation of D•I complexes, thus allowing direct assessment of block. The fraction of channels unblocked, f_u, is plotted against time, and shows that block, which is equivalent to the formation of D•I complexes, occurs with a time constant of 5–6 min at 1 μM. The voltage dependence occurs only over the range where slow inactivation occurs, saturating near −70 mV. This is seen clearly in **Figure 10b**, where f_u is plotted against holding potential. In terms of the model shown above, $f_u = ([R] + [I])/([R] + [D•I] + [I])$. Substituting the above relations for K_I and S_∞ and eqn [1] into this expression, an equation for f_u as a function of [D] and V can be derived:

$$f_u = \cfrac{1}{1 + \cfrac{[D]}{K_I}\left(\cfrac{1}{1 + \exp\left[(V_S - V)/k\right]}\right)} \quad [2]$$

The solid curves in **Figure 10** were calculated from this equation, using a K_I of 140 nM, $V_S = -78.2$ mV and $k = 4.44$ mV, and fit the data quite well.

The quantitative correspondence of the blocking voltage dependence with model predictions supports the hypothesis that SCBIs bind selectively to the slow-inactivated state of the channel, in comparison with the resting state. However, because of the very slow binding reaction, in the equilibrium studies described to this point, only the resting and slow-inactivated states are populated significantly enough to participate in drug binding. In order to test whether SCBIs bind to the fast-inactivated state, the slow inactivation gate was removed by pretreatment of the internal face of the axon membrane with trypsin. After this pretreatment, block by the pyrazoline RH-1211 at −90 mV was comparable to that in intact axons, indicating that RH-1211 can block fast-inactivated channels as well. Treatment of the axon internally with the protein-modifying reagent N-bromoacetamide (NBA) is known to remove the fast inactivation gate. Pretreatment of axons with both trypsin and NBA therefore yields Na⁺ channels that do not inactivate. At depolarized potentials, RH-1211 was just as potent at blocking these noninactivating channels as intact channels, showing that it can also block open channels (Salgado, 1992).

The metabolic bioactivation of DPX-JW062 and indoxacarb to N-decarbomethoxylated metabolites and the similarity of the mode of action of DCJW to pyrazolines was established by Wing *et al.* (1998), who showed that spontaneous activity and action potential conduction in the CNS of *M. sexta* were blocked, consistent with block of Na$^+$ channels. These workers also established that the DCJW Na$^+$ channel block was voltage-dependent. Subsequently, studies in dorsal unpaired median (DUM) neurons isolated from the CNS of *P. americana* with the patch clamp technique have also shown the action of DCJW on Na$^+$ channels (Lapied *et al.*, 2001). Action potentials in these pacemaking neurons were blocked by 100 nM DCJW, and this action was shown to be due to block of voltage-dependent Na$^+$ channels. The block was very potent, with a 50% inhibitory concentration (IC$_{50}$) of 28 nM at −90 mV, in good agreement with the IC$_{50}$ of racemic DCJW of 40 nM in blocking the compound action potential in *M. sexta* CNS determined by Wing *et al.* (1998). In addition, Wing *et al.* (1998) showed that *M. sexta* compound action potential block by DCJW was stereospecific for the *S*-enantiomer, which is also the only enantiomer toxic to insects.

Although the SCBIs enhance slow inactivation, it is important to realize that the slow inactivation of any particular channel is not affected. Instead, slow inactivation of the entire population of channels is enhanced because of the addition to the system of the highly stable SCBI-bound slow-inactivated state. In the presence of the SCBI, time for this very slow equilibration of the drug with the receptors must be allowed in order to measure slow inactivation properly. It is not enough to simply measure slow inactivation with pulses long enough to attain a steady state in the absence of SCBI. While pulses on the order of 1 s are long enough to attain steady-state slow inactivation in control axons, 15 min are required in the presence of insecticide (Salgado, 1992). Direct measurements of the effect of DCJW and indoxacarb on Na$^+$ channels have until now only been carried out with the whole cell voltage clamp method. With this technique, it is difficult to make measurements from a single cell for more than 30 min, so it is difficult to measure the effects of these slow-acting insectides under steady state conditions. The block on washing of DCJW requires at least 15 min to reach a steady state level and appears to be irreversible (Lapied *et al.*, 2001; Tsurubuchi and Kono, 2003; Zhao *et al.*, 2003). However, after equilibration of the cell with DCJW at a negative holding potential, the level of block increases in response to membrane depolarization on a faster time course, with steady state levels being attained within 2–3 min (Zhao *et al.*, 2003). Using 3 min conditioning pulses, Zhao *et al.* (2003) demonstrated that DCJW indeed enhanced slow inactivation of two different Na channel subtypes

studied in isolated neurons; shifts of 12–13 mV in the direction of hyperpolarization were observed.

An additional effect on the *P. americana* DUM neurons observed by Lapied *et al.* (2001) was a strong hyperpolarization of the resting potential, associated with an increase in membrane resistance, indicating the block of a depolarizing conductance, thought to be the background Na$^+$ channels involved in the maintenance of the resting potential in these pacemaking neurons and characterized by Lapied *et al.* (1989, 1999). This finding is extremely interesting and should be investigated in more detail, because it demonstrates the existence of a second potential target site for SCBIs.

2.4.5. SCBIs Act at Site 10 on the Na$^+$ Channel

The voltage-dependent Na$^+$ channel blocking action of pyrazolines was immediately recognized as being similar to that of local anesthetics, class I anticonvulsants, and class I antiarrhythmics (Salgado, 1992), a structurally broad range of drugs (Clare *et al.*, 2000; Anger, 2001) all known to act at a common blocker site within the Na$^+$ channel pore. Consistent with the nomenclature of Soderlund, this will be called site 10. Like the SCBIs, drugs acting at site 10 all exhibit voltage dependence of block, deriving from selective binding to open and inactivated channel states.

Drugs acting at site 10 displace the binding of the radioligand [^{3}H]batrachotoxin (BTX)-B from the BTX binding site in the Na$^+$ channel (Postma and Catterall, 1984; Catterall, 1987). The interaction is clearly allosteric, because local anesthetics can block BTX-modified open channels without displacing the toxin (Wasserstrom *et al.*, 1993; Zamponi *et al.*, 1993), although they speed the dissociation rate of BTX from its binding site. Pyrazolines were also shown to potently displace specific binding of [^{3}H]BTX-B (Deecher *et al.*, 1991; Salgado, 1992). Payne *et al.* (1998) further examined the interaction between the pyrazoline RH-3421 and the local anesthetic dibucaine in BTX binding studies. Each of these compounds decreased the potency of the other as an inhibitor of BTX binding approximately as much as expected from the assumption that they share a common binding site. Furthermore, RH-3421 increased the dissociation rate of [^{3}H]BTX-B from its binding site, also expected from an action at site 10 (Deecher *et al.*, 1991).

Lapied *et al.* (2001) examined the interactions of DCJW with the local anesthetic lidocaine and the guanidinium blocker tetrodotoxin, which is known to block the pore from the external face, at a binding site distinct from the local anesthetics. The IC_{50} for DCJW was not affected by the presence of an IC_{50} concentration of tetrodotoxin (TTX) in the external solution, consistent with independent action of the two compounds at distinct binding sites. In the presence of an IC_{50} concentration of lidocaine, however, the IC_{50} for DCJW was increased about 30-fold. A twofold shift in equilibrium binding would be expected from the hypothesis that both compounds act at the same site (Cheng and Prusoff, 1973), so the mechanism of the observed 30-fold shift is not fully understood. Nevertheless, available evidence is consistent with the action of the SCBIs at site 10.

Studies on permanently charged quaternary derivatives of local anesthetics were crucial in localizing site 10 within the Na^+ channel pore. Being permanently charged, these molecules cannot cross membranes and are therefore not clinically useful. However, when applied within the cell, they are able to enter the channel from the internal mouth, only when the channel is open, giving rise to a phenomenon known as use dependence – increase in the number of blocked channels with stimulation. Because the permanently charged quaternary molecule cannot fit through the ion-selective constriction in the channel known as the selectivity filter, it remains trapped within the channel after activation gate closure. Alternatively, more lipophilic compounds acting at site 10, when caught within the closed channel, can diffuse out laterally through lipophilic pathways in the channel wall. SCBIs behave like local anesthetics with very slow kinetics, and do not display use dependence (Zhao et al., 2003), suggesting that the insecticides can only access the binding site from the lipophilic pathway and not through the internal mouth of the pore.

In addition to modulation of blocker access to the binding site by the activation gate, channel gating also modulates the equilibrium binding affinity of the channel for the molecules. Like the pyrazolines and DCMP/DCJW, there is little affinity of the blocker molecule for the resting state, but strong binding to the open and inactivated states. The bound molecule is thought to lie in a hydrophobic pocket that can be reached via either the aqueous or hydrophobic pathways. When bound, the molecule is thought to lie within the pore and block ion flow through it. Several excellent reviews on the interaction of blocking drugs with Na^+ channels have recently been published (Clare et al., 2000; Anger et al., 2001; Wang and Wang, 2003).

The fact that blockers acting at site 10 bind strongly to a number of states after channel opening but not to the resting state, indicates that the movements of the S6 segments (see Section 2.4.6) associated with channel activation lead to formation of the receptor for these compounds. The subsequent fast and slow inactivation steps are thought to involve processes at the internal membrane face that involve blocking of the pore by a so-called inactivation particle, presumably without further significant changes in the conformation of site 10.

2.4.6. The Molecular Nature of Site 10 in Insects

The primary structure of the Na^+ channel α subunit consists of four homologous repeat domains (D1 to D4), each with six transmembrane segments S1 to S6. The six transmembrane segments of each domain are thought to be arranged into a quasicylindrical pseudosubunit, and all four of these pseudosubunit domains are arranged in pseudotetrametric fashion, forming the Na^+ channel pore in the space between them.

The existence of several mutations in the S6 segments from domains 1, 3, and 4 that affect potency of both tertiary and quaternary local anesthetics, the latter of which have access to the receptor only through aqueous pathways, indicates that the S6 segments line the pore of the Na channel (**Figures 11** and **12**). The voltage sensor for activation gating has been localized, also using site-directed mutagenesis, to the positively charged S4 segments, which are thought to exhibit voltage-dependent outward and/or rotary movements to open the activation gate, which is thought to be located at the cytoplasmic ends of the S6 segments. Although the voltage sensors are located on the S4 segments, their movement appears to be coupled to critical movements of the S6 segments, leading to opening of the activation gate at the cytoplasmic S6 ends, and modulation of three receptor sites known to be associated with the midsections of the S6 segments: the blocker site (site 10), the pyrethroid site (site 7), and the BTX site (site 2) (**Figure 13**). Several amino acid residues clustered around the middles of the S6 segments which, when mutated, affect binding of compounds to these sites, have been identified, and those associated with the binding of Na^+ channel blocking drugs have been the subject of much recent research because they potentially identify the faces of the S6 segments that face the pore (Yarov-Yarovoy et al., 2002). These mutations are colored red in the sequence alignments in **Figure 11**. **Figure 12** shows a helical wheel representation of the four S6 segments arranged according to Kondratiev and Tomaselli (2003), who have only considered the residues

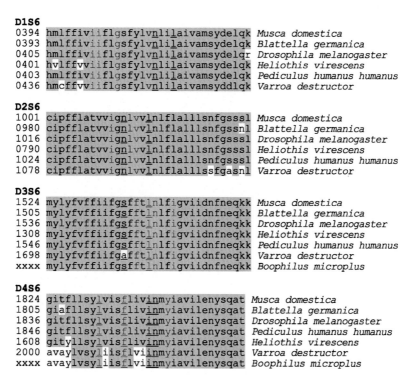

```
D1S6
0394  hmlffiviiiflgsfylvnlilaivamsydelqk  Musca domestica
0393  hmlffiviiiflgsfylvnlilaivamsydelqk  Blattella germanica
0405  hmlffiviiiflgsfylvnlilaivamsydelqr  Drosophila melanogaster
0401  hvlffvviiflgsfylvnlilaivamsydelqk   Heliothis virescens
0403  hmlffiviiiflgsfylvnlilaivamsydelqk  Pediculus humanus humanus
0436  hmcffvviiiflgsfylvnlilaivamsyddlqk  Varroa destructor

D2S6
1001  cipfflatvvignlvvlnlflalllsnfgsssl  Musca domestica
0980  cipfflatvvignlvvlnlflalllsnfgssnl  Blattella germanica
1016  cipfflatvvignlvvlnlflalllsnfgsssl  Drosophila melanogaster
0790  cipfflatvvignlvvlnlflalllsnfgsssl  Heliothis virescens
1024  cipfflatvvignlvvlnlflalllsnfgsssl  Pediculus humanus humanus
1078  cipfflatvvignlvvlnlflalllssfgasnl  Varroa destructor

D3S6
1524  mylyfvffiifgsfftlnlfigviidnfneqkk  Musca domestica
1505  mylyfvffiifgsfftlnlfigviidnfneqkk  Blattella germanica
1536  mylyfvffiifgsfftlnlfigviidnfneqkk  Drosophila melanogaster
1308  mylyfvffiifgsfftlnlfigviidnfneqkk  Heliothis virescens
1546  mylyfvffiifgsfftlnlfigviidnfneqkk  Pediculus humanus humanus
1698  mylyfvffiifgafftlnlfigviidnfneqkk  Varroa destructor
xxxx  mylyfvffiifgsfftlnlfigviidnfneqkk  Boophilus microplus

D4S6
1824  gitfllsylvisflivinmyiavilenysqat  Musca domestica
1805  giafllsylvisflivinmyiavilenysqat   Blattella germanica
1836  gitfllsylvisflivinmyiavilenysqat  Drosophila melanogaster
1846  gitfllsylvisflivinmyiavilenysqat  Pediculus humanus humanus
1608  gityllsylvisflivinmyiavilenysqat  Heliothis virescens
2000  avaylvsyliisflviinmyiavilenysqat  Varroa destructor
xxxx  avaylvsyliisflviinmyiavilenysqat  Boophilus microplus
```

Figure 11 Sequence alignment of the pore-forming S6 transmembrane segments from various arthropod pest species, to show the location of residues known to be important for binding of local anesthetics (red), batrachotoxin (underscored), and pyrethroids (blue). Residues conforming with the consensus sequence are shown with a gray background.

known to affect blocker affinity, without alterations in the voltage dependence or kinetics of gating that would otherwise account for an increase in the IC_{50} for block. The S6 segments can be arranged so that all such residues, shown in red, face into the pore. A total of 10 residues affecting BTX action have also been identified scattered among the middle of all four S6 segments (Wang and Wang, 2003), but it is not yet clear if these are all associated with the binding site.

Residues in the S6 segments of domains 1, 2, and 3 that are known to affect pyrethroid sensitivity are shown in blue in **Figures 11** and **12**. These are on the sides thought to face away from the pore. Mutations conferring pyrethroid resistance have also been identified in the S5 segments, indicating that the pyrethroid receptors may be located at the interfaces between S5 and S6 segments. The number of pyrethroid receptors per channel is not certain, but the locations of the resistance mutations in the S6 segments suggests that there could be as many as four receptors. Interestingly, quantitative studies with tetramethrin isomers on squid axon Na^+ channels indicated that there may be at least three pyrethroid receptors in each channel (Lund and Narahashi, 1982).

The extremely high conservation of the S6 segments among the arthropods compared in **Figure 11** is striking. The residues associated with the binding sites are all absolutely conserved. However, the P loops in the extracellular S5–S6 connecting segments dip deep into the external mouth of the barrel formed by the S6 segments to form the selectivity filter and the outer boundary of the inner chamber containing site 10 (Cronin *et al.*, 2003). The contributions of the P loops to the binding of blockers in site 10 have not yet been explored. **Figure 13** shows a diagram summarizing current knowledge of the locations of the S4, S5, and S6 segments and the three receptors associated with the S6 segments. TTX is also shown, binding to site 1 at the external mouth of the channel.

2.4.7. Biochemical Measurements of the Effects of SCBIs

While electrophysiology and in particular voltage-clamp measurements are essential for studying the detailed interactions of drugs with ion channels, certain receptor binding techniques are also useful, especially when quantitative potency data on many compounds is needed. Displacement of [^{3}H]BTX-B

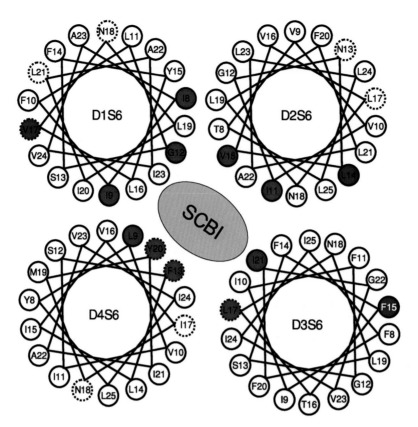

Figure 12 Helical wheel representation of residues 8–25 of the four S6 transmembrane segments arranged so that the residues known to be important for blocker binding (red) face the pore. Pyrethroid-binding residues are blue and BTX-binding residues are shown as broken circles. (Adapted from Kondratiev, A., Tomaselli, G.F., **2003**. Altered gating and local anesthetic block mediated by residues in the I-S6 and II-S6 transmembrane segments of voltage-dependent Na$^+$ channels. *Mol. Pharmacol. 64*, 741–752; and Wang, S.Y., Wang, G.K., **2003**. Voltage-gated Na$^+$ channels as primary targets of diverse lipid-soluble neurotoxins. *Cell Signal. 15*, 151–159.)

has been used effectively to quantify the potency of pyrazolines on Na$^+$ channels. The quantitative correspondence with electrophysiology and biological activity is good (Deecher *et al.*, 1991; Salgado, 1992), and this is the method of choice for structure–activity relationship studies.

Block of the Na$^+$ flux through channels activated by the activator veratridine has also been used to measure effects of dihydropyrazoles. In this case, Na$^+$ flux can be assessed directly as flux of ^{22}Na$^+$ into membrane vesicles (Deecher and Soderlund, 1991), or indirectly from the resulting depolarization as measured by voltage-sensitive dyes (Nicholson, 1992) or depolarization-triggered release of preloaded radiolabeled neurotransmitter (Nicholson and Merletti, 1990). With flux assays, however, veratridine is often used at a high concentration in order to achieve a strong signal, but since this activator acts at the BTX site, it inhibits the action of the pyrazolines (Deecher and Soderlund, 1991) and thereby interferes with accurate potency measurements.

2.4.8. Intrinsic Activity of Indoxacarb on Na$^+$ Channels

The initial mode of action studies with indoxacarb indicated that it was a proinsecticide, being metabolized to DCJW, which potently blocked the compound action potential in *M. sexta* nerve in a voltage-dependent manner (Wing *et al.*, 1998). Indoxacarb itself also had this effect, but was much slower acting; while 1 μM DCJW blocked the compound action potential by 50% after 10 min, 40 min were required for indoxacarb to exert this same effect, and in that preparation one cannot be certain that the compound is not metabolized to DCJW within the ganglion. However, studies on patch-clamped neurons, which cannot metabolize the compound, confirmed that indoxacarb indeed has intrinsic activity on Na$^+$ channels. The effects of the two compounds were comparable, but while indoxacarb block was reversible with about 5 min washing or with hyperpolarization, block by DCJW was not reversible with either washing or hyperpolarization (Tsurubuchi and Kono, 2003; Zhao *et al.*, 2003).

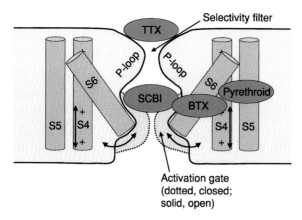

Figure 13 Diagrammatic representation of the location of the binding sites associated with the S6 segments of the Na⁺ channel. The extracellular mouth of the channel is at the top of the figure. Two sets of the S4, S5, and S6 transmembrane helices are shown. Bidirectional arrows show the movements of the S4 and S6 segments associated with activation gating. The thick solid curves show the outline of the channel protein in the activated or open conformation, and the thick dotted line shows the activation gate at the cytoplasmic end of the S6 segments in the closed position. There is one tetrodotoxin (TTX) and one Na⁺ channel blocker insecticide (SCBI) binding site per channel, but the number of batrachotoxin (BTX) and pyrethroid sites is not certain, although only one of each is shown. All three drug sites are in the middle of the S6 segments, and the pyrethroid site also includes residues in segment 5.

Table 2 Action of DCJW and indoxacarb on the abdominal stretch receptor organ of *Spodoptera frugiperda*

Concentration[a] (M)	Average percent block (five preparations, in exposure)	
	DCJW	Indoxacarb
3e-9	0	0
1e-8	40	0
3e-8	100	25
1e-7	100	60
3e-7	100	100

[a] In a saline containing, in mM, 28 NaCl, 16 KCl, 9 CaCl₂, 1.5 NaH₂PO₄, 1.5 mgCl₂, and 175 sucrose, pH 6.5, the stretch receptor fired continuously for several hours between 25 and 75 Hz.

The actions of indoxacarb and DCJW were also compared quantitatively on the abdominal stretch receptor organs of *S. frugiperda* (**Table 2**). In this preparation, the insecticides in solution are perfused rapidly over the body wall containing the stretch receptor, a small, thin organ directly exposed to the saline. Significant metabolism in this preparation is unlikely, but cannot be excluded. The organ fires continually at a rate of between 25 and 75 Hz for many hours, and is very sensitive to block by SCBIs. DCJW blocked activity consistently within 20–40 min at 3×10^{-8} M, but in five tests at 10^{-8} M

the average percent block within 1 h was 40%. Indoxacarb was also intrinsically active, but 3- to 10-fold weaker than DCJW, blocking consistently at 3×10^{-7} M, but only 60% within 1 h at 1×10^{-7} M and 25% at 3×10^{-8} M.

In conclusion, it appears that indoxacarb is weaker (about three to ten times) than DCJW on Na⁺ channels, and that the block is readily reversible. Furthermore, the much slower effect of indoxacarb on the compound action potentials in *M. sexta* ganglia in comparison with DCJW indicates that indoxacarb enters the ganglion much more slowly than does DCJW. Based on the appearance of insect symptoms concurrent with the appearance of high levels of metabolically formed DCJW in larval Lepidoptera (Wing *et al.*, 1998, 2000) and the greater potency and irreversibility of DCJW Na⁺ channel block, it appears that the toxicologically significant compound after indoxacarb or DPX-JW062 administration is DCMP or DCJW, respectively.

2.4.9. Effects of SCBIs on Alternative Target Sites

Other possible target sites for SCBIs have been considered. The voltage-gated Ca²⁺ channel α subunits belong to the same protein superfamily as the voltage-gated Na⁺ channels, and have a similar structure, composed of 24 transmembrane segments arranged into four domains of six segments each. Not unexpectedly, Ca²⁺ channels were also found to be blocked by pyrazolines, although at higher concentrations than Na⁺ channels (Zhang and Nicholson, 1993; Zhang *et al.*, 1996). So far, there is no evidence that Ca²⁺ channel effects are significant in insect poisoning. In the initial work on the pyrazolines, it was observed in paralyzed cockroaches and *M. sexta* larvae that the heart continued to beat normally, and conduction across the cholinergic cercal–giant fiber synapses as well as γ-aminobutyric acid (GABA) and glutamate synapses on muscle, all Ca²⁺ channel-dependent processes, functioned normally (Salgado, 1990). Furthermore, DCJW had no effect on low-voltage activated and high-voltage activated calcium currents in isolated cockroach neurons (Lapied *et al.*, 2001).

A strong hyperpolarization of *M. sexta* muscle, on the order of 20 mV, was observed after poisoning by pyrazolines (Salgado, 1990). The effect was not reversible with prolonged washing and could not be reproduced by treatment with pyrazoline in solution, and therefore appears to be secondary and not related to the hyperpolarization of DUM neurons observed by Lapied *et al.* (2001). Muscles of *Periplaneta* or *Musca* were not hyperpolarized during pyrazoline poisoning (Salgado, 1990).

Table 3 Potency of indoxacarb to pest insects in the laboratory: larval ingestion and contact vs. adult contact

	Insecticidal potency of indoxacarb (LC$_{50}$ values, ppm)		
		Direct spray contact	
Insects	Ingestion	Larvae	Adults[a]
Heliothis virescens	1.47	12.96	56
Helicoverpa zea	0.68	2.55	
Spodoptera frugiperda	0.61	5.25	
Spodoptera exigua	1.96	7.66	
Trichoplusia ni	0.28		45
Plutella xylostella	3.69		30
Empoasca fabae	1.11		
Cydia pomonella	7.68		
Lygus lineolaris	1.46		

[a] Toxicity measured with methylated seed oil (1%) or vegetable oil plus organosilicone surfactant (0.5%) at 7 days posttreatment.

2.5. Biological Potency of Indoxacarb

2.5.1. Spectrum and Potency of Indoxacarb in the Laboratory

Laboratory assays show that lepidopteran, hemipteran, and homopteran pests are inherently very sensitive to indoxacarb, and are significantly more affected when indoxacarb is ingested (Wing et al., 2000). When treated plant leaves were fed to a variety of lepidopterous larvae, this resulted in 50% lethal concentration (LC$_{50}$) values ranging from 0.3 to 7.7 ppm when evaluated 3 days posttreatment (**Table 3**). These values are similar to LC$_{50}$ values published by other researchers (Seal and McCord, 1996; Pluschkell et al., 1998; Liu and Sparks, 1999; Liu et al., 2002; Giraddi et al., 2002; Smirle et al., 2002; Ahmad et al., 2003; Liu et al., 2003), and indicates the high toxicity of this compound to Lepidoptera. Indoxacarb halts insect feeding within 4 h after ingestion. Insect mortality generally takes place within 2–4 days, and it is critical to evaluate both mortality after 72 h and feeding damage to assess the compound's true effectiveness. Trichoplusia, heliothines, and Spodoptera are the most sensitive among the lepidopterans while Plutella and Cydia are somewhat less sensitive. Variation in activity among species may be due to the rate of bioconversion of indoxacarb to the active metabolite DCJW or to the amount of product consumed. Insecticidal activity of indoxacarb against the sucking insects Empoasca fabae and L. lineolaris in the laboratory are similar to that of the lepidoptera (**Table 3**). Lepidopteran adults were also effected when they ingested a sugar water solution with indoxacarb.

The ingestion activity of indoxacarb to lepidopteran larvae is about three to nine times greater than its dermal contact activity (**Table 3**). For some insects however, such as Lobesia botrana, the potency of the two modes of entry is almost equivalent (Anonymous, 2000). Indoxacarb shows negligible activity when insects walk on a dried residue, and has no activity as a fumigant.

Indoxacarb was also significantly less active as a direct contact spray to moths compared to larvae; however, it is still effective. Directly spraying lepidopteran moths with indoxacarb plus methylated seed oil at 1% or a vegetable oil plus organosilicone surfactant at 0.5% resulted in LC$_{50}$ values ranging from 30 to 56 ppm for lepidopteran adults at 7 days post treatment (**Table 3**). The ability of indoxacarb to provide moth control or suppression in the field will contribute to overall pest suppression.

Indoxacarb has inherent toxicity to select coleopteran (Parimi et al., 2003), hemipteran (Lygus), homopteran (Myzus, Empoasca, Nephotettix, Nilaparvata), and dipteran (Chen et al., 2003) pests after feeding on treated leaves or artificial diets (Wing et al., 2000) (**Table 3**). Indoxacarb has low water solubility; however, leaf penetration and translaminar activity are responsible for controlling certain sucking pests and is enhanced by inclusion of oil-based formulations or tank-mixed adjuvants. Indeed, the compound is toxic in the field to plantbugs, fleahoppers, and leafhoppers; however, it is ineffective against species such as aphids because of poor systemicity and low phloem oral bioavailability. Overall, indoxacarb's utility as a field insecticide active against sucking insects is constrained by its very limited systemicity, compared to many neonicotinoids.

Indoxacarb exhibits ovilarvicidal activity on Lepidoptera such as L. botrana, Eupoecilia ambiguella, Heliothis armigera, and Cydia pomonella (Anonymous, 2000). Laboratory data indicates that the probable mechanism of ovilarvicidal activity is due to adsorption of indoxacarb into the chorion and subsequent oral uptake as the neonate chews through the chorion to hatch. Indoxacarb directly sprayed onto the egg is more effective than placed in a dried residual deposit on which the eggs are laid.

2.5.2. Safety of Indoxacarb to Beneficial Insects

Beneficial insects walking over a dried residue of indoxacarb on a leaf are generally unaffected, as shown by Mead-Briggs et al. (1996) with the hoverfly Episyrphus balteatus and predatory mite Typhlodromus pyri. Indoxacarb is safe to the parasitoids Eurytoma pini, Haltichella rhyacioniae, Bracon sp., and Macrocentrus ancylivorus (Nowak et al., 2001), Cotesia marginiventris and Trichogramma pretiosum (Ruberson and Tillman, 1999) and for the predator G. punctipes (Tillman et al., 1998). Williams et al. (2003), Musser and Shelton (2003), Michaud and Grant (2003), Tillman et al. (2001), and Baur et al. (2003) also reported the safety of indoxacarb to beneficial insects. The insecticidal selectivity of indoxacarb against herbivorous pests is explained by Andaloro et al. (2000). There is little movement of product through the tarsi of insects thus rendering carnivorous predators and parasitic wasps largely insensitive to indoxacarb. However, any insect behavior that culminates in ingestion of the product, such as preening (ants), probing (nabids), or lapping exudates from the leaf (apple maggot flies) will impact insect survival. Indoxacarb controls fire ants, Solenopsis spp., on crops (Turnipseed and Sullivan, 2000) as a result of ants preening their antennae and legs, and subsequent ingestion of preened materials. Indoxacarb is formulated on a 5–7 μm silica particle and may readily adhere to the body of the insect through antennal drumming behavior and general locomotory activity.

2.5.3. Sublethal Effects of Indoxacarb

Liu et al. (2002) observed numerous effects of indoxacarb on the development of cabbage looper (Trichoplusia ni) larvae feeding on field-aged leaf residues 18–20 days old as well as on subsequent pupae and emerging adults. These effects were also documented on heliothine and Spodoptera species fed on diet with indoxacarb at 0.1 and 0.01 ppm (Andaloro, unpublished data). Larval development was greatly extended compared to the untreated check; survivors pupated 1–2 weeks later. Treated larvae were much smaller than untreated larvae. Some larvae could not molt or pupate properly, with exuvia remaining attached and sometimes severely restricting the body. Treated lepidopteran

larvae and Colorado potato beetle (Leptinotarsa decemlineata) larvae sometimes are unable to dig into the soil and thus pupate on top of the soil. In the laboratory, lepidopteran adults emerging from treated larvae often have abnormal wings or are unable to fly. Liriomyza adults whose larvae were exposed to leaves treated with sublethal doses of indoxacarb emerged disoriented, stumbling, and unable to fly (Leibee, unpublished data). There are sufficient data indicating indoxacarb has significant growth regulating effects on the larvae, pupae, and adults of pest insect species observed which contribute to overall suppression of the field population.

2.5.4. Spectrum and Insecticidal Potency of Indoxacarb in the Field

Indoxacarb is active against a wide spectrum of insect pests from at least ten orders and well over 30 families (**Table 4**). Lepidopteran insects represent the majority of pests on the indoxacarb label but it is also used to control many species of plant bugs, leafhoppers, fleahoppers, weevils, beetles, flies, cockroaches, and ants. Ahmad et al. (2003), Brickle et al. (2001), and Andaloro et al. (2001) documented the activity of indoxacarb against lepidopteran pests on cotton, while Hammes et al. (1999), Tillman et al. (2001), and Teague et al. (2000) showed indoxacarb activity against hemipteran pests of cotton. The rates of indoxacarb range from approximately 40 to 125 g ha^{-1} active ingredient to control agricultural pests. These rate ranges are very consistent regardless of environmental conditions that can often cause degradation of performance of other registered products. Indoxacarb formulations are extremely stable to ultraviolet radiation, pH, and temperature extremes, and have a positive temperature correlation. Indoxacarb provides excellent rainfastness relative to most labeled products (Flexner et al., 2000) providing growers with additional flexibility and control at labeled rates despite rainfall within hours after application.

Leaf penetration by indoxacarb not only results in activity against leafhoppers and plant bugs, but also pinworm, such as the tomato pinworm (Keiferia lycopersicella) and the tomato leafminer (Tuta absoluta), and provides suppression of lepidopteran leafminers such as the spotted tentiform leafminer (Phyllonorycter blancardella) and the citrus leafminer (Phyllocnistis citrella). Indoxacarb is also active against some dipteran leafminers such as Liriomyza trifolii on tomato and other vegetable crops.

The rate range for nonagricultural pests, such as structural, nuisance, or household pests ranges from 0.01% to 1.0% v/v concentration. Indoxacarb has shown excellent activity when prepared in a bait

Table 4 Spectrum of insects pests controlled by indoxacarb and range of indoxacarb rates for control[a]

Global field use rate range[b]	Insect orders	Selected insect families	Selected representative genera
40–125 g ha^{-1}	Lepidoptera	Arctiidae	*Estigmene*
		Crambidae	*Ostrinia, Diaphania, Hellula, Desmia, Herpetogramma, Loxostege, Pyrausta, Maruca, Neoleucinodes*
		Gelechiidae	*Phthorimaea, Keiferia, Anarsia, Tuta*
		Geometridae	*Operophtera*
		Gracillariidae	*Phyllonorycter, Phyllocnistis*
		Hesperiidae	*Lerodea*
		Lasiocampidae	*Malacosoma*
		Lymantriidae	*Lymantra*
		Noctuidae	*Heliothus/Helicoverpa, Spodoptera, Trichoplusia, Agrotis, Autographa, Alabama, Lithophane, Pseudoplusia, Mamestra, Chrysodeixis, Orthosia, Earias, Plusia, Diaparopsis, Mythimma, Sylepta, Thaumetopea*
		Plutellidae	*Plutella*
		Pieridae	*Pieris, Colias*
		Pyralidae	*Hyphantria, Hellula, Maruca, Ostrinia*
		Sphingidae	*Manduca*
		Tortricidae	*Capua, Lobesia, Endopiza, Cydia, Grapholita, Epiphyas, Cnephasia, Argyrotaenia, Platynota, Eupoecillia, Sparganothis, Archips, Pandemis, Sasemia, Adoxophyes, Diaparopsis*
		Zygaenidae	*Harrisinia*
40–125 g ha^{-1}	Coleoptera	Anobiidae	*Lasioderma*
		Chrysomellidae	*Diabrotica, Leptinotarsa, Chaetocnema*
		Curculionidae	*Hypera, Cryptophlabes, Anthonomus, Conotrachelus, Sitophilus, Phyllobius, Curculio, Diaprepes, Peritelus*
		Tenebrionidae	*Rhizopertha*
0.05–1%	Hymenoptera	Formicidae	*Linepithemia, Camponotus, Solenopsis, Monomorium, Lasius, Atta*
		Diprionidae	*Hoplocampa*
75–125 g ha^{-1}	Hemiptera	Miridae	*Lygus, Pseudatomoscelis, Creontiades, Tydia*
40–125 g ha^{-1}	Homoptera	Cicadellidae	*Empoasca, Typhlocyba, Erythroneura, Scaphoideus*
50–100 g ha^{-1}	Orthoptera	Acrididae	*Schistocerca, Locusta*
		Gryllotalpidae	*Scapteriscus*
0.1–1%	Blattaria	Blattellidae	*Blattella, Supella*
		Blattidae	*Periplaneta, Blatta*
50–125 g ha^{-1}	Diptera	Agromyzidae	*Liriomyza*
0.01–0.1%		Culicidae	*Anopheles*
0.01–0.1%		Muscidae	*Musca*
75–125 g ha^{-1}		Tephritidae	*Rhagoletis*
0.09–0.5%	Isoptera	Rhinotermitidae	*Reticulitermes*
0.1–1%	Thysanura	Lepismatidae	*Lepisma*

[a] The majority of these insects are on indoxacarb labels or development work is ongoing to include them on labels, or sufficient data exists indicating the activity of indoxacarb. The list of insect families and genera is not inclusive of the total insect control activity of indoxacarb.

[b] Indoxacarb rates are expressed in grams of active ingredient per hectare or percent of active ingredient published on global labels or used in laboratory and field experiments to achieve a high level of insect mortality.

against German cockroaches (*Blattella germanica*) (Appel, 2003) and fire ant species (Barr, 2003; Patterson, 2003). However, the suspension concentrate formulation is also effective against certain roach, ant, fly, silverfish, and termite species when the pest is sprayed directly or as a surface or soil residual spray.

2.5.5. Indoxacarb and Insecticide Resistance

Indoxacarb is highly active against insect pest species that are already resistant to other insecticide classes. This has been demonstrated in the housefly (*Musca domestica*) resistant to spinosad (Shono and Scott, 2003) and organophosphates (Sugiyama *et al.*, 2001); tobacco budworm (*Heliothis virescens*) resistant to spinosad (Young *et al.*, 2001); German cockroach resistant to pyrethroids (Appel, 2003; Dong, unpublished data); tortricid leafrollers resistant to organophosphates (Olszak and Pluciennik, 1998); cotton bollworm (*H. armigera*) resistant to cyclodienes, organophosphates, carbamates and pyrethroids (Ahmad *et al.*, 2003); diamondback moth (*Plutella xylostella*) resistant to several conventional chemistries and spinosad (Boyd, 2001); beet armyworm (*Spodoptera exigua*) resistant to pyrethroids, organophosphates, and benzoylurea insect growth regulators (Eng, 1999); and *H. virescens* resistant to multiple conventional chemistries (Holloway *et al.*, 1999).

It is likely that the insects mentioned above are resistant to conventional insecticides due to a variety of mechanisms, including increased metabolism, decreased penetration and target site insensitivity. Because indoxacarb is bioactivated via esterase/amidase enzymes, overproduction of esterases in insects resistant to organophosphates or pyrethroids could lead to faster liberation of the active toxin DCMP than in nonresistant insects. This suggests that resistant insects may in fact develop a negative cross-resistance to indoxacarb, as has been observed in laboratory strains of *H. armigera* (Gunning and Devonshire, 2003). However, it is important to note that we have no clear evidence thus far for negative cross-resistance to indoxacarb in the field, indicating the speed and ease with which susceptible Lepidoptera can bioactivate indoxacarb. The aggregate resistance data also indicate that insects with increased resistance to conventional chemistries also have highly sensitive Na$^+$ channel DCMP binding sites, as would be expected since there has been no previous exposure to commercial SCBIs in the field. Thus, indoxacarb is an excellent rotation partner for alternating modes of action in insect resistance management programs.

As has been mentioned, certain field strains of *C. rosaceana*, which have developed resistance to many conventional insecticides, are also poorly sensitive to indoxacarb (Ahmad *et al.*, 2002). This may be due to an enhanced ability to detoxify indoxacarb; however, this insect is not on the indoxacarb US label and is an extremely unusual example of such poor susceptibility in Lepidoptera.

Insect pests representing eight species (*H. virescens, H. armigera, S. exigua, S. eridania, S. littoralis, P. xylostella,* and *T. absoluta*) in over 20 countries have been targeted for a sustained susceptibility monitoring program by DuPont Agricultural Products (Andaloro, unpublished data). These populations are evaluated by a feeding assay throughout each season in locations that have a history of insecticide resistance, and where a significant number of indoxacarb applications are made, so that early signs of resistance can be detected before widespread field failures occur. This proactive approach to indoxacarb resistance management is beneficial not only for preservation of this product in key markets, but also to maintain its usefulness as a rotation partner, thus promoting the longevity of

other products with different modes of action, and thus ultimately benefiting the agricultural field in general.

2.6. Conclusions

In summary, indoxacarb is the first insecticide product that acts by blocking the voltage-dependent Na$^+$ channel at site 10. The mode of action of this novel SCBI class was determined by sequential observation of poisoned insect symptoms, electrophysiology (starting with intact neuronal preparations and progressing towards more targeted ion channel techniques), and confirmation with Na$^+$ flux and biochemical radioligand binding techniques. It is believed that this approach is the most direct and elegant to determine the mode of action of insect neurotoxins, as opposed to testing the toxin in broad panels of specific target site biochemical assays and trusting that positive responses are indicative of toxicologically significant modes of action. Thus far no qualitative difference has been observed in the mode of action of the subclasses of the SCBIs. Indoxacarb itself is able to bind to Na$^+$ channels, but the main toxicologically active principle is the insect bioactivation metabolite DCMP. SCBIs appear to bind to a novel site, which appears to correspond to site 10 within the channel pore, and have almost no effect on resting Na$^+$ channels, so that in axons that normally have a very negative resting potential, they exert no effect unless the cell is artificially depolarized. Alternatively, sensory neurons and certain pacemaking neurons in the CNS generate action potentials in a region of the cell with a relatively low resting potential, where currents flowing through the cell membrane depolarize it, giving rise to trains of action potentials at a frequency that is dependent on membrane potential. Because the Na$^+$ channels in these regions are continually shuttling through the inactivated state, and are even, because of the low resting potential, spending a significant amount of time in the inactivated state, they are highly susceptible to compounds such as the SCBIs, which bind to the inactivated state. Inactivation is voltage-dependent, and at any given potential there is a dynamic equilibrium between inactivated and resting channels. As inactivated channels become bound to the active compound, they are removed from this equilibrium, with the result that more channels become inactivated to take their place in the equilibrium. In a sense, SCBIs pull channels into the inactivated state. While local anesthetics can access their binding site via an aqueous pathway from the internal solution as well as via a lipid pathway, SCBIs may have access only via the lipid pathway.

At a physiological level, the very slow off-rate of the receptor-bound SCBI (as evidenced by extremely slow washout in electrophysiology) may be largely responsible for the irreversible feeding inhibition seen in poisoned insects. Since this occurs at very low doses of indoxacarb in lepidopteran larvae, it is likely that in pseudoparalyzed insects only a relatively small proportion of total Na^+ channels is poisoned. Thus once DCJW is pseudoirreversibly bound to the slow inactivated state, return of the Na^+ channel to the unpoisoned inactivated state is very slow, which helps explain the high potency of indoxacarb/DCMP in halting insect feeding.

Indoxacarb was designed as a proinsecticide to confer favorable environmental degradation and mammalian safety, while maintaining high levels of insecticidal activity. Proinsecticide design is a common strategy to enhance insecticide selectivity (Prestwich, 1990). The enzymatic cleavage in pest insects via N-decarbomethoxylation is a novel insecticide bioactivation. The divergent indoxacarb metabolic pathways (insects primarily to N-decarbomethoxylated sodium channel blockers, while mammals degrade the compound primarily via other routes leading to multiple inactive metabolites) ensures a high degree of nontarget organism safety, a critical feature for modern insecticides.

Also, since hydrolytic esterases are largely responsible for the bioactivation, it is likely that insects that have become resistant to pyrethroids and organophosphates via induction of detoxifying esterases may in fact liberate DCMP more rapidly than susceptible insects. Because of this feature, and DCMP's novel Na^+ channel target site, accumulated field evidence shows that indoxacarb is generally highly active against insects resistant to these other commercial insecticide classes, and is thus an excellent insect pest management tool. This includes the pyrethroids and DDT, which bind to a different site on the voltage-gated Na^+ channel.

These properties have contributed to indoxacarb's strong potency in the field against Lepidoptera and some other pest orders. The bioavailability characteristics of indoxacarb make it ideal for foliar applications, where a surface residue of a strong chewing insect compound is desired. Its limited leaf penetration allows control of some Homoptera, some Hemiptera and both lepidopteran and dipteran leafminers, and its broad spectrum also confers activity against many nonagricultural pests such as termites, cockroaches, and certain ants. However, as with any new insecticide class, proactive resistance management is key to ensuring the maintenance of this compound and its target site of action as effective pest management tools.

References

Ahmad, M., Hollingworth, R.M., Wise, J.C., 2002. Broad-spectrum insecticide resistance in oblique-banded leafroller *Choristoneura rosaceana* (Lepidoptera: Tortricidae) from Michigan. *Pest Mgt Sci.* 58, 834–838.

Ahmad, M., Arif, M.I., Ahmad, Z., 2003. Susceptibility of *Helicoverpa armigera* (Lepidoptera: Noctuidae) to new chemistries in Pakistan. *Crop Protect.* 22, 539–544.

Andaloro, J.T., Wing, K.D., Green, J.H., Lang, E.B., 2000. Steward® dispersion and cotton leaf interactions: impact on cotton insect pests and safety to beneficial arthropods. *Proc. Beltwide Cotton Conf.* 2, 939–940.

Andaloro, J.T., Edmund, R.M., Castner, E.P., Williams, C.S., Sherrod, D.W., 2001. Steward® SC field performance against heliothines: speed of action, symptmology, and behavior of treated larvae. *Proc. Beltwide Cotton Conf.* 2, 925–926.

Anger, T., Madge, D.J., Mulla, M., Riddall, D., 2001. Medicinal chemistry of neuronal voltage-gated Na^+ channel blockers. *J. Med. Chem.* 44, 115–137.

Anonymous, 2000. Indoxacarb Technical Bulletin for Europe, Middle East and Africa. DuPont de Nemours and Co., Nambsheim, France.

Appel, A.G., 2003. Laboratory and field performance of an indoxacarb bait against German cockroaches (Dictyoptera: Blattellidae). *J. Econ. Entomol.* 96, 863–870.

Barr, C.L., 2003. Fire ant mound and foraging suppression by indoxacarb bait. In: Proc. Red Imported Fire Ant Conf., Palm Springs, CA, pp. 50–51.

Baur, M.E., Ellis, J., Hutchinson, K., Boethel, D.J., 2003. Contact toxicity of selective insecticides for non-target predaceous hemipterans in soybeans. *J. Entomol. Sci.* 38, 269–277.

Boyd, V.L., 2001. A preemptive strike: adopt sound resistance management plan to reduce chances of problems like those in Hawaii. *The Grower* 34(4), 40.

Brickle, D.S., Turnipseed, S.G., Sullivan, M.J., 2001. Efficacy of insecticides of different chemistries against *Helicoverpa zea* (Lepidoptera: Noctuidae) in transgenic *Bacillus thuringiensis* and conventional cotton. *J. Econ. Entomol.* 94, 86–92.

Catterall, W.A., 1987. Common modes of drug action on Na^+ channels: local anesthetics, antiarrhythmics and anticonvulsants. *Trends Pharmacol. Sci.* 8, 57–65.

Chen, H., McPheron, B., Krawczyk, G., Hull, L., 2003. Laboratory evaluation of surface residue toxicity of insecticides to apple maggot flies. *Int. Pest Control March/April 2003* 45, 92–93.

Cheng, Y.-C., Prusoff, W.H., 1973. Relation between the inhibition constant (K_i) and the concentration of inhibitor which causes fifty percent inhibition (I_{50}) of an enzymatic reaction. *Biochem. Pharmacol.* 22, 3099–3108.

Clare, J.J., Tate, S.N., Nobbs, M., Romanos, M.A., 2000. Voltage-gated Na$^+$ channels as therapeutic targets. *Drug Discov. Today 5*, 506–520.

Cronin, N.B., Reilly, A., Duclohier, H., Wallace, B.A., 2003. Binding of the anticonvulsant drug lamotrigine and the neurotoxin batrachotoxin to voltage-gated sodium channels induces conformational changes associated with block and steady-state activation. *J. Biol. Chem. 278*, 10675–10682.

Deecher, D.C., Soderlund, D.M., 1991. RH 3421, an insecticidal dihydropyrazole, inhibits Na$^+$ channel-dependent Na$^+$ uptake in mouse brain preparations. *Pestic. Biochem. Physiol. 39*, 130–137.

Deecher, D.C., Payne, G.T., Soderlund, D.M., 1991. Inhibition of [^{3}H]batrachotoxinin A 20-alpha-benzoate binding to mouse brain Na$^+$ channels by the dihydropyrazole insecticide RH 3421. *Pestic. Biochem. Physiol. 41*, 265–273.

Eng, O., 1999. Indoxacarb: a potent compound for controlling the beet armyworm (*Spodoptera exigua*) on fruit vegetables in Malaysia. *Proc. 3rd Conf. Entomol. Soc. Malaysia* 101.

EPA, 2000. Fact Sheet on Indoxacarb. Issued Oct. 30, 2000. Environmental Protection Agency, Washington, DC.

Finlayson, L.H., Lowenstein, O., 1958. The structure and function of abdominal stretch receptors in insects. *Proc. Roy. Soc. London B 148*, 433–449.

Flexner, L., Green, J., Wing, K., Cameron, R., Saienii, J., et al., 2000. Steward$^®$ 150 SC rainfastness and residual control on cotton. *Proc. Beltwide Cotton Conf. 2*, 940.

Giraddi, R.S., Dasareddy, S.V., Lingappa, S.L., 2002. Bioefficacy of new molecules of insecticides against gram pod-borer (*Helicoverpa armigera*) in pigeonpea (*Cajanus cajan*). *Indian J. Agric. Sci. 72*, 311–312.

Gunning, R.V., Devonshire, A.L., 2003. Negative cross-resistance between indoxacarb and pyrethroids in the cotton bollworm, *Helicoverpa armigera*, in Australia: a tool for resistance management. *Proc. Brighton Crop Protect. Conf. 2*, 788.

Hammes, P., Sherrod, D.W., Hammes, G.G., 1999. Cotton pest and beneficial insect management strategies with Steward insect control product in the Southeast US. *Proc. Beltwide Cotton Conf. 2*, 1224–1225.

Harder, H.H., Riley, S.L., McCann, S.F., Irving, S.N., 1996. DPX-MP062: a novel broad-spectrum, environmentally soft insect control compound. *Proc. Brighton Crop Protects Conf. 1*, 48–50.

Hasan, R., Nishimura, K., Okada, M., Adamatsu, M., Inoue, M., et al., 1996. Stereochemical basis for the insecticidal activity of carbamoylated and acylated pyrazolines. *Pestic. Sci. 46*, 105–112.

Hodgkin, A.L., Huxley, A.F., 1952. A quantitative description of membrane current and its application to conduction and excitation in nerve. *J. Physiol. 117*, 500–544.

Holloway, J.W., Forrester, N.W., Leonard, B.R., 1999. New insecticide chemistry for cotton IPM. *Proc. Beltwide Cotton Conf. 2*, 1086.

Ishaaya, I., 2001. Biochemical processes related to insecticide action: an overview. In: Ishaaya, I. (Ed.), Biochemical Sites of Insecticide Action and Resistance. Springer, New York, pp. 1–16.

Jacobson, R.M., 1985. N-aryl 3-aryl-4,5-dihydro-1H-pyrazole-2-pyrazolines. US Patent 5 798 311.

Jacobson, R.M., 1989. A class of insecticidal dihydropyrazoles. In: Crombie, L.E. (Ed.), Recent Advances in the Chemistry of Insect Control. Royal Society of Chemistry, Cambridge, pp. 206–211.

Kondratiev, A., Tomaselli, G.F., 2003. Altered gating and local anesthetic block mediated by residues in the I-S6 and II-S6 transmembrane segments of voltage-dependent Na$^+$ channels. *Mol. Pharmacol. 64*, 741–752.

Lapied, B., Malecot, C.O., Pelhate, M., 1989. Ionic species involved in the electrical activity of single adult aminergic neurones isolated from the sixth abdominal ganglion of the cockroach *Periplaneta americana*. *J. Exp. Biol. 144*, 535–549.

Lapied, B., Stankiewicz, M., Grolleau, F., Rochat, H., Zlotkin, E., et al., 1999. Biophysical properties of scorpion alpha-toxin-sensitive background Na$^+$ channel contributing to the pacemaker activity in insect neurosecretory cells (DUM neurons). *Eur. J. Neurosci. 11*, 1449–1460.

Lapied, B., Grolleau, F., Sattelle, D.B., 2001. Indoxacarb, an oxadiazine insecticide, blocks insect neuronal Na$^+$ channels. *Br. J. Pharmacol. 132*, 587–595.

Liu, T.-X., Sparks, A.N., 1999. Efficacies of selected insecticides on cabbage looper and diamondback moth on cabbage in south Texas. *Subtrop. Plant Sci. 51*, 56–60.

Liu, T.-X., Sparks, A.N., Chen, W., Liang, G.-M., Brister, C., 2002. Toxicity, persistence and efficacy of Indoxacarb on cabbage looper (Lepidoptera: Noctuidae). *J. Econ. Entomol. 95*, 360–367.

Liu, T.-X., Sparks, A.N., Chen, W., 2003. Toxicity, persistence and efficacy of indoxacarb and two other insecticides on *Plutella xylostella* (Lepidoptera: Plutellidae) immatures in cabbage. *Int. J. Pest Mgt 49*, 235–241.

Lund, A.E., Narahashi, T., 1982. Dose-dependent interaction of the pyrethroid isomer with sodium channels of squid axon membranes. *Neurotoxicology 3*, 11–24.

McCann, S.F., Annis, G.D., Shapiro, R., Piotrowski, D.W., Lahm, G.P., et al., 2001. The discovery of indoxacorb: oxadiazineses a new class of pyrazoline – type insecticides. *Pest Mgt Sci. 57*, 153–164.

McCann, S.F., Annis, G.D., Shapiro, R., Piotrowski, D.W., Lahm, G.P., et al., 2002. Synthesis and biological activity of oxadiazine and triazine insecticides: the discovery of Indoxacarb. In: Baker, D.R., Fenyes, J.G., Lahm, G.P., Selby, T.S., Stevenson, T.M. (Eds.), Synthesis and Chemistry of Agrochemicals, vol. 6. American Chemical Society, Washington, DC, pp. 166–177.

Mead-Briggs, M., Bakker, F.M., Grove, A.J., Primiani, M.M., 1996. Evaluating the effects of multiple-application plant protection products on beneficial arthropods by means of extended laboratory tests: case study with predatory mites and hoverflies, and the insecticides

JW-062 and DPX-MP062. *Proc. Brighton Crop Protect. Conf. 1*, 307–313.

Michaud, J.P., Grant, A.K., 2003. IPM-compatability of foliar insecticides for citrus: indices derived from toxicity to beneficial insects from four orders. *J. Insect Sci. 3*, 18–27.

Mulder, R., Wellinga, K., 1973. Insecticidal 1-(phenylcarbamoyl)-2-pyrazolines. German Patent DE 2 304 584.

Musser, F.R., Shelton, A.M., 2003. *Bt* sweet corn and selective insecticides: impacts on pests and predators. *J. Econ. Entomol. 96*, 71–80.

Nicholson, R.A., 1992. Insecticidal dihydropyrazoles antagonize the depolarizing action of veratridine in mammalian synaptosomes as measured with a voltage-sensitive dye. *Pestic. Biochem. Physiol. 42*, 197–202.

Nicholson, R.A., Merletti, E.L., 1990. The effect of dihydropyrazoles on release of [^{3}H]GABA from nerve terminals isolated from mammalian cerebral cortex. *Pestic. Biochem. Physiol. 37*, 30–40.

Nowak, J.T., McGravy, K.W., Fettig, C.J., Berisford, C.W., 2001. Susceptibility of adult hymenopteran parasitoids of the Nantucket Pine Tip Moth (Lepidoptera: Tortricidae) to broad-spectrum and biorational insecticides in a laboratory study. *J. Econ. Entomol. 94*, 1122–1129.

Olszak, R.W., Pluciennik, Z., 1998. Preliminary investigations on effectiveness of two modern insecticides in controlling codling moth, plum fruit moth and leafrollers. *Proc. Brighton Conf. Crop Protect. 2*, 57.

Parimi, S., Meinke, L.J., Nowatzki, T.M., Chandler, L.D., French, B.W., et al., 2003. Toxicity of insecticide–bait mixtures of insecticide resistant and susceptible western corn rootworms (Coleoptera: Chrysomelidae). *Crop Protect. 22*, 781–786.

Patterson, R.S., 2003. Spring and fall applications of granular indoxacarb baits for fire ant control in Florida. In: Proc. Red Imported Fire Ant Conf., Palm Springs, CA, pp. 51–52.

Payne, G.T., Deecher, D.C., Soderlund, D.M., 1998. Structure–activity relationships for the action of dihydropyrazole insecticides on mouse brain sodium channels. *Pestic. Biochem. Physiol. 60*, 177–185.

Pichon, Y., Ashcroft, F.M., 1985. Nerve and muscle: electrical activity. In: Kerkut, G.A., Gilbert, L.I. (Eds.), Comprehensive Insect Physiology Biochemistry and Pharmacology, vol. 5. Pergamon, Oxford, pp. 85–113.

Pluschkell, U., Horowitz, A.R., Weintraub, P.G., Ishaaya, I., 1998. DPX-MP062: a potent compound for controlling the Egyptian cotton leafworm *Spodoptera littoralis* (Boisd.). *Pestic. Sci. 54*, 85–90.

Postma, S.W., Catterall, W.A., 1984. Inhibition of binding of [^{3}H]batrachotoxin A 20-a-benzoate to Na$^+$ channels by local anesthetics. *Mol. Pharmacol. 25*, 219–227.

Prestwich, G.D., 1990. Proinsecticides: metabolically activated toxicants. In: Hodgson, E., Kuhr, R.J. (Eds.), Safer Insecticides – Development and Use. Marcel, New York, pp. 281–335.

Ruberson, J.R., Tillman, P.G., 1999. Effect of selected insecticides on natural enemies in cotton: laboratory studies. *Proc. Beltwide Cotton Conf. 2*, 1210–1213.

Salgado, V.L., 1990. Mode of action of insecticidal dihydropyrazoles: selective block of impulse generation in sensory nerves. *Pestic. Sci. 28*, 389–411.

Salgado, V.L., 1992. Slow voltage-dependent block of Na$^+$ channels in crayfish nerve by dihydropyrazole insecticides. *Mol. Pharmacol. 41*, 120–126.

Salgado, V.L., 1998. Studies on the mode of action of spinosad: insect symptoms and physiological correlates. *Pestic. Biochem. Physiol. 60*, 91–102.

Schouest, L.P., Salgado, V.L., Miller, T.A., 1986. Synaptic vesicles are depleted from motor nerve terminals of deltamethrin-treated house fly larvae, *Musca domestica. Pestic. Biochem. Physiol. 25*, 381–386.

Scott, M.T., 2000. Comparative degradation and metabolism of indoxacarb, an oxadiazine insecticide. *Am. Chem. Soc. Mtg,* San Francisco, CA, Abstracts.

Seal, D.M., McCord, E., 1996. Management of diamondback moths, *Plutella xylostella* (L.) (Lepidoptera: Plutellidae), in cole crops. *Nucl. Sci. Applic. 7*, 49–55.

Shapiro, R., Annis, G.D., Blaisdell, C.T., Dumas, D.J., Fuchs, J., et al., 2002. Toward the manufacture of indoxacarb. In: Baker, D.R., Fenyes, J.G., Lahm, G.P., Selby, T.P., Stevenson, T.M. (Eds.), Synthesis and Chemistry of Agrochemicals, vol. 6. American Chemical Society, Washington, DC, pp. 178–185.

Shono, T., Scott, J.G., 2003. Spinosad resistance in the housefly, *Musca domestica*, is due to a recessive factor on autosome 1. *Pestic. Biochem. Physiol. 75*, 1–7.

Smirle, M.J., Lowery, D.T., Zurowski, C.L., 2002. Resistance and cross-resistance to four insecticides in populations of oblique-banded leafroller (Lepidoptera: Tortricidae). *J. Econ. Entomol. 95*, 820–825.

Sugiyama, S., Tsurubuchi, Y., Karasawa, A., Nagata, K., Kono, Y., et al., 2001. Insecticidal activity and cuticular penetration of Indoxacarb and its *N*-decarbomethoxylated metabolite in organophosphorus insecticide-resistant and susceptible strains of the housefly *Musca domestica* (L.). *J. Pestic. Sci. 26*, 117–120.

Teague, T.G., Tugwell, N.P., Muthiah, S., Hornbeck, J.M., 2000. New insecticides for control of tarnished plant bug: results from field and cage studies and laboratory bioassays. *Proc. Beltwide Cotton Conf. 2*, 1214–1217.

Tillman, P.G., Hammes, G.G., Sacher, M., Connair, M., Brady, E.A., et al., 2001. Toxicity of a formulation of the insecticide indoxacarb to the tarnished plant bug, *Lygus lineolaris* (Hemiptera: Miridae), and the big-eyed bug, *Geochoris punctipes* (Hemiptera: Lygaeidae). *Pest Mgt Sci. 58*, 92–100.

Tillman, P.G., Mulrooney, J.E., Mitchell, W., 1998. Susceptibility of selected beneficial insects to DPX-MP062. *Proc. Beltwide Cotton Conf. 2*, 1112–1114.

Turnipseed, S.G., Sullivan, M.J., 2000. Activity of Steward® against plant bugs, bollworms, and predacious arthropods in cotton in South Carolina. *Proc. Beltwide Cotton Conf. 2*, 953–954.

Tsurubuchi, Y., Karasawa, A., Nagata, A., Shono, T., Kono, Y., 2001. Insecticidal activity of oxadiazine insecticide indoxacarb and its N-decarbomethoxylated metabolite and their modulations of voltage-gated Na+ channels. *Appl. Entomol. Zool.* 36, 381–385.

Tsurubuchi, Y., Kono, Y., 2003. Modulation of sodium channels by the oxadiazine insecticide indoxacarb and its N-decarbomethoxylated metabolite in rat dorsal root ganglion neurons. *Pest Mgt Sci.* 59, 999–1006.

Wang, S.Y., Wang, G.K., 2003. Voltage-gated Na+ channels as primary targets of diverse lipid-soluble neurotoxins. *Cell Signal.* 15, 151–159.

Wasserstrom, J.A., Liberty, K., Kelly, J., Santucci, P., Myers, M., 1993. Modification of cardiac Na+ channels by batrachotoxin: effects on gating, kinetics, and local anesthetic binding. *Biophys. J.* 65, 386–395.

Wiersma, C.A.G., Furshpan, E., Florey, E., 1953. Physiological and pharmacological observations on muscle organs of the crayfish, *Cambarus clarkii* Girard. *J. Exp. Biol* 30, 136–151.

Williams, L. III, Price, L.D., Manrique, V., 2003. Toxicity of field-weathered insecticide residues to *Anaphes iole* (Hymenoptera: Mymaridae), an egg parasitoid of *Lygus lineolaris* (Heteroptera: Miridae), and implications for inundative biological control in cotton. *Biol. Control* 26, 217–223.

Wing, K.D., Schnee, M.E., Sacher, M., Connair, M., 1998. A novel oxadiazine insecticide is bioactivated in lepidopteran larvae. *Arch. Insect Biochem. Physiol.* 37, 91–103.

Wing, KD, Sacher M., Kagaya, Y., Tsurubuchi, Y., Mulderig, L., et al., 2000. Bioactivation and mode of action of the oxadiazine indoxacarb in insects. *Crop Protect.* 19, 537–545.

Yarov-Yarovoy, V., McPhee, J.C., Idsvoog, D., Pate, C., Scheuer, T., et al., 2002. Role of amino acid residues in transmembrane segments IS6 and IIS6 of the Na+ channel subunit in voltage-dependent gating and drug block. *J. Biol. Chem.* 277, 35393–35401.

Young, H.P., Bailey, W.D., Roe, R.M., 2001. Mechanism of resistance and cross-resistance in a laboratory, Spinosad-selected strain of the tobacco budworm and resistance in laboratory-selected cotton bollworms. *Proc. Beltwide Cotton Conf.* 2, 1167–1170.

Zamponi, G.W., Doyle, D.D., French, R.J., 1993. Fast lidocaine block of cardiac and skeletal muscle Na+ channels: one site with two routes of access. *Biophys. J.* 65, 80–90.

Zhang, A., Nicholson, R.A., 1993. The dihydropyrazole RH-5529 blocks voltage-sensitive calcium channels in mammalian synaptosomes. *Pestic. Biochem. Physiol.* 45, 242–247.

Zhang, A., Towner, P., Nicholson, R.A., 1996. Dihydropyrazole insecticides: interference with depolarization-dependent phosphorylation of synapsin I and evoked release of L-glutamate in nerve-terminal preparations from mammalian brain. *Pestic. Biochem. Physiol.* 54, 24–30.

Zhao, X., Ikeda, T., Yeh, J.Z., Narahashi, T., 2003. Voltage-dependent block of sodium channels in mammalian neurons by the oxadiazine insecticide indoxacarb and its metabolite DCJW. *Neurotoxicol.* 24, 83–96.

Zlotkin, E., 2001. Insecticides affecting voltage-gated ion channels. In: Ishaaya, I. (Ed.), Biochemical Sites of Insecticide Action and Resistance. Springer, New York, pp. 43–76.

A2 Addendum: Indoxacarb and the Sodium Channel Blockers: Chemistry, Physiology, and Biology in Insects

V L Salgado, BASF Corporation, Research Triangle Park, NC, USA

© 2010 Elsevier B.V. All Rights Reserved

Since the publication of the main chapter, a second sodium channel blocker insecticide (SCBI) of the semicarbazone subclass (**Figure 2D** of the original chapter) has reached the market. Optimization of the semicarbazones, which can be thought of as ring-opened pyrazolines, led to metaflumizone (Takagi *et al.*, 2007; **Figure A1**), which is being codeveloped globally by Nihon Nohyaku Co., Ltd, BASF SE, and Pfizer Animal Health, a business unit of Pfizer Inc. Metaflumizone provides good to excellent control of most economically important Lepidoptera pests and certain pests in the orders Coleoptera, Hemiptera, Hymenoptera, Diptera, Isoptera, and Siphonaptera (BASF, 2007). Metaflumizone provides long-lasting control of fleas on companion animals with a single spot-on application and is being marketed for this use under the trade name Promeris by Pfizer Animal Health (Rugg and Hair, 2007). The mechanism of action of metaflumizone as a sodium channel blocker with selectivity for the slow-inactivated state was confirmed by Salgado and Hayashi (2007).

State-dependent block of sodium channels by indoxacarb and its *N*-decarbomethoxyllated metabolite (DCJW) was demonstrated for the first time in insect sodium channels by Zhao *et al.* (2005). State-dependent binding of DCJW and RH-3421 to the rat $Na_v1.4$ sodium channel was confirmed, and it was also shown that the anticonvulsant phenytoin decreased the potency of DCJW and RH-3421. This supports the earlier studies demonstrating competitive interactions between SCBIs and local anesthetics (LAs) in ^{3}H-BTX-B binding assays discussed in Sect. 6.2.4.5 and the hypothesis that the SCBI binding site on the sodium channel overlaps that of local anesthetics and anticonvulsants (Silver and Soderlund, 2005b).

Three recent studies examine how variation in voltage dependence of slow inactivation among sodium channel variants affects sensitivity to SCBIs. Zhao *et al.* (2005) identified two types of TTX-sensitive sodium currents in American cockroach neurons, with large differences in the half-point potentials of fast and slow inactivation and a corresponding difference in the sensitivity to DCJW and indoxacarb. Furthermore, Song *et al.* (2006) identified a variant of the German cockroach $BgNa_v1$ sodium channel with a lysine to glutamate substitution (K1689E) in D4S4 that caused a large negative shift of the half-point potentials of both fast and slow inactivation and a corresponding increase in sensitivity to DCJW. Similarly, a valine to lysine substitution (V787K) in D2S6 of the rat $Na_v1.4$ sodium channel caused a large negative shift in the voltage dependence of slow inactivation and a corresponding increase in sensitivity to DCJW and RH-3421 (von Stein and Soderlund, 2009). In all of these studies, each of the sodium channel variants was sensitive to SCBIs only at potentials in the range where gating occurs, lending support to the state-dependent binding hypothesis.

Significant advances in mapping the SCBI binding site with site-directed mutagenesis have been made in the last 5 years (reviewed by Silver *et al.*, 2009b). These studies have confirmed that the SCBI binding site overlaps the LA binding site and also show that different residues are involved in binding these two groups of compounds. Work so far has focused on residues phenylalanine F13 and tyrosine Y20 in D4S6 (Figs. 11 and 12 of the original chapter), which are particularly important in the binding of local anesthetics. Silver and Soderlund (2007), using the rat $Na_v1.4$ sodium channel expressed in Xenopus oocytes, found that alanine substitution at F13 significantly reduced sensitivity to block by DCJW and RH-3421. Substitution of other amino acids at this position showed that an aromatic side chain is required for high potency of local anesthetics (Li *et al.*, 1999), and this residue may be important for aromatic–aromatic interactions with SCBIs in the rat $Na_v1.4$ sodium channel. Silver and Soderlund (2007) found that alanine substitution of Y20 in rat $Na_v1.4$ channels increased their sensitivity to DCJW and RH-3421 58-fold and 16-fold respectively, and modestly increased the sensitivity to

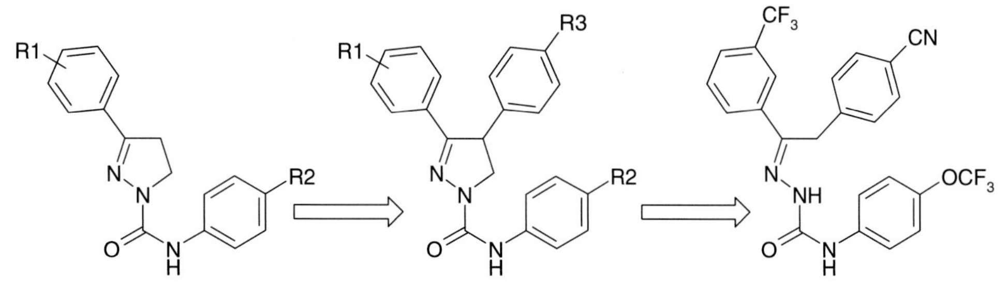

Figure A1 Chemical evolution of the semicarbazone metaflumizone from pyrazolines.

indoxacarb. Thus, while providing important pi-electron interactions for local anesthetics, the aromatic residue at Y20 in D4S6 might hinder access of SCBIs to their binding site. In the German cockroach sodium channel variant BgNa$_v$1-1A, alanine substitution at positions F13 and Y20 of D4S6 had somewhat different effects on the action of SCBIs (Silver et al., 2009a). F13A substitution in the cockroach channel did not change its sensitivity to indoxacarb or DCJW, and made it hypersensitive to metaflumizone. Furthermore, this mutation greatly accelerated recovery of the channels from block by metaflumizone. Therefore, in contrast to its role in the rat Na$_v$1.4 channel in the binding of LAs and SCBIs, the phenyl functional group at position F13 in D4S6 does not appear to participate in the binding of SCBIs in BgNa$_v$1-1A channels, and may in fact hinder the access of metaflumizone (but not DCJW) to its binding site. As mentioned above, alanine substitution of Y20 in D4S6 of rat Na$_v$1.4 sodium channels increased sensitivity to DCJW 58-fold (Silver and Soderlund, 2007). The corresponding mutation increased the sensitivity of BgNa$_v$1-1A channels 13-fold to DCJW and 11-fold to metaflumizone. Thus, the aromatic side chain at position 20 of D4S6, while crucial for local anesthetic binding, does not appear to participate in binding interactions with SCBIs, and in fact seems to hinder access of SCBIs to their binding site in both insect and mammalian sodium channels.

In conclusion, while F13 of D4S6 is important for binding of both LAs and SCBIs in the rat Na$_v$1.4 channel, it does not affect binding of DCJW in the BgNa$_v$1-1A channel, and in fact hinders access of metaflumizone to the binding site. Y20 in D4S6, while crucial for LA binding, hinders access of SCBIs to their site in both insect and mammalian channels. The fact that residues involved in local anesthetic binding sometimes antagonize binding of SCBIs suggests that the SCBI and LA binding sites overlap, but that the binding energy of SCBIs comes in part from different residues, which have yet to be identified.

References

BASF Agricultural Products, 2007. Metaflumizone Worldwide Technical Brochure.

Li, H.-L., Galue, A., Meadows, L., Ragsdale, D.S., 1999. A molecular basis for the different local anesthetic affinities of resting versus open and inactivated states of the sodium channel. Mol. Pharmacol. 55, 134–141.

Rugg, D., Hair, J.A., 2007. Dose determination of a novel formulation of metaflumizone plus amitraz for the treatment and control of fleas (Ctenocephalides felis) and ticks (Rhipicephalus sanguineus) on dogs. Vet. Parasitol. 150, 203–208.

Salgado, V.L., Hayashi, J.H., 2007. Metaflumizone is a novel sodium channel blocker insecticide. Vet. Parasitol. 150, 182–189.

Silver, K.S., Soderlund, D.M., 2005. State-dependent block of rat Nav1.4 sodium channels expressed in Xenopus oocytes by pyrazoline-type insecticides. Neurotoxicology 26, 397–406.

Silver, K.S., Soderlund, D.M., 2007. Point mutations at the local anesthetic receptor site modulate the state-dependent block of rat Na$_v$1.4 sodium channels by pyrazoline-type insecticides. Neurotoxicology 28, 655–663.

Silver, K.S., Nomura, Y., Salgado, V.L., Dong, K., 2009a. Role of the sixth transmembrane segment of domain IV of the cockroach sodium channel in the action of sodium channel blocker insecticides. Neurotoxicology 30, 613–621.

Silver, K.S., Song, W., Nomura, Y., Salgado, V.L., Dong, K., 2009b. Mechanism of action of sodium channel blocker insecticides (SCBIs) on insect sodium channels. Pestic. Biochem. Physiol. in press, doi:10.1016/j.pestbp.2009.09.001.

Song, W., Liu, Z., Dong, K., 2006. Molecular basis of differential sensitivity of insect sodium channels to DCJW, a bioactive metabolite of the oxadiazine insecticide indoxacarb. Neutotoxicol 27, 237–244.

Takagi, K., Hamaguchi, H., Nishimatsu, T., Konno, T., **2007**. Discovery of metaflumizone, a novel semicarbazone insecticide. *Vet. Parasitol. 150*, 177–181.

von Stein, R.T., Soderlund, D.M., **2009**. A point mutation that enhances slow sodium channel inactivation confers resting-state sensitivity to block by sodium channel blocker insecticides (SCBIs). Program No. 314.7/B58.

2009 Neuroscience Meeting Planner. Society for Neuroscience, Chicago, IL, Online.

Zhao, X., Ikeda, T., Salgado, V.L., Yeh, J.Z., Narahashi, T., **2005**. Block of two subtypes of sodium channels in cockroach neurons by indoxacarb insecticides. *Neurotoxicology 26*, 455–465.

3 Neonicotinoid Insecticides

P Jeschke and R Nauen, Bayer CropScience AG,
Monheim, Germany

© 2010, 2005 Elsevier B.V. All Rights Reserved

3.1. Introduction

The discovery of neonicotinoids as important novel pesticides can be considered a milestone in insecticide research of the past three decades. Neonicotinoids represent the fastest-growing class of insecticides introduced to the market since the commercialization of pyrethroids (Nauen and Bretschneider, 2002). Like the naturally occurring nicotine, all neonicotinoids act on the insect central nervous system (CNS) as agonists of the postsynaptic nicotinic acetylcholine receptors (nAChRs) (Bai et al., 1991; Liu and Casida, 1993a; Yamamoto, 1996; Chao et al., 1997; Zhang et al., 2000; Nauen et al., 2001), but, with remarkable selectivity and efficacy against pest insects while being safe for mammals. As a result of this mode of action there is no cross-resistance to conventional insecticide classes, and therefore the neonicotinoids have begun replacing pyrethroids, chlorinated hydrocarbons, organophosphates, carbamates, and several other classes of compounds as insecticides to control insect pests on major crops (Denholm et al., 2002). Today the class of neonicotinoids are part of a single mode of action group as defined by the Insecticide Resistance Action Committee (IRAC; an Expert Committee of Crop Life) for pest management purposes (Nauen et al., 2001).

Neonicotinoids are potent broad-spectrum insecticides possessing contact, stomach, and systemic activity. They are especially active on hemipteran pest species, such as aphids, whiteflies, and planthoppers, but they are also commercialized to control coleopteran and lepidopteran pest species (Elbert et al., 1991, 1998). Because of their physicochemical properties they are useful for a wide range of different application techniques, including foliar, seed treatment, soil drench, and stem application in several crops. Due to the favorable mammalian safety characteristics (Matsuda *et al.*, 1998; Yamamoto *et al.*, 1998; Tomizawa *et al.*, 2000) neonicotinoids like imidacloprid are also important for the control of subterranean pests, and for veterinary use (Mencke and Jeschke, 2002).

3.2. Neonicotinoid History

In the early 1970s, the former Shell Development Company's Biological Research Center in Modesto, California, invented a new class of nitromethylene heterocyclic compounds capable of acting on the nAChR. Starting with a random screening to discover lead structures from university sources, Shell detected the 2-(dibromo-nitromethyl)-3-methyl pyridine (SD-031588) (**Figure 1**) from a pool of chemicals from Prof. Henry Feuer of Purdue University (Feuer and Lawrence, 1969), which revealed an unexpected low-level insecticidal activity against housefly and pea aphid.

Further structural optimization of this insecticide lead structure led to the active six-membered tetrahydro-2-(nitromethylene)-2*H*-1,3-thiazine, nithiazine (SD-03565, SKI-71) (Soloway *et al.*, 1978, 1979; Schroeder and Flattum, 1984; Kollmeyer *et al.*, 1999). The molecular design by Shell appears rational and straightforward. The chemistry had been largely concentrated on the nitromethylene enamine skeleton (Kagabu, 2003a). Today, this early prototype can be considered as the first generation of the so-called neonicotinoid insecticides (**Figure 2**). Nithiazine showed higher activity than parathion against housefly adults (*Musca domestica*), and 1662 times higher activity against the target insect, the lepidopteran corn earworm larvae (*Helicoverpa zea*), combined with good systemic behavior in plants and low mammalian toxicity (Soloway *et al.*, 1978, 1979; Kollmeyer *et al.*, 1999; Tomizawa and Casida, 2003). However, due to

the photochemically unstable 2-nitromethylene chromophore (**Figure 3**) in the field tests, nithiazine was never commercialized for broad agricultural use (Soloway *et al.*, 1978, 1979; Kagabu and Medej, 1995; Kagabu, 1997a; Kollmeyer *et al.*, 1999). Alternatively, photostabilization using the formyl moiety was not adequate for practical application (Kollmeyer *et al.*, 1999). Nevertheless, a knock-down fly product against *M. domestica* containing nithiazine as the active ingredient of a housefly trap device for poultry and animal husbandry has recently been commercialized (Kollmeyer *et al.*, 1999).

In the early 1980s synthesis work was initiated at Nihon Tokushu Noyaku Seizo K. K. (presently Bayer CropScience K. K.) on the basis of this first remarkable neonicotinoid lead structure and the unique insecticidal spectrum of activity (Kagabu *et al.*, 1992; Moriya *et al.*, 1992). Instead of the lepidopteran larva *H. zea*, the target insect for studying structure–activity relationships (SARs) and optimizing biological activity was the green rice leafhopper (*Nephotettix cincticeps*), because it is a major hemipteran pest of rice in Japan (Kagabu, 1997a). At the beginning of the project, a new pesticide screening method using rice seedlings was developed for continuous monitoring of the combined systemic and contact activities of compounds over 2 weeks against leafhoppers and planthoppers (Sone *et al.*, 1995).

The six-membered 2-(nitromethylene)-tetrahydro-1,3-thiazine ring was replaced with different N^1-substituted *N*-heterocyclic ring systems. Starting with 1-methyl-2-(nitromethylene)-imidazolidine, it was found that the activity depends on ring size ($5 > 6 > 7$-ring system). The N^1-benzylated 2-(nitromethylene)-imidazolidine 5-ring was the most active one. Introduction of substituted benzyl residues in the 1-position, like the *para*-chlorobenzyl moiety, and of nitrogen-containing *N*-hetarylmethyl

	Compound		Toxicity index (TI)[a]	
1970–1978 (Shell)	Shell no.	E	Pea aphid	Corn earworm
	SD-031588		~5	0
	SD-033420	CH₂	~3	155
	SD-035347	NH	+	5.2
	SD-035651	S	24	1662
	Nithiazine (SKI-71)			

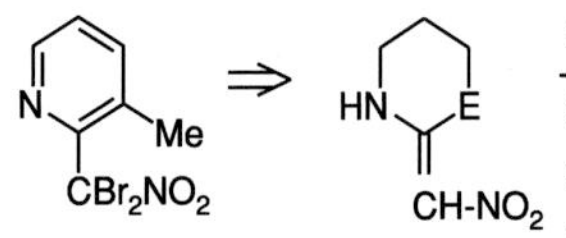

[a]TI = (LC₅₀ parathion/LC₅₀ test compound) × 100.

Figure 1 Development of nithiazine (SKI-71) by Shell. (Data from Kollmeyer, W.D., Flattum, R.F., Foster, J.P., Powel, J.E., Schroeder, M.E., *et al.*, **1999**. Discovery of the nitromethylene heterocycle insecticides. In: Yamamoto, I., Casida, J.E. (Eds.), Neonicotinoid Insecticides and the Nicotinic Acetylcholine Receptor. Springer, New York, pp. 71–89 and Kagabu, S., **2003a**. Molecular design of neonicotinoids: past, present and future. In: Voss, G., Ramos, G. (Eds.), Chemistry of Crop Protection: Progress and Prospects in Science and Regulation. Wiley–VCH, New York, pp. 193–212.)

Pharmacophores [—N—C(E)=X—Y]	Structure type	
	Ring system (R¹—R², R¹—Z—R²)	Noncyclic structure (R¹, R²)
Nitroenamines (E = S, N) (nitromethylenes) [—N—C(E)=CH—NO₂]	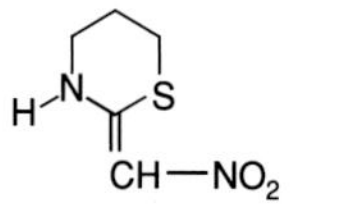 Nithiazine	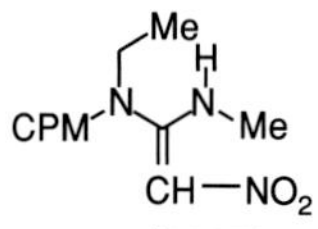 Nitenpyram (1987)ᵃ
Nitroguanidines (E = N) [—N—C(N)=N—NO₂]	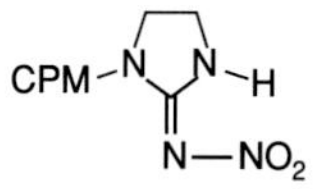 Imidacloprid (1985)ᵃ	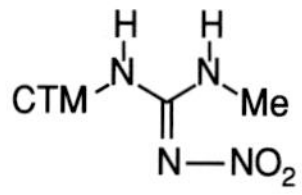 Clothianidinᵇ (1989)ᵃ
	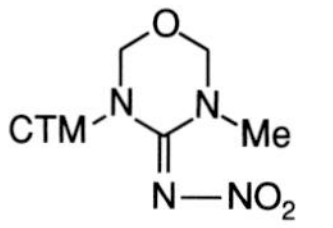 Thiamethoxam (1992)ᵃ	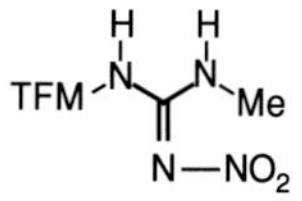 (±)-Dinotefuran (1993)ᵃ
Cyanoamidines (E = S, Me) [—N—C(E)=N—CN]	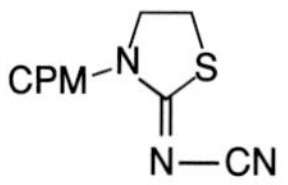 Thiacloprid (1986)ᵃ	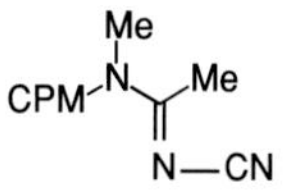 Acetamiprid (1988)ᵃ

ᵃYear of the first patent application (priority date) covering the insecticide.

Figure 2 Chemical structures of the first generation neonicotinoid nithiazine and commercialized neonicotinoids displaying the different types of pharmacophors. CPM, 6-chloro-pyrid-3-ylmethyl; CTM, 2-chloro-1,3-thiazol-5-ylmethyl; TFM, (±)-6-tetrahydro-fur-3-ylmethyl. (Reproduced with permission from Jeschke, P., Schindler, M., Beck, M., 2002. Neonicotinoid insecticides: retrospective consideration and prospects. *Proc. BCPC: Pests and Diseases 1*, 137–144.)

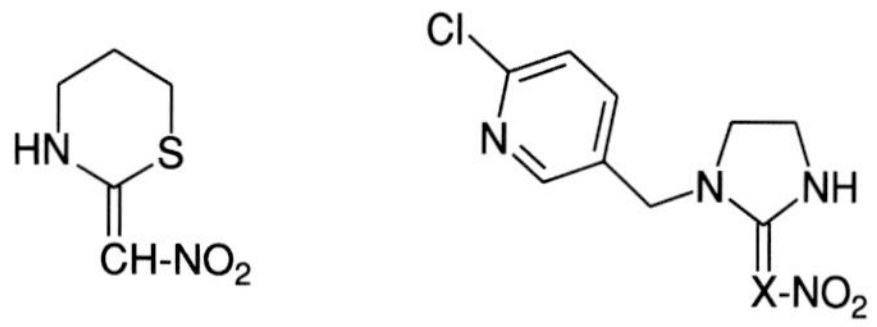

	Nithiazine (SKI-71)	(NTN32692) X = —CH=	(NTN33893) Imidacloprid X = —N=
λ_{max} (nm)	343	323	269
Water solubility (g l⁻¹ at 20 °C)	200	3.4	0.57
LC₉₀ (ppm)	40	0.32	0.32

Figure 3 Electronic absorption (λ_{max} in nm), water solubility (g l⁻¹ at 20 °C), and insecticidal efficacy (ppm) of nithiazine (SKI-71), nitromethylene (NTN32692), and imidacloprid (NTN33893) against green rice leafhopper (*Nephotettix cincticeps*). LC₉₀, concentration at which >90% of insects are killed (Mencke and Jeschke, 2002).

residues, like the pyrid-3-ylmethyl moieties, enhanced the insecticidal activity by a factor of 5 and 25 when compared with the N^1-benzylated 2-(nitromethylene)-imidazolidine (Kagabu, 1997a). Yamamoto had previously recognized that the pyrid-3-ylmethyl residue was an essential moiety for insecticidal activity, by referring to the nAChR agonist model of Beers and Reich (1970), and thus synthesized a number of pyrid-3-yl-amines (Yamamoto and Tomizawa, 1993).

This chemical optimization procedure led to discovery of 1-(6-chloro-pyrid-3-ylmethyl)-2-nitro-methylene-imidazolidine (NTN32692), which had over 100-fold higher activity than nithiazine against *N. cincticeps* (using a strain resistant to organophosphorus compounds, methylcarbamates, and pyrethroides) (Kagabu *et al.*, 1992; Moriya *et al.*, 1992). However, the 2-nitromethylene chromophore of NTN32692 absorbs strongly (**Figure 3**) in sunlight (wavelength 290–400 nm), and rapidly decomposes

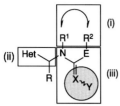

(i) Ring systems (R^1—R^2, R^1—Z—R^2; Z = O, NMe) or noncyclic structures (R^1, R^2)

(ii) Heterocyclic N-substituents [Het—CHR—; R = H > alkyl], e.g., CPM, CTM, and TFM

(iii) Different types of pharmacophores [—N—C(E)=X—Y] e.g., nitroenamines [—N—C(E)=CH—NO$_2$] with E = S, N; nitroguanidines [—N—C(N)=N—NO$_2$] with E = N; and cyanoamidines [—N—C(E)=N—CN] with E = S, Me

Structure of Het-CHR-	Chemical name of this moiety	Abbreviation
Cl—pyridine—CH$_2$—	6-Chloro-pyrid-3-ylmethyl-	CPM
Cl—thiazole—CH$_2$—	2-Chloro-1,3-thiazol-5-ylmethyl-	CTM
O—furan—CH$_2$—	(±)-6-Tetrahydro-fur-3-ylmethyl-	TFM

Figure 4 Structural segments for neonicotinoids.

in the field to noninsecticidal compounds. In order to avoid absorption of sunlight, a more stable functional group was necessary. After preparation of about 2000 compounds, imidacloprid (NTN33893) containing a 2-(N-nitroimino) group (see Section 3.3.1.1), the first member of the second-generation neonicotinoids, emerged from this project (Elbert et al., 1990, 1991; Bai et al., 1991; Nauen et al., 2001). The presence of an N-nitroimino group at the 2-position of the imidazolidine ring makes little difference to insecticidal activity compared with compounds that contain a nitomethylene group at this position, but the presence of =N—NO$_2$ significantly reduces affinity for the receptor (Liu et al., 1993b; Tomizawa and Yamamoto, 1993; Yamamoto et al., 1998). This suggests that the reduction in binding, resulting from the presence of the nitrogen atom at the 2-position, is compensated by the increase in hydrophobicity, which enhances transport to the target sites (Yamamoto et al., 1998; Matsuda et al., 2001). Compared with nithiazine, the biological efficacy of imidacloprid against the green rice leafhopper could be enhanced 125-fold. Furthermore, imidacloprid is about 10 000-fold more insecticidal than (S)-nicotine, a natural insecticide (Tomizawa and Yamamoto, 1993; Yamamoto et al., 1995). This breakthrough to the novel systemic insecticide imidacloprid was achieved by coupling a special heterocyclic group, the 6-chloro-pyrid-3-ylmethyl (CPM) residue (**Figure 4**), to the 2-(N-nitroimino)-imidazolidine building-block, new in this class of chemistry.

With the introduction of the insecticide imidacloprid to the market in 1991 Bayer AG began a successful era of the so-called CNI^{TM}s (chloronicotinyl insecticides syn. neonicotinoids), a milestone in insecticide research. Imidacloprid has, in the last decade, become the most successful, highly effective, and best selling insecticide worldwide used for crop protection and veterinary pest control.

In connection with these excellent results, a parallel change to other electron-withdrawing substituents like 2-N-cyanoimino, having essentially shorter maximum electronic absorptions than 290 nm, led to the discovery of thiacloprid (see Section 3.3.1.2), a second member of the CNI group. Attracted by Bayer's success with imidacloprid, several different companies such as Takeda (now Sumitomo Chemical Takeda Agro), Agro Kanesho, Nippon Soda, Mitsui Toatsu (now Mitsui Chemicals, Inc.), Ciba Geigy (now Syngenta), and others initiated intensive research and developed their own neonicotinoid insecticides. Research in these companies was facilitated because neonicotinoid chemistry showed a relatively broad spectrum of activity (Wollweber and Tietjen, 1999; Roslavtseva, 2000). Since the market introduction of imidacloprid, neonicotinoids have become the fastest-growing class of chemical insecticides. This tremendous success can be explained by their unique chemical and biological properties, such as broad-spectrum insecticidal activity, low application rates, excellent systemic characteristics such as uptake and translocation in plants, new mode of action, and favorable safety profile.

3.3. Chemical Structure of Neonicotinoids

In general, all these commercialized or developed compounds can be divided into ring systems containing neonicotinoids, and neonicotinoids having noncyclic structures which differ in their molecular characteristics. Structural requirements for both ring systems containing neonicotinoids and neonicotinoids having noncyclic structures consist of different segments (Nauen *et al.*, 2001): the bridging fragment $[R^1–R^2]$ (i) and for the noncyclic type compounds, the separate substituents $[R^1, R^2]$ (i), the heterocyclic group [Het] (ii), the bridging chain [–CHR–] and the functional group [=X–Y] as part of the pharmacophore [–N–C(E)=X–Y] (iii) (**Figures 2** and **4, Table 1**).

The methylene group is normally used as the bridging chain (–CHR– with R = H). Other groups such as an ethylene or substituted methylene group decrease the biological activity. The pharmacophore (iii) can be represented by the group [–N–C(E) =X–Y], where =X–Y is an electron-withdrawing group and E is a NH, NR′, CH$_2$, O, or S moiety. It is well known that the pharmacophore type influences the insecticidal activity of the neonicotinoids. In general, the greatest insecticidal activity is observed for the compounds containing the nitroenamine (nitromethylene) [–N–C(E)=CH–NO$_2$; E = S, N], nitroguanidine [–N–C(E)=N–NO$_2$; E = N], or cyanoamidine [–N–C(E)=N–CN; E = S, CH$_3$] pharmacophore (Shiokawa *et al.*, 1995). Besides its influence on biological activity, the pharmacophore is also responsible for some specific properties such as photolytic stability, degradation in soil, metabolism in plants, and toxicity to different animals (Kagabu *et al.*, 1992; Moriya *et al.*, 1992; Minamida *et al.*, 1993; Tomizawa and Yamamoto, 1993; Shiokawa *et al.*, 1994; Tabuchi *et al.*, 1994).

The term "neonicotinoid" (Yamamoto, 1998, 1999) was originally proposed for imidacloprid and related insecticidal compounds with structural similarity to the insecticidal alkaloid (S)-nicotine, which has a similar mode of action (Tomizawa and Yamamoto, 1993; Tomizawa and Casida, 2003). In addition, a variety of terms have been used to subdivide this chemical class based on structural fragments (**Figure 4**): (a) chloronicotinyls (CNIs, heterocyclic group CPM) (Leicht, 1993), (b) thianicotinyls (heterocyclic group CTM) (Maienfisch *et al.*, 1999a), (c) furanicotinyls (heteroalicyclic group, TFM) (Wakita *et al.*, 2003), (d) nitroimines or nitroguanidines, (e) nitromethylenes, (f) cyanoimines (functional group as part of the pharmacophore), (g) first generation with moiety CPM, (h) second generation with moiety CTM or TFM (Maienfisch *et al.*, 1999a, 2001b), and third generation with moiety TFM (regarding the type of the heterocyclic group) (Wakita *et al.*, 2003).

3.3.1. Ring Systems Containing Commercial Neonicotinoids

Ring systems containing neonicotinoids can have as bridging fragment $[R^1–R^2]$ (i) an alkylene group in the building-block, e.g., ethylene for imidacloprid and thiacloprid (five-membered ring systems) which can also be interrupted by heteroatoms like oxygen (six-membered ring systems) as shown in the case of thiamethoxam.

3.3.1.1. 1-[(6-Chloro-3-pyridinyl)-methyl]-N-nitro-2-imidazolidinimine (imidacloprid, NTN 33893)
Imidacloprid was discovered in 1985 as a novel insecticide with a unique structure and with hitherto unrecognized insecticidal performance (Moriya *et al.*, 1992, 1993; Kagabu, 1997a, 2003a). The trade names for soil application/seed treatment are Admire™ and Gaucho® and the trade names for foliar application are Confidor® and Provado®, while for termite control it is Premise® (see Section 3.10).

The first laboratory synthesis was carried out by reduction of 6-chloronicotinoyl chloride to 2-chloro-5-hydroxymethyl-pyridine using an excess NaBH$_4$ in water, and conversion to the chloride by SOCl$_2$. Imidacloprid was obtained by the coupling reaction of 2-chloro-5-chloromethyl-pyridine (CCMP) (Diehr *et al.*, 1991; Wollweber and Tietjen, 1999) with the 5-ring building-block 2-nitro-imino-imidazolidine, in acetonitrile with potassium carbonate as base. This method was also successfully applied to the synthesis of [^{3}H]imidacloprid using NaB[^{3}H]$_4$ (Latli *et al.*, 1996).

Differences in photostability between the 2-nitromethylene group in compound NTN32692 ($\lambda_{max} = 323$ nm) and the 2-(N-nitroimino) group in imidacloprid ($\lambda_{max} = 269$ nm) (**Figure 3**) were examined by quantum chemical methods like AM1 and *ab initio* calculations (Kagabu, 1997a; Kagabu and Akagi, 1997). Crystallographic analysis of imidacloprid revealed a coplanar relationship of the imidazolidine 5-ring to the nitroimino group in position 2 (Born, 1991; Kagabu and Matsuno, 1997). Shorter ring C–N bond lengths and longer exocyclic C=N bond lengths were observed, suggesting the formation of a planar electron delocalized dicnc system, a so-called push–pull olefin (Kagabu and Matsuno, 1997). An intramolecular hydrogen bond between ^{1}NH$\cdots$O$_2$N–N=C^2 was confirmed by characteristic resonances in the nuclear magnetic resonance (NMR) spectra (correlation spectroscopy (COSY), nuclear overhauser enhancement spectroscopy (NOESY)). In addition, the infrared (IR) spectrum showed a highly chelated ^{1}NH absorption by intense bands at 3356 and 3310 cm^{-1} (Kagabu *et al.*, 1998).

Table 1 Chemical classification of nithiazine and commercialized neonicotinoids

Common name	Scientific name		CAS registry no.	Development codes	Empirical formula	Molecular weight (g mol^{-1})
	IUPAC name	Chemical Abstracts name				
Nithiazine	2-Nitromethylene-1,3-thiazinane	Tetrahydro-2-(nitromethylen)-2H-1,3-thiazine	[58842-20-9]	SD-035651^a IN-A0159^b	C$_5$H$_8$N$_2$O$_2$S	160.2
Imidacloprid	1-(6-Chloro-3-pyridylmethyl)-N-nitroimidazolidin-2-ylidenamine	1-[(Chloro-3-pyridinyl)-methyl]-N-nitro-2-imidazolidinimine	[138261-41-3]	NTN 33893	C$_9$H$_{10}$ClN$_5$O$_2$	255.7
Thiacloprid	N-[3-(6-Chloro-pyridin-3-ylmethyl)-thiazolidin-2-ylidene]-cyanamide	Cyanamide, [3-[(6-chloro-3-pyridinyl)methyl]-2-thiazolidinylidene]	[111988-49-9]	YRC 2894	C$_{10}$H$_9$ClN$_4$S	252.8
Thiamethoxam	3-(2-Chloro-thiazol-5-ylmethyl)-5-methyl-1,3,5-oxadiazinan-4-ylidene (nitro)amine	3-[(2-Chloro-5-thiazolyl)methyl]tetrahydro-5-methyl-N-nitro-4H-1,3,5-oxadiazin-4-imine	[153719-23-4]	CGA293'343	C$_8$H$_{10}$ClN$_5$O$_3$S	291.7
Nitenpyram	(E)-N-(6-Chloro-3-pyridylmethyl)-N-ethyl-N'-methyl-2-nitrovinylidenediamine	N-[(6-Chloro-3-pyridinyl)methyl]-N-ethyl-N'-methyl-2-nitro-1,1-ethenediamine	[120738-89-8] [150824-47-8] (E-isomer)	TI-304	C$_{11}$H$_{15}$ClN$_4$O$_2$	270.7
Acetamiprid	(E)-N^1-[(6-Chloro-3-pyridyl)methyl]-N^2-cyano-N^1-methyl-acetamidine	(E)-N-[(6-chloro-3-pyridinyl)methyl]-N'-cyano-N-methyl-ethanimidamide	[135410-20-7]	NI-25	C$_{10}$H$_{11}$ClN$_4$	222.7
Clothianidin	(E)-1-(2-Chloro-1,3-thiazol-5-ylmethyl)-3-methyl-2-nitroguanidine	[C(E)]-N-[(2-Chloro-5-thiazolyl)methyl]-N'-methyl-N''-nitroguanidine	[210880-92-5]	TI-435	C$_6$H$_8$ClN$_5$O$_2$S	249.7
(±)-Dinotefuran	(RS)-1-methyl-2-nitro-3-(tetrahydro-3-furylmethyl)guanidine	N-Methyl-N'-nitro-N''-[(tetrahydro-3-furanyl)methyl]guanidine	[165252-70-0]	MTI 446	C$_7$H$_{14}$N$_4$O$_3$	202.2

^{a}Shell/DuPont.
^{b}DuPont.

Studies were also done using comparative molecular field analysis (CoMFA), a technique for analysis of three-dimensional quantitative structure–activity relationships (3D QSAR) (Cramer *et al.*, 1988; Okazawa *et al.*, 1998). The model showed that the nitrogen atom of the imidacloprid CPM residue interacts with a hydrogen-donating site of nAChR, and the nitrogen atom at the 1-position of the imidazolidine 5-ring interacts with a negatively charged domain (Okazawa *et al.*, 2000). In comparison with (*S*)-nicotine the selective insecticidal mode of action of imidacloprid was well examined in detail by different researchers (Yamamoto *et al.*, 1995; Kagabu, 1997b; Matsuda *et al.*, 1999; Tomizawa, 2000). There studies focused on the chemical structure, state of ionization, and hydrophobicity of both molecules (**Figure 5**).

The principal target insect pests of imidacloprid in agricultural and veterinary medicinal use are sucking insects. Since imidacloprid provides a partial positively charged δ^+ nitrogen atom (not ionized, full or partial) instead of an ammonium head as does (*S*)-nicotine, imidacloprid has little restriction for translocation into the target area compared

to (*S*)-nicotine (Graton *et al.*, 2003). The interatomic distances between the 2-(*N*-nitroimino)-imidazolidine nitrogen and the pyridyl nitrogen are 5.9 Å, which is a suitable value to bind to the electron-rich and hydrogen-donating recognition sites on nAChR (Yamamoto *et al.*, 1995; Matsuda *et al.*, 1999). The electron-donating atom for hydrogen bonding at 5.9 Å apart from the fully or partial positive charged δ^+ nitrogen is required for potent interaction with the anionic subsite of the insect nAChR (Yamamoto *et al.*, 1995; Tomizawa, 2000). In mammals, imidacloprid has weak effects on any nAChR, while nicotine affects mostly peripheral nACRs resulting in higher toxicity.

The deduced electron deficiency of the nitrogen atom of imidacloprid was proved explicitly by ^{15}N NMR spectroscopic measurements (Yamamoto *et al.*, 1995). Tomizawa *et al.* (2000) recently claimed that the *N*-nitro group of imidacloprid is much more important than the bridgehead nitrogen in electrostatic interaction with nAChRs' because a Mulliken charge of the bridgehead nitrogen, calculated by the MNDO method combined with the PM3 method (semi-empirical molecular orbital technique

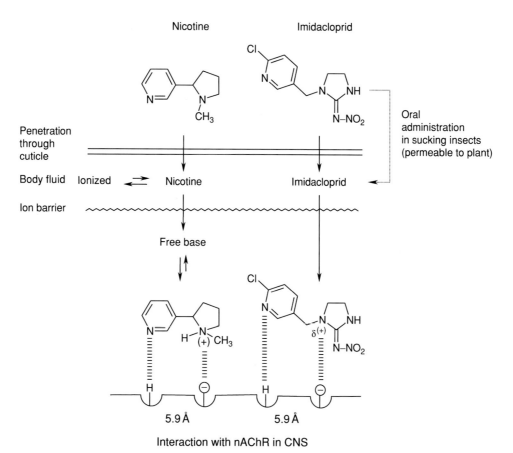

Figure 5 Insecticidal mode of action of (*S*)-nicotine and imidacloprid. (Adapted from Tomizawa, M., **2000**. Insect nicotinic acetylcholine receptors: mode of action of insecticide and functional architecture of the receptor. *Jap. J. Appl. Entomol. Zool. 44*, 1–15.)

for calculating electronic structure), was only marginally positive. However, Matsuda *et al.* (2001) suggested, that the electron-withdrawing power of the *N*-nitro group might be increased by a hydrogen bond involving the *N*-nitro group oxygen and hydrogen-bondable and positively charged amino acid residue, namely lysine or arginine, on the insect receptor, thereby strengthening its interaction with imidacloprid. Kagabu (1996) predicted the important contribution of this *N*-nitro group and its hydrogen-bondable property for insecticidal activity.

3.3.1.2. [3-(6-Chloro-3-pyridinyl)methyl-2-thiazolidinylidene]-cyanamidine (thiacloprid, YRC 2894)

The second member of the CNI[TM] family from Bayer AG launched in 2000 was thiacloprid (Elbert *et al.*, 2000; Yaguchi and Sato, 2001). Similar to imidacloprid, this neonicotinoid thiacloprid (YRC 2894, registered worldwide under the trade name Calypso[TM], and in Japan and Switzerland under the trade names Bariard[TM] and Alanto[TM]), also contains the unique CPM moiety attached to the cyclic 2-(cyanoimino)-thiazolidine (CIT) building-block. Thiacloprid can be synthesized by a convergent one-step technical process starting from two key intermediates, CIT and CCMP (Diehr *et al.*, 1991; Wollweber and Tietjen, 1999; Jeschke *et al.*, 2001). Thiacloprid crystallizes in two different modifications depending on the solvent. The compound crystallizes from dichloromethane as form I (melting point 136 °C) and from isopropanol as form II (melting point 128.3 °C). From its physicochemical data, the technical ingredient is form I.

The results of X-ray crystallographic analysis and an X-ray powder diffraction profile indicate that each form, I ($0.6 \times 0.3 \times 0.2 \, mm^3$) and II ($0.5 \times 0.2 \times 0.1 \, mm^3$), crystallizes with different cell constants in a monoclinic crystal system in the space group $P2_1/c$ (Jeschke *et al.*, 2001). **Figure 6** shows the Ortep plot of the crystal structure of the form I. Interestingly, in the crystal lattice of form II, two symmetric independent molecules in the asymmetric unit cell are fixed (**Figure 6**).

With regard to the configuration of the pharmacophore [−N−C(S)=N−CN] moiety, thiacloprid exists only in the *Z*-configuration in both forms, I and II. In conjunction with the X-ray results, thiacloprid exists in solution exclusively in the stable *Z*-configuration. With respect to the C=N double bond, only one configuration is seen in all NMR spectra (^{1}H, ^{13}C, HMQC, HMBC, NMR spectroscopy), deduced from very sharp proton and carbon signals. In conjunction with X-ray results, there is no evidence for isomerization of the *Z*-isomer in DMSO-d_6 or CDCl$_3$ solution. Thiacloprid shows a single peak maximum at 242 nm in its ultraviolet (UV) spectrum, and it does not absorb natural sunlight. This explains thiacloprid's better photostability in comparison to other neonicotinoids (Jeschke *et al.*, 2001).

Starting from forcefield methods (MMFF94s), the final energies, geometries, and properties were obtained at DFT (Becke and Perdew (BP) functional, triple-zeta valence plus (TZVP) polarization basis, conductor-like screening model − realistic solvent (COSMO-RS) for solvent effects; Perdew, 1986;

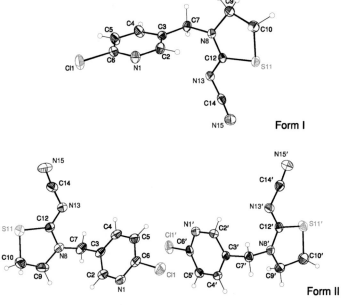

Figure 6 Ortep plot of thiacloprid form I and form II (Jeschke *et al.*, 2001).

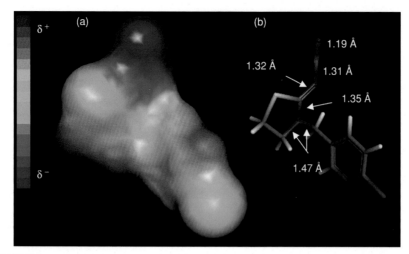

Figure 7 DFT (BR/TZVP) optimized geometry of thiacloprid (Jeschke *et al.*, 2001). (a) Atomic charges projected onto a Conolly surface; (b) some interatomic distances in Å.

Becke, 1988; Klamt, 1995) level of theory (Jeschke *et al.*, 2001) (**Figure 7**). It was found that, in the gas phase, *Z*-configured thiacloprid is about $4 \, \text{kcal mol}^{-1}$ lower in energy than the *E*-isomer. In water, this difference is increased by approximately $1 \, \text{kcal mol}^{-1}$. The preference for the *Z*-configuration stems mainly from steric reasons. The nitrile moiety of the *Z*-isomer has virtually no close contacts to any other atom (sulfur to nitrile-carbon distance $3.1 \, \text{Å}$). In the *E*-isomer, however, the largest possible distance, between the nitrile-carbon and the hydrogen atoms attached to the bridging carbon, is around $2.3 \, \text{Å}$. Quantum chemical calculations show that the strong delocalization of the C=N double bond does not reduce the "double bond character" significantly (i.e., the torsional barrier for rotation around the C=N double bond is not decreased). The molecular dipole moment (derived from the electric field dependence of the DFT wave function) is calculated to be $10.2 \, \text{D}$ in water. This, together with the spatial relation of the pharmacophoric feature (−N−C(S)=N−CN), is very much in line with the pharmacophore model of imidacloprid.

3.3.1.3. 3-[(2-Chloro-5-thiazolyl)methyl]tetrahydro-5-methyl-*N*-nitro-4H-1,3,5-oxadiazin-4-imine (thiamethoxam, CGA 293′343)
Ciba started a research program on neonicotinoids in 1985 (Maienfisch *et al.*, 2001a), and investigated variations of the imidacloprid nitroimino-heterocycle (Maienfisch *et al.*, 2001b). Thiamethoxam (Senn *et al.*, 1998; Maienfisch *et al.*, 1999a, 1999b) was discovered, and developed by Ciba Crop Protection (from 1996 Novartis Crop Protection, now Syngenta Crop Protection) in 1991. Thiamethoxam has been marketed since 1998 under the trademarks Actara®

for foliar and soil treatment, and Cruiser® for seed treatment. Thiamethoxam can be synthesized starting from *N*-methyl-nitroguanidine by treatment with formaldehyde in the presence of formic acid (Göbel *et al.*, 1999). Finally alkylation with 2-chlorothiazol-5-ylmethylchloride in *N*,*N*-dimethylformamide and potassium carbonate as a base afforded the active ingredient in good yields (Maienfisch *et al.*, 1999a).

The SAR profile of this compound demonstrates the rather limited variability of the pharmacophore (−N−C(N)=N−NO₂). Insecticidal activity was only observed if the functional group is a strongly electron-withdrawing group and has a hydrogen accepting head, like *N*-nitroimino and *N*-cyanoimino. As pharmacophore *N*-substituent, a methyl group is clearly superior to hydrogen, an acyl substituent or C₂–C₄ alkyl group. The biological results have led to a general SAR profile for these novel compounds, which can be summarized as follows (Maienfisch *et al.*, 1999b, 2002):

1. All variations of the pharmacophore (N−C(N)=N−NO₂) resulted in a loss of activity.
2. Substitution of the CTM moiety by another N-containing heterocycle reduced the overall insecticidal activity, whereas CPM gave the best results.
3. A perhydro-1,3,5-oxadiazine ring system is clearly better than all other heterocyclic systems like perhydro-1,3,5-thiadiazines, hexahydro-1,3,5-triazines, or hexahydro-1,3-diazines.
4. Introduction of *N*-methyl at the 5-position led to strong increase of insecticidal activity (in contrast to the SAR of imidacloprid), whereas other substituents resulted in clear loss of activity.

3.3.2. Neonicotinoids Having Noncyclic Structures

From the new chemical class of ring system containing neonicotinoids noncyclic structures having the same mode of action can also be deduced. These noncyclic type neonicotinoids can have as separate substituents [R^1, R^2] (i), e.g., for R^1 hydrogen or alkyl like methyl (acetamiprid) and ethyl (nitenpyram) and in the case of E = NH for the substituent R^2 alkyl like methyl (clothianidin and dinotefuran), respectively.

3.3.2.1. N-[(6-chloro-3-pyridinyl)methyl]-N-ethyl-N'-methyl-2-nitro-1,1,-ethenediamine (nitenpyram, TI-304)

Starting from the cyclic nithiazine (Soloway *et al.*, 1978), the acyclic nitenpyram (Bestguard^{TM}) was discovered during optimization of substituents of an acyclic nitroethene (Minamida *et al.*, 1993). The insecticidal activity of 1-(3-pyridylamino)-2-nitroethene compounds against the brown planthopper and the green leafhopper were studied. It was found that the 1-methylamino-1-(pyrid-3-ylmethylamino)-2-nitroethene was more active than the corresponding 1-ethyl, 1-methoxy, or 1-thiomethoxy derivatives. Introduction of sterically large amino groups like ethylamino, isopropylamino, hydrazino, N-pyrrolidino, or N-morpholino decreased the activity against the green rice leafhopper. Furthermore, the methylene bridge between the pyrid-3-yl and nitrogen was replaced with other linkages. However, all attempts at shortening and lengthening the linkage failed to increase the insecticidal activities against the brown planthopper and the green rice leafhopper (Akayama and Minamida, 1999). Heterocyclic aromatic substituents, different from pyrid-3-yl, were incorporated into the nitroethen structure. The replacement of pyrid-3-yl with 6-fluoro-pyrid-3-yl, 6-chloro-pyrid-3-yl, 6-bromo-pyrid-3-yl, and 2-chloro-1,3-thiazol-5-yl enhanced the activity against the brown planthopper.

3.3.2.2. (E)-N-[(6-chloro-3-pyridinyl)methyl]-N'-cyano-N-methyl-ethanimidamide (acetamiprid, NI-25)

Acetamiprid (Takahashi *et al.*, 1992; Matsuda and Takahashi, 1996) has an N-cyanoamidine structure, which contains in analogy to imidacloprid and thiacloprid the CPM moiety. This neonicotinoid was invented during a search for nitromethylene derivatives by Nippon Soda Co. Ltd. in 1989 and was registered in 1995 in Japan. The insecticide is marketed under the trade name Mospilan® for crop protection. The noncyclic acetamiprid was discovered by optimization studies of special 2-N-cyanoimino compounds with an imidazolidine 5-ring obtained from Nihon Bayer (Yamada *et al.*, 1999). The noncyclic 2-nitromethylene and 2-N-nitroimine derivatives are less active than the corresponding ring structures, but numerous noncyclic derivatives show excellent activities against the armyworm and aphid as well as the cockroach. Regarding the substituent on the amino group, the N-methyl group exhibited the highest activity against the diamondback moth, while derivatives with hydrogen, N-methyl, and N-ethyl showed potent activity against the cotton aphid (Yamada *et al.*, 1999).

3.3.2.3. [C(E)]-N-[(2-chloro-5-thiazolyl)methyl]-N'-methyl-N''-nitroguanidine (clothianidin, TI-435)

As a result of continuous investigations of noncyclic neonicotinoids, researchers from Nihon Bayer Agrochem and Takeda in Japan found that these noncyclic neonicotinoids showed high activities against sucking insects. In the optimization process, Takeda researchers were able to demonstrate that compounds containing the nitroguanidine moiety, coupled with the thiazol-5-ylmethyl residue, have increased activity against some lepidopteran pests (Uneme *et al.*, 1999). This is in accordance with the general pharmacophore [−N−C(E)=X−Y], where = X−Y is an electron-withdrawing group such as = N−NO$_2$, and E represents the NH−Me unit. After further optimization in this subclass, clothianidin (TI-435) emerged as the most promising derivative from this program (Ohkawara *et al.*, 2002; Jeschke *et al.*, 2003). Clothianidin has already been registered in Japan for foliar and soil applications under the trade names Dantotsu^{TM} and Fullswing^{TM} (Sumitomo Chemical Takeda Agro), and also in Korea, Taiwan, and other countries. Registrations have also been granted in North America and Europe for seed treatment under the brand name Poncho® (Bayer CropScience).

In the novel noncyclic structure of clothianidin (TI-435) (i.e., R = H; E = NHMe), the N-nitroguanidine pharmacophore [−N−C(N)=N−NO$_2$] is similar to that of imidacloprid, but the CPM group has been replaced by the CTM moiety (Ohkawara *et al.*, 2002) (**Figure 2**). Clothianidin crystallizes under normal laboratory conditions using common solvents as needlelike crystals containing no additional solvent or water molecules. The best-quality crystals for X-ray structure analysis were obtained by slow evaporation of a methanol/H$_2$O (1 : 1) solution at room temperature. **Figure 8** shows a photograph using polarized light of the crystal needle with the dimensions 0.90 × 0.30 × 0.07 mm^3 used for X-ray structure analysis (Jeschke *et al.*, 2003). Clothianidin (TI-435) was subjected to conformational sampling using forcefield methods (MMFF94s). The calculated free energies indicate that at room temperature (RT), the preferred orientation of the N-nitroimino group is in the *trans* position; the Z-isomer with lowest energy is more than 2.6 kcal mol^{-1} above the optimal E-isomer. NMR experiments are in agreement that the N-nitroimino group strongly prefers one orientation only. From calculations as well as the X-ray structure, one can see that the three C–N bonds involving atom C5 have some double bond character. It is worth mentioning that the C5–N3 bond, which is the only formal C=N double bond within the N-nitroimino moiety, is slightly longer than the formal single bonds C5–N5 and C5–N2. This is reflected by the torsional angles found around the C–N bonds during conformational analysis; the respective values are all 180° or −180°. The N-methyl group can flip easily from the anti position into a syn position. The energies of the respective clothianidin (TI-435) conformers, relative to the optimal structure, are below 1.5 kcal mol^{-1}. All these findings are in line with the experimental ^{13}C NMR spectrum, which shows a relatively broad singlet at 138.8 ppm for

the atom C3. This can already be understood qualitatively from the conformational arguments, like rotations of the C–N single bonds and modifications of the positioning of the CTM moiety (Jeschke *et al.*, 2003).

3.3.2.4. (±)-N-methyl-N'-nitro-N''-[(tetrahydro-3-furanyl)methyl]guanidine (dinotefuran, MTI-446)

(±)-Dinotefuran (MTI-446) was discovered by Mitsui Chemicals Inc., and is highly effective as an agonist of nAChRs (Zhang *et al.*, 2000). It has a broad spectrum of activity against insect pests and has low mammalian toxicity (Kodaka *et al.*, 1998, 1999a, 1999b; Hirase, 2003). The trade names for the commercialized product are Starkle® and Albarin®. Similar to imidacloprid and clothianidin, it has a N-nitroguanidine pharmacophore [–N–C(N)=N–NO$_2$], but an alicyclic (±)-tetrahydro-3-furylmethyl (TFM) moiety instead of the halogenated heteroaromatic CPM or CTM moiety in the other neonicotinoids. The discovery of (±)-dinotefuran resulted from the idea of incorporating an N-nitroimino fragment into the structure of acetylcholine (Kodaka *et al.*, 1998; Wakita *et al.*, 2003). (±)-Dinotefuran bears a nonaromatic oxygen atom in the position corresponding to that of the aromatic nitrogen atom of the other neonicotinoids. The potencies of (±)-dinotefuran and a competitive nAChR antagonist (α-bungarotoxin, BGT) in inhibiting [^{3}H]epibatidine binding to *Periplaneta americana* nerve cord membranes were examined (Mori *et al.*, 2001). (±) and (+) Dinotefuran inhibited [^{3}H]epibatidine binding with IC$_{50}$ (inhibitory concentration, where 50% of the tritiated ligand is displaced in membrane preparations from cockroach nerve) values of 890 and 856 nM, respectively. The (−)-enantiomer was about twofold less effective. In contrast the (+)-enantiomer was approximately 50-fold more insecticidal than the (−)-enantiomer of dinotefuran (**Table 2**).

The SAR between the tetrahydrofuran ring moiety and the insecticidal activity was investigated

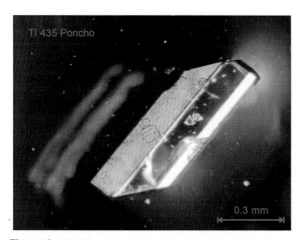

Figure 8 Digital stereo photomicrograph of a single crystal of clothianidin (TI-435) grown from methanol/water at room temperature (Jeschke *et al.*, 2003).

Table 2 Potency of dinotefuran and its isomers in inhibiting [^{3}H]EPI and [^{3}H]α-BGT binding, and their *in vivo* activities

Compound	[^{3}H]EPI binding, IC$_{50}$ (nM)^a	[^{3}H]α-BGT binding, IC$_{50}$ (nM)^a	Knockdown, KD$_{50}$ (nmol g^{-1})	Insecticidal activity, LD$_{50}$ (nmol g^{-1})^a
(±)-Dinotefuran	890 (626–1264)	36.1 (24.9–52.2)	0.351	0.173 (0.104–0.287)
(+)-Dinotefuran	856 (590–1241)	9.58 (6.79–13.52)	0.123	0.0545 (0.0396–0.0792)
(−)-Dinotefuran	1890 (1230–2890)	69.8 (47.1–103.4)	6.70	2.67 (1.55–4.69)

^{a}95% confidence limits in parentheses.
BGT, bungarotoxin; EPI, epibatidine.
Reproduced with permission from Mori, K., Okumoto, T., Kawahara, N., Ozoe, Y., **2001**. Interaction of dinotefuran and its analogues with nicotinic acetylcholine receptors of cockroach nerve cords. *Pest Mgt Sci. 58*, 190–196.

(Kiriyama and Nishimura, 2002), and it was found that:

1. The tetrahydro-fur-2-ylmethyl derivative (shift of oxygen position) reflected diminished activity.
2. The tetrahydro-fur-3-yl moiety is important; the cyclopentanylmethyl (deoxy derivative), the 3-methoxy-N-propyl (open-ring ether) and the N-methyl-pyrrolidine-3-ylmethyl derivatives showed very low or negligible activity.
3. Introduction of a methyl group at the 4- or 5-position of the tetrahydro-fur-3-yl ring gave intermediate activity (Wakita et al., 2003), but introduction of methyl at the 2- or 3-position of the THF ring system and in α-position of the side chain decreases activity. An ethyl group reduced greatly activity even if it was at the 4-position (Wakita et al., 2003).

The structural factors in the N-nitro guanidine moiety of (±)-dinotefuran were described as follows:

1. Deletion or extension to the alkyl group, like ethyl, of the terminal N-methyl group and addition of one more N-methyl group usually decrease the activity.
2. Carbocyclic cyclization to 5- and 6-ring systems maintains the activity.
3. Perhydro-1,3,5-oxadiazine 6-ring cyclization similar to that in thiamethoxam is inactive.
4. Modification of the N-nitroimino chromophore of dinotefuran to the nitromethylene, but not to the N-cyanoimino group maintains the insecticidal activity.

These results indicate that the modification of this moiety of (±)-dinotefuran results in drastic changes in potency and that the incorporation of structural fragments known from previous neonicotinoid insecticides does not necessarily lead to compounds retaining higher activity (Mori et al., 2001). Also in this structural type the distance of three methylene chains between the nitrogen on the imidazolidine ring and the hydrogen acceptor oxygen was important (Wakita et al., 2003).

3.3.3. Bioisosteric Segments of Neonicotinoids

Retrospective considerations regarding bioisosteric segments of the developed and commercial neonicotinoids gave insight into general structural requirements (segments i–iii, **Figure 4**) for all the different ring systems and noncyclic structures (Jeschke et al., 2002). Several common molecular features, when comparing compounds with ring systems, like imidazolidine (imidacloprid), and its isosteric alternatives, like thiazolidine (thiacloprid), perhydro-1,3,5-oxadiazine (thiamethoxam), or hexahydro-1,3,5-triazine (Agro

Kanesho's AKD-1022) and the functional groups like N-nitroimino [=N−NO$_2$], N-cyanoimino [=N−CN], or nitromethylene [=CH−NO$_2$] of these insecticides, have been described (Tomizawa et al., 2000). After superposition of the most active derivatives, it was possible to state the molecular shape similarity of the second generation neonicotinoids. It was found that electrostatic similarity of the most active compounds correlates well with the binding affinity (Nakayama and Sukekawa, 1998; Sukekawa and Nakayama, 1999), and a similar correlation was obtained by CoMFA (Nakayama, 1998; Okazawa et al., 1998).

3.3.3.1. Ring systems versus noncyclic structures
In comparison to the corresponding ring systems, the noncyclic structures exhibit similar broad insecticidal activity by forming a so-called quasi-cyclic conformation when binding to the insect nAChR (Kagabu, 2003a). Thus, the three commercial noncyclic structures – nitenpyram, acetamiprid, and clothianidin – can be regarded as examples, if retrosynthetic considerations are carried out (Jeschke et al., 2002; Kagabu, 2003a). The noncyclic neonicotinoides are generally less lipophilic than the corresponding neonicotinoids with a ring structure (see Section 3.3.4).

Based on the CoMFA results, a binding model for imidacloprid was described. This model clarified that the nitrogen of the CPM moiety interacts with a hydrogen-donating site of the nAChR, and that the nitrogen atom at the 1-position of the imidazolidine ring interacts with the negatively charged domain (Nakayama, 1998; Okazawa et al., 1998). Furthermore, the binding activity of noncyclic structures (e.g., acetamiprid, nitenpyram, and related compounds) to the nAChR of houseflies was measured and the results were analyzed using CoMFA. Superposition of stable conformations of nitenpyram, acetamiprid, and imidacloprid showed that the preferred regions for negative electrostatic potentials near the oxygen atoms of the nitro group, as well as the sterically forbidden regions beyond the imidazolidine 3-nitrogen atom of imidacloprid, were important for binding (Okazawa et al., 2000). The area around the 6-chloro atom of the CPM moiety was described as a sterically permissible region. Apparently the steric interactions were more important for noncyclic neonicotinoids than for the cyclic derivative, imidacloprid. Finally, it was also demonstrated that the noncyclic structures bind to the nAChR recognition site in a manner similar to ring structures like imidacloprid, and that the electrostatic properties of the noncyclic amino and cyclic imidazolidine structures affected their binding affinity (**Figure 9**).

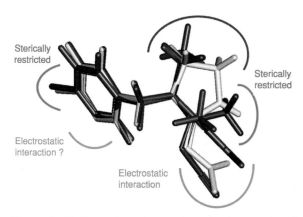

Figure 9 Stable conformations and predicted properties of binding site: imidacloprid (white), nitenpyram (blue), acetamiprid (red). (Adapted from Akazawa *et al.*, 2000.)

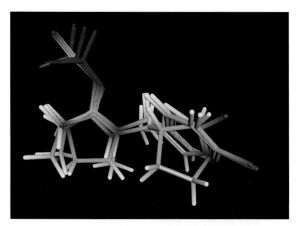

Figure 10 Alignment of DFT/BP/SVP/COSMO optimized geometries of imidacloprid, clothianidin, dinotefuran, and acetamiprid. The alignment was done by minimization of the mutual spatial distance of three pharmacophoric points, namely (i) the positively charged carbon atom connected to the $=N-NO_2$ and $=N-CN$ moiety, respectively, (ii) the nitro/cyano groups themselves, and (iii) the nitrogens of the aromatic rings and the oxygen of the tetrahydrofuran ring.

Figure 10 shows atom-based aligments of imidacloprid, clothianidin, dinotefuran and acetamiprid. The aligment shown does not necessarily reflect the active conformation. However, the following arguments hold true for each of these family of conformers. The ion pairs of the aromatic nitrogen atoms in imidacloprid, clothianidin, and acetamiprid point in the same direction, and that this is also true for one of the two ion pairs of oxygen in dinotefuran. The tetrahydrofuryl ring of dinotefuran is more or less perpendicular to the heteroaromatic ring systems of the other neonicotinoids.

Visual inspection of the nucleophilic Fukui functions (**Figure 11**) shows that the tetrahydrofuryl ring of dinotefuran differs dramatically from the other neonicotinoids under investigation. While the latter all show some large contributions at the aromatic moiety, the tetrahydrofuryl moiety seems to be much less attractive for nucleophilic attack.

3.3.3.2. Isosteric alternatives to the heterocyclic N-substituents The nitrogen-containing hetarylmethyl group as N-substituent (CPM, CTM) has a remarkably strong influence on the insecticidal activity. X-ray crystal structure analysis of imidacloprid and related neonicotinoids indicated that distances between the van der Waals surface of the CPM nitrogen and the atomic center of the pharmacophoric nitrogen are 5.45–6.06 Å (Tomizawa *et al.*, 2000). This range coincides with the distance between the ammonium nitrogen and carbonyl oxygen of acetylcholine, and between the nitrogen atoms of (*S*)-nicotine (Kagabu, 1997a). Alternatively, the CPM and CTM moieties were assumed to be able to participate in hydrogen bonding, like the pyridine ring of (*S*)-nicotine, and that this is important for the insecticidal activity. The CTM substituent generally confers higher potency in the clothianidin and N-desmethyl-thiamethoxam series than the CPM moiety in the imidacloprid, thiacloprid, acetamiprid, and nitenpyram series (Zhang *et al.*, 2000). Surprisingly, replacing both CPM and CTM by an oxygen-containing five-membered heterocycle resulted in a novel N-substituent TFM, that led to the development of the insecticide (±)-dinotefuran. It was found that the TFM structure can be taken as an isoster of the CPM and CTM moiety (Kagabu *et al.*, 2002). In an attempt to understand this, the hydrogen bonding regions of CPM, CTM, and TFM were projected onto their respective Connolly surfaces (Jeschke *et al.*, 2002) (**Figure 12**).

3.3.3.3. Bioisosteric pharmacophors The particularly high potency of the neonicotinoids bearing N-nitroimino, N-cyanoimino, or nitromethylene moieties, which have a negative electrostatic potential, implies a positive electrostatic potential for the corresponding insect nAChR recognition site (Nakayama and Sukekawa, 1998). Therefore, considerable attention has been given to the possible involvement of the pharmacophoric nitrogen in neonicotinoid action. In order to understand better the structural requirements, binding activity was analyzed using CoMFA (Akamatsu *et al.*, 1997). SAR analyses have also been performed for *in vitro* activities (Nishimura *et al.*, 1994). In particular, 3D QSAR

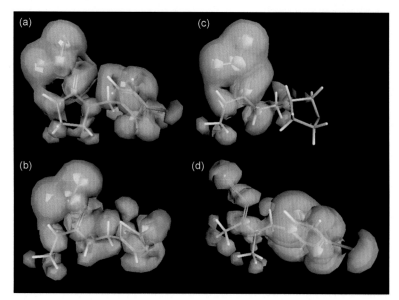

Figure 11 Isosurfaces for Fukui functions for nucleophilic attack of (a) imidacloprid, (b) clothianidin, (c) dinotefuran, and (d) acetamiprid. Three levels of isosurfaces are displayed: 0.005 (green, opaque), 0.001 (yellow, transparent), and 0.0005 (white, transparent).

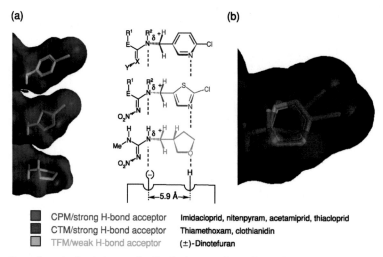

Figure 12 Isosteric alternatives to the heterocyclic *N*-substituents. Connolly surfaces of the CPM moiety from imidacloprid, nitenpyram, acetamiprid (blue), CTM moiety from thiamethoxam and clothianidin (magenta), and TFM from (±)-dinotefuran (green). Atomic charges have been driven via the Mulliken partitioning scheme DFT/BP/TZVP/COSMO Kohn–Sham orbitals. (a) H-bonding potentials is mapped onto the Connolly surface; (b) atomic charges are projected onto the Connolly surface. (Reproduced with permission from Jeschke, P., Schindler, M., Beck, M., **2002**. Neonicotinoid insecticides: retrospective consideration and prospects. *Proc. BCPC: Pests and Diseases 1*, 137–144.)

procedures are helpful to predict the receptor–ligand interaction (Nakayama, 1998; Okazawa *et al.*, 1998, 2000).

As described in an alternative binding model for imidacloprid, the interatomic distance of 5.9 Å between the oxygen of the nitro group (at the van der Waals surface) and the nitrogen in 1-position was also noted as adequate (Kagabu, 1997a). That means the oxygen of the nitro group and the cyano

nitrogen are well suited as acceptors for hydrogen bonding with the nAChR, in place of ring nitrogen atoms in CPM and CTM or the ring oxygen in TFM. Thus the π-conjugated system composed of a *N*-nitroimino or *N*-cyanoimino group and the conjugated nitrogen in 1-position are considered essential moieties for the binding of neonicotinoids to the putative cationic subsite in insect nAChR (**Figure 13**). On the other hand, Maienfisch *et al.*

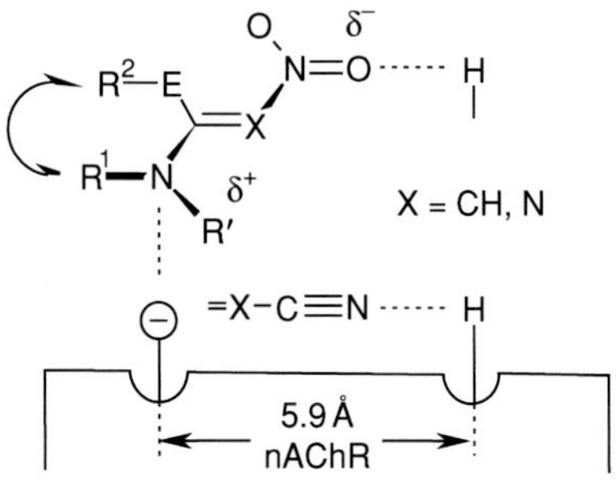

Figure 13 Binding to putative cationic subsite in insect nAChR (Kagabu, 1997a).

(1997) reported that a nitroenamine isomer (N^3-free imidacloprid) posseses similar insecticidal activity to imidacloprid (Boëlle *et al.*, 1998).

3.3.4. Physicochemistry of Neonicotinoids

The physicochemical properties of the ring system and noncyclic neonicotinoids played an important role in their successful development. In addition, photostability is a significant factor in field performance of this class of insecticides (see Section 3.2). As described (Kagabu and Medej, 1995; Kagabu, 1997a; Kagabu and Akagi, 1997), the energy gap for the different functional groups [=X—Y], from the ground state to the single state, excited in the order [=N—CN] > [=N—NO$_2$] > [=CH—NO$_2$]. For practical application of neonicotinoids, such as for soil and seed treatment, as well as for foliar application, the uptake and translocation in plants is crucial for their insecticidal activity. Thereby, not only the bioisosteric segments of neonicotinoids (see Section 3.3.3.3) but also the whole molecular shape (**Figure 12**), including the resultant water solubility, has to be considered. Neonicotinoid insecticides are push–pull olefins made up of conjugated electron donating and accepting groups (Kagabu, 1997a). Such polar, nonvolatile molecules have a high water solubility (**Table 3**) and low P_{OW} values compared with other nonpolar insecticidal classes. The following characteristics were described from Kagabu (1997a):

1. Noncyclic neonicotinoids are less lipophilic than the corresponding ring system neonicotinoids.
2. Concerning the functional group [=X—Y] as part of the pharmacophore [—N—C(E)=X—Y] (iii), water solubility increases in the order of [=CH—NO$_2$] > [=N—CN] > [=N—NO$_2$].
3. Regarding E, the lipophilicity increased in the order of S > C > O > NH (Kagabu, 1996).

Somewhat more lipophilic neonicotinoids should be better for seed treatment application, because uptake by roots is more effective than in the case of more hydrophilic compounds (Briggs *et al.*, 1982) (**Figure 14, Table 3**). Due to the excellent systemic properties of neonicotinoids conferred by the moderate water solubility, these insecticides are effective against their main target pests, which are sucking insects, like aphids, leafhoppers, and whiteflies.

3.3.4.1. Physicochemical properties of commercialized neonicotinoids
The physicochemical properties of a compound are related to its structural formula. However, it is not possible to predict the behavior of a substance, with sufficient certainty, only on the basis of its formula and physicochemical properties (Stupp and Fahl, 2003).

3.3.4.1.1. Imidacloprid Due to the special moieties like the CPM residue and the 2-(N-nitroimino)-imidazolidine 5-ring system, imidacloprid has very weak basic properties under environmental conditions (see Section 3.8.1.2). Water solubility and low partition coefficient in octanol–water are not influenced by pH values between 4 and 9, and at $20\,°C$ (Krohn and Hellpointer, 2002) (**Table 3**). The low partition coefficient of imidacloprid indicates that it has no potential to accumulate in biological tissues and further enrichment in the food chain. The occurrence of imidacloprid in the air is determined by its low vapor pressure of $4 \times 10^{-10}\,Pa$ which excludes volatilization from treated surfaces. The rapid uptake and translaminar transport of imidacloprid from the treated upper leaf side to the lower surface is excellent, as observed in cabbage leaves (Elbert *et al.*, 1991), and in rice and cucumber (Ishii *et al.*, 1994). However, imidacloprid has a considerable acropetal mobility in xylem of plants. In contrast, its penetration and translocation in cotton leaves was less pronounced, as qualitatively reflected by phosphor-imager autoradiography (Buchholz and Nauen, 2001). This xylem mobility makes imidacloprid especially useful for seed treatment and soil application, but it is equally effective for foliar application (Elbert *et al.*, 1991).

Due to its lack of any acidic hydrogen, the pK_a of imidacloprid is >14, and therefore its transport within the phloem is unlikely, as been shown in several studies (Stein-Dönecke *et al.*, 1992; Tröltzsch *et al.*, 1994). Systemic properties were examined using ^{14}C-labeled imidacloprid. The translocation of imidacloprid in winter wheat and its uptake and translocation from treated cotton seeds into the growing parts of the cotton plant is described by Elbert *et al.* (1998). Registered patterns of use of imidacloprid in agriculture now include traditional foliar spray application as well as soil drench application, drip irrigation, trunk (injection) application,

Table 3 Physicochemical properties of commercialized neonicotinoids

Physical and chemical properties	Ring systems			Noncyclic structures			
	Imidacloprid	Thiacloprid	Thiamethoxam	Nitenpyram	Acetamiprid	Clothianidin	(±)-Dinotefuran
Color and physical state	Colorless crystals	Yellow/crystalline powder	Slightly cream/crystalline powder	Pale yellow crystals	Colorless crystals	Clear/coloress solid powder	
Melting point (°C)	144	136	139.1	83–84	98.9	176.8	94.5–101.5
Henry's law constant (Pa $\times$ m^3 mol^{-1}) (at 20°C)	2×10^{-10}				$<5.3 \times 10^{-8}$ (calc)	2.9×10^{-11}	
Density (g ml^{-1}) (at 20°C)	1.54	1.46		1.40 (at 26°C)	1.330	1.61	1.33
Vapor pressure (Pa) (at 25°C)			6.6×10^{-9}		$<1 \times 10^{-6}$	1.3×10^{-10}	
(at 20°C)	4×10^{-10}			1.1×10^{-9}		3.8×10^{-11}	
Solubility in water (g l^{-1}) (at 20°C)	0.61	0.185	4.1 (at 25°C)	840 (pH 7.0)	4.20 (at 25°C)	0.327	54.3 ± 1.3
Solubility in organic solvents (g l^{-1} at 25°C)							
Dichloromethane	67 (at 20°C)	160 (at 20°C)	43.0	nd		1.32	
N-hexane	<0.1	nd	0.00018				
N-heptane	nd	<0.1 (at 20°C)	nd	nd		<0.00104	
Methanol	10 (at 20°C)	nd	10.2	670 (at 20°C)		6.26	
Ethanol	nd						
Xylene	nd	0.3 (at 20°C)	nd	4.5 (at 20°C)		0.0128	
Toluene	0.68		0.63	nd			
1-Octanol	nd	1.4 (at 20°C)	0.63			0.938	
Acetone	50 (at 20°C)	64 (at 20°C)	42.5	290 (at 20°C)		15.2	
Acetonitrile	nd	52 (at 20°C)	78.0	430 (at 20°C)			
Ethyl acetate	nd	9.4 (at 20°C)	5.74	33 (at 20°C)		2.03	
Partition coefficient in octanol–water (at 25°C)	3.72 (at 21°C)						
log P$_{ow}$	0.57 (at 22°C)	1.26 (at 20°C)	−0.13 (at pH 6.8)	−0.64	0.8	0.7	−0.644
log P$_{ow}$ (pH 4.7) (at 20°C)					No significant variations with pH	0.9	
log P$_{ow}$ (pH 9.0) (at 20°C)						0.9	
Dissociation constant pK_a (at 20°C)			No dissociation in range pH 2–12	3.1 and 11.5	0.7 (at 25°C) Weak base	11.09	No dissociation in range pH 1.4–12.3
Flammability		Not highly				Not highly	
Surface tension (mN m^{-1}) (at 20°C)		66				79.6	

nd, not determined.

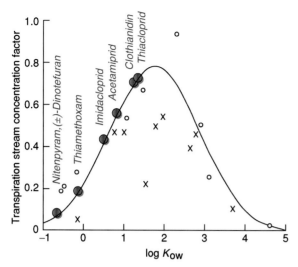

Figure 14 Systemicity of neonicotinoids. Relationship between the translocation of neonicotinoids to barley shoots following uptake by the roots (expressed as the transpiration stream concentration factor), and the octan-1-ol/water partition coefficients (as log K_{ow}). (Diagram and experimental data taken from Briggs, G., Bromilow, R.H., Evans, A.A., **1982**. Relationships between lipophilicity and root uptake and translocation of non-ionized chemicals by barley. *Pestic. Sci. 13*, 495–504). Because of the higher lipophilicity, thiacloprid and clothianidin show the best uptake by the roots from solution, whereas nitenpyram and ($\pm$)-dinotefuran are more xylem mobile than the other neonicotinoids.

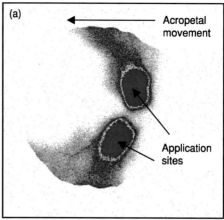

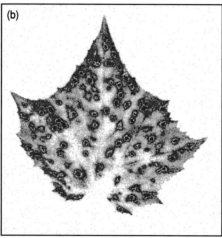

Figure 15 (a) Translocation of [^{14}C]-thiacloprid in cabbage plants 1 day after application of $2 \times 5\,\mu l$ droplets onto the first true leaf. Radioactivity on the surface of the cuticle (residue) was removed by cellulose-acetate stripping; (b) translocation pattern of thiacloprid sprayed to a cucumber leaf.

stem or granular treatments and seed treatment. Because of imidacloprid's systemic properties it is evenly distributed in young, growing plants (Elbert *et al.*, 1991, 1998).

3.3.4.1.2. Thiacloprid

It is stable towards hydrolysis even under conditions of heavy rain. Once applied to leaves, thiacloprid shows good rainfastness and photostability, and remains in or on leaves for a considerable time, thus allowing a continous penetration of the active ingredient into the leaf. Its half-life in water at pH 5, 7, and 9 is over 500 h. Photolysis in water (buffered at pH 7) shows a half-life of >100 days. On soil surfaces thiacloprid is also stable under sunlight irradiation (Jeschke *et al.*, 2001). The penetration and translocation behavior of thiacloprid in cabbage was comparable to those reported for imidacloprid (Buchholz and Nauen, 2002). The amounts of thiacloprid stripped off from cabbage application sites were 23% and 17% at 1 day and 7 days after treatment, respectively. Levels in the true leaf increased from 63% to 77% as measured 1 day and 7 days after application, respectively. These results demonstrate that thiacloprid is readily taken up by cabbage leaves, thus providing good systemic control of leaf-sucking pests. The visualization of the translocation pattern

of [^{14}C]thiacloprid-equivalents by phosphor-imaging technology revealed xylem mobility, i.e., translocation of the active ingredient in the upwards direction, even 1 day after application to cabbage leaves (**Figure 15a**). This xylem mobility is further highlighted in **Figure 15b**, showing excellent distribution of thiacloprid in cucumber leaves after spray application.

3.3.4.1.3. Thiamethoxam

It is a crystalline, odorless compound with a melting point of 139.1 °C. This neonicotinoid has a relatively high water solubility of $4.1\,g\,l^{-1}$ at 25 °C and a low partition coefficient (log $P_{OW} = -0.13$ at pH 6.8). In the range of pH 2 to 12 no dissociation is observed (Maienfisch *et al.*, 2001a). These properties favor a rapid and efficient uptake in plants and xylem transport (Widmer *et al.*, 1999) (**Table 3**). Through this systemic activity, all plant parts situated acropetally from the application can be protected (Maienfisch *et al.*, 1999a, 2001a). Thiamethoxam is hydrolytically very stable at pH 5 with a half-life greater than 1 year at room temperature and stable at pH 7 (estimated half-life at room temperature is 200–300 days) (Maienfisch *et al.*, 2001a). This neonicotinoid is more labile at pH 9, where the half-life is a few days. It is rapidly photolytically degraded with a half-life of about 1 h as a droplet deposit on Teflon. But no decomposition was observed after storage of the active ingredient (a.i.) or formulated product at 54 °C for 2 months; however, at temperatures above 150 °C, exothermic decomposition occurs (Maienfisch *et al.*, 1999a). Under laboratory soils, thiamethoxam degrades at moderate to slow rates. Under field conditions, degradation is generally faster, because of the higher microbial activity of field soils and exposure to light as important degradation pathways (Maienfisch *et al.*, 1999a).

3.3.4.1.4. *Nitenpyram*

It is highly soluble in water ($840 \, \text{g} \, \text{l}^{-1}$), and shows good systemic action (Kashiwada, 1996) and no phytotoxicity, characteristics which enable the various application methods of the neonicotinoid (Akayama and Minamida, 1999). The partition coefficient is low (−0.64) (**Table 3**).

3.3.4.1.5. *Acetamiprid*

It is stable in buffer solutions at pH 4, 5, and 7 and under sunlight. It is degraded slowly at pH 9 at 45 °C. Hydrolysis of acetamiprid in 0.5–1 N sodium hydroxide at 90–95 °C for 2 h gave quantitatively *N*-methyl-(6-chloro-pyrid-3-yl)-methylamine (Tokieda *et al.*, 1997). Because of its high solubility in water (**Table 3**), acetamiprid shows a high systemic and translaminar insecticidal activity (Yamada *et al.*, 1999) and is currently used only foliarly.

3.3.4.1.6. *Clothianidin*

It has no acidic or alkaline properties at the relevant pH (Stupp and Fahl, 2003). Therefore, the pH of the aqueous system has no influence on its physicochemical properties. It is stable to hydrolysis in a pH range of 4 to 9, but photolysis contributes significantly to its degradation in the environment resulting in an elevated mineralization rate. The water solubility of clothianidin is relatively low ($0.327 \, \text{g} \, \text{l}^{-1}$ at 20 °C) in comparison to the other neonicotinoids that have a nitroguanidyl moiety. This is also reflected by the octanol–water partition coefficient, indicating a favorable adsorption to soil (log P_{OW} = 0.7 at 25 °C) (**Table 3**). On basis of these physicochemical properties no bioaccumulation is expected nor any volatilization, and thus no significant amounts of

the substance are to be expected in the atmosphere (Stupp and Fahl, 2003).

3.3.5. Proneonicotinoids

Pharmaceutical conceptions in prodrug design generally focus on its potential in overcoming pharmacokinetic problems and poor oral absorption (Testa and Mayer, 2001). Prodrug activation occurs enzymatically, nonenzymatically, or also sequentially (enzymatic step followed by nonenzymatic rearrangement) (Testa and Mayer, 1998). Depending on both the drug and its prodrug, the therapeutic gain may be modest, marked, or even significant (Balant and Doelker, 1995). The field of neonicotinoid chemistry reflects, e.g., with Mannich adducts, some significant examples for useful proneonicotinoids in crop protection (see Sections 3.3.5.1 and 3.3.5.3).

3.3.5.1. Ring cleavage to noncyclic neonicotinoids

The six-membered perhydro-1,3,5-thiadiazine (Z = S), hexahydro-1,3,5-triazine (Z = NR), and perhydro-1,3,5-oxadiazine (Z = O) containing neonicotinoids can be prepared by a Mannich type cyclization reaction by treatment, for example, cyclization of the *N*-nitroguanidine moiety with two molar equivalents of formaldehyde (HCHO) in the presence of one or more equivalents of hydrogen sulfide (H_2S), an appropriate amine ($R-NH_2$) or water (H_2O) in a suitable solvent. It is known that these cyclic Mannich adducts can cleave into their noncyclic compounds depending on the reaction conditions (pH values). Efficient syntheses of the active ingredient clothianidin, which is readily performed by ring cleavage reaction of the *bis*-aminal structure ($N-CH_2-N$ bonds) within the 6-ring system of the N,N′-dialkylated 2-(*N*-nitroimino)-hexahydro-1,3,5-triazine, have been previously described (Maienfisch *et al.*, 2000; Jeschke *et al.*, 2003).

The hydrolysis (ring cleavage) of different 6-ring intermediates giving noncyclic neonicotinoids were studied in a physiological salt solution at 25 °C (pH 7.34). It was found that the perhydro-1,3,5-oxadiazines (Z = O) have noteably longer half-lives than the appropriate hexahydro-1,3,5-triazines as *bis*-aminals (Kagabu, 2003a) (**Figure 16**).

Agro Kanesho's hexahydro-1,3,5-triazine (Z = NMe) AKD-1022, a six-membered ring compound, is a proinsecticide of clothianidin (Bryant, 2001). Recently, it was demonstrated that thiamethoxam, having a perhydro-1,3,5-oxadiazine (Z = O) structure, is also an *in vivo*, easy-to-cleave neonicotinoid precursor for the highly active noncyclic neonicotinoid clothianidin (Nauen *et al.*, 2003). This suggests a prodrug principle in the mode of action of thiamethoxam, rather than differences in the binding

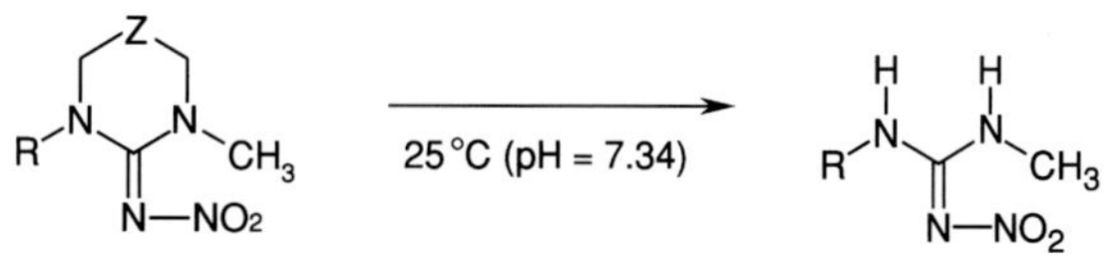

Substituent	Half-life $t_{1/2}$ (in days) at 25 °C (pH = 7.34)		
R	Z = O	Z = S	Z = N—CH₃
2-Chloro-1,3-thiazol-5-ylmethyl- (CTM)	132	12	5.3
6-Chloro-pyrid-3-ylmethyl- (CPM)	72	6	8

Figure 16 Half-life $t_{1/2}$ (days) of neonicotinoids in salt solution at 25 °C (pH 7.34). (Reproduced from Kagabu, S., **2003a**. Molecular design of neonicotinoids: past, present and future. In: Voss, G., Ramos, G. (Eds.), Chemistry of Crop Protection: Progress and Prospects in Science and Regulation. Wiley–VCH, New York, pp. 193–212.)

site (Kayser *et al.*, 2002). Clothianidin is one of the most prominent metabolites in the leaves of thiamethoxam-treated (soil drench) cotton plants. Thiamethoxam is not only rapidly metabolized in plants but also (within minutes) converted to clothianidin in noctuid larvae. The level of clothianidin detected in the hemolymph is quite high, i.e., physiologically relevant, and considering the electrophysiological results in an interaction with the nAChR in the CNS can be assumed to occur (Nauen *et al.*, 2003). Therefore clothianidin is likely to be responsible for the insecticidal potency of thiamethoxam, and not the parent compound itself, an important fact for resistance management strategies (see Section 3.9).

3.3.5.2. Mannich adducts as useful precursors The first patent applications covering a number of substituted 5-nitro-tetrahydro-1,3-pyrimidines, the so-called Mannich adducts of certain insecticidally active nitromethylene compounds, were published in 1988 (e.g., compound Bay T 9992, Bayer AG). Due to their potent toxicity to insects and arthropods, numerous different companies produced compounds of this type (Nauen *et al.*, 2001). These Mannich adducts can be prepared via an aminomethylation reaction by the treatment of the CH-acidic nitromethylenes [−N−C(E)=CH−NO₂; E = NHR²], which react as nitroenamines (Rajappa, 1981), with one or more equivalents of an amine (R³−NH₂), and at least two molar equivalents of the electrophile formaldehyde (HCHO) in a suitable solvent like alcohols, water, polar aprotic solvents such as tetrahydrofuran, and N,N-dimethylformamide, or solid paraformaldehyde (**Figure 17**; Mannich adduct formation, pathway A). In some cases, a small amount of a strong, nonoxidizing acid, such as hydrochloric acid, can be used as a catalyst. The Mannich adducts have a wide range of substituents R³ which inhibit the [³H]imidacloprid

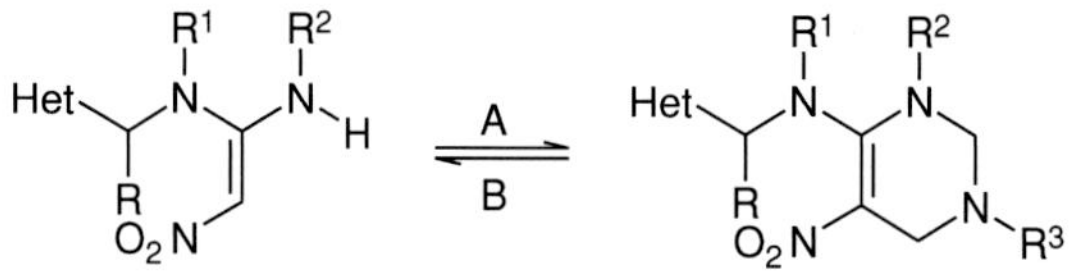

Figure 17 Formation and hydrolysis of Mannich adducts. (Reproduced from Nauen, R., Ebbinghaus-Kintscher, A., Elbert, A., Jeschke, P., Tietjen, K., **2001**. Acetylcholine receptors as sites for developing neonicotinoid insecticides. In: Iishaaya, I., (Ed.), Biochemical Sites of Insecticide Action and Resistance. Springer, New York, pp. 77–105.)

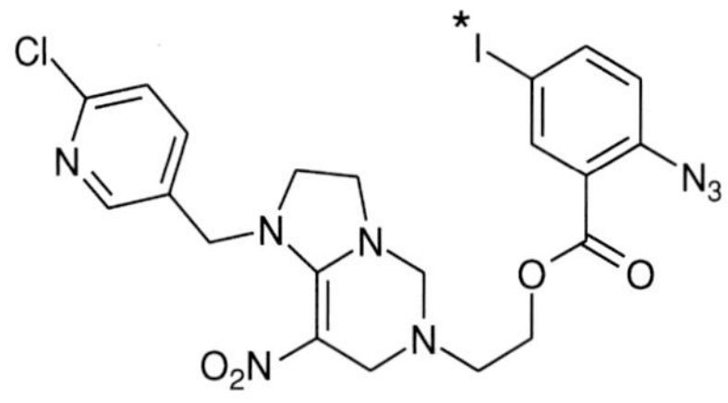

Figure 18 [¹²⁵I] azidoneonicotinoid (AN) photoaffinity radioligand, a useful probe to explore the structure and diversity of insect nAChRs. (Reproduced from Tomizawa, M., Casida, J.E., **2001**. Structure and diversity of nicotinic acetylcholine receptors. *Pest Mgt Sci. 57,* 914–922.)

binding site of *Drosophila* or *Musca* nAChR by 50% at 0.7–24.0 nM (Latli *et al.*, 1997). However, at different pH values, hydrolysis of the Mannich adducts can be observed (**Figure 17**; Mannich adduct cleavage, pathway B).

Application of more stable Mannich adducts in protein biochemistry led to the synthesis of the [¹²⁵I]azidoneonicotinoid (AN) probe by using 1-(6-chloro-pyrid-3-ylmethyl)-2-nitromethylene-imidazolidine (NTN32692) as the photoaffinity radioligand (Latli *et al.*, 1997; Maienfisch *et al.*, 2003) (**Figure 18**). This probe facilitated the identification of insecticide-binding site(s) in insect nAChRs.

In the first photoaffinity labeling of an insect neurotransmitter receptor, [¹²⁵I]AN exhibited high

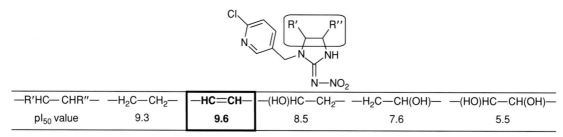

	—R'HC—CHR''—	—H₂C—CH₂—	—HC=CH—	—(HO)HC—CH₂—	—H₂C—CH(OH)—	—(HO)HC—CH(OH)—
pI₅₀ value		9.3	9.6	8.5	7.6	5.5

Figure 19 Modification of the imidacloprid bridging chain and pI₅₀ values for the nicotinic acetylcholine receptor (nAChR) from housefly head membranes. (Reproduced from Nauen, R., Ebbinghaus-Kintscher, A., Elbert, A., Jeschke, P., Tietjen, K., **2001**. Acetylcholine receptors as sites for developing neonicotinoid insecticides. In: Iishaaya, I. (Ed.), Biochemical Sites of Insecticide Action and Resistance. Springer, New York, pp. 77–105.)

affinity (8 nM), and successfully identified a 66 kDa polypeptide in *Drosophila* head membranes. Several cholinergic ligands and neonicotinoid insecticides, like imidacloprid and acetamiprid, strongly inhibited this photoaffinity labeling. Thus, the labeled polypeptide is pharmacologically consistent with the ligand- and insecticide-binding subunit of *Drosophila* nAChR (Tomizawa and Casida, 1997, 2001).

3.3.5.3. Active metabolites of neonicotinoids From the metabolic pathway (see Section 3.7.1) of imidacloprid it is known that hydroxylation of the imidazolidine ring leads in general to the mono- (R' = −H; R'' = −OH) and the bishydroxylated (R', R'' = −OH) derivatives, both of which have reduced affinity (**Figure 19**). Alternatively, the mono (R' = −OH; R'' = −H) derivative reflects a higher level of efficacy (pI₅₀ value 8.5) (Nauen *et al.*, 2001; Sarkar *et al.*, 2001). Interestingly, the olefin metabolite showed a higher pI₅₀ value for nAChR from housefly head membranes than imidacloprid, and provides superior toxicity to some homopterans after oral ingestion (Nauen *et al.*, 1998b, 1999b). This result suggests that for the central ring system the exact rearrangement of the ring atoms, and not the electronic effect of the ring system, seems to be necessary for insecticidal activity. A similar phenomenon has been described for the conjugated pyridone derivatives (Kagabu, 1999).

3.4. Biological Activity and Agricultural Uses

The biological activity and agricultural uses of neonicotinoid insecticides are enormous' and these insecticides are continuing to see new uses (Elbert and Nauen, 2004). It is definitely beyond the scope of this chapter to provide a full overview of the agronomic and horticultural cropping systems that use neonicotinoid insecticides and readers interested in these aspects should refer to many articles

and book chapters published during the past decade, e.g., Elbert *et al.* (1991, 1998), Yamamoto (1999), Kiriyama and Nishimura (2002), and Elbert and Nauen (2004). In order to provide a flavor of the agricultural uses of neonicotinoids, and the affected target pests, a few examples considering imidacloprid are given below. Imidacloprid is presently the most widely used neonicotinoid insecticide worldwide.

3.4.1. Efficacy on Target Pests

Due to their unique properties – high intrinsic acute and residual activity against sucking and some chewing insect species, high efficacy against aphids, whiteflies, leafhoppers and planthoppers, and the Colorado potato beetle, and excellent acropetal translocation – neonicotinoids can be used in a variety of crops (**Figure 20**). These uses include: aphids on vegetables, sugar beet, cotton, pome fruit, cereals, and tobacco; leafhoppers, planthoppers, and water weevil on rice; whiteflies on vegetables, cotton, and citrus; lepidopteran leafminer on pome fruit and citrus; and wireworms on sugar beet and corn (**Table 4**). Termites and turf pests such as white grubs can also be controlled by imidacloprid (Elbert *et al.*, 1990, 1991).

Neonicotinoids such as imidacloprid and thiamethoxam also control important vectors of virus diseases, thereby impairing the secondary spread of viruses in various crops. This control has been observed, e.g., for the persistent barley yellow dwarf virus (BYDV) transmitted by *Rhopalosiphum padi* and *Sitobion avenae* (Knaust and Poehling, 1992). Seed treatments proved highly effective in controlling barley yellow dwarfvirus vectors and the subsequent infection, in a series of field trials in southern England. Sugar beet seed, pelleted with imidacloprid, was well protected especially against infections with beet mild yellow virus transmitted by the peach potato aphid, *Myzus persicae* (Dewar and Read, 1990).

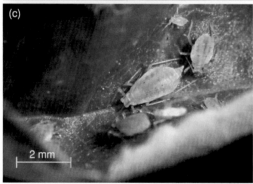

Figure 20 Important pest insects targeted by neonicotinoid insecticides. (a) Sweet-potato whitefly, *Bemisia tabaci*; (b) Colorado potato beetle, *Leptinotarsa decemlineata*; (c) green peach aphid, *Myzus persicae*.

Table 4 Insecticidal efficacy of imidacloprid after foliar application against a variety of pest insects under laboratory conditions

Pest species	Developmental stage	LC$_{95}$, rounded (ppm)
Homoptera		
Aphis fabae	Mixed	8
Aphis gossypii	Mixed	1.6
Aphis pomi	Mixed	8
Brevicoryne brassicae	Mixed	40
Myzus persicae	Mixed	1.6
Myzus persicae (tobacco)	Mixed	8
Phorodon humuli	Mixed	0.32
Laodelphax striatellus	Third instar	1.6
Nephotettix cincticeps	Third instar	0.32
Nilaparvata lugens	Third instar	1.6
Sogatella furcifera	Third instar	1.6
Pseudococcus comstocki	Larvae	1.6
Bemisia tabaci	Second instar	8
Trialeurodes vaporariorum	Adult	40
Hercinothrips femoralis	Mixed	1.6
Coleoptera		
Leptinotarsa decemlineata	Second instar	40
Lema oryzae	Adult	8
Lissorhoptrus oryzophilus	Adult	40
Phaedon cochleariae	Second instar	40
Lepidoptera		
Chilo suppressalis	First instar	8
Helicoverpa armigera	Second instar	200
Plutella xylostella	Second instar	200
Heliothis virescens	Egg	40
Spodoptera frugiperda	Second instar	200

3.4.2. Agricultural Uses

Due to their high systemicity, diverse ways of application are feasible and these methods have been introduced into practice. Soil treatments can be done by incorporation of granules, injection, application with irrigation water, spraying, and use of tablets. Plants or plant parts can be treated by seed dressing, pelleting, implantation, dipping, injection, and painting. These methods have led to a more economic and environmentally friendly use of these products, fitting well into various integrated pest management (IPM) programmes. In this section, imidacloprid has been chosen as an example of a neonicotinoid, and a few examples of its use in various crops is given. Such uses include treatments

for rice in Japan, cotton in the mid-south of the USA, vegetables in Mexico, and cereals in France. In these crops the neonicotinoid insecticides targets mainly early-season pests.

3.4.2.1. Rice In Japan, nursery box application of insecticides is a standard procedure, and it is currently used in 50% of the rice-growing area. This type of treatment is directed against early-season pests like *Lissorhoptrus oryzophilus* (rice water weevil) and *Lema oryzae* (rice leaf beetle). In contrast to commercially available products like carbosulfan, benfuracarb, and propoxur, which control mainly rice water weevil and rice leaf beetle, imidacloprid covers the whole spectrum of early-season pests including *N. cincticeps* (green rice leafhopper), *Laodelphax striatellus* (smaller brown planthopper), and also mid- and late-season hopper species, like *Sogatella furcifera* (white-backed planthopper) and *Nilaparvata lugens* (brown planthopper). Benefits of such nursery box applications include, reduced labor input, longer residual effect, and reduced side effects against nontarget organisms when compared to surface treatments.

Based on 5-year field trials, a use pattern for imidacloprid against the main rice pests has been established. When applied as a granule, 2 granular GR with 1 g a.i. per nursery box (= 200 g a.i. ha^{-1}) 0–3 days before transplanting, the product gives full protection for 50–90 days against different hoppers species. By controlling *L. striatellus*, a 91% reduction of the rice stripe virus was observed. An excellent residual effect was detected up to 50 days after transplanting against *L. oryzophilus*, and up to 60 days against *L. oryzae*.

Apart from the seedling box application, imidacloprid can be used as a granule for water surface application, as a wettable powder (WP) or as a dust formulation for foliar treatment, and also as a seed dressing to control the above mentioned pests on rice. Under field conditions of such use, no apparent effects from imidacloprid treatments were observed against populations of the spiders *Pardosa pseudoannulata* and *Tetragnatha vermiformis*.

Therefore, imidacloprid can be regarded very effective against rice insects and flexible to use, but it is most efficient in nursery box application. It outperforms other conventional insecticides due to its broader efficacy and lower dose rates. Virus diseases are suppressed through effective vector control, and low toxicity has been observed against important beneficials in rice.

3.4.2.2. Cotton

The main cotton pests in the mid South of the USA are *Thrips tabaci*, *Lygus lineolaris* (tarnished plant bug), *Neurocolpus nubilis* (clouded plant bug), *Anthonomus grandis* (boll weevil), *Heliothis virescens* (tobacco budworm), and *Helicoverpa zea* (bollworm). Other pests such as *Pseudatomoscelis seriatus* (cotton fleahopper), *Aphis gossypii* (cotton aphid), *Spodoptera exigua* (beet armyworm) and *S. frugiperda* (fall armyworm) are of lesser importance in this region.

Thrips tabaci attacks the terminal bud and two to four true leaves. The plant is damaged by a reduced stand, retarding growth, killed buds, and delayed fruiting, with significant yield reduction. *Lygus lineolaris* damages by feeding on pinhead squares, which causes young squares to shed. Terminal bud injury leads to multiple branched plants ("crazy cotton"). Symptoms caused by *N. nubilis* are similar to those of *L. lineolaris*.

It is thus evident that damage caused by early-season pests, such as thrips and bugs, can be very important. Loss of first-position squares can have detrimental effects on the yield potential of the plant. If normal fruiting of cotton plants is prevented, this can cause a delay in maturity, which in turn necessitates additional late-season insecticidal applications to control bollworms, tobacco budworms, and boll weevils.

A 480 FS formulation (Gaucho®) is applied for seed dressing, against early infestations of thrips and aphids, at 250 g a.i. per 100 kg cotton seed (37 g a.i. ha^{-1}). Against bugs and later attack of aphids, three foliar split applications at low dose rates are recommended (Provado® 1.6 F, 3 × 15 g a.i. ha^{-1} (**Table 4**). These should be done as band sprays at intervals of 7–10 days. The first application starts 30 days after emergence of the five leaf stage, subsequent to the Gaucho® protection period. The *Heliothis–Helicoverpa* complex is controlled by conventional insecticides in mid-season, whereas boll weevil and armyworms are controlled by azinphos-methyl or cyfluthrin, respectively.

3.4.2.3. Vegetables

Virus diseases have been the main factors affecting the tomato and chilli production in Mexico during the past 20 years, and these diseases have led to significant yield reductions in many regions. These crops are grown on approximately 160 000 ha with constant virus infestation in 30–35% of the fields, with persistent viruses being the most important in Mexican agriculture. In tomato crops, viruses can infest up to 100% of the crop. The damage depends on the type of virus, the population density of vectors, and the developmental stage of the crop. Protection by conventional insecticide treatments under plastic covered greenhouses, however, has not given satisfactory results (Elbert and Nauen, 2004).

3.4.2.4. Cereals

The traditional method in controlling virus vectors in Europe is regular monitoring of aphid attack of young wheat and barley plants after emergence in autumn. If the threshold of 10% infested plants with at least one aphid per plant has been surpassed, usually a pyrethroid will be sprayed. Monitoring has to be continued, because the residual effect of the treatment may be insufficient for vector control. Depending on climatic conditions and the duration of the aphid attack, a second or a third application may be required. This rather time consuming and difficult procedure can be avoided by a seed treatment with imidacloprid.

Barley yellow dwarf virus vectors such as *R. padi* and *S. avenae* are controlled and the secondary spread of the disease is inhibited. In contrast to one or several sprays, seed dressing as the only treatment assures prolonged protection during the critical period, when virus transmission is of importance. Furthermore pyrethroid sprays with a broad spectrum of activity can be substituted by only one application of imidacloprid, which has virtually no effects on beneficials and other nontarget organisms. High yield varieties achieve their optimum yield potential only if they are sown early, which implies an enhanced risk of infestation with aphids and, consequently, with barley yellow dwarf virus. As Gaucho® controls aphids and suppresses the infection with this virus, early-sown cereals are especially well protected against insects and diseases. Therefore, a better tillering is achieved and 15–25% of the seed can be saved.

Both imidacloprid seed treatment and pyrethroid spray application have very good to excellent effects against the vector *R. padi*. The acute effect of the pyrethroid was better due to its quick knockdown activity. Imidacloprid alternatively acts more slowly, so, it takes more time to kill invading aphids, but its residual effect is much more pronounced. Thus imidacloprid use results in long-lasting control of aphids and, as a consequence, considerably better prevention of virus symptoms. A yield increase of 26% over untreated controls was achieved in comparison with, 10% achieved by pyrethroid treatment.

Field trials over 4 years in France confirmed that both wheat and barley are well protected against the above-mentioned pests and diseases. In an average of 76 trials in wheat and 66 trials in barley a yield increase of 3.1 dt ha^{-1} and 4.4 dt ha^{-1}, respectively; was obtained with the Gaucho® treatment when compared with untreated.

3.4.3. Foliar Application

Spray applications are especially used against pests attacking crops such as cereals, maize, rice, potatoes, vegetables, sugar beet, cotton, and deciduous trees. **Table 4** shows the acute activity (estimated LC$_{95}$ (lethal concentration at which 95% of insects are killed) in ppm a.i.) of imidacloprid against a variety of pests, following foliar application (dip and spray treatment) of host plants under laboratory and greenhouse conditions. Imidacloprid was very active to a wide range of aphids. The most susceptible was the damson hop aphid *Phorodon humuli* (LC$_{95}$ 0.32 ppm), which is often highly resistant against conventional insecticides (Weichel and Nauen, 2003). Imidacloprid was highly effective

against some of the most important rice pests, such as leafhoppers and planthoppers, rice leaf beetle *L. oryzae* and rice water weevil *L. oryzophilus*. Although imidacloprid is generally less effective against biting insects, its efficacy against the Colorado potato beetle *Leptinotarsa decemlineata* is relatively high (LC$_{95}$ 40 ppm). The LC$_{95}$ values for the second or third instar larvae of some of the most deleterious noctuid pest species of the order Lepidoptera, *Helicoverpa armigera*, *Spodoptera frugiperda*, and *Plutella xylostella*, were approximately 200 ppm, and higher than those for the species mentioned above (Elbert *et al.*, 1991).

3.4.4. Soil Application and Seed Treatment

Typical soil insect pests such as *Agriotes* sp., *Diabrotica balteata*, or *Hylemyia antiqua* were controlled by incorporation of 2.5–5 ppm a.i. into the soil. Higher concentrations of imidacloprid are necessary to control *Reticulitermes flavipes* (7 ppm) and *Agrotis segetum* (20 ppm). However, imidacloprid activity is much more pronounced against early-season sucking pests, which attack the aerial parts of a wide range of crops. Soil concentrations as low as 0.15 ppm a.i. gave excellent control of *Myzus persicae* and *Aphis fabae* on cabbage in greenhouse experiments (Elbert *et al.*, 1991). A good residual activity is essential for the protection of young plants.

3.5. Mode of Action

The biochemical mode of action of neonicotinoid insecticides has been studied and characterized extensively in the past 10 years. They act selectively on insect nAChRs, a family of ligand-gated ion channels located in the CNS of insects and responsible for rapid neurotransmission. The nAChR is a pentameric transmembrane complex, and each subunit consists of an extracellular domain containing the ligand binding site and four transmembrane domains (Nauen *et al.*, 2001; Tomizawa and Casida, 2003). Neonicotinoid insecticides bind to the acetylcholine binding site located on the hydrophilic extracellular domain of α-subunits. Their ability to displace tritiated imidacloprid from its binding site correlates well with their insecticidal efficacy (Liu and Casida, 1993a; Liu *et al.*, 1993b). [^{3}H]imidacloprid binds with nanomolar affinity to nAChR preparations from insect tissues, and next to the less specific α-bungarotoxin, it is the preferred compound in radioligand competition studies (Lind *et al.*, 1998; Nauen *et al.*, 2001). Furthermore, electrophysiological studies revealed that neonicotinoid insecticides act agonistically on nAChR, and this interaction is again very well correlated with their potential to control target pest species. In contrast antagonistic compounds of mammalian nAChR were shown to be far less active as insecticides (Nauen *et al.*, 1999a). All neonicotinoid insecticides act with nanomolar affinity against housefly and other insect nAChRs, except thiamethoxam, which exhibits a comparatively low affinity for the [^{3}H]imidacloprid binding site. This low affinites was attributed to its proneonicotinoid nature (see Sections 3.3.5 and 3.3.5.1), as it was shown to be activated to clothianidin in insects and plants (Nauen *et al.*, 2003).

The mode of action of neonicotinoid insecticides has been described in several excellent reviews which should serve as a primary source for further informations (Kagabu, 1997a; Matsuda *et al.*, 2001; Nauen *et al.*, 2001; Tomizawa and Casida, 2003). These reviews describe our current knowledge of the structure and function of insect nAChR, characterized by receptor binding studies, phylogenetic considerations regarding receptor homologies between orthologs from different animal species, and electrophysiological investigations, as well as the description of a vast range of a structurally diverse receptor ligands, of course with an emphasis on neonicotinoid insecticides.

3.6. Interactions of Neonicotinoids with the Nicotinic Acetylcholine Receptor

3.6.1. Selectivity for Insect over Vertebrate nAChRs

Electrophysiological measurements reported in numerous studies have revealed that nAChRs are widely expressed in the insect nervous system on both postsynaptic and presynaptic nerve terminals, on the cell bodies of interneurons, motor neurons, and sensory neurons (Goodman and Spitzer, 1980; Harrow and Sattelle, 1983; Sattelle *et al.*, 1983; Breer, 1988; Restifo and White, 1990). Schröder and Flattum (1984) using extracellular electrophysiological recordings were the first to identify that the site of action of the nitromethylene nithiazine was on the cholinergic synapse. A number of subsequent electrophysiological and biochemical binding studies revealed that the primary target of the neonicotinoids were the nAChRs (Benson, 1989; Sattelle *et al.*, 1989; Bai *et al.*, 1991; Leech *et al.*, 1991; Cheung *et al.*, 1992; Tomizawa and Yamamoto, 1992, 1993; Zwart *et al.*, 1992; Liu and Casida, 1993a; Tomizawa *et al.*, 1996). Recent electrophysiological studies indicate that imidacloprid acts as an agonist on two distinct nAChR subtypes on

cultured cockroach dorsal unpaired motoneuron (DUM) neurons (Buckingham *et al.*, 1997), an α-bungarotoxin (α-BGTx) sensitive nAChR with "mixed" nicotinic/muscarinic pharmacology and an α-BGTx insensitive nAChR. Such electrophysiological observations were supported by binding studies with [^{3}H]imidacloprid in membrane preparations from *M. persicae*. These ligand competition studies revealed the presence of high and low-affinity nAChR binding sites for imidacloprid in *M. persicae* (Lind *et al.*, 1998).

The identification of multiple putative nAChR subunits by molecular cloning is consistent with a substantial diversity of insect nAChRs (Gundelfinger, 1992). At present, at least five different subunits have been cloned from *Drosophila melanogaster* (Schulz *et al.*, 1998), from the locust *Locusta migratoria* (Hermsen *et al.*, 1998), and from the aphid *M. persicae* (Huang *et al.*, 1999). Despite the considerable number of subunits identified, only a few functional receptors were obtained after expression of different subunit combinations in *Xenopus* oocytes or cell lines. Initial work suggested that some subunits can form homooligomeric functional receptors when expressed in *Xenopus* oocytes. This was shown for Lα1 from the locust *Schistocerca gregaria* (Marshall *et al.*, 1990; Amar *et al.*, 1995), and for the Mpα1 and Mpα2 from *M. persicae* (Sgard *et al.*, 1998). However, the expression of these subunits was not very effective and generated only small inward currents (5–50 nA) following application of nicotine or acetylcholine. Alternatively, all three *Drosophila* α subunits (ALS, SAD, and Dα2) can form functional receptors in *Xenopus* oocytes when coexpressed with a chicken neuronal β2 subunit (Bertrand *et al.*, 1994; Matsuda *et al.*, 1998; Schulz *et al.*, 1998), suggesting that additional insect nAChR subunits remain to be cloned. Radioligand binding studies using several *M. persicae* α subunits coexpressed with a rat β2 subunit in the *Drosophila* S2 cell line also indicate pharmacological diversity in *M. persicae* (Huang *et al.*, 1999). In these binding studies it was shown that imidacloprid selective targets were formed by Mpα2 and Mpα3, but not Mpα1 subunits. These examples indicate that our understanding of the complexity of insect nAChR is still limited, and that electrophysiology will play an essential role in determining the significance of certain subunit combinations in the mode of action of neonicotinoid and other insecticidally active ligands.

3.6.2. Whole Cell Voltage Clamp of Native Neuron Preparations

The use of isolated neurons from insect CNS for electrophysiological studies is a suitable tool to investigate the mode of action of new insecticidal compounds, which act on a range of neuronal target sites. Primary neuronal cell cultures from *H. virescens* larvae, one of the most important lepidopteran pest species, is one of these suitable tools. *H. virescens* neurons respond to the application of ACh, with a fast inward current of up to 5 nA at a holding potential of −70 mV. The current reversed at a holding potential close to 0 mV, indicating the activation of nonspecific cation channels, i.e., nAChRs.

Figure 21 illustrates the whole cell currents elicited by application of 1 μM nithiazine and the second generation neonicotinoids, acetamiprid, nitenpyram, and clothianidin. All of these compounds act as agonists on the nAChR, but the potency and agonistic efficacy of each of these compounds were quite different. Imidacloprid and clothianidin were the most potent compounds in this *Heliothis* preparation with an EC_{50} of 0.3 μM (**Table 5**). In the case of imidacloprid there was good agreement with electrophysiological measurements

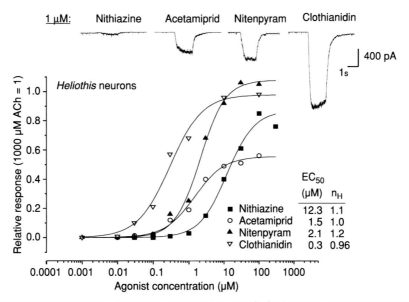

Figure 21 Whole cell current responses of a neuron isolated from the CNS of *Heliothis virescens* after application of different neonicotiniods. The dose-response curve was fitted by the Hill equation. All currents were first normalized to mean amplitudes elicited by 10 μM ACh before and after each test concentration was applied, and then normalized to the relative amplitude elicited by 1000 μM ACh. EC_{50} values given correspond to the half maximal activation of nAChR by each agonist. The Hill coefficient (n_H) of all tested compounds was close to 1. The upper inset shows the corresponding responses for the neonicotinoids at 1 μM (holding potential −70 mV). All currents were obtained from the very same neuron. (Reproduced with permission from Nauen, R., Ebbinghaus-Kintscher, A., Elbert, A., Jeschke, P., Tietjen, K., **2001**. Acetylcholine receptors as sites for developing neonicotinoid insecticides. In: Iishaaya, I. (Ed.), Biochemical Sites of Insecticide Action and Resistance. Springer, New York, pp. 77–105.)

Table 5 Comparison between electrophysiological and [³H]imidacloprid displacement potencies for different neonicotinoids and epibatidine on insect nAChRs. Electrophysiological data (EC_{50} and relative (agonist) efficacy) were obtained from neuron cell bodies isolated from the CNS of *Heliothis virescens*. EC_{50} and relative efficacy values represent the mean of n separate experiments on different neurons. The inhibition of [³H]imidacloprid binding to nAChR in housefly head membrane preparations by the compounds is expressed as pI_{50} value (pI_{50} values (= −log M) correspond to the concentration of cold ligand displacing 50% of bound [³H]imidacloprid from housefly head membranes)

Compound	n	EC_{50} (μM ± SD)	Relative efficacy (1 mM ACh = 1)	[³H]imidacloprid pI_{50}
Imidacloprid	4	0.31 ± 0.15	0.14 ± 0.02	9.3
Clothianidin	3	0.33 ± 0.03	0.99 ± 0.08	9.2
Acetamiprid	3	1.07 ± 0.37	0.56 ± 0.05	8.7
Nitenpyram	3	1.66 ± 0.38	0.98 ± 0.07	8.6
(±)-Epibatidine	3	1.69 ± 0.79	0.20 ± 0.05	6.2
Nithiazine	4	9.60 ± 3.20	0.79 ± 0.06	6.8

Reproduced with permission from Nauen, R., Ebbinghaus-Kintscher, A., Elbert, A., Jeschke, P., Tietjen, K., **2001**. Acetylcholine receptors as sites for developing neonicotinoid insecticides. In: Iishaaya, I. (Ed.), Biochemical Sites of Insecticide Action and Resistance. Springer, New York, pp. 77–105.

recorded from isolated cockroach neurons, where imidacloprid exhibited an EC_{50} of 0.36 μM (Orr et al., 1997). Acetamiprid, nitenpyram, and the natural toxin epibatidine exhibited an EC_{50} of between 1 and 2 μM (**Table 5**). Similar values were also observed for the cockroach preparation, with an EC_{50} between 0.5 and 0.7 μM for epibatidine and acetamiprid (Orr et al., 1997). Nithiazine had the lowest potency, with an EC_{50} of about 10 μM. The neonicotinoids, clothianidin, and nitenpyram were full agonists, whereas acetamiprid, epibatidine, and imidacloprid were partial agonists. The maximal response elicited by saturable concentrations of imidacloprid (100 μM) was only 15% of the maximal current obtained by 1000 μM ACh. In earlier work on isolated cockroach neurons (Orr et al., 1997) and isolated locust neurons (Nauen et al., 1999b), it was also found that imidacloprid acted as a partial agonist on insect nAChRs. The partial agonistic action of imidacloprid was also observed with chicken α4β2 nAChRs, and on a hybrid nAChR formed by the coexpression of a Drosophila α subunit (SAD) with the chicken β2 subunit in Xenopus oocytes (Matsuda et al., 1998). Imidacloprid activates very small inward currents in clonal rat phaeochromocytoma (PC 12) cells, thus also indicating partial agonistic actions (Nagata et al., 1996).

Furthermore, single-cell analysis revealed that imidacloprid activates predominatly a subconductance of approximately 10 pS, whereas acetylcholine activated mostly the high conductance state with 25 pS. Multiple conductance states were also observed in an insect nAChR reconstituted into planar lipid bilayers (Hancke and Breer, 1986) and on locust neurons (van den Breukel et al., 1998).

3.6.3. Correlation between Electrophysiology and Radioligand Binding Studies

There is a good correlation between electrophysiological measurements, using isolated Heliothis neurons, and radioligand binding studies on housefly head membranes regarding the affinity of different ligands to nAChRs (**Figure 22**). This good correlation for the neonicotinoids may indicate that houseflies (binding data) and tobacco budworms (electrophysiology) have similar binding sites for imidacloprid and related compounds. Biochemical investigations using [3H]imidacloprid as a radioligand in a number of different insect membrane preparations, e.g., from Periplaneta americana, Lucilia sericata, D. melanogaster, Manduca sexta, H. virescens, Ctenocephalides felis, M. persicae, and N. cincticeps, indicate that many (if not all) insects have high specific imidacloprid binding sites with nearly identical K_d values of ~1–10 nM (Lind et al.,

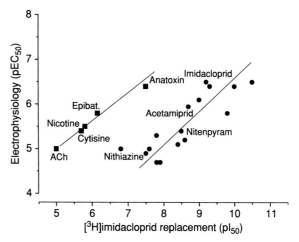

Figure 22 Comparison between electrophysiological and binding potencies of different neonicotinoids and nicotinoids. Electrophysiological data were obtained from neuron cell bodies isolated from the CNS of Heliothis virescens. pEC_{50} values (= −log M) correspond to the half maximal activation of nAChR by each agonist. Binding data: pI_{50} values (= −log M) correspond to the concentration of cold ligand displacing 50% of bound [3H]imidacloprid from housefly head membranes. (Reproduced from Nauen, R., Ebbinghaus-Kintscher, A., Elbert, A., Jeschke, P., Tietjen, K., **2001**. Acetylcholine receptors as sites for developing neonicotinoid insecticides. In: Iishaaya, I. (Ed.), Biochemical Sites of Insecticide Action and Resistance. Springer, New York, pp. 77–105.)

1998). However, it was also found that only the homopteran species seems to have an additional very high affinity binding site.

In general, the pI_{50} values obtained by the displacement of specifically bound [3H]imidacloprid from housefly head membranes were two to four orders of magnitude higher than the electrophysiologically determined pEC_{50} values obtained from isolated Heliothis neurons. Similar differences in biochemical binding and functional assay studies were also observed for different vertebrate nAChRs (review: Holladay et al., 1997). A possible explanation for this phenomenon might be provided by considering the following: it is generally accepted that each nAChR can exist in multiple states, i.e., a resting state, an active (open) state, and one or more desensitized state(s), each of which has different affinities for ligands. The active state has a low affinity for ACh (K_d ranging from about 10 μM to 1000 μM), whereas the desensitized state(s) has a higher affinity (K_d ranging from about 10 nM to 1 μM) for nicotinic ligands (Lena and Changeux, 1993).

The kinetics of the transitions between these states have been resolved for Torpedo nAChR in vitro. The rate of isomerization between the resting and active state lies in the μs to ms timescale, and within the desensitized state over a timeframe of seconds to minutes (Changeux, 1990, cited in Lena and Changeux, 1993). Because binding studies are conducted over a timescale of minutes to hours they may reflect interaction with the desensitized state(s), whereas electrophysiological studies measure the interaction of ligands with the active state. Considering this, it is surprising that there is a direct correlation between electrophysiological and biochemical binding studies for natural alkaloids such as (−)-nicotine, cytisine, (±)-epibatidine and anatoxin A (**Figure 22**). For these compounds the pI_{50} and pEC_{50} values were in the same range, with a good correlation. It is a general observation that natural alkaloids such as nicotine and epibatidine exhibit an agonistic potency in electrophysiological assays on isolated cockroach neurons (Bai *et al.*, 1991; Buckingham *et al.*, 1997; Orr *et al.*, 1997), *Heliothis* neurons (**Figure 22**), and locust neurons (Nauen *et al.*, 1999b) (see below). This potency is comparable to highly insecticidal neonicotinoids like imidacloprid. Matsuda *et al.* (1998), using a hybrid receptor formed by the coexpression of the *Drosophila* α subunit SAD with the chicken ß2 observed a comparable agonistic potency for both (−)-nicotine and imidacloprid. However, the agonistic potency of (+)-epibatidine was about two orders of magnitude greater. In contrast, all binding studies using [^{3}H]imidacloprid on housefly head membranes (Liu and Casida, 1993a, 1993b; Liu *et al.*, 1995; Wollweber and Tietjen, 1999), whitefly preparations (Chao *et al.*, 1997; Nauen *et al.*, 2002; Rauch and Nauen, 2003), and *Myzus* preparations (Nauen *et al.*, 1996, 1998a; Lind *et al.*, 1998), indicate that imidacloprid has a considerably higher potency in replacing specifically bound [^{3}H]imidacloprid than (−)-nicotine.

3.7. Pharmacokinetics and Metabolism

Neonicotinoids have moderate and different water solubility. They are nonionized and not readily hydrolyzed at physiological pH values (**Table 3**). The substances are biodegradable, therefore no accumulation was observed in mammals or through food chains (Roberts and Hutson, 1999). Neonicotinoid metabolism involves mostly detoxification reactions evident in three pathways (Tomizawa and Casida, 2003): (1) the brief period for poisoning at sublethal doses in both insects and mammals, (2) the synergistic effects of detoxification inhibitors on insecticidal action, and (3) the potency of metabolites relative to the parent compounds on nAChRs. The pharmacokinetics of neonicotinoids

and imidacloprid have been investigated in *S. littoralis*, and the obtained profiles were consistent with high metabolism and an important role of non-target tissues as sinks in this less responsive species (Greenwood *et al.*, 2002).

3.7.1. Metabolic Pathways of Commercial Neonicotinoids

3.7.1.1. Imidacloprid The metabolism of imidacloprid is strongly influenced by the method of application (Nauen *et al.*, 1998b). Depending on time and plant species, imidacloprid is degraded more or less completely, as comparative studies in many field crops have revealed (Araki *et al.*, 1994). The uptake and translocation of [^{14}C]imidacloprid after foliar application was studied in cotton. These studies showed that imidacloprid is taken up readily when mixed with the adjuvant Silwet L-77, then translocated acropetally and rapidly metabolized (Nauen *et al.*, 1999b). Experiments with seed-treated cotton plants revealed that only 5% of the applied imidacloprid was taken up by the young plant, and that 27 days after sowing, approximately 95% of the parent compound was metabolized (Tröltzsch *et al.*, 1994).

From results of soil metabolism studies it was found that imidacloprid is completely degradable and will not persist in soil. Under standard laboratory conditions the aerobic degradation of imidacloprid has a half-life (DT$_{50}$) of 156 days. The a.i. is degradated to carbon dioxide, and any residues are firmly bound to the soil matrix (Krohn and Hellpointer, 2002). Radioactively labeled imidacloprid shows that other degradation products, >10% of the applied radioactivity, do not occur. Sunlight and anaerobic conditions accelerated the degradation of imidacloprid. The degradation pathway shows that the neonicotinoid itself, and not the individual degradation products of the parent compound, is present in soil (Krohn and Hellpointer, 2002). Imidacloprid is not stabile in an aqueous environment, and in such an environment sunlight further accelerates the degradation.

The metabolism of imidacloprid was studied in tomatoes, potatoes, and maize (Roberts and Hutson, 1999) (**Figure 23**). In experiments using tomatoes, [^{14}C-*pyridinyl*]imidacloprid was applied to fruits. The main metabolites were the *N*-desnitro-imidacloprid and the 4- and 5-hydroxy derivatives (see Section 3.3.5.3). Small amounts of the cyclic urea, the *N*-nitroso compound were identified. A similar pattern of metabolites to the above mentioned was described for vines, together with a trace of the bicyclic triazinone. Traces of parent imidacloprid

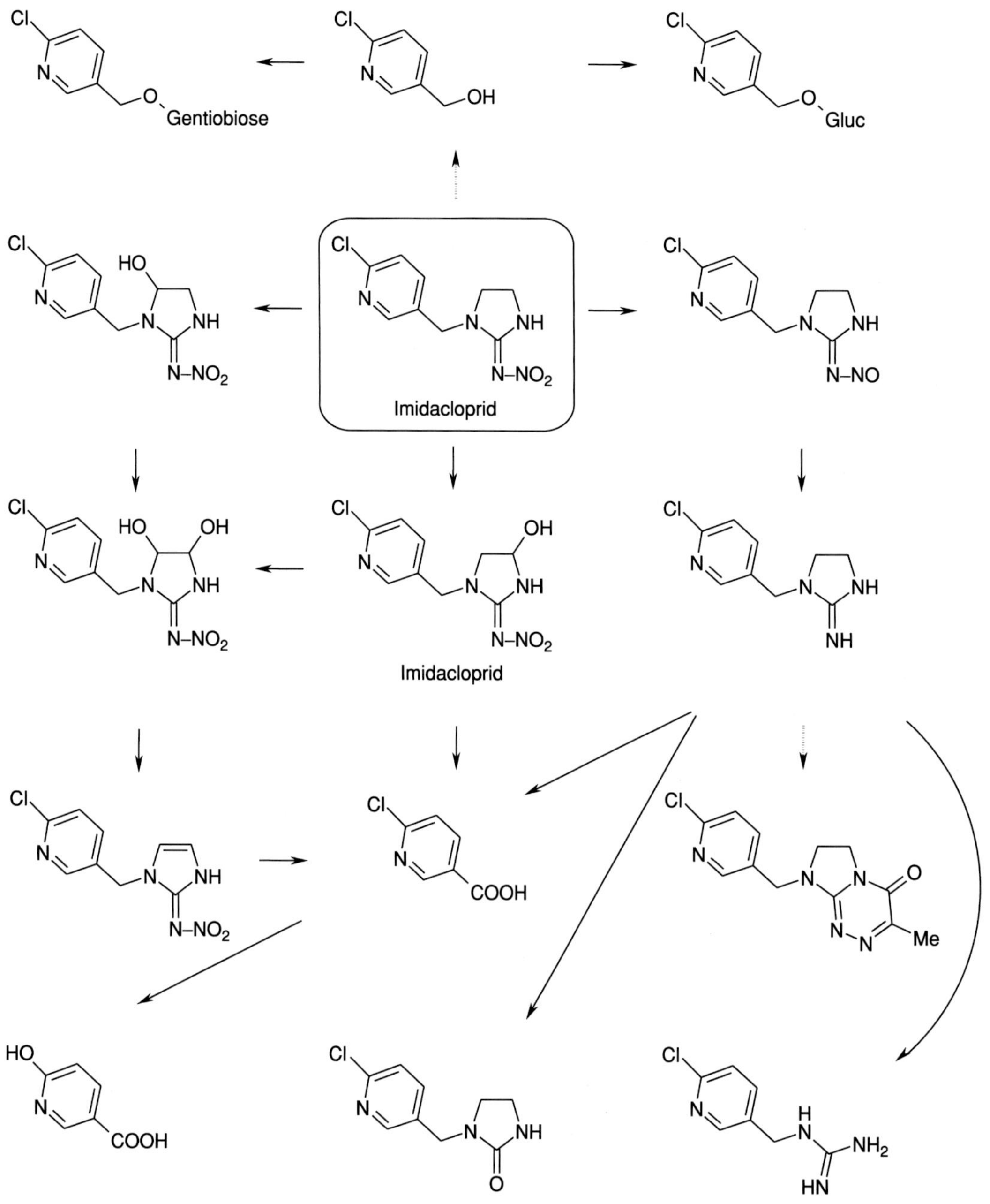

Figure 23 Proposed pathway of imidacloprid metabolism in plants. (Reproduced with permission from Roberts, T.R., Hutson, D.H. (Eds.), **1999**. Metabolic Pathways of Agrochemicals, Part 2, Insecticides and Fungicides. Cambridge University Press, Cambridge.)

and 6-chloro-nicotinic acid (6-CNA) were detected in the tubers. Koester (1992) studied the comparative metabolism of imidacloprid in suspension cultures of cells of several plant species including soybean and cotton, and found that the initial pathway of degradation proceeded mainly by monohydroxylation of the imidazolidine ring and dehydration to the olefin (see Section 3.3.5.3). Small amounts of 6-CNA and the 6-chloro-picolyl alcohol (6-CPA) and its glucopyranosid were formed. These identified

metabolites are consistent with those found in whole plants.

3.7.1.2. Thiacloprid The metabolic pathway of the systemic thiacloprid, in quantitative and also qualitative terms, is similar in all crops (fruiting crops, cotton) investigated (Klein, 2001). The parent compound labeled in [methylene-^{14}C]-position is always the major component. The presence of high amounts of 6-CNA in cotton seed is due to the accumulation of this acid in the seed as a phloem sink, after being secreted from the apoplasm into the phloem as a trap compartment for weak acids (Klein, 2001). It originated from leaf metabolism, where it was also detected (Klein, 2001). All main metabolites identified in plants were also detected in the rat and livestock metabolism study. By using two different labeling positions ([methylene-^{14}C] and [thiazolidine-^{14}C]), besides the unchanged parent compound, 25 metabolites were isolated and identified in the rat excreta. In goats and hens the residue levels in tissues and organs, as well as in milk and eggs, were low due to the fast elimination. There was no bioaccumulation of residues in any of the tissues. Generally, the metabolism of thiacloprid proceeded via the following pathways (Klein, 2001):

- hydroxylation of the parent compound at the 4-position of 1,3-thiazolidine ring and oxidative cleavage at the methylene bridge leading to 6-CPA and 6-CNA followed by conjugation of these two aglycones with sugars
- the corresponding 1,3-thiazolidine ring metabolites mainly comprised of the free 1,3-thiazolidine, the 4-hydroxylated 1,3-thiazolidine and its 4,5-dehydro (olefinic) derivative, as well as conjugates
- the N-nitrile moiety was hydrolyzed to the corresponding N-amide derivative, which was then N-hydroxylated
- the 1,3-thiazolidine ring system was cleaved in 1-position and the sulfur oxidized or methylated, forming the sulfonic acid or the S-methy-sulfoxyl and S-methyl-sulfonyl group, respectively.

3.7.1.3. Thiamethoxam

In metabolism studies, the bulk of thiamethoxam (84–95%) was excreted in the urine with a small amount (2.5–6%) in the feces within 24 h, primarily as unchanged parent compound. This was also observed in rats, mice, ruminants, and poultry. Hydrolysis of the perhydro-1,3,5-oxadiazine ring system in thiamethoxam followed by N-demethylation is described as the major metabolic pathway; two other compounds are formed by loss of the N-nitro group from either metabolite. Reduction of the N-nitro group to a hydrazine, and subsequent conjugation with 2-oxopropionic or acetic acids results in several major metabolites. Thiamethoxam metabolism was well investigated in plants like maize, rice, pears, and cucumbers. The active ingredient exhibits a systemic behavior, i.e., it is translocated via the roots into the whole plant. An accumulation was observed at the leaf borders, with very low residue levels. The same metabolites were found in cereals and fruits.

An aqueous photolysis study with radiolabeled thiamethoxam indicated that the neonicotinoid degrades significantly under photolytic conditions (Schwartz et al., 2000).

3.7.1.4. Nitenpyram

The polar nitenpyram has a relatively short persistence in soil (DT$_{50}$ 1–15 days), which probably offsets the relatively weak sorption that might otherwise lead to mobility through soils. In plants the metabolites 2-[N-(6-chloro-3-pyridylmethyl)-N-ethyl]amino-2-methyliminoacetic acid (CPMA) and N-(6-chloro-3-pyridylmethyl)-N-ethyl-N-methylformamidine (CPMF) were formed. Presumably CPMF was produced in the field by decarboxylation of CPMA (Tsumura et al., 1998). Animal metabolism was investigated by oral administration of nitenpyram labeled with ^{14}C to rats. As a result, 95–98% of the ^{14}C administered was excerted into urine within 2 days after treatment and no accumulation of ^{14}C in the internal organs was observed. The results reflect the high water solubility (840 g l^{-1}), and the low partition coefficient (log P$_{OW}$ = −0.64 at 25 °C) of nitenpyram may contribute to its low toxicity against mammals (Akayama and Minamida, 1999).

3.7.1.5. Acetamiprid

Acetamiprid is a mobile, rapidly biodegradable compound in most soil. The primary degradation pathway is aerobic soil metabolism. Acetamiprid is relatively stable to hydrolysis at environmental temperatures, and photodegrades relatively slowly in water, whereas it is rapidly degraded in soils to form N-(6-chloro-pyrid-3-yl-methyl)-N-methyl-amine and 6-CNA. The DT$_{50}$ values of acetamiprid in clay loam or light clay soils were in range of 1–2 days. In soil from the field or that used in container studies, acetamiprid degraded very rapidly (half-life 12 days). The main pathway was nitrile oxidation of acetamiprid into the N-amidino derivative, which cleaved, affording the N-(6-chloro-pyrid-3-ylmethyl)-N-methyl-amine and finally to 6-CNA. Mineralization to carbon dioxide followed.

Acetamiprid accounted for the majority of the residue in the crops, and metabolites occurred in negligible amounts. The main component found in plant metabolism studies was the parent acetamiprid. Small amounts of N-demethylacetamiprid were formed which cleaved to give 6-CNA and to a lesser extent oxidized to 6-CPA. The latter in turn either formed a conjugate with glucose or was also thought to form a N-(6-chloro-pyrid-3-ylmethyl)-N-methylamine as a minor metabolite (Tokieda et al., 1998; Roberts and Hutson, 1999) (**Figure 24**).

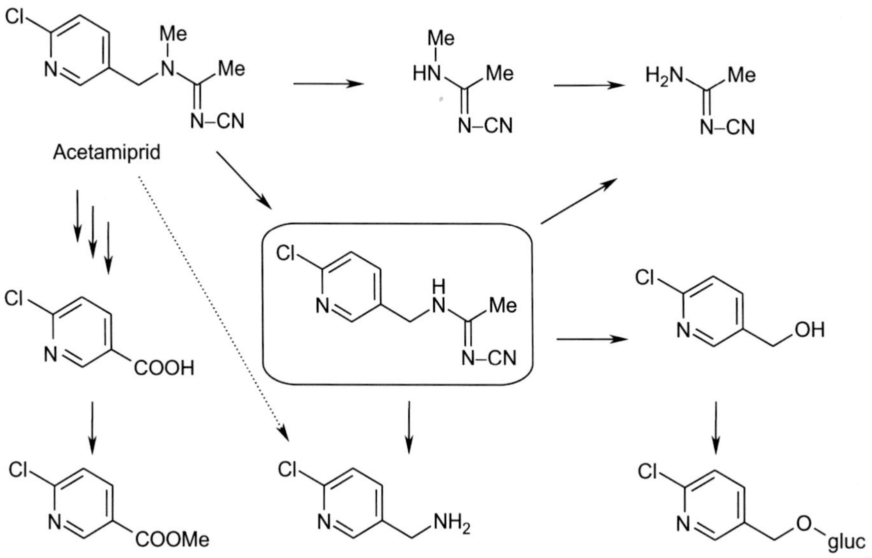

Figure 24 Proposed pathway of acetamiprid metabolism in plants. (Reproduced with permission from Roberts, T.R., Hutson, D.H. (Eds.), **1999**. Metabolic Pathways of Agrochemicals, Part 2, Insecticides and Fungicides. Cambridge University Press, Cambridge.)

The metabolic fate of [2,6-^{14}C-*pyridine*]acetamiprid was investigated in eggplants and apples following application to foliage or fruit, and in cabbage plants after application to foliar and soil. Parent acetamiprid was slowly metabolized to similar products in all plants. Acetamiprid was absorbed by and penetrated the foliage well but it was poorly translocated (Roberts and Hutson, 1999).

3.7.1.6. Clothianidin Studies of the metabolic pathway of clothianidin in maize and sugar beet show a clear and consistent picture (Klein, 2003). The parent compound is always the major component of the residue in the edible parts of the crops. Up to seven metabolites were identified, but most of them did not exceed 10% of the total radioactive residues. Furthermore, all the main metabolites identified in plants were also detected in rat metabolism studies (Klein, 2003; Yokota *et al.*, 2003). Generally, the following reactions are involved in the metabolic pathway of clothianidin in maize and sugar beets after seed treatment application:

- *N*-demethylation to form the *N*-demethyl clothianidin and subsequent hydrolysis to produce the *N*-(2-chloro-1,3-thiazol-5-ylmethyl)urea;
- hydrolysis of the *N'*-nitroimino moiety to form *N*-(2-chloro-1,3-thiazol-5-yl)-*N'*-methylurea and subsequent *N*-demethylation to form *N*-(2-chloro-1,3-thiazol-5-yl)urea;
- denitrification (reduction) to form *N*-(2-chloro-1,3-thiazol-5-yl)-*N*-methylguanidine and C—N bond cleavage within the compound to form *N*-methylguanidine;

- C—N bond cleavage (loss of the 1,3-thiazol-5-ylmethyl moiety) to form *N*-methyl-*N'*-nitroguanidine;
- *N*-demethylation of *N*-methyl-*N'*-nitroguanidine to form *N'*-nitroguanidine;
- denitrification (reduction of *N*-methyl-*N'*-nitroguanidine to form *N*-methylguanidine);
- C—N bond cleavage of *N*-(2-chloro-1,3-thiazol-5-yl)-*N'*-nitroguanidine to form nitroguanidine; and
- C—N bond cleavage (loss of the *N'*-nitroimino moiety) and subsequent oxidation of the intermediate to form 2-chloro-1,3-thiazol-5-carboxylic acid.

Unchanged parent compound is the major constituent of the residue at harvest. The major pathway of metabolism involves *N*-demethylation of clothianidin to produce *N*-(2-chloro-1,3-thiazol-5-ylmethyl)-*N'*-nitroguanidine. In some crop parts *N*-methylguanidine made up some 20% of the residue at harvest (Klein, 2003).

3.8. Pharmacology and Toxicology

Because of the target site selectivity, neonicotinoids are more toxic to aphids, leafhoppers, and other sensitive insects than to mammals and aquatic life (Matsuda *et al.*, 1998; Yamamoto *et al.*, 1998; Tomizawa *et al.*, 2000; Tomizawa and Casida, 2003). Comparative binding studies on insects nAChRs indicate that their ability to displace tritiated imidacloprid from its binding site correlates well with their insecticidal efficacy (Kagabu, 1997a; Nauen *et al.*, 2001).

3.8.1. Safety Profile

The introduction of imidacloprid as the first CNI was a milestone in the general effort to enhance the environmental safety of crop protection agents (Krohn and Hellpointer, 2002) (**Tables 6** and **7**).

3.8.1.1. Mammalian toxicity
The safety factor of imidacloprid, that means the ratio of the lethal doses for mammals to insects, can be estimated as 7300 from the LD_{50} value of $0.062\,mg\,kg^{-1}$ for *M. persicae* by topical application and an oral LD_{50} value for rats of $450\,mg\,kg^{-1}$ (**Table 6**) (Elbert *et al.*, 1990). Therefore imidacloprid ranks as an insecticide with high selective toxicity.

Thiacloprid possesses a low acute toxicity. Birds and mammals are at neither acute nor chronic risk when thiacloprid is applied at the recommended dose under practical conditions.

After oral dosing to rats, thiamethoxam is rapidly and completely absorbed, and up to 90% of the applied material is readily eliminated as parent compound through the urine. Thiamethoxam has a low acute toxicity when applied to rats either orally, dermally, or by inhalation (**Table 6**). The neonicotinoid is not irritant to skin and eyes, and has no sensitizing potential (Maienfisch *et al.*, 1999a, 2001a). Acute and absorption subchronic neurotoxicity studies in rats showed that thiamethoxam is devoid of any neurotoxic potential. However, mechanistic studies showed thiamethoxam demonstrated a phenobarbital-like induction of enzymes in mouse liver. In repeat dose feeding studies in rodents and dogs, the liver and kidneys (rat only) were the main target organs. In life-time rodent feeding studies, mice showed an increased incidence of liver tumors, which was specific to this species since a chronic rat study showed thiamethoxam had no tumorigenic potential (Maienfisch *et al.*, 2001a). A 12-month dog feeding study produced some alterations in blood chemistry, and a minimal change was seen in testes weight in two males at 1500 ppm, and tubular atrophy was observed at 750 and 1500 ppm. Dermal absorption studies in rats, and *in vitro* studies with rat and human skin, resulted in very low absorption of thiamethoxam. After an exposure time of 6 h the systemic absorption of thiamethoxam was between 0.4% and 2.7% within 6 h, and between 0.8% and 2.9% within 48 h. Oral reproductive toxicity studies showed no evidence of developmental impairment or teratogenic potential (Maienfisch *et al.*, 1999a).

Nitenpyram has a minimal adverse effect against mammals (Akayama and Minamida, 1999) (**Table 6**).

Acetamiprid is of relatively low toxicity to mammals, and it has been classified as an "unlikely" human carcinogen. In mammals, acetamiprid caused generalized, nonspecific toxicity and did not appear to have specific target organ toxicity. Acetamiprid has relatively low acute and chronic toxicity in mammals, and there was no evidence of carcinogenicity, neurotoxicity, and/or endocrinic disruption. Aggregate risk estimates for acetamiprid for food and water do not raise concern for acute and chronic levels of exposure. The acute oral toxicity to male rat (LD_{50} $217\,mg\,kg^{-1}$) and mouse (LD_{50} $198\,mg\,kg^{-1}$), as well as female rat (LD_{50} $146\,mg\,kg^{-1}$) and mouse (LD_{50} $184\,mg\,kg^{-1}$) is noncritical. The acute dermal toxicity to male and female rats is low ($>2000\,mg\,kg^{-1}$). Toxicological tests for acetamiprid, e.g., irritation to skin and eyes (rabbit), sensitization to skin (guinea pig), and mutagenicity, as evaluated by the Ames test, were all negative.

3.8.1.2. Environmental fate
The fate of crop protection agents such as neonicotinoids in the environment includes its behavior in the soil, water, and air compartments (Stupp and Fahl, 2003). After application of a crop protection agent (as a formulation) to crops or in/on soil, a part of the active substance is not taken up by the plants remains, e.g., on the soil. Other parts of the agent can be transported to neighboring water bodies via drift, wind erosion, or runoff, or can enter the air by evaporation from the plant surface (Stupp and Fahl, 2003). Therefore, different physicochemical and biological processes are involved in the environmental behavior of an active substance, i.e., hydrolytic and photochemical degradation, microbial transformation and decomposition like metabolism, sorption to the soil, and translocation of the active substance and its degradation products.

Formulated imidacloprid is stable at room temperature in the dark under usual storage conditions; however, the active ingredient is labile in alkaline media (Kagabu and Medej, 1995, 2003b). Numerous field trials carried out in Europe and the USA on the degradation of imidacloprid demonstrate that the active ingredient does not accumulate in the soil (Krohn and Hellpointer, 2002). There is no likelihood of an accumulatiom occurring following repeated yearly applications. Long-term studies in the UK indicate that the maximum concentrations in the soil reaches a plateau, having a relatively low residual level ($\leq 0.030\,mg\,kg^{-1}$) (Krohn and Hellpointer, 2002). Imidacloprid is not stable in an aqueous environment, where sunlight further accelerates its degradation. Those characteristics influence the environmental behavior of imidacloprid on surface water and in water sediment such as in rivers and lakes, as well as in fog and rain.

Table 6 Toxicological properties of commercialized neonicotinoids (selected relevant data)

Study type	Species	Ring systems			Noncyclic structures			
		Imidacloprid[a]	Thiacloprid[b]	Thiamethoxam[c]	Nitenpyram[d]	Acetamiprid[d]	Clothianidin[e]	(±)-Dinotefuran[d]
Acute oral (LD_{50} mg a.i. kg^{-1} bw)	Rat	~450 (m/f)	836 (m) 444 (f)	1563	1680 (m) 1575 (f)	217 (m) 146 (f)	>5000 (m/f)	2804 (m) 2000 (f)
Acute dermal (24 h) (LD_{50} mg a.i. kg^{-1} bw)	Rat	>5000 (m/f)	>2000 (m/f)	>2000	>2000 (m/f)		>2000 (m/f)	
Acute inhalation (4 h, aerosol) (LC_{50} mg a.i. m^{-3} air)	Rat	>5323 (dust)	>2535 (m) ~1223 (f)	3720	>5800 (m/f)	>0.29 (m/f)	>6141 (m/f)	
Skin irritation (4 h)	Rabbit	No irritation	No irritation	No irritation	No irritation	No irritation	No irritation	Acute percutaneous
Eye irritation	Rabbit	None	No irritation	No irritation	Very slight	No irritation	No irritation	Acute percutaneous
Skin sensitation	Guinea pig	No skin sensitation	No skin sensitation	No skin sensitation	No skin sensitation		No skin sensitation	No skin sensitation

[a]Elbert et al. (1990).
[b]Elbert et al. (2000).
[c]Maienfisch et al. (1999a).
[d]Tomlin (2000).
[e]Ohkawara et al. (2002).
m, males; f, females; a.i., active ingredient; bw, body weight.

Table 7 Environmental profile of commercialized neonicotinoids (selected relevant data)

Test species (acute toxicity test)	Ring systems			Noncyclic structures			
	Imidacloprid[a]	Thiacloprid[b]	Thiamethoxam[c]	Nitenpyram[d,g]	Acetamiprid[e]	Clothianidin[f]	(±)-Dinotefuran[e]
Birds							
Mallard duck (LC$_{50}$ mg a.i. kg^{-1} diet)	>5000 (food)	>5000 (food)[g]	>5200 (food)	>5620 (food)		>5200 (food)	1000
Bobwhite quail (LD$_{50}$ mg a.i. kg^{-1} bw)	152	2716	1552	2250 (capsule)	180	>2000	>2000
Japanese quail (LD$_{50}$ mg a.i. kg^{-1} bw)	31	49					
Fish							
Rainbow trout (LC$_{50}$ mg a.i. l^{-1})	211 (96 h)	29.6 (96 h)	>125 (96 h)	>10 (96 h)		>100 (96 h)	>40 (48 h)
Bluegill sunfish (LC$_{50}$ mg a.i. l^{-1})		24.5 (96 h)	>114 (96 h)			>120 (96 h)	
Carp (LC$_{50}$ mg a.i. l^{-1})	280 (96 h)			>1000 (96 h)	>100		>1000 (96 h)
Invertebrates							
Water flea, *Daphnia magna* (EC$_{50}$ mg a.i. l^{-1})	85 (48 h)	>85.1 (48 h)	>100 (48 h)	>100 (48 h)[h]	>1000	>120 (48 h)	1000 (48 h)
Earthworm, *Eisenia foetida* (LC$_{50}$ mg a.i. kg^{-1} dry soil)	10.7 (14 days)	105 (14 days)[g]	>1000 (14 days) (soil)	32.2 (14 days)	>1000	13.21 (14 days) (dry soil)	

[a]Pflüger and Schmuck (1991).
[b]Schmuck (2001).
[c]Maienfisch *et al.* (1999a).
[d]Akayama and Minamida (1999).
[e]Tomlin (2000).
[f]Ohkawara *et al.* (2002).
[g]Technical material.
[h]Unpublished data: Sumitomo Chemical Takeda Agro, 1999.

The occurrence of imidacloprid in the air is determined by its low vapor pressure of 4×10^{-10} Pa, which excludes volatilization from treated surfaces (Krohn and Hellpointer, 2002).

Thiacloprid possesses a favorable environmental profile and will disappear from the environment after having been applied as a crop protection chemical. The effect of technical and formulated (SC 480) thiacloprid on respiratory (carbon turnover) and nitrogen mineralization rates in soil were examined in a series of 28-day studies on soil microflora (Schmuck, 2001). Thiacloprid had no adverse effect either on the microbial mineralization of nitrogen in different soil types like silty sand (0.6% org.C, pH_{KCl} 5.7) or loamy silt soil (2.3% org.C, pH_{KCl} 7.1) after addition of lucerne–grass–green meal ($5\,g\,kg^{-1}$) and at both the maximum and 10 × application rates (Schmuck, 2001). The half-lives in soil measured under field conditions of northern Europe ranged from 9 to 27 days, and in southern Europe from 10 to 16 days (Krohn, 2001). The major metabolites formed are only intermediates before complete mineralization to carbon dioxide occurs. Under practical field conditions thiacloprid is not toxic to earthworms (*Eisenia foetida*) or soil microorganisms (**Table 7**).

Furthermore, thiacloprid and its formulations pose a favorably low toxicity hazard to honeybees and bumblebees. Consequently, the compound can be applied to flowering crops at the recommended rates without posing a risk to bees (Calypso™ up to $200\,g\,a.i.\,ha^{-1}$). In addition, Calypso™ does not affect the pollination efficacy of pollinators (Elbert *et al.*, 2000). Thiacloprid must be classified as being only slightly mobile in soil, and hence it has low potential for leaching into groundwater. In the aquatic environment thiacloprid will also undergo rapid biotic degradation, with a half-life of 12–20 days (Krohn, 2001). Because of its low vapor pressure (3×10^{-10} Pa) and its water solubility of $0.185\,g\,l^{-1}$, the potential for its volatilization is negligible. Acute and chronic toxicity testing in fish and water-fleas (*Daphnia magna*) indicates a very low to negligible risk to these species following accidental contamination with thiacloprid by spray drift. Reproduction of water-fleas and fish was also not affected at environmentally relevant water concentrations of thiacloprid. Other aquatic invertebrates revealed a higher sensitivity to thiacloprid, especially insect larvae and amphipod species such as *Chironomus* and *Hyalella*. Insect larvae were also identified as the most sensitive species in a pond study. However, the rapid dissipation of thiacloprid in water bodies allows for a rapid recovery of affected aquatic insect populations. Algae, represented by *Scenedesmus subspicatus* and *Pseudokirchneriella subcapitata*, are not sensitive to thiacloprid. Toxicity exposure ratio values indicate a high margin of safety for algal species. Harmful concentrations of thiacloprid or its metabolites are most unlikely in aquatic ecosystems. There is no potential for bioconcentration in fish due to the log P_{OW} of 1.26 for thiacloprid. The metabolites occurring in groundwater are of no concern as noted by toxicological, ecotoxicological, and biological testing.

In laboratory soils, thiamethoxam degrades at moderately slow rates (Maienfisch *et al.*, 1999a). The compound is photolyzed rapidly in water. In natural aquatic systems, e.g., rice paddies, degradation also occurs in the absence of light by microbial degradation. Based on the low vapor pressure (**Table 3**) and the results of soil volatility studies, significant volatilization is not expected. Thiamethoxam has a low toxicity by ingestion to birds, and is practically nontoxic to fish, *Daphnia* and molluscs. In addition, the earthworm species E. *foetida* and green algae were also found to be insensitive (Maienfisch *et al.*, 2001a). Thiamethoxam can be classified as slightly to moderately harmful to most beneficial insects, but safe to predatory mites in the field (Maienfisch *et al.*, 1999a). Alternatively, thiamethoxam and all other nitroguanidines have to be considered toxic to bees. However, thiamethoxam showed no bioaccumulation potential, and is moderately mobile in soil and degrades at fast to moderate rates under field conditions (Maienfisch *et al.*, 1999a).

Nitenpyram rapidly decomposed in water, under the conditions of 12 000 lux irradiation intensity using a 500 W xenon lamp (half-life at pH 5, 18–21 min; in purified water, 16 min), and in soil under aerobic conditions (half-lives flooded <1 day; upland 1–3 days) (Akayama and Minamida, 1999).

Toxicity of acetamiprid to fish and *Daphnia* is very low, with LD_{50} values for carp and *Daphnia* of $>100\,mg\,l^{-1}$ and $>1000\,mg\,l^{-1}$, respectively (Yamada *et al.*, 1999). Acetamiprid shows less adverse effects on both natural and enemies, such as predaceous mites, and beneficial insects, as well as honeybee and bumblebee (Takahashi *et al.*, 1998).

Chlothianidin degraded moderately under field conditions and was found to be a neonicotinoid of medium mobility based on laboratory experiments (Stupp and Fahl, 2003). With regard to its behavior in aquatic environments, photolysis contributes significantly to degradation, and clothianidin finally mineralizes to carbondioxide. Its degradation in water/sediment systems under aerobic conditions is moderate. The main part of the applied active substance is bound irreversibly to the sediment, thus avoiding subsequent contamination (Stupp and Fahl, 2003). In addition, in a water/sediment study degradation of clothianidin was observed to be significantly faster (factor 2–3) under anaerobic conditions than in aerobic conditions (Stupp and Fahl, 2003). Based on the physicochemical properties of clothianidin (see Section 3.3.4) no volatilization, and thus, no significant amounts of this compound can be expected in the atmosphere. Therefore clothianidin does not appear to represent a risk for the environment.

3.9. Resistance

Resistance in arthropod pest species comprises a change in the genetic composition of a population in response to selection by pesticides, such that control of the pest species in the field may be impaired at the recommended application rates. Selection for resistance in pest populations (including weeds, fungi, mites, and insects) due to frequent applications of agrochemicals has been one of the major problems in modern agriculture. Resistance to insecticides and acaricides appeared very early, and the extent and economic impact remain greater than for other agrochemicals. Between the beginning of the last century and the mid-1950s, the number of resistant arthropod species grew gradually with only a few resistant species described per decade. However, this rate increased markedly after the introduction of the organophosphates to more than 30 new resistant species every 2 years through the early 1980s. Ten years later, more than 500 arthropod species were known to be resistant to at least one insecticide or acaricide (Georghiou, 1983; Green et al., 1990). Among them are 46 hopper and aphid species, which show in some cases very high levels of resistance to conventional types of insecticide chemistry, i.e., organophosphates, and carbamate and pyrethroid insecticides. The invention and subsequent commercial development of neonicotinoid insecticides, such as imidacloprid, has provided agricultural producers with invaluable new tools for managing some of the world's most destructive crop pests, primarily those of the order Hemiptera (aphids, whiteflies, and planthoppers) and Coleoptera (beetles), including species with a long history of resistance to earlier-used products (**Figure 20**). However, the speed and scale with which imidacloprid, the commercial forerunner of neonicotinoids, was incorporated into control strategies around the world prompted widespread concern over the development of imidacloprid resistance (Cahill et al., 1996; Denholm et al., 2002).

To a large extent these pessimistic forecasts have not been borne out in practice. Imidacloprid has proved remarkably resilient to resistance, and cases of resistance that have been reported are still relatively manageable and/or geographically localized. The existence of strong resistance in some species has nonetheless demonstrated the potential of pests to adapt and resist field applications of neonicotinoids. The ongoing introduction of new molecules (e.g., acetamiprid, thiamethoxam, nitenpyram, thiacloprid, and clothianidin), unless carefully regulated and coordinated, seems bound to increase exposure to neonicotinoids, and to enhance conditions favoring resistant phenotypes.

3.9.1. Activity on Resistant Insect Species

The incidence and management of insect resistance to neonicotinoid insecticides was recently reviewed by Denholm et al. (2002). Resistance to neonicotinoid insecticides is still rare under field conditions and baseline susceptibility data have been provided in the past and more recently especially for imidacloprid in order to monitor for early signs of resistance in some of the most destructive pest insects (Cahill et al., 1996; Elbert et al., 1996; Elbert and Nauen, 1996; Foster et al., 2003; Nauen and Elbert, 2003; Rauch and Nauen, 2003; Weichel and Nauen, 2003). Baseline susceptibility data and derived diagnostic concentrations to monitor resistance are usually calculated from composite log-dose probit-mortality lines, including the combined curves of several strains of a certain pest collected from all over the world or at least different parts of the world where the compound is supposed to be used. A compilation for imidacloprid of such data is given for some of the most important pests, i.e., *Myzus persicae, Aphis gossypii, Phorodon humuli, Bemisia tabaci, Trialeurodes vaporariorum*, and *Leptinotarsa decemlineata* (**Table 8**).

There have been only low levels of tolerance detected in European and Japanese samples of *M. persicae* (Devine et al., 1996; Nauen et al., 1996), but it was not possible to link this tolerance to specific biochemical markers (Nauen et al., 1998a). In most cases the lower susceptibility to imidacloprid and other neonicotinoids was correlated with a decreased efficacy of nicotine (Devine et al., 1996; Nauen et al., 1996). Furthermore differences in hardiness were sometimes observed in field strains bioassayed directly upon receipt, which allow strains to survive longer in comparison to susceptible laboratory populations, when exposed to imidacloprid-treated leaves (Nauen and Elbert, 1997). However, such effects disappeared when the exposure time (especially in systemic bioassays) was extended

Table 8   Baseline susceptibility of some high risk pests to imidacloprid

Species	Diagnostic dose	Bioassay system/assessment time	Reference
Myzus persicae	15 ppm	Leaf dip (6-well plate)	Nauen and Elbert (2003)
Myzus persicae	2.25 ng per aphid	Topical	Foster et al. (2003)
Aphis gossypii	13 ppm	Leaf dip (6-well plate)	Nauen and Elbert (2003)
Phorodon humuli	13 ppm	Leaf dip (6-well plate)	Weichel and Nauen (2003)
Bemisia tabaci	16 ppm	Systemic bioassay	Cahill et al. (1996)
Bemisia tabaci	1 ppm	Leaf dip	Rauch and Nauen (2003)
Leptinotarsa decemlineata	8 ppm	Artificial diet (larvae)	Olson et al. (2000)
Leptinotarsa decemlineata	0.2 μg per beetle	Topical (adult)	Nauen (unpublished data)

from 48 h to 72 h, and also after maintaining such strains under laboratory conditions for some weeks (Nauen and Elbert, 1997). More recently Foster et al. (2003b) demonstrated that tolerance to imidacloprid in M. persicae from different regions in Europe also provided cross-tolerance to acetamiprid. The authors were able to show a clear correlation between ED_{50} values of acetamiprid and imidacloprid for strains with a different degree of tolerance. However, tolerance factors compared to a susceptible reference population never exceeded factors of 20, and field failures were not seen (Foster et al., 2003b).

One species of major concern over the last decade is the tobacco or cotton whitefly, B. tabaci; this is a serious pest in many cropping systems worldwide and several biotypes of this species have been described (Perring, 2001). The most widespread biotype is the B-type, which is also known as B. argentifolii Bellows & Perring. The B-type whitefly is a common pest, particularly in cotton, vegetables, and ornamental crops, both by direct feeding and as a vector of numerous plant pathogenic viruses. In southern Europe, it coexists with another biotype, the Q-type, which was originally thought to be restricted to the Iberian peninsula, but which is now also known to occur in some other countries throughout the Mediterranean area, including Italy and Israel (Brown et al., 2000; Palumbo et al., 2001; Nauen et al., 2002; Horowitz et al., 2003). The biotypes B and Q can easily be distinguished by randomly amplified polymorphic DNA polymerase chain reaction (RAPD-PCR) or native polyacrylamide gel electrophoresis (PAGE) and subsequent visualization of their nonspecific esterase banding pattern (Guirao et al., 1997; Nauen and Elbert, 2000).

As a consequence of extensive exposure to insecticides, B. tabaci has developed resistance to a wide range of chemical control agents (Cahill et al., 1996). The need for a greater diversity of chemicals for whitefly control in resistance management programs has been met by the introduction of several insecticides with new modes of action, which are unaffected by mechanisms of resistance to organophosphates or pyrethroids. Since the introduction of imidacloprid, the neonicotinoids have been the fastest-growing class of insecticides. Imidacloprid exhibits an excellent contact and systemic activity and therefore has been largely responsible for the sustained management of B. tabaci in horticultural and agronomic production systems worldwide. Beside imidacloprid, there are other neonicotinoids with good efficacy against whiteflies, e.g., acetamiprid and thiamethoxam.

In Israel, monitoring of resistance in B. tabaci to imidacloprid and acetamiprid was initiated in 1996 in cotton and greenhouse ornamental crops. After 2 years of use in cotton, no apparent resistance to imidacloprid and acetamiprid was reported (Horowitz et al., 1998). However, 3 years of acetamiprid use in greenhouses in Israel resulted in a 5–10-fold decrease in susceptibility of B. tabaci to acetamiprid (Horowitz et al., 1999). In the past only a few cases of lowered neonicotinoid susceptibility in B-type B. tabaci have been described, among them strains from Egypt and Guatemala that were recently reported (El Kady and Devine, 2003; Byrne et al., 2003). In Arizona, where imidacloprid has been used since 1993, monitoring of B. tabaci populations from cotton fields, melon fields, and greenhouse vegetables suggested reduced susceptibility to imidacloprid from 1995 to 1998, but subsequent monitoring showed that these populations had actually regained and sustained susceptibility to imidacloprid in 1999 and 2000 (Li et al., 2000, 2001). Furthermore, imidacloprid use in Arizona and California remains high, but no signs of reduced control in the field have been reported yet (Palumbo et al., 2001). B-type whiteflies have been shown to develop resistance to imidacloprid under selection pressure in the laboratory (Prabhaker et al., 1997). There was a moderate increase of resistance of up to 17-fold in the first 15 generations, but 82-fold resistance after 27 generations. However, resistance was not stable and disappeared after a few generations without insecticide pressure. Resistance to imidacloprid conferring a high level of cross-resistance to thiamethoxam and acetamiprid was first demonstrated, and best studied in Q-type *B. tabaci* from greenhouses in the Almeria region of southern Spain, but was also detected in single populations from Italy and recently Germany as well (Nauen and Elbert, 2000; Nauen *et al.*, 2002; Rauch and Nauen, 2003). Neonicotinoid resistance seem to remain stable in all field-collected Q-type strains maintained in the laboratory without further selection pressure (Nauen *et al.*, 2002). Neonicotinoid cross-resistance was also reported in B-type whiteflies from cotton in Arizona but at lower levels (Li *et al.*, 2000). More recently a high level of cross-resistance between neonicotinoids was also described in a B-type strain of *B. tabaci* from Israel, and resistance factors detected in a leaf-dip bioassay exceeded 1000-fold (Rauch and Nauen, 2003).

The Colorado potato beetle, *L. decemlineata* has a history of developing resistance to virtually all insecticides used for its control. The first neonicotinoid, i.e., imidacloprid was introduced for controlling Colorado potato beetles in North America in 1995. Concerns over resistance development were reinforced when extensive monitoring of populations from North America showed about a 30-fold variation in LC_{50} values from ingestion and contact bioassays against neonates (Olsen *et al.*, 2000). Much of this variation appeared unconnected with imidacloprid use, and was probably a consequence of cross-resistance from chemicals used earlier. Lowest levels of susceptibility occurred in populations from Long Island, New York, an area that has experienced the most severe resistance problems of all with *L. decemlineata*. Zhao *et al.* (2000) and Hollingworth *et al.* (2002) independently studied single strains collected from different areas in Long Island, both treated intensively with imidacloprid between 1995 and 1997. In the first study, grower's observations of reduced control were supported by resistance ratios for imidacloprid of 100-fold and 13-fold in adults and larvae, respectively. The second study reported 150-fold resistance from topical application bioassays against adults. In this case the strain was also tested with thiamethoxam, which had not been used for beetle control at the time of collection. Interestingly, resistance to thiamethoxam (about threefold) was far lower than to imidacloprid.

Other reports referring to resistance to neonicotinoid insecticides were on species of lesser importance, including species either from field-collected populations or artificially selected strains. Among these were the small brown planthopper, *Laodelphax striatellus* (Sone *et al.*, 1997), western flower thrips, *Franklienella occidentalis* (Zhao *et al.*, 1995), houseflies, *Musca domestica* and German cockroach, *Blattella germanica* (Wen and Scott, 1997), *Drosophila melanogaster* (Daborn *et al.*, 2001), *Lygus hesperus* (Dennehy and Russell, 1996), and brown planthoppers, *Nilaparvata lugens* (Zewen *et al.*, 2003).

3.9.2. Mechanisms of Resistance

Many pest insects and spider mites have developed resistance to a broad variety of chemical classes of insecticides and acaricides, respectively (Knowles, 1997; Soderlund, 1997). One of the three major classes of mechanisms of resistance to insecticides in insects is allelic variation in the expression of target proteins with modified insecticide binding sites, e.g., acetylcholinesterase insensitivity towards organophosphates and carbamates, voltage-gated sodium channel mutations responsible for knock-down resistance to pyrethroids, and a serine to alanine point mutation (*rdl* gene) in the γ-aminobutyric acid (GABA)-gated chloride channel (GABA$_A$-R) at the endosulfan/fipronil/dieldrin binding site (ffrench-Constant *et al.*, 1993; Williamson *et al.*, 1993, 1996; Mutero *et al.*, 1994; Feyereisen, 1995; Dong, 1997; Soderlund, 1997; Zhu and Clark, 1997; Bloomquist, 2001; Gunning and Moores, 2001; Siegfried and Scharf, 2001). The second – and often most important – class of resistance mechanisms in insect pest species is metabolic degradation involving detoxification enzymes such as microsomal cytochrome P-450 dependent monooxygenases, esterases, and glutathione S-transferases (Hodgson, 1983; Armstrong, 1991; Hemingway and Karunarantne, 1998; Bergé *et al.*, 1999; Devonshire *et al.*, 1999; Feyereisen, 1999; Hemingway, 2000; Field *et al.*, 2001; Scott, 2001; Siegfried and Scharf, 2001). The third, least important mechanism is an altered composition of cuticular waxes which affects penetration of toxicants. Reduced penetration of insecticides through the insect cuticle has often been described as a contributing factor, in combination with target site insensitivity or metabolic detoxification (or both), rather than functioning as a major mechanism on its own (Oppenoorth, 1985). Most of the mechanisms mentioned above affect in many cases the efficacy of more than one class of insecticides, i.e., constant selection pressure to one chemical class could to a greater or lesser extent confer cross-resistance to compounds from other chemical classes (Oppenoorth, 1985; Soderlund, 1997).

When the first neonicotinoid insecticide was introduced to the market in 1991, aphids were considered to be high risk pests with regard to their potential to develop resistance to this class of chemicals. They have a high reproductive potential, and extremely short life cycle allowing for numerous generations in a growing season. Combined with frequent applications of insecticides that are usually required to maintain aphid populations below economic thresholds, resistance development is facilitated in these species, resulting in control failures. Such control failures have been reported for organophosphorus compounds for many decades, and more recently also for pyrethroids (Foster et al., 1998, 2000; Devonshire et al., 1999; Foster and Devonshire, 1999).

One of the major aphid pests is the green peach aphid, M. persicae. Resistance of M. persicae to insecticides is conferred by increased production of a carboxylesterase, named E4 or FE4, which provides cross-resistance to carbamates, organophosphorus, and pyrethroid insecticides (Devonshire and Moores, 1982; Devonshire, 1989). This esterase overproduction was shown to be due to gene amplification (Field et al., 1988, 2001). It was the sole resistance mechanism reported in M. persicae for more than 20 years, and only recently an insensitive (modified) acetylcholinesterase was described as a contributing factor in carbamate resistance in M. persicae (Moores et al., 1994a, 1994b; Nauen et al., 1996). The insensitive acetylcholinesterase in M. persicae confers strong resistance to pirimicarb, and a little less to triazamate (Moores et al., 1994a, 1994b; Buchholz and Nauen, 2001). Due to the improvement of molecular biological techniques, Martinez-Torres et al. (1998) recently showed that knockdown resistance to pyrethroids, caused by a point mutation in the voltage-gated sodium channel, is also present in M. persicae. In summary, resistance to all major classes of aphicides occurs in M. persicae; however, the only class of insecticides not yet affected by any of the mechanisms described above are the neonicotinoids, including its most prominent member imidacloprid (Elbert et al., 1996, 1998a; Nauen et al., 1998a; Horowitz and Denholm, 2001).

The most comprehensive studies on the biochemical mechanisms of resistance to neonicotinoid insecticides using an agriculturally relevant pest species were performed in whiteflies, B. tabaci (Nauen and Elbert, 2000; Nauen et al., 2002; Rauch and Nauen, 2003; Byrne et al., 2003). Biochemical examinations revealed that neonicotinoid resistance in Q-type B. tabaci collected in 1999 was not associated with a lower affinity of imidacloprid to nAChRs in whitefly membrane preparations (Nauen et al., 2002). This was confirmed more recently by testing strains ESP-00, GER-01, and ISR-02 obtained in the years 2000–2002 by Rauch and Nauen

(2003). Although neonicotinoid resistance was very high in these strains (up to 1000-fold), the authors found just a 1.3-fold and 1.7-fold difference in binding affinity between strains, and concluded that target site resistance is not involved in neonicotinoid resistance in those strains investigated. Piperonyl butoxide, a monooxygenase inhibitor, is generally used as a synergist to suppress insecticide resistance conferred by microsomal monooxygenases. Experiments with whiteflies pre-exposed to piperonyl butoxide suggested a possible involvement of cytochrome P-450 dependent monooxygenases in neonicotinoid resistance (Nauen et al., 2002). Rauch and Nauen (2003) biochemically confirmed that whiteflies resistant to neonicotinoid insecticides showed a high microsomal 7-ethoxycoumarin O-deethylase activity, i.e., up to eightfold higher compared with neonicotinoid susceptible strains. Furthermore, this enhanced monooxygenase activity could be correlated with imidacloprid, thiamethoxam, and acetamiprid resistance. Significant differences between glutathione S-transferase and esterase levels were not found between neonicotinoid resistant and susceptible strains of B. tabaci (Rauch and Nauen, 2003). Several metabolic investigations in plants and vertebrates showed that imidacloprid and other neonicotinoids undergo oxidative degradation, which may lead to insecticidally toxic and nontoxic metabolites (Araki et al., 1994, Nauen et al., 1999b; Schulz-Jander and Casida, 2002). Metabolic studies in B. tabaci in vivo revealed that the main metabolite in neonicotinoid-resistant strains is 5-hydroxy-imidacloprid, whereas no metabolism could be detected in the susceptible strain (Figure 23) (see Section 3.7.1.1). One can therefore suggest that oxidative degradation is the main route of imidacloprid detoxification in neonicotinoid resistant Q-type whiteflies (Rauch and Nauen, 2003). Compared to imidacloprid, the 5-hydroxy metabolite showed a 13-fold lower binding affinity to whitefly nAChR. This result was in accordance with previous studies with head membrane preparations from the housefly (Nauen et al., 1998). The binding affinity expressed as the IC_{50} was highest with olefine (0.25 nM) > imidacloprid (0.79) > 5-hydroxy (5 nM) > 4-hydroxy (25 nM) > dihydroxy (630 nM) > guanidine and urea (>5000 nM). The biological efficacy in feeding bioassays with aphids correlated also with the relative affinities of the metabolites towards the housefly nAChR (Nauen et al., 1998b). The lower binding affinity of 5-hydroxy-imidacloprid compared to imidacloprid coincides with its lower efficacy against B. tabaci in the sachet test (17-fold). These data show that differences between binding to the nAChR are comparable to the efficacy differences in adult bioassays (Rauch and Nauen, 2003).

Daborn et al. (2001) described D. melanogaster mutants exhibiting an eight- and sevenfold resistance to imidacloprid and DDT, respectively, which showed an overexpression of the P-450 gene Cyp6g1. However, it was not tested whether recombinant Cyp6g1 itself can metabolize imidacloprid; potentially it could metabolize a trans-acting factor which then upregulates presently uncharacterized P-450 genes. However, it was demonstrated that the recombinant cytochrome P-450 isozyme CYP3A4 from human liver can selectively metabolize the imidazolidine moiety of imidacloprid, resulting in 5-hydroxy-imidacloprid as the major metabolite (Schulz-Jander and Casida, 2002). It would therefore be fruitful to investigate if homologs of Cyp6g1 and Cyp3a4, or similar genes, are present in the B. tabaci genome, and if overexpression of these genes is associated with neonicotinoid resistance.

Although considerable resistance to neonicotinoid insecticides was also reported for populations of the Colorado potato beetle from Long Island, NY, the underlying mechanisms associated have not been very well studied. Pharmacokinetic investigations on susceptible and resistant strains did not reveal any differences in the uptake and excretion of ^{14}C-labeled imidacloprid, nor were there any differences in metabolic conversion found (Hollingworth et al., 2002). The authors suggested a modification of the target site to be responsible for the observed resistance. But, preliminary investigations revealed no differences between resistant and susceptible strains in receptor binding studies using [^{3}H]imidacloprid (Nauen and Hollingworth, unpublished data).

Resistance development to imidacloprid would be disastrous in many regions, so there is a demand for an effective resistance management program, and interest in finding and developing new active ingredients for pest control.

3.9.3. Resistance Management

Monitoring and detection of insecticide resistance in order to recommend and implement effective resistance management strategies is currently one of the most important areas of applied entomology. This program is necessary to sustain the activity of as many active ingredients with different modes of action as possible by using alternate spray regimes, rotation, and sophisticated application techniques (Denholm and Rowland, 1992; McKenzie, 1996; Denholm et al., 1998, 1999; Horowitz and Denholm, 2001). Historically, the problem of insecticide resistance was tackled by continuously introducing new active ingredients for pest insect control. The number of insecticides available is high, but the biochemical mechanisms targeted by all these compounds is rather limited (Casida and Quistad, 1998; Nauen and Bretschneider, 2002). A pronounced decline in the introduction of new insecticides since the late 1970s revealed the limitations of a strategy relying on the continuous introduction of new active compounds (Soderlund, 1997).

This problem has been recognized by the regulatory authorities; and the European Plant Protection Organization (EPPO) recently published guidelines outlining the background work on resistance required for registration of a new active ingredient (Heimbach et al., 2000). These requirements, which are provided by agrochemical companies as an essential part of the registration dossier, include

- baseline susceptibility studies (testing several strains of a target species known to be prone to resistance development);
- monitoring (continuous studies on the development of resistance of target species by simple bioassays after the launch of a new compound); and
- possible resistance management strategies (how the new compound should be combined with others in order to expand its life time in the field).

3.10. Applications in Nonagricultural Fields

Because of the high insecticidal efficacy together with its nonvolatility and stability under storage conditions, imidacloprid (Premise®) has been successfully applied as a termiticide (Jacobs et al., 1997a; Dryden et al., 1999), and has also been used for control of turf pests such as white grubs (Elbert et al., 1991). Furthermore, imidacloprid is the active ingredient of the insecticide Merit®, and is commonly incorporated into fertilizers for early control of grubs in turf (Armbrust and Peeler, 2002). In addition, imidacloprid was the first neonicotinoid to be used in a gel bait formulation for cockroach control. The imidacloprid gel showed outstanding activity even after 27 months' under various conditions (Pospischil et al., 1999). As an endoparasiticide imidacloprid exerted activity against the gastrointestinal nematode Haemonchus contortus in sheep only at higher concentrations (Mencke and Jeschke, 2002). Synergistic mixtures containing imidacloprid are patented and these are useful against textile-damaging insects, such as moth (*Tineola bisselliella, Tinea pellionella*) and beetles (*Attagenus, Anthrenus*) (Mencke and Jeschke, 2002). Salmon-parasitizing crabs are controlled by addition of 100 ppm imidacloprid to seawater (Mencke and Jeschke, 2002). Due to its toxicological properties – favorable mammalian safety characteristics (Yamamoto *et al.*, 1995) (**Table 6**), the absence of eye/skin irritation and skin sensitization potential – imidacloprid has been developed for control of lice in humans (Mencke and Jeschke, 2002) and veterinary medicine (Werner *et al.*, 1995). Imidacloprid (worldwide trademark: Advantage®) is the first neonicotinoid to have been developed for topical application in animals (Griffin *et al.*, 1997) (see Section 3.10.1).

Nitenpyram (Capstar®), a fast-acting, orally administered flea treatment, is absorbed into blood of the host animal, and is thus readily available for uptake by feeding fleas (Rust *et al.*, 2003). Therefore, administration of nitenpyram is effective in eliminating adult fleas for up to 48 h after treatment.

3.10.1. Imidacloprid as a Veterinary Medicinal Product

The cat flea (*Ctenocephalides felis*), the primary ectoparasite of companion animals worldwide, will feed on a wide variety of animals in addition to cats and dogs (Rust and Dryden, 1997), although it is not equally well adapted to all hosts (Williams, 1993). Fleas threaten the health of humans and animals due to bite reactions and transmission of diseases (Krämer and Mencke, 2001), and in addition are major nuisance pests. Therefore, flea control, is necessary.

In veterinary medicine, fleas are the primary cause for flea allergic dermatitis. This dermatitis results when the adult flea injects saliva into the host during blood feeding, which accelerates immunological response, leading to secondary infections of the skin.

The last aspect of veterinary importance is the role of fleas in disease, for example, transmission of the cestode *Dipylidium caninum*. The role of fleas in human disease transmission has been known since historic times. The Oriental rat flea (*Xenopsylla cheopis*) is the major transmitter of *Yersinia pestis*, the bacterium that causes the bubonic plague in humans. The cat flea is also capable of transmitting *Y. pestis*, and human plague cases and even deaths associated with infected cats and dogs have been occasionally reported (Rust *et al.*, 1971).

Furthermore, a variety of bacteria and viruses have been reported to be transmitted by the dog flea (*C. canis*) as well as the cat flea (*C. felis*) (Krämer and Mencke, 2001). Moreover, cat owners have a high incidence of cat scratch fever, a zoonotic disease caused by the Gram-negative bacterium *Bartonella henselae*. Recent research showed that flea feces are the major means of disease transmission between cats, and from cats to humans (Malgorzata *et al.*, 2000).

3.10.1.1. Insecticidal efficacy in veterinary medicine
Recommendations for the treatment of fleas on companion animals, and the selection of an insecticide and its formulation, are generally based upon the species and age of the animal to be treated, the level of infestation, the rate of potential reinfestation, and the thoroughness of environmental treatment. However, the selection of an insecticide formulation by a pet owner is actually based on economics and the product's ease of use (Williams, 1993). Another factor in the choice of a flea treatment by pet owners is the safety and toxicology of the insecticide (Krämer and Mencke, 2001).

Imidacloprid, 10% spot-on, was designed to offer a dermal treatment, which means it is applied externally onto a small dorsal area (a spot) of the animal's skin. Criteria for the selection of an appropriate topical flea formulation are good solubility of the compound, good adhesion to the skin, good spreading properties, good local and systemic tolerance, stability, and compatibility with legal standards. The imidacloprid spot-on formulation, which meets all these requirements, contains 10 g a.i. in 100 ml nonaqueous solution. The efficacy of this formulation for flea control on cats and dogs has been reported (Krämer and Mencke, 2001). Imidacloprid applied at the target therapeutic dosage of 10.0 mg kg^{-1} killed 99% of the fleas within 1 day of treatment, and continued to provide 99–100% control of further flea infestation for at least 4 weeks (Hopkins *et al.*, 1996; Arther *et al.*, 1997) (**Figure 25**).

Studies using flea-infested cats (Jacobs *et al.*, 1997b) proved that imidacloprid possess considerable potency against adult fleas on cats, and retains a high level of activity for 4–5 weeks. Imidacloprid was effective for both immediate relief from an existing flea burden (the therapeutic effect), and for longer-term flea control (the prevention or prophylactic effect) (**Table 9**). Imidacloprid spreads and acts using animal skin as the main carrier. The compound was shown to be localized in the water-resistant lipid layer of the skin surface, produced by sebaceous glands, and spread over the body surface and onto the hair (Mehlhorn *et al.*, 1999). Insecticide spread over the skin surface was also reported from a clinical study that observed fast onset of flea control, as early as 6 h posttreatment (Everett *et al.*, 2000). Therefore if the superficial fatty layer of

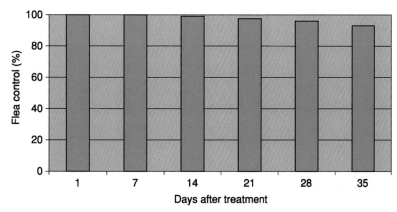

Figure 25 Flea control achieved with imidacloprid 10% spot-on as confirmed by a dose-confirmation study in flea infested dogs (Hopkins *et al.*, 1996).

Table 9 Geometric mean flea counts of two groups of cats, an untreated control and a group treated with imidacloprid 10% spot-on at a dosage of 10.0 mg kg^{-1} bw at day 0 of the study

Weeks after treatment	After 1 day			After 2 days		
	Control	Treated	Reduction (%)	Control	Treated	Reduction (%)
0	36.7	0.01	99.8	31.7	0.0	100.0
1	34.1	0.0	100.0			
2	31.0	0.2	99.3	27.3	0.0	100.0
3	32.1	1.0	97.0	28.9	0.1	99.7
4	36.2	5.0	86.1	27.8	0.9	96.7
5	33.8	9.5	71.9	26.8	3.1	88.3
6	28.1	20.1	28.4	24.8	10.6	57.3

Reproduced from Jacobs, D.E., Hutchinson, M.J., Krieger, K.J., **1997b**. Duration of activity of imidacloprid, a novel adulticide for flea control, against *Ctenocephalides felis* on cats. *Vet. Rec. 140*, 259–260.

the skin is removed by repeated swabbing with alcohol fleas consumed the same amount of blood in treated and untreated dogs. Thus imidacloprid localized in the lipid layer of the skin acts on adult fleas by contact. It was reported that imidacloprid is not taken up by the flea during blood feeding, but is absorbed via the flea's smooth, nonsclerotized intersegmental membranes that are responsible for the insect's mobility. Mehlhorn *et al.* (1999) concluded that this seems reasonable because of imidacloprid's lipophilicity renders it incapable of passing through the sclerotized cuticle. Moreover, initial damage was seen in the ganglia close to the ventral body side (i.e., in subesophageal and thoracic ganglia). Fleas affected by imidacloprid treatment showed characteristic pathohistological changes (Mehlhorn *et al.*, 1999). Muscle fibers and tissue around the subesophageal ganglion were damaged, with the mitochondria and axons showing intensive vacuolization (**Figure 26**). Imidacloprid's mode of action corresponded with the structural findings (Mehlhorn *et al.*, 1999), which showed overall destruction of

the mitochondria, damage of the nerve cells, and disintegration of the insect muscles (**Figure 26**).

Imidacloprid's activity on ectoparasitic insects results from its presence within the lipid layer of the host body surface. Since this lipid layer is always present, imidacloprid remains available for a prolonged time (Hopkins *et al.*, 1996; Mehlhorn *et al.*, 2001a), and reduces the likelihood of its removal during water exposure (Mehlhorn *et al.*, 1999).

The spectrum of imidacloprid activity to ectoparasites is not confined to fleas. It has also proven to be highly effective against both sucking lice (*Linognathus setosus*), and biting or chewing lice (*Trichodectes canis*) (Hanssen *et al.*, 1999). Furthermore imidacloprid acted rapidly on all motile stages of sheep keds (Mehlhorn *et al.*, 2001b). Sheep keds of the species *Melophagus ovinus* are wingless parasitic insects belonging to the dipteran family Hippobiscidae. Besides skin infection, which results in the loss wool quality and meat production, sheep keds are also known to transmit diseases, such as trypanosomiasis. However, ticks did not prove to

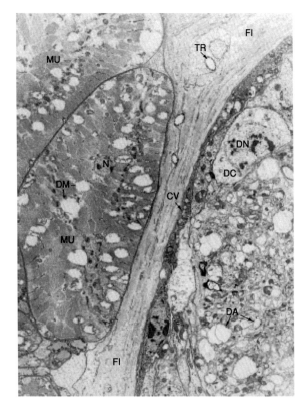

Figure 26   Transmission electron micrograph of a section through an adult cat flea exposed to imidacloprid for 1 h *in vitro*. Note the extensive damage at the level of muscle fibers and the subesophagal ganglion (vacuoles). CV, cellular cover of the ganglion; DA, degenerating axon; DC, degenerating nerve cell; DM, degenerating mitochondrion; FI, Fibrillar layer of connective tissue; Mu, Muscle fiber; TR, tracheole. Magnification ×25 000. (Reproduced with permission from Mehlhorn, H., Mencke, N., Hansen, O., **1999**. Effects of imidacloprid on adult and larval stages of the cat flea *Ctenocephalides felis* after *in vivo* and *in vitro* experiments: a light- and electronmicroscopy study. *Parasitol. Res. 85*, 625–637.)

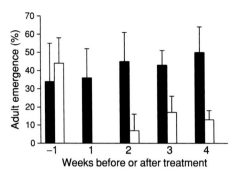

Figure 27   Emergence of adult fleas from flea eggs incubated on blankets used by untreated (control) or imidacloprid treated cats. (Reproduced from Jacobs, D.E., Hutchinson, M.J., Ewald-Hamm, D., **2000**. Inhibition of immature *Ctenocephalides felis felis* (Siphonaptera: Pulicidae) development in the immediate environment of cats treated with imidacloprid. *J. Med. Entomol. 37*, 228–230.)

be sensitive to complete control by imidacloprid (Young and Ryan, 1999).

3.10.1.2. Larvicidal activity   Apart from the fast-acting adulticidal activity of imidacloprid on flea populations, its larvicidal effects have also been investigated. Reinfestation of animals, from earlier deposited eggs, larvae, and preemerged adult fleas, can be overcome by the use of effective larvicidal compounds or insecticides acting on both the adult and the immature stages of fleas. In early studies using imidacloprid, larvicidal activity was observed in the immediate surrounding of treated dogs (Hopkins *et al.*, 1997). Skin debris collected from treated dogs, when mixed into flea rearing media, showed high flea larva mortality. In repeated tests using the skin debris samples, collected at day 7 post-treatment, the larval mortality remained high at

100% even after 51 days (Arther *et al.*, 1997). In *in vitro* studies, flea larvae survived for only 6 h when placed on clipped hair from imidacloprid treated dogs (Mehlhorn *et al.*, 1999). Similar results have also been reported for cats (Jacobs *et al.*, 2000). Adult flea emergence was reduced by 100%, 84%, 60%, and 74% in the first, second, third, and fourth week postimidacloprid treatment, in comparison to untreated controls (**Figure 27**). This persistent larvicidal activity is important, because in the absence of any larvicidal effect of an applied adulticide, reinfestation would occur from eggs deposited prior to treatment (Jacobs *et al.*, 2001). Furthermore, cats wander freely outdoors, and thus may visit sites shared with other flea-infested domestic or wild animals.

3.10.1.3. Flea allergy dermatitis (FAD)   FAD is a disease in which a hypersensitive state is produced in a host in response to the injection of antigenic material from flea salivary glands (Carlotti and Jacobs, 2000). Synonyms for FAD include flea bite allergy and flea bite hypersensitivity. In cats, the disease is also known as feline miliary dermatitis and feline eczema. FAD is one of the most frequent causes of skin conditions in small animals, and a major clinical entity in dogs. FAD is the commonest nonroutine reason for pet owners to seek veterinary advice. Hypersensitivity to flea bites is not only of importance to domestic pets, but is also an important cause of the common skin disease in humans, termed papular urticaria. Detailed investigations carried out on patients exposed to flea-infested pets have shown that the incidence of such reactions is quite high.

Several field studies have been conducted, focusing on the efficacy of imidacloprid on cats and dogs with clinical signs of FAD (Krämer and Mencke, 2001). The efficacy of imidacloprid in flea removal, and the resolution of FAD was tested in dogs and cats from single- and multiple-animal households (Genchi *et al.*, 2000). Flea infestation was examined and FAD dermatitis lesions were ranked according to severity of typical clinical signs. Flea numbers dropped significantly after treatment of animals from both single- and multiple-animal households. In dogs clinical signs of FAD prior to treatment, decreased from 38% to 16% by day 14, and 6% by day 28, thus verifying a rapid adulticidal and high residual activity that lasted at least 4 weeks. There was an effective control of parasites, with rapid improvement of allergy until almost complete remission up to 28 days following the first application. Recently studies on the effect of imidacloprid on cats with clinical signs of FAD confirmed field data published by Genchi *et al.* (2000). Clinical signs of FAD, especially alopecia and pruritus were resolved after monthly treatment using imidacloprid (Keil *et al.*, 2002) (**Figure 28**). Furthermore, controlling FAD is enhanced when blood feeding of fleas is reduced. This so-called "sublethal effect" or "antifeeding effect" was reported using very low concentrations of imidacloprid (Rust *et al.*, 2001, 2002).

3.10.1.4. Imidacloprid as combination partner in veterinary medicinal products   The ability of acaricides to repel or kill ticks, before they attach to a host and feed, is important for the prevention of transmission of tick born pathogens (Young *et al.*, 2003). K9 Advantix™, an effective tick control agent (Spencer *et al.*, 2003, Mehlhorn *et al.*, 2003), is a spot-on product containing 8.8% (w/w) imidacloprid and 44% (w/w) permethrin. The mixture repels and kills four species of ticks, including *Ixodes scapularis*, for up to 4 weeks. It also repels and kills mosquitoes, and kills flea adults and larvae.

Furthermore, a combination containing imidacloprid 10% (w/v) and permethrin 50% (w/v) in a spot-on formulation, has repellent and insecticidal efficacy, against the sand fly species (*Phlebotomus papatasi*) (Mencke *et al.*, 2003), ticks (*I. ricinus, Rhipicephalus sanguineus*), and flea (*C. felis felis*) (Epe *et al.*, 2003) on dogs.

Another combination product, (Advantage Heart™ (10% w/v imidacloprid plus 1% w/v moxidectin), a macrolide antihelmintic, has been developed as a spot-on for dermal application to kittens and cats (Arther *et al.*, 2003). It is intended for monthly application for control of flea infestations and intestinal nematodes, and for prevention of feline heartworm disease. It controls and treats not only established adult gastrointestinal parasites, but also developmental stages, including fourth instar larvae and immature adults of *Toxocara cati* in cats (Hellmann *et al.*, 2003; Reinemeyer and Charles, 2003). Furthermore, the spot-on combination is safe and highly efficacious against *T. canis* and Ancylostomatidae in naturally infested dogs (Hellmann *et al.*, 2003), as well as against *Sarcoptes scabiei* var. *canis* on dogs (Fourie *et al.*, 2003).

3.11. Concluding Remarks and Prospects

The discovery of neonicotinoids as a new class of nAChR ligands can be considered a milestone in insecticide research, and permits an understanding of the functional properties of insect nAChRs. Up to now the most current information regarding nAChRs originated from research with vertebrate receptors. The world market for insecticides is still dominated by compounds irreversibly inhibiting acetylcholinesterase, an important enzyme in the CNS of insects. The market share of these inhibitors,

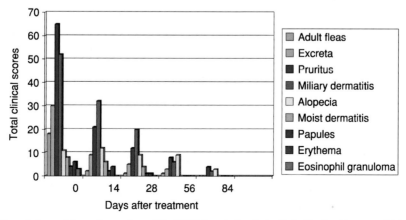

Figure 28   Resolution of signs of flea allergy dermatitis (FAD) in cats treated with imidacloprid over a 84-day study period (Keil *et al.*, 2002).

Table 10 Modes of action of the top selling 100 insecticides/acaricides and their world market share (excluding fumigants, endotoxins and those insecticides with unknown mode of action)

Mode of action	Market share (%) 1987	1999	Change (%)
Acetylcholinesterase	71	52	−20.0
Voltage-gated sodium channel	17	18	+1.4
Acetylcholine receptor	1.5	12	+10.0
GABA-gated chlorine channel	5.0	8.3	+3.3
Chitin biosynthesis	2.1	3.0	+0.9
NADH dehydrogenase	0	1.2	+1.2
Uncouplers	0	0.7	+0.7
Octopamine receptor	0.5	0.6	+0.1
Ecdysone receptor	0	0.4	+0.4

Reproduced with permission from Nauen, R., Bretschneider, T., **2002**. New modes of action of insecticides. *Pestic. Outlook 12*, 241–245. © The Royal Society of Chemistry.

and those insecticides acting on the voltage-gated sodium channel, account for approximately 70% of the world market (Nauen and Bretschneider, 2002) (**Table 10**).

Today, the neonicotinoids are the fastest-growing group of insecticides (estimated marked share in 2005: about 15%), with widespread use in most countries in many agronomic cropping systems, especially against sucking pests but also against ectoparasitic insects. The relatively low risk and target specificity of the products, combined with their suitability for a range of application methods, also will ensure their success as important insecticides in integrated pest management (IPM) strategies.

References

Akamatsu, M., Ozoe, Y., Ueno, T., Fujita, T., Mochida, K., et al., **1997**. Sites of action of noncompetitive GABA antagonists in housflies and rats: three-dimensional QSAR analysis. *Pestic. Sci. 49*, 319–332.

Akayama, A., Minamida, I., **1999**. Discovery of a new systemic insecticide, nitenpyram and its insecticidal properties. In: Yamamoto, I., Casida, J.E. (Eds.), Neonicotinoid Insecticides and the Nicotinic Acetylcholine Receptor. Springer, New York, pp. 127–148.

Amar, M., Thomas, P., Wonnacott, S., Lunt, G.G., **1995**. A nicotinic acetylcholine receptor subunit from insect brain forms a non-desensitizing homo-oligomeric nicotine acetylcholine receptor when expressed in *Xenopus* oocytes. *Neurosci. Lett. 199*, 107–110.

Araki, Y., Bornatsch, W., Brauner, A., Clark, T., Dräger, G., et al., **1994**. Metabolism of imidacloprid in plants. In: Proc. IUPAC Congress, Washington 2B–157.

Armbrust, K.I., Peeler, H.B., **2002**. Effects of formulation on the run-off of imidacloprid from turf. *Pest Mgt Sci. 58*, 702–706.

Armstrong, W.N., **1991**. Glutathione S-transferases: reaction, metabolism, structure, and function. *Chem. Res. Toxicol. 4*, 131–142.

Arther, R.G., Bowman, D.D., McCall, J.W., Hansen, O., Young, D.R., **2003**. Feline Advantage Heart™ (imidacloprid and moxidectin) topical solution as monthly treatment for prevention of heartworm infection (*Dirofilaria immitis*) and control of fleas (*Ctenocephalides felis*) on cats. *Parasitol. Res. 90*, 137–139.

Arther, R.G., Cunningham, J., Dorn, H., Everett, R., Herr, L.G., et al., **1997**. Efficacy of imidacloprid for removal and control of fleas (*Ctenocephalides felis*) on dogs. *Am. J. Vet. Res. 58*, 848–850.

Bai, D., Lummis, S.C.R., Leicht, W., Breer, H., Satelle, D.B., **1991**. Actions of imidacloprid and related nitromethylen on cholinergic receptors of an identified insect motor neuron. *Pestic. Sci. 33*, 197–204.

Balant, L.P., Doelker, E., **1995**. Metabolic considerations in prodrug design. In: Wolff, M.E. (Ed.), Burger's Medicinal Chemistry and Drug Discovery, 5th edn., vol. 1. Wiley-Interscience, New York, pp. 949–982.

Becke, A.D., **1988**. Density-functional exchange-energy approximation with correct asymtotic behavior. *Phys. Rev. A 38*, 3098–3100.

Beers, W.H., Reich, E., **1970**. Structure and activity of acetylcholine. *Nature 228*, 917–922.

Benson, J.A., **1989**. Insect nicotinc acetylcholine receptors as targets for insecticides. In: McFarlane, N.R., Farnham, A.W. (Eds.), Progress and Prospects in Insect Control, BCPC Monograph, vol. 43. British Crop Protection Council, pp. 59–70.

Bergé, J.B., Feyereisen, R., Amichot, M., **1999**. Cytochrome P-450 monooxygenases and insecticide resistance in insects. In: Denholm, I., Pickett, J.A., Devonshire, A.L. (Eds.), Insecticide Resistance: From Mechanisms to Management. CAB International, Wallingford, pp. 25–31.

Bertrand, D., Ballivet, M., Gomez, M., Bertrand, S., Phannavong, B., et al., **1994**. Physiological properties of neuronal nicotinic receptors reconstituted from the vertebrate β2 subunit and *Drosophila* α subunits. *Eur. J. Neurosci. 6*, 869–875.

Bloomquist, J.R., **2001**. GABA and glutamate receptors as biochemical sites for insecticide action. In: Ishaaya, I. (Ed.), Biochemical Sites Important in Insecticide Action and Resistance. Springer, Berlin, pp. 17–41.

Boëlle, J., Schneider, R., Géradin, P., Loubinoux, B., Maienfisch, P., et al., **1998**. Synthesis and insecticidal evaluation of imidacloprid analogs. *Pestic. Sci. 54*, 304–307.

Born, L., **1991**. The molecular and crystal structure of imidacloprid (phase 2). *Pflanzenschutz-Nachrichten Bayer 44*, 137–144.

Breer, H., **1988**. Receptors for acetylcholine in the nervous system of insects. In: Lund, G.G. (Ed.), Neurotox 88: Molecular Basis of Drug and Pesticide Action. Excerpta Medica, Amsterdam, pp. 301–309.

Briggs, G., Bromilow, R.H., Evans, A.A., **1982**. Relationships between lipophilicity and root uptake and translocation of non-ionized chemicals by barley. *Pestic. Sci. 13*, 495–504.

Brown, J.K., Perring, T.M., Cooper, A.D., Bedford, I.D., Markham, P.G., 2000. Genetic analysis of *Bemisia* (Hemiptera: Aleyrodidae) populations by isoelectric focusing electrophoresis. *Biochem. Genet. 38*, 13–25.

Bryant, R., 2001. Chemistries for the newer groups of agrochemical active ingredients. *Agro-Food-Industry Hi-Tech 9/10*, 58–63.

Buchholz, A., Nauen, R., 2001. Organophosphate and carbamate resistance in *Myzus persicae* Sulzer and *Phorodon humuli* Schrank (Homoptera: Aphididae): biological and biochemical considerations. *Mitteil. Deutschen Gesell. Allg. Angew. Entomol. 13*, 227–232.

Buchholz, A., Nauen, R., 2002. Translocation and translaminar bioavailability of two neonicotinoid insecticides after foliar application to cabbage and cotton. *Pest Mgt Sci. 58*, 10–16.

Buckingham, S.D., Lapied, B., Le Corronc, H., Grolleau, F., Sattelle, D.B., 1997. Imidacloprid actions on insect neuronal acetylcholine receptors. *J. Exp. Biol. 200*, 2685–2692.

Byrne, F.J., Castle, S., Prabhaker, N., Toscano, N.C., 2003. Biochemical study of resistance to imidacloprid in B biotype *Bemisia tabaci* from Guatemala. *Pest Mgt Sci. 59*, 347–352.

Cahill, M., Gorman, K., Day, S., Denholm, I., Elbert, A., et al., 1996. Baseline determination and detection of resistance to imidacloprid in *Bemisia tabaci* (Homoptera: Aleyrodidae). *Bull. Entomol. Res. 86*, 343–349.

Carlotti, D.N., Jacobs, D.E., 2000. Therapy, control and prevention of flea allergy dermatitis in dogs and cats. *Vet. Dermatol. 11*, 83–98.

Casida, J.E., Quistad, G.B., 1998. Golden age of insecticide research: past, present, or future? *Annu. Rev. Entomol. 43*, 1–16.

Chao, S.L., Dennehy, T.J., Casida, J.E., 1997. Whitefly (Hemiptera: Aleyrodidae) binding site for imidacloprid and related insecticides: a putative nicotinic acetylcholine receptor. *J. Econ. Entomol. 90*, 879–882.

Cheung, H., Clarke, B.S., Beadle, D.J., 1992. A patch-clamp study of the action of a nitromethylene heterocycle insecticide on cockroach neurons growing *in vitro*. *Pestic. Sci. 34*, 187–193.

Cramer, R.D., Paterson, D.E., Bunce, J.D., 1988. Comparative molecular field analysis (CoMFA). 1. Effect of shape on binding of steroids to carrier proteins. *J. Am. Chem. Soc. 110*, 5959–5967.

Daborn, P., Boundy, S., Yen, J., Pittendrigh, B., ffrench-Constant, R., 2001. DDT resistance in *Drosophila* correlates with Cyp6g1 over-expression and confers cross-resistance to the neonicotinoid imidacloprid. *Mol. Genet. Genom. 266*, 556–563.

Denholm, I., Cahill, M., Dennehy, T.J., Horowitz, A.R., 1999. Challenges with managing insecticide resistance in agricultural pests, exemplified by the whitefly *Bemisia tabaci*. In: Denholm, I., Pickett, J.A., Devonshire, A.L. (Eds.), Insecticide Resistance: From Mechanisms to Management. CAB International, Wallingford, pp. 81–93.

Denholm, I., Devine, G., Foster, S., Gorman, K., Nauen, R., 2002. Incidence and management of insecticide resistance to neonicotinoids. *Proc. Brighton Crop Protection Conference: Pests and Diseases 1*, 161–168.

Denholm, I., Horowitz, A.R., Cahill, M., Ishaaya, I., 1998. Management of resistance to novel insecticides. In: Ishaaya, I., Degheele, D. (Eds.), Insecticides with Novel Modes of Action, Mechanism and Application. Springer, New York, pp. 260–282.

Denholm, I., Rowland, M.W., 1992. Tactics for managing pesticide resistance in arthropods: theory and practice. *Annu. Rev. Entomol. 37*, 91–112.

Dennehy, T.J., Russell, J.S., 1996. Susceptibility of *Lygus* bug populations in Arizona to acephate (Orthene) and bifenthrin (Capture) with related contrasts of other insecticides. In: Proceedings Beltwide Cotton Conferences, Nashville, TN, USA, vol. 2. pp. 771–776.

Devine, G., Harling, Z., Scarr, A.W., Devonshire, A.L., 1996. Lethal and sublethal effects of imidacloprid on nicotine-tolerant *Myzus nicotianae* and *Myzus persicae*. *Pestic. Sci. 48*, 57–62.

Devonshire, A.L., 1989. Insecticide resistance in *Myzus persicae*: from field to gene and back again. *Pestic. Sci. 26*, 375–382.

Devonshire, A.L., Denholm, I., Foster, S., 1999. Insecticide resistance in the peach potato aphid, *Myzus persicae*. In: Denholm, I., Ioannidis, P.M. (Eds.), Combating Insecticide Resistance. AgroTypos SA, Athens, pp. 79–85.

Devonshire, A.L., Moores, G.D., 1982. A carboxylesterase with broad substrate specificity causes organophosphorus, carbamate and pyrethroid resistance in peach potato aphids (*Myzus persicae*). *Pestic. Biochem. Physiol. 18*, 235–246.

Dewar, A.M., Read, L.A., 1990. Evaluation of an insecticidal seed treatment, imidacloprid, for controlling aphids on sugar beet. *Proc. Brighton Crop Protection Conference: Pests and Diseases 2*, 721–726.

Diehr, H.-J., Gallenkamp, B., Jelich, K., Lantzsch, R., 1991. Synthesis and chemical-physical properties of the insecticide imidacloprid (NTN33893). *Pflanzenschutz-Nachrichten Bayer 44*, 107–112.

Dong, K., 1997. A single amino acid change in the para sodium channel protein is associated with knockdown resistance (kdr) to pyrethroid insecticides in German cockroaches. *Insect Biochem. Mol. Biol. 27*, 93–100.

Dryden, M.W., Perez, H.R., Ulitchny, D.M., 1999. Control of fleas on pets and in homes by use of imidacloprid or lufenuron and pyrethrin spray. *J. Am. Vet. Med. Assoc. 215*, 36–39.

El Kady, H., Devine, G.J., 2003. Insecticide resistance in Egyptian populations of the cotton whitefly, *Bemisia tabaci* (Hemiptera: Aleyrodidae). *Pest Mgt Sci. 59*, 865–871.

Elbert, A., Becker, B., Hartwig, J., Erdelen, C., 1991. Imidacloprid: a new systemic insecticide. *Pflanzenschutz-Nachrichten Bayer 44*, 113–136.

Elbert, A., Erdelen, C., Kühnhold, J., Nauen, R., Schmidt, H.W., *et al.*, 2000. Thiacloprid, a novel neonicotinoid insecticide for foliar application. *Proc. Brighton Crop Protection Conference: Pests and Diseases 1*, 21–26.

Elbert, A., Nauen, R., 1996. Bioassays for imidacloprid for a resistance monitoring against the whitefly *Bemisia tabaci*. *Proc. Brighton Crop Protection Conference: Pests and Diseases 2*, 731–738.

Elbert, A., Nauen, R., 2004. New applications for neonicotinoid insecticides using imidacloprid as an example. In: Horowitz, A.R., Ishaaya, I. (Eds.), Novel Approaches to Insect Pest Management in Field and Protected Crops. Springer, New York, pp. 29–44.

Elbert, A., Nauen, R., Cahill, M., Devonshire, A.L., Scarr, A.W., *et al.*, 1996. Resistance management with chloronicotinyl insecticides using imidacloprid as an example. *Pflanzenschutz-Nachrichten Bayer 49*, 5–53.

Elbert, A., Nauen, R., Leicht, W., 1998. Imidacloprid, a novel chloronicotinyl insecticide, biological activity and agricultural importance. In: Ishaaya, I., Degheele, D. (Eds.), Insecticides with Novel Modes of Action, Mechanism, and Application. Springer, New York, pp. 50–73.

Elbert, A., Overbeck, H., Iwaya, K., Tsuboi, S., 1990. Imidacloprid, a novel systemic nitromethylene analog insecticide for crop protection. *Proc. Brighton Crop Protection Conference: Pests and Disease 1*, 21–28.

Epe, C., Coati, N., Stanneck, D., 2003. Efficacy of the compound preparation imidacloprid 10% (w/v)/permethrin 50% (w/v) spot-on against ticks (*I. ricinus, R. sanguineus*) and fleas (*C. felis*) on dogs. *Parasitol. Res. 90*, 122–124.

Everett, R., Cunningham, J., Arther, R., Bledsoe, D.L., Mencke, N., 2000. Comparative evaluation of the speed of flea kill of imidacloprid and selamectin in dogs. *Vet. Therap. 1*, 229–234.

Feuer, H., Lawrence, J.P., 1969. The alkyl nitrate nitration of active methylene compounds. 6. A new synthesis of α-nitroalkyl heterocyclics. *J. Am. Chem. Soc. 91*, 1856–1857.

Feyereisen, R., 1995. Molecular biology of insecticide resistance. *Toxicol. Lett. 82*, 83–90.

ffrench-Constant, R.H., Rocheleau, T.A., Steichen, J.C., Chalmers, A.E., 1993. A point mutation in *Drosophila* GABA receptor confers insecticide resistance. *Nature 363*, 449–451.

Field, L.M., Blackmann, R.L., Devonshire, A.L., 2001. Evolution of amplified esterase genes as a mode of insecticide resistance in aphids. In: Ishaaya, I. (Ed.), Biochemical Sites Important in Insecticide Action and Resistance. Springer, Berlin, pp. 209–219.

Field, L.M., Devonshire, A.L., Forde, B.G., 1988. Molecular evidence that insecticide resistance in peach potato aphids, *Myzus persicae* (Sulz), results from amplification of an esterase gene. *Biochem. J. 251*, 309–315.

Foster, S.P., Denholm, I., Devonshire, A.L., 2000. The ups and downs of insecticide resistance in peach potato aphids (*Myzus persicae*) in the UK. *Crop Protect. 19*, 873–879.

Foster, S.P., Denholm, I., Harling, Z.K., Moores, G.D., Devonshire, A.L., 1998. Intensification of resistance in UK field populations of the peach potato aphid, *Myzus persicae* (Homoptera: Aphididae) in 1996. *Bull. Entomol. Res. 88*, 127–130.

Foster, S.P., Denholm, I., Thompson, R., 2003. Variation in response to neonicotinoid insecticides in peach potato aphids, *Myzus persicae* (Hemiptera: Aphididae). *Pest Mgt Sci. 59*, 166–173.

Foster, S.P., Devonshire, A.L., 1999. Field-simulator study of insecticide resistance conferred by esterase-, MACE-, and kdr-based mechanisms in the peach potato aphid, *Myzus persicae* (Sulzer). *Pestic. Sci. 55*, 810–814.

Fourie, L.J., Du Rand, C., Heine, J., 2003. Evaluation of the efficacy of an imidacloprid 10%/moxidectin 2.5% spot-on against *Sarcoptes scabiei* var. *canis* on dogs. *Parasitol. Res. 90*, 135–136.

Genchi, C., Traldi, G., Bianciardi, P., 2000. Efficacy of imidacloprid on dogs and cats with natural infestations of fleas, with special emphasis on flea hypersensitivity. *Vet. Therap. 2*, 71–80.

Georghiou, G.P., 1983. Management of resistance in arthropods. In: Georghiou, G.P., Saito, T. (Eds.), Pest Resistance to Pesticides. Plenum, New York, pp. 769–792.

Göbel, T., Gsell, L., Huter, O.F., Maienfisch, P., Naef, R., *et al.*, 1999. Synthetic approaches towards CGA 293′343: a novel broad-spectrum insecticide. *Pestic. Sci. 55*, 355–357.

Goodman, C.S., Spitzer, N.C., 1980. Embryonic development of neurotransmitter receptors in grasshopers. In: Sattelle, D.B., Hall, L.M., Hidebrand, J.G. (Eds.), Receptors for Neurotransmitters, Hormones and Pheromones in Insects. Elsevier, Amsterdam, pp. 195–307.

Graton, J., Berthelot, M., Gal, J.-F., Laurence, C., Lebreton, J., *et al.*, 2003. The nicotinic pharmacophore: thermodynamics of the hydrogen-bonding complexation of nicotin, nornicotine, and models. *J. Organ. Chem. 68*, 8208–8221.

Green, M.B., LeBaron, H.M., Moberg, W.K., 1990. Managing Resistance to Agrochemicals: From Fundamental to Practical Strategies. American Chemical Society, Washington, DC.

Greenwood, R., Ford, M.G., Scarr, A., 2002. Neonicotinoid pharmacokinetics. *Proc. Brighton Crop Protection Conference: Pest and Diseases 1*, 153–160.

Griffin, L., Krieger, K., Liege, P., 1997. Imidacloprid: a new compound for control of fleas initiated dermatitis. *Suppl. Comp. Cont. Educ. Prac. Vet. 19*, 17–20.

Guirao, P., Beitia, F., Cenis, J.L., 1997. Biotype determination of Spanish populations of *Bemisia tabaci* (Hemiptera: Aleyrodidae). *Bull. Entomol. Res. 87*, 587–593.

Gundelfinger, E.D., 1992. How complex is the nicotinic receptor system of insects? *Trends Neurosci. 15*, 206–211.

Gunning, R.V., Moores, G.D., 2001. Insensitive acetylcholinesterase as sites for resistance to organophosphates and carbamates in insects: insensitive acetylcholinesterase confers resistance in Lepidoptera.

In: Ishaaya, I. (Ed.), Biochemical Sites Important in Insecticide Action and Resistance. Springer, Berlin, pp. 221–238.

Hancke, W., Breer, H., **1986**. Channel properties of an insect neuronal acetylcholine receptor protein reconstituted in planar lipid bilayers. *Nature 321*, 171–175.

Hanssen, I., Mencke, N., Asskildt, H., Ewald-Hamm, D., Dorn, H., **1999**. Field study on the insecticidal efficacy of Advantage against natural infestans of dogs with lice. *Parasitol. Res. 85*, 347–348.

Harrow, I.D., Sattelle, D.B., **1983**. Acetylcholine receptors on the cell body membrane of giant interneurone 2 in the cockroach *Periplaneta americana. J. Exp. Biol. 105*, 339–350.

Heimbach, U., Kral, G., Niemann, P., **2000**. Implementation of resistance risk analysis of plant protection products in the German authorisation procedure. In: Proc. Brighton Crop Protection Conference: Pests and Diseases, vol. 7B-2. pp. 771–776.

Hellmann, K., Knoppe, T., Radeloff, I., Heine, J., **2003**. The anthelmintic efficacy and the safety of a combination of imidacloprid and moxidectin spot-on in cats and dogs under field conditions in Europe. *Parasitol. Res. 90*, 142–143.

Hemingway, J., **2000**. The molecular basis of two contrasting metabolic mechanisms of insecticide resistance. *Insect Biochem. Mol. Biol. 30*, 1009–1015.

Hemingway, J., Karunarantne, S.H.P.P., **1998**. Mosquito carboxylesterases: a review of the molecular biology and biochemistry of a major insecticide resistance mechanism. *Med. Vet. Entomol. 12*, 1–12.

Hermsen, B., Stetzer, E., Thees, R., Heiermann, R., Schrattenholz, A., et al., **1998**. Neuronal nicotinic receptors in the locust *Locusta migratoria. J. Biol. Chem. 273*, 18394–18404.

Hirase, K., **2003**. Dinotefuran: a novel neonicotenoid insecticide. *Agrochem. Japan 82*, 14–16.

Hodgson, E., **1983**. The significance of Cytochrome P-450 in insects. *Insect Biochem. 13*, 237–246.

Holladay, M.W., Dart, M.J., Lynch, J.K., **1997**. Neuronal nicotinic acetylcholine receptors as targets for drug discovery. *J. Med. Chem. 40*, 4169–4194.

Hollingworth, R.M., Mota-Sanchez, D., Whalon, M.E., Graphius, E., **2002**. Comparative pharmokinetics of imidacloprid in susceptible and resistant Colorado potato beetles. In: Proc. 10th IUPAC Int. Congr. Chem. Crop Protection, Abstracts 1, p. 312.

Hopkins, T.J., Kerwick, C., Gyr, P., Woodley, I., **1996**. Efficacy of imidacloprid to remove and prevent *Ctenocephalides felis* infestations on dogs and cats. *Austral. Vet. Practit. 26*, 150–153.

Hopkins, T.J., Woodley, I., Gyr, P., **1997**. Imidacloprid topical formulation: larvicidal effect against *Ctenocephalides felis* in the surrounding of treated dogs. *Suppl. Comp. Cont. Educ. Pract. Vet. 19*, 4–10.

Horowitz, A.R., Denholm, I., **2001**. Impact of insecticide resistance mechanisms on management strategies. In: Ishaaya, I. (Ed.), Biochemical Sites Important in

Insecticide Action and Resistance. Springer, Berlin, pp. 323–338.

Horowitz, A.R., Denholm, I., Gorman, K., Cenis, J.L., Kontsedalov, S., et al., **2003**. Biotype Q of *Bemisia tabaci* identified in Israel. *Phytoparasitica 31*, 94–98.

Horowitz, A.R., Denholm, I., Gorman, K., Ishaaya, I., **1999**. Insecticide resistance in whiteflies: current status and implication for management. In: Denholm, I., Ioannidis, P. (Eds.), Combating Insecticide Resistance. AgroTypos SA, Athens, pp. 8996–8998.

Horowitz, A.R., Weintraub, P.G., Ishaaya, I., **1998**. Status of pesticide resistance in arthropod pest in Israel. *Phytoparasitica 26*, 31–240.

Huang, Y., Williamson, M.S., Devonshire, A.L., Windass, J.D., Lansdell, S.J., et al., **1999**. Molecular characterization and imidacloprid selectivity of nicotinic acetylcholine receptor subunits from the peach potato aphid *Myzus persicae. J. Neurochem. 73*, 380–389.

Ishii, Y., Kobori, I., Araki, Y., Kurogochi, S., Iwaya, K., et al., **1994**. HPLC determination of the new insecticide imidacloprid and its behavior in rice and cucumber. *J. Agric. Food Chem. 42*, 2917–2921.

Jacobs, D.E., Hutchinson, M.J., Ewald-Hamm, D., **2000**. Inhibition of immature *Ctenocephalides felis felis* (Siphonaptera: Pulicidae) development in the immediate environment of cats treated with imidacloprid. *J. Med. Entomol. 37*, 228–230.

Jacobs, D.E., Hutchinson, M.J., Fox, M.T., Krieger, K.J., **1997a**. Comparison of flea control strategies using imidacloprid or lufenuron in a controlled simulated home environment. *Am. J. Vet. Res. 58*, 1260–1262.

Jacobs, D.E., Hutchinson, M.J., Krieger, K.J., **1997b**. Duration of activity of imidacloprid, a novel adulticide for flea control, against *Ctenocephalides felis* on cats. *Vet. Rec. 140*, 259–260.

Jacobs, D.E., Hutchinson, M.J., Stannek, D., Mencke, N., **2001**. Accumulation and persistence of flea larvicidal activity in the immediate environment of cats treated with imidacloprid. *Med. Vet. Entomol. 15*, 342–345.

Jeschke, P., Moriya, K., Lantzsch, R., Seifert, H., Lindner, W., et al., **2001**. Thiacloprid (Bay YRC2894): a new member of the chloronicotinyl insecticide (CNI) *Pflanzenschutz-Nachrichten Bayer 54*, 147–160.

Jeschke, P., Schindler, M., Beck, M., **2002**. Neonicotinoid insecticides: retrospective consideration and prospects. *Proc. Brighton Crop Protection Conference: Pests and Diseases 1*, 137–144.

Jeschke, P., Uneme, H., Benet-Buchholz, J., Stölting, J., Sirges, W., et al., **2003**. Clothianidin (TI-435): the third-member of the chloronicotinyl insecticide (CNI^{TM}) family. *Pflanzenschutz-Nachrichten Bayer 56*, 5–24.

Kagabu, S., **1996**. Studies on the synthesis and insecticidal activity of neonicotinoid compounds. *J. Pestic. Sci. 21*, 237–239.

Kagabu, S., **1997a**. Chloronicotinyl insecticides: discovery, application, and future perspective. *Rev. Toxicol. 1*, 75–129.

Kagabu, S., **1997b**. Molecular structures and properties of imidacloprid and analogous compounds. *Phytoparasitica 25*, 347.

Kagabu, S., **1999**. Discovery of chloronicotinylinsecticides. In: Yamamoto, I., Casida, J.E. (Eds.), Neonicotinoid Insecticides and the Nicotinic Acetylcholine Receptor. Springer, New York, pp. 91–106.

Kagabu, S., **2003a**. Molecular design of neonicotinoids: past, present and future. In: Voss, G., Ramos, G. (Eds.), Chemistry of Crop Protection: Progress and Prospects in Science and Regulation. Wiley–VCH, New York, pp. 193–212.

Kagabu, S., **2003b**. Insecticides: imidacloprid. In: Plimmer, J.R. (Ed.), Encyclopedia of Agrochemicals. Wiley-Interscience, New York, pp. 933–944.

Kagabu, S., Akagi, T., **1997**. Quantum chemical consideration of photostability of imidacloprid and related compounds. *J. Pest. Sci. 22*, 84–89.

Kagabu, S., Matsuda, K., Komai, K., **2002**. Preparation of dinotefuran related compounds and agonistic action on SADβ2 hybrid nicotinic acetylcholine receptors expressed in *Xenopus laevis* oocytes. *J. Pestic. Sci. 27*, 374–377.

Kagabu, S., Matsuno, H., **1997**. Chloronicotinyl insecticides. 8. Crystal and molecular structures of imidacloprid and analogous compounds. *J. Agric. Food Chem. 45*, 276–281.

Kagabu, S., Medej, S., **1995**. Stability comparison of imidacloprid and related compounds under simulated sunlight, hydrolysis conditions, and to oxygen. *Biosci. Biotechnol. Biochem. 59*, 980–985.

Kagabu, S., Moriya, K., Shibuya, K., Hattori, Y., Tsuboi, S., *et al.*, **1992**. 1-(6-Halonicotinyl)-2-nitromethylene-imidazolidines as potential new insecticides. *Biosci. Biotechnol. Biochem. 56*, 362–363.

Kagabu, S., Yokoyama, K., Iwaya, K., Tanaka, M., **1998**. Imidacloprid and related compounds: structure and water solubility of *N*-alkyl derivatives of imidacloprid. *Biosci. Biotechnol. Biochem. 62*, 1216–1224.

Kashiwada, Y., **1996**. Bestguard (nitenpyram, TI-304) – a new systemic insecticide. *Agrochem. Japan 68*, 18–19.

Kayser, H., Lee, C., Wellmann, H., **2002**. Thiamethoxam and imidacloprid bind to different sites on nicotinic receptors: conserved pharmacology among aphids. In: Proc. 10th IUPAC Int. Congr. Chem. Crop Protection, Abstracts 1, p. 305.

Keil, K., Wellington, J., Ciszewski, D., **2002**. Efficacy of Advantage[TM] in controlling flea allergy dermatitis in cats. *Suppl. Comp. Cont. Educ. Pract. Vet. 24*, 6–9.

Kiriyama, K., Nishimura, K., **2002**. Structural effects of dinotefuran and analogues in insecticidal and neural activities. *Pest Mgt Sci. 58*, 669–676.

Klamt, A., **1995**. Conductor-like screening model for real solvents. A new approach to the quantitative calculation of solvation phenomena. *J. Phys. Chem. 99*, 2224–2235.

Klein, O., **2001**. Behaviour of thiacloprid (YRC 2894) in plants and animals. *Pflanzenschutz-Nachrichten Bayer 54*, 209–240.

Klein, O., **2003**. Behaviour of clothianidin (TI-435) in plants and animals. *Pflanzenschutz-Nachrichten Bayer 56*, 75–101.

Knaust, H.J., Poehling, H.M., **1992**. Studies of the action of imidacloprid on grain aphids and their efficiency to transmit BYD virus. *Pflanzenschutz-Nachrichten Bayer 45*, 381–408.

Knowles, C.O., **1997**. Mechanisms of resistance to acaricides. In: Sjut, V. (Ed.), Molecular Mechanisms of Resistance to Agrochemicals, vol. 13. Springer, Berlin, pp. 57–77.

Kodaka, K., Kinoshita, K., Wakita, T., Yamada, E., Kawahara, N., *et al.*, **1998**. MTI-446: a novel systemic insect control compound. In: Proc. Brighton Crop Protection Conference: Pests and Diseases, vol. 1. pp. 21–26.

Kodaka, K., Kinoshita, K., Wakita, T., Yamada, E., Kawahara, N., *et al.*, **1999a**. Synthesis and activity of novel insect control compound MTI-446 and its derivatives. In: Proc. 24th Annu. Mtg Pestic. Sci. Soc. Japan Abstracts, p. 85.

Kodaka, K., Kinoshita, K., Wakita, T., Yamada, E., Kawahara, N., *et al.*, **1999b**. Synthesis and activity of novel insect control compound MTI-446 and its derivatives. In: Proc. 15th Ass. Pestic. Design Res. Abstracts, p. 27.

Koester, J., **1992**. Comparative metabolism [pyridinyl-14C-methyl]imidacloprid in plant cell suspension cultures. In: Proc. Brighton Crop Protection Conference: Pests and Diseases, vol. 2. pp. 901–906.

Kollmeyer, W.D., Flattum, R.F., Foster, J.P., Powel, J.E., Schroeder, M.E., *et al.*, **1999**. Discovery of the nitromethylene heterocycle insecticides. In: Yamamoto, I., Casida, J.E. (Eds.), Neonicotinoid Insecticides and the Nicotinic Acetylcholine Receptor. Springer, New York, pp. 71–89.

Krämer, F., Mencke, N., **2001**. Flea biology and control: the biology of the cat flea, control and prevention with imidacloprid in small animals. Springer, New York.

Krohn, J., **2001**. Behaviour of thiacloprid in the environment. *Pflanzenschutz-Nachrichten Bayer 54*, 281–290.

Krohn, J., Hellpointer, E., **2002**. Environmental fate of imidacloprid. *Pflanzenschutz-Nachrichten Bayer 55* (special edn.), 3–26.

Latli, B., Than, C., Morimoto, H., Williams, P.G., Casida, J.E., **1996**. [6-Chloro-3-pyridylmethyl-^{1}H]neonicotinoids as high affinity radioligands for the nicotinic acetylcholine receptor: preparation using NaB[1]H[4], and LiB[1]H[4]. *J. Labeled Compd. Radiopharm. 38*, 971–978.

Latli, B., Tomizawa, M., Casida, J.E., **1997**. Synthesis of a novel [^{125}I]neonicotinoid photoaffinity probe for the *Drosophila* nicotinic acetylcholine receptor. *Bioconjugate Chem. 8*, 7–14.

Leech, C.A., Jewess, P., Marshall, J., Sattelle, D.B., **1991**. Nitromethylene actions on *in situ* and expressed insect nicotinic acetylcholine receptors. *FEBS Lett. 290*, 90–94.

Leicht, W., **1993**. Imidacloprid: a chloronicotinyl insecticide. *Pestic. Outlook 4*, 17–24.

Léna, C., Changeux, J.-P., **1993**. Allosteric modulations of the nicotinic acetylcholine receptor. *Trends Neurosci.* 16, 181–186.

Li, Y., Dennehy, T.J., Li, S., Wigert, M.E., Zarborac, M., *et al.*, **2001**. Sustaining Arizona's fragile success in whitefly resistance management. In: Proc. Beltwide Cotton Conferences, National Cotton Council, Memphis, TN, pp. 1108–1114.

Li, Y., Dennehy, T.J., Li, X., Wigert, M.E., **2000**. Susceptibility of Arizona whiteflies to chloronicotinyl insecticides and IGRs: new developments in the 1999 season. In: Proc. Beltwide Cotton Conferences, National Cotton Council, Memphis, TN, pp. 1325–1330.

Lind, R.J., Clough, M.S., Reynolds, S.E., Earley, F.G.P., **1998**. [^{3}H]Imidaclorid labels high- and low-affinity nicotinic acetylcholine receptor-like binding sites in the aphid *Myzus persicae* (Hemiptera: Aphididae). *Pestic. Biochem. Physiol.* 62, 3–14.

Liu, M.-Y., Casida, J.E., **1993a**. High affinity binding of [^{3}H]imidacloprid in the insect acetylcholine receptor. *Pestic. Biochem. Physiol. 1993*, 40–46.

Liu, M.-Y., Lanford, J., Casida, J.E., **1993b**. Relevance of [^{3}H]imidacloprid binding site in housefly head acetylcholine receptor to insecticidal activity of 2-nitromethylenen- and 2-nitroimino-imidazolidines. *Pestic. Biochem. Physiol.* 46, 200–206.

Liu, M.-Y., Latli, B., Casida, J.E., **1995**. Imidacloprid binding site in *Musca* nicotinic acetylcholine receptor: interactions with physostigmine and a varity of nicotinic agonists with chloropyridyl and chlorothiazolyl substituents. *Pestic. Biochem. Physiol.* 52, 170–181.

Maienfisch, P., Angst, M., Brandl, F., Fischer, W., Hofer, D., *et al.*, **2001a**. Chemistry and biology of thiamethoxam: a second generation neonicotinoid. *Pest Mgt Sci.* 57, 906–913.

Maienfisch, P., Brandl, F., Kobel, W., Rindlisbacher, A., Senn, R., **1999a**. CGA 293,343: a novel, broad-spectrum neonicotinoid insecticide. In: Yamamoto, I., Casida, J.E. (Eds.), Neonicotinoid insecticides and the Nicotinic Acetylcholine Receptor. Springer, New York, pp. 177–209.

Maienfisch, P., Gonda, J., Jacob, O., Kaufmann, L., Pitterna, T., *et al.*, **1997**. In: 214th Am. Chem. Soc. Natl Mtg Abstracts, Agro-018.

Maienfisch, P., Gsell, L., Rindlisbacher, A., **1999b**. Synthesis and insecticidal activity of CGA 293343: a novel broad-spectrum insecticide. *Pestic. Sci.* 55, 351–355.

Maienfisch, P., Haettenschwiler, J., Rindlisbacher, A., Decock, A., Wellmann, H., *et al.*, **2003**. Azido-neonicotinoids as candidate photoaffinity probes for insect nicotinic acetylcholine receptors. 1. *Chimia* 57, 710–714.

Maienfisch, P., Huerlimann, H., Haettenschwiler, J., **2000**. A novel method for the preparation of N,N'-disubstituted-N''-nitroguanidines, including a practical synthesis of the neonicotinoid insecticide clothianidin. *Tetrahedron Lett.* 41, 7187–7191.

Maienfisch, P., Huerlimann, H., Rindlisbacher, A., Gsell, L., Dettwiler, H., *et al.*, **2001b**. The discovery of thiamethoxam: a second generation neonicotinoid. *Pest Mgt Sci.* 57, 165–176.

Maienfisch, P., Rindlisbacher, A., Huerlimann, H., Haettenschwiler, J., Desai, A.K., *et al.*, **2002**. 4-Nitroimino-1,3,5-oxadiazines: a new type of neonicotinoid insecticides. *ACS Symp. Series 800*, 219–230.

Malgorzata, G., Mikolajczyk, M.S., O'Reilly, K.L., **2000**. Clinical disease in kittens inoculated with a pathogenic strain of *Bartonella henselae*. *Am. J. Vet. Res.* 61, 375–379.

Marshall, J., Buckingham, S.D., Shingai, R., Lunt, G.G., Goosey, M.W., *et al.*, **1990**. Sequence and functional expression of a single α subunit of an insect nicotinic acetylcholine receptor. *EMBO J.* 9, 4391–4398.

Martinez-Torres, D., Foster, S.P., Field, L.M., Devonshire, A.L., Williamson, M.S., **1998**. A sodium channel point mutation is associated with resistance to DDT and pyrethroid insecticides in the peach potato aphid, *Myzus persicae* (Sulzer) (Hemiptera: Aphididae). *Insect Mol. Biol.* 8, 1–8.

Matsuda, K., Buckingham, S.D., Freeman, J.C., Squire, M.D., Baylis, H.A., *et al.*, **1998**. Effects of the α subunit on imidacloprid sensitivity of recombinant nicotinic acetylcholine receptors. *Br. J. Pharmacol.* 123, 518–524.

Matsuda, K., Buckingham, S.D., Freeman, J.C., Squire, M.D., Baylis, H.A., *et al.*, **1999**. Role of the α subunit of nicotinic acetylcholine receptor in the selective action of imidacloprid. *Pestic. Sci.* 55, 211–213.

Matsuda, K., Buckingham, S.D., Kleier, D., Rauh, J.J., Grauso, M., *et al.*, **2001**. Neonicotinoids: insecticides acting on insect nicotinic acetylcholine receptors. *Trends Pharmacol. Sci.* 22, 573–578.

Matsuda, M., Takahashi, H., **1996**. Mospilan (acetamiprid, NI-25): a new systemic insecticide. *Agrochem. Japan 68*, 12–20.

McKenzie, J.A., **1996**. Ecological and Evolutionary Aspects of Insecticide Resistance. Academic Press, San Diego, CA.

Mehlhorn, H., D'Haese, J., Mencke, N., Hanssen, O., **2001b**. *In vivo* and *in vitro* effects of imidacloprid on sheep keds (*Melophagus ovinus*): a light and electron microscopic study. *Parasitol. Res.* 87, 331–336.

Mehlhorn, H., Hansen, O., Mencke, N., **2001a**. Comparative study on the effects of three insecticides (fipronil, imidacloprid, selamectin) on the developmental stages of the cat flea (*Ctenocephalides felis* Buche 1835): a light and electron microscopic analysis of *in vivo* and *in vitro* experiments. *Parasitol. Res.* 87, 198–207.

Mehlhorn, H., Mencke, N., Hansen, O., **1999**. Effects of imidacloprid on adult and larval stages of the cat flea *Ctenocephalides felis* after *in vivo* and *in vitro* experiments: a light- and electronmicroscopy study. *Parasitol. Res.* 85, 625–637.

Mehlhorn, H., Schmahl, G., Mencke, N., Bach, T., **2003**. The effect of an imidacloprid and permethrin combination against developmental stages of *Ixodes ricinus* ticks. *Parasitol. Res.* 90, 119–121.

Mencke, N., Jeschke, P., **2002**. Therapy and prevention of parasitic insects in veterinary medicine using imidacloprid. *Curr. Topics Med. Chem.* 2, 701–715.

Mencke, N., Volf, P., Volvofa, V., Stanneck, D., 2003. Repellent efficacy of a combination containing imidacloprid and permethrin against sand flies (*Phlebotomus papatasi*) on dogs. *Parasitol. Res.* 90, 108–111.

Minamida, I., Iwanaga, K., Tabuchi, T., Aoki, I., Fusaka, T., et al., 1993. Synthesis and insecticidal activity of acylic nitroethene compounds containing a heteroarylmethylamino group. *J. Pestic. Sci.* 18, 41–48.

Moores, G.D., Devine, G.J., Devonshire, A.L., 1994a. Insecticide-insensitive acetylcholinesterase can enhance esterase-based resistance in *Myzus persicae* and *Myzus nicotianae*. *Pestic. Biochem. Physiol.* 49, 114–120.

Moores, G.D., Devine, G.J., Devonshire, A.L., 1994b. Insecticide resistance due to insensitive acetylcholinesterase in *Myzus persicae* and *Myzus nicotianae*. In: Proc. Brighton Crop Protection Conference: Pests and Diseases, vol. 1. pp. 413–418.

Mori, K., Okumoto, T., Kawahara, N., Ozoe, Y., 2001. Interaction of dinotefuran and its analogues with nicotinic acetylcholine receptors of cockroach nerve cords. *Pest Mgt Sci.* 58, 190–196.

Moriya, K., Shibuya, K., Hattori, Y., Tsuboi, S., Shiokawa, K., et al., 1992. 1-(6-Chloronicotinyl)-2-nitroimino-imidazolidines and related compounds as potential new insecticides. *Biosci. Biotechnol. Biochem.* 56, 364–365.

Moriya, K., Shibuya, K., Hattori, Y., Tsuboi, S., Shiokawa, K., et al., 1993. Structural modification of the 6-chloropyridyl moiety in the imidacloprid skeleton: introduction of a five-membered heteroaromatic ring and the resulting insecticidal activity. *Biosci. Biotech. Biochem.* 57, 127–128.

Mutero, A., Pralavorio, M., Bride, J.M., Fournier, D., 1994. Resistance-associated point mutations in insecticide insensitive acetylcholinesterase. *Proc. Natl Acad. Sci. USA* 91, 5922–5926.

Nagata, K., Aistrup, G.L., Song, J.-H., Narahashi, T., 1996. Subconductance-state currents generated by imidacloprid at the nicotinic acetylcholine receptor in PC 12 cells. *NeuroReport* 7, 1025–1028.

Nakayama, A., 1998. Molecular similarity and structure–activity relationship of neonicotinoid insecticides. *J. Pestic. Sci.* 23, 336–343.

Nakayama, A., Sukekawa, M., 1998. Quantitative correlation between molecular similarity and receptor-binding activity of neonicotinoid insecticides. *Pestic. Sci.* 52, 104–110.

Nauen, R., Bretschneider, T., 2002. New modes of action of insecticides. *Pestic. Outlook* 12, 241–245.

Nauen, R., Ebbinghaus, U., Tietjen, K., 1999a. Ligands of the nicotinic acetylcholine receptor as insecticides. *Pestic. Sci.* 55, 608–610.

Nauen, R., Ebbinghaus-Kintscher, A., Elbert, A., Jeschke, P., Tietjen, K., 2001. Acetylcholine receptors as sites for developing neonicotinoid insecticides. In: Iishaaya, I. (Ed.), Biochemical Sites of Insecticide Action and Resistance. Springer, New York, pp. 77–105.

Nauen, R., Ebbinghaus-Kintscher, U., Salgado, V.L., Kaussmann, M., 2003. Thiamethoxam is a neonicotinoid precursor converted to clothianidin in insects and plants. *Pest. Biochem. Physiol.* 76, 55–69.

Nauen, R., Elbert, A., 1997. Apparent tolerance of a field-collected strain of *Myzus nicotianae* to imidacloprid due to strong antifeeding response. *Pestic. Sci.* 49, 252–258.

Nauen, R., Elbert, A., 2000. Resistance of *Bemisia* spp. (Homoptera: Aleyrodidae) to insecticides in southern spain with special reference to neonicotinoids. *Pest Mgt Sci.* 56, 60–64.

Nauen, R., Elbert, A., 2003. European monitoring of resistance to common classes of insecticides in *Myzus persicae* and *Aphis gossypii* (Homoptera: Aphididae) with special reference to imidacloprid. *Bull. Entomol. Res.* 93, 47–54.

Nauen, R., Hungenberg, H., Tollo, B., Tietjen, K., Elbert, A., 1998a. Antifeedant-effect, biological efficacy and high affinity binding of imidacloprid to acetylcholine receptors in tobacco associated *Myzus persicae* (Sulzer) and *Myzus nicotianae* Blackman (Homoptera: Aphididae). *Pestic. Sci.* 53, 133–140.

Nauen, R., Reckmann, U., Armborst, S., Stupp, H.P., Elbert, A., 1999b. Whitefly-active metabolites of imidacloprid: biological efficacy and translocation in cotton plants. *Pestic. Sci.* 55, 265–271.

Nauen, R., Strobel, K., Tietjen, K., Otsu, Y., 1996. Aphicidal activity of imidacloprid against a tobacco feeding strain of *Myzus persicae* (Homoptera: Aphididae) from Japan closely related to *Myzus nicotianae* and highly resistant to carbamates and organophosphates. *Bull. Entomol. Res.* 86, 165–171.

Nauen, R., Stumpf, N., Elbert, A., 2002. Toxicological and mechanistic studies on neonicotinoid cross resistance in Q-type Bemisia tabaci (Hemiptera: Aleyrodidae). *Pest Mgt Sci.* 58, 868–875.

Nauen, R., Tietjen, K., Wagner, K., Elbert, A., 1998b. Efficacy of plant metabolites of imidacloprid against *Myzus persicae* and *Aphis gossypii* (Homoptera: Aphididae). *Pestic. Sci.* 52, 53–57.

Nishimura, K., Kanda, Y., Okazawa, A., Ueno, T., 1994. Relationship between insecticidal and neurophysiological activities of imidacloprid and related compounds. *Pestic. Biochem. Physiol.* 50, 51–59.

Ohkawara, Y., Akayama, A., Matsuda, K., Andersch, W., 2002. Clothianidin: a novel broad-spectrum neonicotinoid insecticide. In: Proc. Brighton Crop Protection Conference: Pests and Diseases, vol. 1. pp. 51–58.

Okazawa, A., Akamatsu, M., Nishiwaki, H., Nakagawa, Y., Miyagawa, H., et al., 2000. Three-dimensional quantitative structure–activity relationship analysis of acyclic and cyclic chloronicotinyl insecticides. *Pest Mgt Sci.* 56, 509–515.

Okazawa, A., Akamatsu, M., Ohoka, A., Nishiwaki, H., Cho, W.-J., Nakagawa, Y., et al., 1998. Prediction of the binding mode of imidacloprid and related compounds to housefly head acetylcholine receptors using three-dimensional QSAR analysis. *Pestic. Sci.* 54, 134–144.

Olsen, E.R., Dively, G.P., Nelson, J.O., 2000. Baseline susceptibility to imidacloprid and cross-resistance patterns in Colorado potato beetle (Coleoptera: Chrysomelidae) populations. *J. Econ. Entomol. 93*, 447–458.

Oppenoorth, F.J., 1985. Biochemistry and genetics of insecticide resistance. In: Kerkut, G.A., Gilbert, L.I. (Eds.), Comprehensive Insect Physiology, Biochemistry, and Pharmacology, vol. 12. Pergamon, Oxford, pp. 731–773.

Orr, N., Shaffner, J., Watson, G.B., 1997. Pharmacological characterization of an epibatidine binding site in the nerve cord of *Periplaneta americana*. *Pestic. Biochem. Physiol. 58*, 183–192.

Palumbo, J.C., Horowitz, A.R., Prabhaker, N., 2001. Insecticidal control and resistance management for *Bemisia tabaci*. *Crop Protect. 20*, 739–765.

Perdew, J.P., 1986. Density functional approximation for the correlation energy of the inhomogeneous electron gas. *Phys. Rev. B 33*, 8822–8824.

Perring, T.M., 2001. The *Bemisia tabaci* species complex. *Crop Protect. 20*, 725–737.

Pflüger, W., Schmuck, R., 1991. Ecotoxicological profile of imidacloprid. *Pflanzenschutz-Nachrichten Bayer 44*, 145–158.

Pospischil, R., Schneider, U., Böcker, T., Junkersdorf, J., Nentwig, G., *et al.*, 1999. Efficacy of imidacloprid for cockroach control in a gel bait formulation. *Pflanzenschutz-Nachrichten Bayer 52*, 386–400.

Prabhaker, N., Toscano, N.C., Castle, S.J., Henneberry, T.J., 1997. Selection for imidacloprid resistance in silverleaf whiteflies from the Imperial Valley and development of a hydroponic bioassay for resistance monitoring. *Pestic. Sci. 51*, 419–428.

Rajappa, S., 1981. Nitroenamines: preparation, structure, and synthetic potential. *Tetrahedron Lett. 37*, 1453–1480.

Rauch, N., Nauen, R., 2003. Biochemical markers linked to neonicotinoid cross-resistance in *Bemisia tabaci* (Hemiptera: Aleyrodidae). *Arch. Insect Biochem. Physiol. 54*, 165–176.

Reinemeyer, C.R., Charles, S., 2003. Evaluation of the efficacy of a combination of imidacloprid and moxidectin against immature *Toxocara cati* in cats. *Parasitol. Res. 90*, 140–141.

Restifo, L.L., White, K., 1990. Molecular and genetic approaches to neurotransmitter and neuromodulator systems in *Drosophila*. *Adv. Insect. Physiol. 22*, 115–219.

Roberts, T.R., Hutson, D.H. (Eds.), 1999. Metabolic Pathways of Agrochemicals, Part 2, Insecticides and Fungicides. Cambridge University Press, Cambridge.

Roslavtseva, S.A., 2000. Neonicotinoids as a new promising group of insecticides. *Agrokhimiya 1*, 49–52.

Rust, J.H., Jr., Cavanaugh, D.C., O'Shita, R., Marshall, J.D., Jr., 1971. The role of domestic animals in the epidemiology of plague. 1. Experimental infection of dogs and cats. *J. Infect. Dis. 124*, 522–526.

Rust, M.K., Dryden, M.W., 1997. The biology, ecology, and management of the cat flea. *Annu. Rev. Entomol. 42*, 451–473.

Rust, M.K., Hinkle, N.C., Waggoner, M., Mencke, N., Hansen, O., *et al.*, 2001. The influence of imidacloprid on adult cat flea feeding. *Suppl. Comp. Cont. Educ. Vet. Pract. 23*, 18–21.

Rust, M.K., Waggoner, M., Hinkle, N.C., Mencke, N., Hansen, O., *et al.*, 2002. Disruption of adult cat flea feeding by imidacloprid. *Suppl. Comp. Cont. Educ. Vet. Pract. 24*, 4–5.

Rust, M.K., Waggoner, M.M., Hinkle, N.C., Stansfield, D., Barnett, S., 2003. Efficacy and longevity of nitenpyram against adult cat fleas (Siphonaptera: Pulicidae). *J. Med. Entomol. 40*, 678–681.

Sarkar, M.A., Roy, S., Kole, R.K., Chowdhury, A., 2001. Persistence and metabolism of imidacloprid in different soils of West Bengal. *Pest Mgt Sci. 57*, 598–602.

Sattelle, D.B., Buckingham, S.D., Wafford, K.A., Sherby, S.M., Barkry, N.M., *et al.*, 1989. Actions of the insecticide 2(nitromethylene)tetrahydro-1,3-thiazine on insect and vertebrate nicotinic acetylcholine recptors. *Proc. Roy. Soc. B 237*, 501–514.

Sattelle, D.B., Harrow, I.D., Hue, B., Pelhate, M., Gepner, J.I., *et al.*, 1983. α-Bungarotoxin blocks excitatory synaptic transmission between cercal sensory neurons and giant interneurons. *J. Exp. Biol. 107*, 473–489.

Schmuck, R., 2001. Ecotoxicological profile of the insecticide thiacloprid. *Pflanzenschutz-Nachrichten Bayer 54*, 161–184.

Schröder, M.E., Flattum, R.F., 1984. The mode of action and neurotoxic properties of the nitromethylene heterocycle insecticides. *Pestic. Biochem. Physiol. 22*, 148–160.

Schulz, R., Sawruk, E., Mülhardt, C., Bertrand, S., Baumann, A., *et al.*, 1998. Dα3, a new functional α subunit of nicotinic acetylcholine receptors from *Drosophila*. *J. Neurochem. 71*, 853–862.

Schulz-Jander, D.A., Casida, J.E., 2002. Imidacloprid insecticide metabolism: human cytochrome P-450 isozymes differ in selectivity for imidazolidine oxidation versus nitroimine reduction. *Toxicol. Lett. 132*, 65–70.

Schwartz, B.J., Sparrow, F.K., Heard, N.E., Thede, B.M., 2000. Simultaneous derivatization and trapping of volatile products from aqueous photolysis of thiamethoxam insecticide. *J. Agric. Food Chem. 48*, 4671–4675.

Scott, J.G., 2001. Cytochrome P-450 monooxygenases and insecticide resistance: lessons from CYP6D1. In: Ishaaya, I. (Ed.), Biochemical Sites Important in Insecticide Action and Resistance. Springer, Berlin, pp. 255–267.

Senn, R., Hofer, D., Hoppe, T., Angst, M., Wyss, P., *et al.*, 1998. CGA 293343: a novel broad-spectrum insecticide supporting sustainable agriculture worlwide. In: Proc. Brighton Crop Protection Conference: Pests and Diseases, vol. 1. pp. 27–36.

Sgard, F., Fraser, S.P., Katkowska, M.J., Djamgoz, M.B.A., Dunbar, S.J., *et al.*, 1998. Cloning and functional characterization of two novel nicotinic acetylcholine receptor α subunits from the insect pest *Myzus persicae*. *J. Neurochem. 71*, 903–912.

Shiokawa, K., Tsuboi, S., Iwaya, K., Moriya, K., **1994**. Development of a chloronicotinyl insecticide, imidacloprid. *J. Pestic. Sci. 19*, 209–217; 329–332.

Shiokawa, K., Tsuboi, S., Moriya, K., Kagabu, S., **1995**. Chloronicotinyl insecticides: development of imidacloprid. In: 8th Int. Congr. Pestic. Chemi.: Option 2000, pp. 49–59.

Siegfried, B.D., Scharf, M.E., **2001**. Mechanisms of organophosphate resistance in insects. In: Ishaaya, I. (Ed.), Biochemical Sites Important in Insecticide Action and Resistance. Springer, Berlin, pp. 269–291.

Soderlund, D.M., **1997**. Molecular mechanisms of insecticide resistance. In: Sjut, V. (Ed.), Molecular Mechanisms of Resistance to Agrochemicals. Springer, Berlin, pp. 21–56.

Soloway, S.B., Henry, A.C., Kollmeyer, W.D., Padgett, W.M., Powell, J.E., *et al.*, **1978**. Nitromethylene heterocycles as insecticides. In: Shankland, D.L., Hollingworth, R.M., Smyth, T., Jr. (Eds.), Pesticide and Venom Neurotoxicology. Plenum, New York, pp. 153–158.

Soloway, S.B., Henry, A.C., Kollmeyer, W.D., Padgett, W.M., Powell, J.E., *et al.*, **1979**. Nitromethylene insecticides. *Adv. Pestic. Sci. 2*, 206–217.

Sone, S., Hattori, Y., Tsuboi, S., Otsu, Y., **1997**. Difference in suceptibility to imidacloprid of the population of the small brown planthopper, *Laodelphax striatellus* Fallen, from various localities in Japan. *J. Pestic. Sci. 20*, 541–543.

Sone, S., Yamada, Y., Tsuboi, S., **1995**. New pesticide screening method for leaf hoppers and plant hoppers using rice seedlings. *Jap. J. Appl. Entomol. Zool. 39*, 171–173.

Spencer, J.A., Butler, J.M., Stafford, K.C., Pough, M.B., Levy, S.A., *et al.*, **2003**. Evaluation of permethrin and imidacloprid for prevention of *Borrelia burgdorferi* transmission from blacklegged ticks (*Ixodes scapularis*) to *Borrelia burgdorferi*-free dogs. *Parasitol. Res. 90*, 106–107.

Stein-Dönecke, U., Führ, F., Wieneke, J., Hartwig, J., Leicht, W., **1992**. Influence of soil moisture on the formulation of dressing zones and uptake of imidacloprid after seed treatment of winter wheat. *Pflanzenschutz-Nachrichten Bayer 45*, 327–368.

Stupp, H.-P., Fahl, U., **2003**. Environmental fate of clothianidin (TI-435; Poncho®). *Pflanzenschutz-Nachrichten Bayer 56*, 59–74.

Sukekawa, M., Nakayama, A., **1999**. Application of molecular similarity analysis in 3D-QSAR of neonicotinoid insecticides. *J. Pestic. Sci. 24*, 38–43.

Tabuchi, T., Fusaka, T., Iwanaga, K., **1994**. Studies on acyclic nitroethene compounds. 3. Synthesis and insecticidal activity of acyclic nitroethene compounds containing (6-substituted)-3-pyridylamino group. *J. Pestic. Sci. 19*, 119–125.

Takahashi, H., Mitsui, J., Takakusa, N., Matsuda, M., Yoneda, H., *et al.*, **1992**. NI-25, a new type of systemic and broad sprectrum insecticide. In: Proc. Brighton Crop Protection Conference: Pests and Diseases, vol. 1. pp. 89–96.

Takahashi, H., Takakusa, N., Suzuki, J., Kishimoto, T., **1998**. Development of a new insecticide, acetamiprid. In: Abstracts 23rd Ann. Mtg Pestic. Sci. Soc. Japan, p. 23.

Testa, B., Mayer, J.M., **1998**. Design of intramolecularly activated prodrugs. *Drug Metab. Rev. 30*, 787–807.

Testa, B., Mayer, J.M., **2001**. Concepts in prodrug design to overcome pharmacokinetic problems. In: Testa, B., Waterbeemd, H. van de, Folkers, G., Guy, R. (Eds.), Pharmacokinetic Optimization in Drug Research: Biological, Physicochemical, and Computational Strategies. Verlag Helvetica Chimica Acta, Zurich, Switzerland, pp. 85–95.

Tokieda, M., Ozawa, M., Kobayashi, S., Gomyo, T., **1997**. Method for determination of total residues of the insecticide acetamiprid and its metabolites in crops by gas chromatography. *J. Pestic. Sci. 22*, 77–83.

Tokieda, M., Ozawa, M., Kobayashi, S., Gomyo, T., **1998**. In: Proc. 9th Int. Congr. Pestic. Chem., 2, Poster 7C-029.

Tomizawa, M., **2000**. Insect nicotinic acetylcholine receptors: mode of action of insecticide and functional architecture of the receptor. *Jap. J. Appl. Entomol. Zool. 44*, 1–15.

Tomizawa, M., Casida, J.E., **1997**. [^{125}I]azidoneonicotinoid photoaffinity labeling of insecticide-binding subunit of *Drosophila* nicotinic acetylcholine receptor. *Neurosci. Lett. 237*, 61–64.

Tomizawa, M., Casida, J.E., **2001**. Structure and diversity of nicotinic acetylcholine receptors. *Pest Mgt Sci. 57*, 914–922.

Tomizawa, M., Casida, J.E., **2003**. Selective toxicity of neonicotinoids attributable to specificity of insect and mammalian nicotinic receptors. *Annu. Rev. Entomol. 48*, 339–364.

Tomizawa, M., Latli, B., Casida, J.E., **1996**. Novel neonicotinoid–agarose affinity column for *Drosophila* and *Musca* nicotinic acetylcholine receptors. *J. Neurochem. 67*, 1669–1676.

Tomizawa, M., Latli, B., Casida, J.E., **1999**. Structure and function of insect nicotinic acetylcholine receptors studied with nicotinoid insecticide affinity probes. In: Yamamoto, I., Casida, J.E. (Eds.), Nicotinoid Insecticides and Nicotinic Acetylcholine Receptor. Springer, New York, pp. 271–292.

Tomizawa, M., Lee, D.L., Casida, J.E., **2000**. Neonicotinoid insecticides: molecular feature conferring selectivity for insect versus mammalian nicotinic receptors. *J. Agric. Food Chem. 48*, 6016–6024.

Tomizawa, M., Yamamoto, I., **1992**. Binding of nicotinoids and the related compounds to the insect nicotinic acetylcholine receptor. *J. Pestic. Sci. 17*, 231–236.

Tomizawa, M., Yamamoto, I., **1993**. Structure activity relationships of neonicotinoids and imidacloprid analogs. *J. Pestic. Sci. 18*, 91–98.

Tomlin, C.D.S., **2000**. The pesticide Manual, 12th edn. British Crop Protection Council.

Tröltzsch, C.M., Führ, F., Wieneke, J., Elbert, A., **1994**. Einfluss unterschiedlicher Bewässerungsverfahren auf die aufnahme von Imidacloprid durch Baumwolle nach Saatgutbeizung. *Pflanzenschutz-Nachrichten Bayer 47*, 249–303.

Tsumura, Y., Nakamura, Y., Tonagai, Y., Kamimoto, Y., Tanaka, Y., *et al.*, **1998**. Determination of neonicotinoid pesticide nitenpyram and its metabolites in agricultural products. *Shokuhin Eiseigaku Zasshi 39*, 127–134.

Uneme, H., Iwanaga, K., Higuchi, N., Kando, Y., Okauchi, T., *et al.*, **1999**. Syntheses and insecticidal activity of nitroguanidine derivatives. *Pestic. Sci. 55*, 202–205.

van den Breukel, I., van Kleef, R.G.D.M., Zwart, R., Oortgiesen, M., **1998**. Physiostigmine and acetylcholine differentially activate nicotinic receptor subpopulations in *Locusta migratoria* neurons. *Brain Res. 789*, 263–273.

Wakita, T., Kinoshita, K., Yamada, E., Yasui, N., Kawahara, N., *et al.*, **2003**. The discovery of dinotefuran: a novel neonicotinoid. *Pest Mgt Sci. 59*, 1016–1022.

Weichel, L., Nauen, R., **2003**. Monitoring of insecticide resistance in damson hop aphids, *Phorodon humuli* Schrank (Hemiptera: Aphididae) from German hop gardens. *Pest Mgt Sci. 59*, 991–998.

Werner, G., Hopkins, T., Shmidl, J.A., Watanabe, M., Krieger, K., **1995**. Imidacloprid, a novel compound of the chloronicotinyl group with outstanding insecticidal activity in the on-animal treatment of pests. *Pharmacol. Res. 31* (Suppl.), 136.

Wen, Z.M., Scott, J.G., **1997**. Cross-resistance to imidacloprid in strains of German cockroach (*Blattella germanica*) and housefly (*Musca domestica*). *Pestic. Sci. 49*, 367–371.

Widmer, H., Steinemann, A., Maienfisch, P., **1999**. Chemical and physical properties of thiamethoxam (CGA 293343). In: Abstracts 218th Am. Chem. Sci. Natl Mtg, p. 134.

Williams, B., **1993**. Reproductive success of cat fleas, *Ctenocephalides felis*, on calves as unusual host. *Med. Vet. Entomol. 7*, 94–98.

Williamson, M.S., Denholm, I., Bell, C.A., Devonshire, A.L., **1993**. Knockdown resistance (kdr) to DDT and pyrethroid insecticides maps to a sodium channel gene locus in the housefly (*Musca domestica*). *Mol. Gen. Genet. 240*, 17–22.

Williamson, M.S., Martinez-Torres, D., Hick, C.A., Devonshire, A.L., **1996**. Identification of mutations in the housefly *para*-type sodium channel gene associated with knock-down resistance (kdr) to pyrethroid insecticides. *Mol. Gen. Genet. 252*, 51–60.

Wollweber, D., Tietjen, K., **1999**. Chloronicotinyl insecticides: a success of the new chemistry. In: Yamamoto, I., Casida, J.E. (Eds.), Neonicotinoid Insecticides and the Nicotinic Acetylcholine Receptor. Springer, New York, pp. 109–125.

Yaguchi, Y., Sato, T., **2001**. Thiacloprid (Bariard), a novel neonicotinoid insecticide for foliar application. *Agrochem. Japan 79*, 14–16.

Yamada, T., Takahashi, H., Hatano, R., **1999**. A novel insecticide, acetamiprid. In: Yamamoto, I., Casida, J.E. (Eds.), Neonicotinoid Insecticides and Nicotinic Acetylcholine Receptor. Springer, New York, pp. 149–175.

Yamamoto, I., **1996**. Neonicotinoids: mode of action and selectivity. *Agrochem. Japan 68*, 14–15.

Yamamoto, I., **1998**. Nicotine: old and new topics. *Rev. Toxicol. 2*, 61–69.

Yamamoto, I., **1999**. Nicotine to neonicotinoids: 162 to 1997. In: Yamamoto, I., Casida, J.E. (Eds.), Neonicotinoid Insecticides and the Nicotinic Acetylcholine Receptor. Springer, New York, pp. 3–27.

Yamamoto, I., Tomizawa, M., **1993**. New development of nicotinoid insecticides. In: Mitsui, T., Matsumura, F., Yamaguchi, I. (Eds.), Pesticide Development: Molecular Biological Approaches. Pesticide Sci. Soc. Japan, Tokyo, pp. 67–83.

Yamamoto, I., Tomizawa, M., Saito, T., Miyamoto, T., Walcott, E.C., *et al.*, **1998**. Structural factors contributing to insecticidal and selective actions of neonicotinoids. *Arch. Insect. Biochem. Physiol. 37*, 24–32.

Yamamoto, I., Yabuta, G., Tomizawa, M., Saito, T., Miyamoto, T., *et al.*, **1995**. Molecular mechanism for selective toxicity of nicotinoids and neonicotinoids. *J. Pestic. Sci. 20*, 33–40.

Yokota, T., Mikata, K., Nagasaki, H., Ohta, K., **2003**. Absorption, tissue distribution, excretion, and metabolism of clothianidin in rats. *J. Agric. Food Chem. 51*, 7066–7072.

Young, D.R., Arther, R.G., Davis, W.L., **2003**. Evaluation of K9 Advantix™ vs. Frontline Plus® topical treatments to repel brown dog ticks (*Rhipicephalus sanguineus*) on dogs. *Parasitol. Res. 90*, 116–118.

Young, D.R., Ryan, W.G., **1999**. Comparison of Frontline® Top Spot™, Preventic® collar alone or combined with Advantage® in control of flea and tick infestations in water immersed dogs. In: Proc. 5th Int. Symp. Ectoparas. Pets, Fort Collins, Colorado, April, p. 24.

Zewen, L., Zhaojun, H., Yinchang, W., Lingchun, Z., Hongwei, Z., *et al.*, **2003**. Selection for imidacloprid resistance in *Nilaparvata lugens*: cross-resistance patterns and possible mechanisms. *Pest Mgt Sci. 59*, 1355–1359.

Zhang, A., Kayser, H., Maienfisch, P., Casida, J.E., **2000**. Insect nicotinic acetylcholine receptor: conserved neonicotinoid specifity of [³H]imidacloprid binding site. *J. Neurochem. 75*, 1294–1303.

Zhao, J.Y., Liu, W., Brown, J.M., Knowles, C.O., **1995**. Insecticide resistance in field and laboratory strains of Western flower thrips (Thysanoptera: Thripidae). *J. Econ. Entomol. 88*, 1164–1170.

Zhao, J.Z., Bishop, B.A., Graphius, E.J., **2000**. Inheritance and synergism of resistance to imidacloprid in the Colorado potato beetle (Coleoptera: Chrysomelidae). *J. Econ. Entomol. 93*, 1508–1514.

Zhu, K.Y., Clark, J.M., **1997**. Validation of a point mutation of acetylcholinesterase in Colorado potato beetle by polymerase chain reaction coupled to enzyme inhibition assay. *Pestic. Biochem. Physiol. 57*, 28–35.

Zwart, R., Oortgiesen, M., Vijverberg, H.P., **1992**. The nitromethylene heterocycle 1-(pyridin-3-yl-methyl)-2-nitromethylene-imidazolidine distinguishes mammalian from insect nicotinic receptor subtypes. *Eur. J. Pharmacol. 228*, 165–169.

A3 Addendum: The Neonicotinoid Insecticides

P Jeschke, Bayer CropScience AG, Monheim am
Rhein, Germany
R Nauen, Bayer CropScience AG, Monheim am
Rhein, Germany

© 2010 Elsevier B.V. All Rights Reserved

During the last 5 years neonicotinoid insecticides
and their biochemical target-site, the insect nicotinic
acetylcholine receptor (nAChR), have been described in several excellent reviews, which should
serve as additional sources for information
(Tomizawa et al., 2005; Thany et al., 2007; Jeschke,
2007a, b; Jeschke and Nauen, 2008; Elbert et al.,
2008). These reviews and other numerous articles
describe the current knowledge of seven commercially marketed active ingredients such as:
(1) the ring systems containing neonicotinoids
like imidacloprid (Hopkins et al., 2005; Thielert,
2006), thiacloprid (Jeschke and Nauen, 2008) and
thiamethoxam (Maienfisch, 2007), as well as (2)
neonicotinoids having non-cyclic structures like
acetamiprid, nitenpyram, clothianidin, and dinotefuran (Jeschke and Nauen, 2008; Wakita, 2008). In
addition, transformation of neonicotinoids (1) into
(2), exemplified by a partial cleavage of thiamethoxam into clothianidin in-vivo in insects and
plant tissues, was further discussed (Kayser et al.,
2007; Jeschke and Nauen, 2007).

There has been rapid progress in research on
nAChRs. In 2001, the crystal structures of the first
soluble homopentameric acetylcholine binding protein (AChBP) was resolved (Brejc et al., 2001; Smit
et al., 2001). This AChBP subtype is secreted by glia
cells of a mollusk, the freshwater snail *Lymnaea
stagnalis* (L-AChBP). Some years later a second
AChBP subtype (A-AChBP), isolated from the
saltwater mollusk *Aplysia californica*, was characterized. A-AChBP shares only 33% amino acid identity with L-AChBP, but it possesses all the functional
residues identified in L-AChBP (Hansen et al., 2004;
Celie et al., 2005). Whereas A-AChBP has a similar
high sensitivity for both electronegative neonicotinoids and cationic nicotinoids and the L-AChBP
demonstrates lower neonicotinoid and higher
nicotinoid sensitivities (Tomizawa et al., 2008;
Tomizawa and Casida, 2009). Thus, these two
AChBPs from mollusks have distinct pharmacology

suggestive of the nAChRs from species as divergent
as mammals and insects (Tomizawa et al., 2008).

Furthermore, a refined 4 Å resolution electron
microscopy structure of the heteropentameric verte-
brate muscle type $(\alpha1)_2\beta\gamma\delta$ nAChR showed considerable structural similarity to the L-AChBP ligand
binding domain (LBD) (Unwin, 2005). Because of
this similarity, L-AChBP is now considered a structural and functional surrogate for the extracellular
LBD of insect nAChRs (Talley et al., 2008). Therefore, the ACh binding site of the crystalline structures of AChBPs can be used as an example for the
N-terminal domain of an α-subunit of insect
nAChRs as a template for docking simulations
(CoMFA, 3D QSAR) of ACh ligands such as neonicotinoid insecticides by so-called manual or automated modeling methods.

Because of the identical pharmacological profile of
AChBPs for neonicotinoids and nicotinoid radioligands (Tomizawa et al., 2007a, b), a deeper insight
into molecular determinants of neonicotinoid agonists interacting with their respective binding sites
on insect nAChRs is possible (Gao et al., 2006). It
was found that the chlorine atom in 6-chloropyrid-3-
ylmethyl (CPM) of imidacloprid and thiacloprid contacts isoleucine (Ile) 106/methionine (Met) 116 and
the pyridine nitrogen is directed to Ile 118/tryptophan
(Trp) 147 via a solvent bridge (Casida and Tomizawa,
2008). The guanidine/amidine plane undergoes
π-stacking with tyrosine (Tyr) 188 and the pharmacophore group (e.g., [=N–NO$_2$], [=CH–NO$_2$] or
[=N–CN]) interacts with serine (Ser) 189/cysteine
(Cys) 190, thereby defining the binding pocket that
models AChBPs and possibly insect nAChRs potency
and selectivity. Characteristic of several agonists such
as imidacloprid and thiacloprid, loop C largely envel-
ops the ligand, positioning aromatic side chains to
interact optimally with conjugated and hydrophobic
regions of the neonicotinoid insecticides, which is
consistent with the results of solution-based photoaffinity labelings (Talley et al., 2008).

Today, photoaffinity labeling (e.g., with optimized 5-azido-6-chloropyridin-3-ylmethyl photoaffinity probes) combined with mass spectrometry (MS) provides a direct and physiologically relevant chemical biology method for 3D structural investigations of ligand–receptor interactions (Tomizawa *et al.*, 2007a, b). This is an important approach for studying subtype-selective agonists (Tomizawa *et al.*, 2008) and mapping the elusive neonicotinoid binding site (Tomizawa *et al.*, 2007b). The results were used to establish structural models of the two AChBP subtypes. In A-AChBP, neonicotinoids and nicotinoids are nestled in similar bound conformations (Tomizawa *et al.*, 2008). Surprisingly, for L-AChBP, the neonicotinoid insecticides have two bound conformations that are inversely relative to each other. Accordingly, the subtype selectivity is based on two disparate bound conformations of nicotinic agonists (Tomizawa *et al.*, 2008). Recently, the nicotinic agonist interactions with A-AChBP have been precisely defined by scanning 17 Met and Tyr mutants within the binding site by photoaffinity labeling with 5-azido-6-chloropyridin-3-ylmethyl probes that have similar affinities to their non-azido counterparts (Tomizawa *et al.*, 2009).

On the other hand, agonist actions of commercial neonicotinoid insecticides were studied by single electrode voltage-clamp electrophysiology on *n*AChRs expressed by neurons isolated from thoracic ganglia of the American cockroach, *Periplaneta americana* (Ihara *et al.*, 2006; Tan *et al.*, 2007). Based on maximal inward currents, neonicotinoid insecticides could be divided into the two subgroups defined above: (1) ring systems containing neonicotinoids and *(S)*-nicotine were relatively weak partial agonists causing only 20–25% of the maximal ACh current and (2) neonicotinoids having non-cyclic structures were much more effective agonists producing 60–100% of the maximal ACh current (Miyagi *et al.*, 2006; Ihara *et al.*, 2006; Tan *et al.*, 2007). Thiamethoxam, even at 100 µM, failed to cause an inward current and showed no competitive interaction with other neonicotinoids on *n*AChRs, indicating that it is not a direct-acting agonist or antagonist (Tan *et al.*, 2007).

In addition, in most insect non-α subunits, lysine (Lys) or arginine (Arg) moieties are found. These basic residues may interact with the *N*-nitro group of neonicotinoid insecticides through electrostatic forces and *H*-bonding, strengthening the *n*AChR–insecticide interaction (Shimomura *et al.*, 2006; Wang *et al.*, 2007). On the other hand, high-resolution crystal structures of L-AChBP with neonicotinoid insecticides such as imidacloprid and clothianidin suggested that the guanidine moiety in

both stacks with Tyr 185, while the *N*-nitro group of imidacloprid but not of clothianidin makes a *H*-bond with Gln 55. The *H*-bond of NH at position 1 with the backbone carbonyl group of Trp 143, offers for clothianidin an explanation for the diverse actions of neonicotinoids on insect *n*AChRs (Ihara *et al.*, 2008).

Since 2004, several research groups have published further evidence related to the submolecular basis for the mechanism of target-site selectivity of commercial neoncotinoids for insect *n*AChRs (e.g., structural features of agonist binding loops C and D; Toshima *et al.*, 2009) over vertebrate *n*AChRs, based on the *n*AChR subunit composition (Tomizawa and Casida, 2005; Matsuda *et al.*, 2005). The activity of neonicotinoids on wild-type and mutant α7 nicotinic receptors was also investigated using voltage-clamp electrophysiology. It was found, that when neonicotinoids bind to the receptor, the *N*-nitro group is located close to loops D and F, which was supported by the models of the agonist binding domain of the α7 nicotinic receptor (Shimomura *et al.*, 2006). Recently, similar electrophysiological studies on native *n*AChRs and on wild-type and mutagenized recombinant *n*AChRs have shown that basic residues particular to loop D of insect *n*AChRs are likely to interact electrostatically with the *N*-nitro group of neonicotinoid insecticides (Matsuda *et al.*, 2009).

Ongoing design of active novel ingredients and their optimization has to consider in its process conformational transitions of *n*AChRs (Jeschke, 2007b). Molecular interactions of neonicotinoid insecticides containing different pharmacophore variants with *n*AChRs have been mapped by chemical and structural neurobiological approaches, thereby encouraging the biorational and receptor structure-guided design of novel nicotinic ligands (Ohno *et al.*, 2009). Recently, replacement of the nitromethylene pharmacophore with nitro-conjugated systems was described (Jeschke, 2007b; Shao *et al.*, 2009). The methyl [1-(2-trifluoromethyl-pyridin-5-yl)ethyl]-*N*-cyano-sulfoximine (Sulfoxaflor; common name ISO-proposed) was prepared by Dow AgroSciences, which has a new *N*-cyano-sulfoximine [–S(O)=N–CN] pharmacophore variant (Loso *et al.*, 2007).

After 16 years of use, insects pests such as whiteflies *Bemisia tabaci* (Gennadius) and *Trialeurodes vaporariorum* (Westwood) (Gorman *et al.*, 2007), the brown planthopper *Nilaparvata lugens* (Stål) (Nauen and Denholm, 2005; Gorman *et al.*, 2008), the Colorado potato beetle *Leptinotarsa decemlineata* (Say) (Nauen and Denholm, 2005), and a few others like the mango leafhopper *Idioscopus clypealis* (Lethierry) have developed resistance to neonicotinoids in some parts of the world (Elbert et al., 2008). A laboratory-selected nAChR subunit Nlα1 point mutation (Y151S) observed is most likely responsible for conferring target-site resistance to neonicotinoid insecticides in the brown planthopper N. lugens (Liu et al., 2007). The Y151S mutation has a significantly reduced effect on neonicotinoid agonist activity when the subunit Nlα1 is coassembled with Nlα2 than when expressed as the sole α-subunit in a heteromeric nAChR (Liu et al., 2009). A mutation at this site (Y151M) when introduced into Nlα1 and coexpressed with rat β2 in Xenopus oocytes also showed lower agonist activity (Zhang et al., 2008). In this context, the possible role of the subunit Nlβ1 in neonicotinoid sensitivity was investigated by A-to-I RNA editing as well (Yao et al., 2009).

One area of major concern over the last decade is resistance development in B. tabaci (Nauen and Denholm, 2005). Owing to the obvious advantage of B. tabaci Q-biotypes in neonicotinoid use environments, resistance started to spread all over the world and is now no longer restricted to intense European cropping systems such as in southern Spain. Cross-resistance studies with both acetamiprid and thiamethoxam revealed that the strain that had been laboratory selected with thiamethoxam for 12 generations exhibited almost no cross-resistance to acetamiprid (original strain collected in the Ayalon Valley late in the cotton-growing season 2002), whereas the acetamiprid-selected strain exhibited high cross-resistance of >500-fold to thiamethoxam (Horowitz et al., 2004). A possible explanation for the lack of cross-resistance to acetamiprid in thiamethoxam-selected whiteflies is selection for different resistant traits (Horowitz et al., 2004). Resistance in thiamethoxam-selected whiteflies might be associated with an activation mechanism, as it has been shown that thiamethoxam is most probably a pro-drug easily converted to clothianidin (Jeschke and Nauen, 2007), whereas in acetamiprid-selected whiteflies the activated compound itself (clothianidin) is the primary target for detoxification and results in broad cross-resistance to all neonicotinoids (Horowitz et al., 2004).

One interesting finding is that resistance to imidacloprid is age specific in both B- and Q-type strains of B. tabaci (Nauen et al., 2008). The authors showed that in contrast to adults exhibiting high levels of resistance to imidacloprid, young nymphs are still susceptible. The highest observed resistance ratio at LD_{50} expressed in prepupal nymphs was 13, compared with atleast 580 in their adult counterparts (Nauen et al., 2008). This has considerable implications for resistance management strategies, since targeted nymphs are still well controlled and

selection pressure is released from adults. Neonicotinoid resistance in B. tabaci is most likely mediated by CYP6CM1(vQ), a cytochrome P_{450} monooxygenase described to be highly over-expressed in female adults (Karunker et al., 2008), and shown to hydroxylate imidacloprid at the 5-position of the imidazolidine ring system when heterologously expressed in Escherichia coli (Karunker et al., 2009). Another whitefly species reported to have developed resistance to neonicotinoids is the greenhouse whitefly, Trialeurodes vaporariorum (Gorman et al., 2007; Karatolos et al., 2009). Similar to B. tabaci, this species also shows cross-resistance to pymetrozine (Karatolos et al., 2009).

Furthermore, field populations of brown planthoppers, N. lugens and Sogatella furciera Horváth, resistant to neonicotinoids have been described in East and South-East Asia (Gorman et al., 2008; Matsumura et al., 2008). Mechanistic studies showed a clear correlation between resistance ratios to imidacloprid and the extent of O-deethylation of ethoxycoumarin, indicating that resistance is likely conferred by monooxygenases similar to that observed in whiteflies (Puinean et al., in press). Finally, target-site resistance due to mutations in nAChR subunits has been suggested to occur in the Colorado potato beetle L. decemlineata (Tan et al., 2008). More information on general aspects of insecticide resistance can be found on the website of the Insecticide Resistance Action Committee (IRAC, www.irac-online.org; McCaffery and Nauen, 2006).

The increasing success of neonicotinoids as an insecticide class also relies on a high degree of versatile application methods (foliar, seed, or soil treatment), not seen to the same extent in other chemical classes (Elbert et al., 2008). New formulations have been developed for neonicotinoid insecticides (examples from Bayer CropScience) to optimize their bioavailability through improved rain fastness, better retention, and spreading of the spray deposit on the leaf surface, combined with higher leaf penetration through the cuticle and translocation within the plant (Baur et al., 2007; Elbert et al., 2008). The new formulation technology O-TEQ® (oil dispersion, OD) for foliar application was developed for imidacloprid (Confidor®) and thiacloprid (Calypso®) (Baur et al., 2007). The O-TEQ® formulations facilitate leaf penetration, particularly under suboptimal conditions for foliar uptake.

Combined formulations of neonicotinoids with pyrethroids such as Confidor S® (imidacloprid and cyfluthrin for control of tobacco pests in South America), Muralla® (imidacloprid and deltamethrin for vegetable and rice in Central America and Chile), or Connect® (imidacloprid and β-cyfluthrin for control of stinkbugs in soybean) and other insecticides are also being developed with an aim of broadening the insecticidal spectrum of neonicotinoids and to replace WHO Class I products from older chemical classes (Elbert *et al.*, 2008). In addition to the insecticidal properties of imidacloprid, a stress shield mode of action was identified. It supports plants in moderating the effects of abiotic and biotic stress. Therefore, Trimax® (Hopkins *et al.*, 2005), an optimized imidacloprid formulation was developed, which moderates water stress in plants with an average lint yield increase in cotton of 10% (Thielert, 2006; Gonias *et al.*, 2008).

In addition to crop protection, applications of neonicotinoid insecticides in non-agricultural fields is also expanding in the last 5 years, e.g., household sectors, lawn and garden, and for controlling ectoparasites in animal health (Rust, 2005). New bait gel products containing imidacloprid like Maxforce Prime® (Bayer EnvironmentalSciences), for the control of cockroaches, and thiamethoxam like OptigardTM (Syngenta), for broad-spectrum control of ants, are now on the market. Other products include Agita® 10 WG (Novartis Animal Health), a water-soluble granule containing thiamethoxam and the fly pheromone *(Z)*-9-tricosen for use against the house fly *Musca domestica* and against synanthropic flies (Nurita *et al.*, 2008).

In the field of animal health, AdvantageMulti™ (or Advocate®; Bayer Animal Health) is a spot-on formulation of the macrolactone moxidectin and imidacloprid which shows efficacy against ear mites (*Otodectes cynotis*) and fleas (*Ctenocephalides felis*) (Wenzel *et al.*, 2008). K9 Advantix® (Bayer Animal Health) is a spot-on topical solution of imidacloprid and permethrin for dogs, that work synergistically against the most common and important external parasites such as the long star tick, like *Amblyomma americanum*. Finally, Vectra 3D™ (Summit Vet-Pharm) is a new topical spot-on ectoparasiticide containing a combination of dinotefuran, permethrin, and pyriproxyfen against *A. americanum* and the Gulf Coast tick *A. maculatum* on dogs (Coyne, 2009).

Over the last 5 years, sales of the neonicotinoids have nearly trebled. Future expansion will be driven by growth of the established commercial neonicotinoids. The chemical class will further benefit from organophosphate and carbamate replacements. Generic competition will lead to price erosions, which also will open new opportunities for neonicotinoids in low-price markets (Elbert *et al.*, 2008). Combined with active life-cycle management such as optimized formulations and new combinations, neonicotinoids will be the most important chemical class in the next few years in modern crop protection.

References

Baur, P., Arnold, R., Giessler, S., Mansour, P., Vermeer, R., 2007. Bioavailability of insecticides from O-TEQ® formulations: overcoming barriers for systemic active ingredients. *Pflanzenschutz-Nachrichten Bayer (English edition) 60*, 27–42.

Brejc, K., van Dijk, W.J., Klaassen, R.V., Schurmans, M., van der Oost, J., Smit, A.B., Sixma, T.K., 2001. Crystal structure of an ACh-binding protein reveals the ligand-binding domain of nicotinic receptors. *Nature 411*, 269–276.

Casida, J.E., Tomizawa, M., 2008. Insecticide interactions with γ-aminobutyric acid and nicotinic receptors: predictive aspects of structural models. *J. Pesticide Sci. 33*, 4–8.

Celie, P.H., Kasheverow, I.E., Mordvintsev, D.Y., Hogg, R.C., van Nierop, P., van Elk, R., van Rossum-Fikkert, S.E., Zhamak, M.N., Bertrand, D., Tsetlin, V., Sixma, T.K., Smit, A.B., 2005. Crystal structure of nicotinic acetylcholine receptor homologue AChBP in complex with an α-conotoxin PnIA variant. *Nat. Struct. Mol. Biol. 12*, 1–7.

Coyne, M.J., 2009. Efficacy of a topical ectoparasiticide containing dinotefuran, pyriproxyfen, and permethrin against *Amblyomma americanum* (Lone Star Tick) and *Amblyomma maculatum* (Gulf Coast tick) on dogs. *Vet. Ther. Res. Appl. Vet. Med. 10*, 17–23.

Elbert, A., Haas, M., Springer, B., Thielert, W., Nauen, R., 2008. Applied aspects of neonicotinoid uses in crop protection. *Pest Manage. Sci. 64*, 1099–1105.

Gao, F., Mer, G., Tonelli, M., Hansen, S.B., Burghard, T.P., Taylor, P., et al., 2006. Solution NMR of acetylcholine binding protein reveals agonist-mediated conformational change of the C-loop. *Mol. Pharmacol. 70*, 1230–1235.

Gonias, E.D., Oosterhuis, D.M., Bibi, A.C., 2008. Physiologic response of cotton to the insecticide imidacloprid under high-temperature stress. *J. Plant Growth Regul. 27*, 77–82.

Gorman, K., Devine, D., Bennison, J., Coussons, P., Punchard, N., Denholm, I., 2007. Report of resistance to the neonicotinoid insecticide imidacloprid in *Trialeurodes vaporariorum* (Hemiptera: Aleyrodidae). *Pest Manage. Sci. 63*, 555–558.

Gorman, K., Liu, Z., Brüggen, K.U., Nauen, R., 2008. Neonicotinoid resistance in rice brown planthopper, *Nilaparvata lugens*. *Pest Manage. Sci. 64*, 1122–1125.

Hansen, S.B., Talley, T.T., Radic, Z., Taylor, P., 2004. Structural and ligand recognition characteristics of an acetylcholine-binding protein from *Aplysia californica*. *J. Biol. Chem. 279*, 24197–24202.

Hopkins, J.A., Vodrazka, K., Rudolph, R., Bell, J., 2005. Trimax: assisting cotton growers in yield maximization. *Proc. Beltwide Cotton Conf.* 2644/1–2644/5.

Horowitz, A.R., Kontsedalov, S., Ishaya, I., 2004. Dynamics of resistance to the neonicotinoids acetamiprid and thiamethoxam in *Bemisia tabaci* (Homoptera: Aleyrodidae). *J. Econ. Entomol. 97*, 2051–2056.

Ihara, M., Brown, L.A., Ishida, C., Okuda, H., Sattelle, D.B., Matsuda, K., 2006. Actions of imidacloprid, clothianidin and related neonicotinoids on nicotinic acetylcholine receptors of American cockroach neurons and their relationships with insecticidal potency. *J. Pestic. Sci. 31*, 35–40.

Ihara, M., Shimomura, M., Ishida, C., Nishiwaki, H., Akamatsu, M., Sattelle, D.B., Matsuda, K., 2007. A hypothesis to account for the selective and divers actions of neonicotinoid insecticides at their molecular targets, nicotinic acetylcholine receptors: catch and release in hydrogen bond networks. *Invert. Neurosci. 7*, 47–51.

Jeschke, P., 2007a. Chemical structural features of commercial neonicotinoids. In: Krämer, W., Schirmer, U. (Eds.), Modern Crop Protection Compounds. Wiley-VCH, Weinheim, Germany, pp. 958–961.

Jeschke, P., 2007b. Nicotinic acetylcholine receptors as a continous source for rational insecticides. In: Ishaaya, I., Nauen, R., Horowitz, A.R. (Eds.), Insecticides Design Using Advanced Technologies. Springer, Berlin-Heidelberg, Germany, pp. 151–195.

Jeschke, P., Nauen, R., 2007. Thiamethoxam: a neonicotinoid precursor converted to clothianidin in insects and plants. In: Lyga, J.W., Theodoridis, G. (Eds.), Synthesis and Chemistry of Agrochemicals VII. ACS Symposium Series 948, Washington, DC, pp. 51–65.

Jeschke, P., Nauen, R., 2008. Neonicotinoids – from zero to hero in insecticide chemistry. *Pest Manage. Sci. 64*, 1084–1098.

Karatolos, N., Gorman, K., Williamson, M., Nauen, R., Ffrench-Constant, R., Denholm, I., 2009. Resistance to neonicotinoid insecticides in the greenhouse whitefly, *Trialeurodes vaporariorum. Intergr. Control Protect. Crops Mediterr. Clim. 49*, 103–106.

Karunker, I., Benting, J., Lueke, B., Ponge, T., Nauen, R., Roditakes, E., Vontas, J., Gorman, K., Morin, D.I.S., 2008. Over-expression of cytochrome P_{450} CYP6CM1 is associated with high resistance to imidacloprid in the B and Q biotypes of *Bemisia tabaci* (Hemiptera: Aleyrodidae). *Insect. Biochem. Mol. Biol. 38*, 634–644.

Karunker, I., Morou, E., Nikou, D., Nauen, R., Sertchook, R., Stevenson, B.J., Paine, M.J.I., Morin, S., Vontas, J., 2009. Structural model and functional characterization of the Bemisia tabaci CYP6CM1vQ, a cytochrome P_{450} associated with high levels of imidacloprid resistance. *Insect. Biochem. Mol. Biol. 39*, 697–706.

Kayser, H., Wellmann, H., Lee, C., Decock, A., Gomes, M., Cheek, B., Lind, R., Baur, M., Hattenschwiler, J., Maienfisch, P., 2007. Thiamethoxam: high-affinity binding and unusual mode of interference with other neonicotinoids at aphid membranes. In: Lyga, J.W., Theodoridis, G. (Eds.), Synthesis and Chemistry of Agrochemicals VII. ACS Symposium Series 948, Washington, DC, pp. 67–81.

Liu, Z., Williamson, M.S., Lansdell, S.J., Han, Z., Denholm, I., Millar, N.S., 2007. Target-site resistance to neonicotinoid insecticides in the brown planthopper *Nilaparvata lugens*. In: Ohkawa, H., Miyagawa, H.,

Lee, P.W. (Eds.), Pesticide Chemistry. Wiley-VCH, Weinheim, Germany, pp. 271–274.

Liu, Z., Han, Z., Zhang, Y., Song, F., Yao, X., Liu, S., Gu, J., Millar, N.S., 2009. Heteromeric co-assembly of two insect nicotinic acetylcholine receptor α subunits: influence on sensitivity to neonicotinoid insecticides. *J. Neurochem. 108*, 498–506.

Loso, M.R., Nugent, B.M., Huang, J.X., Rogers, R.B., Zhu, Y., Renga, J.M., Hegde, V., Demark, J.J., 2007. Insecticidal N-substituted (6-haloalkylpyridin-3-yl)alkyl-sulfoximines. *Pat. Appl. WO 095229 A2* (Dow AgroSciences LLC).

Maienfisch, P., 2007. Six-membered heteroxycles (thiamethoxam, AKD 1022). In: Krämer, W., Schirmer, U. (Eds.), Modern Crop Protection Compounds. Wiley-VCH, Weinheim, Germany, pp. 994–1013.

Matsuda, K., Shimomura, M., Ihara, M., Akamatsu, M., Sattelle, D.B., 2005. Neonicotinoids show selective and diverse actions on their nicotinic receptor targets: Electrophysiology, molecular biology, and receptor modeling studies. *Biosci. Biotechnol. Biochem. 69*, 1442–1452.

Matsuda, K., Kanaoka, S., Akamatsu, M., Sattelle, D.B., 2009. Diverse actions and target-site selectivity of neonicotinoids: structural insights. *Mol. Pharmacol. 76*, 1–10.

Matsumura, M., Takeuci, H., Satoh, M., Sanada-Morimura, S., Otuka, A., Watanabe, T., Van Thanh, D., 2008. Species-specific insecticide resistance to imidacloprid and fipronil in the rice planthoppers *Nilaparvata lugens* and *Sogatella furcifera* in East and South-east Asia. *Pest Manage. Sci. 64*, 1115–1121.

McCaffery, A., Nauen, R., 2006. Insecticide resistance action committee (IRAC): public responsibility and enlightened industrial self-interest. *Outlook Pest Manage. 17*, 11–14.

Miyagi, S., Komaki, I., Ozoe, Y., 2006. Identification of a high-affinity binding site for dinotefuran in the nerve cord of the American cockroach. *Pest Manage. Sci. 62*, 293–298.

Nauen, R., Denholm, I., 2005. Resistance of insect pests to neonicotinoid insecticides: current status and future prospects. *Arch. Insect Biochem. Physiol. 58*, 200–215.

Nauen, R., Bielza, P., Denholm, I., Gorman, K., 2008. Age specific expression of resistance to a neonicotinoid insecticide in the whitefly *Bemisia tabaci*. *Pest Manage. Sci. 64*, 1106–1110.

Nurita, A.T., Abu Hassan, A., Nur Aida, H., Norasmah, B., 2008. Field evaluations of the granular fly bait, Quick Bayt and the paint-on fly bait, Agita against synanthropic flies. *Trop. Biomed. 25*, 126–133.

Ohno, I., Tomizawa, M., Durkin, K.A., Naruse, Y., Casida, J., Kagabu, S., 2009. Molecular features of neonicotinoid pharmacophore variants interacting with insect nicotinic receptor. *Chem. Res. Toxicol. 22*, 476–482.

Puinean, A.M., Denholm, I., Millar, N.S., Nauen, R., Williamson, M.S., 2009. Characterization of imidacloprid resistance mechanisms in the brown planthopper, *Nilaparvata lugens* Stal (Hemiptera: Delphacidae). *Pestic.*

Biochem. Physiol., doi: 10.1016/j.pestbp.2009.06.008 (Still in press).

Rust, M.K., 2005. Advances in the control of *Ctenocephalides felis* (cat flea) on cats and dogs. *Trends Parasitol. 21*, 232–236.

Shao, X., Li, Z., Qian, X., Xu, X., 2009. Design, synthesis, and insecticidal activities of novel analogues of neonicotinoids: replacement of nitromethylene with nitroconjugated system. *J. Agric. Food Chem. 57*, 951–957.

Shimomura, M., Yokota, M., Ihara, M., Akamatsu, M., Sattelle, D.B., Matsuda, K., 2006. Role in the selectivity of neonicotinoids of insect-specific basic residues in loop D of the nicotinic acetylcholine receptor agonist binding site. *Mol. Pharmcol. 70*, 1255–1263.

Smit, A.B., Syed, N.I., Schaap, D., van Minnen, J., Klumpermann, J., Kits, K.S., Lodder, H., van der Schors, R.C., van Elk, R., Sorgedrager, B., Brejc, K., Sixma, T.K., Geraerts, W.P.M., 2001. A glia-derived acetylcholine binding protein that modulates synaptic transmission. *Nature 411*, 261–278.

Talley, T.T., Harel, M., Hibbs, R.E., Radic, Z., Tomizawa, M., Casida, J.E., Taylor, P., 2008. Atomic interactions of neonicotinoid agonists with AChBP: molecular recognition of the distinctive electronegative pharmacophore. *Proc. Natl. Acad. Sci. USA 105*, 7606–7611.

Tan, J., Galligan, J.J., Hollingworth, R.M., 2007. Agonist actions of neonicotinoids on nicotinic acetylcholine receptors expressed by cockroach neurons. *NeuroToxicology 28*, 829–842.

Tan, J., Salgado, V.L., Hollingworth, L.W., 2008. Neural actions of imidacloprid and their involvement in resistance in the colorado potato beetle, *Leptinotarsa decemlineata* (Say). *Pest Manage. Sci. 64*, 37–47.

Thany, S.H., Lenaers, G., Raymond-Delpech, V., Sattelle, D.B., Lapied, B., 2007. Exploring the pharmacological properties of insect nicotinic acetylcholine receptors. *Trends Pharmacol. Sci. 28*, 14–22.

Thielert, W., 2006. A unique product: the story of the imidacloprid stress shield. *Pflanzenschutz-Nachrichten Bayer (English Edition) 59*, 73–86.

Tomizawa, M., Casida, J.E., 2005. Neonicotinoid insecticide toxicology: mechanisms and selective action. *Annu. Rev. Pharmacol. Toxicol. 45*, 247–268.

Tomizawa, M., Casida, J.E., 2009. Molecular recognition of neonicotinoid insecticides: The determinants of life or death. *Acc. Chem. Res. 42*, 260–269.

Tomizawa, M., Millar, N.S., Casida, J.E., 2005. Pharmacological profiles of recombinant and native insect nicotinic acetylcholine receptors. *Insect Biochem. Mol. Biol. 35*, 1347–1355.

Tomizawa, M., Maltby, D., Medzihradszky, K.F., Casida, J.E., 2007a. Defining nicotinic agonist binding surface through photoaffinity labeling. *Biochemistry 46*, 8798–8806.

Tomizawa, M., Talley, T.T., Maltby, D., Durkin, K.A., Medzihradszky, K.F., Burlingame, A.L., Taylor, P., Casida, J.E., 2007b. Mapping the exclusive neonicotinoid binding site. *Proc. Natl. Acad. Sci. USA 104*, 9075–9080.

Tomizawa, M., Maltby, D., Talley, T.T., Durkin, K.A., Medzihradszky, K.F., Burlingame, A.L., Taylor, P., Casida, J.E., 2008. Atypical nicotinic agonist bound conformations conferring subtype selectivity. *Proc. Natl. Acad. Sci. USA 105*, 1728–1732.

Tomizawa, M., Talley, T.T., Park, J.F., Maltby, D., Medzihradszky, K.F., Durkin, K.A., Cornejo-Bravo, J.M., Burlingame, A.L., Casida, J.E., Taylor, P., 2009. Nicotinic agonist binding site mapped by methionine- and thyrosine-scanning coupled with azidochloropyridinyl photoafinity labeling. *J. Med. Chem. 52*, 3735–3741.

Toshima, K., Kanaoka, S., Yamada, A., Tarumoto, K., Akamatsu, M., Sattelle, D.B., Matsuda, K., 2009. Combined roles of loops C and D in the interactions of a neonicotinoid insecticide imidacloprid with the alpha4-beta2 nicotinic acetylcholine receptor. *Neuropharmacology 56*, 264–272.

Unwin, N., 2005. Refined structure of the nicotinic acetylcholine receptor at 4 Å resolution. *J. Mol. Biol. 346*, 967–989.

Wakita, T., 2008. Development of new insecticide, dinotefuran. *Yuki Gosei Kagaku Kyokaishi 66*, 716–720.

Wang, Y., Cheng, J., Qian, X., Zhong, L., 2007. Actions between neonicotinoids and key residues of insect *n*AChR based on an ab initio quantum chemistry study: Hydrogen bonding and cooperative π-π interaction. *Bioorg. Med. Chem. 15*, 2624–2630.

Wenzel, U., Heine, J., Mengel, H., Erdmann, F., Schaper, R., Heine, S., Daugschiess, A., 2008. Efficacy of imidacloprid 10% moxidectin 1% (Advocate/Advantage Multi) against fleas (*Ctenocephalides felis felis*) on ferrets (*Mustela putorius furo*). *Parasitol. Res. 103*, 231–234.

Yao, X., Song, F., Zhang, Y., Shao, Y., Li, J., Liu, Z., 2009. Nicotinic acetylcholine receptor β1 subunit from the brown planthopper, *Nilaparvata lugens*: A-to-I RNA editing and its possible roles in neonicotinoid sensitivity. *Insect Biochem. Mol. Biol. 39*, 348–354.

Zhang, Y., Liu, S., Gu, J., Song, F., Yao, X., Liu, Z., 2008. Imidacloprid acts as an antagonist on insect nicotinic acetylcholine receptor containing the Y151M mutation. *Neurosci. Lett. 446*, 97–100.

4 Insect Growth- and Development-Disrupting Insecticides

T S Dhadialla, Dow AgroSciences LLC,
Indianapolis, IN, USA
A Retnakaran, Canadian Forest Service,
Sault Ste. Marie, ON, Canada
G Smagghe, Ghent University, Ghent, Belgium

© 2010 Elsevier B.V. All Rights Reserved

4.1. Introduction

In 1967 Carrol Williams proposed that the term "third generation pesticide" be applied to the potential use of the insect juvenile hormone (JH) as an insecticide, and suggested that it would not only be environmentally benign but that the pest insects would also be unable to develop resistance. However, it took several years before the first commercial juvenile hormone analog (JHA) made its debut (reviews: Retnakaran *et al.*, 1985; Staal, 1975). Since then, several compounds that adversely interfere with the growth and development of insects have been synthesized, and have been collectively referred to as "insect growth regulators (IGRs)" (review: Staal, 1982). Concerns over eco-toxicology and mammalian safety have resulted in a paradigm shift from the development of neurotoxic, broad-spectrum insecticides towards softer, more environmentally friendly pest control agents such as IGRs. This search has led to the discovery of chemicals that: (1) interfere with physiological and biochemical systems that are unique to either insects in particular or arthropods in general; (2) have insect-specific toxicity based on either molecular target site or vulnerability to a developmental stage; and (3) are safe to the environment and nontarget species. At the physiological level, opportunities to develop insecticides with such characteristics existed, *inter alia*, in the endocrine regulation of growth, development, reproduction, and metamorphosis. The rationale was that if the pest insect is treated with a chemical analog, which mimics the action of hormones like JHs and ecdysteroids, at an inappropriate stage the hormonal imbalance would force the insect to go through abnormal development leading to mortality. It was also thought that because such chemicals would in many instances work via the receptor(s) of these hormones, it was less likely for the affected population to develop target site resistance. Yet other target sites are the biosynthetic steps of cuticle formation, which insects share with other arthropods. Adversely interfering with this process would result in the inability of the intoxicated insect to molt and undergo further development. Unlike neurotoxic insecticides that are fast acting, IGRs are in general slow acting, which might result in more damage to the crop. However, some IGRs, such as ecdysone agonists, induce feeding inhibiton, significantly reducing the damage to below acceptable levels. The slow mode of action of some IGRs, such as chitin synthesis inhibitors, can be an advantage in controlling social insects, such as termites, where the material has to be carried to the brood and spread to other members.

It was in the early seventies that the first chitin synthesis inhibitor, a benzoylphenylurea, was discovered by scientists at Philips-Duphar BV, (now Crompton Corp., Weesp, The Netherlands), and marketed as Dimilin® by the Uniroyal Chemical Company (Crompton Corp. Middlebury, CT) in the USA. Since then several new analogs that interfere with one or more steps of cuticle synthesis have been synthesized and are being marketed for controlling various pests. These products are reviewed in Section 4.4, with emphasis on recent developments.

After the initial success with the synthesis of methoprene by Zoecon (Palo Alto, CA, USA) very few new JHAs with good control potential other than pyriproxyfen and fenoxycarb have been developed. These compounds have been particularly useful in targeting the eggs and embryonic development as well as larvae. The synthesis of pyriproxifen and fenoxycarb was a departure from the terpenoid structure of JHs and earlier JHAs, and is reviewed in Section 4.3. For extensive reviews on earlier JHAs the reader is referred to Retnakaran *et al.* (1985) and Staal *et al.* (1975).

Early attempts in the 1970s to synthesize insecticides with molting hormone (20-hydroxyecdysone) activity failed because they were based on a cholesterol backbone, which resulted in chemical and metabolic instability of the steroid nucleus (Watkinson and Clarke, 1973). It took nearly two more decades before the first nonsteroidal ecdysone agonist, based on the bisacylhydrazine class of compounds, was synthesized (Hsu, 1991). Structure activity optimization of the first such compound over the subsequent few years led to the synthesis of four highly effective compounds that have since been commercialized (see Section 4.2.1). The reader is also referred to earlier reviews on the ecdysone agonist insecticides (Oberlander *et al.*, 1995; Dhadialla *et al.*, 1998).

For more basic information, the reader is referred to the chapters in this series on insect hormones (ecdysteroids, JHs) and cuticle synthesis for further details on hormone physiology and chitin synthesis to augment the brief overviews presented in this chapter.

4.1.1. Physiological Role and Mode of Action of the Insect Molting Hormone

Amongst the animal kingdom, arthropods display a remarkable adaptability and diversity in inhabiting very different ecological niches. Between larval and adult stages of a given insect, an insect undergoes distinct developmental and morphological changes that help it to survive in different environments. For example, mosquito larvae are perfectly suited to surviving in an aqueous environment, whereas the adult mosquitos inhabit terrestrial and aerial environments. Larval or nymphal stages of most of the agricultural pests develop on host plant species, and the adult assumes an aerial space to look for food and a mate, returning to the host plant only to oviposit and start another life cycle.

The growth and development from one stage to another is regulated by two main hormones, the steroidal insect molting hormone, 20-hydroxyecdysone (20E; **Figure 1, 1**) and the sesquiterpenoid JH, of which there are five types (**Figure 15**). Even as an insect embryo grows, it undergoes embryonic molts, where each molt is regulated by 20E. The molting process continues through the larval and pupal stages culminating in the adult stage. While molting to accommodate growth is regulated by 20E, the development from an egg to a larva to a pupa to an adult is regulated by the timing and titers of JH (**Figure 17**; also reviewed extensively by Riddiford, 1996). In the adult stage, both these hormones, being pleiotropic, change their roles to regulating reproductive processes (Wyatt and Davey, 1996).

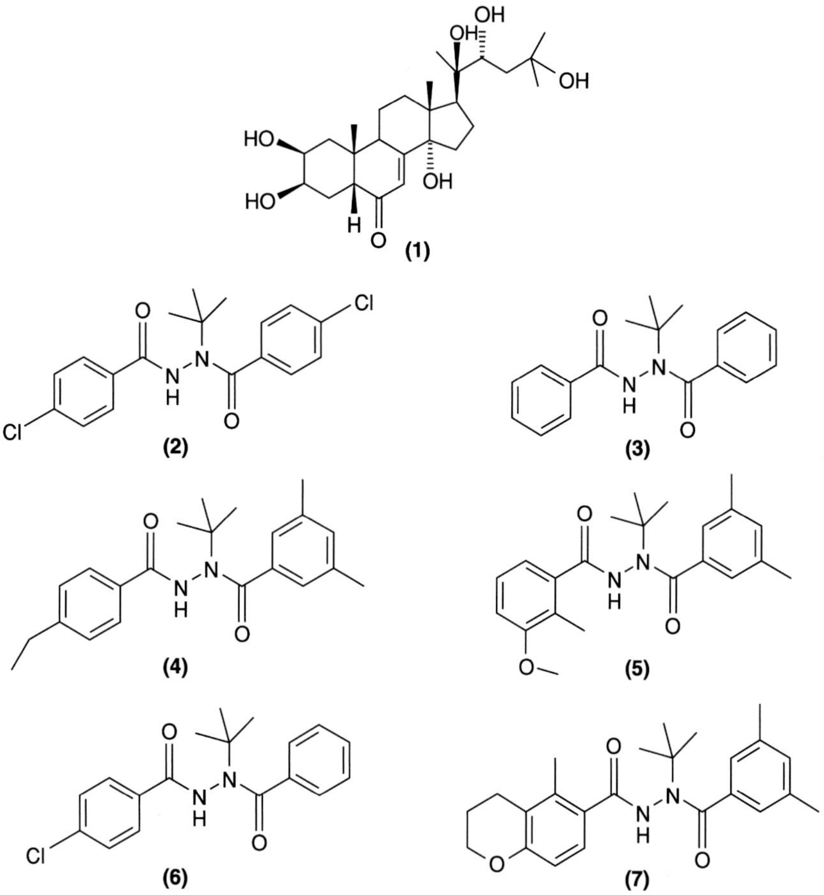

Figure 1 Chemical structures of 20-hydroxyecdysone (**1**), the first discovered symmetrically substituted dichloro-dibenzoylhydrazine (**2**), RH-5849 (**3**), tebufenozide (**4**), methoxyfenozide (**5**), halofenozide (**6**), and chromofenozide (**7**).

The molting process is initiated by an increase in the titer of 20E, and is completed following its decline and the release of eclosion hormone. In preparation for a molt and as the 20E titers increase, the larva stops feeding and apolysis of the epidermis from the old cuticle takes place leaving an ecdysial space that is filled with molting fluid containing inactive chitinolytic enzymes. During this time, the epidermal cells also reorganize in order that large quantities of protein can be synthesized for deposition of a new cuticle. This happens with up- and downregulation of a number of genes encoding a number of epidermal proteins and enzymes (Riddiford, 1994). The fall in the 20E titer triggers the activation of enzymes in the molting fluid for digestion of the procuticle underlying the old cuticle. This event is followed by the resorption of the molting fluid and preparation for ecdysis (Retnakaran et al., 1997; Locke, 1998). Finally, when the 20E titer is cleared from the system, the eclosion hormone is released, which in turn results in the release of the ecdysis-triggering hormone, and all

these events lead to the ecdysis of the larva leaving behind the remnants of the old cuticle (Truman et al., 1983; Zitnanova et al., 2001). With the completion of ecdysis, feeding resumes and endocuticular deposition continues during the intermolt period.

4.1.1.1. Molecular basis of 20E action Of the two hormones 20E and JH, the molecular basis of 20E action is much better understood. The pioneering work of Clever and Karlson (1960) and Ashburner (1973) on the action of 20E to induce/repress "puffs" in the polytene chromosomes of *Chironomus* and *Drosophila* led Ashburner and co-workers to propose a model for ecdysteroid action (Ashburner et al., 1974). According to this model, 20E binds to an ecdysone receptor to differentially regulate several classes of "early" and "late" genes. While the "early" genes are activated by the 20E–receptor complex, the "late" genes are repressed. The protein products of the "early" genes such as E75 and CHR3 derepress the expression of the "late" genes such as DOPA decarboxylase

(DDC) and at the same time repress their own expression.

4.1.1.2. Ecdysone receptors

The ecdysone receptor complex is a heterodimer of two proteins, ecdysone receptor (EcR) and ultraspiracle (USP), which is a homolog of the mammalian retinoic acid receptor (RXR) (Yao *et al.*, 1992, 1995; Thomas *et al.*, 1993). In several insects, both EcR and USP exist in several transcriptional and splice variants, presumably for use in a stage- and tissue-specific way (review: Riddiford *et al.*, 2001). Both EcR and USP are members of the steroid receptor superfamily that have characteristic DNA and ligand binding domains. Ecdysteroids have been shown to bind to EcR only when EcR and USP exist as heterodimers (Yao *et al.*, 1993), although additional transcriptional factors are required for ecdysteroid dependent gene regulation (Arbeitman and Hogness, 2000; Tran *et al.*, 2000). Moreover, EcR can heterodimerize with RXR to form a functional ecdysteroid receptor complex in transfected cells (Yao *et al.*, 1992; Tran *et al.*, 2000). cDNAs encoding both EcR and USPs from a number of dipteran (Koelle *et al.*, 1991; Imhof *et al.*, 1993; Cho *et al.*, 1995; Kapitskaya *et al.*, 1996; Hannan and Hill, 1997, 2001; Veras *et al.*, 1999), lepidopteran (Kothapalli *et al.*, 1995; Swevers *et al.*, 1995), coleopteran (Mouillet *et al.*, 1997; Dhadialla and Tzertzinis, 1997), homopteran (Zhang *et al.*, 2003; Dhadialla *et al.*, unpublished data; Ronald Hill, personal communication), and orthopteran (Saleh *et al.*, 1998; Hayward *et al.*, 1999, 2003) insects, tick (Guo *et al.*, 1997) and crab (Chung *et al.*, 1998) have been cloned. Some of the EcRs and USPs have been characterized in ligand binding (Kothapalli *et al.*, 1995; Kapitskaya *et al.*, 1996; Dhadialla *et al.*, 1998) and cell transfection assays (Kumar *et al.*, 2002; Toya *et al.*, 2002). In all cases, the DNA binding domains (DBDs) of EcRs show a very high degree of homology and identity. However, homology between the ligand binding domains (LBDs) of EcRs varies from 70% to 90%, although all EcRs studied so far bind 20E and other active ecdysteroids. The DBDs of USPs are also highly conserved. The USP LBDs, however, show very interesting evolutionary dichotomy: the LBDs from the locust, *Locusta migratoria*, the mealworm beetle, *Tenebrio molitor*, the hard tick, *Amblyoma americanum*, and the fiddler crab, *Uca puglitor*, show about 70% identity with their vertebrate homolog, but the same sequences from dipteran and lepidopteran USPs show only about 45% identity with those from other arthropods and vertebrates (Guo *et al.*, 1997; Hayward *et al.*, 1999; Riddiford

et al., 2001). The functional significance of RXR-like LBDs in USPs of primitive arthropods is not well understood, because EcRs from the same insects still bind ecdysteroids (Guo *et al.*, 1997; Chung *et al.*, 1998; Hayward *et al.*, 2003; Dhadialla, unpublished data for *Tenebrio molitor* EcR and USP (TmEcR/TmUSP)).

The crystal structures of USPs from both *Heliothis virescens* and *Drosophila melanogaster* have been elucidated by two groups (Billas *et al.*, 2001; Clayton *et al.*, 2001). The crystal structure of USP is similar to its mammalian homolog RXR, except that USP structures show a long helix-1 to helix-3 loop and an insert between helices 5 and 6. These variations seem to lock USP in an inactive conformation by displacing helix 12 from the agonist conformation. Both groups found that crystal structures of the two USPs had large hydrophobic cavities, which contained phospholipid ligands.

Finally, the crystal structures of *Heliothis viresens* EcR/USP (HvEcR/HuUSP) heterodimers liganded with an ecdysteroid or a nonsteroidal ecdysone agonist have been determined (Billas *et al.*, 2003; see Section 4.2.2.2 for more details). The crystal structure of liganded EcR/USP from the silverleaf whitefly, *Bemesia tabaci*, has also been determined and awaits publication (Ronald Hill, personal communication).

4.2. Ecdysteroid Agonist Insecticides

4.2.1. Discovery of Ecdysone Agonist Insecticides and Commercial Products

Although attempts to discover insecticides with an insect molting hormone activity were made in the early 1970s (Watkinson and Clarke, 1973), it was not until a decade later that the first bisacylhydrazine ecdysone agonist ((2) in **Figure 1**) was serendipitously discovered by Hsu (1991) at Rohm and Haas Company, Springs House, PA, USA. Several years later, after several chemical iterations of this early lead, a simpler, unsubstituted, but slightly more potent analog, RH-5849 ((3) in **Figure 1**), was discovered (Aller and Ramsay, 1988). Further work on the structure and activity of RH-5849, which had commercial-level broad spectrum activity against several lepidopteran, coleopteran, and dipteran species (Wing, 1988; Wing and Aller, 1990), resulted in more potent and cost-effective bisacylhydrazines with a high degree of selective pest toxicity (review: Dhadialla *et al.*, 1998). Of these, three bisacylhydrazine compounds, all substituted analogs of RH-5849, coded as RH-5992 (tebufenozide (**4**); **Figure 1**), RH-2485 (methoxyfenozide (**5**); **Figure 1**), and RH-0345 (halofenozide (**6**); **Figure 1**) have been

Table 1   Bisacylhydrazine insecticides with ecdysone mode of action

Structure	Common name	Coded as	Industry	Registered names	Pest spectrum
	Tebufenozide	RH-5992	Rohm and Haas Co.,[a] Dow AgroSciences LLC[b]	MIMIC®, CONFIRM®, ROMDAN®	Lepidoptera
	Methoxyfenozide	RH-2485	Rohm and Haas Co.,[a] Dow AgroSciences LLC[b]	INTREPID®, RUNNER®, PRODIGY®, FALCON®	Lepidoptera
	Chromofenozide	ANS-118, CM-001	Nippon Kayaku, Saitame, Japan and Sankyo, Ibaraki, Japan	MATRIC®, KILLAT®	Lepidoptera
	Halofenozide	RH-0345	Rohm and Haas Co.,[a] Dow AgroSciences LLC[b]	MACH 2®	Lepidoptera, Coleoptera

[a]Discovered and commercialized by Rohm and Haas Company, Spring House, PA, USA.
[b]Now owned and commercialized by Dow AgroSciences, LLC., Indianapolis, IN, USA.

commercialized (**Table 1**). Both tebufenozide and methoxyfenozide are selectively toxic to lepidopteran larvae (Hsu, 1991). However, methoxyfenozide is more potent than tebufenozide, and is toxic to a wider range of lepidopteran pests of cotton, corn, and other agronomic pests (Ishaaya et al., 1995; Le et al., 1996; Trisyono and Chippendale, 1997). Halofenozide has a broader and an overall insect control spectrum somewhat similar to that of RH-5849, but with significantly higher soil-systemic efficacy against scarabid beetle larvae, cutworms, and webworms (RohMid LLC, 1996). Another bisacylhydrazine, chromafenozide (**7**) coded as ANS-118; **Figure 1**) was discovered and developed jointly by Nippon Kayaku Co., Ltd., Saitama, Japan and Sankyo Co., Ltd. Ibaraki, Japan (Yanagi et al., 2000; Toya et al., 2002; **Table 1**). Chromafenozide, registered under the trade names, MATRIC® and KILLAT®, is commercialized for the control of lepidopteran larval pests of vegetables, fruits, vines, tea, rice, and ornamentals in Japan (Yanagi et al., 2000; Reiji et al., 2000).

4.2.1.1. Synthesis and structure–activity relationships (SAR)   The synthesis of symmetrical and asymmetrical 1,2-dibenzoyl-1-*t*-butyl hydrazines can be achieved by the following two step reaction shown in **Figure 2**. In 1984, during the process of synthesizing compound number (**1** in **Figure 2**), which was to be used for the synthesis of another class of compounds, Hsu (Rohm and Haas Company) obtained an additional undesired product (**2** in **Figure 2**), 1,2-di (4-chlorobenzoyl)-1-*t*-butyl-hydrazine (A = 4-Cl). Testing of this undesired by-product revealed that it had ecdysteroid activity, and eventually led to the development of a bisacylhydrazine class of compounds. This serendipitous discovery can be attributed to the inquiring mind of Hsu, who decided to test this compound for biological activity even though it was a contaminant! By treating equivalent amounts of different benzoyl chlorides with **1** (benzoyl hydrazide), unsubstituted analogs (**3**) can be prepared as shown in **Figure 2**.

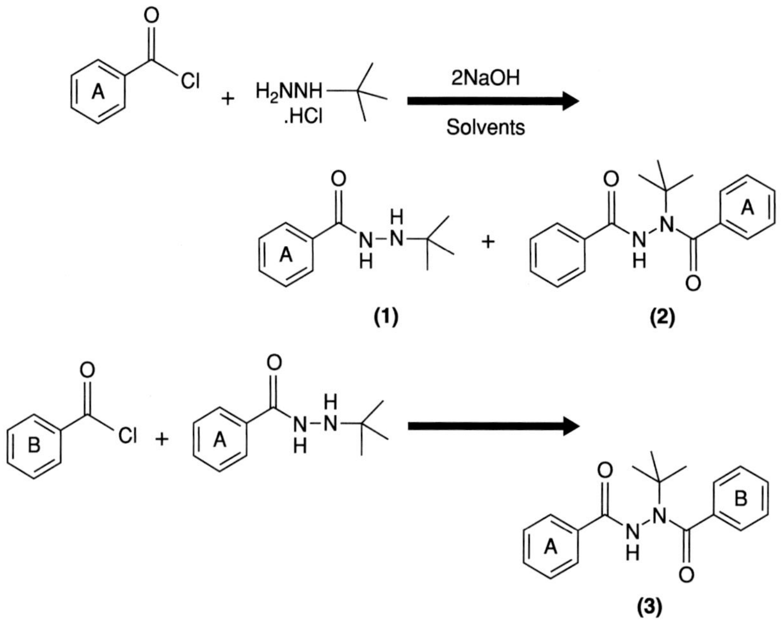

Figure 2 Synthetic routes for unsubstituted and substituted bisacylhydrazines. See text for details. (Reprinted with permission from Hsu, A.C.-T., Fujimoto, T.T., Dhadialla, T.S., **1997**. Structure-activity study and conformational analysis of RH-5992, the first commercialized nonsteroidal ecdysone agonist. In: Hedin, P.A., Hollingworth, R.M., Masler, E.P., Miyamoto, J., Thompson, D.G. (Eds.), Phytochemicals for Pest Control, ACS Symposium Series, vol. 658. American Chemical Society, pp. 206–219; © American Chemical Society.)

The discovery of the first ecdysone agonist (1) offered 18 possible sites for possible structural variation. Between 1985 and 1999, approximately 4000 structural bisacylhydrazine analogs were synthesized at the Rohm and Haas Company. This resulted in the selection of 22 candidate compounds for field testing; three of these were selected for commercial development.

Extensive SAR research was carried out around the bisacylhydrazine chemistry to discover additional more potent nonsteroidal agonists with different pest spectrum activity (Oikawa et al., 1994a, 1994b; Nakagawa et al., 1995a, 1995b, 1998, 2000, 2001a, 2001b; Hsu et al., 1997; Shimuzu et al., 1997; Smagghe et al., 1999b). Some general conclusions for the most active compounds can be drawn, in addition to those drawn by Hsu et al. (1997), as illustrated below:

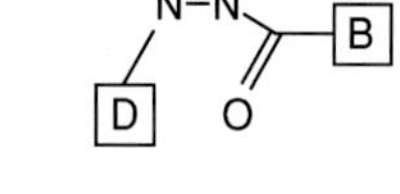

A and B rings: phenyl, heterocyclic, or alkyl. The most active analogs are those that have different substituents on two phenyl rings (e.g., chromofenozide,

halofenozide, tebufenozide, and methoxyfenozide). However, the phenyl ring furthest away from the t-butyl group (A region) can be replaced with alkyl chains without greatly affecting the biological activities (Shimuzu et al., 1997). Phenyl ring (A region) can have various 2, 3, 4, 5 methylene or ethylene dioxan substituents and maintain lepidopteran specific activity. Analogs derived from symmetrically substituted phenyl rings are generally less active than asymmetrically substituted compounds.

D region: In most cases, hydrogen is necessary for activity. When substituted with an unhydrolyzable group, the analogs tend not to be active.

C region: A bulky alkyl group like t-butyl is important for activity. The t-butyl group can be modified to a benzodioxan group without significantly decreasing the affinity of the compound.

Based on the X-ray crystal structures of RH-5849, a number of models have been developed that demonstrate superimposition of RH-5849 on the 20E structure (Chan et al., 1990; Mohammed-Ali et al., 1995; Cao et al., 2001; Kasuya et al., 2003). More recent X-ray crystallography studies of liganded (20E or diacylhydrazine) HvEcR/HvUSP show that the two compounds share common sites occupied by the side chain of 20E and the B ring of bisacylhydrazine (Billas et al., 2003).

4.2.2. Bisacylhydrazines as Tools of Discovery

The discovery of bisacylhydrazine has been a catalyst in stimulating research on the biological activity of ecdysteroids and nonsteroidal ecdysone agonists in whole insects, tissues, and in cells (reviews: Dhadialla *et al.*, 1998; Oberlander *et al.*, 1995). Because bisacylhydrazines have a greater metabolic stability than 20E *in vivo*, it was possible to confirm and further understand the effects of rising and declining titers of 20E during growth and development (Retnakaran *et al.*, 1995).

With the cloning and sequencing of cDNAs encoding EcRs from different insects it became clear that there were homology differences in the LBDs of the cloned EcRs but the functional significance of these differences was not apparent. In almost all insects, the effect of 20E is manifested by an interaction with EcR in the ecdysone receptor complex. Therefore, it is paradoxical that tebufenozide and methoxyfenozide, which also manifest insect toxicity via interaction with EcR, are predominantly toxic to lepidopteran insects and to a lesser degree to other insects. Through the use of these ecdysone agonists it became clear that while 20E can bind to different LBDs of EcRs from different orders of insects, the same LBDs bound bisacylhydrazines with unequal affinities, which reflected their insect specificity (reviewed by Dhadialla *et al.*, 1998). It also became clear that while a relatively low but similar affinity (K_d = 10–100 nM) of 20E to different EcRs was functional, only high-affinity (nM range) interactions of tebufenozide and methoxyfenozide with EcR were functionally productive (**Table 2**). On the other hand, the low affinity of halofenozide was most likely compensated for by its high metabolic stability in susceptible insects.

4.2.2.1. Molecular modeling
Molecular modeling studies on both EcR LBDs (Wurtz *et al.*, 2000; Kasuya *et al.*, 2003) and bisacylhydrazines (Chan *et al.*, 1990; Mohammed-Ali *et al.*, 1995; Hsu *et al.*, 1997; Cao *et al.*, 2001) were conducted, perhaps partly prompted by the molecular modeling and X-ray crystal structure data of LBDs of steroid receptors from vertebrates and the search for new classes of nonsteroidal ecdysone agonists/antagonists. The results suggested that the tertiary protein structure of EcR LBDs is similar to that of vertebrate steroid receptor LBDs, and that helix 12 forms a "mouse trap" for a ligand in the binding cavity (**Figure 3**; see also description of crystal structures of EcR and USP below). Results of molecular modeling and *in silico* docking experiments with 20E and RH-5849 have provided conflicting information about the orientation of 20E in the binding pocket. However, the recent description of the X-ray crystal structure of ligand bound HvEcR/HvUSP LBDs has provided definite answers to this (see below).

4.2.2.2. Mutational analysis
In a very elegant study, Kumar *et al.* (2002) extended the *in silico* molecular modeling approach to construct a homology model of *Choristoneura fumiferana* EcR (CfEcR) LBD with 20E docked in to predict amino acid residues involved in interaction with 20E. The authors identified 17 amino acid residues as possibly being critical for 20E binding. In order to confirm this prediction, all 17 of these amino acids were mutated to alanine, except for A110 (where the 110 residue is numbered starting from helix one, which otherwise is A393 in the full-length CfEcR), which was mutated to proline. When the single point mutated LBDs were tested in ligand binding assays and transactivation assays (in insect and mammalian cells and in mice *in vivo*) the A110P mutant was ineffective in the two assays for a response to ecdysteroids (ponasterone A and 20E). While there was a 30% decrease in transactivation assays for the nonsteroidal bisacylhydrazines, their

Table 2 Relative binding affinities of ponasterone A and bisacylhydrazine insecticides to EcR/USPs in cellular extracts of dipteran, lepidopteran, and coleopteran cells and (biological activities)

Ligand	K_d (nM)		
	Drosophila Kc cells	Plodia interpunctella[a]	Leptinotarsa decemlineata[b]
Ponasterone A	0.7	3	3
Tebufenozide	192	3 (predominantly lepidoptera active)	218
Methoxyfenozide	124	0.5 (predominantly lepidoptera active)	ND
Halofenozide	493	129 (lepidoptera active)	2162 (coleoptera active)

[a]Imaginal wing disc cell line.
[b]Embryonic cell line.
ND, not determined.

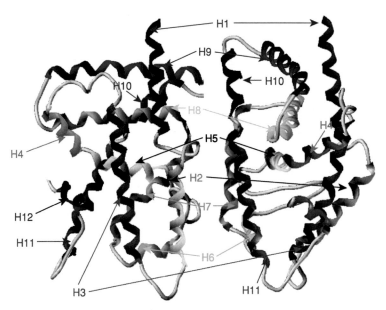

Figure 3   Different elevations of computer model(s) of the three-dimensional structure of the LBD of insect ecdysone receptors. The different helices are represented in different colors, and indicated by arrows and numbers. After a ligand has docked into the ligand binding pocket of EcR heterodimerizd with USP, helix 12 closes on the binding cavity like a "mouse trap."

ability to bind was unaffected. Changing A110 to leucine was as effective as the change to proline. Further work along these lines should make it possible not only to design new effective chemistries as ecdysone agonists/antagonists for pest control, but also to design EcR LBDs that result in productive binding with different chemistries, thus allowing their use in several gene switch applications.

The necessity of EcR and USP forming a heterodimer for ligand interaction coupled with the difficulty in producing large enough quantities of EcR and USP LBDs required for X-ray diffraction studies have been overcome by Billas *et al.* (2003). These authors reported the crystal structures of the LBDs of the moth *H. virescens* EcR-USP heterodimer in complex with the ecdysteroid ponasterone A and with a nonsteroidal, lepidopteran specific agonist, BY106830 (a bisacylhydrazine). Comparison of the crystal structures liganded with ponasterone A and BY106830 revealed that the two ligands occupy very different but slightly overlapping spaces in the ligand binding pockets. The overlap of the two ligands was observed on the side chain of ponasterone A with the *t*-butyl group and the benzoyl ring closest to it in BY106830. The presence of the *t*-butyl group and its occupation in a hydrophobic groove of the HvEcR LB pocket confers lepidopteran specificity on it. Further examination of the residues in this hydrophobic cavity revealed that V384 in helix 5, which was conserved in lepidopteran insect EcR LBDs and is replaced by methionine in other

insects, is essential for the lepidopteran specificity. The crystal structure observations supported the results of Kumar *et al.* (2002).

4.2.2.3.   Ligand-dependent conformational changes

Yao *et al.* (1995) used limited proteolysis protection to demonstrate that binding of muristerone A, a potent ecdysteroid, to *Drosophila melanogaster* EcR and USP (DmEcR/DmUSP) induces a conformation change in EcR that can be detected by sodium dodecylsulfate polyacrylamide gel electrophoresis (SDS–PAGE) and autoradiography. In these experiments, incubation of muristerone A with DmEcR-^{35}S-methionine labeled/DmUSP proteins produced by transcription and translation of corresponding cDNAs in rabbit reticulocyte cell-free mixtures, afforded partial protection to EcR from trypsin as the protease. In an extension of this approach, Dhadialla *et al.* (unpublished data) used EcR-^{35}S-methionine labeled/USP proteins produced *in vitro* to demonstrate that muristerone A bound to DmEcR/DmUSP, *Aedes aegypti* EcR and USP (AeEcR/AeUSP), and CfEcR/CfUSP protected an EcR fragment of 36 kDa from limited proteolysis with trypsin, chymotrypsin, and proteinase K. On the other hand, the ability of tebufenozide (RH-5992) to induce a similar conformational change upon binding to EcR, which affords limited proteolytic protection as with muristerone A, correlated with its affinity to the various EcRs (**Table 3**). Interestingly, peptide fragments of a similar size (36 kDa)

Table 3 Size distribution of protease protected fragments of ^{35}S-labeled EcRs incubated in the presence of homologous USPs and an ecdysteroid, muristerone A, or a bisacylhydrazine, tebufenozide

| | Protected fragment size (kDa) | | | | | |
| | DmEcR | | AaEcR | | CfEcR | |
Protease	Muristerone A	Tebufenozide	Muristerone A	Tebufenozide	Muristerone A	Tebufenozide
Chymotrypsin	36		45	45	ND	ND
Trypsin	40		50	50	ND	ND
Proteinase K	36		36	36	36	36
K_d (nM)		336		28		0.5

ND, not determined.
Cloned EcR and USP cDNAs from the three insects, *Drosophila melanogaster* (Dm), *Aedes aegypti* (Aa), and *Choristoneura fumiferana* (Cf) were expressed using *in vitro* transcription and translations of the rabbit reticulocyte system. However, EcRs were *in vitro* labeled with ^{35}S-methionine. Both EcR and USPs from the same species were incubated in the presence of 1 μm muristerone A or 1 μm tebufenozide and binding allowed to reach equilibrium. At that point the binding reactions were aliquoted and subjected to increasing protease concentrations for 20 min at RT (limited proteolysis). The digests were analyzed by SDS-PAGE, and the labeled EcR bands visualized by fluorography. The K_d values are the same as those given in **Table 2**.

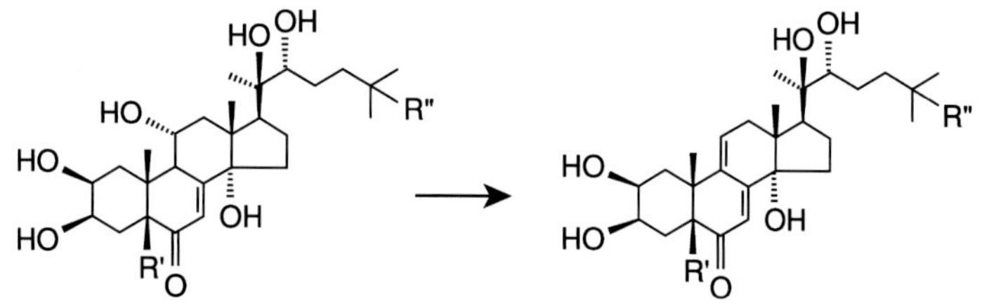

Ajugasterone C (R′ = H, R″ = H) ⟶ Dacryhainansterone (R′ = H, R″ = H)

Muristerone A (R′ = OH, R′ = H) ⟶ Kaladasterone (R′ = OH, R″ = H)

Turkesterone (R′ = H, R″ = OH) ⟶ 25-Hydroxydacryhainansterone (R′ = H, R″ = OH)

Figure 4 Schematic derivatization of ajugasterone C, muristerone A, and turkesterone to photo affinity ecdysteroid 7,9(11)-dien-6-ones, dacryhainansterone, kaladasterone, and 25-hydroxydacryhainansterone, respectively. (Adapted from Bourne, P.C., Whiting, P., Dhadialla, T.S., Hormann, R.E., Girault, J.-P., *et al.*, 2002. Ecdysteroid 7,9(11)-dien-6-ones as potential photoaffinity labels for ecdysteroid binding proteins. *J. Insect Sci. 2*, 1–11.

to the three EcRs with either muristerone A or tebufenozide (in the cases of AaEcR and CfEcR) bound were protected from proteolysis with proteinase K, suggesting similar ligand-induced conformational changes. Such a conformational change *in vivo* may be necessary for ligand-dependent transactivation of a gene.

With respect to the hypothesis of Billas *et al.* (2003) that in the EcR/USP heterodimer the EcR ligand binding pocket may conform to accommodate an active ligand, it is not clear if the changes seen in limited proteolysis experiments (above) are a result of just ligand-induced changes in the binding pocket or a result of ligand-induced changes in the whole of the EcR LBD.

4.2.2.4. Photoaffinity reagents for studying receptor–ligand interactions Before the X-ray

crystal structures of the EcR/USP heterodimer were available, alternative approaches to understanding the interaction of ecdysteroid and nonsteroidal bisacylhydrazine ligands with residues in the EcR ligand binding pocket were explored. One such approach was to synthetically produce photoaffinity analogs of ecdysteroids (Bourne *et al.*, 2002) and bisacylhydrazines (Dhadialla *et al.*, unpublished data) as has been achieved for vertebrate steroids and their receptors (Katzenellenbogen and Katzenellenbogen, 1984). An ideal affinity labeling reagent should have: (1) a high affinity for the binding protein with low nonspecific binding to other proteins; (2) a photo-reactive functional group; and (3) a radiolabeled moiety for detection purposes.

Bourne *et al.* (2002) synthesized three ecdysteroid 7,9(11)-dien-7-ones (**Figure 4**; dacryhainansterone, 25-hydroxydacryhainansterone, and kaladasterone) as photoaffinity ecdysteroids by dehydration of the corresponding 11α-hydroxy ecdysteroids (ajugasterone, turkesterone, and muristerone A, respectively). Of the three photoaffinity ecdysteroids, dacryhainansterone and kaladasterone showed the greatest potential for use in determining contact amino acid residues in EcR LBDs. All three dienone ecdysteroids retained their biological activity in the *D. melanogaster* B_{II} cell assay developed by Clement *et al.* (1993). However, upon irradiation with UV at 350 nM in the presence of bacterially expressed DmEcR/DmUSP and CfEcR/CfUSP, both dacryhainansterone and kaladasterone blocked subsequent specific binding of tritiated ponasterone A by >70%. These results clearly showed the potential of using radiolabeled dienones for further characterization of their interaction within EcR LBDs.

Photoaffinity analogs of bisacylhydrazines have also been produced and used to characterize their binding to EcRs from dipterans and lepidopterans (Dhadialla *et al.*, unpublished data). A tritiated benzophenone analog of methoxyfenozide (**Figure 5**), coded as RH-131039, displays the above-mentioned characteristic requirements of a photoaffinity reagent as well as the specificity to bind to lepidopteran EcR (CfEcR) with high affinity. Binding to dipteran EcR (DmEcR) could not be detected. Binding of either tritiated ponasterone A or tritiated RH-131039 to CfEcR/CfUSP could be competitively displaced using excess unlabeled RH-131039, tebufenozide, ponasterone A, or 20E. These results indicate that the ecdysteroids and the two nonsteroidal ecdysone agonists share a common binding site in the ligand binding pocket of CfEcR/CfUSP (**Figure 6**), an observation that has been confirmed by Billas *et al.* (2003) in their work on the crystal structure of liganded HvEcR/HvUSP. However, further experiments, which can now be done with the ecdysteroid and bisacylhydrazine photoaffinity compounds, need to be done to determine the ligand–receptor contact residues. Similar results were obtained for AaEcR/AaUSP, but not for DmEcR/DmUSP, indicating that even within an insect order, differences in the homology and folding of EcR LBDs is sufficient to discriminate between affinities of nonsteroidal ecdysone agonists. The bisacylhydrazine and ecdysteroid photoaffinity reagents described are additional useful tools that can be combined with data from chemical structure–activity relationships (SAR), analytical (liquid chromatography/mass spectrometery (LC/MS) and mass laser desorption ionization (MALDI)), and mutational analysis approaches to hypothesize and/or define the three-dimensional space of non-lepidopteran EcR LBDs. However, the availability of the liganded HvEcR/HvUSP LBD

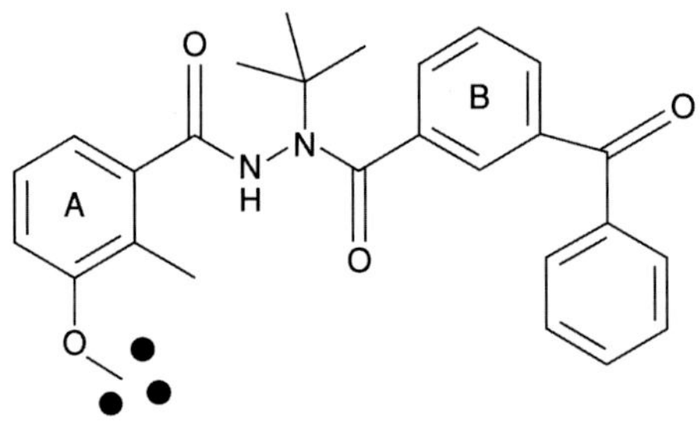

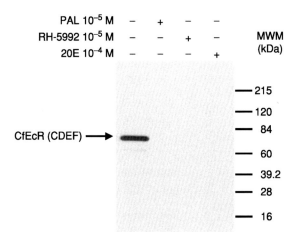

RH-131039

Figure 5   Structure of the tritiated photoaffinity benzophenone analog, RH-131039, of methoxyfenozide. The three dots on the methoxy substitution on the A-phenyl ring indicate tritiation of the methyl group. The benzophenone group on the B-phenyl ring can be activated with UV light for it to cross-link to the nearest amino acid residue.

Figure 6   Fluorograph to show specific binding of tritiated RH-131039 (PAL; photo affinity ligand) to CDEF domain of *Choristoneura fumiferana* EcR (CfEcR) expressed by *in vitro* transcription and translation using rabbit reticulocyte system. CfEcR (CDEF domain) plus CfUSP was incubated with tritiated PAL in the presence of excess cold PAL, RH-5992, or 20E at the indicated concentrations under equilibrium binding conditions after which the reaction mixture was irradiated with $UV_{350\,nm}$ for 20 min at 4 °C. The reaction mixture was then mixed with SDS-PAGE sample buffer containing 1 mM DTT, heated for 5 min at 95 °C and then subjected to SDS-PAGE. The gel was dried and fluorographed to get the above image. The numbers on the right indicate relative migration of molecular weight markers (MWM).

crystal structure provides the needed information for defining the three-dimensional ligand receptor interaction for homology modeling of such interactions with non-lepidopteran EcR LBDs, and for the design and discovery of new biologically potent ligands as insecticides for different insects orders.

4.2.3. Mode of Action of Bisacylhydrazines

In the earlier reviews on insecticides with ecdysteroid and JH activity (Oberlander *et al.*, 1995; Dhadialla *et al.*, 1998), much of the focus on mode of action of ecdysone agonists was from data using RH-5849 and RH-5992. RH-5849, the unsubstituted bisacylhydrazine that was not commercialized, had a broader spectrum of insect toxicity than the four that have been commercialized. In this chapter, the focus will be on work published since 1996 on the commercialized bisacylhydrazine insecticides tebufenozide, methoxyfenozide, chromofenozide, and halofenozide. References will be made to earlier publications where relevant.

4.2.3.1. Bioassay
A number of *in vitro* and *in vivo* assays have been used to study the effects and mode of action of the bisacylhydrazine insecticides (reviews: Oberlander *et al.*, 1995; Dhadialla *et al.*, 1998). Larvae of susceptible and nonsusceptible insects (described below), and a number of insect cell lines and dissected tissues have been used *in vitro* to understand the mode of action of bisacylhydrazines. Wing *et al.* (1988) were the first to use *D. melanogaster* embryonic Kc cells to demonstrate that like 20E, RH-5849 also induced aggregation and clumping of otherwise confluent cultures of Kc cells. Similar morphological effects of tebufenozide, methoxyfenozide, and halofenozide have also been demonstrated for cell lines derived from embryos or tissues of *Drosophila* (Clement *et al.*, 1993), the mosquito, *Aedes albopictus* (Smagghe *et al.*, 2003a), the midge, *Chironomus tentans* (Spindler-Barth *et al.*, 1991), the forest tent caterpillar, *Malacosoma disstria*, the spruce budworm, *Choristoneura fumiferana* (Sohi *et al.*, 1995), the Indian meal moth, *Plodia interpunctella* (Oberlander *et al.*, 1995), the cotton boll weevil, *Anthonomus grandis* (Dhadialla and Tzertzinis, 1997), the European corn borer, *Ostrinia nubilalis*, the Southwestern corn borer, *Diatraea grandiosella*, and the cotton bollworm, *Helicoverpa zea* (Trisyono *et al.*, 2000).

Some of these cell lines, imaginal wing discs, and larval claspers from susceptible insects have also been used to study the relative binding affinities, and biochemical and molecular effects of tebufenozide (Mikitani, 1996b), methoxyfenozide, or halofenozide (**Table 4**) and other ecdysone agonists

(Nakagawa *et al.*, 2002). Cytosolic and or nuclear extracts from 20E responsive cells and tissues containing functional ecdysone receptors, as well as bacterially expressed EcRs and USPs from different insects, have also been used to determine the relative binding affinities of bisacylhydrazines or to screen for new chemistries with similar mode of action in radiometric competitive receptor binding assays (**Table 5**).

Finally, in order to increase the throughput for screening a higher number of compounds, Tran *et al.* (2000) reconstituted ligand-dependent transactivation of *C. fumiferana* EcR in yeast. Reconstitution of the system required transforming yeast with four plasmids carrying cDNAs encoding CfEcR, CfUSP, or RXRγ, an activation factor, a glucocorticoid receptor interacting protein (GRIP), and a plasmid with four copies of an ecdysone response element (hsp27) in tandem fused to a reporter gene (**Figure 7**). The potency of a number of ecdysone agonists in transactivating the reporter gene in the reconstituted EcR ligand dependent assay correlated well with the known activities of the tested bisacylhydrazines in whole insect assays (**Figure 8**), indicating the utility of such assays for high throughput screening of new and novel chemistries with ecdysone mode of action. A similar yeast-based strain was also reconstituted using AaEcR/AaUSP (Tran *et al.*, 2001).

4.2.3.2. Whole organism effects
The commercialized bisacylhydrazine insecticides are toxic to susceptible insects mainly as a result of ingestion. Toxicity via topical application is expressed only when very high doses are applied. The effects of bisacylhydrazines in susceptible insects have been studied in a number of insects (Slama, 1995; Smagghe *et al.*, 1996b, 1996c, 2002; Retnakaran *et al.*, 1997; Dhadialla *et al.*, 1998; Carton *et al.*, 2000). In general, because bisacylhydrazines are more metabolically stable *in vivo* than ecdysteroids and because they are true ecdysone agonists at the receptor level, ingestion of bisacylhydrazines creates "hyperecdysonism" in susceptible insects,

Table 4   Relative affinities of 20E and bisacylhydrazines in imaginal wing evagination assays

Species	20E	RH-5849	RH-5992	RH-2485	RH-0345	Reference
Drosophilla melanogaster	$0.011\,\mu mol\,l^{-1}$	$1\,\mu mol\,l^{-1}$	$1.1\,\mu mol\,l^{-1}$	ND	ND	Farkas and Slama (1999)
Chironomus tentans	278 nM	27 nM	7.3 nM	ND	230 nM	Smagghe *et al.* (2002)
Spodoptera littoralis	291 nM	44500 nM	403 nM	10.9 nM	472 nM	Smagghe *et al.* (2000a)
Galleria mellonella	321 nM	865 nM	8.9 nM	ND	ND	Smagghe *et al.* (1996a)
Leptinotarsa decemlineata	60.7 nM	461 nM	757 nM	ND	ND	Smagghe *et al.* (1996a)

ND, not determined.

Table 5   Relative binding affinities of bisacylhydrazine compounds to cellular extracts or *in vitro* expressed EcR and USP proteins from insects of different orders

	20E	Pon A	RH-5849	RH-5992	RH-2485	RH-0345	Reference
Drosophila melanogaster	145 nM	0.9 nM		336 nM			Dhadialla, (unpublished data), Cherbas *et al.* (1988)
Aedes aegypti	28 nM	2.8 nM		30 nM			Dhadialla, unpublished data; Kapitskaya *et al.* (1996)
Chironomus tentans		0.35 ± 0.28 and 6.5 ±2.4 nM					Grebe *et al.* (2000), Smagghe *et al.* (2001)
Spodoptera littoralis	158 nM			86.7 nM	24.3 nM		Smagghe *et al.* (2000a)
Plodia interpunctella	210 nM	3 nM		3 nM			Dhadialla, unpublished data
Galleria mellonella	106 nM		911 nM	22 nM			Smagghe *et al.* (1996a)
Spodoptera frugiperda (Sf9 cells)	166 nM	8.9 nM	363 nM	1.5 nM	3.5 nM		Nakagawa *et al.* (2000)
Anthonomus grandis	247 nM	6.1 nM		12 000 nM			Dhadialla and Tzertzinis (1997)
Leptinotarsa decemlineata	425 nM		740 nM	1316 nM			Smagghe *et al.* (1996a)
Tenebrio molitor		6 nM	>10 μM	>10 μM	>10 μM	>10 μM	Dhadialla, unpublished data
Locusta migratoria	1000 nM	1.18 nM	>10 μM	>10 μM	>10 μM	>10 μM	Hayward *et al.* (2003)
Bamesia argentfolli		8 nM	>10 μM	>10 μM	>10 μM	>10 μM	Dhadialla, unpublished data

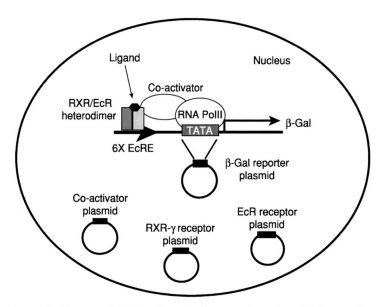

Figure 7   Schematic diagram showing a yeast cell transformed with plasmids carrying cDNAs encoding ecdysone receptor, RXR-γ, co-activator glucocorticoid receptor interaction protein (GRIP), and β-galactosidase (β-Gal) reporter gene fused to 6 heat shock protein 27 (hsp27) ecdysone response elements in a reconstituted and ecdysteroid (muristerone A) inducible high throughput screening system for discovery of new nonsteroidal ecdysone agonists.

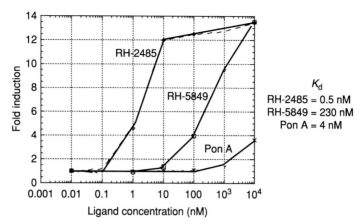

Figure 8 Dose-dependent induction of reporter (β-gal) gene in an ecdysone inducible, transformed yeast cell assay system, with RH-2485 (methoxyfenozide), RH-5849, and the phytoecdysteroid, ponasterone A. While the fold induction achieved with RH-2485 and RH-5849 correlate with their relative affinities (indicated on the right of the graph) for the ecdysone receptor (CfEcR) used in this study, the activation achieved by ponasterone does not correlate with its determined binding affinity to ecdysone receptor.

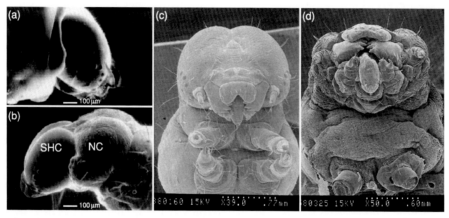

Figure 9 Scanning electron micrographs of control treated (a and c) and tebufenozide intoxicated (b and d) by ingestion (fava bean leaves dipped in a 10 ppm aqueous solution and 100 ng, respectively) second instar southern armyworm (a and b) and third instar spruce budworm (c and d) larvae 48 h after intoxication. In both control armyworm and budworm larvae (a and c), well formed sclerotized cuticle and mouthparts can be seen. In the intoxicated armyworm (b), the larvae is tricked into undergoing a precocious molt and the slipped head capsule (SHC) can be seen over the mouthparts. The new cuticle (NC) covering the head is soft, malformed, and unsclerotized. The soft unsclerotized mouthparts of the intoxicated spruce budworm (d) are shown after manually removing the slipped head capsule. In both cases, tebufenozide intoxicated larvae initiate a molt, slip head capsule, but are unable to complete a molt, and hence die of starvation, desiccation, and hemorrhage from the malformed unsclerotized new cuticle. (Adapted and composed from Dhadialla, T.S., Carlson, G.R., Le, D.P., **1998**. New insecticides with ecdysteroidal and juvenile hormone activity. *Annu. Rev. Entomol. 43*, 545–569; as well as other unpublished electron micrographs.)

thus inducing effects and symptomology of a molt event. One of the first effects of bisacylhydrazine ingestion by susceptible larvae is feeding inhibition within 3–14 h (Slama, 1995; Smagghe *et al.*, 1996a; Retnakaran *et al.*, 1997), which prevents further plant damage. During this time, synthesis of a new cuticle begins and apolysis of the new cuticle from the old one takes place. Subsequently, the intoxicated larvae become moribund, slip their head capsule (**Figures 9** and **10**), and, in extreme cases, the hind gut may be extruded (**Figure 11**). The new cuticle is not tanned or sclerotized. One resulting consequence is that the mouth parts under the slipped

head capsule remain soft and mushy, preventing any crop damage even if the head capsule came off from mechanical or physical force. The larvae ultimately die as a result of their inability to complete a molt, starvation, and desiccation due to hemorrhage.

The reasons for the lethal precocious molt effects of bisacylhydrazines have been investigated at the ultrastructural level in *C. fumiferana* (Retnakaran *et al.*, 1997), in beet armyworm, *Spodoptera exigua* (Smagghe *et al.*, 1996c), in tomato looper, *Chrysodeixis chalcites* (Smagghe *et al.*, 1997), in the Colorado potato beetle, *Leptinotarsa*

Figure 10 Control (left) and tebufenozide intoxicated by ingestion (right) larvae of the white tussock moth. While the control larva continues normal growth and development, the tebufenozide intoxicated larva undergoes a precocious lethal molt. In this case the intoxicated larva is in the slipped head capsule stage.

Figure 11 Photomicrograph showing the severe growth inhibitory effects of ingested tebufenozide in spruce budworm larvae. While both larvae show slipped head capsules, one larva (bottom) shows an additional effect, extrusion of the gut.

decemlineata (Smagghe *et al.*,1999c; Dhadialla and Antrium, unpublished observations), and in cultured abdominal sternites of the mealworm, *Tenebrio molitor* (Soltani *et al.*, 2002). Some general conclusions can be drawn from these studies. Examination of the cuticle following intoxication with any of the three bisacylhydrazines revealed

that the larvae synthesize a new cuticle that is malformed (**Figure 12**; Dhadialla and Antrim, unpublished data). Unlike during normal cuticle synthesis, the lamellate endocuticle deposition in bisacylhydrazine intoxicated larvae is disrupted and incomplete. The epidermal cells in intoxicated larvae have fewer microvilli, show hypertrophied Golgi complex and an increased number of vesicles compared to normal epidermal cells active in cuticle synthesis. The visual observations of precocious production of new cuticle have also been demonstrated by *in vivo* and *in vitro* experiments to demonstrate the inductive effects of tebufenozide and 20E on the amount of chitin in *S. exigua* larval cuticle and chitin synthesis in cultured claspers of *O. nubilalis*, respectively (Smagghe *et al.*, 1997).

At the physiological level the state of "hyperecdysonism," coined by Williams (1967), manifested by bisacylhydrazines in intoxicated susceptible larvae is achieved by various mechanisms. Blackford and Dinan (1997) demonstrated that while larvae of the tomato moth, *Lacnobia oleracea*, detoxified ingested 20E as expected, it remained susceptible to ecdysteroid agonists RH-5849 and RH-5992. This suggested that the metabolic stability of the ecdysone agonists induced the "hyperecdysonism" state in the tomato moth larvae. In another study, RH-5849 was shown to repress steroidogenesis in the larvae of the blowfly, *Caliphora vicina*, as a result of its action on the ring gland (Jiang and Koolman, 1999). However, the production of ecdysteroids in abdominal sterintes of *T. molitor*, cultured *in vitro* in the presence of RH-0345, increased compared to that in control cultured abdominal sternites (Soltani *et al.*, 2002). Ecdysteroid production, measured by an ecdysteroid enzyme immunoassay, by sternites cultured *in vitro* in the presence of 1–10 μM RH-0345 increased with increasing incubation times and concentrations of the bisacylhydrazine. Topical application of 10 μg RH-0345 to newly ecdysed pupae also caused a significant increase in hemolymph ecdysteroid amount as compared to control treated pupal hemolymph ecdysteroid levels. However, there was no effect in the timing of the normal ecdysteroid release in the hemolymph. Contrary to the *in vitro* and *in vivo* effects observed in the above *Tenebrio* study, Williams *et al.* (1997) observed that injection of RH-5849 into larvae of the tobacco hornworm, *Manduca sexta*, induced production of midgut cytosolic ecdysone oxidase and ecdysteroid phosphotransferase activities, which are involved in the inactivation of 20E. In addition, both 20E and RH-5849 caused induction of ecdysteroid 26-hydroxylase activity in the midgut mitochondria

(a)

Endocuticle

Epidermal
cells

(b)

Old
cuticle

Molting
fluid

New
malformed
cuticle

Epidermal
cells

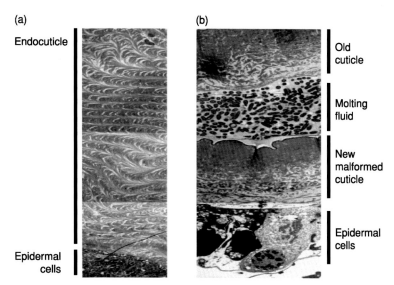

Figure 12 Transmission electron micrographs of the integument of control (a) and halofenozide (fava bean leaves dipped in a 100 ppm solution) ingested (b) second instar Colorado potato beetle larvae (same magnification). In the control integument, the epidermal cells appear normal (not clear in this figure) and the new cuticle is deposited in an ordered lamellate manner. In the integument of halofenozide intoxicated larva, the epidermal cells are highly vacuolated and the new cuticle is malformed, lacking ordered lamellate deposition. Although apolysis of the old cuticle has taken place, it is not shed, and electron dense granulated material (bacterial?) is seen in the ecdysial space. (Both micrographs taken at same magnification).

and microsomes. Subsequently, Williams *et al.* (2002) demonstrated that induction of ecdysteroid 26-hydroxylase by RH-5849, RH-5992 (tebufenozide), and RH-0345 (halofenozide) may be a universal action of bisacylhydrazines in susceptible lepidopteran larvae. These effects were not observed in nonsusceptible larvae of the waxmoth, *Galleria mellonella*, whose ecdysteroid receptor is capable of binding RH-5992 (Williams *et al.*, 2002). The binding in this case may not be sufficient for transactivation of genes that are involved in the molting process, and induced or repressed by 20E.

It is interesting to compare what happens at the molecular level in response to changing hormone levels in control insects compared to bisacylhydrazine intoxicated larvae as shown in **Figure 13**. Bisacylhydrazines, by virtue of their action via the ecdysone receptor, activate genes that are dependent upon increasing titers of 20E. However, those genes that are normally activated either by decreasing titers or absence of 20E are not expressed in the presence of ingested bisacylhydrazine in the hemolymph. During a normal molt cycle, the release of eclosion hormone to initiate the eclosion behavior in the molting larvae is dependent upon clearance of 20E titers in the hemolymph (Truman *et al.*, 1983). The presence of bisacylhydrazines in the hemolymph inhibits the release of eclosion hormone, thus resulting in an unsuccessful lethal molt.

The expression and repression of several genes during a molt cycle in lepidopteran (*M. sexta* and

C. fumiferana) larvae treated with RH-5992 have been investigated (Retnakaran *et al.*, 1995, 2001; Palli *et al.*, 1995, 1996, 1999). In control larvae, at the time of the rise in titer of 20E, as well as those treated with RH-5992, early genes such as MHR3 (in *Manduca*), CHR3 (in *Choristoneura*), and E75, (in *Drosophila*) are expressed. As 20E titers decline, DDC, which requires a transient exposure to 20E followed by its clearance, is expressed. In RH-5992, intoxicated larvae genes like DDC are not expressed, which prevents tanning and hardening of the already malformed new cuticle. During the intermolt period, when 20E is absent, genes like those for the 14 kDa larval cuticle protein (LCP14) that are normally repressed in the presence of 20E, are expressed. Once again, the prolonged presence of RH-5992 in the hemolymph prevents the expression of LCP14, and perhaps other genes that would normally be depressed in the absence of 20E. As a consequence of all the changes induced or repressed by RH-5992, and other bisacylhydrazine insecticides, the molting process is completely derailed at the ultrastructural, physiological and molecular level leading to precocious lethal molt in susceptible larvae.

Farkas and Slama (1999) studied the effects of RH-5849 and RH-5992 on chromosomal puffing, imaginal disc proliferation, and pupariation in larvae of *D. melanogaster*. In this study, they demonstrated that the two bisacylhydrazine compounds acted like 20E in the induction of early

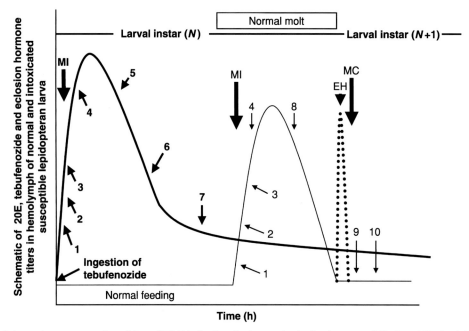

Figure 13 Schematic representation of titers 20E (thin line) and release of eclosion hormone (EH; dotted line), which triggers the ecdysis of the larva to complete a normal molt. The solid bold line represents relative titers of ingested tebufenozide in a susceptible lepidopteran larva (adapted from Dhadialla, unpublished data). Owing to the metabolic stability and amount of tebufenozide in the insect hemolymph, eclosion hormone, the release of which is normally dependent upon the complete decline of 20E, is not released and the intoxicated insect is not able to complete the molt, which leads to premature death. Both methoxyfenozide and halofenozide undergo similar metabolic fate, which is detrimental to intoxicated insect stage. The ecdysone agonists trigger a molt attempt anytime during the feeding stage of a susceptible larval instar. Events that take place during the molt and are dependent upon the increasing and decreasing titers of 20E are also shown. The numbers in bold and regular font represent different events triggered by tebufenozide and 20E, respectively. **1**, 1: Inhibition of feeding; **2**, 2: initiation of new cuticle synthesis; **3**, 3: apolysis of old cuticle from new cuticle resulting in an ecdysial space filled with molting fluid; **4**, 4: head capsule slippage; MI: molt initiated; MC: molt completed; **5**: derailment of the molting process; **6**: eclosion hormone is not released, and the larva stays trapped in its old cuticle and slipped head capsule covering the mouth parts (refer to Figure 9) causing it to starve; **7**: molt attempt is lethal and the ecdysone agonist intoxicated larvae dies of starvation, hemorrhage, and desiccation; 8: cuticle formation continues and molting fluid starts to be resorbed; 9: molt attempt is completed after release of EH and larva ecdyses into the next larval stage; 10: new cuticle hardens and the mouth parts are sclerotized so that the larva may continue its growth and development into the next stage.

chromosomal puffs (74EF and 75B), and regression of pre-existing puffs (25AC, 68C) in larval salivary glands. In the chromosomal puff assay, the ED_{50} for bisacylhydrazine compounds were, however, an order of magnitude less than for 20E. Results of additional experiments in this study using different assays, i.e., glycoprotein glue secretion, imaginal disc evagination, and rescue of phenotypic expression in ecdysone deficient mutants showed the potencies of the bisacylhydrazines were two orders of magnitude lower than for 20E. In spite of the quantitative differences in the assays used, the results confirmed the ecdysone-mimetic action of the two bisacylhydrazines.

In a study to understand the role of ecdysteroids in the induction and maintenance of the pharate first instar diapause larvae of the gypsy moth, *Lymantria dispar*, Lee and Denlinger (1997) demonstrated that a diapause specific 55 kDa gut protein could be induced in ligature isolated larval abdomens

with injections of either 20E or RH-5992. They also demonstrated that the effect of KK-42, an imidazole derivative known to inhibit ecdysteroid biosynthesis, to prevent prediapausing pharate first instar larvae from entering diapause could be reversed by application of 20E or RH-5992. In this case, RH-5992 was two orders of magnitude more active than 20E.

Using 1,5-disubstituted imidazoles, Sonoda *et al.* (1995) demonstrated that the inductive effects of these compounds on precocious metamorphosis of *Bombyx mori* larvae could be reversed by tebufenozide.

4.2.4. Basis for Selective Toxicity of Bisacylhydrazine Insecticides

Unlike halofenozide, which is toxic to both coleopteran and lepidopteran larvae, both tebufenozide and methoxyfenozide are selectively toxic to lepidopteran larvae with a few exceptions of toxicity to dipteran insects, like the midge, C. tentans (Smagghe et al., 2002) and the mosquito species (Darvas et al., 1998). This, at first, was a surprise, especially after the discovery that bisacylhydrazines are true ecdysone agonists, the activity of which, like that of 20E, is manifested via interaction with the ecdysone receptor complex. There are three basic reasons why an ingested insecticide may be differentially toxic to different insects: (1) metabolic differences between susceptible and nonsusceptible insects; (2) lack of transport to the target site in nonsusceptible insects; and (3) target site differences between susceptible and nonsusceptible insects.

Smagghe and Degheele (1994) found no differences in the pharmacokinetics and metabolism of ingested RH-5992 in two susceptible species, Spodoptera exigua and S. exempta, and a nonsusceptible insect, Leptinotarsa decemlineata. Similar results were obtained when the pharmacokinetics and metabolism of RH-5992 were investigated in larvae of susceptible S. exempta and the nonsusceptible Mexican bean beetle, Epilachna verivesta (Dhadialla and Thompson, unpublished results).

In competitive displacement EcR binding assays, it became apparent that one of the reasons for the differential selectivity of tebufenozide and methoxyfenozide might be due to differences in the LBDs of EcRs from different insects. While 20E binds to EcR complexes from insects from different orders with similar affinities, the affinity of tebufenozide and methoxyfenozide is very disparate for insects even within an insect order (**Table 2**). Hence, the very high affinity of tebufenozide and methoxyfenozide to lepidopteran EcR complexes correlates very well with their toxicity to larvae within this order, as well as those few amongst the Diptera. However, the same logic does not explain the toxicity of halofenozide and lack of toxicity of tebufenozide to coleopteran larvae, especially when the affinity of tebufenozide is higher than that of halofenozide for binding to EcRs from a coleopteran insect. Perhaps, in this case, the low binding affinity of halofenozide is compensated for by its greater metabolic stability (Smagghe and Degheele, 1994; Farinos et al., 1999).

To determine if there are additional reasons at the cellular level for the selective toxicity of tebufenozide, Sundaram et al. (1998) made use of four lepidopteran and dipteran cell lines all of which are responsive to 20E indicating the presence of functional EcRs. A lepidopteran cell line from C. fumiferana midgut cells (CF-203) and a dipteran cell line (DM-2) derived from D. melanogaster embryos responded to 20E in a similar dose-dependent induction of the transcriptional factor, hormone

receptor 3 (HR3) CHR3 and DHR3 recspectively, which represents one of the early 20E inducible genes in the ecdysone signaling pathway. The lepidopteran CF-203 cells could produce CHR3 mRNA at 10^{-10} M RH-5992 concentration, as opposed to 10^{-6} M of this compound required for induction of very low amounts of DHR3 transcripts. In further experiments using ^{14}C-RH-5992, the authors were able to show that CF-203 cells accumulated and retained sixfold higher amounts of the radiolabeled compound than DM-2 cells over the same incubation period. Similar results were obtained with another lepidopteran cell line (MD-66) derived from Malacosomma disstria, and a dipteran cell line (Kc) from D. melanogaster embryos. This significantly higher retention of radiolabeled RH-5992 in lepidopteran cells compared to dipteran cells was in contrast to similar levels of retention of tritiated ponasterone A, a potent phytoecdysteroid. The extremely low amounts of RH-5992 in dipteran cells was due to an active efflux mechanism that was temperature-dependent and could be blocked with 10^{-5} M oubain, an inhibitor of Na$^+$, K$^+$-ATPase.

In developing a yeast based ecdysteroid responsive reporter gene assay, Tran et al. (2001) had to use a yeast strain that was deficient in the pdr5 and snq2 genes, both of which are involved in multiple drug resistance, to increase the sensitivity of the transformed strain to both steroidal and nonsteroidal ecdysone agonists. Subsequently, Retnakaran et al. (2001) demonstrated that while the wild-type yeast Saccharomyces cervisiae actively excluded tebufenozide, the pdr5 deleted mutant strain retained significantly higher amounts of radiolabeled tebufenozide (Hu et al., 2001; Retnakaran et al., 2001). Retransformation of the mutant strain with the pdr5 gene enabled the active exclusion of radiolabeled tebufenozide, thus once again suggesting a role for an ATP binding cassette transporter.

Some general conclusions can be drawn from the above studies regarding the selective insect toxicity of tebufenozide and methoxyfenozide. While the absence of significant differences in the pharmacokinetics and metabolism of these compounds in susceptible and nonsusceptible insects is not directly responsible for selective toxicity, it is indirectly, because the hormone levels of tebufenozide, after its ingestion by lepidopteran larvae, are about 2000-fold higher than the K_d equivalent for the lepidopteran EcR. The selective toxicity of tebufenozide and methoxyfenozide to lepidopteran larvae is probably due to the clearance of these agonists as shown in resistant cell lines where they are actively pumped out. It would be interesting to investigate if cells derived from C. tentans and mosquito tissues, like lepidopteran cells, are devoid of an active mechanism to eliminate tebufenozide and methoxyfenozide, because ecdysone receptors from both these dipterans have high affinities for bisacylhydrazines (Kapitskaya et al., 1996; Smagghe et al., 2002) and are susceptible to tebufenozide (Smagghe et al., 2002).

4.2.5. Spectrum of Activity of Commercial Products

4.2.6.1. Chromofenozide (MATRIC®; KILLAT®; ANS-118; CM-001)   Jointly developed by Nippon Kayaku Co. Ltd (Saitama, Japan), and Sankyo, Co. Ltd. (Ibaraki, Japan), chromofenozide is the latest amongst the nonsteroidal ecdysone agonist insecticides registered for control of lepidopteran

pests on vegetables, fruits, vines, tea, rice, arboriculture, ornamentals, and other crops in Japan (**Table 6**). Chromofenozide has relatively safe mammalian, avian, and aquatic organism toxicology (**Table 7**). It also has no adverse effects on nontarget arthropods, including beneficials.

4.2.5.2. Halofenozide (MACH II; RH-0345)   Unlike tebufenozide, methoxyfenozide, and chromofenozide, halofenozide is more soil systemic and has a broader pest spectrum of activity (Lepidoptera and Coleptera; **Table 6**). It was developed for control of beetle grub and lepidopteran larval pests of turf in lawns and on golf courses.

Halofenozide gave excellent control of the Japanese beetle (*Popillia japonica*) and the oriental

Table 6   Spectrum of pest activity of the four commercial ecdysone agonist insecticides

Tebufenozide and methoxyfenozide (Lepidoptera specific)	Methoxyfenozide (Lepidoptera specific)	Chromofenozide (Lepidoptera specific)	Halofenozide (Lepidoptera and Coleoptera specific)
Adoxophyes spp.	Clysia ambiguella	Spodoptera littura	Agrotis ipsilon
Anticarcia gemmatalis	Grapholitha molesta	Spodoptera exigua	Spodoptera frugiperda
Boarmia rhombodaria	Heliothis spp.	Spodoptera littoralis	Sod webworm (Crambinae)
Chilo suppressalis	Helicoverpa zea	Plutella xylostella	Anomala orientalis
Choristoneura spp.	Ostrinia nubilalis	Cnaphalocrosis medinalis	Maladera canstanea
Cnapholocrocis medinalis	Phyllonorychter spp.	Ostrinia furnacalis	Ataenius spretulus
Cydia spp.	Tuta absoluta	Adoxophyes orana	Rhizotrogus majalis
Diatraea spp.	Keiferia lycopersicella	Heliothis virescens	Popillia japonica
Diaphania spp.		Lobesia botrana	Cotinis nitida
Hellula spp., Homona magnamima		Cydia pomonella	Cyclocephala spp.
Lobesia botrana		Chilo suppressalis	Phyllophaga spp.
Lymantria dispar		Anticarcia gemmatalis	Hyperodes near anthricinus
Planotortrix spp.,			
Platynota idaeusalis			
Plusia spp., Pseudoplusia includens			
Rhicoplusia nu, Sparganothis spp.			
Spodoptera spp.			
Trichoplusia ni			
Epiphyas postvittana, Grapholitha prunivora			
Memestra brassicae, Operophthera brumata			

Table 7   Ecotoxicological profile of bisacylhydrazine insecticides

	Tebufenozide	Methoxyfenozide	Chromofenozide	Halofenozide
Avian: mallard duck, LC_{50} (8-day dietary)	>5000 mg kg^{-1}	>5620 mg kg^{-1}		>5000 mg kg^{-1}
Bobwhite quail, LC_{50} (8-day dietary)	>5000 mg kg^{-1}	>5620 mg kg^{-1}	>5000 mg kg^{-1} (Japanese quail, 14-day)	4522 mg kg^{-1}
Aquatic: bluegill sunfish, acute LC_{50} (96 h)	3.0 mg l^{-1}	>4.3 mg l^{-1}		>8.4 mg l^{-1}
Daphnia magna, acute EC_{50} (48 h)	3.8 mg l^{-1}	3.7 mg l^{-1}	>189 mg l^{-1} (3 h)	3.6 mg l^{-1}
Honeybee (oral and contact), acute			>100 μg bee^{-1} (contact)	
LD_{50}	>234 μg bee^{-1}	>100 μg bee^{-1}	>133 μg bee^{-1} (oral)	>100 μg bee^{-1}
Earthworm, LC_{50} (14 days)	1000 mg kg^{-1}	>1213 mg kg^{-1}	>1000 mg kg^{-1}	980 mg kg^{-1}

beetle (*Exomala orientalis*) when applied at 1.1 kg active ingredient (a.i.) ha^{-1} to a mixed field population on turf plots (Cowles and Villani, 1996; Cowles *et al.*, 1999). In the same study, higher rates (1.7–2.2 kg a.i. ha^{-1}) were required to control the European chafer, *Rhizotrogus (Amphimallon) majalis (Razoumowsky)*, population. The Asiatic garden beetle, *Maladera castanaea* (Arrow), was not sensitive to any of the doses of halofenozide tested.

The recommended rates for the control of lepidopteran larvae (such as cutworms, sod webworms, armyworms, and fall armyworms) and beetle grubs (Japanese beetle, northern and southern masked chafer, June beetle, Black turfgrass ataenius beetle, green June beetle, annual bluegrass weevil larvae, bill bugs, aphodius beetles, European chafer, and oriental beetle) are 12 kg a.i. ha^{-1} and 18–24 kg a.i. ha^{-1}, respectively.

4.2.5.3. Tebufenozide (MIMIC; CONFIRM; ROMDAN; RH-5992) and methoxyfenozide (RUNNER; INTREPID; PRODIGY; RH-2485)

Tebufenozide, at low use rates, shows remarkable insect selectivity for the control of lepidopteran larvae, including many members of the families Pyralidae, Pieridae, Tortricidae, and Noctuidae, which are pests of tree fruit, vegetables, row crops, and forests (**Table 6**; Le *et al.*, 1996; Dhadialla *et al.*, 1998; Carlson *et al.*, 2001). Laboratory studies have shown the following: tebufenozide administered through artificial diet or by leaves treated with tebufenozide is efficacious against *C. fumiferana* (Retnakaran *et al.*, 1999), corn earworm and fall armyworm (Chandler *et al.*, 1992), and beet armyworm (Chandler, 1994); both tebufenozide and methoxyfenozide are efficacious against laboratory reared and field populations of codling moth, *Cydia pomonella* (Pons *et al.*, 1999; Knight *et al.*, 2001), laboratory reared and multi-resistant cotton leafworm, *Spodoptera littoralis* (Smagghe *et al.*, 2000b), field populations of filbertworm, *Cydia latiferreana*, on hazelnut, and against *D. grandiosella* (Trisyono and Chippendale, 1998).

Tebufenozide has been used very successfully for the control of *C. fumiferana*, which is the most destructive insect defoliator of spruces, *Picea* sp., and balsam fir, *Abies balsamea* (L.) Mill, in North American forests (Sundaram *et al.*, 1996; Cadogan *et al.*, 1997, 1998, 2002). Cadogan *et al.* (1998) used a 240 g l^{-1} formulation of commercial tebufenozide (MIMIC® 240LV) mixed with water and a tracer dye (Rhodamine WT), which was sprayed onto the trees using a light-wing aircraft, to determine the foliar deposit of the insecticide and its

efficacy against *C. fumiferana*. Single applications of 1 or 2 L ha^{-1} resulted in a 61.4–93.6% or 85.6–98.3% reduction, respectively, of the spruce budworm populations. Following double applications, the mean population reductions ranged from 93.6 to 98.3% compared to mean population reductions of 61% in untreated control fields. Defoliation in untreated control plots was about 92% and in plots with single or double spray applications, defoliation was 40–62% or 31–62%, respectively. These results indicated that a double spray application was most efficacious.

MIMIC 240LV has also provided effective control of Jackpine budworm, *Choristaneura pinus*, and *Lymantria dispar* in North American forests, as well as control of processionary caterpillar, *Thaumetopomea pityocampa*, and European pine shoot moth, *Rhyaciona buoliana*, in forests in Spain and Chile, respectively (Lidert and Dhadialla, 1996; Butler *et al.*, 1997).

Two of the major lepidopteran pests of tree-fruits are codling moth and leaf rollers, which have been controlled by frequent applications of azinphosmethyl and other organophosphate insecticides (Riedl *et al.*, 1999). With increasing requirements to use more selective insecticides without effecting biological control of other pests, much attention has been paid to the use of IGRs like chitin synthesis inhibitors, JHAs, and agonists of 20E. The effectiveness of tebufenozide against larvae of the codling moth, *C. pomonella*, has been tested both under laboratory and field conditions (Pons *et al.*, 1999; Knight, 2000). Pons *et al.* (1999) found that tebufenozide has ovicidal and larvicidal but not adulticidal activity against *C. pomonella*. Tebufenozide was about 73 times more effective when presented to neonate larvae in artificial diet as compared to treated apples; possibly because once the neonates have burrowed their way into the apples they are no longer exposed to the insecticides. However, in tebufenozide treated artificial diets, the susceptibility to the insecticide decreased with increasing age. Because of the feeding behavior of codling moth larvae, which generally do not feed and do not ingest large amounts of foliage or fruit as they burrow into the core of the fruit, the exposure to the insecticides is very limited. However, the reproductive and ovicidal effects of tebufenozide may be more important than its larvicidal effects (Knight, 2000). Knight (2000) found that in 30 field trials, tebufenozide exhibited both residual and topical ovicidal effects for codling moth egg masses laid on leaves. Similar results were obtained by Pons *et al.* (1999).

These studies indicated that the timing of tebufenozide sprays was important in not only preventing codling moth injury to the apples, but also in controlling oblique banded leaf rollers for which both tebufenozide and methoxyfenozide have been shown to have lethal and sublethal effects (Riedl and Brunner, 1996; Sun et al., 2000). Knight (2000) proposed that delaying the first application of tebufenozide until after petal fall in apple for *Choristoneura rosaceana* (Harris) would correspond to an approximate codling moth timing of 30-degree days at which time this insect would be susceptible as well. This strategy, which targets the adults and eggs, was effective in the control of codling moth.

Both tebufenozide and methoxyfenozide are active primarily by ingestion, but also exhibit selective contact and ovicidal activity (Trisyono and Chippendale, 1997; Sun and Barrett, 1999; Sun et al., 2000).

4.2.6. Ecotoxicology and Mammalian Reduced Risk Profiles

Both tebufenozide and methoxyfenozide have been classified by the US Environmental Agency (EPA) as reduced risk pesticides because of their low acute and chronic mammalian toxicity, low avian toxicity, and their safety to most beneficial arthropods and compatibility with integrated pest management programs. Both tebufenozide and methoxyfenozide are toxic primarily to lepidopteran pests, but they are also toxic to a few dipteran (*C. tentans*, mosquitos) pests. Of the 150 insect species tested, both tebufenozide and methoxyfenozide were found to be safe to members of the insect orders Hymenoptera, Coleoptera, Diptera, Heteroptera, Hemiptera, Homoptera, and Neuroptera (Glenn Carlson, unpublished data). When used at rates that are 18- to 1500-fold greater than that producing 90% mortality in lepidopteran larvae, both tebufenozide and methoxyfenozide had little or no effect on a panel of non-lepidopteran pests (Coleoptera, Homoptera, mites, and nematodes; Trisyono et al., 2000; Carlson et al., 2001; Medina et al., 2001). Similarly, laboratory and/or field studies have shown that the two insecticides have little or

no adverse effect on a wide range of non-lepidopteran beneficial insects, like honeybees and predatory insects/mites, when applied at rates used for control of lepidopteran larvae.

In 1998, methoxyfenozide became one of the only four pesticide products to be awarded the "Presidential Green Chemistry Award," which was established by the US Government during 1995 to recognize outstanding chemical processes and products that reduce negative impact on human health and the environment relative to the currently available technology.

The ecotoxicological and mammalian toxicity data for the four commercial ecdysone agonist insecticides are shown in **Tables 7** and **8**. These data clearly show the low mammalian, avian, and aquatic toxicity of these insecticides.

4.2.7. Resistance, Mechanism, and Resistance Potential

Tebufenozide is the first nonsteroidal ecdysone agonist with commercial application and has been used in Western Europe since the mid 1990s. In the USA, the first uses occurred concurrently in 1994 in Alabama and Mississippi under Section 18 exemptions (Walton et al., 1995; Dhadialla et al., 1998).

Resistance to tebufenozide was documented for the first time in the codling moth *C. pomonella* around Avignon in southern France (Sauphanor and Bouvier, 1995; Sauphanor et al., 1998b) and in the greenheaded leafroller, *Planotortrix octo*, in New Zealand (Wearing, 1998). A small survey with the beet armyworm, *S. exigua*, collected in different greenhouses in Western Europe (Spain, Belgium, The Netherlands) showed lack of resistance development to tebufenozide (Smagghe et al., 2003b). In this same period, Smagghe et al. (1999a) kept a laboratory strain of *S. exigua* under continuous pressure with sublethal concentrations of tebufenozide over 12 generations, and in the sixth generation a significantly lower toxicity (about fivefold) was observed. In addition, retention and

Table 8   Mammallian reduced risk profile of bisacylhydrazine insecticides

	Tebufenozide	Methoxyfenozide	Chromofenozide	Halofenozide
Acute oral LD$_{50}$ (rat, mouse)	>5000 mg kg^{-1}	>5000 mg kg^{-1}	>5000 mg kg^{-1}	2850 mg kg^{-1}
Acute dermal LD$_{50}$	>5000 mg kg^{-1}	>2000 mg kg^{-1}	>2000 mg kg^{-1}	>2000 mg kg^{-1}
Eye irritation (rabbit)	Nonirritating	Nonirritating	slightly irritating	Moderately irritating, positive for contact
Dermal sensitization (guinea pig)	Nonsensitizer	Negative	Mildly sensitizing	Allergy
Ames assay	Negative	Negative	Negative	Negative
Acute inhalation	>4.3 mg l^{-1}	>4.3 mg l^{-1}		>2.7 mg l^{-1}
Reproduction (rat)	No effect	No effect	No effect	No effect

metabolic fate studies showed the importance of oxidative metabolism, leading to a rapid clearance from the body. Interestingly, a decrease in oviposition was noted with the lower toxicity indicating a possible fitness cost related to development of resistance (Smagghe and Degheele, 1997; Smagghe *et al.*, 1998). More recently, Moulton *et al.* (2002) reported resistance to tebufenozide in third instar larvae, which reached levels of 150-fold in a selected strain of *S. exigua* from Bangbuathong (Thailand) as compared with a laboratory reference strain. In this region of Thailand, many insecticides including organophosphates (OPs), pyrethroids, benzoylphenylurea (BPUs), and *Bacillus Thurigiensis* (Bt) formulations, and even new IGRs have been rendered ineffective due to ill-advised agricultural practices, most notably dilution of insecticide residues on leaves by overhead drench irrigation (Moulton *et al.*, 2002). This practice is likely to be responsible for the high incidence of insecticide multiresistance in this area and the highly accelerated rate of resistance development. When this Thailand strain was dosed with methoxyfenozide, 120-fold lower toxicity was observed. These selection assays with tebufenozide and methoxyfenozide showed a reduction in toxicity for both compounds, suggesting at least some commonality of resistance mechanism (Moulton *et al.*, 2002). A greenhouse raised strain of *S. exigua*, in southern Spain was also found to be resistant to tebufenozide and methoxyfenozide. Although the level of resistance was lower in this second strain, it was high enough to allow studies on the mechanism(s) of resistance. In general, a higher breakdown activity leads to lower levels of the parent toxophore. For tebufenozide and methoxyfenozide the major first phase route of detoxification was through oxidation (Smagghe *et al.*, 1998, 2003). In addition, piperonyl butoxide, a P450 inhibitor, significantly synergized the toxicity of tebufenozide and methoxyfenozide, whereas DEF, an esterase inhibitor, was ineffective (Smagghe *et al.*, 1998, 1999), indicating that a lower toxicity was more likely from an increase in oxidative activity, rather than in esterase activity. At the cellular level, Retnakaran *et al.* (2001) reported that tebufenozide accumulated selectively in lepidopteran Cf-203 (*C. fumiferana*) cells in contrast to dipteran Dm-2 cells (*D. melanogaster*), which actively excluded the compound (see Section 4.2.4 for more details). Perhaps such exclusion systems may also account for the fact that older instars of the white-marked tussock moth (*Orgyia leucostigma*) are resistant to tebufenozide (Retnakaran *et al.*, 2001). The characterization of all such possible resistance processes is essential to provide information that is helpful

to prevent resistance from developing towards this valuable group of IGRs. But more information is needed with additional strains collected from the field, especially where growers have severe pest control problems, before a general interpretation can be formulated on resistance risks for this new type of IGRs in the field.

While the available data suggest oxidative metabolism as the primary reason for development of resistance to tebufenozide and methoxyfenozide, there is no evidence so far that suggests that target site modification could be involved as another route to resistance development in field or laboratory insect populations. However, at the cellular level, there is evidence that there could be alterations in the target site(s). Using insect cell lines, Cherbas and co-workers were the first to report that *in vitro* cultured Kc cells of *D. melanogaster* did not respond to 20E after continuous exposure (Cherbas, personal communication). Similarly, Spindler-Barth and Spindler (1998) reported that cells of another dipteran, *C. tentans*, maintained in the continuous presence of increasing concentrations of 20E or tebufenozide over a period of 2 years, developed resistance to both compounds. In these resistant subclones, all 20E regulated responses that are known to occur in sensitive cells were no longer detectable, suggesting that the hormone signalling pathway itself was interrupted (Spindler-Barth and Spindler, 1998; Grebe *et al.*, 2000). Further ligand binding experiments with extracts containing ecdysone receptors from susceptible and resistant cells indicated that ligand binding to EcR from resistant clones was significantly decreased. Moreover, an increase in 20E metabolism and a reduction in receptor concentration were noted in some clones. Similar effects have also been observed in another study using imaginal discs of *S. littoralis* selected for resistance to tebufenozide (Smagghe *et al.*, 2001).

While resistance to ecdysone agonist insecticides is inevitable, it can be prevented from occurring sooner with good resistance monitoring programs.

4.2.8. Other Chemistries and Potential for New Ecdysone Agonist Insecticides

After the discovery of the first three bisacylhydrazine insecticides by scientists at Rohm and Haas Company, Spring House, PA, USA, a number of laboratories initiated such efforts to discover new and novel chemistries with ecdysone mode of action and different pest selectivity. Mikitani (1996) at Sumitomo Company, Japan, discovered a benzamide, 3,5,-di-*t*-butyl-4-hydroxy-N-isobutyl-benzamide

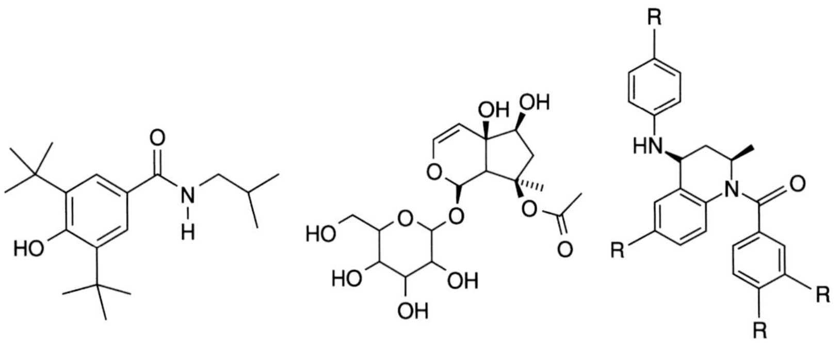

Figure 14 Chemical structures of new ecdysone agonists reported in literature: (**1**) 3,5-di-*t*-butyl-4-hydroxy-*N*-isobutyl-benzamide (DTBHIB) from Sumitomo; (**2**) 8-*O*-acetylhrapagide from Merck Research Laboratories, Westpoint, PA, USA and (**3**) tetrahydroquino-line from FMC Corporation, Princeton, NJ, USA. In (3) R = halide.

(DTBHIB; **Figure 14**) and Elbrecht *et al.* (1996) at Merck Research Laboratories, Westpoint, PA, USA, reported the isolation of an iridoid glycoside, 8-O-acetylharpagide (**Figure 14**), from *Ajuga reptans*. Both these compounds were reported to induce 20E-like morphological changes in *Drosophila* Kc cells, as well as competitively displace tritiated ponasterone A from *Drosophila* ecdysteroid receptors with potencies similar to that of RH-5849, the unsubstituted bisacylhydrazine. However, the insecticidal activity of these compounds was not described. Attempts to replicate the results of Mikitani (1996) using DTBHIB and analogs failed to demonstrate that these compounds were competitive inhibitors of tritiated ponasterone A binding to DmEcR/DmUSP produced by *in vitro* transcription and translation (Dhadialla, unpublished observations). On the other hand, Dinan *et al.* (2001) demonstrated that the results obtained by Elbrecht *et al.* (1996) were due to co-purification of ecdysteroids in their 8-O-acetylharpagide preparation. When used as a highly purified preparation, Dinan *et al.* (2001) found that 8-O-acetylharpagide was not active as an agonist or an antagonist in *D. melanogaster* BII cell bioassay, and neither did it compete with tritiated ponasterone A for binding to the lepidopteran ecdysteroid receptor complex from *C. fumiferana*.

Finally, scientists at FMC discovered a new tetrahydroquinoline (THQ) class of compounds (**Figure 14**) that competitively displaced tritiated ponasterone A from both dipteran (*D. melanogaster*) and lepidopteran (*H. virescens*) EcR and USP heterodimers (unpublished data). Interestingly, the most active analogs of this class of compounds bound the DmEcR/DmUSP with much higher affinity than the HvEcR/HvUSP. This is the reverse of what was observed with bisacylhydrazines. When tested for ligand binding to EcR/USP proteins from

L. migratoria, B. argentifoli, and T. molitor, to which bisacylhydrazines show no measurable affinity, members of the THQ were found to bind with measurable affinity (μM range; Dhadialla and colleagues, unpublished results). Further work on this chemistry was continued at Rohm and Haas Company, Spring House, PA, USA, and its subsidiary, RheoGene LLC, Malvern, PA, USA, which resulted in the synthesis of a number of analogs that were active in transactivating reporter genes fused to different insect EcRs and heterodimeric partners (*L. migratoria* RXR, and human RXR) in mouse NIH3T3 cells (Michelotti *et al.*, 2003).

The discovery of THQs with ligand binding activities to various EcRs (Michelotti *et al.*, 2003), some of which do not interact with bisacylhydrazines, and the interpretation of X-ray crystal structure results of liganded HvEcR/HvUSP (Billas *et al.*, 2003) provides good evidence for the potential to discover new chemistries with ecdysone agonist activities. It should also be possible to design chemistries that specifically interact with EcR/USPs from a particular insect order.

4.2.9. Noninsecticide Applications of Nonsteroidal Ecdysone Agonists; Gene Switches in Animal and Plant Systems

A number of researchers started to explore the utilization of the ecdysone receptor as an inducible gene switch due to the knowledge that neither mammals nor plants have ecdysone receptors, and the discovery of bisacylhydrazine as true ecdysone agonists with reduced risk ecotoxicology and mammalian profiles. Gene switches are inducible gene regulation systems that can be used to control the expression of transgenes in cells, plants, or animals. There is a recognized need for tightly regulated eukaryotic molecular gene switch applications, such as for gene therapy, and in understanding the role played by specific proteins in signaling pathways, cell differentiation, and development (Allgood and Eastman, 1997). Other applications include large-scale production of proteins in cells, cell-based high-throughput screening assays, and regulation of traits in both plants and animals. Unlike all other previous and current insecticides, commercial nonsteroidal ecdysone agonists are the first class of insect toxic chemistry that has led to such a high level of interest and utility outside the insecticide arena. The following is a brief overview of the use of bisacylhydrazine insecticides and EcRs as gene switches in mammalian and plant systems.

4.2.9.1. Gene switch in mammalian systems

Christopherson et al. (1992) demonstrated that DmEcR could function as an ecdysteroid-dependent transcription factor in human 293 cells cotransfected with an RSV-based expression vector that encoded DmEcR and a reporter gene, which contained four copies of Drosophila Hsp27 linked to a MTV promoter-CAT construct. They demonstrated that the activity of DmEcR could not be activated by any of the mammalian steroid hormones tested. Even amongst the ecdysteroids, the phytoecdysteroid muristerone A was the best transactivator of the reporter gene, while 20E was not as effective. These authors further demonstrated that chimeric receptors containing the LBD of DmEcR and the DBD of glucocorticoid receptor could also function when cotransfected with plasmids containing glucocorticoid receptor response elements fused to a reporter gene. Subsequently, No et al. (1996) demonstrated that an EcR-based gene switch consisting of DmEcR, Homo sapiens RXR, and appropriate response elements fused to a reporter gene in mammalian cells or transgenic mice could be transactivated with micromolar levels of muristerone A. Muristerone was shown to maintain its activity when injected into mice and was not found to be toxic, teratogenic, or inactivated by serum binding proteins. Moreover, overexpression of modified DmEcR and RXR did not appear to to be toxic, at least in transfected cells. This study, like an earlier one (Yao et al., 1995), demonstrated that HsRXR could substitute for its insect homolog and heterodimer partner for EcR, USP. However, in all these studies, the range of activating ligands was limited to ecdysteroids and, in particular, to muristerone A. In addition, RXR, which is present in almost all mammalian cells, was found not to be as good a partner for EcR as USP. Hence, very high endogenous levels were necessary for stimulation with muristerone A to occur.

With the availability of EcR and USPs from other insects, especially lepidopteran insects, and understanding the high affinity of nonsteroidal ecdysone agonists like tebufenozide and methoxyfenozide for lepidopteran EcRs/USPs, Suhr et al. (1998) demonstrated that the B. mori EcR (BmEcR), unlike DmEcR, could be transactivated to very high levels in mammalian cells with tebufenozide without adding an exogenous heterodimer partner like RXR. The endogenous levels of RXR were enough to provide the high transactivation levels. The work by Suhr et al. (1998) further demonstrated that while the D domain (hinge region) of EcR was necessary for high-affinity heterodimerization with USP, both D and E (LBD) domains were necessary for high-affinity interaction with RXR. By creating chimeric EcR using parts of E-domains from BmEcR and DmEcR, these investigators showed that a region in the middle of the E-domain (amino acids 402 to 508) in BmEcR constituted the region conferring high specificity for tebufenozide. Subsequent research by others has mainly focused on using EcRs and USPs from other insects as well as using different promoters for expression of the transfected genes (Kumar et al., 2002; Dhadialla et al., 2002; Palli and Kapitskaya, 2002; Palli et al., 2002, 2003). Based on EcR homology modeling studies and site-directed mutagenesis of selected amino acid residues, Kumar et al. (2002) found that a CfEcR mutant, which had an A110P mutation in the LBD, was responsive only to nonsteroidal ecdysone agonists, including methoxyfenozide, but not to any of the ecdysteroids tested. This demonstrated that the affinity and functional specificity of an EcR for a ligand can be altered, thus offering the possibility of using multiplexed EcR-based gene switches that could regulate different traits with different ligands. Palli et al. (2003) developed a two-hybrid EcR based gene switch, which consisted of constructs containing DEF domains of CfEcR fused to the GAL4 DNA binding domain, and CfUSP or Mus musculus retinoid X receptor (MsRXR) EF domains fused to the VP16 activation domain. These constructs were tested in mammalian cells for their ability to drive a luciferase gene placed under the control of GAL4 response elements and synthetic TATAA promoter. This combination gave very low level basal activation of the reporter gene in the absence of the inducer. In the presence of the inducer, there was a rapid increase in the expression of the luciferase reporter gene, reaching levels as high as 8942-fold greater than basal level by 48 h. Withdrawal of the ligand resulted in 50% and 80% reduction of the reporter gene by 12 h and 24 h, respectively.

These studies clearly demonstrate the potential of utilizing EcR based gene switches that can be activated by ecdysteroids and nonsteroidal ecdysone agonists like tebufenozide, methoxyfenozide, and others. The fact that both tebufenozide and methoxyfenozide are registered as commercial insecticides, and have proven reduced risk mammalian and ecotoxicology profiles, makes them very attractive as inducers of the EcR based gene switches. The work of Kumar et al. (2002) clearly demonstrates the potential of mutating EcR to change its ligand specificity, thus opening additional possibilities of extending the use of the EcR gene switch in a multiplexed manner.

4.2.9.2. Gene switch for trait regulation in plants The reader is referred to very good recent reviews on this topic that not only describe the EcR-based chemically inducible gene regulation systems, but also other systems that have utility in plants (Jepson et al., 1998; Zuo and Chua, 2000; Padidam, 2003). This section is restricted to descriptions of EcR-based gene switch systems.

In the mid-1990s, a number of agricultural companies initiated research to exploit the use of the EcR-based gene switch and nonsteroidal ecdysone agonists like tebufenozide as chemical inducers for regulation of traits (for example, fertility, flowering, etc.) in plants. Initial work was done using DmEcR and DmUSP as components of the gene switch (Goff et al., 1996). In this case, the researchers used chimeric polypeptides (GAL4 DBD fused to LBD of DmEcR and VP16 activation domain fused to DmUSP) to activate the luciferase reporter gene fused to GAL4 response element in maize cells. In the presence of 10 µM tebufenozide, about 20- to 50-fold activation of luciferase expression was obtained. Subsequently, a number of researchers developed variants of EcR gene switches using different chimeric combinations of heterologous DBD, LBDs from different lepidopteran EcRs, and transactivation domains that could interact with appropriate response elements to transactivate a reporter gene in a ligand-dependent manner (Martinez et al., 1999a; Unger et al., 2002; Padidam et al., 2003). For example, Jepson et al. (1996) and Martinez et al. (1999b) used chimeric H. virescens EcR (HvEcR) composed of glucocorticoid receptor transactivation and DBD fused to LBD of HvEcR and GUS reporter gene fused to glucocorticoid receptor response element for transfection of maize and tobacco protoplast. In both cases, weak transactivation of the GUS reporter gene was obtained with tebufenozide and muristerone A, though the response with the latter was much

lower than with tebufenozide. However, 10 µM and higher concentrations of the two ligands were required to transactivate the reporter genes.

Padidam et al. (2003) used a chimeric C. fumiferana EcR (CfEcR) composed of a CfEcR LBD, GAL4 and LexA DBDs, and VP16 activation domains that could be activated with methoxyfenozide in a dose-dependent manner from a GAL4- or LexA-response element to express a reporter gene. These researchers used Arabidopsis and tobacco plants for transformation with the gene switch components and obtained several transgenic plants that had little or no basal level of expression in the absence of methoxyfenozide. In the presence of methoxyfenozide, reporter expression was several fold higher than in the absence of methoxyfenozide. The above studies provided ample evidence of the utility of EcR gene switch, especially those that utilize EcR from a lepidopteran species, and tebufenozide and methoxyfenozide as chemical inducers for trait regulation in plants.

Demonstration of the utility of EcR-based gene switch for trait regulation in maize was demonstrated by Unger et al. (2002). A mutation in maize (MS45), which results in male-sterile phenotype, could be reversed by complementation to gain fertility using methoxyfenozide-dependent chimeric receptor gene switch to express the wild-type MS45 protein in tapetum and anthers. These researchers used the EcR LBD from, O. nubialis to generate a chimeric receptor. The chimeric receptor was introduced into MS45 maize with the MS45 gene fused to the GAL4 response element, which in the absence of methoxyfenozide were male sterile. However, application of methoxyfenozide to plants containing either a constitutive promoter or anther specific promoter resulted in the restoration of fertility to MS45 plants grown in either the greenhouse or the field.

It is interesting to note that in all the above studies, except those by Goff et al. (1996), reporter transactivation response to tebufenozide, methoxyfenozide, or muristerone A via the EcR gene switch was obtained without the requirement of an exogenous heterodimeric partner (USP or RXR), suggesting that there may be other factor(s) in plants that can substitute for USP as a partner for EcR, or that EcR can function as a homodimer. However, as far as is known, there is no evidence of EcR binding an ecdysteroid or nonsteroid ligand in the absence of USP or RXR. Irrespective, these studies provide ample demonstration of the utility of EcR-based gene switch, which can be regulated with an ecdysteroid or a nonsteroidal ecdysone agonist. The use of nonsteroidal ecdysone agonists, like any of the commercialized products, is attractive because their reduced risk for the environment and mammals, birds, aquatic animals, and beneficial arthropods has been found to be acceptable. The specificity and the high affinity with which the bisacylhydrazines bind ecdysteroid receptors also offers the opportunity to utilize different ligands and ecdysone receptors to regulate more than one trait in a plant or an animal system.

4.2.10. Conclusions and Future Prospects of Ecdysone Agonists

The discovery of bisacylhydrazine insecticides with a new mode of action, very high degree of selective insect toxicity and reduced risk to the environment, has spurred a lot of interest in this mode of action and chemistry. The biochemical and molecular information on insect EcRs from different orders of insects, crystal structures of liganded HvEcR/HvUSP, and of DmUSP and HvUSP, and the discovery of new non-bisacylhydrazine chemistries, offers potential for the discovery of other new and novel chemistries with ecdysone mode of action and those that would be selectively toxic to nonlepidopteran and coleopteran pests. Although the existing bisacylhydrazine insecticides are most efficacious when ingested, the use of this type of control method could be extended to act against other types of pests if ways could be found for these insecticides to be effective via topical or systemic applications. These bisacylhydrazine insecticides, because of their new mode of action, insect selectivity, and reduced risk ecotoxicology and mammalian profiles, are ideally suited for use in integrated insect management programs.

An altogether different benefit from the discovery of nonsteroidal ecdysone agonists as insecticides has been in the area of gene switch applications in plants and animals. In this particular application, it may be possible to alter the molecular structure of ecdysone receptor LBDs to accommodate new chemistries as gene switch ligands. This is an exciting active area of research that, if successfully applied, offers a number of opportunities in trait and gene regulation in both plant and animal systems. The search for new and novel chemistries that target ecdysone receptors will continue to be a fruitful area of research for several years.

4.3. Juvenile Hormone Analogs

Kopec (1922) extirpated the brain of a caterpillar and demonstrated that this prevented pupation; he attributed this phenomenon to a humoral factor produced by the brain. This discovery heralded the study of hormonal regulation of metamorphosis in insects. Wigglesworth (1936) later described the secretion of a hormone that prevents metamorphosis from a pair of glands, the corpora allata (CA) attached to the base of the brain; he called it the "status quo" or juvenile hormone (JH). A few years later, Fukuda (1994) described the prothoracic glands as the source of the molting hormone or ecdysone, which was later characterized as a steroid hormone (Butenandt and Karlson, 1954). At about the same time, the physiology of JH was elegantly worked out by Williams (1952). The chemical characterization of its structure, however, eluded researchers until 1967 when Röller and co-workers finally showed that it was a sesquiterpene, epoxy farnesoic acid methyl ester. At this time, Williams (1967) made the now famous statement, "third generation pesticides" in describing the use of JHs as environmentally safe control agents to which the insect will be unable to develop resistance, the first and second generation pesticides being the inorganic and the chlorinated hydrocarbons, respectively. The development of highly potent synthetic analogs of JH, which were several fold more active than the native hormone, gave credence to William's claim (Henrick et al., 1973).

This section describes the biological activity of JH and its analogs as well as their modes of action as understood today. Some of the major uses of JHAs for pest management are cataloged and finally prospects for future development are mentioned.

4.3.1. Juvenile Hormone Action

Juvenile hormone has a unique terpenoid structure and is the methyl ester of epoxy farnesoic acid. This sesquiterpene exists in at least six different forms (**Figure 15**). JH III is the most common type and is present in most insects. Five different JHs, JH 0, I, II, III, and 4-methyl-JH I, have been described in lepidopterans. The bis-epoxide of JH, JH B$_3$, is found along with JH III in higher Diptera (Cusson and Palli, 2000). Various studies show that JH III is the main JH in most insects whereas JH I and II are the principal ones in Lepidoptera. Biosynthesis of JH proceeds through the mevalonate pathway with acetyl coenzyme A (acetyl CoA) serving as the building block. Propionyl CoA is used wherever ethyl side chains occur. Epoxidation and esterification are the terminal steps in the biosynthetic process (Schooley and Baker, 1985; Brindle et al., 1987).

JH secretion by the corpus allatum is regulated by two neurohormones, the allatotropins that stimulate secretion and the allatostatins that inhibit production. Severing the nerve connections or extirpating the neurosecretory cells results in the loss of control (Tobe and Stay, 1985; Tobe *et al.*, 1985; Lee *et al.*, 2002). JH is cleared from the hemolymph by JH esterase, which selectively cleaves the methyl ester inactivating the hormone (Wroblewski *et al.*,

1990; Feng *et al.*, 1999). JH being highly lipophilic is made soluble by JH binding proteins (JHB) to facilitate transportation to the target sites as well as protect it from degradation from nonspecific esterases (Goodman, 1990). The receptor rich structure of the corpus allatum from the female cockroach, *Diploptera punctata*, is illustrated in **Figure 16** (Chiang *et al.*, 2002).

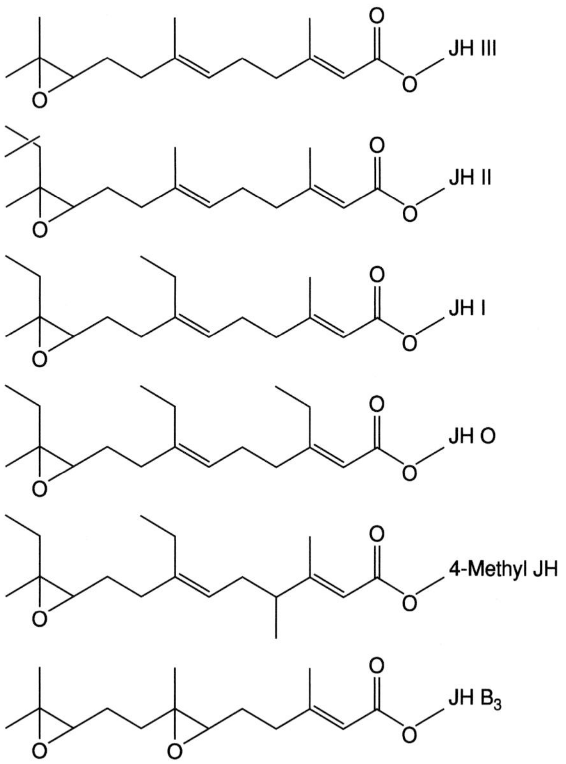

JH is perhaps one of the most pleiotropic hormones known and functions in various aspects of metamorphosis, reproduction, and behavior (Riddiford, 1994, 1996; Wyatt and Davey, 1996; Cusson and Palli, 2000; Palli and Retnakaran, 2000; Hiruma, 2003; Riddiford *et al.*, 2003).

The major function of JH is the maintenance of the larval status or the so-called juvenilizing effect. During the last larval instar in lepidoptera there is an absence of JH during the commitment peak of 20E, which results in the reprogramming of metamorphosis towards pupation (**Figure 17**). In the absence of JH, this ecdysone peak induces the expression of the broad complex gene (*BrC* or *Broad*), which is a transcription factor that initiates the larval–pupal transformation through several microRNAs (miRNAs) (Zhou and Riddiford, 2002; Sempere *et al.*, 2003). In adults, JH secretion resumes and is responsible for yolk protein (vitellogenin) synthesis and transport into the ovaries (Wyatt and Davey 1996). In addition, JH is also responsible for adult diapause where the ovaries do not develop due to the absence of JH and this effect

Figure 15 Chemical structure of naturally occurring JHs.

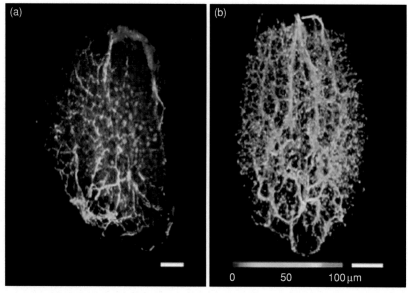

Figure 16 Corpus allatum of female cockroach, *Diploptera punctata*. (a) *N*-methyl-D-aspartate subtype of glutamate receptors (NMDAR) in the nerve fibers (green) involved in JH synthesis and nuclei (red) of parenchyma cells counterstained with propidium iodide. (b) Allatostatin-immunoreactive nerve fibers in the gland (depth code with different colors indicates distance of an object from the surface). Scale bar = 50 μm. (Reproduced with permission from Chiang, A.-S., *et al.*, **2002**, Insect NMDA receptors mediate juvenile hormone biosynthesis. *Proc. Natl Acad. Sci. USA 99*, 37–42; © National Academy of Sciences, USA.)

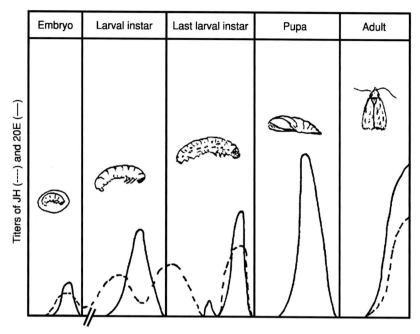

Figure 17   Juvenile hormone and 20-hydroxyecdysone (20E) titers in the various stages of a typical lepidopteran. MP, molt peak of 20E; CP, commitment peak of 20E; RD, reproduction peak of 20E.

can be reversed by applying JH. JH has also been shown to play a role in caste determination, pheromone production, polyphenism, migration, antifreeze protein production, female sexual behavior, male accessory gland secretion, etc. (Wyatt and Davey, 1996).

4.3.2. Putative Molecular Mode of Action of JH

The mode of action at the molecular level has not been completely understood at present. Three types of action have been hypothesized and varying degrees of support for each of these modes have been reported. First is the direct action where it directly induces the secretion of a product (Iyengar and Kunkel, 1995; Feng *et al.*, 1999). Second is the indirect action where it modulates the action of 20E (Dubrovski *et al.*, 2000; Zhou and Riddiford, 2002). In these two instances, the action is at the genomic level where JH has to bind to nuclear receptors to evoke its action. The third involves the role of JH in the transport of vitellogenin into the oocytes through a membrane receptor (Sevala and Davey, 1989). A search for a nuclear as well as a membrane receptor by many investigators over several years has not yielded any concrete results. A nuclear protein isolated from *M. sexta* epidermis using a human retinoic acid receptor as a probe turned out to be an ecdysone-induced transcription factor, a homolog of *Drosophila* hormone receptor 3 (DHR3). A 29 kDa epidermal nuclear protein was found to be a low-affinity JHB (Palli *et al.*, 1994; Charles *et al.*, 1996).

A *D. melanogaster* mutant tolerant to the JHA, methoprene (Met) contained an 85 kDa protein that showed high affinity to JH III (Wilson and Fabian, 1986; Shemshedini *et al.*, 1990). The gene was cloned and was identified as a member of the basic helix–loop–helix (bHLH) PAS family of transcriptional regulators (Ashok *et al.*, 1998). Attempts to demonstrate the *Met* gene as the JH receptor have only added more confusion to the picture. The ultraspiracle (*USP*) gene, a homolog of the human retinoid-X receptor (RXR) and the heterodimeric partner of the EcR, was demonstrated to be involved in signal transduction of JH III in both yeast (Jones and Sharp, 1997) and Sf9 (Jones *et al.*, 2001; Xu *et al.*, 2002) cell based assays. In both cases, the cells were transformed with DNA sequences encoding DmUSP, in which JH III application resulted in transactivation. Isolated, recombinant DmUSP specifically bound JH III, and the authors report that *USP* homodimerizes and changes tertiary conformation, including the movement of the LBD α-helix 12 (Jones *et al.*, 2001; Xu *et al.*, 2002). Xu *et al.* (2002) also reported that ligand binding pocket point mutants of *USP* that did not bind JH III also acted as dominant negative suppressors of JH III activation of reporter promoter, and addition of wild-type *USP* rescued this activation. However, Hayward *et al.* (2003) were not able to demonstrate binding of tritiated JH III to USP from *L. migratoria*. In contrast to results obtained from Jones *et al.* (2001) and Xu *et al.* (2002), the crystal structure of both HvUSP and DmUSP with a lipophilic ligand revealed the LBD α-helix 12 in an open, antagonist position (Billas *et al.*, 2001; Clayton *et al.*, 2001). While it is unclear whether *USP* is the true nuclear JH receptor, the evidence so far is at best equivocal.

Another approach to eluciding the molecular target(s) of JH action has been the use of JH responsive genes. Several have been identified, such as *jhp21* from *L. migratoria* (Zhang *et al.*, 1996). A response element in the promoter region of this gene binds to a protein, which is thought to be a transcription factor activated by JH (Zhou and Riddiford, 2002). Unfortunately, many of these JH responsive genes appear to be induced by JH at a slow rate, suggesting that they may not be the primary site of regulation. However, there are some genes, such as the JH esterase gene (*JHE*) from a *C. fumiferana* cell line, CF-203, that are induced by JH I within 1 h of administration (Feng *et al.*, 1999). This Cf *jhe* appears to be a primary JH responsive gene, which is induced by JH I and suppressed by 20E in a dose-dependent manner. Work in authors' laboratory has shown that a 30 bp minimal promoter region upstream of the Cf *jhe* transcription start site appears to bind to a nuclear protein isolated from

the CF-203 cells (Retnakaran *et al.*, unpublished data). Whether or not it needs to be a high-affinity, low-abundance protein to qualify as a putative JH nuclear receptor is currently under investigation. It has been suggested that JH could act through a membrane receptor as well, as shown by the work of Davey on vitellogenin transport into the oocytes (Wyatt and Davey, 1996). JH acts on the follicle cells of the ovary and makes them contract, causing large spaces to appear between the cells; this process has been termed "patency." This intercellular space formation (patency) permits the vitellogenin from the hemolymph to enter the ovary and gain access to the surface of the oocyte and subsequently enter these cells. This phenomenon was originally described in *R. prolixus* but was later shown in many other species including locusts (Wyatt and Davey, 1996). This membrane-bound receptor appears to be coupled to a G protein and the signal is transduced through a protein kinase system activating an ATPase that permits, among other things, the transport of vitellogenin into the cells (**Figure 18**).

And so the search goes on to unambiguously identify the nuclear and membrane receptors of JH. Characterizing such a receptor would open the

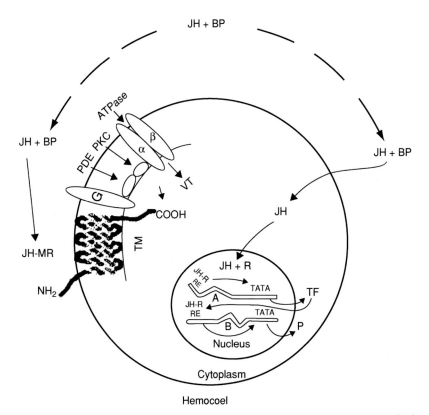

Figure 18 Putative dual receptor mechanism for JH action. A and B are two DNA transcripts; ATPase, Na^+K^+ ATPase with α and β subunits; BP, JH binding protein; G, G protein; JH, juvenile hormone; JH-BP, JH bound to binding protein; JH-MR, JH bound to membrane receptor; PDE, phosphodiesterase; PKC, protein kinase C; RE, hormone response element; TATA, transcription initiation site; TF, transcription factor; TM, transmembrane region. (Reproduced with permission from Wyatt, G.R., Davey, K.G., **1996**. Cellular and molecular actions of JH. II. Roles of JH in adult insects. *Adv. Insect Physiol. 26*, 1–155.)

door for a more rational approach to developing target specific control agents.

4.3.3. Pest Management with JHAs

The idea that insects would be unable to develop resistance against JH if it was used as a control agent was one of the driving principles behind the impetus to develop this hormone as an insecticide (Williams, 1967). The difficulty and cost of synthesizing such a complex molecule with a labile epoxide moiety and susceptibility to degradation delayed the realization of this concept. However, it soon became apparent that several synthetic analogs of JH, many of them several fold more active than the native hormone, could be used as control agents. The discovery that some synthetic synergists of insecticides have intense JH activity (Bowers, 1968) was the

harbinger for the development of various new analogs with diverse chemical structures. Consequently, the modes of action of these analogs may not necessarily be similar.

Over the years, numerous JHAs have been synthesized and their relative potencies, structure–activity relationships, and differential effects on various species have been studied (Slama *et al.*, 1974; Romanuk, 1981). Many naturally occurring JHAs, also called juvenoids, have been isolated from plants such as the "paper factor" from the balsam fir tree (*Abies balsamea*), and juvocimenes from the sweet basil plant (*Ocimum basilicum*) (Bowers and Nishida, 1980). During coevolution, the plants probably developed these JHAs to defend themselves against insects. Some representative examples of these compounds are shown in **Figure 19**.

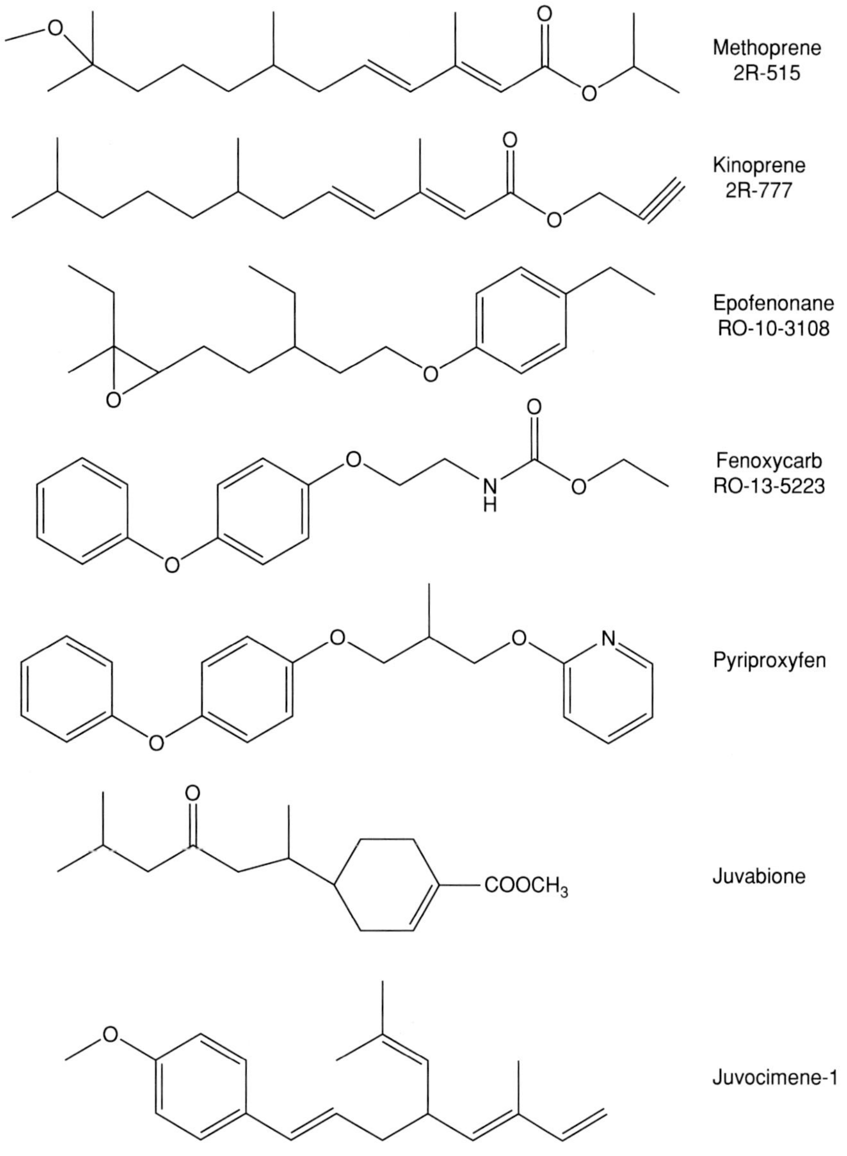

Figure 19 Some synthetic and naturally occurring JHAs.

JHAs can be broadly classified into two groups: the terpenoid JHAs such as methoprene and kinoprene and the phenoxy JHAs such as fenoxycarb and pyriproxyfen, with some such as epofenonane straddling the two (**Figure 19**).

4.3.3.1. Biological action of JHAs JHAs mimic the action of natural JH and, as such, can theoretically interfere with all the functions of JH. However, only a few such functions have been capitalized for control purposes (Retnakaran et al., 1985). The organismal effects that have been exploited for control are relatively few, the most common being that seen on the last larval instar. Treatment of the larva in its last instar with JHA severely interferes with normal metamorphosis and results in various larval–pupal intermediates that do not survive (**Figure 20**). This effect, however, is not observed if the last instar larva is treated with the JHA methoprene during the first day or at the pharate stage (prepupal stage) of the stadium. If treated on day 1, the larva molts into a supernumerary larval instar and if it receives JHA at the pharate stage there is no effect (Retnakaran 1973a, 1973b; Retnakaran et al., 1985). Morphogenetic control is ideally suited for controlling insects that are pests as adults, such as mosquitos and biting flies.

JHAs block embryonic development at blastokinesis and act as ovicides (Riddiford and Williams, 1967; Retnakaran, 1970). In some instances, when older eggs are treated, there is a delayed effect that is manifested at the last larval instar when it molts into a supernumerary instar (Riddiford, 1971). This type of ovicidal effect is useful especially in controlling fleas. In some cases where adults are treated, the material gets transferred to the eggs where they block embryonic development (Masner et al., 1968). JHAs induce sterility in both sexes of adult tsetse flies (Langley et al., 1990).

Adult or reproductive diapause is caused by the absence of JH secretion and this condition can be terminated by JHA treatment (De Wilde et al., 1971; Retnakaran, 1974).

4.3.3.2. Properties and mode of action of selected JHAs
4.3.3.2.1. Methoprene (ALTOSID®, APEXSE®, DIANEX®, PHARORID®, PRECOR®, VIODAT®) Methoprene is perhaps one of the best known terpenoid JHAs developed for pest control. A mutant strain of D. melanogaster that was tolerant to methoprene, the so-called Met flies, was generated by Wilson and Fabian (1986). These flies were also found to be tolerant to JH III, JH B₃, and several JHAs but not to many classes of insecticides. It was therefore tempting to speculate that this would be an elegant source to discover the JH receptor. An 85 kDa protein isolated from the fat body of wild flies was found to bind with high affinity to JH III. The same 85 kDa protein from Met flies showed a sixfold lower affinity for JH III (Shemshedini et al., 1990). When the Met gene was cloned, it became

Figure 20 Morphogenetic effects of a JHA, methoprene, on the last larval instar of some lepidopteran insects. (a) A spruce budworm (*Choristoneura fumiferana*) pupa showing larval head and legs; (b) a forest tent caterpillar (*Malacosoma disstina*) showing aberrant mouthparts; (c) a pupa-like larva of the Eastern hemlock looper, *Lambdina fiscellaria*; (d) deformed Eastern hemlock looper larva with everted wing discs.

apparent that it was a bHLH-PAS and belonged to a family of transcriptional regulators, and this gene was not vital for the survival of the flies (Ashok *et al.*, 1998; Wilson and Ashok, 1998). This raises the possibility that JH activity could be exhibited by compounds that may interfere at any step during the synthesis, transportation, and target-site activity. Being extremely pleiotropic, the target-site activity could easily span a wide spectrum of functions. The methoprene-tolerant *Met* gene probably encodes a nonvital insecticide target protein of one type or another (Wilson and Ashok, 1998).

Methoprene is by far the most thoroughly studied JHA. Extensive EPA data collected over several years have shown that this JHA is relatively nontoxic to most nontarget organisms. Methoprene is rapidly broken down and excreted; its half-life in the soil is about 10 days. Mild toxicity to birds and some aquatic organisms has been observed. It has been used as a mosquito larvicide and for controlling many coleopterans, dipterans, homopterans, and siphonopterans (Harding, 1979).

4.3.3.2.2. Kinoprene (ENSTAR®)
This is also an extremely mild JHA with little or no toxic effects. It breaks down in the environment and is relatively nonpersistent. It is relatively nonhazardous to bees and is nontoxic to birds, fish, and beneficial insects. It is decomposed by sunlight. It has been effectively used for controlling scales, mealy bugs, aphids, whiteflies, and fungal gnats. Its effects are morphological, ovicidal, and it acts as a sterilant (Harding, 1979).

4.3.3.2.3. Fenoxycarb (INSEGAR®, LOGIC®, TORUS®, PICTYL®, VARIKILL®)
Fenoxycarb was the first phenoxy JHA found to be effective (Grenier and Grenier, 1993). It is unusual in having a carbamate moiety but does not show any carbamate-like toxicity. This JHA is considered slightly toxic to nontarget species, especially aquatic crustaceans, and carries the signal word "caution" on its label. It has been used as an effective control agent against fire ants (as bait), fleas, mosquitos, cockroaches, scale insects, and sucking insects. It is considered a general use insecticide. Unlike methoprene and kinoprene, fenoxycarb can be toxic to some of the beneficial insects such as neuropterans (Liu and Chen, 2001).

4.3.3.2.4. Pyriproxifen (KNACK®, SUMILARV®, ADMIRAL®)
This JHA is a phenoxy analog similar to fenoxycarb. It is by far one of the most potent JHAs currently available. It is mildly toxic to some aquatic organisms but is nontoxic to bees. Its effects are similar to the other JHAs and causes morphogenetic effects as well as sterility. It has been used effectively in the form of bait for controlling the red imported fire ant (*Solenopsis invicta*) in California. It has been used for controlling aphids, scales, whiteflies, and the Pear psylla, and the tsetse fly (*Glossina morsitans*) in Zimbabwe. In the latter, treatment of both sexes results in pyriproxifen being transferred to the female uterus where the embryo is killed (Langley *et al.*, 1990).

4.3.4. Use of JHAs for Pest Management

JJHAs have been successfully used for controlling certain types of pests, especially where the pest is the adult stage. It has not been very effective against most lepidopteran agricultural pests because the larval stage is responsible for crop destruction. While some of the control measures tested since the previous edition of this series (Retnakaran *et al.*, 1985) are described below, **Table 9** summarizes the effects of the various JHA insecticides on agricultural and stored insect pests.

4.3.4.1. Control of public health and veterinary insects
In most instances, the adult form is the active pest and, therefore, these species can be managed by JHA insecticides. Mosquitos, fleas, cockroaches, fire ants, and tsetse flies are some of the major pests that are vulnerable to JHAs (**Table 10**).

4.3.4.2. Anti-JH for pest management
Compounds that prevent JH production, facilitate JH degradation, or destroy the corpus allatum all belong to this group. It is a catchall of various compounds that negate the activity of JH. Ideally, such compounds should be very effective pest control agents. Treatment of newly emerged larvae with such a compound would theoretically create miniature pupae, thus abbreviating the destructive part of the insect's life history. Many of the anti-JH compounds turned out to be highly toxic and did not pan out as good control agents. Once the JH receptor is characterized, it can be a target for control. Some of the effects that have been studied with anti-JH compounds are shown in **Table 11**.

4.3.5. Effects on Nontarget Invertebrates and Vertebrates

The effects of JHA insecticides have been studied and extensively reviewed (Grenier and Grenier, 1993; Miyamoto *et al.*, 1993; Palli *et al.*, 1995). The phenoxy JHAs, fenoxycarb, and pyriproxifen have been shown to be toxic to a number of dipteran, coleopteran, and hemipteran predators and parasitoids of scale insects (Mendel *et al.*, 1994).

Table 9   Control of agricultural and stored product insects with JHA insecticides

Insect species	Host	Hormone analog	Activity	Reference
Nephotettix cincticeps N. nigropictus N. virescens Recilia dorsalis (leafhoppers)	Rice	NC-170 4-chloro-5-(6-chloro-3-pyridylmethoxy)-2-(3,4-dichlorophenyl)-pyridazin-3(2H)-one	Good activity, 100 ppm suppressed population for more than 6 weeks	Miyake et al. (1991)
Adoxophyes orana Pandemis heparana Archips podana, Archips rosana (leafrollers)	Apple Orchards	Fenoxycarb and epofenonane	Adequate reduction, reinfestation never resulted in an increase to a harmful population level; good activity	de Reede et al. (1985)
Lycoriella mali (mushroom pest)	Compost	Methroprene	Resistance to this compound does not seem to be developing rapidly	Keil et al. (1988)
Heliothis virescens (tobacco budworm)	Cotton	Fenoxycarb	Good activity, opens new aspects for the control of lepidopterous pests with this compound	Masner et al. (1986)
Rhagoletis pomonella (apple maggot)	Apple	Pyriproxyfen	Good activity	Duan et al. (1995)
Ceratitis capitata (Mediterranean fruit fly)	Fruits in general	Methoprene	Good activity prevents adult eclosion	Saul et al. (1987)
Adoxophyes orana Pandemis heparana (leafrollers)	Apple orchards	Epofenonane and fenoxycarb	Good activity, the foliar residue remained active for at least 4 weeks and parasites seem to be less susceptible to it than the host itself	de Reede et al. (1984)
Manduca sexta (tobacco hornworm)	Tobacco plants	JHAs such as 1,3- and 1,4-bisNCS	Good activity, can disrupt insect growth and development	Ujváry et al. (1989)
Ceroplastes floridensis (Florida wax scale)	Guava plants	Fenoxycarb	Insects on treated plants deposited only a small amount of wax	Eisa et al. (1991)
Cacopsylla pyricola (Pear psylla)	Pear	Fenoxycarb	May still be a viable strategy for managing this	Krysan (1990)
Melanoplus sanaguinipes, M. differentialis (grasshopper)	Rangelands and crops	Fenoxycarb	Good activity; eliminates egg production and reduces oviposition	Capinera et al. (1991)
Spodoptera littura (tobacco cutworm)	Vegetable crops	Pyriproxyfen	Good activity, reduces the number of eggs	Hastakoshi (1992)
Christoneura rosaceana (leafroller)	Fruit trees	Fenoxycarb	Good activity, abnormal development is noted in all treatments	Aliniazee and Arshad (1998)
Cydia pomonella (codling moth)	Apple and pear	Phenoxyphenol pyridine and pyrazine carboxamides N-(4-phenoxyphenol) pyridinecarboxamides and N-(4-phenoxyphenyl) pyrazinecarboxamides	Good activity, prolongs development	Balcells et al. (2000)
Laodelphax striatellus (brown planthopper)	Rice	NC-170 4-chloro-5-(6-chloro-3-pyridylmethoxy)-2-(3,4-dichlorophenyl)-pyridazine-3(H)-one	Good activity, 100 ppm prevents larvae from entering diapause	Miyake et al. (1991)
Dacus dorsalis (oriental fruit fly)	Papayas	Methoprene	Good activity, higher doses showed no survival	Saul et al. (1987)
Trichoplusia ni (cabbage looper)	Cabbage	Methoprene	Good activity, amount of eggs is reduced	Campero and Haynes (1990)

Table 9 Continued

Insect species	Host	Hormone analog	Activity	Reference
Bemisia tabaci (sweet potato whitefly)	Cotton and vegetable crops	Pyriproxyfen	Good activity, strong translaminar effect and acts on all stages	Ishaaya and Horowitz (1992)
Bemisia tabaci (sweet potato whitefly)	Cotton and vegetable crops	insect growth regulators such as buprofezin and pyriproxyfen	activity not well seen	Palumbo *et al.* (2001)
Trialeurodes vaporariorum (greenhouse whitefly)	Greenhouse crops	Pyriproxyfen	Good activity, affects all stages	Ishaaya *et al.* (1994)
Reticulitermes santonensis *Reticulitermes flaviceps* *Coptotermes formosanus* (termites)	Wood	Carbamate derivative of 2-(4-hydroxybenzyl) cyclokexanone (W-328)	Good activity, 500 ppm concentration results in significant soldier differentiation	Hrdý *et al.* (2001)
Reticulitermes speratus (termites)		Ethyl(2-(*p*-phenoxy-phenoxy)ethyl)carbamate	Good activity, results in the collapse of the whole termite colony	Tsunoda *et al.* (1986)
Ips paraconfusus Lanier (California beetle)	Logs of ponderosa pine	Fenoxycarb	Good activity, acts as an effective chemosterilant	Chen and Border (1989)
Lipaphis erysimi (mustard aphid)	Leafy vegetables	Pyriproxyfen	Good activity, causes direct mortality, reduces longevity and inhibits progeny formation	Liu and Chen (2000)
Cacopsylla pyricola (Pear psylla)	Pear	Fenoxycarb	Good activity	Lyoussoufi *et al.* (1994)
Reticulitermes speratus (Japanese subterranean termite)	Wood in general	Ethyl (2-(*p*-phenoxy-phenoxy)ethyl)carbamate	Good activity, disturbance of caste differentiation results in collapse of the colony	Tsunoda *et al.* (1986)
Rhyzopertha dominica (lesser grain borer)	Wheat grain	2',7',-epoxy-3,7'-dimethyl-undec-2'-enyl6-thyl-3-pyridyl ether	Good activity, 4 ppm resulted in insect mortality	Mkhize (1991)
Tribolium castaneum (red flour beetle), *Rhyzopertha dominica* (lesser grain borer), *Sitophilus oryzae* (rice weevil)–red flour beetle	Wheat flour	Methoprene and pyriproxyfen	Good activity, 20 ppm reduced the population by 80–99%	Kostyukovsky *et al.* (2000)

However, the ectoparasite of the California red scale and Florida wax scale, *Aphytis holoxanthus*, was insensitive to pyriproxifen (Peleg, 1988). Fenoxycarb was also found to be toxic to *Colpoclypeus florus*, a parasitoid of *Adoxophyes orana* and *Pandemis heparana*. Pupation in the predatory coccinellid, *Chilocorus bipustulatus* L., was inhibited when larvae fed on the scale insect, *Chrysomphalus aonidum*, were dipped into a 0.025% solution of fenoxycarb (cited in Grenier and Grenier, 1993). Additional effects of JHA insecticides have been summarized in **Table 12**.

The toxicological profiles of JHA insecticides for mammalian (rat), aquatic (rainbow trout and Daphnia), predatory, and parasitoid insects have been summarized in **Table 13**. In general, JHA insecticides have low acute toxicity to fish, birds, and mammals. Both fenoxycarb and pyriproxifen have very low toxicity to adult bees. However, effects on brood development have been observed as a result of worker bees feeding pollen containing fenoxycarb residues to larvae (Wildboltz, 1988).

Last instars of certain aquatic insects are susceptible to JHA insecticides. Morphogenetic abnormalities have been observed in the heteropteran, *Notonecta unifaciata*, and in the dragon flies, *Anax junius* and *Pantala hymenaea*, with applications of fenoxycarb (Miura and Takahashi, 1987). Similarly, pyriproxifen also produced morphogenetic effects when applied to last instars of the

Table 10 Control of some insect pests of public health and veterinary importance with JHA insecticides

Insect species	Host	Hormone analog	Activity	Reference
Monomorium pharaonis (pharaoh ant)	Large office buildings, houses, apartment buildings, food establishments and hospitals	Pyriproxyfen	Good activity when used at concentrations of 0.25%, 0.5%, and 1%	Vail et al. (1995)
Culex quinquefasciatus Aedes aegypti (mosquitos)	Humans	VCRC/INS/A-23	Activity was good and the compound was found cost effective	Tyagi et al. (1985)
Chironomus fusciceps (midge)	Sulfur containing pods near residential areas	Pyriproxyfen	Good activity, at 0.00177 and 0.05369 ppm caused 50 to 90% emergence inhibition	Takagi et al. (1995)
Glossian morsitans, G. pallidipes (testse flies)	Cattle and humans	Pyriproxyfen	Good activity, it can replace pesticides	Hargrove and Langley (1993)
Blattella germanica (german cockroach)	Residences	Fenyxocarb and hydroprene	It can reduce the population	King and Bennett (1988)
Anopheles balabacensis (mosquitos)	Humans	Pyriproxyfen	Good activity, 0.005 ppb reduce the egg and sperm production and blood feeding and copulating activity	Iwanaga and Kanda (1988)
Ctenocephalides felis (cat flea)	Cat	Methoprene	Activity is 12.7 and 127 ng cm^{-2}, resulted in 15–5.2% adult emergence. Good control	Moser et al. (1992)
Solenopsis invicta (red imported fire ant)		Pyriproxyfen	Good activity, cause 80–85% reduction in colony size	Banks and Lofgren (1991)
Blattella orientalis (oriental cockroach)	Industrial, food manufacturing, and domestic premises	S-hydroprene	Good activity	Edwards and Short (1993)
Ctenocephalides felis (cat flea)	Cat	Methoprene	Good activity when cat fleas less than 24 h old were treated	Olsen (1985)
Culex pipiens pallens Culex tritaeniorhynchus	Humans	Pyriproxyfen	Activity remains unclear	Kamimura and Arakawa (1991)
Ctenocephalides felis and Ctenocephalides canis (dog and cat fleas)	Cat and dog	Methoprene and pyriproxifen	Good activity	Taylor (2001)
Culex pipiens (mosquitos)	Humans	4-alkoxyphenoxy- and 4-(alkylphenoxy) alkanaldoxime o-ethers	Good activity	Hayashi et al. (1989)
Culex pipiens	Humans	(Phenoxyphenoxy) and (benzylphenoxy) propyl ethers	Good activity	Niwa et al. (1989)
Alphitobius diaperinus (mealworm)	Poultry production	Methoprene and fenoxycarb	Good activity	Edwards and Abraham (1985)
Ctenocephalides felis (cat fleas)	Cats	CGA-2545'728	Good activity	Rasa et al. (2000)
Ctenocephalides felis (cat fleas)	Cats	Methoprene, pyriproxifen and fenoxycarb	Good activity	Miller et al. (1999)

Table 11 Insect control with some anti-JH compounds

Insect species	Host	Hormone analog	Activity	Reference
Manduca sexta (tobacco hornworm)	Tobacco, tomato	ETB (ethyl 4-[2-{tert-butyl carbonyloxy} butoxy]benzoate)	Some activity, 50 μg of ETB caused formation of larval–pupal intermediates after the fourth instar	Beckage and Riddiford (1983)
Bombyx mori (silkworm)	Mulberry	1-Citronellyl-5-phenylimidazole (KK-22)	Some activity, induction of precocious metamorphosis seems to correlate with prolongation of the larval developmental period in the third instar	Kuwano and Eto (1984)
Blattella germanica (German cockroach)	Household pest	Precocenes	Short-term activity seems to be effective and long-term activity shows that damage to CA is irreversible	Belles et al. (1985)
Bombyx mori (silkworm)	Mulberry	Ethoxy (KK-110) and 4-chlor-phenyl (KK-135)	Activity can be inhibited by stimultaneous application of methoprene	Kuwano et al. (1990)
Nilaparvata lugens (brown planthopper)	Rice	Precocene II	Activity with an increase in good ovarian growth	Ayoade et al. (1995)
Spodoptera littoralis (Egyptian cotton worm)	Cotton	Precocene I and II	Good activity but should be undertaken with care in order to minimize disruptive effects on parasitoids	Hagazi et al. (1998)
Hyphantria cunea (Fall webworm)	Hardwood trees	Fluoromevalonate (Fmev, ZR-3516)	Activity was shown as evoking varying degrees of ecdysial disturbance, which resulted in death of the insect	Farag and Varjas (1983)
Oxycarenus lavaterae (Lygaeid)	European forests	Precocenes	Activity shows a high degreee of uniformity	Belles and Baldellou (1983)
Bombyx mori (silkworm)	Mulberry	1-Benzyl-5-[(E)-2,6-dimethyl-1,5-heptadienyl] imidazole (KK-42)	Good activity, 100% precocious pupation	Kuwano et al. (1985)
Locusta migratoria migratorioides; Oncopeltus fasciatus; Schistocerca gergaria (locusts)	Various plants	Precocene II	Good activity, ovipositor protrusion, and death at higher concentrations	Degheele et al. (1986)
Nilaparvata lugens (Brown planthopper)	Rice	Precocene II (anti-allatin)	Good activity with high rate of mortality	Pradeep and Nair (1989)
Spodoptera mauritia (lawn army worm)	Agricultural crops	Fluoromevalonate (Fmev)	Good activity, induced various morphogenetic abnormalities, death before pupation occurred	Nair and Rajalekshmi (1989)
Drosophila melanogaster (fruit fly)	Various fruits	5-Ethoxy-6-(4-methoxyphenyl)methyl-1,3-benzodioxole J2581	Activity can be reversed by application of JH of methoprene	Song et al. (1990)
Diploptera punctata (cockroach)	Various types of diet	1,5-disubstituted imidaxoles	Good activity, powerful inhibitors of the last step of juvenile synthesis	Unnithan et al. (1995)
Manduca sexta (tobacco hornworm)	Tomato and tobaco	Farnesol dehydrogenase	Good activity, unique dehydrogenase that should be examined further	Sen and Garvin (1995)
Oncopeltus fasciatus (milkweed bug)	Milkwood	6,7-Dimethoxy-2,2-dimethyl 2H-chromene	Activity shows defiency of JH	Bowers and Unnithan (1995)

Continued

Table 11   Continued

Insect species	Host	Hormone analog	Activity	Reference
Heliothis virescens and Trichoplusia ni (cabbage looper)	Field crops and vegetables	Virus AcJHE-SG (transgenic virus with JH esterase gene)	Good activity, death occurred from contraction-paralysis and disruption of the normal sequence of events at the molt	Bonning et al. (1995)
Sogatella furcifera (whitebacked rice planthopper)	Rice	Precocene II	Activtiy showed inductio metathetely by itself occurred	Miyake and Mitsui (1995)
Zaprionus paravittiger (banana fruitfly)	Bananas	Precocene	Good activity, with 0.076, 0.1 ppm in 1 μL of acetone, adult longevity and fecundity were reduced	Rup Pushpinder and Baniwal (1985)
Musca domestica (housefly) Diploptera punctata (cockroach)	Various diets	2-(1-Imidazolyl)-2-methyl-1-phenyl-1-propanone and 2-methyl-1-phenyl-2-(1,2,4,-triazol-1-yl)-1-propanone	Activity was seen at 0.2 mM	Bélai et al. (1988)
Aphis craccivora (ground nut aphid)	Legumes	Precocene II	Good activity, with 2.0 μg/aphid, precocious metamorphosis was seen	Srivastava and Jaiswl (1989)
Pieris brassicae (cabbage butterfly) and Leptimotarsa decemlineata (Colorado potato beetle)	Vegetables and potato	2,2-Dimethylchromene derivatives	Good activity	Darvas et al. (1989)
Heliothis zea (corn earworm)	Corn	Precocene II	Good activity, growth and development were inhibited	Binder and Bowers (1991)

Table 12   Effects of JH analog insecticides on beneficial and aquatic species

Parasite/predator/aquatic	Host/habitat	Hormone analog	Activity	Reference
Chilocorus nigritus	Citrus red scale	Pyriproxyfen	Harmful to nontargeted species	Magaula and Samways (2000)
Daphnia pulex	Aquatic	Methoprene	Decrease in the incidence of all-male broods and an increase in the incidence of all-female broods	Peterson et al. (2000)
Aphanteles congregatus	Tobacco hornworm	Methoprene	Low dose would allow the insect to pupate normally	Beckage and Riddiford (1981)
Copidosoma floridanum	Cabbage looper	Methoprene	Slight activity, inhibited morphogenesis	Strand et al. (1991)
Encarsia pergandiella E.transvena, E.formosa	White fly	Pyriproxyfen	Relatively safe to parasitoids	Liu and Stansly (1997)
Moina macrocopa	Aquatic	Methoprene	Low activity on reproduction	Chu et al. (1997)
Ceroplastes destructor	Citrus orchards	Pyriproxyfen and fenoxycarb	Arrested the first and second instar	Wakgari and Giliomee (2001)
Chrysoperla rufilabris	Aphids	Fenoxycarb and pyriproxyfen	Ovicidal effect and delay in development	Liu and Chen (2001)
Rhithropanopeus harrishii (mud crab)	Aquatic	Methoprene	No activity was noticed	Celestial and McKenney (1993)

Table 13 Ecotoxicological profile of JHA insecticides (from various EPA reports)

Compound	Mammals (rat) LD_{50}	Fish (rainbow trout) 96h LC_{50}	Crustacea (Daphnia) 48h LC_{50}	Honey bees LC_{50}	Predators	Parasitoids
Methoprene[a]	>34 g kg^{-1}	>3.3. mg l^{-1}	>0.51 mg l^{-1}	>1000 mg bee^{-1}	Minimal effects	Minimal effects
Kinoprene[b]	>5 g kg^{-1}	>20 mg l^{-1}	>0.113 mg l^{-1}	Nontoxic	Minimal effects	Minimal effects
Fenoxycarb[c]	>10 g kg^{-1}	>1.6 mg l^{-1}	>1.6 mg l^{-1}	>1000 ppm	Low toxicity	Low toxicity
Pyriproxifen[d]	>5 g kg^{-1}	>325 mg l^{-1}	>400 mg l^{-1}	>100 mg bee^{-1}	Low toxicity	Low toxicity

[a]ALTOSID®, APEXSE®, DIANEX®, PHARORID®, VIODAT®.
[b]ENSTAR®.
[c]INSEGAR®, LOGIC®, TORUS®, PICTYL®, VARIKILL®.
[d]KNACK®, SUMILARV®, ADMIRAL®.

dragonfly, *Orthetrum albistrum* and the midge, *Chironomus yoshimatsui* (Miyamoto *et al.*, 1993).

4.3.6. Conclusions and Future Research of JHAs

JHAs have not proven to be the wonderful control agents they were purported to be. However, they have advantages for controlling pests of public health. Many of them are environmentally attractive. The anti-JH compounds have, for the most part, remained at the experimental stage. Interfering with JH action will become an attractive option, once the JH receptor is characterized.

4.4. Insecticides with Chitin Synthesis Inhibitory Activity

If the insect cuticle, which provides the exoskeletal structure, is disrupted during its formation, it is lethal to the insect. The cuticle needs to be waterproof for protection, soft and flexible to allow movement, extensible in between segments to allow for increase during feeding and growth, and also rigid to provide firm points of muscle attachment as well as for mandibles and claws. The cuticle is the product of a single layer of epidermal cells, and consists of different layers of which the procuticle contains chitin, a major component (30–60%) of insect cuticle. Chitin is a β-1,4-linked aminopolysaccharide homopolymer of *N*-acetylglucosamine (GlcNAc), and is by far one of the most abundant biological materials found on earth. This nitrogenous amino sugar polysaccharide is cross-linked to proteins via biphenyl linkages to form a protective matrix, which consists of a chitin microfibers–protein complex. Chitin biosynthesis, which is not fully understood, is a complex process consisting of a series of enzymatic steps beginning with a glucose unit, which is converted to GlcNAc that is linked with UTP, transported within the cell in combination with dolichol phosphate, and polymerized into chitin. The newly polymerized chitin is covalently linked to proteins to form chitin microfibrils in the cuticle. Polymerization to form chitin is catalyzed by the enzyme chitin synthase (CHS or chitin-UDP-glucosamine-transferase), which occurs in several forms (CHS 1, 2 and 3; EC 2.4.1.16). This enzyme has probably undergone sequential gene duplication and divergence during evolution, resulting in its expression in different forms in diverse species, one of which is the human hyaluronan synthase. It has a conserved amino acid sequence essential for chitin biosynthesis in yeasts, and it has evolved into two types: the fungal form, which occurs as an inactive zymogen requiring proteolysis for activation, and the arthropod form, which is membrane bound. Over the past three decades, the chitin biosynthetic pathway has proven to be important for developing insect control agents that selectively inhibit any of the chitin synthetic steps in insects. Two types of insect regulatory chitin synthesis inhibitors (CSI) have been developed and used as commercial compounds for controlling agricultural pests: the benzoylphenyl ureas (BPUs), and buprofezin and cyromazine (Spindler *et al.*, 1990; Retnakaran and Oberlander, 1993; Londerhausen, 1996; Palli and Retnakaran, 1999).

4.4.1. Brief Review of Old Chitin Synthesis Inhibitors

The insecticidal activity of the first BPUs was discovered around 1970 by scientists at Philips-Duphar BV, (now Crompton Corp., Weesp, The Netherlands) and the first commercial compound of this series was diflubenzuron, Dimilin® (Van Daalen *et al.*, 1972; Grosscurt, 1978) marketed by Uniroyal Chemical (**Table 14**). Diflubenzuron and older CSIs have already been extensively reviewed by Retnakaran *et al.* (1985), Grosscurt *et al.* (1987), and Retnakaran and Wright (1987).

Table 14 Chemical structures of the chitin synthesis inhibitors (CSI), 11 benzoylphenyl ureas (BPUs), and buprofezin and cyromazine, their biological effects and their target pest species

Compound	Chemical structure	Biological effects	Uses to control pest species
Bistrifluron		Inhibition of larval molting leading to death	Whiteflies in tomato (*Trialeurodes vaporariorum* and *Bemisia tabaci*), and lepidopterous insects in vegetables, cabbage, persimmon, apples, and other fruits (e.g., *Spodoptera exigua*, *Plutella xylostella*, *Stathmopoda masinissa*, and *Phyllonorycter ringoniella*) at 75–400 g ha^{-1}
Chlorfluazuron		Acts as an antimolting agent, leading to death of the larvae and pupae	*Heliothis*, *Spodoptera*, *Bemisia tabaci*, and other chewing insects on cotton, and *Plutella*, *Thrips* and other chewing insects on vegetables. Also used on fruit, potatoes, ornamentals, and tea. Applied at 2.5 g hl^{-1}
Diflubenzuron		Nonsystemic insect growth regulator with contact and stomach action. Acts at time of insect molting, or at hatching of eggs	Wide range of leaf-eating insects in forestry, woody ornamentals, and fruit. Controls certain major pests in cotton, soybeans, citrus, tea, vegetables, and mushrooms. Also controls larvae of flies, mosquitos, grasshoppers, and migratory locusts. Used as an ectoparasiticide on sheep for control of lice, fleas, and blowfly larvae. Is suitable for inclusion in integrated control programs. Effective at 25–75 g a.i./ha against most leaf-feeding insects in forestry; in concentrations of 0.01–0.015% a.i. against codling moth, leaf miners, and other leaf-eating insects in top fruit; in concentrations of 0.0075–0.0125% a.i. against citrus rust mite in citrus; and at a dosage of 50–150 g a.i./ha against a number of pests in cotton (cotton boll weevil, armyworms, leafworms), soya beans (soya bean looper complex) and maize (armyworms). Also for control of larvae of mushroom flies in mushroom casing (1 g a.i. m^{-2}); mosquito larvae (25–100 g a.i. ha^{-1}); fly larvae in animal housings (0.5–1 g a.i. m^{-2} surface); and locusts and grasshoppers (60–67.5 g a.i. ha^{-1})
Fluazuron		Systemic acarine growth regulator with contact and stomach action inhibiting chitin formation	Ixodicide for strategic control of the cattle tick *Boophilus microplus* (including all known resistant strains) on beef cattle

Name	Structure	Mode of action	Uses
Flucycloxuron	(structure)	Nonsystemic acaricide and insecticide inhibiting the molting process in mites and insects. It is only active against eggs and larval stages. Adult mites and insects are not affected	Eggs and larval stages of Eriophyid and Tetranychid mite species on a variety of crops, including fruit crops, vegetables, and ornamentals. Also controls larvae of a number of insects. Because of its relative selectivity, it is well-suited for integrated control programs. Recommended concentration for control of mites on fruit crops is 0.01–0.015% a.i. On insects, good activity has been found against codling moth, leaf miners, and some leaf rollers in pome fruit, at the same concentration. In ornamentals grown under glass/plastic, lower concentrations can be used. In grapes, the recommended dosage is 125–150 g a.i. ha^{-1}
Flufenoxuron	(structure)	Insect and acarid growth regulator with contact and stomach action. Treated larvae die at the next molt or during the ensuing instar. Treated adults lay nonviable eggs	Control of immature stages of many phytophagous mites (*Aculus, Brevipalpus, Panonychus, Phyllocoptruta, Tetranychus* spp.) and insect pests on pome fruit, vines, citrus fruit, tea, cotton, maize, soybeans, vegetables, and ornamentals
Hexaflumuron	(structure)	Ingested, systemic insecticide	In agriculture for control of larvae of Lepidoptera, Coleoptera, Homoptera, and Diptera on top fruit, cotton, and potatoes. Major use now is in bait, for control of subterranean termites
Lufenuron	(structure)	Insecticide/acaricide acting mostly by ingestion; larvae are unable to molt, and also cease feeding	Lepidoptera and Coleoptera larvae on cotton, maize, and vegetables; and citrus whitefly and rust mites on citrus fruit, at 10–50 g ha^{-1}. Also for the prevention and control of flea infestations on pets
Novaluron	(structure)	Insecticide acts by ingestion and contact, affecting molting. Causes abnormal endocuticular deposition and abortive molting	Under development by Makhteshim Chemical Works for control of Lepidoptera, whitefly, and agromyzid leaf miners in top fruit, vegetables, cotton, and maize
Teflubenzuron	(structure)	Antimolting agent, leading to death of the larvae and pupae	*Heliothis, Spodoptera, Bemisia tabaci* and other chewing insects on cotton; and *Plutella, Thrips* and other chewing insects on vegetables. Also used on fruit, potatoes, ornamentals, and tea. Applied at 2.5 g hl^{-1}.
Triflumuron	(structure)	Ingested insecticide, acting by inhibition of molting	Lepidoptera, Psyllidae, Diptera, and Coleoptera on fruit, soybeans, vegetables, forest trees, and cotton. Also used against larvae of flies, fleas, and cockroaches in public and animal health

Continued

Table 14 Continued

Compound	Chemical structure	Biological effects	Uses to control pest species
Buprofezin		Probable chitin synthesis and prostaglandin inhibitor. Hormone disturbing effect, leading to suppression of ecdysis. Persistent insecticide and acaricide with contact and stomach action; not translocated in the plant. Inhibits molting of nymphs and larvae, leading to death. Also suppresses oviposition by adults; treated insects lay sterile eggs	Insecticide with persistent larvicidal action against Homoptera, some Coleoptera, and also Acarina. Effective against Cicadellidae, Deltocephalinae (leafhoppers), and Delphacidae (planthoppers) in rice, at $50-250$ g ha^{-1}; Cicadellidae (lady beetle) in potatoes; Aleyrodidae (whitefly) in citrus, cotton, and vegetables, at $0.025-0.075$ g ha^{-1}; Coccidae, Diaspididae (scale insects) and Pseudococcidae (mealybugs) in citrus and top fruit, at $25-50$ g hl^{-1}; Tarsonemidae in vegetables, at $250-500$ g ha^{-1}. Suitable for IPM programs
Cyromazine		Contact action, interfering with molting and pupation. When used on plants, action is systemic: applied to the leaves, it exhibits a strong translaminar effect; applied to the soil, it is taken up by the roots and translocated acropetally	Control of Diptera larvae in chicken manure by feeding to poultry or treating the breeding sites. Also used to control flies on animals. Used as a foliar spray to control leaf miners (*Liriomyza* spp.) in vegetables (e.g., celery, melons, tomatoes, lettuce), mushrooms, potatoes, and ornamentals, at $75-225$ g ha^{-1}; also used at $190-450$ g ha^{-1} in drench or drip irrigation

Data compiled from Palli, S.R., Retnakaran, A., **1999**. Molecular and biochemical aspects of chitin synthesis inhibition. In: Jollés, P., Muzzarelli, R.A.A. (Eds.), Chitin and Chitinases. Birkhäuser-Verlag, pp. 85–98; Tomlin, C.D.S., Ed., **2000**. The Pesticide Manual, 12th edn. British Crop Protection Council Publications; and Ishaaya, I., **2001**. Biochemical processes related to insecticide actions: an overview. In: Ishaaya, I. (Ed.), Biochemical Sites of Insecticide Action and Resistance. Springer, pp. 1–16.

The discovery of diflubenzuron spawned the discovery and development of a whole array of new analogs by different agricultural companies (**Table 14**). These include: chlorfluazuron (AIM®, ATABRON®, HELIX®, JUPITER®; Ishihara Sangyo), flucycloxuron (ANDALIN®; Uniroyal Chemical), flufenoxuron (CASCADE®; BASF), hexaflumuron (CONSULT®, RECRUIT®, TRUENO®; Dow AgroSciences LLC), lufenuron (MATCH®; Syngenta AG), teflubenzuron (NOMOLT®, DART®, DIARACT®; BASF) and triflumuron (ALSYSTIN®, BAYCIDAL®; Bayer AG) (Zoebelein *et al.*, 1980; Haga *et al.*, 1982, 1992; Becher *et al.*, 1983; Sbragia *et al.*, 1983; Retnakaran *et al.*, 1985; Anderson *et al.*, 1986; Retnakaran and Wright, 1987; Grosscurt *et al.*, 1987; Sheets *et al.*, 2000; Tomlin, 2000).

4.4.2. New Chemistries and Products

In the last decade, some newer BPU analogs were discovered: novaluron (RIMON®), bistrifluron, fluazuron (ACATAK®), and noviflumuron (RECRUIT® III) by Makhteshim Chemicals, Dongbu Hannong Chemical Co., Syngenta AG, and Dow AgroSciences LLC, respectively (Bull *et al.*, 1996; Ishaaya *et al.*, 1996; Kim *et al.*, 2000; Tomlin, 2000; Karr *et al.*, 2004) (**Table 14**).

Two other IGRs, buprofezin (APPLAUD®) and cyromazine (NEOPREX®, TRIGARD®, VETRAZIN®), with chemistries different from BPUs, but which also interfere with molting and chitin biosynthesis were developed by Nihon Nohyaku and Ciba-Geigy AG (now Syngenta AG), respectively (**Table 14**; Hall and Foehse 1980; Williams and Berry, 1980; Kanno *et al.*, 1981; Reynolds and Blakey, 1989; Tomlin, 2000).

4.4.3. Mode of Action and SAR

Typically, the chemistry and symptoms of the CSIs are unique, and not similar to other insecticides. Studies with diflubenzuron, the most thoroughly investigated compound of the BPUs, revealed that it alters cuticle composition, especially inhibition of chitin, resulting in abnormal endocuticular deposition that affects cuticular elasticity and firmness, and causes abortive molting. The reduced chitin levels in the cuticle seem to result from inhibition of biochemical processes leading to chitin formation. To date, it is not clear whether inhibition of chitin synthetase is the primary biochemical site for the reduced level of chitin, since in some studies BPUs do not inhibit chitin synthetase in cell-free systems (Retnakaran *et al.*, 1985; Grosscurt and Jongsma, 1987; Retnakaran and Wright, 1987; Londerhausen, 1996; Oberlander and Silhacek, 1998; Palli and Retnakaran, 1999). In addition to

chitin synthetase inhibition, other modes of action have been suggested for BPUs, such as: (1) inhibit the transport of UDP-GlcNAc across biomembranes; (2) block the binding of chitin to cuticular proteins resulting in inhibition of cuticle deposition and fibrillogenesis; (3) inhibit the formation of chitin due to an inhibition of the protease that activates chitin synthase, and activation of chitinases and phenoloxidases, which are both connected with chitin catabolism; (4) affect ecdysone metabolism, resulting in ecdysone accumulation that stimulates chitinase, which in turn digests nascent chitin; (5) block the conversion of glucose to fructose-6-phosphate; and (6) inhibit the DNA synthesis (Soltani *et al.*, 1984; Retnakaran *et al.*, 1985; Cohen, 1985; Retnakaran and Wright, 1987; Retnakaran and Oberlander, 1993; Mikólajczyk *et al.*, 1994; Zimowska *et al.*, 1994; Oberlander and Silhacek, 1998; Palli and Retnakaran, 1999; Oberlander and Smagghe, 2001). In addition, recent studies using imaginal discs and cell-free systems indicated that BPUs inhibit the 20-hydroxecdysone dependent GlcNAc incorporation into chitin, suggesting that BPUs affect ecdysone dependent biochemical sites, which lead to chitin inhibition (Mikolajczyl *et al.*, 1994; Zimowska *et al.*, 1994; Oberlander and Silhacek, 1998). BPUs act mainly by ingestion, and for the most part they are effective as larvicides, but in some species they also suppress oviposition. They act as ovicides, reducing the egg-laying rate or hindering the hatching process by inhibiting embryonic development. In most cases, the embryo was fully developed in the egg but the larva failed to hatch (Retnakaran and Wright, 1987; Ishaaya, 2001). BPUs were also found to reduce cyst and egg production in two free-living plant nematodes (Veech, 1978; Evans, 1985). In addition to being toxic by ingestion, some BPUs exhibit contact toxicity. Moreover, most BPUs, except hexaflumiron, have no systemic or translaminar activity (Retnakaran and Wright, 1987; Tomlin, 2000; Ishaaya, 2001).

The mode of action of buprofezin resembles that of the BPUs, although its structure is not analogous (Uchida *et al.*, 1985; De Cock and Degheele, 1998; Ishaaya, 2001). The compound strongly inhibits the incorporation of ^{3}H-glucose and GlcNAc into chitin. As a result of chitin deficiency, the procuticle of treated larval stages loses its elasticity and the insect is unable to molt. In addition, buprofezin may work as a prostaglandin inhibitor that may lead to suppression of ecdysis and slightly reduced DNA synthesis. It also exerts its effect on egg-hatch and on the larval stages in the rice planthopper, *Nilaparvata lugens*, and the whiteflies, *Bemesia tabaci and*

Trialeurodes vaporariorum (De Cock *et al.*, 1995; De Cock and Degheele, 1998; Ishaaya, 2001). It has no ovicidal activity but suppresses embryogenesis through adults (**Table 14**).

Although the exact mode of action of the IGR cyromazine, which is predominantly active against dipteran larvae, is not known, evidence has been presented to suggest that its target site for interference with sclerotization is different from that of BPUs (Biddington, 1985). In larvae intoxicated with cyromazine, the cuticle rapidly becomes less extensible and unable to expand compared with the cuticle of untreated larvae. The cuticle may be stiffer because of increased cross-linking between the various cuticle components, the nature of which remains unknown. As summarized in **Table 14**, cyromazine is an IGR with contact action interfering with molting and pupation. It has good systemic activity. When applied to the leaves, it exhibits a strong translaminar activity, and when applied to the soil it is taken up by the roots and translocated acropetally (Hall and Foehse, 1980; Awad and Mulla, 1984; Reynolds and Blakey, 1989; Viñuela and Budia, 1994; Tomlin, 2000).

With classical SAR (Hansch-Fujita), the effects of different substituents on the benzene rings in BPUs were analyzed for larvicidal activity against *Chilo suppressalis*, *S. littoralis*, and *B. mori* (Nakagawa *et al.*, 1987, 1989a, 1989b). For the benzoyl moiety, the toxicity was higher with a higher total hydrophobicity, a higher electron withdrawal from the side chain, and a lower steric bulkiness of *ortho*-substituents. In addition, introduction of electron-withdrawing and hydrophobic substituents at the *para*-position of the phenyl (aniline) moiety enhanced the larvicidal activity, whereas bulkier groups were unfavorable. Stacking did not occur between the two aromatic moieties along the urea moiety (Sotomatsu *et al.*, 1987). To ascertain the above results, the relative activity of BPUs was determined by measuring the incorporation of [^{14}C]-GlcNAc in larval integuments cultured *in vitro* (Nakagawa *et al.*, 1989b). There was a colinear relationship between *in vitro* activities and *in vivo* larvicidal toxicities if the hydrophobic factor(s) participating in transport were considered separately. In addition, integuments of the three Lepidoptera were incubated in conditions with and without the synergists piperonylbutoxide (PB) and *S,S,S*-tributylphosphorotrithioate (DEF). In the Qualitative Structure–activity Relation (QSAR) equation measured without synergist, an electron-withdrawing effect was favorable to the activity, but an electron-donating group was favorable to the activity in the presence of PB. These results mean that electron-withdrawing groups are playing a role in suppressing the oxidative metabolism. DEF had no significant effect, suggesting that hydrolytic degradation of the phenyl moiety was not of significant consequence as compared to its oxidative degradation. In summary, the SAR results suggested that for BPUs the specific larvicidal spectrum is due to inherent differences in metabolism in addition to differences in the physiochemical substituent effects (Nakagawa *et al.*, 1987, 1989a, 1989b; Sotomatsu *et al.*, 1987).

4.4.4. Spectrum of Activity

Most CSI compounds are very potent against a variety of different pests with the highest activity towards lepidopterous insects and whiteflies. The main commercial applications are in field crops/agriculture, forestry, horticulture, and in the home as summarized in **Table 14** (Retnakaran and Wright, 1987; Tomlin, 2000; Ishaaya, 2001). Novaluron acts by ingestion and contact against lepidopterans (*S. littoralis*, *S. exigua*, *S. frugiperda*, *Tuta absoluta* and *Helicoverpa armigera*), whitefly, *B. tabaci*, eggs and larvae, and different stages of the leafminer, *Liriomyza huidobrensis*. Bistrifluron is active against various lepidopteran pests and whiteflies an apple, Brassica leafy vegetables, tomato, persimmon, and other fruits (Kim *et al.*, 2000). Hexaflumuron and the newer noviflumuron are now used in bait for control of subterranean termites (Sheets *et al.*, 2000; Karr *et al.*, 2004). As a specialty in the series of CSIs, fluazuron is the only ixodicide with a strong activity against cattle ticks (Kim *et al.*, 2000) (**Table 14**).

Buprofezin has a persistent larvicidal action against sucking Homoptera, such as the greenhouse whitefly, *T. vaporariorum*, the sweet potato whitefly, *B. tabaci*, both of which are important pests of cotton and vegetables, the brown planthopper, *N. lugens* in rice, the citrus scale insects, *Aonidiella aurantii* and *Sassetia oleae*, and some Coleoptera and Acarina. In contrast, cyromazine is used for control of dipteran larvae in chicken manure. It is also used as a foliar spray to control leafminers (*Liriomyza* sp.) in vegetables and ornamentals, and to control flies on animals (Hall and Foehse, 1980; Williams *et al.*, 1980; Kanno *et al.*, 1981; Reynolds and Blakey, 1989; Tomlin, 2000) (**Table 14**).

4.4.5. Ecotoxicology and Mammalian Safety

Overall, CSIs have selective insect toxicities and as such are considered "soft insecticides." For diflubenzuron, the harbinger of all BPUs, its environmental fate has been extensively investigated and was

Table 15 Environmental effects of some CSIs

Compound	Mammals	Fish	Crustacae	Bees	Predators	Parasitoids
Bistrifluron	Rat LD$_{50}$ >5 g kg^{-1}	Carp LC$_{50}$ (48 h) >0.5 mg l^{-1}	Daphnia LC$_{50}$ (48 h) 0.9 µg l^{-1}	LD$_{50}$ >100 µg bee^{-1}	Highly toxic	No – very little effects
Chlorfluazuron	Rat LD$_{50}$ >8.5 g kg^{-1}	Carp LC$_{50}$ (48 h) >300 mg l^{-1}	Daphnia LC$_{50}$ (48 h) 7.1 µg l^{-1}	LD$_{50}$ >100 µg bee^{-1}	Safe – little effects	Safe-little adverse effects
Diflubenzuron	Rat LD$_{50}$ >4.6 g kg^{-1}	Zebra fish, rainbow trout LC$_{50}$ (96 h) >0.2 mg l^{-1}		Not hazardous LD$_{50}$ >100 µg bee^{-1}		
Flucycloxuron	Rat LD$_{50}$ >5 g kg^{-1}	Rainbow trout, sunfish LC$_{50}$ (96 h) >0.1 mg l^{-1}	Daphnia LC$_{50}$ (48 h) 4.4 µg l^{-1}	LD$_{50}$ >100 µg bee^{-1}	No – moderate effects	No – harmful effects
Flufenoxuron	Rat LD$_{50}$ >3 g kg^{-1}	Rainbow trout LC$_{50}$ (96 h) >100 mg l^{-1}			No adverse effects – harmful	No – little adverse effects
Hexaflumuron	Rat LD$_{50}$ >5 g kg^{-1}	rainbow trout: not lethal, Tilapia LC$_{50}$ >3 µg l^{-1}	Daphnia LC$_{50}$ (96 h) 0.1 µg l^{-1}	LD$_{50}$ >100 µg bee^{-1}	Very little effects	Harmless, but ectoparasites strongly affected
Lufenuron	Rat LD$_{50}$ >2 g kg^{-1}	Carp, rainbow trout, sunfish LC$_{50}$ (96 h) >30–70 mg l^{-1}		LC$_{50}$ >38 µg/bee^{-1} LD$_{50}$ >8 µg/bee^{-1}	Safe – harmful	Moderately harmful
Novaluron	Rat LD$_{50}$ >5 g kg^{-1}	Rainbow trout, sunfish LC$_{50}$ (96 h) >1 mg l^{-1}	Daphnia LC$_{50}$ (48 h) 58 µg l^{-1}	Nontoxic at recommended rates; LC$_{50}$ >100 µg bee^{-1} LD$_{50}$ >1 mg bee^{-1}	Low toxicity – harmful	Little–harmful effects
Teflubenzuron	Rat LD$_{50}$ >5 g kg^{-1}	Trout, carp LC$_{50}$ (96 h) >0.5 mg l^{-1}		LD$_{50}$ >1 mg bee^{-1}	Harmless – highly toxic	No – very little effects
Triflumuron	Rat LD$_{50}$ >5 g kg^{-1}	Rainbow trout LC$_{50}$ (96 h) >320 mg l^{-1}	Daphnia LC$_{50}$ (48 h) 225 µg l^{-1}	Toxic	Harmless – moderately toxic	Moderately harmful
Buprofezin	Rat LD$_{50}$ 2.4 g kg^{-1}	Carp, rainbow trout LC$_{50}$ (48 h) >2.7–1.4 mg l^{-1}	Daphnia LC$_{50}$ (3 h) 50.6 mg l^{-1}	No effect at 2 g l^{-1}	Harmless – moderately toxic	Very little effects
Cyromazine	Rat LD$_{50}$ 3.4 mg kg^{-1}	Carp, rainbow trout, sunfish LC$_{50}$ (96 h) >90 mg l^{-1}	Daphnia LC$_{50}$ (48 h) >9.1 mg l^{-1}	No effect up to 5 µg	Harmless – moderately toxic	Harmless – moderate effects

Data compiled from Darvas, B., Polgár, L.A., **1998**. Novel-type insecticides: specificity and effects on non-target organisms. In: Ishaaya, I., Degheele, D. (Eds.), Insecticides with Novel Modes of Action: Mechanism and Application. Springer, pp. 188–259; Tomlin, C.D.S. (Ed.), **2000**. The Pesticide Manual, 12th edn. British Crop Protection Council Publications; and Ishaaya, I., **2001**. Biochemical processes related to insecticide actions: an overview. In: Ishaaya, I. (Ed.), Biochemical Sites of Insecticide Action and Resistance. Springer, pp. 1–16.

shown to be broken down by various microbial agents without accumulation in soil and water. The fear that spray run-off into streams may cause widespread mortality of nontarget species has not been realized. Moreover, its chitin synthesis inhibitory action is quite specific, and related biochemical processes, such as chitin synthesis in fungi and biosynthesis of hyaluronic acid and other mucopolysaccharides in chickens, mice, and rats, are not affected. A representative list of the salient environmental effects of CSIs against mammals (rat), vertebrates (fish), crustaceans, bees, predators, and parasitoids is given in **Table 15**. However, owing to selective toxicity towards arthropods, BPUs have, varying degrees of effects on crustaceans as well as beneficial insects, which requires their use with care. In this case, the IOBC/WPRS (International Organization for Biological and Integrated control of Noxious Animals and Plants, West Palaearctic Regional Section, 2004) sequential testing scheme has proven its value for testing the side effects on a species-by-species basis and under (semi-)field conditions (Hassan, 1992). However, since most of the BPU formulations need to be ingested to be effective, topical effects on parasites, predators, and pollinators are minimal. Retnakaran and Wright (1987), Perez-Farinos *et al.* (1998), and Medina *et al.* (2002) claimed that the selectivity of BPUs may result from a relatively low cuticular absorption in residual and direct contact assays. As summarized in **Table 15**, the majority of the CSIs are harmless or exert little adverse effect on bees, predators, or parasitoids, which renders these BPUs acceptable for inclusion in integrated pest management (IPM) programs (Elzen, 1989; Hassan *et al.*, 1991, 1994; Vogt, 1994; Van de Veire *et al.*, 1996; Darvas and Polgar, 1998; Sterk *et al.*, 1999; Tomlin, 2000).

4.4.6. Resistance, Mechanism for Resistance and Resistance Potential

Pimprikar and Georghiou (1979) were the first to report very high levels of resistance (>1,000-fold) to diflubenzuron in a housefly population stressed with the compound. Since then, several research groups in different parts of the world have reported resistance to CSIs in Lepidoptera, Diptera, and whiteflies.

In Southeast Asia, selection of diamondback moth, *Plutella xylostella,* from Thailand during six generations with chlorfluazuron and teflubenzuron resulted in resistance levels of 109-fold and 315-fold, respectively (Ismail and Wright, 1991, 1992). While marked cross-resistance between chlorfluazuron and teflubenzuron was demonstrated, there was no evidence of cross-resistance to diflubenzuron, and little or no cross-resistance to flufenoxuron and

hexaflumuron. Pretreatment with the synergists PB and DEF increased the toxicity of chlorfluazuron and teflubenzuron up to 34-fold and 28-fold, respectively, suggesting that microsomal monooxygenases and esterases are involved in resistance. In another study with *P. xylostella* from Malaysia, selection with diflubenzuron increased resistance to teflubenzuron, but had no effect on chlorfluazuron (Furlong and Wright, 1994). In this study, while use of PB and DEF suggested the involvement of microsomal monooxygenases and esterases in teflubenzuron resistance, glutathion-S-transferase (GSTs) had limited involvement.

As documented by Moffit *et al.* (1988) and Sauphanor *et al.* (1998), the codling moth *C. pomonella* shows very high levels of resistance to diflubenzuron in the USA and France. In southern France, failure in *C. pomonella* control was observed for several years, revealing a 370-fold resistance for diflubenzuron and cross-resistance with two other BPUs, teflubenzuron (7-fold) and triflumuron (102-fold), as well as to the ecdysone agonist, tebufenozide (26-fold) (Sauphanor and Bouvier, 1995). Interestingly, resistance to diflubenzuron was linked to cross-resistance to deltamethrin. In both cases, enhanced mixed-function oxidase and GST activities are involved in resistance, rather than target site modification. In addition, a fitness cost described in both resistant strains was mainly associated with metabolic resistance (Boivin *et al.*, 2001). Finally, a lack of relationship between ovicidal and larvicidal resistance for diflubenzuron in *C. pomonella* may be due to different transport properties as well as differential enzymatic metabolism (Sauphanor *et al.*, 1998).

In some strains of *S. littoralis*, resistance to diflubenzuron was also increased by a factor of about 300 (El-Guindy *et al.*, 1983). El Saidy *et al.* (1989) reported that diflubenzuron and teflubenzuron were hydrolyzed rapidly by all tissues tested, and the gut wall was the most active tissue reaching 61% hydrolytic breakdown for diflubenzuron and 16% for teflubenzuron. Interestingly, profenofos and DEF could inhibit degradation of both BPUs tested under optimal conditions. Ishaaya and Casida (1980) had earlier reported that these organophosphorous compounds can inhibit insect esterases in larval gut integument. The strong synergism activity of DEF and profenofos indicated that the major route of detoxification in *S. littoralis* was through hydrolysis, while oxidative metabolism was found to be of minor importance for resistance.

In the Australian sheep blowfly, *Lucilia cuprina*, resistance to diflubenzuron was inherited in a codominant (S male × R female) or incompletely recessive (R male × S female) manner, and there was also some maternal influence on inheritance of monooxygenase activities, suggesting that diflubenzuron resistance is polygenic and involves mechanisms additional to monooxygenases (Kotze and Sales, 2001).

McKenzie and Batterman (1998) reported the development of resistance to cyromazine in populations of *L. cuprina*, which, however, had remained susceptible for almost 20 years of exposure. This resistance in *L. cuprina* was controlled by a single gene in each variant and two resistance loci were identified: one (Cyr4) closely linked to the marker "reduced eyes" on chromosome IV, and the other, Cyr5, closely linked to the "stubby bristles" marker on chromosome V (Yen et al., 1996). Similarly, in *D. melanogaster*, cyromazine resistance was due to a mutation in a single, but different gene than the one identified in resistant *L. cuprina*. The resistance genes, designated *Rst(2)Cyr* and *Rst(3)Cyr*, were localized to puff positions 64 on chromosome II and 47 on chromosome III, respectively (Adcock et al., 1993).

With respect to resistance in *Musca domestica*, Keiding (1999) reviewed the period 1977 to 1994. In this period, only two IGRs, diflubenzuron and cyromazine, were widely used for fly control, either by direct contact application to breeding sites, manure, garbage, etc., or as an admixture in feed for poultry or pigs. Widespread use of diflubenzuron or cyromazine by direct treatment of manure has not led to resistance of practical importance except for resistance to diflubenzuron on a few farms in the Netherlands. However, there are reports showing that the use of either compound mixed with feed has resulted in the development of moderate to high resistance to these compounds. High resistance in adult flies to organophosphates and pyrethroids does not usually confer cross-resistance to the larvicidal effects of diflubenzuron or cyromazine, but cross-resistance between organophosphorous compounds and JHAs (methoprene) has been reported. In Denmark, Keiding et al. (1992) came to the same conclusions where housefly control with both compounds was carried out on farms for 1–9 years. On none of the farms was any general increase of tolerance to diflubenzuron found. In the same study, two strains were selected with diflubenzuron and cyromazine where moderate to high resistance developed if the selection pressure was strong, especially when used as feed-through applications on poultry farms where all feed contains diflubenzuron or cyromazine. If the treatment of fly breeding sources is less complete, resistance problems may not develop.

More recently, Kristensen and Jespersen (2003) observed resistance in houseflies to diflubenzuron for the first time in Denmark, and some field populations with resistance to cyromazine. A fivefold cyromazine resistant strain was established and this was 3-, 5-, and 90-fold resistant to diflubenzuron, triflumuron, and methoprene, respectively. In reviewing the mechanism of resistance in houseflies to diflubenzuron and cyromazine, Ishaaya (1993) indicated that resistance was due to increased levels of detoxifying enzymes, decreased penetration, and enhanced excretion of the compounds. Cross-resistance between cyromazine and diflubenzuron was observed but the mechanism seems more target site related than metabolic.

For the control of whiteflies, novel compounds like buprofezin were introduced at the beginning of the 1990s (Horowitz et al., 1994; De Cock and Degheele, 1998). In Israel, a survey in cotton fields over 4 years (1989–1992) with two or three applications of buprofezin per season indicated a fivefold increase in tolerance in the *B. tabaci*, and the resistance ratio for resistance to pyriproxifen was >500-fold (Horowitz and Ishaaya, 1994). Based on the success of the program in Israel, the usefulness of buprofezin was once again demonstrated in a resistance management program in Arizona to control resistance to pyriproxifen in populations of the whitefly, *Bermisia argentifolii* in cotton (Dennehy and Williams, 1997). In another study in southern Spain where highly multiresistant strains of *B. tabaci* showed lower efficacy to buprofezin, pyriproxyfen resistance was not obvious (Elbert and Nauen, 2000). For the greenhouse whitefly, *T. vaporariorum*, which is a major pest problem in protected crops in Western Europe, a >300-fold resistance was scored in a strain collected in northern Belgium (De Cock et al., 1995). No cross-resistance was recorded for pyriproxyfen and diafenthiuron, indicating their potential in a resistance management program. In UK greenhouses, Gorman et al. (2002) also found very strong resistance to buprofezin in *T. vaporariorum*, and these strains were cross-resistant to teflubenzuron.

The chitin synthesis inhibition site has proven to be important for developing control agents that act against important groups of insect pests and many of these CSIs are environmentally safe. However, details of the exact mode of action of CSIs have not been elucidated so far. Although there were already some incidences of resistance to CSIs in the last two decades in different pest insects, these IGR compounds remain suitable for use as rotation partners in integrated and resistance management programs based on their new and selective mode of action in contrast to broad-spectrum neurotoxins.

At present, it is believed that chitin synthesis has by no means been fully exploited as an attractive target and that opportunities exist to further discover new CSIs. A better understanding of the biosynthetic pathway for chitin synthesis and cuticle deposition, as well as precise mode of action of CSIs would allow for development of better and more efficient high throughput assays for discovery of new and novel CSIs.

4.5. Conclusions and Future Prospects of Insect Growth- and Development-Disrupting Insecticides

The three classes of insecticides reviewed in this chapter are much slower acting than those acting on neural target sites. The end user has, of course, been used to seeing insects die within a very short time following application of neuroactive insecticides. The discovery and availability of insecticides that inhibit growth and development of insects brought a paradigm shift from the faster acting neurotoxic insecticides. This change has necessitated educating the distributors and the users on the mode of action of these new insecticides. The bisacylhydrazine insecticides are generally faster acting than the JHA and CSI insecticides. Moreover, an attractive feature of the bisacylhydrazine insecticides is their ability to prevent crop damage by inhibition of feeding within 3–12 h after application.

Of the three classes of insecticides that disrupt growth and development in insects, the mode of action of nonsteroidal ecdysone agonist bisacylhydrazine insecticides is the best understood at the molecular level. This detailed understanding has been possible, both with the cloning and expression of cDNAs encoding EcR and USPs from several insects, and the availability of stable and easy to synthesize bisacylhydrazines. Unlike the JHAs and CSI, the molecular targets for bisacylhydrazine insecticides are not only known but the interaction of some of these insecticides with specific amino acid residues in the ligand binding pocket of the target ecdysone receptor are also known. Moreover, reasons for the selective insect toxicity of bisacylhydrazine insecticides are also well understood. Publication of the crystal structure of unliganded and liganded (with ecdysteroid and bisacylhydrazine) HvEcR/HvUSP, and the discovery of new nonsteroidal ecdysone agonist chemistries like tetrahydroquinolines, provides new tools, and suggests possibilities of discovering new ecdysone agonist

insecticides. It should now be possible to use combinatorial chemistry approaches, around leads generated either via *in silico* screening or rational design (based on the three-dimensional structures and interactions of ligand in ligand binding pocket of EcR), to discover new and novel chemistries that target EcRs from specific insect orders. The discovery of additional new, novel, selective, and potent nonsteroidal ecdysone agonists will create opportunities to extend their applications in individually or simultaneously regulating different ecdysone receptor gene switches in plants and animals. Further, the discovery of non steroidal ecdysone agonist bisacylhydrazine insecticides has opened tremendous opportunities to explore both basic and applied biology.

The molecular basis of action of JHA and CSI insecticides is not well understood, although both classes of chemistries were discovered long before the bisacylhydrazine insecticides. The discovery of the JH receptor(s) has been elusive. To complicate matters further, it is not clear if the JHAs use the same molecular target/site as the natural JHs do (Dhadialla *et al.*, 1998). Dhadialla *et al.* (1998) alluded to the possibility that the JH receptor may be a complex of two or more proteins, and that JH and JHAs could manifest their action by interacting either with different proteins in the complex or by interacting at different, but effective, sites. Consequently, the research to discover new JH agonist or antagonist chemistries has been slow. An antagonist of JH (different from a precocene type of mode of action) that acts either by inhibiting one of the JH biosynthesis steps or antagonizes the action of JH at the receptor level would be useful. With that, it may be possible to have agonists/antagonists of JH with insect selective toxicity, as has been possible for the ecdysone agonists.

Even though the precise mode of action of CSI insecticides is unknown, many analogs and variants of the original diflubenzuron, Dimilin®, have been synthesized and registered as insecticides. A better understanding of the biosynthetic pathway for chitin synthesis and cuticle deposition, and molecular characterization of the various enzymes involved, as well as the precise mode of action of CSIs, would allow for development of better and more efficient high-throughput assays for discovery of new and novel CSIs.

Finally, in spite of their slower speed of kill of insect pests than the faster acting neurotoxic insecticides, the three classes of insect growth and development disrupting insecticides are well suited for use in insect pest control. They are also suited for resistance management programs due to their novel and different modes of action. The bisacylhydrazine insecticides are particularly attractive, due to their selective insect toxicity and high degree of mammalian and ecotoxicological reduced risk profiles.

Acknowledgments

The authors would like to thank Natalie Filion, CFS, Ontario, Canada, and Karan Dhadialla, University of Chicago, IL, USA, for sorting and typing the references in the required format. TSD also expresses his deep appreciation to Dow Agrociences LLC., Indianapolis, IN, USA (especially Drs. W. Kleshick and S. Evans) for support and encouragement in undertaking this writing. Part of the research was supported by grants from Gerome Canada and the Canadian Forest Service to AR and colleagues. GS also acknowledges the Fund for Scientific Research Flanders (FWO, Brussels, Belgium).

References

Adcock, G.J., Batterham, P., Kelly, L.E., McKenzie, J.A., 1993. Cyromazine resistance in Drosophila melanogaster (Diptera, Drosophilidae) generated by ethyl methanesulfonate mutagenesis. J. Econ. Entomol. 86, 1001–1008.

Aliniazee, M.T., Arshad, M., 1998. Susceptibility of immature stages of the obliqued banded leafroller, Choristoneura rosaceana (Lepidoptera: Tortricidae) to fenoxycarb. J. Entomol. Soc. Br. Columbia 95, 59–63.

Allgood, V.E., Eastman, E.M., 1997. Chimeric receptors as gene switches. Curr. Opin. Biotechnol. 8, 474–479.

Aller, H.E., Ramsay, J.R., 1988. RH-5849 – A novel insect growth regulator with a new mode of action. Brighton Crop Prot. Conf. 2, 511–518.

Anderson, M., Fisher, J.P., Robinson, J., Debray, P.H., 1986. Flufenoxuron – an acylurea acaricide/insecticide with novel properties. Brighton Crop Prot. Conf. 1, 89–96.

Arbeitman, M.N., Hogness, D.S., 2000. Molecular chaperones activate the Drosophila ecdysone receptor, an RXR heterodimer. Cell 101, 67–77.

Asano, S., Kuwano, E., Eto, M., 1986. Precocious metamorphosis induced by an anti-juvenile hormone compound applied to 3rd instar silkworm larvae, Bombyx mori L. (Leopidoptera: Bombycidae). App. Entomol. Zool. 21(2), 305–312.

Ashburner, M., 1973. Sequential gene activation by ecdysone in polytene chromosomes of Drosophila melangaster. I. Dependence upon ecdysone concentration. Devel. Biol. 35, 47–61.

Ashburner, M., Chiara, C., Meltzer, P., Richards, G., 1974. Temporal control of puffing activity in polytene chromosomes. Cold Spring Harbor Symp. Quant. Biol. 38, 655–662.

Ashok, M., Turner, C., Wilson, T.G., 1998. Insect juvenile hormone resistance gene homology with the bHLH-PAS family of transcriptional regulators. Proc. Natl Acad. Sci. USA 95(6), 2761–2766.

Awad, T.I., Mulla, M.S., 1984. Morphogenetic and histopathological effects induced by the insect growth regulator cyromazine in Musca domestica (Diptera, Muscidae). J. Med. Entomol. 21, 419–426.

Ayoade, O., Morooka, S., Tojo, S., 1995. Induction of macroptery, precocious metamorphosis, and retarded ovarian growth by topical application of precocene II, with evidence for its non-systemic allaticidal effects in the brown planthopper, Nilaparvata lugens. J. Insect Physiol. 42(6), 529–540.

Balcells, M., Avilla, J., Profitos, J., Canela, R., 2000. Synthesis of phenoxyphenyl pyridine and pyrazine carboxamides activity against Cydia pomonella (L.) eggs. J. Agric. Food Chem. 48, 83–87.

Banks, W.A., Lofgren, C.S., 1991. Effectiveness of the insect growth regulator pyriproxyfen against the red imported fire ant Hymenoptera formicidae. J. Entomol. Sci. 26(3), 331–338.

Becher, H.M., Becker, P., Prokic-Immel, R., Wirtz, W., 1983. CME, a new chitin synthesis inhibiting insecticide. Brighton Crop Prot. Conf. 1, 408–415.

Beckage, N.E., Riddiford, L.M., 1981. Effects of methoprene and juvenile hormone on larval ecdysis, emergence, and metamorphosis of the endoparasitic wasp, Apanteles Congregatus. J. Insect Physiol. 28(4), 329–334.

Beckage, N.E., Riddiford, L.M., 1983. Lepidopteran antijuvenile hormones: effects on development of Apanteles congregatus in Manduca sexta. J. Insect Physiol. 29(8), 633–637.

Bélai, I., Matolcsy, G., Farnsworth, D.E., Feyereisen, R., 1988. Inhibition of insect cytochrome P-450 by some metyrapone analogues and compounds containing a cyclopropylamine moiety and their evaluation as inhibitors of juvenile hormone biosynthesis. Pestic. Sci. 24, 205–219.

Bellés, X., Baldellou, M.I., 1983. Precocious metamorphosis induced by precocenes on Oxycarenus lavaterae. Entomol. Exp. Appl. 34(2), 129–133.

Bellés, X., Messegues, A., Piulachs, M.B., 1985. Sterilization induced by precocenes on females of Blatella gexmanica(L): short and long term effects. Zeitschrift fur Angewandte Entomologie 100, 409–417.

Biddington, K.C., 1985. Ultrastructural changes in the cuticle of the sheep blowfly, Lucilia cuprina, induced by certain insecticides and biological inhibitors. Tissue Cell 17, 131–140.

Billas, I.M.L., Moulinier, L., Rochel, N., Moras, D., 2001. Crystal structure of the ligand-binding domain of the ultraspiracle protein USP, the ortholog of retinoid X receptors in insects. J. Biol. Chem. 276, 7465–7474.

Billas, I.M.L., Twema, T., Garnier, J.-M., Mitschler, A., Rochel, N., et al., 2003. Structural adaptability in the ligand-binding pocket of the ecdysone receptor. Nature 426, 91–96.

Binder, B.F., Bowers, W.S., **1991**. Behavioral changes and growth inhibition in last instar larvae of *Heliothis zea* induced by oral and topical application of precocene II. *Entomol. Exp. Appl.* 59, 207–217.

Blackford, M., Dinan, L., **1997**. The tomato moth *Lacnobia oleracea* (Lepidoptera: Noctuidae) detoxifies ingested 20-hydroxyecdysone, but is susceptible to the ecdysteroid agonists RH-5849 and RH-5992. *Insect Biochem. Mol. Biol.* 27, 167–177.

Boivin, T., Chabert d'Hieres, C., Bouvier, J.C., Beslay, D., Sauphanor, B., **2001**. Pleiotropy of insecticide resistance in the codling moth, *Cydia pomonella. Entomol. Exp. Appl.* 99, 381–386.

Bonning, B.C., Hoover, K., Booth, T.F., Duffey, S., Hammock, B.D., **1995**. Development of a recombinant baculovirus expressing a modified juvenile hormone esterase with potential for insect control. *Arch. Insect Biochem. Physiol.* 30, 177–194.

Bourne, P.C., Whiting, P., Dhadialla, T.S., Hormann, R.E., Girault, J.-P., *et al.*, **2002**. Ecdysteroid 7,9(11)-dien-6-ones as potential photoaffinity labels for ecdysteroid binding proteins. *J. Insect Sci.* 2, 1–11 (available online at http://www. insectscience.org/2.11).

Bouvier, J.C., Boivin, T., Beslay, D., Sauphanor, B., **2002**. Age-dependent response to insecticides and enzymatic variation in susceptible and resistant codling moth larvae. *Arch. Insect Biochem. Physiol.* 51, 55–66.

Bowers, W.S., **1968**. Juvenile hormone: activity of natural and synthetic synergists. *Science* 161, 895–897.

Bowers, W.S., Nishida, R., **1980**. Juvocimences L potent juvenile hormone mimics from sweet basil. *Science* 209, 1030–1032.

Bowers, W.S., Unnithan, G.C., Fukushima, J., Toda, J., Sugiyama, T., **1995**. Synthesis and biological activity of furanyl anti-juvenile hormone compounds. *Pestic. Sci.* 43, 1–11.

Brindle, P.A., Baker, F.C., Tsai, L.W., Reuter, C.C., Schooley, D.A., **1987**. Sources of proprionate for the biogenesis of ethyl-branched insect juvenile hormones: role of isoleucine and valine. *Proc. Natl Acad. Sci. USA* 84, 7906–7910.

Bull, M.S., Swindale, S., Overend, D., Hess, E.A., **1996**. Suppression of *Boophilus microplus* populations with fluazuron – an acarine growth regulator. *Austral. Vet. J.* 74, 468–470.

Butenandt, A., Karlson, P., **1954**. Uber die isolierung eines metamorphose-hormons des insekten in kristallisierter from *Z. Naturforsch. Teil B* 9, 389–391.

Butler, L., Kondo, V., Blue, D., **1997**. Effects of tebufenozide (RH-5992) for gypsy-moth (Lepidoptera, Lymantriidae) suppression on nontarget canopy arthropods. *Env. Entomol.* 26(5), 1009–1015.

Cadogan, B.L., Retnakaran, A., Meating, J.H., **1997**. Efficacy of RH5992, a new insect growth regulator against spruce budworm (Lepidoptera: Tortricidae) in a boreal forest. *J. Econ. Entomol.* 90, 551–559.

Cadogan, B.L., Thompson, D., Retnakaran, A., Scharbach, R.D., Robinson, A., *et al.*, **1998**. Deposition of aerially applied tebufenozide (RH5992) on balsam

fir (*Abies balsamea*) and its control of spruce budworm (*Choristoneura fumiferana* [Clem.]). *Pestic. Sci.* 53, 80–90.

Cadogan, B.L., Scharbach, R.D., Krause, R.E., Knowles, K.R., **2002**. Evaluation of tebufenozide carry-over and residual effects on spruce budworm (Lepidoptera: Tortricidae). *J. Econ. Entomol.* 95, 578–586.

Campero, D.M., Haynes, K.F., **1990**. Effects of methoprene on chemical communication, courtship, and oviposition in the cabbage looper (Lepidoptera: Noctuidae). *J. Econ. Entomol.* 83(6), 2263–2268.

Cao, S., Qian, X., Song, G., **2001**. N'-tert-butyl-N'-aroyl-N-(alkoxycarbonylmethyl)-N-aroylhydrazines, a novel nonsteroidal ecdysone agonist: synthesis, insecticidal activity, conformational, and crystal structure analysis. *Can. J. Chem.* 79, 272–278.

Capinera, J.L., Epsky, N.D., Turick, L.L., **1991**. Responses of *Malonoplus sanguinipes* and *M. differentialis* (Orthoptera: Acrididae) to fenoxycarb. *J. Econ. Entomol.* 84(4), 1163–1168.

Carlson, G.R., Dhadialla, T.S., Hunter, R., Jansson, R.K., Jany, C.S., *et al.*, **2001**. The chemical and biological properties of methoxyfenozide, a new insecticidal ecdysteroid agonist. *Pest Mgt. Sci.* 57, 115–119.

Carton, B., Heirman, A., Smagghe, G., Tirry, L., **2000**. Relationship between toxicity, kinetics and *in vitro* binding of nonsteroidal ecdysone agonists in the cotton leafworm and the Colorado potato beetle. *Med. Fac. Landbouww. Univ. Gent.* 65(2a), 311–322.

Celestial, D.M., McKenney, C.L.Jr., **1993**. The influence of an insect growth regulator on the larval development of the mud crab *Rhithropanopeus harrisii. Environ. Pollut.* 85(2), 169–173.

Chan, T.H., Ali, A., Britten, J.F., Thomas, A.W., Strunz, G.M., *et al.*, **1990**. The crystal structure of 1,2-dibenzoyl-1-tert-butylhydrazine, a non-steroidal ecdysone agonist, and its effects on spruce bud worm (*Choristoneura fumiferana*). *Can. J. Chem.* 68, 1178–1181.

Chandler, L.D., Pair, S.D., Harrison, W.E., **1992**. RH-5992, a new insect growth regulator active against corn earworm and fall armyworm (Lepidoptera: Noctuidae). *J. Econ. Entomol.* 85, 1099–1103.

Chandler, L.D., **1994**. Comparative effects of insect growth regulators on longevity and mortality of beet armyworm (Lepidoptera: Noctuidae) larvae. *J. Entomol. Sci.* 29, 357–366.

Charles, J.P., Wojtasek, H., Lentz, A.J., Thomas, B.A., Bonning, B.C., *et al.*, **1996**. Purification and reassessment of ligand binding by the recombinant, putative juvenile hormone receptor of the tobacco hornworm, *Manduca sexta. Arch. Insect. Biochem. Physiol.* 31, 371–393.

Cherbas, P., Cherbas, L., Lee, S.-S., Nakanishi, K., **1988**. [125I]iodo-ponasterone A is a potent ecdysone and a sensitive radioligand for ecdysone receptors. *Proc. Natl Acad. Sci., USA* 85, 2096–2100.

Chen, N.M., Borden, J.H., **1989**. Adverse effect of fenoxycarb on reproduction by the california fivespined IPS,

IPS *Paraconfusus lanier* (Coleoptera: Scolytidae). *Can. Entomol. 121*, 1059–1068.

Chiang, A.-S., Lin, W-Y., Liu, H-P., Pszezalkowski, A., Fu, T-F., *et al.*, 2002. Insect NMDA receptors mediate juvenile hormone biosynthesis. *Proc. Natl Acad. Sci. USA 99*, 37–42.

Cho, W.-L., Kapitskaya, M.Z., Raikhel, A.S., 1995. Mosquito ecdysteroid receptor: analysis of the cDNA and expression during vitellogenesis. *Insect Biochem. Mol. Biol. 25*, 19–27.

Christopherson, K.S., Mark, M.R., Bajaj, V., Godowski, P.J., 1992. Ecdysteroid-dependent regulation of genes in mammalian cells by a *Drosophila* ecdysone receptor and chimeric transactivators. *Proc. Natl Acad. Sci. USA 89*, 6314–6318.

Chu, K.H., Wong, C.K., Chiu, K.C., 1997. Effects of the insect growth regulator (S)-methoprene on survival and reproduction of the freshwater cladoceran, *Moina macrocopa. Environ. Pollut. 96*(2), 173–178.

Chung, A.C.-K., Durica D.S., Clifton, W., Roe, A., Hopkins, P.M., 1998. Cloning of crustacean ecdysteroid receptor and retinoid-X-receptor gene homologs and elevation of retinoid-X-receptor mRNA by retinoic acid. *Mol. Cell. Endocrinol. 139*, 209–227.

Clayton, G.M., Peak-Chew, S.Y., Evans, R.M., Schwabe, J.W.R., 2001. The structure of the ultraspiracle ligand-binding domain reveals a nuclear receptor locked in an inactive conformation. *Proc. Natl Acad. Sci. USA 98*, 1549–1554.

Clement, C.Y., Bradbrook, D.A., Lafont, R., Dinan, L., 1993. Assessment of a microplate-based bioassay for the detection of ecdysteroid-like or antiecdysteroid activities. *Insect Biochem. Mol. Biol. 23*, 187–193.

Clever, U., Karlson, P., 1960. Induktion von puff-veränderungen in den speicheldriisenchromosomen von *Chironomus tentans* durch ecdysone. *Exp. Cell. Res. 20*, 623–662.

Cohen, E., 1985. Chitin synthetase activity and inhibition in different insect microsomal preparations. *Experientia 41*, 470–472.

Cowles, R.S., Villani, M.G., 1996. Susceptibility of Japanese beetle, oriental beetle, and european chafer (Coleoptera: Scarabaedae) to halofenozide, an insect growth regulator. *J. Econ. Entomol. 89*(6), 1556–1565.

Cowles, R.S., Alm, S.R., Villani, M.G., 1999. Selective toxicity of halofenozide to exotic white grubs (Coleoptera: Scarabaeidae). *J. Econ. Entomol. 92*(2), 427–434.

Cusson, M., Palli, S.R., 2000. Can juvenile hormone research help rejuvenate integrated pest management? *Can. Entomol. 132*, 263–280.

Darvas, B., Pap, L., Kelemen, M., Laszlo, P., 1998. Synergistic effects of verbutin with dibenzoylhydrazine-type ecdysteroid agonists on larvae of *Aedes aegypti* (Diptera: Culicidae). *J. Econ. Entomol. 91*, 1260–1264.

Darvas, B., Polgár, L.A., 1998. Novel-type insecticides: specificity and effects on non-target organisms. In: Ishaaya, I., Degheele, D. (Eds.), Insecticides with Novel Modes of Action: Mechanism and Application. Springer, Berlin, pp. 188–259.

Darvas, B., Timár, T., Varjas, L., Kulcsár, P., Hosztafi, S., Bordás, B., 1989. Synthesis of novel 2,2-dimethylchromene derivatives and their toxic activity on larvae of *Pieris brassicae* (Lep., Pieridae), and *Leptinotarsa decemlineata* (Col., Chrysomelidae)., *Acta Phytopathol. Entomol. Hungarica 24*(3–4), 455–472.

De Cock, A., Degheele, D., 1998. Buprofezin: a novel chitin synthesis inhibitor affecting specifically planthoppers, whiteflies and scale insects. In: Ishaaya, I., Degheele, D. (Eds.), Insecticides with Novel Modes of Action: Mechanism and Application. Springer, pp. 74–91.

De Cock, A., Ishaaya, I., Van de Veire, M., Degheele, D., 1995. Response of buprofezin-susceptible and -resistant strains of *Trialeurodes vaporariorum* (Homoptera: Aleyrodidae) to pyriproxyfen and diafenthiuron. *J. Econ. Entomol. 88*, 763–767.

De Reede, R.H., Alkema, P., Blommers, L.H.M., 1985. The use of the insect growth regulator fenoxycarb and epofenonane against leafrollers in integrated pest management in apple orchards. *Entomol. Exp. Appl. 39*, 265–272.

De Reede, R.H., Groendijk, R.F., Wit, A.K.M., 1984. Field tests with the insect growth regulators, epofenonane and fenoxycarb, in apple orchards against leafrollers and side-effects on some leafroller parasites. *Entomol. Exp. Appl. 35*, 275–281.

Dedos, S.G., Asahina, M., Fugo, H., 1993. Effect of fenoxycarb application on the pupal–adult development of the silkworm, *Bombyx mori. J. Sericult. Sci. Japan 62*(4), 276–285.

Degheele, D., Fontier, H., Auda, M., De Loof, A., 1986. Preliminary study on the mode of action of precocene II on *Musca domestica* L. *Lucilia caesar* L. *Med. Fac. Landbouww. Univ. Gent. 51*(1), 101–108.

Dennehy, T.J., Williams, L., 1997. Management of resistance in *Bemisia* in Arizona cotton. *Pestic. Sci. 51*, 398–406.

DeWilde, J., De Kort, C.A.D., DeLoof, A., 1971. The significance of juvenile hormone titers. *Mitt. Schweiz. Ent. Ges. 44*, 79–86.

Dhadialla, T.S., Carlson, G.R., Le, D.P., 1998. New insecticides with ecdysteroidal and juvenile hormone activity. *Annu. Rev. Entomol. 43*, 545–569.

Dhadialla, T.S., Cress, D.E., Carlson, G.R., Hormann, R.E., Palli, S.R., *et al.*, 2002. Ecdysone receptor, retinoid X receptor and ultraspiracle protein based dual switch inducible gene expression modulation system. *PCT Int. Appl. WO 2002029075.* pp. 79.

Dhadialla, T.S., Tzertzinis, G., 1997. Characterization and partial cloning of ecdysteroid receptor from a cotton boll weevil embryonic cell line. *Arch. Insect Biochem. Physiol. 35*, 45–57.

Dinan, L., Whiting, P., Bourne, P., Coll, J., 2001. 8-O-Acetylharpagide is not an ecdysteroid agonist. *Insect Biochem. Mol. Biol. 31*, 1077–1082.

Duan, J.J., Prokopy, R.J., Yin, C.-M., Bergwiler, C., Oouchi, H., 1995. Effects of pyriproxyfen on ovarian development and fecundity of *Rhagoletis pomonella* flies. *Entomol. Exp. Appl. 77*, 17–21.

Dubrovski, E.B., Dubrovskaya, V.A., Bilderback, A.L., Berger, E.M., 2000. The isolation of two juvenile hormone-inducible genes in *Drosophila melanogaster*. *Devel. Biol. 224(2)*, 486–495.

Edwards, J.P., Abraham, L., 1985. Laboratory evaluation of two insect juvenile hormone analogues against *Alphitobius diaperinus* (Panzer) (Coleoptera: Tenebrionidae). *J. Stored Prod. Res. 21(4)*, 189–194.

Edwards, J.P., Short, J.E., 1993. Elimination of a population of the oriental cockroach (Dictyoptera: Blatidae) in a simulated domestic environment with the insect juvenile hormone analog (S)-hydropene. *J. Econ. Entomol. 86(2)*, 436–443.

Eisa, A.A., El-Fatah, M.A., El-Nabawi, A., El-Dash, A.A., 1991. Inhibitory effects of some insect growth regulators on developmental stages, fecundity and fertility of the florida wax scale, *Ceroplastes floridensis*. *Phytoparasitica 19(1)*, 49–55.

El Saidy, M.F., Auda, M., Degheele, D., 1989. Detoxification mechanism of diflubenzuron and teflubenzuron in the larvae of *Spodoptera littoralis* (Boisd.). *Pestic. Biochem. Physiol. 35*, 211–222.

Elbert, A., Nauen, R., 2000. Resistance of *Bemisia tabaci* (Homoptera: Aleyrodidae) to insecticides in southern Spain with special reference to neonicotinoids. *Pest Manag. Sci. 56*, 60–64.

Elbrecht, A., Chen, Y., Jurgens, T., Hensens, O.D., Zink, D.L., *et al.*, 1996. 8-O-acetyl-harpagide is a nonsteroidal ecdysteroid agonist. *Insect Biochem. Mol. Biol. 26*, 519–523.

El-Guindy, M.A., El-Rafai, A.R.M., Abdel-Satter, M.M., 1983. The pattern of cross-resistance to insecticides and juvenile hormone analogues in a diflubenzuron-resistant strain of the cotton leafworm, *Spodoptera littoralis*. *Pestic. Sci. 14*, 235–245.

Elzen, G.W., 1989. Sublethal effects of pesticides on beneficial parasitoids. In: Jepson, P.C. (Ed.), Pesticides and Non-target Invertebrates. Intercept, pp. 29–159.

Evans, K., 1985. An approach to control of *Globodera rostochiensis* using inhibitors of collagen and chitin synthesis. *Nematologica 30*, 247–250.

Farag, A.I., Varjas, L., 1983. Precocious metamorphosis and moulting deficiencies induced by an anti-JH compound FMEV in the fall webworm, *Hyphantria cunea*. *Entomol. Exp. Appl. 34*, 65–70.

Farinos, G.P., Smagghe, G., Tirry, L., Castañera, P., 1999. Action and pharmakokinetics of a novel insect growth regulator, halofenozide, in adult beetles of *Aubeonymus mariaefranciscae* and *Leptinotarsa decemlineata*. *Arch. Insect Biochem. Physiol. 41*, 201–213.

Farkas, R., Slama, K., 1999. Effect of bisacylhydrazine ecdysteroid mimics (RH-5849 and RH-5992) on chromosomal puffing, imaginal disc proliferation and pupariation in larvae of *Drosophila melanogaster*. *Insect Biochem. Mol. Biol. 29*, 1015–1027.

Feng, Q.L., Ladd, T.R., Tomkins, B.L., Sundaram, M., Sohi, S.S., *et al.*, 1999. Spruce budworm (*Christoneura fumiferana*) juvenile hormone esterase: hormonal regulation, developmental expression and cDNA cloning. *Mol. Cell. Endocrinol. 148*, 95–108.

Fukuda, S., 1994. The hormonal mechanism of larval moulting and metamorphosis in the silkworm. *J. Fac. Sci. Imp. Univ. Tokyo 4*, 477–532.

Furlong, M.J., Wright, D.J., 1994. Examination of stability of resistance and cross-resistance patterns to acylurea insect growth regulators in field populations of the diamondback moth, *Plutella xylostella*, from Malaysia. *Pestic. Sci. 42*, 315–326.

Goff, S.A., Crossland, L.D., Privalle, L.S., 1996. Control of gene expression in plants by receptor mediated transactivation in the presence of a chemical ligand. *PCT Int. Appl. WO 96/27673*. pp. 59.

Goodman, W.G., 1990. A simplified method for synthesizing juvenile hormone–protein conjugates. *J. Lipid Res. 31(2)*, 354–357.

Gordon, R., 1995. Toxic effects of a commercial formulation of fenoxycarb against adult and egg stages of the eastern spruce budworm, *Choristoneura fumiferana* (Clemens) (Lepidoptera: Tortricidae). *Can. Entomol. 127*, 1–5.

Gorman, K., Hewitt, F., Denholm, I., Devine, G.J., 2002. New developments in insecticide resistance in the glasshouse whitefly (*Trialeurodes vaporariorum*) and the two-spotted spider mite (*Tetranychus urticae*) in the UK. *Pest Manag. Sci. 58*, 123–130.

Grebe, M., Rauch, P., Spindler-Barth, M., 2000. Characterization of subclones of the epithelial cell line from *Chironomus tentans* resistant to the insecticide RH 5992, a non-steroidal moulting hormone agonist. *Insect Biochem. Mol. Biol. 30*, 591–600.

Grenier, S., Grenier, A.-M., 1993. Fenoxycarb, a fairly new insect growth regulator: a review of its effects on insects. *Ann. Appl. Biol. 122*, 369–403.

Grosscurt, A.C., 1978. Diflubenzuron: some aspects of its ovicidal and larvicidal mode of action and an evaluation of its practical possibilities. *Pestic. Sci. 9*, 373–386.

Grosscurt, A.C., ter Haar, M., Jongsma, B., Stoker, A., 1987. PH 70–23: a new acaricide and insecticide interfering with chitin deposition. *Pestic. Sci. 22*, 51–59.

Guo, X., Harmon, M.A., Laudet, V., Manglesdorf, D.J., Palmer, M.J., 1997. Isolation of a functional ecdysteroid receptor homologue from the ixodid tick *Amblyomma americanum* (L). *Insect Biochem. Mol. Biol. 27*, 945–962.

Haga, T., Toki, T., Koyanagi, T., Nishiyama, R., 1982. Structure-activity relationships of a series of benzoyl-pyridyloxyphenyl-urea derivatives. *Int. Congr. Pestic. Chem. 2*, 7.

Haga, T., Toki, T., Tsujii, Y., Nishaymar, R., 1992. Development of an insect growth regulator, chlorfluazuron. *J. Pestic. Sci. 17*, S103–S113.

Hagazi, E.M., El-Singaby, N.R., Khafagi, W.E., 1998. Effects of precocenes (I and II) and juvenile hormone I on *Spodoptera littoralis* (Boisd.) (Lep., Noctuidae) larvae parasitized by *Microplitis rufiventris Kok.* (Hym., Braconidae). *J. Appl. Entomol.* 122(8), 453–456.

Hall, R.D., Foehse, M.C., **1980.** Laboratory and field tests of CGA-72662 for control of the house fly and face fly in poultry, bovine, or swine manure. *J. Econ. Entomol.* 73, 564–569.

Hannan, G.N., Hill, R.J., **1997.** Cloning and characterization of LcEcR: a functional ecdysone receptor from the sheep blowfly, *Lucillia cupirina. Insect Biochem. Mol. Biol.* 27, 479–488.

Hannan, G.N., Hill, R.J., **2001.** *Lcusp,* an ultraspiracle gene from the sheep blowfly, *Lucilia cupirina:* cDNA cloning, developmental expression of RNA and confirmation of function. *Insect Biochem. Mol. Biol. 31,* 771–781.

Harding, W.C., **1979.** Pesticide profiles, part one: insecticides and miticides. *Univ. Maryland, Coop. Ext. Serv. Bull. 267,* 1–30.

Hargrove, J.W., Langley, P.A., **1993.** A field trial of pyriproxyfen-treated targets as an alternative method for controlling tsetse (Diptera: Glossinidae). *Bull. Entomol. Res. 83,* 361–368.

Hassan, S.A., **1992.** Guidelines for testing the effects of pesticides on beneficial organisms: description of test methods. *IOBC/WPRS Bull. 15,* 1–186.

Hassan, S.A., Bigler, F., Bogenschultz, H., Boller, E., Brun, J., *et al.,* **1991.** Results of the 5th joint pesticide testing program carried out by the IOBC/WPRS working group "Pesticides and beneficial organisms." *Entomophaga 36,* 55–67.

Hassan, S.A., Bigler, F., Bogenschultz, H., Boller, E., Brun, J., *et al.,* **1994.** Results of the 6th joint pesticide testing program of the IOBC/WPRS working group "Pesticides and beneficial organisms." *Entomophaga 39,* 107–119.

Hastakoshi, M., **1992.** An inhibitor mechanism for oviposition in the tobacco cutworm: *Spodoptera litura* by juvenile hormone analogue pyriproxyfen. *J. Insect Physiol. 38*(10), 793–801.

Hayashi, T., Iwamura, H., Nakagawa, Y., Fujita, T., **1989.** Development of (4-alkoxyphenoxy) alkanaldoxime O-ethers as potent insect juvenile hormone mimics and their structure-activity relationships. *J. Agric. Food Chem. 37,* 467–472.

Hayward, D.C., Bastiani, M.J., Truman, J.W.H., Riddiford, L.M., Ball, E.E., **1999.** The sequence of *Locusta* RXR, homologous to *Drosophila* ultraspiracle, and its evolutionary implications. *Devel. Genes Evol. 209,* 564–571.

Hayward, D.C., Dhadialla, T.S., Zhou, S., Kuiper, M.J., Ball, E.E., *et al.,* **2003.** Ligand specificity and developmental expression of RXR and ecdysone receptor in the migratory locust. *J. Insect Physiol.* 49, 1135–1144.

Henrick, C.A., Staal, G.B., Siddal, J.B., **1973.** Alkyl 3,7,1,1-trimethyl-2,4-dodecadienoates, a new class of potent insect growth regulators with juvenile hormone activity. *J. Agric. Food Chem.* 21, 354–359.

Hicks, B.J., Gordon, R., **1992.** Effects of the juvenile hormone analog fenoxycarb on various developmental stages of the eastern spruce budworm, *Choristoneura fumiferana* (Clemens) (Lepidoptera: Tortricidae). *Can. Entomol.* 124, 117–123.

Hicks, B.J., Gordon, R., **1994.** Effect of the juvenile hormone analog fenoxycarb on post-embryonic development of the eastern spruce budworm, *Choristoneura fumiferana,* following treatment of the egg stage. *Entomol. Exp. Appl.* 71, 181–184.

Hill, T.A., Foster, R.E., **2000.** Effect of insecticides on the diamondback moth (Lepidoptera: Plutellidae) and its parasitoid *Diadegma insulare* (Hymenoptera: Ichneumonidae). *J. Econ. Entomol.* 93(3), 763–768.

Hiruma, K., **2003.** Juvenile hormone action in insect development. In: Henry, H.L., Norman, A.W. (Eds.), Encyclopedia of Hormones. Elsevier Science, Amsterdam, pp. 528–535.

Horowitz, A.R., Ishaaya, I., **1994.** Managing resistance in insect growth-regulators in the sweet-potato whitefly (Homoptera, Aleyrodidae). *J. Econ. Entomol.* 87, 866–871.

Horowitz, A.R., Forer, G., Ishaaya, I., **1994.** Managing resistance in Bemisia tabaci in Israel with emphasis on cotton. *Pestic. Sci.* 42, 113–122.

Hrdý, I., Kuldová, J., Wimmer, Z., **2001.** A juvenile hormone analogue with potential for termite control: laboratory test with *Reticulitermes santonensis, Reticulitermes flaviceps* and *Coptotermes formosanus* (Isopt., Rhinotermitidae). *J. Appl. Entomol.* 125(7), 403–411.

Hsu, A.C.-T., Fujimoto, T.T., Dhadialla, T.S., **1997.** Structure-activity study and conformational analysis of RH-5992, the first commercialized nonsteroidal ecdysone agonist. In: Hedin, P.A., Hollingworth, R.M., Masler, E.P., Miyamoto, J., Thompson, D.G. (Eds.), Phytochemicals for Pest Control, ACS Symposium Series, vol. 658. American Chemical Society, pp. 206–219.

Hsu, AC.-T., **1991.** 1,2-Diacyl-1-alkyl-hydrazines; a novel class of insect growth regulators. In: Baker, D.R., Fenyes, J.G., Moberg, W.K. (Eds.), Synthesis and Chemistry of Agrochemicals, II. ACS Symposium Series, vol. 443. American Chemical Society, pp. 478–490.

Hu, W., Feng, Q., Palli, S.R., Krell, P.J., Arif, B.M., *et al.,* **2001.** The ABC transporter Pdr5p mediates the efflux of nonsteroidal ecdysone agonists in *Sacchromyces cerevisiae. Eur. J. Biochem.* 268, 3416–3422.

Imhof, M.O., Rusconi, S., Lezzi, M., **1993.** Cloning of a *Chironomus tentans* cDNA encoding a protein c(EcRH) homologous to the *Drosophila melanogaster* ecdysone receptor (dEcR). *Insect Biochem. Mol. Biol.* 23, 115–124.

Ishaaya, I., **1993.** Insect detoxifying enzymes – their importance in pesticide synergism and resistance. *Arch. Insect Biochem. Physiol.* 22, 263–276.

Ishaaya, I., **2001.** Biochemical processes related to insecticide actions: an overview. In: Ishaaya, I. (Ed.), Biochemical Sites of Insecticide Action and Resistance. Springer, Berlin, pp. 1–16.

Ishaaya, I., Casida, J.E., **1980.** Properties and toxicological significance of esterases hydrolyzing permethrin and cypermethrin in *Trichoplusia ni* larval gut integument. *Pestic. Biochem. Physiol. 14*, 178–184.

Ishaaya, I., De Cock, A., Degheele, D., **1994**. Pyriproxyfen, a potent suppressor of egg hatch and adult formation of the greenhouse whitefly (Homoptera: Aleyrodidae). *J. Econ. Entomol. 87(5)*, 1185–1189.

Ishaaya, I., Horowitz, A.R., **1992**. Novel phenoxy juvenile hormone analog (pyriproxyfen) suppresses embryogenesis and adult emergence of sweetpotato whitefly (Homoptera: Aleyrodidae). *J. Econ. Entomol. 85(6)*, 2113–2117.

Ishaaya, I., Yablonski, S., Horowitz, A.R., **1995**. Comparative toxicology of two ecdysteroid agonists, RH-2485 and RH-5992, on susceptible and pyrethroid resistant strains of the Egyptian cotton leafworm, *Spodoptera littoralis*. *Phytoparasitica 23*, 139–145.

Ishaaya, I., Yablonski, S., Mendelson, Z., Mansour, Y., Horowitz, A.R., **1996**. Novaluron (MCW-275), a novel benzoylphenyl urea, suppressing developing stages of lepidopteran, whitefly and leafminer pests. *Brighton Crop Prot. Conf. 2*, 1013–1020.

Ismail, F., Wright, D.J., **1991**. Cross-resistance between acylurea insect growth regulators in a strain of *Plutella xylostella* L. (Lepidoptera, Yponomeutidae). *Pestic. Sci. 33*, 359–370.

Ismail, F., Wright, D.J., **1992**. Synergism of teflubenzuron and chlorfluazuron in an acylurea-resistant field strain of *Plutella xylostella* L. (Lepidoptera, Yponomeutidae). *Pestic. Sci. 34*, 221–226.

Iwanaga, K., Kanda, T., **1988**. The effects of a juvenile hormone active oxime ether compound on the metamorphosis and reproduction of an anopheline vector, *Anopheles balabacensis* (Diptera: Culicidae). *Appl. Entomol. Zool. 23(2)*, 186–193.

Iyengar, A.R., Kunkel, J.G., **1955**. Follicle cell calmodulin in *Blatella gexmanica*: transcript accumulation during vitellogenesis is regulated by juvenile hormone. *Devel. Biol. 170*, 314–320.

Jepson, I., Martinez, A., Greenland, A.J., **1996**. A gene switch comprising an ecdysone receptor or fusion product allows gene control by external chemical inducer and has agricultural and pharmaceutical applications. *PCT Int. Appl. WO 9637609*. pp. 121.

Jepson, I., Martinez, A., Sweetman, J.P., **1998**. Chemical inducible gene expression systems for plants. A review. *Pestic. Sci. 54*, 360–367.

Jiang, R.-J., Koolman, J., **1999**. Feedback inhibition of ecdysteroids: evidence for a short feedback loop repressing steroidogenesis. *Arch. Insect Biochem. Physiol. 41*, 54–59.

Jones, G., Sharp, P.A., **1997**. Ultraspiracle: an invertebrate nuclear receptor for juvenile hormones. *Proc. Natl Acad. Sci. USA 94*, 13499–13503.

Jones, G., Wozniak, M., Chu, Y.-X., Dhar, S., Jones, D., **2001**. Juvenile hormone III-dependent conformational changes of the receptor ultraspiracle. *Insect Biochem. Mol. Biol. 32*, 33–49.

Kamimura, K., Arakawa, R., **1991**. Field evaluation of an insect growth regulator, pyriproxyfen, against *Culex*

pipiens and *Culex tritaeniorhynchus*. *Jap. J. San. Zool. 42(3)*, 249–254.

Kanno, H., Ikeda, K., Asai, T., Maekawa, S., **1981**. 2-Tert-butylimino-3-isopropyl-5-perhydro-1,3,5-thiodiazin 4-one (NNI 750), a new insecticide. *Brighton Crop Prot. Conf. 1*, 59–69.

Kapitskaya, M., Wang, S., Cress, D.E., Dhadialla, T.S., Raikhel, A.S., **1996**. The mosquito *ultraspiracle* homologue, a partner of ecdysteroid receptor heterodimer: cloning and characterization of isoforms expressed during vitellogenesis. *Mol. Cell. Endocrinol. 121*, 119–132.

Karr, L.I., Sheets, J.J., King, J.E., Dripps, J.E., **2004**. Laboratory performance and pharmacokinetics of the benzoylphenylurea noviflumuron in eastern subterranean termites (Isoptera: Rhinotermitidae). *J. Econ. Entomol 97*, 593–600.

Kasuya, A., Sawada, Y., Tsukamoto, Y., Tanaka, K., Toya, T., et al., **2003**. Binding mode of ecdysone agonists to the receptor: comparative modeling and docking studies. *J. Mol. Model. 9*, 58–65.

Katzenellenbogen, J.A., Katzellenbogen, B.S., **1984**. Affinity labeling of receptors for steroid and thyroid hormones. *Vit. Horm. 41*, 213–274.

Keiding, J., **1999**. Review of the global status and recent development of insecticide resistance in field populations of the housefly, *Musca domestica* (Diptera: Muscidae). *Bull. Entomol. Res. 89*, S9–S67.

Keiding, J., Elkhodary, A.S., Jespersen, J.B., **1992**. Resistance risk assessment of 2 insect development inhibitors, diflubenzuron and cyromazine, for control of the housefly *Musca domestica* L. 2. Effect of selection pressure in laboratory and field populations. *Pestic. Sci. 35*, 27–37.

Keil, C.B., Othman, M.H., **1988**. Effects of methoprene on *Lycoriella mali* (Diptera: Sciaridae). *J. Econ. Entomol. 81(6)*, 1592–1597.

Kim, K.S., Chung, B.J., Kim, H.K., **2000**. DBI-3204: A new benzoylphenyl urea insecticide with a particular activity against whitefly. *Brighton Crop Prot. Conf. 1*, 41–46.

King, J.E., Bennett, G.W., **1988**. Mortality and developmental abnormalities induced by two juvenile hormone analogs on nymphal german cockroaches (Dictyoptera: Blattellidae). *J. Econ. Entomol. 81(1)*, 225–227.

Knight, A.L., **2000**. Tebufenozide targeted against codling moth (Lepidoptera: Tortricidae) adults, eggs, and larvae. *J. Econ. Entomol. 93(6)*, 1760–1767.

Knight, A.L., Dunley, J.E., Jansson, R.K., **2001**. Baseline monitoring of codling moth (Lepidoptera: Tortricidae) larval response to benzoylhydrazine insecticides. *J. Econ. Entomol. 94*, 264–270.

Koelle, M.R., Talbot, W.S., Segraves, W.A., Bender, M.T., Cherbas, P., et al., **1991**. The *Drosophila* EcR gene encodes an ecdysone receptor, a new member of the steroid receptor superfamily. *Cell 67*, 59–77.

Kopec, S., **1922**. Studies on the necessity of the brain for the inception of insect metamorphosis. *Biol. Bull. 42*, 323–341.

Kostyukovsky, M., Chen, B., Atsmi, S., Shaaya, E., 2000. Biological activity of two juvenoids and two ecdysteroids against three stored product insects. *Insect Biochem. Mol. Biol. 30*, 891–897.

Kothapalli, R., Palli, S.R., Ladd, T.R., Sohi, S.S., Cress, D., et al., 1995. Cloning and developmental expression of the ecdysone receptor gene from the spruce budworm, *Choristoneura fumiferana*. *Devel. Genet. 17*, 319–330.

Kotze, A.C., Sales, N., 2001. Inheritance of diflubenzuron resistance and monooxygenase activities in a laboratory-selected strain of *Lucilia cuprina* (Diptera: Calliphoridae). *J. Econ. Entomol. 94*, 1243–1248.

Kristensen, M., Jespersen, J.B., 2003. Larvicide resistance in *Musca domestica* (Diptera: Muscidae) populations in Denmark and establishment of resistant laboratory strains. *J. Econ. Entomol. 96*, 1300–1306.

Krysan, J.L., 1990. Fenoxycarb and diapause: a possible method of control for pear psylla (Homoptera: Psyllidae). *J. Econ. Entomol. 83*(2), 293–299.

Kumar, M.B., Fujimoto, T., Potter, D.W., Deng, Q., Palli, S.R., 2002. A single point mutation in ecdysone receptor leads to increased ligand specificity: implications for gene switch applications. *Proc. Natl Acad. Sci. USA 99*, 14710–14715.

Kuwano, E., Kikuchi, M., Eto, M., 1990. Synthesis and insect growth regulatory activity of 1-neopentyl-5-substituted imidazoles. *J. Fac. Agric. Kyushu Univ. 35*(1–2), 35–41.

Kuwano, E., Takeya, R., Eto, M., 1985. Synthesis and anti-juvenile hormone acitivty of 1-substituted-5-[(E)-2,6-dimethyl-1,5-heptadienyl]imidazoles. *Agric. Biol. Chem. 49*(2), 483–486.

Kuwano, E., Takeya, R., Eto, M., 1984. Synthesis and anti-juvenile hormone activity of 1-citronellyl-5-substituted imidazoles. *Agric. Biol. Chem. 48*(12), 3115–3119.

Langley, P.A., Felton, T., Stafford, K., Oouchi, H., 1990. Formulation of pyriproxyfen, a juvenile hormone mimic, for tsetse control. *Med. Vet. Entomol. 4*, 127–133.

Le, D.P., Thirugnanam, M., Lidert, Z., Carlson, G.R., Ryan, J.B., 1996. RH-2485: A new selective insecticide for caterpillar control. *Proc. Brighton Crop Prot. Conf. 2*, 481–486.

Lee, K.-Y., Denlinger, D.L., 1997. A role for ecdysteroids in the induction and maintenance of the pharate first instar diapause of the gypsy moth, *Lymantria dispar*. *J. Insect Physiol. 43*, 289–296.

Lee, K.-Y., Chamberlin, M.E., Horodyski, F.M., 2002. Biological activity of *Manduca sexta* allatotropin-like peptides, predicted products of tissue-specific and developmentally regulated alternatively spliced mRNAs. *Peptides 23*, 1933–1941.

Legaspi, J.C., Legaspi, B.C., Jr., Saldana, R.R., 1999. Laboratory and field evaluations of biorational insecticides against the Mexican rice borer (Lepidoptera: Pyralidae) and a parasitoid (Hymenoptera: Braconidae). *J. Econ. Entomol. 92*(4), 804–810.

Lidert, Z., Dhadialla, T.S., 1996. MIMIC insecticide an ecdysone agonist for the global forestry market. In: Proceedings of the International Conference on Integrated Management of Forestry Lymantridae, Warsaw-Sekocin, pp. 59–68.

Liu, T.-X., Chen, T.-Y., 2000. Effects of a juvenile hormone analog, pyriproxyfen, on the apterous form of *Lipaphis erysimi*. *Entomol. Exp. Appl. 98*, 295–301.

Liu, T.-X., Chen, T.-Y., 2001. Effects of the insect growth regulator fenoxycarb on immature *Chrysoperla rufilabris* (Neuroptera: Chyrsopidae). *Fl. Entomol. 84*(4), 628–633.

Liu, T.-X., Stansly, P.A., 1997. Effects of pyriproxyfen on three species of *Encarsia* (Hymenoptera: Aphelinidae), endoparasitoids of *Bemisia argentifolii* (Homoptera: Aleyrodidae). *Biol. Microbiol Cont. 90*(2), 404–411.

Locke, M., 1998. Epidermis. In: Harrison, F.W., Locke, M. (Eds.), Microscopic Anatomy of Invertebrates, vol. 11A: Insecta. Wiley-Liss, New York, pp. 75–138.

Londerhausen, M., 1996. Approaches to new parasiticides. *Pestic. Sci. 48*, 269–292.

Lopez, J.D., Jr., Latheef, M.A., Meola, R.W., 1999. Effect of insect growth regulators on feeding response and reproduction of adult bollworm. *Proc. Beltwide Cotton Conf. 2*, 1214–1221.

Lyoussoufi, A., Gadenne, C., Rieus, R., d'Arcier, F.F., 1994. Effects of an insect growth regulator, fenoxycarb, on the diapause of the pear psylla, *Cacopsylla pyri*. *Entomol. Exp. Appl. 72*(3), 239–244.

Magaula, C.N., Samways, M.J., 2000. Effects of insect growth regulators on *Chilocorus nigritus* (Fabricius) (Coleoptera: Coccinellidae), a non-target natural enemy of citrus red scale, *Aonidiella aurantii* (Maskell) (Homoptera: Diaspididae), in southern Africa: evidence from laboratory and field trials. *African Entomol. 8*(1), 47–56.

Martinez, A., Sparks, C., Drayton, P., Thompson, J., Greenland, A., Jepson, I., 1999a. Creation of ecdysone receptor chimeras in plants for controlled regulation of gene expression. *Mol. Gen. Genet. 261*, 546–552.

Martinez, A., Sparks, C., Hart, C.A., Thompson, J., Jepson, I., 1999b. Ecdysone agonist inducible transcription in transgenic tobacco plants. *Plant J. 19*, 97–106.

Mascarenhas, V.J., Leonard, B.R., Burris, E., Graves, J.B., 1996. Beet armyworm (Lepidoptera: Noctuidae) control on cotton in Louisiana. *Fl. Entomol. 79*, 336–343.

Masner, P., Angst, M., Dorn, S., 1986. Fenoxycarb, an insect growth regulator with juvenile hormone activity: a candidate for *Heliothis virescens* (F.) control on cotton. *Pestic. Sci. 18*, 89–94.

Masner, P., Salama, K., Landa, V., 1968. Natural and synthetic materials with insect hormone activity. IV. Specific female sterility effects produced by a juvenile hormone analogue. *J. Embryol. Exp. Morphol. 20*, 25–31.

McKenzie, J.A., Batterham, P., 1998. Predicting insecticide resistance: mutagenesis, selection and response. *Phil. Trans. Roy. Soc. Lond. B Biol. Sci. 353*, 1729–1734.

Medina, P., Budia, F., Tirry, L., Smagghe, G., Viñuela, E.,
2001. Compatibility of spinosad, tebufenozide and
azadirachtin with eggs and pupae of the predator
Chrysoperla carnea (Stephens) under laboratory
conditions. *Biocontrol Sci. Technol. 11*, 597–610.

Medina, P., Smagghe, G., Budia, F., del Estal, F., Tirry, L.,
et al., 2002. Significance of penetration, excretion,
transovarial uptake to toxicity of three insect growth
regulators in predatory lacewing adults. *Arch. Insect
Biochem. Physiol. 51*, 91–101.

Mendel, Z., Blumberg, D., Ishaaya, I., 1994. Effects of
some insect growth regulators on natural enemies of
scale insects (Hom.: Cocoidea). *Entomophaga 39*,
199–209.

Michelotti, E.L., Tice, C.M., Palli, S.R., Thompson, C.S.,
Dhadialla, T.S., 2003. Tetrahydroquinolines for modulating the expression of exogenous genes via an
ecdysone receptor complex. *PCT Int. Appl. WO
2003105849.* pp. 129.

Mikitani, K., 1996a. A new nonsteroidal class of ligand
for the ecdysteroid receptor 3,5-di-*tert*-butyl-4-
hydroxy-N-isobutyl-benzamide. *Biochem. Biophys.
Res. Commun. 227*, 427–432.

Mikitani, K., 1996b. Ecdysteroid receptor binding activity
and ecdysteroid agonist activity at the level of gene
expression are correlated with the activity of dibenzoyl
hydrazines in larvae of Bombyx mori. *J. Insect Physiol.
42*, 937–941.

Mikólajczyk, P., Oberlander, H., Silhacek, D.L., Ishaaya,
I., Shaaya, E., 1994. Chitin synthesis in *Spodoptera
frugiperda* wing imaginal discs. I. Chlorfluazuron,
diflubenzuron, and teflubenzuron inhibit incorporation but not uptake of [^{14}C]-N-acetyl-D-glucosamine.
Arch. Insect Biochem. Physiol. 25, 245–258.

Miller, R.J., Broce, A.B., Dryden, M.W., Hopkins, T.,
1999. Susceptibility of insect growth regulators and
cuticle deposition of the cat flea (Siphonaptera: Pulicidae) as a function of age. *J. Med. Entomol. 36*(6),
780–787.

Miura, T., Takahashi, R.M., 1987. Impact of fenoxycarb,
a carbamate insect growth regulator, on some aquatic
invertebrates abundant in mosquito breeding habitats.
J. Am. Mosq. Control Assoc. 3, 476–480.

Miyake, T., Mitsui, T., 1995. Multiple physiological
activity of an anti-juvenile hormone, precocene 2 on
the whitebacked rice planthopper. *J. Pestic. Sci. 20*,
17–24.

Miyake, T., Haruyama, H., Mitsui, T., Sakurai, A., 1991.
Effects of a new juvenile hormone mimic, NC-170, on
metamorphosis and diapause of the small brown
planthopper, *Laodelphax striatellus. J. Pestic. Sci. 17*,
75–82.

Miyake, T., Haruyama, H., Ogura, T., Mitsui, T., Sakurai,
A., 1990. Effects of insect juvenile hormone active NC-
170 on metamorphosis, oviposition and embryogenesis
in leafhoppers. *J. Pest. Sci. 16*, 441–448.

Miyamoto, J., Hirano, M., Takimoto, Y., Hatakoshi, M.,
1993. Insect growth regulators for insect control,
with emphasis on juvenile hormone analogs: present
and future prospects. In: Duke, S.O., Menn, J.J.,
Plimmer, J.R. (Eds.), Pest Control with Enhanced
Environmental Safety, ACS Symposium Series, vol.
524. American Chemical Society, Washington, DC,
pp. 144–168.

Mkhize, J.N., 1991. Activity of a juvenile hormone analog
as a protectant against the lesser grain borer, *Rhizo-
pertha dominica* (F.) (Coleoptera: Bostrichidae).
Insect Sci. Appl. 13(2), 183–187.

Moffit, H.R., Westgrad, P.H., Mantey, K.D., Van de Baan,
H.E., 1988. Resistance to diflubenzuron in codling
moth (Lepidoptera: Tortricidae). *J. Econ. Entomol.
81*, 1511–1515.

Mohammed-Ali, A.D., Chan, T.-H., Thomas, A.W.,
Strunz, G.M., Jewett, B., 1995. Structure–activity relationship study of synthetic hydrazines as ecdysone agonists in the control of spruce budworm (*Choristoneura
fumiferana*). *Can. J. Chem. 73*, 550.

Moser, B.A., Koehler, P.G., Patterson, R.S., 1992. Effect of
methoprene and diflubenzuron on larval development
of the cat flea (Siphonaptera: Pulicidae). *J. Econ.
Entomol. 85*(1), 112–116.

Mouillet, J.F., Delbecque, J.P., Quennedey, B.,
Delachambre, J., 1997. Cloning of two putative
ecdysteroid receptor isoforms from *Tenebrio molitor*
and their developmental expression in the epidermis during metamorphosis. *Eur. J. Biochem. 248*,
856–863.

Moulton, J.K., Pepper, D.A., Jansson, R.K., Dennehy, T.J.,
2002. Pro-active management of beet armyworm
(Lepidoptera: Noctuidae) resistance to tebufenozide
and methoxyfenozide: baseline monitoring, risk assessment, and isolation of resistance. *J. Econ. Entomol. 95*,
414–424.

Muyle, H., Gordon, R., 1989. Effects of selected juvenile
hormone analogs on sixth instar larvae of the eastern
spruce budworm, *Choristoneura fumiferana* Clemens
(Lepidoptera: Tortricidae). *Can. Entomol. 121*,
1111–1116.

Nair, V.S.K., Rajalekshmi, E., 1989. Effects of fluorome-
valonate on penultimate and last instar larvae of *Spodoptera mauritia* Boisd. *Indian J. Expt. Biol. 27*(2),
170–173.

Nakagawa, Y., Akagi, T., Iwamura, H., Fujita, T., 1989a.
Quantitative structure-activity studies of benzoylphenylurea larvicides. VI. Comparison of substituent
effects among activities against different insect species.
Pestic. Biochem. Physiol. 33, 144–157.

Nakagawa, Y., Hattori, K., Minakuchi, C., Kugimiya, S.,
Ueno, T., 2000a. Relationships between structure
and molting hormonal activity of tebufenozide, methoxyfenozide, and their analogs in cultured integument
system of *Chilo suppressalis* Walker. *Steroids 65*,
117–123.

Nakagawa, Y., Hattori, K., Shimizu, B., Akamatsu, M.,
Miyagawa, H., *et al.*, 1998. Quantitative structure-
activity studies of insect growth regulators XIV.

Three-dimensional quantitative structure-activity relationship of ecdysone agonists including dibenzoylhydrazine analogs. *Pestic. Sci. 53*, 267–277.

Nakagawa, Y., Minakuchi, C., Takahashi, K., Ueno, T., 2002. Inhibition of [^{3}H]ponasterone A binding by ecdysone agonists in the intact Kc cell line. *Insect Biochem. Mol. Biol. 32*, 175–180.

Nakagawa, Y., Minakuchi, C., Ueno, T., 2000b. Inhibition of [^{3}H]ponasterone A binding by ecdysone agonists in the intact Sf-9 cell line. *Steroids 65*, 537–542.

Nakagawa, Y., Matsutani, M., Kurihara, N., Nishimura, K., Fujita, T., 1989b. Quantitative structure-activity studies of benzoylphenylurea larvicides. VIII. Inhibition of N-acetylglucosamine incorporation into the cultured integument of *Chilo suppressalis* Walker. *Pestic. Biochem. Physiol. 43*, 141–151.

Nakagawa, Y., Shimuzu, B., Oikawa, N., *et al.*, 1995a. Three-dimensional quantitative structure-activity analysis of steroidal and dibenzoylhydrazine-type ecdysone agonists. In: Hanch, C., Fujita, T. (Eds.), Classical and Three-Dimensional QSAR in Agrobiochemistry ACS Symposium, vol. 606. American Chemical Society, Washington, DC, pp. 288–301.

Nakagawa, Y., Smagghe, G., Kugimiya, S., *et al.*, 1999. Quantitative structure-activity studies of insect growth regulators: XVI. Substituent effects of dibenzoylhydrazines on the insecticidal activity to Colorado potato beetle *Leptinotarsa decemlineata*. *Pestic. Sci. 55*, 909–918.

Nakagawa, Y., Smagghe, G., Tirry, L., Fujita, T., 2001a. Quantitative structure–activity studies of insect growth regulators: XIX. Effects of substituents on the aromatic moiety of dibenzoylhydrazines on larvicidal activity against the beet armyworm *Spodoptera exigua*. *Pest Manag. Sci. 58*, 131–138.

Nakagawa, Y., Smagghe, G., Van Paemel, M., Tirry, L., Fujita, T., 2001b. Quantitative structure-activity studies of insect growth regulators: XVIII. Effects of substituents on the aromatic moiety of dibenzoylhydrazines on larvicidal activity against the Colorado potato beetle *Leptinotarsa decemlineata*. *Pest Mgt. Sci. 57*, 858–865.

Nakagawa, Y., Sotomatsu, T., Irie, K., Kitahara, K., Iwamura, H., *et al.*, 1987. Quantitative structure-activity studies of benzoylphenylurea larvicides. III. Effects of substituents at the benzoyl moiety. *Pestic. Biochem. Physiol. 27*, 143–155.

Nakagawa, Y., Soya, Y., Nakai, K., *et al.*, 1995b. Quantitative structure-activity studies of insect growth regulators. XI. Stimulation and inhibition of N-acetylglucosamine incorporation in a cultured integument system by substituted *N-tert*-butyl-*N,N'*-dibenzoylhydrazines. *Pestic. Sci. 43*, 339–345.

Niwa, A., Iwamura, H., Nakagawa, Y., Fujita, T., 1989. Development of (phenoxyphenoxy)- and (benzylphenoxy) propyl ethers as potent insect juvenile hormone mimetics. *J. Agric. Food Chem. 37*, 462–467.

No, D., Yao, T.-P., Evans, M.E., 1996. Ecdysone-inducible gene expression in mammalian cells and transgenic mice. *Proc. Natl Acad. Sci. USA 93*, 3346–3351.

Oberlander, H., Silhacek, D.L., Porcheron, P., 1995. Non-steroidal ecdysteroid agonists: Tools for the study of hormonal action. *Arch. Insect Biochem. Physiol. 28*, 209–223.

Oberlander, H., Silhacek, D.L., 1998. New perspectives on the mode of action of benzoylphenyl urea insecticides. In: Ishaaya, I., Degheele, D. (Eds.), Insecticides with Novel Modes of Action: Mechanism and Application. Springer, pp. 92–105.

Oberlander, H., Smagghe, G., 2001. Imaginal discs and tissue cultures as targets for insecticide action. In: Ishaaya, I. (Ed.), Biochemical Sites of Insecticide Action and Resistance. Springer, Berlin, pp. 133–150.

Oikawa, N., Nakagawa, Y., Nishimura, K., Ueno, T., Fujita, T., 1994a. Quantitative structure-activity analysis of larvicidal 1-(substituted benzoyl)-2-benzoyl-1-*tert*-butylhydrazines against *Chilo suppressalis*. *Pestic Sci. 41*, 139–148.

Oikawa, N., Nakagawa, Y., Nishimura, K., Ueno, T., Fujita, T., 1994b. Quantitative structure-activity studies of insect growth regulators. X. Substituent effects on larvicidal activity of 1-*tert*-butyl-1-(2-chlorobenzoyl)-2-(substituted benzoyl)hydrazines against *Chilo suppressalis* and design synthesis of potent derivatives. *Pestic. Biochem. Physiol. 48*, 135–144.

Olsen, A., 1985. Ovicidal effect on the cat flea, *Ctenocephalides felis* (Bouché), of treating fur of cats and dogs with methoprene. *Int. Pest Control.* January/February, 10–14.

Padidam, M., 2003. Chemically regulated gene expression in plants. *Curr. Opin. Plant Biol. 6*, 169–177.

Padidam, M., Gore, M., Lu, D.L., Smirnova, O., 2003. Chemical-inducible, ecdysone receptor-based gene expression system for plants. *Transgenic Res. 12*, 101–109.

Palli, S.R., Kapitskaya, M.Z., 2002. Novel substitution variants of nuclear receptors and their use in a dual switch inducible system for regulation of gene expression. *PCT Int. Appl. WO 2002066615.* pp. 110.

Palli, S.R., Kapitskaya, M.Z., Kumar, M.B., Cress, D.E., 2003. Improved ecdysone receptor-based inducible gene regulation system. *Eur. J. Biochem. 270*, 1308–1315.

Palli, S.R., Kumar, M.B., Cress, D.E., Fujimoto, T.T., 2002. Substitution variants of nuclear receptors and their use in a dual switch inducible system for regulation of gene expression. *PCT Int. Appl. WO 2002066612.* pp. 148.

Palli, S.R., Ladd, T.R., Sohi, S.S., Cook, B.J., Retnakaran, A., 1996. Cloning and developmental expression of *Choristoneura* hormone receptor 3, an ecdysone-inducible gene and a member of the steroid receptor superfamily. *Insect Biochem. Mol. Biol. 26*, 485–499.

Palli, S.R., Ladd, T.R., Tomkins, W., Primavera, M., Sundaram, M.S., *et al.*, 1999. Biochemical and biological modes of action of ecdysone agonists on the spruce budworm. *Pestic. Sci. 55*, 656–657.

Palli, S.R., Primavera, M., Tomkins, W., Lambert, D., Retnakaran, A., 1995. Age-specific effects of a non-steroidal ecdysteroid agonist, RH-5992, on the spruce budworm, Choristoneura fumiferana (Lepidoptera: Tortricidae). Eur. J. Entomol. 92, 325–332.

Palli, S.R., Retnakaran, A., 1999. Molecular and biochemical aspects of chitin synthesis inhibition. In: Jollès, P., Muzzarelli, R.A.A. (Eds.), Chitin and Chitinases. Birkhäuser Verlag, pp. 85–98.

Palli, S.R., Retnakaran, A., 2000. Ecdysteroid and juvenile hormone receptors: properties and importance in developing novel insecticides. In: Ishaaya, I. (Ed.), Biochemical Sites of Insecticide Action and Resistance. Springer, pp. 107–132.

Palli, S.R., Touchara, K., Charles, J., Bonning, B.C., Atkinson, J.K., et al., 1994. A nuclear juvenile hormone-bonding protein from larvae of Manduca sexta: A putative receptor for the metamorphic action of juvenile hormone. Proc. Natl Acad. Sci. USA 91(13), 6191–6195.

Palumbo, J.C., Horowitz, A.R., Prabhaker, N., 2001. Insecticidal control and resistance management for Bemisia tabaci. Crop Protect. 20, 739–765.

Peleg, B.A., 1988. Effect of a new phenoxy juvenile hormone analog on California red scale (Homoptera: Diaspididae), Florida wax scale (Homoptera: Coccidae) and the ectoparasite Aphytis holoxanthus DeBache (Hymenoptera: Aphelinidae). J. Econ. Entomol. 81, 88–92.

Perera, S.C., Ladd, T.R., Dhadialla, T.S., Krell, P.J., Sohi, S.S., et al., 1999. Studies on two ecdysone receptor isoforms of the spruce budworm, Choristoneura fumiferana. Mol. Cell. Endocrinol. 152, 73–84.

Perez-Farinos, G., Smagghe, G., Marco, V., Tirry, L., Castañera, P., 1998. Effects of topical application of hexaflumuron on adult sugar beet weevil Aubeonymus mariaefranciscae on embryonic development: pharmacokinetics in adults and embryos. Pestic. Biochem. Physiol. 61, 169–182.

Peterson, J.K., Kashian, D.R., Dodson, S.I., 2000. Methoprene and 20-OH-ecdysone affect male production in Daphnia pulex. Environ. Toxicol. Chem. 20(3), 582–588.

Pimprikar, G.D., Georghiou, G.P., 1979. Mechanisms of resistance to diflubenzuron in the house fly, Musca domestica L. Pestic. Biochem. Phsyiol. 12, 10–22.

Pons, S., Riedl, H., Avilla, J., 1999. Toxicity of the ecdysone agonist tebufenozide to codling moth (Lepidoptera: Tortricidae). J. Econ. Entomol. 92, 1344–1351.

Pradeep, A.R., Nair, V.S.K., 1989. Morphogenetic effects of precocene II in the brown planthopper Nilaparvata lugens Stal. (Homoptera: Delphacidae). Acta Entomol. Biochem. 86(3), 172–178.

Rasa, Cordellia G., Meola, Roger W., Shenker, R., 2000. Effects of a new insect growth regulator, CGA-255'728, on the different stages of the cat flea (Siphonaptera: Pulicidae). Entomol. Soc. Am. 37(1), 141–145.

Reiji, I., Shinya, N., Takashi, O., Yasushi, T., Keiji, T., et al., 2000. Ecdysone mimic insecticide, chromafenozide: field efficacy against the apple tortrix. Annu. Rep. Sankyo Res. Lab. 52, 59–62.

Retnakaran, A., 1970. Blocking of embryonic development in the spruce budworm, Choristoneura fumiferana (Lepidoptera: Tortricidae), by some compounds with juvenile hormone activity. Can. Entomol. 102, 1592–1596.

Retnakaran, A., 1973a. Hormonal induction of supernumerary instars in the spruce budworm, Choristoneura fumiferana (Lepidoptera: Tortricidae). Can. Entomol. 105, 459–461.

Retnakaran, A., 1973b. Ovicidal effect in the white pine weevil, Pissodes strobi (Coleoptera: Curculionidae), of a synthetic analogue of juvenile hormone. Can. Entomol. 105, 591–594.

Retnakaran, A., 1974. Induction of sexual maturity in the white pine weevil, Pissodes strobi (Coleoptera: Curculionidae), by some analogue of juvenile hormone. Can. Entomol. 106, 831–834.

Retnakaran, A., Gelbic, I., Sundaram, M., Tomkins, W., Ladd, T., et al., 2001. Mode of action of the ecdysone agonist tebufenozide (RH-5992), and an exclusion mechanism to explain resistance to it. Pest Mgt. Sci. 57, 951–957.

Retnakaran, A., Granett, J., Ennis, T., 1985. Insect growth regulators. In: Kerkut, G.A., Gilbert, L.I. (Eds.), Comprehensive Insect Physiology, Biochemistry and Pharmacology, vol. 12. Pergamon, Oxford, pp. 529–601.

Retnakaran, A., Hiruma, K., Palli, S.R., Riddiford, L.M., 1995. Molecular analysis of the mode of action of RH-5992, a lepidopteran-specific, non-steroidal ecdysteroid agonist. Insect Biochem. Mol. Biol. 25, 109–117.

Retnakaran, A., Macdonald, A., Tomkins, W.L., Davis, C.N., Brownright, A.J., et al., 1997. Ultrastructural effects of a non-steroidal agonist, RH-5992, on the sixth instar larvae of spruce budworm, Choristoneura fumiferana. J. Insect Physiol. 43, 55–68.

Retnakaran, A., Oberlander, H., 1993. Control of chitin synthesis in insects. In: Muzzarelli, R.A.A. (Ed.), Chitin Enzymology. European Chitin Society, pp. 89–99.

Retnakaran, A., Tomkins, W.L., Primavera, M.J., Palli, S.R., 1999. Feeding behavior of the first-instar Choristoneura fumiferana and Choristoneura pinus pinus (Lepidoptera: Tortricidae). Can. Entomol 131, 79–84.

Retnakaran, A., Wright, J.E., 1987. Control of insect pests with benzoylphenylureas. In: Wright, J.E., Retnakaran, A. (Eds.), Chitin and Benzoylphenyl Ureas. Dr. W. Junk Publishers, Dordrecht, Netherlands, pp. 205–282.

Reynolds, S.E., Blakey, J.K., 1989. Cyromazine causes decreased cuticle extensibility in larvae of the tobacco hornworm, Manduca sexta. Pestic. Biochem. Physiol. 35, 251–258.

Riddiford, L.M., 1971. Juvenile hormone and insect embryogenesis. Mitt. Schweiz. Ent. Ges. 44, 177–186.

Riddiford, L.M., **1994**. Cellular and molecular actions of juvenile hormone. I. General considerations and premetamorphic actions. *Adv. Insect Physiol. 24*, 213–274.

Riddiford, L.M., **1996**. Juvenile hormone: the status of its "status quo" action. *Arch. Insect Biochem. Physiol. 32*, 271–286.

Riddiford, L.M., Cherbas, P., Truman, J.W., **2001**. Ecdysone receptors and their biological functions. *Vit. Horm. 60*, 1–73.

Riddiford, L.M., Hiruma, K., Zhou, X., Nelson, C.A., **2003**. Insights into the molecular basis of the hormonal control of molting and metamorphosis from *Manduca sexta* and *Drosophila melanogaster*. *Insect Biochem. Mol. Biol. 33*, 1327–1338.

Riddiford, L.M., Williams, C.M., **1967**. The effects of juvenile hormone analogues on the embryonic development of silkworms. *Proc. Natl Acad. Sci. USA 57*, 595–601.

Riedl, H., Blomefield, T.L., Gilomee, J.H., **1999**. A century of codling moth control in South Africa. II. Current and future status of codling moth management. *J. S. Afr. Soc. Hortic. Sci. 8*, 32–54.

Riedl, H., Brunner, J.F., **1996**. Insect growth regulators provide new pest control tools for Pacific Northwest orchards. In: Proc. 92nd Annu. Mtg. Washington State Hortic. Assoc., pp. 189–200.

RohMid, L.L.C., **1996**. RH-0345, Turf and Ornamental Insecticide. *Technical Infor. Bull.*, 9.

Roller, H., Dahm, K.H., Sweely, C.C., Trost, B.M., Ange, W., **1967**. The structure of the juvenile hormone. *Chem. Int. Ed. Engl. 6*, 179.

Romanuk, M., **1981**. Structure-activity relationships in selected groups of juvenoids. In: Sehnal, F., Zabra, A., Menn, J.J., Cymborowsti, B. (Eds.), Regulation of Insect Development and Behaviour, Part 1. Wroclaw Technical Univeristy Press, Wroclaw, Poland, pp. 247–260.

Rup, P.J., Baniwal, A., **1985**. Responses of *Zaprionus paravittiger* (Drosophilidae: Diptera) to anti-juvenile hormone, precocene. *Insect Sci. Appl. 6(6)*, 671–675.

Saleh, D.S., Zhang, J., Wyatt, G.R., Walker, V.K., **1998**. Cloning and characterization of an ecdysone receptor cDNA from *Locusta migratoria*. *Mol. Cell. Endocrinol. 143*, 91–99.

Sariaslani, F.S., McGee, L.R., Ovenall, D.W., **1987**. Microbial transformation of Precocene II: Oxidative reactions by *Streptomyces griseus*. *Appl. Environ. Microbiol. 53(8)*, 1780–1784.

Saul, S.H., Mau Ronald, F.L., Oi, D., **1995**. Laboratory trials of methoprene-impregnated waxes for disinfesting papayas and peaches of the Mediterranean fruit fly (Diptera: Tephritidae). *J. Econ. Entomol. 78*, 652–655.

Saul, S.H., Mau Ronald, F.L., Kobayashi, R.M., Tsuda, D.M., *et al.*, **1987**. Laboratory trials of methoprene-impregnated waxes for preventing survival of adult oriental fruit flies (Diptera: Tephritidae) from infested papayas. *J. Econ. Entomol. 80*, 494–496.

Sauphanor, B., Bouvier, J.C., **1995**. Cross-resistance between benzoylureas and benzoylhydrazines in the codling moth, *Cydia pomonella* L. *Pestic. Sci. 45*, 369–375.

Sauphanor, B., Brosse, V., Monier, C., Bouvier, J.C., **1998a**. Differential ovicidal and larvicidal resistance to benzoylureas in the codling moth, *Cydia pomonella*. *Entomol Exp. Appl. 88*, 247–253.

Sauphanor, B., Bouvier, J.C., Brosse, V., **1998b**. Spectrum of insecticide resistance in *Cydia pomonella* (Lepidoptera: Tortricidae) in southeastern France. *J. Econ. Entomol. 91*, 1225–1231.

Saxena, R.C., Dixit, O.P., Sukumaran, P., **1992**. Laboratory assessment of indigenous plant extracts for antijuvenile hormone activity in *Culex quinquefasciatus*. *Indian J. Med. Res. 95*, 204–206.

Sbragia, R., Bisabri-Ershadi, B., Rigterink, R.H., **1983**. XRD-473, a new acylurea insecticide effective against *Heliothis*. *Brighton Crop Prot. Conf. 1*, 417–424.

Schooley, D.A., Baker, F.C., **1985**. Juvenile hormone biosynthesis. In: Kerkut, G.A., Gilbert, K.I. (Eds.), Comprehensive Insect Physiology, Biochemistry and Pharmacology, vol. 7. Pergamon, Oxford, pp. 363–389.

Sempere, L.F., Sokol, N.S., Dubrovsky, E.B., Berger, E.M., Ambros, V., **2003**. Temporal regulation of microRNA expression in *Drosophila melanogaster* mediated by hormonal signals and Broad-Complex gene activity. *Devel. Biol. 259*, 9–18.

Sen, S.E., Garvin, G.M., **1995**. Substrate requirements for lepidopteran farnesol dehydrogenase. *J. Agric. Food Chem. 43*, 820–825.

Sevala, V.L., Davey, K.G., **1989**. Action of juvenile hormone on the follicle cells of *Rhodnius prolixus*: Evidence for a novel regulatory mechanism involving protein kinase c. *Experientia. 45*, 355–356.

Shaw, P.W., Walker, J.T.S., **1996**. Biological control of wooly apple aphid by *Aphelinus mali* in an integrated fruit production programme in Nelson. In: Proc. 49th N.Z. Plant Prot. Conf., pp. 59–63.

Sheets, J., Karr, L.L., Dripps, J.E., **2000**. Kinetics and uptake, clearance, transfer, and metabolism of hexaflumuron by eastern subterranean termites (Isoptera: Rhinotermitidae). *J. Econ. Entomol. 93*, 871–877.

Shemshedini, L., Wilson, T.G., **1990**. Resistance ot juvenile hormone and an insect growth regulator in *Drosophila* is associated with an altered cytosolic juvenile hormone-binding protein. *Proc. Natl Acad. Sci. USA 87(6)*, 2072–2076.

Shimuzu, B., Nakagawa, Y., Hattori, K., Nishimura, K., Kurihara, N., *et al.*, **1997**. Molting hormone and larvicidal activities of aliphatic acyl analogs of dibenzoylhydrazine insecticides. *Steroids 62*, 638–642.

Singh, G., Sidhu, H.S., **1992**. Effect of juvenile hormone analogues on the morphogenesis of apteriform mustard sphid, *Lipaphis Erysimi* (Kalt.). *J. Insect Sci. 5(1)*, 73–74.

Slama, K., **1995**. Hormonal status of RH-5849 and RH-5992 synthetic ecdysone agonists (ecdysoids) examined on several standard bioassays for ecdysteroids. *Eur. J. Entomol. 92*, 317–323.

Slama, K., Romanuk, M., Sorm, F., **1974**. Insect Hormones and Bioanalogues. Springer, New York, NY, 477 pp.

Slama, K., Williams, C.M., **1996**. The juvenile hormone. V. The sensitivity of the bug, *Pyrrhocoris apterus*, to a hormonally active factor in american paper-pulp. *Biol. Bull. 130*, 235–246.

Smagghe, G., Braeckman, B.P., Huys, N., Raes, H., **2003a**. Cultured mosquito cells *Aedes albopictus* C6/36 (Dip., Culicidae) responsive to 20-hydroxyecdysone and non-steroidal ecdysone agonist. *J. Appl. Entomol. 127*, 167–173.

Smagghe, G., Carton, B., Decombel, L., Tirry, L., **2001**. Significance of absorption, oxidation, and binding to toxicity of four ecdysone agonists in multi-resistant cotton leafworm. *Arch. Insect Biochem. Physiol. 46*, 127–139.

Smagghe, G., Carton, B., Heirman, A., Tirry, L., **2000a**. Toxicity of four dibenzoylhydrazine correlates with evagination-induction in the cotton leafworm. *Pest. Biochem. Physiol. 68*, 49–58.

Smagghe, G., Carton, B., Wesemael, W., Ishaaya, I., Tirry, L., **1999a**. Ecdysone agonists – mechanism of action and application on *Spodoptera* species. *Pestic. Sci. 55*, 343–389.

Smagghe, G., Degheele, D., **1994**. The significance of pharmacokinetics and metabolism to the biological activity of RH-5992 (Tebufenozide) in *Spodoptera exempta, Spodoptera exigua*, and *Leptinotarsa decemlineata*. *Pestic. Biochem. Physiol. 49*, 224–234.

Smagghe, G., Degheele, D., **1997**. Comparative toxicity and tolerance for the ecdysteroid mimic tebufenozide in a laboratory and field strain of cotton leafworm (Lepidoptera: Noctuidae). *J. Econ. Entomol. 90*, 278–282.

Smagghe, G., Dhadialla, T.S., Derycke, S., Tirry, L., Degheele, D., **1998**. Action of the ecdysteroid agonist tebufenozide in susceptible and artificially selected beet armyworm. *Pestic. Sci. 54*, 27–34.

Smagghe, G., Dhadialla, T.S., Lezzi, M., **2002**. Comparative toxicity and ecdysone receptor affinity of non-steroidal ecdysone agonists and 20-hydroxyecdysone in *Chironomus tentans*. *Insect Biochem. Mol. Biol. 32*, 187–192.

Smagghe, G., Eelen, H., Verschelde, E., Richter, K., Degheele, D., **1996a**. Differential effects of nonsteroidal ecdysteroid agonists in Coleoptera and Lepidoptera: Analysis of evagination and receptor binding in imaginal discs. *Insect Biochem. Mol. Biol. 26*, 687–695.

Smagghe, G., Gelman, D., Tirry, L., **1997**. *In vivo* and *in vitro* effects of tebufenozide and 20-hydroxyecdysone on chitin synthesis. *Arch. Insect Biochem. Physiol. 41*, 33–41.

Smagghe, G., Medina, P., Schuyesmans, S., Tirry, L., Viñuela, E., **2000b**. Insecticide resistant monitoring of tebufenozide for managing *Spodoptera exigua* (Hübner [1808]). *Bol. San. Veg. Plagas 26*, 475–481.

Smagghe, G., Nakagawa, Y., Carton, B., Mourad, A.K., Fujita, T., *et al.*, **1999b**. Comparative ecdysteroid action of ring-substituted dibenzoylhydrazines in *Spodoptera exigua*. *Arch. Insect Biochem. Physiol. 41*, 42–53.

Smagghe, G., Pineda, S., Carton, B., Del Estal, P., Budia, F., Viñuela, E., **2003b**. Toxicity and kinetics for methoxyfenozide in greenhouse-selected *Spodoptera exigua* (Lepidoptera: Noctuidae). *Pest Mgt. Sci. 59*, 1203–1209.

Smagghe, G., Salem, H., Tirry, L., Degheele, D., **1996b**. Action of a novel insect growth regulator tebufenozide against different developmental stages of four stored product insects. *Parasitica 52*, 61–69.

Smagghe, G., Viñuela, E., Budia, F., Degheele, D., **1996c**. *In vivo* and *in vitro* effects of the nonsteroidal ecdysteroid agonist tebufenozide on cuticle formation in *Spodoptera exigua*: An ultrastructural approach. *Arch. Insect Biochem. Physiol. 32*, 121–134.

Smagghe, G., Viñuela, E., Van Limbergen, H., Budia, F., Tirry, L., **1999c**. Nonsteroidal moulting hormone agonists: effects on protein synthesis and cuticle formation in Colorado potato beetle. *Entomol. Exp. Appl. 93*, 1–8.

Sohi, S.S., Palli, S.R., Retnakaran, A., **1995**. Forest insect cell lines responsive to 20-hydroxyecdysone and two nonsteroidal ecdysone agonists, RH-5849 and RH-5992. *J. Insect Physiol. 41*, 457–464.

Soltani, N., Aribi, N., Berghiche, H., Lakbar, S., Smagghe, G., **2002**. Activity of RH-0345 on ecdysteroid production and cuticle secretion in *Tenebrio molitor* pupae *in vivo* and *in vitro*. *Pestic. Biochem. Physiol. 72*, 83–90.

Soltani, N., Besson, M.T., Delachambre, J., **1984**. Effect of diflubenzuron on the pupal-adult development of *Tenebrio molitor* L. (Coleoptera: Tenebrionidae): growth and development, cuticle secretion, epidermal cell density and DNA synthesis. *Pestic. Biochem. Physiol. 21*, 256–264.

Song, Q., Ma, M., Ding, T., Ballarino, J., Wu, S.-J., **1990**. Effects of a benzodioxole, J2581 (5-ethoxy-6-(4-methoxyphenyl)methyl-1,3-benzodioxole), on vitellogenesis and ovarian development of *Drosophila melanogaster*. *Pestic. Biochem. 37*(1), 12–23.

Sonoda, M., Kuwano, E., Taniguchi, E., **1995**. Precocious metamorphosis induced by 1,5-disubstituted imidazoles is counteracted by tebufenozide (RH-5992), an ecdysteroid agonist. *Nippon Noyaku Gakkaishi 20*, 325–327.

Sotomatsu, T., Nakagawa, Y., Fujita, T., **1987**. Quantitative structure-activity studies of benzoylphenylurea larvicides. IV. Benzoyl ortho substituent effects and molecular conformation. *Pestic. Biochem. Physiol. 27*, 156–164.

Spindler, K.D., Spindler-Barth, M., Londershausen, M., **1990**. Chitin metabolism: a target for drugs against parasites. *Parasitol. Res. 76*, 283–288.

Spindler-Barth, M., Spindler, K.-D., **1998**. Ecdysteroid resistant subclones of the epithelial cell line from *Chironomus tentans* (Insecta, Diptera). I. Selection and characterization of resistant clones. *In Vitro Cell. Devel. Biol. Animal 34*, 116–122.

Spindler-Barth, M., Turberg, A., Spindler, K.-D., **1991**. On the action of RH 5849, a nonsteroidal ecdysteroid agonist, on a cell line from *Chironomus tentans. Arch. Insect Biochem. Physiol. 16*, 11–18.

Srivastava, U.S., Jaiswal, A.K., **1989**. Precocene II Induced effects in the aphid *Aphid craccivora* Koch. *Insect Sci. Appl. 10*(4), 471–475.

Sterk, G., Hassan, S.A., Baillod, M., Bakker, F., Bigler, F., *et al.*, **1999**. Results of the seventh joint pesticide testing programme carried out by the IOBC/WPRS-Working Group Pesticides and Beneficial Organisms. *BioControl 44*, 99–117.

Staal, G.B., **1975**. Insect growth regulators with juvenile hormone activity. *Annu. Rev. Entomol. 20*, 417–460.

Staal, G.B., **1982**. Insect control with growth regulators interfering with the endocrine system. *Entomol. Exp. Appl. 31*, 15–23.

Stevens, P.S., Stevens, D., **1994**. An insect growth regulator for controlling leafrollers in kiwifruit. In: Proc. 47th N.Z. Plant Prot. Conf., pp. 310–313.

Strand, M.R., Goodman, W.G., Baehrecke, E.H., **1991**. The juvenile hormone titer of *Trichoplusia ni* and its potential role in embryogenesis of the polyembryonic wasp *Copidosoma floridanum. Insect Biochem. 21*(2), 205–214.

Suh, C.P.-C., Orr, D.B., Van Duyn, J.W., **2000**. Effect of insecticides on *Trichogramma exiguum* (Trichogrammatidae: Hymenoptera) preimaginal development and adult survival. *J. Econ. Entomol. 93*(3), 577–583.

Suhr, S.T., Gil, E.B., Senut, M.-C., Gage, F.H., **1998**. High level transactivation by a modified *Bombyx* ecdysone receptor in mammalian cells without exogenous retinoid X receptor. *Proc. Natl Acad. Sci. USA 95*, 7999–8004.

Sun, X., Barrett, B.A., **1999**. Fecundity and fertility changes in adult codling moth (Lepidoptera: Tortricidae) exposed to surfaces treated with tebufenozide and methoxyfenozide. *J. Econ. Entomol. 92*, 1039–1044.

Sun, X., Barrett, B.A., Biddinger, D.J., **2000**. Fecundity and fertility changes in adult redbanded leafroller and obliquebanded leafroller (Lepidoptera: Tortricidae) exposed to surfaces treated with tebufenozide and methoxyfenozide. *Entomol. Exp. Applic. 94*, 75–83.

Sundaram, K.M.S., Sundaram, A., Sloane, L., **1996**. Foliar persistence and residual activity of tebufenozide against spruce budworm larvae. *Pestic. Sci. 47*, 31–40.

Sundaram, M., Palli, S.R., Ishaaya, I., Krell, P.J., Retnakaran, A., **1998a**. Toxicity of ecdysone agonists correlates with the induction of CHR3 mRNA in the spruce budworm. *Pestic. Biochem. Physiol. 62*, 201–208.

Sundaram, M., Palli, S.R., Krell, P.J., Sohi, S.S., Dhadialla, T.S., *et al.*, **1998b**. Basis for selective action of a synthetic molting hormone agonist, RH-5992 on lepidopteran insects. *Insect Biochem. Mol. Biol. 28*, 693–704.

Sundaram, M., Palli, S.R., Smagghe, G., Ishaaya, I., Feng, Q.L., *et al.*, **2002**. Effect of RH-5992 on adult development in the spruce budworm, *Choristoneura fumiferana. Insect Biochem. Mol. Biol. 32*, 225–231.

Swevers, L., Drevet, J.R., Lunke, M.D., Iatrou, K., **1995**. The silkmoth homolog of the *Drosophila* ecdysone receptor (B1 Isoform): cloning and analysis of expression during follicular cell differentiation. *Insect Biochem. Mol. Biol. 25*, 857–866.

Takagi, M., Tsuda, Y., Wada, Y., **1995**. Laboratory evaluation of a juvenile hormone mimic, pyriproxyfen, against *Chironomus fusciceps* (Diptera: Chironomidae). *J. Am. Mosquito Control Assoc. 11*(4), 474–475.

Tatar, M., Yin, C.-M., **2001**. Slow aging during insect reproductive diapause: Why butterflies, grasshoppers and flies are like worms. *Exp. Geront. 36*, 723–738.

Taylor, M.A., **2001**. Recent developments in ectoparasiticides. *Vet. J. 161*, 253–268.

Thomas, H.E., Stunnenberg, H.G., Stewart, A.F., **1993**. Heterodimerization of the *Drosophila* ecdysone receptor with retinoid X receptor and ultraspiracle. *Nature 362*, 471–475.

Tobe, S.S., Ruegg, R.P., Stay, B.A., Baker, F.A., Miller, C.A., *et al.*, **1985**. Juvenile hormone titer and regulation in the cockroach *Diploptera punctata. Experientia 41*, 1028–1034.

Tobe, S.S., Stay, B., **1985**. Structure and regulation of the corpus allatum. *Adv. Insect Physiol. 18*, 305–432.

Tomlin, C.D.S. (Ed.) **2000**. The Pesticide Manual, 12th edn. British Crop Protection Council Publications.

Toya, T., Fukasawa, H., Masui, A., Endo, Y., **2002**. Potent and selective partial ecdysone agonist activity of chromafenozide in sfg cells. *Biochem. Biophys. Res. Comm. 292*, 1087–1091.

Tran, H.T., Askari, H.B., Shaaban, S., Price, L., Palli, S.R., *et al.*, **2000**. Reconstruction of ligand-dependent transactivation of *Choristoneura fumiferana* ecdysone receptor in yeast. *J. Mol. Endocrinol. 15*, 1140–1153.

Tran, H.T., Shaaban, S., Askari, H.B., Walfish, P.G., Raikhel, A.S., *et al.*, **2001**. Requirement of co-factors for the ligand-mediated activity of the insect ecdysteroid receptor in yeast. *J. Mol. Endocrinol. 27*, 191–209.

Trisyono, A., Chippendale, M., **1997**. Effect of the nonsteroidal ecdysone agonists, methoxyfenozide and tebufenozide, on the European corn borer (Lepidoptera: Pyralidae). *J. Econ. Entomol. 90*(6), 1486–1492.

Trisyono, A., Chippendale, M., **1998**. Effect of the ecdysone agonists, RH-2485 and tebufenozide, on the Southwestern corn borer, *Diatraea grandiosella. Pestic. Sci. 53*, 177–185.

Trisyono, A., Goodman, C.L., Grasela, J.J., McIntosh, A.H., Chippendale, G.M., **2000**. Establishment and characterization of an *Ostrinia nubilalis* cell line, and its response to ecdysone agonists. *In Vitro Cell. Devel. Biol. Animal 36*, 400–404.

Truman, J.W., Rountree, D.B., Reiss, S.E., Schwartz, L.M., **1983**. Ecdysteroids regulate the release and action of eclosion hormone in the tobacco hornworm, *Manduca sexta (L). J. Insect Physiol. 29*, 895–900.

Tsunoda, K., Doki, H., Nishimoto, K., **1986**. Effect of developmental stages of workers and nymphs of *Reticulitermes speratus* (Kolbe) (Isoptera: Rhinotermitidae) on caste differentiation induced by JHA treatment. *Mat. Organismen Berlin 21*(1), 47–62.

Tyagi, B.K., Kalyanasundaram, M., Das, P.K., Somachary, N., **1985**. Evaluation of a new compound (VCRC/INS/A-23) with juvenile hormone activity against mosquito vectors. *Ind. J. Med. Res. 82*, 9–13.

Uchida, M., Asai, T., Sugimoto, T., **1985**. Inhibition of cuticle deposition and chitin biosynthesis by a new insect growth regulator buprofezin in *Nilaparvata lugens* Stål. *Agric. Biol. Chem. 49*, 1233–1234.

Ujváry, I., Matolcsy, G., Riddiford, L.M., Hiruma, K., Horwath, K.L., **1989**. Inhibition of spiracle and crochet formation and juvenile hormone activity of isothiocyanate derivatives in the tobacco hornworm, *Manduca sexta*. *Pestic. Biochem. Physiol. 35*, 259–274.

Unger, E., Cigan, A.M., Trimnell, M., Xu, R.-J., Kendall, T., *et al.*, **2002**. A chimeric ecdysone receptor facilitates methoxyfenozide-dependent restoration of male fertility in ms45 maize. *Transgenic Res. 11*, 455–465.

Unnithan, G.C., Andersen, J.F., Hisano, T., Kuwano, E., Feyereisen, R., **1995**. Inhibition of juvenile hormone biosynthesis and methyl fernesoate epoxidase activity by 1,5-disubstituted imidazoles in the cockroach, *Diploptera punctata*. *Pestic. Sci. 43*, 13–19.

Vail, K.M., Williams, D.F., **1995**. Pharaoh ant (Hymenoptera: Formicidae) colony development after consumption of pyriproxyfen baits. *J. Econ. Entomol. 88*(6), 1695–1702.

Van Daalen, J.J., Meltzer, J., Mulder, R., Wellinga, K., **1972**. A selective insecticide with a novel mode of action. *Naturwissenschaften 59*, 312–313.

Van de Veire, M., Smagghe, G., Degheele, D., **1996**. Laboratory test method to evaluate the effect of 31 pesticides on the predatory bug, *Orius laevigatus* (Het.: Anthocoridae). *Entomophaga 41*, 235–243.

Van Laecke, K., Degheele, D., **1991**. Detoxification of diflubenzuron and teflubenzuron in the larvae of the beet armyworm (*Spodoptera exigua*) (Lepidoptera: Noctuidae). *Pestic. Biochem. Physiol. 40*, 181–190.

Van Laecke, K., Smagghe, G., Degheele, D., **1995**. Detoxifying enzymes in greenhouse and laboratory strain of beet armyworm (Lepidoptera: Noctuidae). *J. Econ. Entomol. 88*, 777–781.

Veech, J.A., **1978**. The effect of diflubenzuron on the reproduction of free living nematodes. *Nematologica 24*, 312–320.

Veras, M., Mavroidis, M., Kokolakis, G., Gourzi, P., Zacharopoulou, A., *et al.*, **1999**. Cloning and characterization of CcEcR: An ecdysone receptor homolog from the Mediterranean fruit fly *Ceratitis capitata*. *Eur. J. Biochem. 265*, 798–808.

Viñuela, E., Budia, F., **1994**. Ultrastructure of *Ceratitis capitata* Wiedemann larval integument and changes induced by cyromazine. *Pestic. Biochem. Physiol. 48*, 191–201.

Vogt, H. (Ed.) **1994**. Pesticides and Beneficial Organisms, IOBC/WPRS Bulletin. vol. 17. IOBC/WPRS Publications.

Wakgari, W., Giliomee, J., **2001**. Effects of some conventional insecticides and insect growth regulators on different phenological stages of the white – scale, *Ceroplastes destructor* Newstead (Hemiptera: Coccidae), and its primary parasitoid, *Aprostocetus ceroplastae* (Girault) (Hymenoptera: Eulophidae). *Intern. J. Pest Manag. 47*(3), 179–184.

Walton, L., Long, J.W., Spivey, J.A., **1995**. Use of Confirm insecticide for control of beet armyworm in cotton under Section 18 in MS and AL. *Proc. Beltswide Cotton Conf. 2*, 46–47.

Watkinson, I.A., Clarke, B.S., **1973**. The insect moulting hormone system as a possible target site for insecticidal action. *PANS 14*, 488–506.

Wearing, C.H., **1998**. Cross-resistance between azinphosmethyl and tebufenozide in the greenheaded leafroller, *Planotortrix octo*. *Pestic. Sci. 54*, 203–211.

Wigglesworth, V.B., **1936**. The function of the corpus allatum in the growth and reproduction of *Rhodnius prolixus* (Hemiptera). *Q. J. Microsc. Sci. 79*, 91–121.

Wildboltz, T., **1988**. Integrated pest management in Swiss apple orchards: stability and risks. *Entomol. Exp. Appl. 49*, 71–74.

Williams, C.M., **1952**. Physiology of insect diapause. IV. The brain and prothoracic glands as an endocrine system in the *Cecropia* silkworm. *Biol. Bull. 103*, 120–138.

Williams, C.M., **1967**. The juvenile hormone. II. Its role in the endocrine control of molting, pupation, and adult development in the *Cecropia* silkworm. *Biol. Bull. 121*, 572–585.

Williams, D.R., Chen, J.-H., Fisher, M.J., Rees, H.H., **1997**. Induction of enzymes involved in molting hormone (ecdysteroid) inactivation by ecdysteroids and an agonist, 1,2-dibenzoyl-1-*tert*-butylhydrazine (RH-5849). *J. Biol. Chem. 272*, 8427–8432.

Williams, D.R., Fisher, M.J., Smagghe, G., Rees, H.H., **2002**. Species specificity of changes in ecdysteroid metabolism in response to ecdysteroid agonists. *Pestic. Biochem. Physiol. 72*, 91–99.

Williams, R.E., Berry, J.G., **1980**. Evaluation of CGA 72662 as a topical spray and feed additive for controlling house fly breeding in chicken manure. *Poultry Sci. 59*, 2207–2212.

Wilson, T.G., Ashok, M., **1998**. Insecticide resistance resulting from an absence of target-site gene product. *Proc. Natl Acad. Sci. USA 95*(24), 14040–14044.

Wilson, T.G., Fabian, J., **1986**. A *Drosophila melanogaster* mutant resistant to a chemical analog of juvenile hormone. *Devel. Biol. 118*(1), 190–201.

Wing, K.D., **1988**. RH 5849, a non-steroidal ecdysone agonist: effects on a *Drosophila* cell line. *Science 241*, 467–469.

Wing, K.D., Aller, H.E., **1990**. Ecdysteroid agonists as novel insect growth regulators. In: Cassida, J.E. (Ed.), Pesticides and Alternatives: Innovative Chemical and

Biological Approaches to Pest Control. Elsevier, pp. 251–257.

Wing, K.D., Slawecki, R., Carlson, G.R., 1988. RH-5849, a nonsteroidal ecdysone agonist: effects on larval Lepidoptera. *Science 241*, 470–472.

Wroblewski, V.J., Harshman, L.G., Hanzlik, T.N., Hammock, B.D., 1990. Regulation of juvenile hormone esterase gene expression in the tobacco budworm (*Heliothis virescens*). *Arch. Biochem. Biophys. 278*, 461–466.

Wurtz, *et al.*, 2000. A new model for 20-hydroxyecdysone and dibenzoylhydrazine binding: A homology modeling and docking approach. *Protein Sci. 9*, 1073–1084.

Wyatt, G.R., Davey, K.G., 1996. Cellular and molecular actions of juvenile hormone. II. Roles of juvenile hormone in adult insects. *Adv. Insect Physiol. 26*, 1–155.

Xu, Y., Fang, F., Chu, Y.-X., Jones, D., Jones, G., 2002. Activation of transcription through the ligand-binding pocket of the orphan nuclear receptor spiracle. *Eur. J. Biochem. 269*, 6026–6036.

Yanagi, M., Watanabe, T., Masui, A., Yokoi, S., Tsukamoto, Y., *et al.*, 2000. ANS-118: A novel insecticide. *Proc. Brighton Crop Prot. Conf. 2*, 27–32.

Yanagi, M., 2000. Development of a novel lepidopteran insect control agent, chromofenozide (MATRIC), and prospect of success. *Agrochem. Jpn. 76*, 16–18.

Yao, T.-P., Forman, B.M., Jiang, Z., Cherbas, L., Chen, J.-D., *et al.*, 1995. Functional ecdysone receptor is the product of *EcR* and *Ultraspiracle* genes. *Nature 366*, 476–479.

Yao, T.-P., Segraves, W.A., Oro, A.E., McKeown, M., Evans, R.M., 1992. *Drosophila* ultraspiracle modulates ecdysone receptor function via heterodimer formation. *Cell 71*, 63–72.

Yen, J.L., Batterham, P., Gelder, B., McKenzie, J.A., 1996. Predicting resistance and managing susceptibility to cyromazine in the Australian sheep blowfly *Lucilia cuprina*. *Aust. J. Exp. Agri. 36*, 413–420.

Zhang, J., Cress, D.E., Palli, S.R., Dhadialla, T.S., 2003. cDNAs encoding ecdysteroid receptors of whitefly and their use in screening for pesticides. *PCT Int. Appl. WO 2003027266*. pp. 85.

Zhang, J., Wyatt, G.R., 1996. Cloning and upstream sequence of a juvenile hormone-regulated gene from the migatory locust. *Gene 175*(1–2), 193–197.

Zhou, X., Riddiford, L.M., 2002. Broad specifies pupal development and mediates the 'status quo' action of juvenile hormone on the pupal-adult transformation in *Drosophila* and *Manduca*. *Development 129*, 2259–2269.

Zimowska, G., Mikólajczyk, P., Silhacek, D.L., Oberlander, H., 1994. Chitin synthesis in *Spodoptera frugiperda* wing discs. II. Selective action of chlorfluazuron on wheat germ agglutinin binding and cuticle ultrastructure. *Arch. Insect Biochem. Physiol. 27*, 89–108.

Zitnanova, I., Adams, M.E., Zitnan, D., 2001. Dual ecdysteroid action on the epitracheal glands and central nervous system preceding ecdysis of *Manduca sexta*. *J. Exp. Biol. 204*, 3483–3495.

Zoebelein, G., Hammann, I., Sirrenberg, W., 1980. BAY-SIR-8514, a new chitin synthesis inhibitor. *J. Appl. Entomol. 89*, 289–297.

Zuo, J., Chua, N.-H., 2000. Chemical-inducible systems for regulated expression of plant genes. *Curr. Opin. Biotech. 11*, 146–151.

Relevant Website

http://www.iobc.wprs.org – International Organization for Biological and Integrated Control of Noxious Animals and Plants/West Palaearctic Regional Section **2004**.

A4 Addendum: Recent Progress on Mode of Action of 20-Hydroxyecdysone, Juvenile Hormone (JH), Non-Steroidal Ecdysone Agonist and JH Analog Insecticides

T S Dhadialla, Dow AgroSciences LLC, Indianapolis, IN, USA

© 2010 Elsevier B.V. All Rights Reserved

A4.1. Recent Progress on Mode of Action of 20-Hydroxyecdysone and Non-Steroidal Ecdysone Agonist Insecticides

The reader is also referred to a more recent publication on nonsteroidal ecdysone agonist insecticides (Dhadialla and Ross, 2007). Since the last publication of this chapter, the crystal structure of the ecdysone receptor ligand binding domain (LBD) from the whitefly, *Bemesia tabacci* (Bt), heterodimeric ecdysone receptor complex has been published (Carmichael *et al.*, 2005). Analysis of the amino acid residues lining the hormone binding pocket within the EcR subunit indicates that 22 of 25 residues making hydrogen bonds or nonpolar contacts with the hormone are highly conserved both in nature and side-chain rotameric conformation (Carmichael *et al.*, 2005). The residues that are in contact with the hormone and that vary between insects cause remarkably little change in pocket topography. For example, the effect of replacing BtEcR Met-272 by the smaller HvEcR Val-384 is minimized in the region encompassing the hormone by a change in the rotameric conformation of the adjacent conserved, non-ligand-interacting residue BtEcR Leu-308 – HvEcRLeu-420 (see **Figure A1**). In fact, regions of the ligand binding pocket surface in contact with the hormone are remarkably well conserved, not only in shape, but also in the overall hydrophobic and polar character.

The same is not true for those parts of the pocket formed by residues that do not contribute to hormone interaction. For example, the change in BtEcR Leu-308 – HvEcRLeu-420 rotameric conformation described earlier also contributes to a loosening of residue packing in a region of the *H. virescens* pocket adjacent to, but not in contact with, the steroid side chain (**Figure A1**). This loosening is further enhanced by the orientation of HvEcR Gln-503 toward the surface of the protein, in contrast to the conformation of the corresponding BtEcR Met-389, which is directed toward BtEcR Leu-308 (**Figure A1**). Calculation of the molecular

surface of the pocket with a 1.2 Å probe reveals an extension of the *H. virescens* pocket in this region, which is absent in the BtEcR protein. It is precisely here that the bisacylhydrazine inserts itself into the lepidopteran protein atomic volume, suggesting a mechanism contributing to differential binding of the insecticide across taxonomic orders.

In an effort to develop a high-throughput nonradiometric ligand binding assay for insect EcRs, Graham *et al.* (2007) found that fluorescein can be attached to the end of the ecdysteroid side chain with little or no effect on binding to the receptor protein. This led to the development of a recombinant receptor-based fluorescence polarisation ligand binding assay, which was readily automated for screening of a chemical library (Graham *et al.*, 2009). This high-throughput assay facilitated the discovery of compounds killing sheep body lice down to 0.5 ppm (Ronald Hill, personal communication).

Hannan *et al.* (2009) demonstrated that it is possible to selectively inhibit the synthesis of ecdysone receptor subunit proteins, using RNAi approaches. RNAi knock-down of ecdysone receptor synthesis may find practical applications, for example for the control of sucking insect pests of agricultural importance.

Finally, considerable progress has been made in use of nonsteroidal ecdysone agonist insecticides as ligands for EcR-based gene switches in plants for regulated gene expression (Tavva *et al.*, 2006, 2007a, 2007b, 2008). These investigators improved the sensitivity of the EcR-based gene switch to nanomolar concentrations of methoxyfenozide compared to micromolar amounts needed to activate earlier versions of EcR gene switches. This was achieved by using gene constructs containing LBD of CfEcR fused to GAL4 DNA binding domain and VP16 activation domain fused to LBD of *Locusta migratoria* retinoid-X-Receptor (LmRXR) to transform plants. Earlier

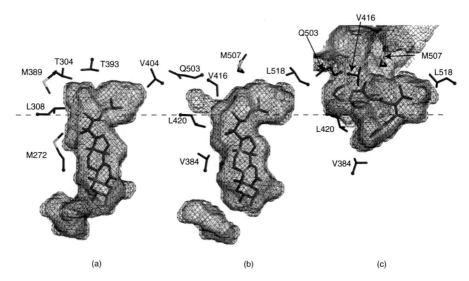

(a)   (b)   (c)

Figure A1 The ligand-binding pockets of (a) BtEcR-LBD, showing PonA bound, (b) HvEcR-LBD, with PonA bound and (c) HvEcR-LBD, showing the bisacylhydrazine BYI06830 bound. Pockets were generated both with a 1.4 Å radius probe (transparent green), as well as with a 1.2 Å probe (black mesh to locate more loosely-packed regions of the pocket wall. The looser packing of the walls of the HvEcR-LBD ligand-binding pocket in the vicinity of residues L420, V416 and Q503 is evidenced both by the orientation of the side chains of these residues and by the intrusion into the protein atomic volume of the ecdysteroid-binding pocket wall calculated with the 1.2 Å probe. Opening of this region upon BYI06830 binding is evident in (c), with the pocket volume now opening outwards to solvent. No such region of looser packing is evident in the walls of the ligand-binding pocket of BtEcR-LBD (see panel a). Reproduced with permission from Carmichael *et al.* (2005).

versions of the gene switch used only the EcR LBDs and were 1000 times less sensitive than the gene switch reported by Tavva *et al.* (2007a).

A4.1. Recent Progress on Mode of Action of 20-Hydroxyecdysone and Non-Steroidal Ecdysone Agonist Insecticides

Prior research on JH or its analogs (JHA) and 20-E had assumed that these two growth and developmental hormones have independent pathways and targets for action. Grace Jones and her collaborators were the first to indicate that one of the EcR complex proteins, ultraspiracle, is also a receptor for JH (Jones and Sharp, 1997; Jones *et al.*, 2001; Xu *et al.*, 2002). More recently, Konopova and Jindra (2007) used RNAi-aided knock-down of the *Drosophila Methoprene tolerant gene (Met)* ortholog in the flour beetle, *Tribolium castaneum*, to demonstrate precocious metamorphosis of early-stage beetle larvae and that the TcMet RNAi insects' phenotypes were similar to those observed for allatectomized lepidopteran larvae. Results from more recent work on the honeybee, *Apis melifera*, the mosquito, *Aedes aegypti*, tobacco budworm, *Heliothis virescens*, and *T. castaneum* show that the target proteins involved in ecdysteroid (EcR and USP) and JH (Met) action interact and mediate cross-talk between these two important hormones (Bitra and Palli, 2009;

Li *et al.*, 2007; Wu *et al.*, 2006; Parthasarathy and Palli, 2007, 2008a, 2008b; Franssens *et al.*, 2006a; 2006b; Flatt *et al.*, 2008). Li *et al.* (2007) identified a JH response element (JHRE1) common to 16 genes regulated by JH III. These researchers showed that two proteins (FKBP39 and Chd64) that were identified using DNA affinity column prepared using conserved JHRE1 interact with EcR, USP, and Met proteins, suggesting that 20E and JHIII may function through multiple protein complexes that include proteins involved in signal transduction of both these hormones. Some of the aforementioned studies have been aided by the use of RNAi approaches to knocking out the function of target genes in both *T. castaneum* and *Drosophila* L57 cells (Parthasarathy *et al.*, 2008; Li *et al.*, 2007). In spite of these studies, major gaps in our understanding of the biochemical and molecular details of the mode of action of JH and its analogs remain. At the same time, it is studies like these that use diverse systems, approaches, and tools, which will help us gain a better understanding of the elusive mode of action of JH and its analogs.

Studies like these, in combination with the ability to use RNAi gene knock down approaches, not only enhance our understanding of hormone action, but also open up possibilities for intervening in new protein targets that are critical to the growth and development of insects.

Acknowledgments

We sincerely thank Dr. Ronald J. Hill, CSIRO Food and Nutritional Sciences, Australia, and Professor Subha Reddy Palli, University of Kentucky, for providing updates on the mode of action of 20E and nonsteroidal ecdysone agonists and JH and JHAs, respectively.

References

Bitra, K., Palli, S.R., 2009. Interaction of proteins involved in ecdysone and juvenile hormone signal transduction. *Arch. Insect Biochem. Physiol. 70*, 90–105.

Carmichael, J.A., Lawrence, M.C., Graham, L.D., Pilling, P.A., Epa, V.C., Noyce, L., Ho, V., Hannan, G.N., Hill, R.J., 2005. The X-ray structure of a hemipteran ecdysone receptor ligand-binding domain – comparison with a lepidopteran ecdysone receptor ligand-binding domain and implications for insecticide design. *J Biol. Chem. 280*, 22258–22269.

Dhadialla, T.S., Ross, R.Jr., 2007. Bisacylhydrazines: Novel Chemistry for Insect Control. In: Kramer, W., Shirmer, U. (Eds.), Modern Crop Protection Compounds. Wiley, KGaA, Weinheim, pp. 773–796.

Flatt, T., Heyland, A., Rus, F., Porpiglia, E., Sherlock, C., Yamamoto, R., Garbuzov, A., Palli, S.R., Tatar, M., Silverman, N., 2008. Hormonal regulation of the humoral innate immune response in Drosophila melanogaster. *J. Exp. Biol. 211*, 2712–2724.

Franssens, V., Smagghe, G., Simonet, G., Claeys, I., Breugelmans, B., De Loof, A., Vanden, B.J., 2006a. 20-Hydroxyecdysone and juvenile hormone regulate the laminarin-induced nodulation reaction in larvae of the flesh fly, *Neobellieria bullata*. *Dev. Comp. Immunol. 30*, 735–740.

Franssens, V., Smagghe, G., Simonet, G., Claeys, I., Breugelmans, B., De Loof, A., Vanden, B.J., 2006b. Determination of the effects of ecdysteroids and JH on nodulation responses. *In Vitro Cell. Dev. Biol. Anim. 42*, 6A.

Graham, L.D., Johnson, W.M., Pawlak-Skrzecz, A., Eaton, R.E, Bliese, M., Howell, L., Hannan, G.N., Hill, R.J., 2007. Ligand binding by recombinant domains from insect ecdysone receptors. *Insect Biochem. Mol. Biol. 37*, 611–626.

Graham, L.D., Johnson, W.M., Tohidi-Esfahani, D., Pawlak-Skrzecz, A., Bliese, M., Lovrecz, G.O., Lu, L., Howell, L., Hannan, G.N., Hill, R.J., 2009. Ecdysone receptors of pest insects – molecular cloning, characterisation, and a ligand binding domain-based fluorescent polarization screen. In: Smagghe, G. (Ed.), Ecdysone: Structures and Functions. Springer, Berlin, pp. 447–474.

Hannan, G.N., Hill, R.J., Dedos, S.G., Swevers, L., Iatrou, K., Tan, A., Parthasarathy, R., Bai, H., Zhang, Z., Palli, S.R., 2009. Applications of RNA interference in ecdysone research. In: Smagghe, Guy (Ed.), Ecdysone: Structures and Functions. Springer, Berlin, pp. 205–227.

Jones, G., Sharp, P.A., 1997. Ultraspiracle: an invertebrate nuclear receptor for juvenile hormones. *PNAS. 94*, 13499–13503.

Jones, G., Wozniak, M., Chu, Y.-X., Dhar, S., Jones, D., 2001. Juvenile hormone III-dependent conformational changes of the nuclear receptor ultraspiracle. *Insect Biochem. Molec. Biol. 32*, 33–49.

Konopova, B., Jindra, M., 2007. Juvenile hormone resistance gene Methoprene- tolerant controls entry into metamorphosis in the beetle *Tribolium castaneum*. *Proc. Natl. Acad. Sci. USA 104*, 10488–10493.

Li, Y., Zhang, Z., Robinson, G.E., Palli, S.R., 2007. Identification and characterization of a juvenile hormone response element and its binding proteins. *J. Biol. Chem. 282*, 37605–37617.

Parthasarathy, R., Palli, S.R., 2007. Developmental and hormonal regulation of midgut remodeling in a lepidopteran insect, *Heliothis virescens*. *Mech. Dev. 124*, 23–34.

Parthasarathy, R., Palli, S.R., 2008a. Molecular analysis of juvenile hormone analog action in controlling the metamorphosis of the red flour beetle, *Tribolium castaneum*. *Arch. Insect Biochem. Physiol.* (in press).

Parthasarathy, R., Palli, S.R., 2008b. Proliferation and differentiation of intestinal stem cells during metamorphosis of the red flour beetle, *Tribolium castaneum*. *Dev. Dyn. 237*, 893–908.

Parthasarathy, R., Tan, A., Palli, S.R., 2008. bHLH-PAS family transcription factor methoprene-tolerant plays a key role in JH action in preventing the premature development of adult structures during larval-pupal metamorphosis. *Mech. Dev. 125*, 601–616.

Tavva, V.S., Dinkins, R.D., Palli, S.R., Collins, G.B., 2006. Development of a highly sensitive ecdysone receptor gene switch for applications in plants. *Plant J. 45*, 457–469.

Tavva, V.S., Palli, S.R., Dinkins, R.D., Collins, G.B., 2007a. Applications of EcR gene switch technology in functional genomics. *Arch. Insect. Biochem. Physiol. 65*, 164–179.

Tavva, V.S., Dinkins, R.D., Palli, S.R., Collins, G.B., 2007b. Development of a tightly regulated and highly inducible ecdysone receptor gene switch for plants through the use of retinoid X receptor chimeras. *Transgenic Res. 16*, 599–612.

Tavva, V.S., Palli, S.R., Dinkins, R.D., Collins, G.B., 2008. Improvement of a monopartite ecdysone receptor gene switch and demonstration of its utility in regulation of transgene expression in plants. *FEBS J. 275*, 2161–2176.

Wu, Y., Parthasarathy, R., Bai, H., Palli, S.R., 2006. Mechanisms of midgut remodeling: juvenile hormone analog methoprene blocks midgut metamorphosis by modulating ecdysone action. *Mech. Dev. 123*, 530–547.

Xu, Y., Fang, F., Chu, Y.X., Jones, D., Jones, G., 2002. Activation of transcription through the ligand-binding pocket of the orphan nuclear receptor ultraspiracle. *Eur. J. Biochem. 269*, 6026–6036.

Further Reading

Hatakoshi, M., 2007. A new juvenoid – Pyriproxifen. In: Kramer, W., Shirmer, U. (Eds.), Modern Crop Protection Compounds. Wiley, KGaA, Weinheim, pp. 797–811.

Sheets, J.J., 2007. Chitin Synthesis. In: Kramer, W., Shirmer, U. (Eds.), Modern Crop Protection Compounds. Wiley, KGaA, Weinheim, pp. 813–824.

5 Azadirachtin, a Natural Product in Insect Control

A J Mordue (Luntz), University of Aberdeen,
Aberdeen, UK
E D Morgan, Keele University, Keele, UK
A J Nisbet, University of Melbourne, Werribee,
Australia

© 2010, 2005 Elsevier B.V. All Rights Reserved

5.1. Introduction

The dominance of synthetic pesticides, which began with the introduction of DDT in the 1940s, almost displaced completely the natural pesticides used before that time. Only pyrethrum and derris (the active compound is rotenone) survived in specialized uses. In the following half-century the problems of persistence and toxicity of the synthetics grew and regulation followed them. Slowly, one after the other was restricted in use, until we have reached the point where there are no longer available synthetic crop protectants for certain uses. At the same time fundamental research has uncovered entirely new ways of dealing with pests that range from use of pheromone traps through parasitoids to feeding deterrents. Unique among these new discoveries is the compound azadirachtin.

Azadirachtin is a colorless crystalline compound, found in the seeds of the neem tree, *Azadirachta indica* A. Juss. 1830. The compound is of great interest because it is a powerful feeding deterrent for a number of phytophagous insects, and has the power, at parts per million concentration, to inhibit growth and development of all insects ingesting it. This substance, together with a few compounds of closely related structure in the same seeds, provides the most powerful natural pesticide discovered since the era of synthetic pesticides. Its spectrum of activity includes many other arthropods and even annelids and nematodes. Yet it appears to have little or no toxicity for vertebrates. Its isolation, structure determination, synthesis, toxicity, mode of action, spectrum of activity, and practical application have all absorbed the interest of scientists of many disciplines during the past four decades. Exploration of the properties and uses of azadirachtin continues, with a steady flow of about 45 papers per year appearing on various aspects of the compound.

A resurgence of interest in azadirachtin may be at hand. With the elucidation of its mode of action, which is drawing close, there will be the possibility of studying that site in detail with high-throughput screening methods to search for other molecules with greater field efficacy. Additionally, azadirachtin may prove to be an exciting tool for research scientists studying cellular processes and the cell cycle in invertebrate versus vertebrate cells.

5.1.1. The Neem Tree

The modern investigation of the properties of neem seeds began, as often happens, simultaneously in two directions, with the field work of Schmutterer in the Sudan, and laboratory work (unpublished) at the Anti-Locust Research Centre in London. In 1966, Morgan and Butterworth began a systematic search for the substance or substances in neem seeds responsible for the feeding deterrent effect on the desert locust *Schistocerca gregaria*, based on an antifeedant bioassay developed at the Anti-Locust Research Centre (Haskell and Mordue (Luntz), 1969). This work resulted in the isolation of the compound called azadirachtin (Butterworth and Morgan, 1968, 1971). It was soon shown to be effective systemically (Gill and Lewis, 1971), and to have growth-disrupting effects as well (Ruscoe, 1972). Neem products were slow to come to market while powerful synthetics dominated sales, but products are now available in several countries, including the USA, but elsewhere, the toxicity testing rules made for single-compound synthetics still make the licensing of neem products very difficult (Isman, 1997) (see Section 5.4).

Other compounds of related chemical structure and similar antifeedant properties have since been isolated from the seeds in smaller quantities. Therefore consideration must be given to the group when discussing the properties of azadirachtin. However, azadirachtin itself remains the most accessible and most abundant compound of the group.

The neem tree appears to be a native of Burma or the Indian subcontinent (Schmutterer, 2002). Indian sources say parts of the neem tree have been used since ancient times as protection against pests, citing the use of neem leaves for protection against book-worm and pests in stored rice, as well as in Ayurve-dic medicine (Puri, 1999). The tree belongs to the family Meliaceae, which includes various species of mahogany, but the wood is not of high quality.

There are only two other species recognized in the genus: *A. excelsa* (= *A. integrifolia*, locally known as marrango), a native of the Philippine Islands (Schmutterer and Doll, 1993), and the Thai neem tree, *A. siamensis*, very close in appearance to *A. indica*, and formerly thought to be a subspecies, but now awarded full species status. The latter forms hybrids with neem. All contain azadirachtin-like substances with similar properties (Kalinowski et al., 1997). Today neem is grown widely across the drier tropics from Australia to Africa, and South and Central America. It is easily grown, widely distributed, exists in large numbers (an estimated 14 million in India alone), so the seeds and their extracts are readily available.

5.2. Azadirachtin Research up to 1985

Research on azadirachtin from its discovery to 1985 is best summarized in three international symposia, two held in Germany, in 1980 (Schmutterer et al., 1981) and in 1983 (Schmutterer and Ascher, 1984), and one in Kenya in 1986 (Schmutterer and Ascher, 1987). In the 15–20 years following the isolation of azadirachtin, the major biological effects of neem seed extracts were established and shown to be due largely to the active ingredient azadirachtin. The antifeedant effects, insect growth regulatory effects, and sterility effects were estab-lished for a wide variety of phytophagous insects, and field trials were carried out to show the effec-tiveness of neem as a crop protection agent. Early promise of this, together with its large safety margin to vertebrates and potential effectiveness against other pests such as nematodes, led to an upsurge in neem research (Schmutterer and Ascher, 1984, 1987). By 1985 neem pesticides were promoted for use in developing countries by the Deutsche Gesellschaft für Technische Zusamenarbeit (GTZ) as a real alternative in biological and integrated pest control where sufficient raw material was available. In the Western world neem seed, neem oil, and several of their components were being tested in the USA as insect antifeedants and disruptants of normal growth and development. The first commer-cial neem seed formulation was patented in the USA (Larson, 1985) and granted registration by the En-vironment Protection Agency. Neem tree planta-tions for seed supplies were cultivated in the USA, Central and South America, and Australia.

By this time also the broad areas of its mode of action were established although the details were still not fully understood. Insect antifeedancy was shown to be particularly effective in lepidopteran pests compared with others such as the Coleoptera and Hemiptera and was related to effects on the mouthpart chemoreceptors (Simmonds and Blaney, 1984). Studies on the insect growth regulatory (IGR) mode of action of azadirachtin in, for example, the blowfly *Calliphora vicinia*, the milkweed bug *Oncopeltus fasciatus*, the tobacco hornworm *Manduca sexta*, and the African migratory locust *Locusta migratoria* revealed that ecdysone biosynthesis and catabolism were affected by azadirachtin treatment to cause the major insect growth regulatory effects (Redfern *et al.*, 1982; Sieber and Rembold, 1983; Schlüter *et al.*, 1985; Dorn *et al.*, 1986; Bidmon *et al.*, 1987). In adult insects, such as *L. migratoria*, effects of azadirachtin on egg development were shown to be related to changes in both juvenile hormone (JH) and ecdysone titers (Rembold *et al.*, 1986). In addition to the effects of azadirachtin on JH and ecdysone and the interactions between both hormones, Schlüter (1987) was able to demonstrate in the Mexican bean beetle *Epilachna varivestis* that azadirachtin also affected cells in the process of proliferation and differentiation. Mitosis in the wing disc was blocked from metaphase onwards by azadirachtin injections.

By 1986 azadirachtin was established as an antifeedant, insect growth regulator, and sterilant with a novel mode of action. The basis for its mode of action was known to involve the neurosecretory–neuroendocrine axis and perhaps other sites including particular stages in cell division. Although the full structure had not yet been elucidated or synthetic routes explored, the use of neem extracts to produce insecticides with good market potential was thought to be feasible. Studies to overcome a number of problems preventing commercial use of neem relating to standardization and regulatory matters were still required, as were studies on its possible role in integrated pest management strategies.

5.3. Chemistry of Neem Products

Azadirachtin, formula $C_{35}H_{44}O_{16}$, m.p. 160 °C, $[\alpha]_D^{25}$-66°(CHCl$_3$, $c = 0.5$) (**Figure 1**) belongs to the large group of plant triterpenoids, and to the narrower group of limonoids, which are frequently found in plants of the Meliaceae. Limonoids are triterpenoids that have their side chains shortened by four carbon atoms and the remaining four atoms cyclized as a furan ring (e.g., nimbin and salannin, and their desacetyl-derivatives; **Figure 1**). The molecule of azadirachtin is unusually highly oxygenated, and in a molecular model the surface is covered by functional

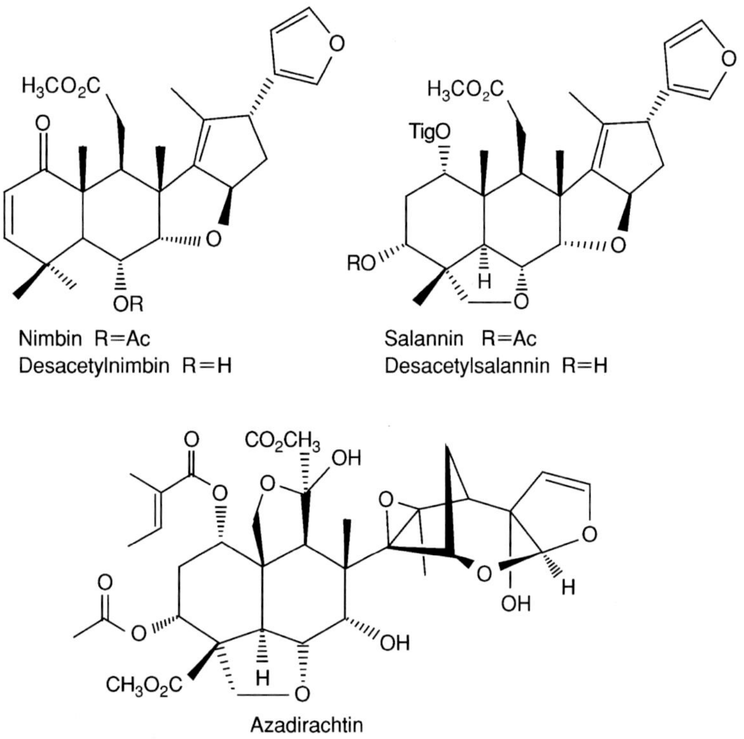

Figure 1 The structure of azadirachtin, together with the principal limonoids (nimbin and salannin, and their desacetyl-derivatives) from the seeds of neem (*Azadirachta indica*). Nimbin possesses almost no antifeedant effect and that of salannin is small compared to azadirachtin.

groups. The molecule contains an acetate, a tiglate ester, two methyl esters, a secondary and a tertiary alcohol, an epoxide, a vinyl ether, which is part of an acetal, and a hemiacetal. This overabundance of reactive groups makes its chemistry very difficult (Ley et al., 1993). Just how these contribute to its insecticidal properties is not yet understood. Another locust antifeedant called meliantriol of lower activity than azadirachtin was announced shortly before the isolation of azadirachtin (Lavie et al., 1967), but its isolation has never been repeated.

Structural studies on azadirachtin began immediately after its isolation, but because of its complexity and the sensitivity of some of the functional groups, progress was difficult. The structure was finally solved 18 years after isolation by a combination of modern nuclear magnetic resonance (NMR) techniques and X-ray crystallography (Bilton et al., 1987; Kraus et al., 1987; Turner et al., 1987). Synthesis of azadirachtin presents a great challenge to organic chemists. Great progress has been made towards it by the groups of Ley (Durand-Reville et al., 2002) and others (Nicolaou et al., 2002; Fukuzaki et al., 2002), but as yet it has not been completed.

5.3.1. Neem Limonoids

Largely because of the interest in azadirachtin, the triterpenoids of neem have been studied intensively. About 150 such compounds have now been described (Akhila and Rani, 1999), most of them found in very small quantities in various parts of the tree. Only about one-third of them have been tested for biological activity, and none has shown greater activity than the azadirachtin group.

Unfortunately many writers speak of "neem" as if that were a single commodity. The different parts of the neem tree all have different properties and contain different substances. The lack of more accurate description and the frequent lack of analytical data on what is contained in seed extracts have reduced the value of some of the published work on such extracts. The seeds are the only practical source of azadirachtin and its group of compounds. Any of this substance in other parts of the tree is in marginally small concentration.

5.3.2. Isolation of Azadirachtin

The triterpenoid present in greatest quantity in the seeds is usually salannin (Figure 1), a simpler triterpenoid with only weak activity as a feeding deterrent. Azadirachtin, representing 0.1–1.0% (mean 0.6%) of the weight of the seed kernels is next in quantity. The other compounds of azadirachtin-like structure and biological properties occur in progressively smaller amounts. Chemically, they divide into three structural types (Figure 2): the azadirachtins, with a hemiacetal group at carbon atom number 11, the azadirachtols, without the hemiacetal, and the meliacarpins, in which the methoxycarbonyl ester at C29 is replaced by methyl. Not all compounds found fit easily into these types and new trace constituents continue to be found (e.g., Luo et al., 1999; Malathi et al., 2003). Rearrrangement products, known as azadirachtinins (Figure 3), of much lower biological activity, are found in the seeds and are also formed during isolation of the limonoids. The names azadirachtin A, azadirachtin B, etc., are sometimes used. These are incorrect names. For the correct chemical names for all these triterpenoids and limonoids, see Kraus (2002). Also present in lesser quantities are nimbin, 3-desacetylnimbin, and 6-desacetylsalannin (Johnson et al., 1996) (Figure 1), simpler limonoids of lesser activity and interest. Compounds of similar structure are found in the related Melia genus. For example, Persian lilac or chinaberry, Melia azedarach, contains meliacarpins (Figure 2), but the seeds are extremely hard and their extracts are toxic to mammals.

A number of isolation procedures have been described, but all require extraction with solvent from the ground seeds, followed by solvent partition to separate the more polar triterpenoids from the oil. After that, methods differ more, but all require some form of chromatography to separate the individual compounds, either gravity column chromatography (Johnson and Morgan, 1997a), flash columns (Jarvis et al., 1999), or preparative high-performance liquid chromatography (HPLC) (Lee and Klocke, 1987; Govindachari et al., 1990, 1996). Extraction by supercritical carbon dioxide has been examined, but is not as efficient as solvent extraction (Johnson and Morgan, 1997b; Ambrosino et al., 1999). There is a lot of current interest in microwave-assisted chemistry, and a microwave-assisted extraction of seeds has been described (Dai et al., 2001).

5.3.3. Analysis

The number of triterpenoids in neem seeds and their similarity in physical properties makes analysis for the important pesticidal compounds both important and difficult. The standard method in use for some time is reverse-phase HPLC with ultraviolet (UV) absorption at short wavelength (214–218 nm) (e.g., Deota et al., 2000). Normal-phase supercritical fluid chromatography offers advantages where the equipment is available (Johnson and Morgan, 1997a). A colorimetric method for the whole group of triterpenoids, using vanillin–sulfuric acid has been

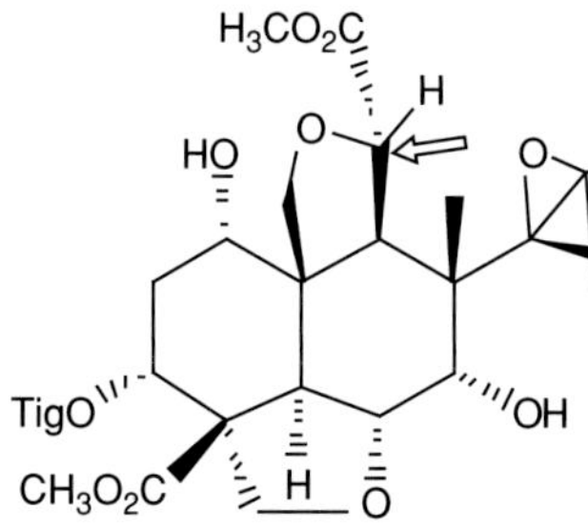

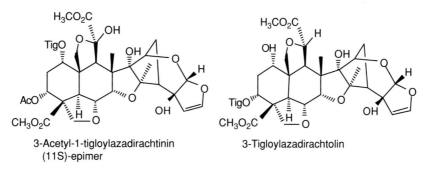

Figure 2 Examples of minor products related to azadirachtin from neem seeds. The essential difference of structure that makes an azadirachtol or meliacarpin is indicated by an arrow. Marrangin is the chief insecticidal compound of the marrango tree (*Azadirachta excelsa*). The names azadirachtin A, azadirachtin B, etc., are sometimes used. These are incorrect names. For the correct chemical names for all teiterpernoids and limonoids, see Kraus (2002).

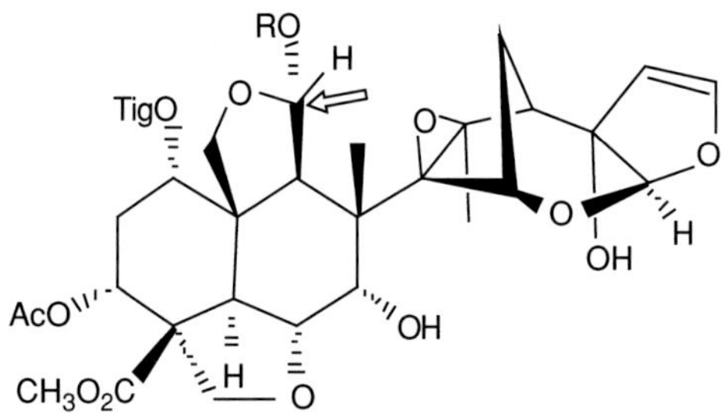

3-Tigloylazadirachtol

11-Demethoxycarbonylazadirachtin
(azadirachtin H) R=H
11-Acetyl-11-demethoxycarbonylazadirachtin
(marrangin) R=OAc

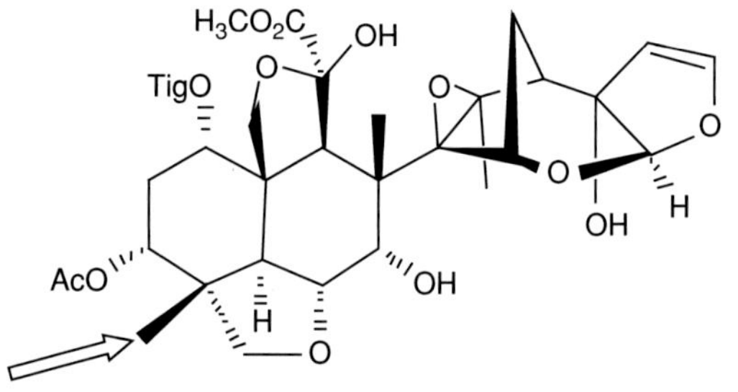

3-Acetyl-1-tigloyl-11-hydroxy-meliacarpin
(azadirachtin D)

3-Acetyl-1-tigloylazadirachtinin
(11S)-epimer

3-Tigloylazadirachtolin

Figure 3 Rearrangement products from azadirachtin and 3-tigloylazadirachtol. These compounds, of much lower activity than the others illustrated, are formed by reaction between the epoxide ring and the hydroxyl group at C7.

reported, but this is rather nonspecific (Dai *et al.*, 1999), and more specific enzyme-linked immunosorbent assays (ELISAs) have been developed that will give detection from 0.5 to 1000 ppb for azadirachtin and close relatives (Schutz *et al.*, 1997, Hemalatha *et al.*, 2001). Additionally, monoclonal antibodies to azadirachtin produced by phage display techniques are currently being assessed (Robertson, Mordue (Luntz), Mordue, and Porter, unpublished data). The most accurate and sensitive

analytical method, but also the most costly in equipment, is liquid chromatography–mass spectrometry (Schaaf *et al.*, 2000; Pozo *et al.*, 2003).

5.3.4. Stability and Persistence

Azadirachtin is soluble in polar organic solvents (methanol, ethanol, acetone, chloroform, ethyl acetate) and slightly soluble in water $(1.29\,g\,l^{-1})$. Aqueous solutions are usually made by dissolving first in ethanol or acetone and carefully diluting with water to avoid formation of the glassy mass when the solid is wetted with water, and which then dissolves very slowly. Solutions in organic solvents are stable almost indefinitely (Jarvis *et al.*, 1998) but solutions in water at pH 7 or higher are unstable. Aqueous solutions are most stable from pH 4 to 6 (Jarvis *et al.*, 1998).

While degradability is one great advantage in agricultural application of neem triterpenoids, the rate of disappearance is a little too high according to some authors. One report says azadirachtin is stable for less than 1 week under ambient conditions (Sundaram, 1996), another says its stability is less than 3 months (Brooks *et al.*, 1996). Scott and Kaushik (2000) gave a half-life of 36–48 h in water exposed to sunlight. The DT_{50} (time for 50% of azadirachtin to disappear) in soil at 25 °C was 20 and 31.5 days in unautoclaved and autoclaved soil respectively (Stark and Walter, 1995). Probably the most comprehensive studies on the environmental behavior and stability of azadirachtin have been carried out by the Canadian Forest Service for its use on young trees (Sundaram, 1996). Its persistence on balsam fir and red oak foliage; its dissipation in forest nursery soils; absorption, leaching, and desorption from sandy loam forest soil; photostability on foliage; and rate of hydrolysis in natural waters have all been measured (Sundaram, 1996; Sundaram *et al.*, 1997). They have also made efforts to formulate azadirachtin against hydrolysis and UV degradation, but no large-scale study of formulation for stability appears to have been made.

5.4. Neem Insecticides in Pest Control

5.4.1. Background

Today, after 40 years of discovery, research, and development of neem as a natural pesticide there are well-established products in the organic agriculture and niche markets in both North America and Western Europe. Azadirachtin, the main active ingredient of neem seeds, is an extremely effective antifeedant to many phytophagous insects, and an IGR and sterilant to all insects tested. Other constituents of neem seeds, including the oil, have been shown also to be effective control agents against plant nematodes and fungal pathogens.

The complexity of the azadirachtin molecule and our inability, as yet, to synthesize it has precluded its production as a synthetic pesticide, hence use of the natural product is the only choice. Neem as a botanical pesticide has many excellent attributes

that include its broad-spectrum IGR effects, systemic action in some plants, minimal disruption of natural enemies and pollinators, rapid breakdown in the environment, and lack of toxicity to vertebrates. However, its initial promise has not been realized. Problems have arisen during the research and development phase and these relate to its mode of action, where (as with most IGRs) azadirachtin acts slowly and although feeding and crop damage ceases shortly after treatment, pest insects remain alive on treated crops, reducing acceptance by growers and consumers. Neem pesticides have limited persistence on plants and require multiple applications against certain pests when used as a stand-alone treatment. This, together with the cost of production of a consistent, high-quality product with 1% or more of azadirachtin by weight from annual supplies of seeds, results in the cost of neem pesticides to be some three times greater than synthetic pyrethroids (Isman, 2004). Regulatory issues have also featured large in the development of neem and have been a formidable barrier to the successful commercialization of neem-based pesticides. The Environmental Protection Agency of the USA chose to simplify the approval process of neem insecticides by recognizing azadirachtin as the sole active ingredient and deeming the remaining chemical components present in neem kernels to be "inert ingredients" (Isman, 2004). However, Europe and Japan, for example, have required identification and toxicological studies for all major constituents of neem, processes that have greatly delayed its introduction into these and other countries.

Despite these problems neem insecticides are established as valuable components in integrated crop management systems and accepted by organic producers. In California neem insecticides are used on over 60 food crops, particularly lettuce and tomato, and account for 0.25% of insecticide use on tomato (California EPA, 2001, in Isman, 2004). In Europe, Neemazal insecticides (Trifolio-M GmbH) have been cleared for use in Germany after the parent company completed toxicological studies on all compounds present in their neem products. Neem is expected to show continued sales growth in highly developed countries as new markets continue to be developed. Growth in organic food production and the gradual reduction, and ultimate elimination, of synthetic neurotoxic insecticides will both favor expansion of the market for neem. The introduction of neem products specifically formulated for particular crops and crop pests will also aid expansion, as would intensive education of growers.

In developing countries, simple inexpensive formulation of neem extracts must be encouraged as an important source of pest control at the level of subsistence farming, as long as such unregulated products are not confused with legitimate regulated products of high technical standard.

5.4.2. Use in Integrated Pest Management

Azadirachtin exhibits good efficacy against key pests such as whiteflies, leaf miners, fungus gnats, thrips, aphids, and many leaf-eating caterpillars (Immaraju, 1998). Its use is recommended for insecticide resistance control, integrated pest control, and organic pest control programs. Research in community-level interactions with a view to the use of neem-based insecticides in IPM strategies are revealing subtle and complex effects that may or may not be useful. For example, the ability of azadirachtin to inhibit protein synthesis is demonstrated in the obliquebanded leafroller, *Choristoneura rosaceana* by an inhibition of midgut esterases (Smirle *et al.*, 1996; Lowery and Smirle, 2000). This is seen also in resistant Colorado potato beetles (*Leptinotarsa decemlineata*) treated with both neem and *Bacillus thuringiensis* (Trisyono and Whalon, 1999) and may be a useful tool in the management of insect populations where insecticide resistance has developed as a result of elevated esterase activity.

Neem has been suggested for use in maintaining beneficial insect levels, as predators and parasitoids are not exposed directly to the insecticide on the plant and are not harmed by it, as shown in the brown citrus aphid and its parasitoid *Lysiphlebus testaceipes* (Tang *et al.*, 2002) and the glasshouse whitefly *Trialeurodes vaporariorum* and its paraitoid *Encarsia formosa* (Simmonds *et al.*, 2002). However, the indirect exposure of predators or parasitoids by feeding on or developing within prey exposed to azadirachtin may result in classical IGR effects on the beneficial if concentrations are high (Beckage *et al.*, 1988; Osman and Bradley, 1993; Raguraman and Singh, 1998; Qi *et al.*, 2001). Overall, however, if concentrations of neem insecticides are kept to the recommended levels no harm should ensue, for example, for the diamondback moth *Plutella xylostella* no adverse effects were apparent on the survival and foraging behavior of *Diadegma* parasitoids (Akol *et al.*, 2002). There is evidence that larval honeybees have some resistance to azadirachtin in neem, and may also be safe from its effects by a lack of translocation back to the hive in nectar or pollen (Naumann and Isman, 1996).

5.4.3. Toxicity to Nontarget Organisms

Azadirachtin has displayed remarkable selectivity in toxicity studies, consistently showing no toxicity to vertebrates until very high concentrations are used. Investigations into mammalian toxicology of pure azadirachtin demonstrated a "no-observed-effect level" of $1500\,mg\,kg^{-1}\,day^{-1}$ (the highest quantity tested) when azadirachtin was administered orally to rats for 90 days (Raizada *et al.*, 2001). In addition to whole-animal toxicological data, a number of *in vitro* assays of the effects of azadirachtin on several mammalian cell lines have established that the compound is not effective against mammalian cells (measured as effects on proliferation, direct toxicity, and morphological alterations, e.g., micronucleation) until concentrations above $10\,\mu M$ are used (Reed and Majumder, 1998; Akudugu *et al.*, 2001; Salehzadeh *et al.*, 2002; Robertson, 2004) (see Section 5.7.5).

Similarly with plants, phytotoxicity has been demonstrated particularly at the seedling stage, however, again not until high concentrations are used. Using tobacco seedlings Nisbet *et al.* (1996a) demonstrated toxic effects seen as short-term wilting and long-term deformation at 500 ppm pure azadirachtin. No effects on seedlings were seen at lower doses (1–200 ppm). Using turnip and barley seedlings no significant phytotoxicity was seen with 1–200 ppm azadirachtin (Stamatis, Allen, and Mordue (Luntz), unpublished data), although the dry weight of turnip seedlings 14 days post-treatment with 200 ppm azadirachtin was significantly lower than controls. Cell suspension cultures of *Agrostis stolonifera* when treated with azadirachtin at 150, 250, 500, and 1000 ppm during exponential growth showed significant loss in cell viability (measured as % viable cells in $10\,\mu l$ cell suspension aliquots using fluorescein diacetate) within 2 h of treatment. The EC_{50} (concentration at which 50% of cells lost viability) at 2 h post-treatment was approximately 300 ppm ($400\,\mu M$) (Stamatis, Zounos, Allen, and Mordue (Luntz), unpublished data).

5.4.4. Resistance

One of many advantages of the azadirachtin group of compounds in practical use is that development of resistance, which has become such a problem with single-compound synthetic pesticides, is delayed by use of the mixture. Resistance and desensitization can be avoided with the mixture of neem triterpenoids. Feng and Isman (1995) showed that the peach potato aphid *Myzus persicae* developed resistance to pure azadirachtin over 40 generations, but the same did not happen with neem seed extract. In another study with *Spodoptera litura* larvae on cabbage, the larvae became desensitized to pure azadirachtin in both choice and nonchoice tests, but did not desensitize to the seed extract, even though the latter contained the same amount of azadirachtin (Bomford and Isman, 1996).

5.4.5. Systemic Action

The discovery by Gill and Lewis (1971) that azadirachtin was effective systemically was largely ignored for some years, but recently, that valuable aspect of its properties has again received attention. The uptake of azadirachtin from solution by cabbage leaves was demonstrated with the cabbage white butterfly *Pieris brassicae* (Arpaia and van Loon, 1993). Tobacco (*Nicotiana clevelandii*) seedlings also absorbed azadirachtin from solution through their roots and its subsequent translocation to leaves caused disturbances in the feeding behavior of the aphid *M. persicae*, resulting in a reduced ability of the aphids to acquire phloem-located viruses from infected seedlings (Nisbet *et al.*, 1993, 1996a). Sundaram *et al.* (1995) showed that neem extract watered around the roots of young aspen (*Populus tremuloides*) was taken up within 3 h and translocated to the stem and foliage within 3 days. Soil drenching of bean plants (*Phaseolus vulgaris*) with neem extract controlled the pea leaf miner *Lyriomyza huidobrensis* (Weintraub and Horowitz, 1997). More recently it has been shown that systemic dispersion of neem extract can be used to control *Ips pini* bark beetle in lodgepole pine (*Pinus contorta*) logs (Duthie-Holt *et al.*, 1999). No works appears to have been done to show whether all the azadirachtin-like compounds are translocated, or whether some more readily than others.

5.5. Antifeedant Effects of Azadirachtin

Azadirachtin has marked antifeedant activity against a large number of insect species, its effects being mediated through contact chemoreception (primary antifeedancy) and internal feedback mechanisms (secondary antifeedancy); the latter related to toxic effects on the gut, e.g., enzyme production, cell proliferation, and motility (Timmins and Reynolds, 1992; Nasiruddin and Mordue (Luntz), 1993a; Trumm and Dorn, 2000). Insects vary markedly in their behavioral sensitivity to azadirachtin. The desert locust *Schistocerca gregaria* and many species of Lepidoptera, are among the most sensitive, being deterred by as little as 0.007 ppm (in diets) whereas the Hemiptera and Coleoptera are much less sensitive with EC_{50} values of around 100 ppm or more (e.g., Isman, 1993; Mordue (Luntz) and Blackwell, 1993). Whereas *Schistocerca gregaria* prefers to starve to death rather than ingest azadirachtin, the primary antifeedant effect on aphids of azadirachtin applied

systemically in plants occurs at levels far higher than those causing IGR and sterility effects (Lowery and Isman, 1993; Nisbet *et al.*, 1993, 1994; Mordue (Luntz) *et al.*, 1996). Hematophagous insects such as *Culex* mosquitoes are also less sensitive to the antifeedant effects of azadirachtin than to its IGR effects (Su and Mulla, 1998, 1999).

Whereas azadirachtin alone shows toxic IGR and sterilant actions against insects, several compounds in the plant biosynthetic pathway leading to azadirachtin have antifeedant effects against phytophagous insects (Aerts and Mordue (Luntz), 1997). The antifeedant role is an important phenomenon in the overall toxic effects of azadirachtin in those insects that are sensitive to it behaviorally and this phenomenon has been utilized in bioassays to explore the structure–activity relationships of the azadirachtin molecule. Both the decalin and the dihydrofuranacetal fragments of azadirachtin are important in eliciting antifeedant activity. Methylation of the hydroxy substitutions on the molecule and the addition of bulky groups to the dihydrofuran ring decrease antifeedant activity (Blaney *et al.*, 1990; Blaney and Simmonds, 1994; Govindachari *et al.*, 1995; Simmonds *et al.*, 1995).

The primary behavioral antifeedant response of insects to azadirachtin is mediated via neural input from the contact chemoreceptors. Inhibition of feeding behavior results from stimulation of deterrent receptors by azadirachtin often coupled with an inhibition of sugar receptors (Simmonds and Blaney, 1984). In *Spodoptera littoralis*, *Helicoverpa armigera*, and *H. assulta*, cells sensitive to sucrose and azadirachtin are mainly in the lateral sensillum styloconicum with azadirachtin evoking high impulse discharges (Simmonds *et al.*, 1995; Tang *et al.*, 2000). The firing of sucrose sensitive cells is reduced in the presence of azadirachtin and in some species the firing of the deterrent cells to azadirachtin is also reduced in the presence of sucrose (Simmonds and Blaney, 1996). This phenomenon, known as peripheral interaction, varies among insects and occurs in *S. littoralis* and *S. gregaria* but not in *L. migratoria* (Simmonds and Blaney, 1996). Investigations of the antifeedant mode of action by both electrophysiological recordings and behavioral analysis have shown that both polyphagous and oligophagous insects are behaviorally responsive to azadirachtin with the most responsive species being able to differentiate extremely small changes in the parent molecule (Blaney *et al.*, 1990, 1994; Nasiruddin and Mordue (Luntz), 1993b; Simmonds *et al.*, 1995). In Lepidoptera the antifeedant response is also correlated with increased neural activity of the chemoreceptors. A comparison of antifeedant effects with physiological effects as measured by electrophysiological responses of taste receptors in lepidopteran larvae, antifeedant behavioral responses in locusts, and [^{3}H]dihydroazadirachtin-binding studies using insect cell lines and homogenates of locust testes is interesting. Results show that all of the effects are linked in terms of the potency of synthetic analogs of azadirachtin. This suggest a causal link between specific binding to membrane proteins and the ability of the azadirachtin molecule to exert biological effects (Mordue (Luntz) et al., 1998).

5.6. Neuroendocrine Effects of Azadirachtin

5.6.1. Insect Growth Regulation

Over 400 species of insect to date have been shown to be susceptible to the neuroendocrine effects of azadirachtin, exhibiting the now classic symptoms of abnormal molts, larval–adult intermediates, mortality at ecdysis, and delayed molts, often resulting in greatly extended instar lengths. Many papers on a wide variety of insect taxa have allowed detailed descriptions of symptoms to be recorded that show consistency between all insects tested and an ED$_{50}$ (effective dose for 50% of the population) of between 1 and 4 μg g^{-1} bodyweight (Koul and Isman, 1991; Mordue (Luntz) and Blackwell, 1993; Mordue (Luntz) et al., 1995; Riba et al., 2003). In O. fasciatus, Epilachna varivestis, and L. migratoria complete molt inhibition has been shown, by radioimmunoassay, to be due to a blockage of ecdysteroid synthesis and release, a delay in appearance of the last ecdysteroid peak, and a slow, abnormal decline in this peak (Redfern et al., 1982; Sieber and Rembold, 1983; Schlüter et al., 1985; Mordue (Luntz) et al., 1986). In Rhodnius prolixus a single dose of azadirachtin reduced hemolymph ecdysteroid titers to levels too low for induction of ecdysis (Garcia et al., 1990). This blockage of ecdysteroid production is not due to a direct action of azadirachtin on the prothoracic glands (Koolman et al., 1988) but is caused by two specific actions of azadirachtin. These actions are in the cascade of events leading from brain activation and competency to molt to 20-hydroxyecdysone (20E) production in the hemolymph and its effects on target tissues. First, the release of the morphogenetic brain peptide prothoracicotropic hormone (PTTH) responsible for inducing synthesis and release of ecdysone from the prothoracic glands is blocked in azadirachtin-treated insects. In vitro cultures of brain gland complexes or prothoracic glands of Calliphora vicinia, Bombyx mori, Heliothis virescens, or Manduca sexta in the presence of azadirachtin and PTTH will produce and release normal levels of ecdysone (Bidmon et al., 1987; Pener et al., 1988; Barnby and Klocke, 1990). However, decapitation and ligation experiments in B. mori, R. prolixus, and Peridroma saucia have shown that release of PTTH is deficient in azadirachtin-treated insects (Koul et al., 1987; Garcia et al., 1990; Koul and Isman, 1991). This deficiency is due to effects on the corpus cardiacum, in terms of release of PTTH and indeed other peptide hormones related to molting such as eclosion hormone or bursicon (Sieber and Rembold, 1983; Mordue (Luntz) et al., 1986; Bidmon et al., 1987; Rembold et al., 1989). Staining of neurosecretion with paraldehyde fuchsin revealed a build-up of material in the corpora cardiaca and neuropilar areas of the brain neurosecretory system in L. migratoria (Subrahmanyam et al., 1989).

Experiments on L. migratoria with [^{3}H]dihydroazadirachtin revealed heavy labeling within the corpora cardiaca but not in the brain beyond the neurilemma barrier membrane (Subrahmanyam et al., 1989). This suggests, that azadirachtin directly affects the corpora cardiaca by inhibiting the release of hormones, perhaps by operating via effects on neurotransmitters such as acetylcholine, γ-aminobutyric acid (GABA), serotonin, and octopamine (Bidmon et al., 1987; Käuser et al., 1987; Koolman et al., 1988; Banerjee and Rembold, 1992). Also through that axis, azadirachtin has been shown to slow down the rate of synthesis and transport of PTTH by the brain neurosecretory cells, for example in H. virescens (Barnaby and Klocke, 1990). Additionally, studies on the ultrastructure of the ring complex of Lucilia cuprina after azadirachtin treatment revealed degenerative structural changes of the nuclei of all endocrine glands (prothoracic glands, corpus allatum, corpus cardicum), suggesting that effects at the transcriptional–translational level were the key to the generalized disruption of neuroendocrine function (Meurant et al., 1994).

Second, in addition to the effect of azadirachtin on PTTH synthesis, transport, and release, azadirachtin has direct effects on the production of ecdysone 20-monooxygenase from the midgut and fat body; the insect cytochrome P450-dependent hydroxylase responsible for the conversion of ecdysone to its more active metabolite, 20E. In C. vicinia, Drosophila melanogaster, Aedes aegypti, and M. sexta, 20-monooxygenase levels were affected within 1 h of azadirachtin exposure at levels varying between 10^{-4} and 10^{-5} M azadirachtin

(Bidmon *et al.*, 1987; Smith and Mitchell, 1988; Mitchell *et al.*, 1997). Such effects are also clearly demonstrated in *Tenebrio molitor* pupae where ecdysone hemolymph levels were shown to remain relatively constant after azadirachtin treatment (1 μg per insect), whereas there was a highly significant depletion of immunoreactive ecdysteroids (Marco *et al.*, 1990). Variations in successful recovery of development by treatment with exogenous 20E (Pener *et al.*, 1988; Barnby and Klocke, 1990; Marco *et al.*, 1990) support the fact that different target sites for azadirachtin exist and are manifested during growth and molting processes at different times.

In parallel with molting hormone deficiencies in azadirachtin-treated insects, changes in JH also play a significant role in the overall syndrome of azadirachtin poisoning at the molt. That JH hemolymph titers are reduced or delayed is demonstrated in a variety of ways. Azadirachtin inhibits JH-stimulated supernumerary molts in chilled *Galleria mellonella* larvae (Malczewska *et al.*, 1988) perhaps by a blockage of synthesis and release of JH. In the last instar larvae of *M. sexta*, azadirachtin causes supernumerary molts by delaying and extending the JH peak into the critical period for commitment to larval rather than pupal cuticle (Schlüter *et al.*, 1985; Beckage *et al.*, 1988). The importance of timing of hemolymph JH levels with ecdysteroid levels is demonstrated by the selective destruction of larval crochets or hooked setae on the prolegs in *M. sexta* (Reynolds and Wing, 1986; Beckage *et al.*, 1988). The epidermal cells producing such structures require the presence of both ecdysteroids and JH in order to make cuticle and in the absence of JH (e.g., on day 2 of the last larval instar) (Riddiford, 1981) the epidermis loses this ability. Other examples of an imbalance in the presence or absence of ecdysteroid and JH levels are seen in the occurrence of uneverted tanned pupal wing discs in *E. varivestis* (Schlüter, 1987), effects linked with high JH titers such as green hemolymph, brown and green cuticle, lack of black pigment, and supernumerary molts in locusts, *S. littoralis*, and *G. mellonella* (Schmutterer and Freres, 1990; Nicol and Schmutterer, 1991; von Friesewinkel and Schmutterer, 1991; Gelbic and Némec, 2001) and cuticular melanization resulting in black spots related to the absence of JH and low levels of ecdysteroids (Hori *et al.*, 1984; Koul *et al.*, 1987; Malczewska *et al.*, 1988).

It is of interest that azadirachtin potentiates the action of the ecdysteroid agonist RH-2485 in *S. littoralis* (Adel and Sehnal, 2000). There was no effect of azadirachtin on the hyperecdysonic action of high concentrations of diacylhydrazine RH-2485 to induce precocious molting (immediate and fatal molts) thus substantiating the view that azadirachtin does not bind to epidermal ecdysteroid receptors (Koolman *et al.*, 1988). However, azadirachtin was shown to potentiate supernumerary molts in early instar larvae of *S. littoralis* treated with low concentrations of RH-2485, perhaps explained by a slow decline in hemolymph JH levels. Azadirachtin also shows synergy with RH-2485 with regard to deaths at the molt and sterility (Adel and Sehnal, 2000).

5.6.2. Reproduction

Adverse effects of azadirachtin on ovarian and testes development, fecundity, and fertility are an important component of overall toxicity and insect control (Karnavar, 1987; Mordue (Luntz) and Blackwell, 1993; Mordue (Luntz), 2000). Sterility effects in females due to interference with vitellogenin synthesis and its uptake into oocytes has been demonstrated in many insects including *L. migratoria*, *O. fasciatus*, *S. exempta*, and *R. prolixus* (Rembold and Sieber, 1981; Dorn *et al.*, 1987; Feder *et al.*, 1988; Tanzubil and McCaffery, 1990). Reduced fecundity has been recorded in the fruit fly *Ceratitis capitata*, the leaf miner *Liriomyza trifolii*, peach potato aphid *Myzus persicae*, *S. exempta*, *S. littoralis*, and the southern green stink bug, *Nezara viridula* (Parkman and Pienkowski, 1990; Stark *et al.*, 1990; Tanzubil and McCaffrey, 1990; Nisbet *et al.*, 1994; Adel and Sehnal, 2000; Riba *et al.*, 2003); such fertility effects relate to resorption of yolk proteins in developing eggs (Rembold and Sieber, 1981) and lower viability of emerging larvae after treatment of the parental generation (Tanzubil and McCaffery, 1990). In males, fertility reduction is seen in a reduced potency in *O. fasciatus* (Dorn, 1986), a degeneration of spermatocytes in *Mamestra brassicae* (Shimizu, 1988) and blocked cell division in developing spermatocytes in *L. migratoria* leading to significantly smaller mature testes (Linton *et al.*, 1997).

The endocrine events causing reproductive malfunction are very similar to those seen with growth and molting and relate to disruptions in ecdysteroid production by the ovaries and in JH levels in the hemolymph. The latter are most probably caused by alterations to the brain neurosecretory peptides, the allatostatins and allatotropins that control synthesis and release of JH.

Whether production of JH esterase levels, controlling JH catabolism, is affected by azadirachtin is not yet known. In adult female *L. migratoria* ecdysteroid synthesis by the ovaries, together with oogenesis, is inhibited by 10 μg azadirachtin per insect (Rembold and Sieber, 1981), and JH titers are decreased (Rembold *et al.*, 1986), these effects being associated with an accumulation of neurosecretory material in the corpora cardiaca and brain (Subrahmanyam *et al.*, 1989). In *R. prolixus* reduction in oocyte growth and egg production are caused also by reduced ecdysteroid levels in hemolymph and ovaries (Feder *et al.*, 1988; Garcia *et al.*, 1991). In the cotton stainer bug *Dysdercus koenigii* and the peach potato aphid *Myzus persicae* azadirachtin applied topically or in the diet inhibits and delays embryogenesis, resulting either in many of the embryos remaining at a late stage of development or in the viviparous aphid being born dead with undeveloped appendages (Koul, 1984; Nisbet *et al.*, 1994). It is thought that JH control of embryogenesis and the rate of development of embryos in aphids are important here.

In adult female earwigs, *Labidura riparia*, azadirachtin has been shown to have marked cytological effects on ovaries resulting in severely reduced ovarian development. Due to cyclical reproductive changes in *L. riparia* throughout adult life, relating to the production of eggs (vitellogenesis) and a nonvitellogenic period when the female cares for her eggs, there was the possibility of investigating the specific role of azadirachtin in vitellogenesis control. In this insect JH controls both vitellogenin production in the fat body and stimulates vitellogenin uptake by the ovaries. In addition, ovarian ecdysteroids play a role in vitellogenesis and, at the peak of production, oviposition. They also provide a feedback loop to the brain pars lateralis neurosecretory cells causing release of allatostatins, which block JH synthesis and release enabling successive follicle degenerations during the nonvitellogenic period. Azadirachtin treatment blocks vitellogenesis and corpus allatum activity as seen by ultrastructural studies and concomitantly increases allatostatin build-up in the pars lateralis as measured by antibodies to allatostatin 3 (*Blattella germanica* BLAST-3) (Sayah *et al.*, 1996, 1998). This phenomenon together with the fact that JH treatment rescues the effect of azadirachtin on vitellogenesis suggest that azadirachtin affects those peptides controlling corpus allatum activity in a similar manner to the azadirachtin effects on PTTH and hence ecdysone production during the molt.

Two major conclusions can be drawn regarding the mode of action of azadirachtin on neuroendocrine activity in insects during the molt and reproductive development. First, in both molting and sterility, the processes of synthesis, transport, and release of morphogenetic peptide hormones in the brain are at the center of the mode of action of azadirachtin. Second, in the molt (but not yet proved during reproductive development) the production of important enzymes for controlling hemolymph levels of active hormones (20-hydroxy-monooxygenase levels that convert ecdysone to 20E) are inhibited by azadirachtin.

5.7. Studies on the Mode of Action of Azadirachtin

5.7.1. Studies using Insect Cell Lines

The analysis of direct cytological effects of azadirachtin in whole insects, or even in isolated insect tissues, is complicated by the multiplicity of physiological consequences resulting from the compound (see Section 5.6). This complexity has been addressed by analyzing the pharmacology of azadirachtin uptake and binding in cultured *S. frugiperda* Sf9 cells (derived from pupal ovarian tissue) and *D. melanogaster* Kc167 cells (derived from embryonic cells). Such cells rapidly accumulate [^{3}H]azadirachtin from culture medium and are enriched with binding sites compared with many tissue extracts derived from whole insects (Nisbet *et al.*, 1995, 1997; Mordue (Luntz) *et al.*, 1999; Mordue (Luntz), 2002; Robertson, 2004). The effects of azadirachtin on the proliferation and cell cycle events in these cells (Rembold and Annadurai, 1993; Salehzadeh *et al.*, 2002, 2003; Robertson, 2004), accompanied by the above binding studies have given detailed insights into the mode of action of the compound.

Initial studies examining the effects of azadirachtin on the proliferation of Sf9 cells used relatively high concentrations (1 μM) in the culture medium to produce profound effects on cell proliferation (Rembold and Annadurai, 1993; Mordue (Luntz) and Nisbet, 2000). Robertson (2004) has also shown significant reductions in the number of viable cells using the trypan blue exclusion assay within 48 h of treatment with 1 μM azadirachtin for both Sf9 and Kc167 cells. In addition to this effect on cell proliferation, azadirachtin evoked a characteristic cytotoxic effect on cultured cells. Reed and Majumdar (1998) used azadirachtin concentrations of about 0.1 μM in cell culture to demonstrate increases in cell volume, blebbings, blisters, and holes in cell membranes resulting in cytoplasmic extrusions. Since these studies much more refined work has been carried out to discover where in the cell azadirachtin is acting and at what concentration.

5.7.2. Effects on Cell Cycle Events

The effects of azadirachtin in blocking cell division have already been established in whole tissues such as imaginal wing discs of *E. varivestis* (Schlüter,

1985, 1987) and in locust testes, where meiosis was blocked at prometaphase I (Linton *et al.*, 1997). Salehzadeh *et al.* (2002) examined the effects of azadirachtin on the proliferation of Sf9 cells using dehydrogenase activity to estimate cell numbers and found an EC_{50} of 0.15 nM after 96 h incubation. Cell proliferation studies using [^{3}H]thymidine up-take during cell division revealed also an inhibition of mitosis in Sf9 cells 24 h post-treatment with an EC_{50} value of 1.5 nM azadirachtin (Robertson, 2004). This nuclear effect was confirmed by Comet assays that directly measure DNA damage. Significant damage was revealed in Sf9 cells within 6 h of treatment; with an EC_{50} value of 5 nM (Robertson, 2004). The reduction in proliferation was related to a time- and concentration-dependent antimitotic effect of azadirachtin on the cells, with concentrations between 10 nM and 5 μM causing the accumulation of mitotic figures, incomplete mitosis, cytokinesis, and an increased proportion of the cells in the G_2/M phase of the cell cycle (Salehzadeh *et al.*, 2003). The "G_2" (GAP2) and the "M" (mitosis) stages occur after DNA synthesis and are the stages when the chromosomes separate and cytoplasmic division occurs. These events are dependent upon multiple cellular processes, including the synthesis and functioning of microtubules; a number of compounds (e.g., colchicine) interfere with these processes and inhibit mitosis. Azadirachtin and colchicine inhibit the cell cycle at similar points, an observation that encouraged Salehzadeh *et al.* (2003) to examine the ability of azadirachtin to prevent colchicine-flouroscein from binding to cells previously incubated for 4 h with 10 μM azadirachtin. A 45% reduction in fluorescence was recorded in azadirachtin-treated cells. This led to the conclusion that azadirachtin was binding either directly to tubulin or causing conformational changes that prevented colchicine binding (Salehzadeh *et al.*, 2003). While [^{3}H]dihydroazadirachtin is rapidly incorporated by Sf9 cells and binds to the nuclei, it does so in a highly specific, high-affinity, saturable, and essentially irreversible manner (Nisbet *et al.*, 1997). Based on these kinetic data, saturation of the azadirachtin-binding element of Sf9 cells requires much lower concentrations or shorter incubation times than 10 μM for 4 h. Thus, it seems unlikely that colchicine and azadirachtin share the same binding site in Sf9 cells; see Schlüter (1987) for a similar conclusion in wing bud development of *E. varivestis* and Billker *et al.* (2002) in microgametogenesis of *Plasmodium berghei*. An alternative explanation for the reduction in colchicine labeling in azadirachtin-treated cells is that the concentration of azadirachtin used, 10 μM, caused cytotoxic damage and reduced the availability of colchicine-binding sites.

5.7.3. Binding Studies

There is clearly substantial evidence to show that azadirachtin inhibits cell proliferation in Sf9 cells and that it does so by inhibiting mitotic events that rely on the correct synthesis and assembly of microtubules. [^{3}H]dihydroazadirachtin binds to the nuclei of Sf9 cells (Nisbet *et al.*, 1997) and to fractions of homogenates of locust testes with very similar binding properties (Nisbet *et al.*, 1995; Mordue (Luntz) *et al.*, 1999). The binding sites on locust testes were determined by microautoradiography (Nisbet *et al.*, 1996b) and shown to be associated with developing sperm tails, i.e., a region of intense microtubule synthesis. In addition azadirachtin interferes with spermatogenic meiosis in locust testes, arresting meiosis at metaphase I (Linton *et al.*, 1997), an effect that could be due to disruption in microtubule synthesis and assembly. Treatment of homogenates of locus testes with nocodazole, which disrupts polymerized microtubules, or prolonged cold-stirring in buffers to promote microtubule depolymerization failed to reduce the binding of [^{3}H]dihydroazadirachtin (Nisbet *et al.*, unpublished data).

In systems other than insect cell lines there is good evidence that azadirachtin interferes with microtubule function. In the malarial parasite *P. berghei* azadirachtin does not interfere with the assembly of microtubules during microgametogenesis but confocal laser microscopy and transmission electron microscopy have shown that the formation of mitotic spindles and axonemes was disrupted, i.e., the patterning of microtubules into more complex structures (Billker *et al.*, 2002).

In the absence of a good insect cell model for analyzing protein secretion, mammalian cell lines have been used to attempt to understand a possible link between the effects of azadirachtin on protein synthesis, microtubule assembly, and secretory events. Incubation of mouse mammary acini with 0.5 mM azadirachtin reduced overall protein synthesis by about 75% but failed to prevent the secretion of [^{35}S]methionine-labeled proteins in pulse-chase experiments (Nisbet *et al.*, unpublished data). The secretion of proteinaceous vesicles from murine mammary epithelial cells is dependent upon the structural integrity of a microtubular network and the existence of tubulins in their polymerized forms. In contrast to the lack of effect of azadirachtin on protein secretion, colchicine and nocodazole treatments of these cells disrupted microtubule assembly and inhibited protein secretion (Patton, 1974; Rennison *et al.*, 1992). Salehzadeh *et al.* (2003) prepared tubulin from pig brains and repolymerized the protein *in vitro* in the presence of 0.1 mM azadirachtin, but this treatment only reduced polymerization by about 20%. In addition, tubulin and repolymerized microtubules prepared from sheep brains failed to specifically bind [³H]dihydroazadirachtin *in vitro* (Nisbet *et al.*, unpublished data).

5.7.4. Effects on Protein Synthesis

Azadirachtin directly inhibits protein synthesis in a variety of tissues where cells are producing enzymes: midgut cells producing trypsin (Timmins and Reynolds, 1992), midgut and fat body cells producing 20-monooxygenases for ecdysone catabolism (Bidmon *et al.*, 1987; Smith and Mitchell, 1988), and midgut and fat body cells producing detoxification enzymes in insecticide-resistant insects (Lowery and Smirle, 2000). These effects in whole insects are seen also in insect cell lines where both cell division and protein synthesis are inhibited in a dose-dependent manner. Using *S. gregaria* injected with azadirachtin, Annadurai and Rembold (1993) were able to study polypeptide profiles of brain, corpora cardiaca, subesophageal ganglion, and hemolymph using high-resolution two-dimensional gel electrophoresis. Using Sf9 cells Robertson (2004) also carried out polypeptide separation with two-dimensional gels with spot analysis by hierarchical and K means clustering. Both groups concluded that differential effects on protein synthesis occurred although overall protein synthesis was reduced. Robertson (2004) showed that in Sf9 cells out of 381 protein spots 52.2% were decreased significantly in intensity, 28.9% were increased in size, and 18.9% were unchanged.

5.7.5. Studies Using Mammalian Cell Lines

Azadirachtin is not effective against mammalian cells (measured as effects on proliferation, direct toxicity, and morphological alterations, e.g., micronucleation) until concentrations above 10 μM are reached (Reed *et al.*, 1998; Akudugu *et al.*, 2001; Salehzadeh *et al.*, 2002; Robertson, 2004). On cultured rat dorsal root ganglion neurons, neuronal excitability was altered by modulation of potassium conductances, but only after azadirachtin treatment at 10 μM or higher (Scott *et al.*, 1999). In contrast, the EC_{50} for the effect of the compound on the viability of insect cells (Sf9) *in vitro* is several orders of magnitude lower than that for mammalian cell lines, e.g., 150 pM for Sf9 cells, >10 μM for mouse L929 cells (Salehzadeh *et al.*, 2002). Using mammalian cell lines to investigate azadirachtin mode of action it was shown that azadirachtin inhibited

protein synthesis in mouse mammary cells at concentrations above 100 μM but had no effects on protein secretion from these cells (Nisbet *et al.*, unpublished data). Cytosolic concentrations of Ca^{2+} in mouse mammary cells were unaffected by the presence of 500 μM azadirachtin. It is therefore apparent that azadirachtin does not inhibit protein synthesis in mammalian cells as a secondary effect of instantaneous mobilization of calcium stores from the endoplasmic reticulum, as has been demonstrated for other protein synthesis inhibitors, e.g., vasopressin (Reilly *et al.*, 1998), arachidonate, carbachol, or 1,2-bis(*o*-aminophenoxy)ethane-*N*,*N*,*N'*,*N'*-tetraacetic acid tetra(acetoxy-methyl) ester (BAPTA/AM) (Wong *et al.*, 1993). It is also clear from the lack of effect on resting calcium levels or protein secretion that azadirachtin does not induce reductions in protein synthesis through general toxicity or structural cellular damage and that the molecule may interfere with a specific cellular event involved in protein synthesis.

5.7.6. Resolving the Mode of Action of Azadirachtin

It would appear from current knowledge that azadirachtin may have more than one mode of action. First, it alters or prevents the formation of new assemblages of organelles or cytoskeleton resulting in the disruption of cell division, blocked transport and release of neurosecretory peptides, and inhibition of spermatozoa formation. Second, it inhibits protein synthesis in cells that are metabolically active and have been switched on to produce large amounts of protein such as digestive enzymes by midgut cells, cultured cells with high rates of proliferation, or cell production of mixed function oxidases for detoxification of xenobiotics. At the molecular level azadirachtin may act by altering or preventing transcription or translation of proteins expressed at particular stages of the cell cycle. Future work to elucidate this novel mode of action must now concentrate on the characterization and identification of binding sites using proteomic, microarray, and differential display techniques.

The mode of action of azadirachtin is unlikely to be resolved through the use of mammalian cell lines or proteins. The differences between homologous proteins in mammals and insects may in fact be the key to the specificity of azadirachtin. The use of modern protein purification and production methods should assist in resolving this question. The rapidly expanding range of proteomic tools available for synthesizing short peptides or recombinant polypeptides (e.g., phage display) could provide a high-throughput method of determining the amino acid residues and conformation required for azadirachtin binding, giving leads to the identity of the binding protein in insects.

Azadirachtin has profound effects on protein synthesis in insect cells and tissues and its antimitotic actions may result from the absence of proteins involved directly in microtubule assembly and function, or proteins involved in posttranslation modifications of these elements (e.g., protein kinases and phosphatases). Thorough stepwise analysis of the protein profile of cells treated with azadirachtin is possible by two-dimensional electrophoresis, such as in the studies initiated by Robertson (2004). This will provide valuable information on the time course of the effects of azadirachtin on the synthesis of individual proteins and even on the posttranslational modification of proteins. Modern molecular techniques to investigate the effects of azadirachtin on gene transcription will be invaluable to investigate this phenomenon. Differential gene expression in azadirachtin-treated cells at time points after treatment could be investigated using RNA arbitrarily primed-polymerase chain reaction (RAP-PCR) or cDNA subtraction to give an indication of which genes are upregulated or downregulated early in the process of the cytotoxic action of azadirachtin. With the advent of high-throughput expressed sequence tag (EST) sequencing and microarray analysis, EST databases could be generated from untreated insect cell lines, printed on microarrays, and hybridized with cDNA derived from treated and untreated cell lines over a wide range of time points after treatment. Ideally, these time-course analyses would be performed in azadirachtin-susceptible *Drosophila* cell lines (Robertson 2004; Mordue, Dhadialla, and Mordue (Luntz), unpublished data), to take advantage of the availability of the full annotated genome for this organism rather than attempting to determine homology with genes from Sf9 cells for example. Recent work has identified a putative Azadirachtin binding proteins with an apparent molecular mass of 591 kD. The protein was identified, by native polyacrylamide gel electrophoresis ligand overlay blots, in nuclear protein extracts from *Drosophila* KC167 cells using BSA-conjugated azadirachtin as a ligand (S.L. Robertson, T.S. Dhandialla, N. Weiting, S.V. Ley, W. Mordue, A.J. Mordue (Luntz), unpublished data).

References

Adel, M.M., Sehnal, F., **2000**. Azadirachtin potentiates the action of ecdysteroid agonist RH-2485 in *Spodoptera littoralis*. *J. Insect Physiol.* 46, 267–274.

Aerts, R.J., Mordue (Luntz), A.J., **1997**. Feeding deterrence and toxicity of neem triterpenoids. *J. Chem. Ecol.* 23, 2117–2132.

Akhila, A., Rani, K., **1999**. Chemistry of the neem tree (*Azadirachta indica* A. Juss.). *Prog. Chem Org. Nat. Prod. 78*, 47–149.

Akol, A.M., Sithanantham, S., Njagi, P.G.N., Varela, A., Mueke, J.M., **2002**. Relative safety of sprays of two neem insecticides to *Diadegma mollipla* (Holmgren), a parasitoid of the diamondback moth: effects on adult longevity and foraging behaviour. *Crop Protect.* 21, 853–859.

Akudugu, J., Gäde, G., Böhm, L., **2001**. Cytotoxicity of azadirachtin A in human glioblastoma cell lines. *Life Sci.* 68, 1153–1160.

Ambrosino, P., Fresa, R., Fogliano, V., Monti, S.M., Ritieni, A., **1999**. Extraction of azadirachtin A from neemseed kernels by supercritical fluid and its evalution by HPLC and LC/MS. *J. Agric. Food Chem.* 47, 5252–5256.

Annadurai, R.S., Rembold, H., **1993**. Azadirachtin A modulates the tissue specific 2D polypeptide patterns of the desert locust, *Schistocerca gregaria*. *Naturwissenschaften 80*, 127–130.

Arpaia, S., van Loon, J.J.A., **1993**. Effects of azadirachtin after systemic uptake into *Brassica oleracea* on larvae of *Pieris brassicae*. *Entomol. Exp. Applic.* 66, 39–45.

Banerjee, S., Rembold, H., **1992**. Azadirachtin-A interferes with control of serotonin pools in the neuroendocrine system of locusts. *Naturwissenschaften 79*, 81–84.

Barnby, M.A., Klocke, J.A., **1990**. Effects of azadirachtin on levels of ecdysteroids and prothoracicotropic hormone-like activity in *Heliothis virescens* (Fabr.) larvae. *J. Insect Physiol.* 36, 125–131.

Beckage, N.E., Metcalf, J.S., Nielsen, B.D., Nesbit, D.J., **1988**. Disruptive effects of azadirachtin on development of *Cotesia congregata* in host tobacco hornworm larvae. *Arch. Insect Biochem. Physiol.* 9, 47–65.

Bidmon, H.J., **1986**. Ultrastructural changes of the prothorax glands of untreated and with azadirachtin treated *Manduca sexta* larvae (Lepidoptera, Spingidae). *Entomol. Gen.* 12, 1–17.

Bidmon, H.J., Käuser, G., Möbus, P., Koolman, J., **1987**. Effect of azadirachtin on blowfly larvae and pupae. In: Proceedings of the 3rd International Neem Conference, pp. 253–271.

Billker, O., Shaw, M.K., Jones, I.W., Ley, S.V., Mordue, A.J., et al., **2002**. Azadirachtin disrupts formation of organised microtubule arrays during microgametogenesis of *Plasmodium berghei*. *J. Eukaryot. Microbiol.* 49, 489–497.

Bilton, J.N., Broughton, H.B., Jones, P.S., Ley, S.V., Lidert, Z., et al., **1987**. An X-ray crystallographic, mass spectrometric and NMR study of the limonoid insect antifeedant azadirachtin and related derivatives. *Tetrahedron 43*, 2805–2815.

Blaney, W.M., Simmonds, M.S.J., Ley, S.V., Anderson, J.C., Smith, S.C., et al., **1994**. Effect of azadirachtin-derived decalin (perhydronaphthalene) and dihydrofuranacetal (furo[2,3-*b*]pyran) fragments on the feeding behavior of *Spodoptera littoralis* (perhydronaphtha-lene). *Pest. Sci. 40*, 169–173.

Blaney, W.M., Simmonds, M.S.J., Ley, S.V., Anderson, J.C., Toogood, P.L., **1990**. Antifeedant effects of azadirachtin and structurally related compounds on lepidopterous larvae. *Entomol. Exp. Applic. 55*, 149–160.

Bomford, M.K., Isman, M.B., **1996**. Desensitization of fifth instar *Spodoptera litura* to azadirachtin and neem. *Entomol. Exp. Applic. 81*, 307–313.

Brooks, M.W., Roy, S., Uden, P.C., Vittum, P., **1996**. The extraction and determination of azadirachtin from soil and insects by supercritical fluid extraction. *J. Testing Eval. 24*, 149–153.

Butterworth, J.H., Morgan, E.D., **1968**. Isolation of a substance that suppresses feeding in locusts. *J. Chem. Soc., Chem. Commun.* 23–24.

Butterworth, J.H., Morgan, E.D., **1971**. Investigation of the locust feeding inhibition of the seeds of the neem tree, *Azadirachta indica*. *J. Insect Physiol. 17*, 969–977.

Dai, J.M., Yaylayan, V.A., Raghavan, G.S.V., Pare, J.R., **1999**. Extraction and colorimetric determination of azadirachtin-related limonoids in neem seed kernels. *J. Agric. Food Chem. 47*, 3738–3742.

Dai, J.M., Yaylayan, V.A., Raghavan, G.S.V., Pare, J.R.J., Liu, Z., et al., **2001**. Influence of operating parameters on the use of microwave-assisted process (MAP) for the extraction of azadirachtin-related limonoids from neem (*Azadirachta indica*) under atmospheric pressure conditions. *J. Agric. Food Chem. 49*, 4584–4588.

Deota, P.T., Upadhay, P.R., Patel, K.B., Mehta, K.J., Kamath, B.V., et al., **2000**. Estimation and isolation of azadirachtin from neem *Azadirachta indica* A. Juss. seed kernels using high performance liquid chromatography. *J. Liqu. Chromatogr. 23*, 2225–2235.

Dorn, A., **1986**. Effects of azadirachtin on reproduction and egg development of the heteropteran *Oncopeltus fasciatus* Dallas. *J. Appl. Entomol. 102*, 313–319.

Dorn, A., Rademacher, J.M., Sehn, E., **1986**. Effects of azadirachtin on the moulting cycle, endocrine system and ovaries in last-instar larvae of the milkweed bug, *Oncopeltus fasciatus*. *J. Insect Physiol. 32*, 321–328.

Dorn, A., Rademacher, J.M., Sehn, E., **1987**. Effects of azadirachtin on reproductive organs and fertility in the large milkweed bug, *Oncopeltus fasciatus*. In: Proceedings of the 3rd International Neem Conference, pp. 273–288.

Durand-Reville, T., Gobbi, L.B., Gray, B.L., Ley, S.V., Scott, J.S., **2002**. Highly selective entry to the azadirachtin skeleton via a Claisen rearrangement/radical cyclization sequence. *Org. Lett. 4*, 3847–3850.

Duthie-Holt, M.A., Borden, J.H., Rankin, L.J., **1999**. Translocation and efficiency of a neem-based insecticide in lodgepole pine using *Ips pini* (Coleoptera: Scolytidae) as an indicator species. *J. Econ. Entomol. 92*, 180–186.

Feder, D., Valle, D., Rembold, H., Garcia, E.S., **1988**. Azadirachtin-induced sterilization in mature females of *Rhodnius prolixus*. *Z. Naturforsch. C 43*, 908–913.

Feng, R., Isman, M.B., **1995**. Selection for resistance to azadirachtin in the green peach aphid *Myzus persicae*. *Experientia 51*, 831–833.

Fukuzaki, T., Kobayashi, S., Hibi, T., Ikuma, Y., Ishihara, J., et al., **2002**. Studies aimed at the total synthesis of azadirachtin: a modeled connection of C-8 and C-14 in azadirachtin. *Org. Lett. 4*, 2877–2880.

Garcia, E.S., Gonzales, M.S., Azambuja, P., **1991**. Effects of azadirachtin in *Rhodnius prolixus*: data and hypotheses. *Mem. Inst. Oswaldo Cruz, Rio de Janiero 86*, 107–111.

Garcia, E.S., Luz, N., Azambuja, P., Rembold, H., **1990**. Azadirachtin depresses the release of prothoracicotropic hormone in *Rhodnius prolixus* larvae: evidence from head transplantations. *J. Insect Physiol. 36*, 679–682.

Gelbic, I., Némec, V., **2001**. Developmental changes caused by metyrapone and azadirachtin in *Spodoptera littoralis* (Boisd.) (Lep., Noctuidae) and *Galleria mellonella* (L.) (Lep., Pyralidae). *J. Appl. Entomol. 125*, 417–422.

Gill, J.S., Lewis, C.T., **1971**. Systemic action of an insect feeding deterrent. *Nature 232*, 402–403.

Govindachari, T.R., Gopalakrishnan, G., Suresh, G., **1996**. Isolation of various azadirachtins from neem oil by preparative high performance liquid chromatography. *J. Liqu. Chromatogr. 19*, 1729–1733.

Govindachari, T.R., Narasimhan, N.S., Suresh, G., Partho, P.D., Gopalakrishnan, G., et al., **1995**. Structure-related insect antifeedant and growth-regulating activities of some limonoids. *J. Chem. Ecol. 21*, 1585–1600.

Govindachari, T.R., Sandhya, G., Ganeshraj, S.P., **1990**. Simple method for the isolation of azadirachtin by preparative high-performance liquid chromatography. *J. Chromatogr. 513*, 389–391.

Haskell, P.T., Mordue (Luntz), A.J., **1969**. The role of mouthpart receptors in the feeding behaviour *Schistocerca gregaria*. *Entomol. Exp. Applic. 12*, 423–440.

Hemalatha, K., Venugopal, N.B.K., Rao, B.S., **2001**. Determination of azadirachtin in agricultural matrixes and commercial formulations by enzyme-linked immunosorbent assay. *J. AOAC Int. 84*, 1001–1010.

Hori, M., Hiruma, K., Riddiford, L. M., **1984**. Cuticular melanization in the tobacco hornworm larva. *Insect Biochem. 14*, 267–274.

Immaraju, J.A., **1998**. The commercial use of azadirachtin and its integration into viable pest control programmes. *Pest. Sci. 54*, 285–289.

Isman, M.B., **1993**. Growth-inhibitory and antifeedant effects of azadirachtin on six noctuids of regional economic importance. *Pest. Sci. 38*, 57–63.

Isman, M.B., **1997**. Neem and other botanical insecticides: barriers to commercialization. *Phytoparasitica 25*, 339–344.

Isman, M.B., **2004**. Factors limiting commercial success of neem insecticides in North America and Western Europe. In: Koul, O., Wahab, S. (Eds.), Neem Today and in the New Millennium. Kluwer, London, pp. 33–41.

Jarvis, A.P., Johnson, S., Morgan, E.D., 1998. Stability of the natural insecticide azadirachtin in aqueous and organic solvents. *Pest. Sci.* 53, 217–222.

Jarvis, A.P., Morgan, E.D., Edwards, C., 1999. Rapid separation of triterpenoids from neem seed extracts. *Phytochem. Anal.* 10, 39–43.

Johnson, S., Morgan, E.D., 1997a. Comparison of chromatographic systems for triterpenoids from neem (*Azadirachta indica*) seeds. *J. Chromatogr. A* 761, 53–63.

Johnson, S., Morgan, E.D., 1997b. Supercritical fluid extraction of oil and triterpenoids from neem seeds. *Phytochem. Anal.* 8, 228–232.

Johnson, S., Morgan, E.D., Peiris, C.J., 1996. Development of the major triterpenoids and oil in the fruit and seeds of neem (*Azadirachta indica*). *Ann. Bot.* 78, 383–388.

Kalinowski, H.O., Krack, C., Ermel, K., Chriathamjaree, C., 1997. Isolation and characterization of 1-tigloyl-3-acetylazadirachtol from the seed kernels of the Thai neem *Azadirachta siamensis* Valeton. *Z. Naturforsch. B-J. Chem Sci.* 52, 1413–1417.

Karnavar, G.K., 1987. Influence of azadirachtin on insect nutrition and reproduction. *Proc. Indian Acad. Sci.* 96, 341–347.

Kaüser, G., Brandtner, H.M., Bidmon, H.J., Koolman, J., 1987. Ecdysone synthesis and release by the brain–ring gland complex of blowfly larvae. *J. Insect Physiol.* 34, 563–569.

Koolman, J., Bidmon, H.J., Lehmann, M., Kaüser, G., 1988. On the mode of action of azadirachtin in blowfly larvae and pupae. In: Sehnal, F., Zabza, A., Denlinger, D.L. (Eds.), Endocrinological Fronteirs in Physiological Insect Ecology, vol. 1. Wrocław Technical University Press, Wrocław, pp. 55–67.

Koul, O., 1984. Azadirachtin. 1. Interaction with the development of red cotton bugs. *Entomol. Exp. Applic.* 36, 85–88.

Koul, O., Amanai, K., Ohtaki, T., 1987. Effect of azadirachtin on the endocrine events of *Bombyx mori*. *J. Insect Physiol.* 33, 103–108.

Koul, O., Isman, M.B., 1991. Effects of azadirachtin on the dietary utilization and development of the variegated cutworm *Peridroma saucia*. *J. Insect Physiol.* 37, 591–598.

Kraus, W., 2002. Azadirachtin and other triterpenoids. In: Schmutterer, H. (Ed.), The Neem Tree, 2nd edn. Neem Foundation, Mumbai, pp. 39–111.

Kraus, W., Bokel, M., Bruhn, A., Cramer, A., Klaiber, I., et al., 1987. Structure determination by nmr of azadirachtin and related compounds from *Azadirachta indica*. *Tetrahedron Lett.* 26, 6435–6438.

Larson, R.O., 1985. Stable anti-pest neem seed extract. US Patent No. 4,556,562.

Lavie, D., Jain, M.K., Shpan-Gabrielith, S.R., 1967. A locust phago-repellent from two *Melia* species. *J. Chem. Soc., Chem. Commun.* 910–911.

Lee, S.M., Klocke, J.A., 1987. Combined Florisil, droplet counter-current and high performance liquid chromatography for preparative isolation and purification of azadirachtin from neem (*Azadirachta indica*) seeds. *J. Liqu. Chromatogr.* 10, 1151–1163.

Ley, S.V., Denholm, A.A., Wood, A., 1993. The chemistry of azadirachtin. *Nat. Prod. Rep.* 10, 109–157.

Linton, Y.M., Nisbet, A.J., Mordue (Luntz), A.J., 1997. The effects of azadirachtin on the testes of the desert locust, *Schistocerca gregaria*. *J. Insect Physiol.* 43, 1077–1084.

Lowery, D.T., Isman, M.B., 1993. Antifeedant activity of extracts from neem, *Azadirachta indica*, to strawberry aphid, *Chaetosiphon fragaefolii*. *J. Chem. Ecol.* 19, 1761–1773.

Lowery, D.T., Smirle, M.J., 2000. Toxicity of insecticides to obliquebanded leafroller, *Choristoneura rosaceana*, larvae and adults exposed previously to neem seed oil. *Entomol. Exp. Applic.* 95, 201–207.

Luo, X.-D., Ma, Y.-B., Wu, S.-H., Wu, D.-G., 1999. Two novel azadirachtin derivatives from *Azadirachta indica*. *J. Nat. Prod.* 62, 1022–1024.

Malathi, R., Kabaleeswaran, V., Rajan, S.S., Gopalakrishnan, G., Govindachari, T.R., 2003. Structure of azadirachtin-I, 11-β epimer. *J. Chem. Crystallogr.* 33, 229–232.

Malczewska, M., Gelman, D.B., Cymborowski, B., 1988. Effect of azadirachtin on development, juvenile hormone and ecdysteroid titres in chilled *Galleria mellonella* larvae. *J. Insect Physiol.* 34, 725–732.

Marco, M-P., Pascual, N., Bellés, X., Camps, F., Messeguer, A., 1990. Ecdysteroid depletion by azadirachtin in *Tenebrio molitor* pupae. *Pestic. Biochem. Physiol.* 38, 60–65.

Meurant, K., Sernia, C., Rembold, H., 1994. The effects of azadirachtin-A on the morphology of the ring complex of *Lucilia cuprina* (Wied.) larvae (Diptera, Insecta). *Cell Tissue Res.* 275, 247–254.

Mitchell, M.J., Smith, S.L., Johnson, S., Morgan, E.D., 1997. Effects of neem tree compounds azadirachtin, salannin, nimbin and 6-desacetylnimbin on ecdysone 20-monooxygenase activity. *Arch. Insect Biochem. Physiol.* 35, 199–209.

Mordue (Luntz), A.J., 2002. The cellular actions of azadirachtin. In: Schmutterer, H. (Ed.), The Neem Tree *Azadirachta indica* A. Juss. and other Meliaceous Plants. VCH Verlagsgesellschaft, Weinheim, pp. 266–274.

Mordue (Luntz), A.J., Blackwell, A., 1993. Azadirachtin: an update. *J. Insect Physiol.* 39, 903–924.

Mordue (Luntz), A.J., Evans, K.A., Charlet, M., 1986. Azadirachtin, ecdysteroids and ecdysis in *Locusta migratoria*. *Comp. Biochem. Physiol. C* 85, 297–301.

Mordue (Luntz), A.J., Nisbet, A.J., 2000. Azadirachtin from the neem tree (*Azadirachta indica*): its actions against insects. *Ann. Soc. Entomol. Brasil* 29, 615–632.

Mordue (Luntz), A.J., Nisbet, A.J., Jennens, L., Ley, S.V., Mordue, W., 1999. Tritiated dihydroazadirachtin binding to *Schistocerca gregaria* testes and *Spodoptera* Sf9 cells suggests a similar mechanism of action for azadirachtin. In: Singh, R.P., Saxena, R.C. (Eds.), *Azadirachta indica* A. Juss. IBH Publishing Company Pvt Ltd, New Delhi, pp. 247–258.

Mordue (Luntz), A.J., Nisbet, A.J., Nasiruddin, M., Walker, E., 1996. Differential thresholds of azadirachtin for feeding deterrence and toxicity in locusts and an aphid. *Entomol. Exp. Applic. 80*, 69–72.

Mordue (Luntz), A.J., Simmonds, M.S.J., Ley, S.V., Blaney, W.M., Mordue, W., *et al.*, 1998. Actions of azadirachtin, a plant allelochemical, against insects. *Pest. Sci. 54*, 277–284.

Mordue (Luntz), A.J., Zounos, A., Wickramananda, I.R., Allan, E.J., 1995. Neem tissue culture and the production of insect antifeedant and growth regulatory compounds. *BCPC Symp. Proceed. 63*, 187–194.

Nasiruddin, M., Mordue (Luntz), A.J., 1993a. The effect of azadirachtin on the midgut histology of the locusts, *Schistocerca gregaria* and *Locusta migratoria. Tissue Cell. 25*, 875–884.

Nasiruddin, M., Mordue (Luntz), A.J., 1993b. The protection of barley seedlings from attack by *Schistocerca gregaria* using azadirachtin and related analogs. *Entomol. Exp. Applic. 70*, 247–252.

Naumann, K., Isman, M.B., 1996. Toxicity of a neem (*Azadirachta indica* A. Juss) insecticide to larval honey bees. *Am. Bee J. 136*, 518–520.

Nicol, C.M.Y., Schmutterer, H., 1991. Contact effects of seed oil from the Neem Tree, *Azadirachta indica* (A. Juss), on nymphs of the gregarious phase of the desert locust, *Schistocerca gregaria* (Forsk.). *J. Appl. Entomol. 111*, 197–205.

Nicolaou, K.C., Poecker, A.J., Follmann, M., Baati, R., 2002. Model studies towards azadirachtin. 2. Construction of the crowded C8–C14 bond by transition metal chemistry. *Angew. Chem. Int. Edn. 41*, 2107–2110.

Nisbet, A.J., Mordue (Luntz), A.J., Grossman, R.B., Jennens, L., Ley, S.V., *et al.*, 1997. Characterisation of azadirachtin binding to Sf9 nuclei *in vitro. Arch. Insect Biochem. Physiol. 34*, 461–473.

Nisbet, A.J., Mordue (Luntz), A.J., Mordue, W., 1995. Detection of [22,23-³H]dihydroazadirachtin binding sites on membranes from *Schistocerca gregaria* (Forskal) testes. *Insect Biochem. Mol. Biol. 25*, 551–558.

Nisbet, A.J., Mordue (Luntz), A.J., Mordue, W., Williams, L.M., Hannah, L., 1996b. Autoradiographic localisation of [22,23-³H]dihydroazadirachtin binding sites in desert locust testes. *Tissue Cell 28*, 725–729.

Nisbet, A.J., Woodford, J.A.T., Strang, R.H.C., 1994. The effects of azadirachtin-treated diets on the feeding behaviour and fecundity of the peach potato aphid, *Myzus persicae. Entomol. Exp. Applic. 71*, 65–72.

Nisbet, A.J., Woodford, J.A.T., Strang, R.H.C., 1996a. The effects of azadirachtin on the acquisition and inoculation of potato leafroll virus by *Myzus persicae. Crop Protect. 15*, 9–14.

Nisbet, A.J., Woodford, J.A.T., Strang, R.H.C., Connolly, J.D., 1993. Systemic antifeedant effects of azadirachtin on the peach potato aphid *Myzus persicae. Entomol. Exp. Applic. 68*, 87–98.

Osman, M.Z., Bradley, J., 1993. Effects of neem seed extracts on *Pholeastor (Apantales) glomeratus* L (Hym. Braconidae), a parasitoid of *Pieris brassicae* L. (Lep. Pieridae). *J. Appl. Entomol.–Z. Angew. Entomol. 115*, 259–265.

Parkman, P., Pienkowski, R.L., 1990. Sublethal effects of neem seed extract on adults of *Liriomyza trifolii* (Diptera, Agromyzidae). *J. Econ. Entomol. 83*, 1246–1249.

Patton, S., 1974. Reversible suppression of lactation by colchicine. *FEBS Lett. 48*, 85–87.

Pener, M.P., Rowntree, D.B., Bishoff, S.T., Gilbert, L.I., 1988. Azadirachtin maintains prothoracic gland function but reduces ecdysteroid titres in *Manduca Sexta* pupae: *in vivo* and *in vitro* studies. In: Sehnal, F., Zabza, A., Denlinger, D.L. (Eds.), Endocrinological Frontiers in Physiological Insect Ecology. Wrocław Technical University Press, Wrocław, pp. 41–54.

Pozo, O.J., Marin, J.M., Sancho, J.V., Hernandez, F., 2003. Determination of abamectin and azadirachtin residues in orange samples by liquid chromatography-electrospray tandem mass spectrometry. *J. Chromatogr. A. 992*, 133–140.

Puri, H.S., 1999. Neem the Divine Tree. Harwood Academic, Amsterdam.

Qi, B.Y., Gordon, G., Gimme, W., 2001. Effects of neem-fed prey on the predacious insects *Harmonia conformis* (Boisduval) (Coleoptera: Coccinellidae) and *Mallada signatus* (Schneider) (Neuroptera: Chrysopidae). *Biol. Control 22*, 185–190.

Raguraman, S., Singh, R.P., 1998. Behavioral and physiological effects of neem (*Azadirachta indica*) seed kernel extract on larval parasitoid, *Bracon hebetor. J. Chem. Ecol. 24*, 1241–1250.

Raizada, R.B., Srivastava, M.K., Kaushal, R.A., Singh, R.P., 2001. Azadirachtin, a neem biopesticide: subchronic toxicity assessment in rats. *Food Chem. Toxicol. 39*, 477–483.

Redfern, R.E., Kelly, T.J., Borkovec, A.B., Hayes, D.K., 1982. Ecdysteroid titres and moulting aberrations in last stage *Oncopeltus* nymphs treated with insect growth regulators. *Pest. Biochem. Physiol. 18*, 351–356.

Reed, E., Majumdar, S.K., 1998. Differential cytotoxic effects of azadirachtin on *Spodoptera frugiperda* and mouse cultured cells. *Entomol. Exp. Applic. 89*, 215–221.

Reilly, B.A., Brostrom, M.A., Brostrom, C.O., 1998. Regulation of protein synthesis in ventricular myocytes by vasopressin: the role of sarcoplasmic/endoplasmic reticulum Ca^{2+} stores. *J. Biol. Chem. 273*, 3747–3755.

Rembold, H., Annadurai, R.S., 1993. Azadirachtin inhibits proliferation of Sf9 cells in monolayer culture. *Z. Naturforsch. C 38*, 495–499.

Rembold, H., Sieber, K.P., 1981. Inhibition of oogenesis and ovarian ecdysteroid synthesis by azadirachtin in *Locusta migratoria migratorioides* (R and F). *Z. Naturforsch. C 36*, 466–469.

Rembold, H., Schlagintweit, B., Ulrich, G.M., 1986. Activation of juvenile hormone synthesis *in vitro* by a corpus cardiacum factor from *Locusta migratoria*. *J. Insect Physiol. 32*, 91–94.

Rembold, H., Subrahmanyam, B., Müller, T., **1989**. Corpus cardiacum: a target for azadirachtin. *Experientia 45*, 361–363.

Rennison, M.E., Handel, S.E., Wilde, C.J., Burgoyne, R.D., **1992**. Investigation of the role of microtubules in protein secretion from lactating mouse mammary epithelial cells. *J. Cell Sci. 102*, 239–247.

Reynolds, S.E., Wing, K.D., **1986**. Interactions between azadirachtin and ecdysteroid-dependent systems in the tobacco hornworm, *Manduca sexta*. Abstracts 5th Int. Congr. Pest. Chem., IUPAC, Ottawa, Canada, 2 D/E-08.

Riba, M., Martí, J., Sans, A., **2003**. Influence of azadirachtin on development and reproduction of *Nezara viridula* L. (Het., Pentatomidae). *J. Appl. Entomol. 127*, 37–41.

Riddiford, L.M., **1981**. Hormonal control of epidermal cell development. *Am. Zool. 21*, 751–762.

Robertson, S.L., **2004**. Studies on the mode of action of azadirachtin from the neem tree *Azadirachta indica*, using insect cell lines. PhD thesis. University of Aberdeen, UK.

Ruscoe, C.N.E., **1972**. Growth disrupting effects of an insect antifeedant. *Nature New Biol. 236*, 159–160.

Salehzadeh, A., Akhkha, A., Cushley, W., Adams, R.L.P., Kusel, J.R., *et al.*, **2003**. The antimitotic effect of the neem terpenoid azadirachtin on cultured insect cells. *Insect Biochem. Mol. Biol. 33*, 681–689.

Salehzadeh, A., Jabbar, A., Jennens, L., Ley, S.V., Annadurai, R.S., *et al.*, **2002**. The effects of phytochemical pesticides on the growth of cultured invertebrate and vertebrate cells. *Pest Mgt Sci. 58*, 268–276.

Sayah, F., Fayet, C., Idaomar, M., Karlinsky, A., **1996**. Effect of azadirachtin on vitellogenesis of *Labidura riparia* (Insecta, Dermaptera). *Tissue and Cell 28*, 741–749.

Sayah, F., Idaomar, M., Soranzo, L., Karlinsky, A., **1998**. Endocrine and neuroendocrine effects of azadirachtin in adult females of the earwig *Labidura riparia*. *Tissue and Cell 30*, 86–94.

Schaaf, O., Jarvis, A.P., van der Esch, S.A., Giagnacovo, G., Oldham, N.J., **2000**. Rapid and sensitive analysis of azadirachtin and related triterpenoids from neem (*Azadirachta indica*) by high performance liquid chromatography-atmospheric pressure chemical ionization mass spectrometry. *J. Chromatogr. A 886*, 89–97.

Schlüter, U., **1985**. Occurrence of weight-gain reduction and inhibition of metamorphosis and storage protein-formation in last larval instars of the mexican bean beetle, *Epilachna varivestis*, after injection of azadirachtin. *Entomol. Exp. Applic. 39*, 191–195.

Schlüter, U., **1987**. Effects of azadirachtin on developing tissues of various insect larvae. In: Proceedings of the 3rd International Neem Conference, pp. 331–348.

Schlüter, U., Bidmon, H.J., Grewe, S., **1985**. Azadirachtin affects growth and endocrine events in larvae of the tobacco hornworm, *Manduca sexta*. *J. Insect Physiol. 31*, 773–777.

Schmutterer, H. (Ed.) **2002**. The Neem Tree, 2nd edn. Neem Foundation, Mumbai.

Schmutterer, H., Ascher, K.R.S. (Eds.) **1984**. Natural Pesticides from the Neem Tree (*Azadirachta indica* A. Juss) and Other Tropical Plants. German Agency for Technical Cooperation, Eschborn.

Schmutterer, H., Ascher, K.R.S. (Eds.) **1987**. Natural Pesticides from the Neem Tree and Other Tropical Plants. German Agency for Technical Cooperation, Eschborn.

Schmutterer, H., Ascher, K.R.S., Rembold, H. (Eds.) **1981**. Natural Pesticides from the Neem Tree (*Azadirachta indica* A. Juss). German Agency for Technical Cooperation, Eschborn.

Schmutterer, H., Doll, M., **1993**. The marrango or Philippine neem tree, *Azadirachta excelsa* (= *A integrifolia*): a new source of insecticides with growth-regulating properties. *Phytoparasitica 21*, 79–86.

Schmutterer, H., Freres, T., **1990**. Influence of neem-seed oil on metamorphosis, color and behavior of the desert locust, *Schistocerca gregaria* (Forsk), and of the African migratory locust, *Locusta migratoria migratorioides* (R and F). *Z. Pflanzenkrank. Pflanzenschutz. 97*, 431–438.

Schutz, S., Wengatz, I., Goodrow, M.H., Gee, S.J., Hummel, H.E., *et al.*, **1997**. Development of an enzyme-linked immunosorbent assay for azadirachtins. *J. Agric. Food Chem. 45*, 2363–2368.

Scott, I.M., Kaushik, N.K., **2000**. The toxicity of a neem insecticide to populations of culicidae and other aquatic invertebrates as assessed in *in situ* microcosms. *Arch. Environ. Contam. Toxicol. 39*, 329–336.

Shimizu, T., **1988**. Suppressive effects of azadirachtin on spermiogenesis of the diapausing cabbage armyworm, *Mamestra brassicae, in vitro*. *Entomol. Exp. Applic. 46*, 197–199.

Sieber, K.P., Rembold, H., **1983**. The effects of azadirachtin on the endocrine control of molting in *Locusta migratoria*. *J. Insect Physiol. 29*, 523–527.

Simmonds, M.S.J., Blaney, W.M., **1984**. Some effects of azadirachtin on lepidopterous larvae. In: Proceedings of the 2nd International Neem Conferences, pp. 163–180.

Simmonds, M.S.J., Blaney, W.M., **1996**. Azadirachtin: advances in understanding its activity as an antifeedant. *Entomol. Exp. Applic. 80*, 23–26.

Simmonds, M.S.J., Blaney, W.M., Ley, S.V., Anderson, J.C., Banteli, R., *et al.*, **1995**. Behavioral and neurophysiological responses of *Spodoptera littoralis* to azadirachtin and a range of synthetic analogs. *Entomol. Exp. Applic. 77*, 69–80.

Simmonds, M.S.J., Manlove, J.D., Blaney, W.M., Khambay, B.P.S., **2002**. Effect of selected botanical insecticides on the behavior and mortality of the glasshouse whitefly *Trialeurodes vaporariorum* and the parasitoid *Encarsia formosa*. *Entomol. Exp. Applic. 102*, 39–47.

Smirle, M.J., Lowery, D.T., Zurowski, C.L., **1996**. Influence of neem oil on detoxication enzyme activity in the obliquebanded leafroller, *Choristoneura rosaceana*. *Pest. Biochem. Phys. 56*, 220–230.

Smith, S.L., Mitchell, M.J., 1988. Effects of azadirachtin on insect cytochrome-P450 dependent ecdysone 20-monooxygenase activity. *Biochem. Biophys. Res. Comm. 154*, 559–563.

Stark, J.D., Vargas, R.I., Thalman, R.K., 1990. Azadirachtin: effects on metamorphosis, longevity, and reproduction of three tephritid fruit-fly species (Diptera, Tephritidae). *J. Econ. Entomol. 83*, 2168–2174.

Stark, J.D., Walter, J.F., 1995. Persistence of azadirachtin A and azadirachtin B in soil: effects of temperature and microbial activity. *J. Environ. Sci. Health B 30*, 685–698.

Su, T.Y., Mulla, M.S., 1998. Antifeedancy of neem products containing azadirachtin against *Culex tarsalis* and *Culex quinquefasciatus* (Diptera: Culicidae). *J. Vector Ecol. 23*, 114–122.

Su, T.Y., Mulla, M.S., 1999. Effects of neem products containing azadirachtin on blood feeding, fecundity, and survivorship of *Culex tarsalis* and *Culex quinquefasciatus* (Diptera: Culicidae). *J. Vector Ecol. 24*, 202–215.

Subrahmanyam, B., Müller, T., Rembold, H., 1989. Inhibition of turnover of neurosecretion by azadirachtin in *Locusta migratoria*. *J. Insect Physiol. 35*, 493–497.

Sundaram, K.M.S., 1996. Azadirachtin biopesticide: a review of studies conducted on its analytical chemistry, environmental behaviour and biological effects. *J. Environ. Sci. Health B 31*, 913–948.

Sundaram, K.M.S., Campbell, R., Sloane, L., Studens, J., 1995. Uptake, translocation, persistence and fate of azadirachtin in aspen plants (*Populus tremuloides* Michx) and its effect on pestiferous two-spotted spider-mite (*Tetranychus urticae* Koch). *Crop Protect. 14*, 415–421.

Sundaram, K.M.S., Sundaram, A., Curry, J., Sloane, L., 1997. Formulation selection, and investigation of azadirachtin-A persistence in some terrestrial and aquatic components of a forest environment. *Pest. Sci. 51*, 74–90.

Tang, D.L., Wang, C.Z., Luo, L., Qin, J.D., 2000. Comparative study on the responses of maxillary sensilla styloconica of cotton bollworm *Helicoverpa armigera* and oriental tobacco budworm *H. assulta* larvae to phytochemicals. *Sci. China Series C Life Sci. 43*, 606–612.

Tang, Y.Q., Weathersbee, A.A., Mayer, R.T., 2002. Effect of neem seed extract on the brown citrus aphid (Homoptera: Aphididae) and its parasitoid *Lysiphlebus testaceipes* (Hymenoptera: Aphidiidae). *Environ. Entomol. 31*, 172–176.

Tanzubil, P.B., McCaffery, A.R., 1990. Effects of azadirachtin and aqueous neem seed extracts on survival, growth and development of the African armyworm, *Spodoptera exempta*. *Crop Protect 9*, 383–386.

Timmins, W.A., Reynolds, S.E., 1992. Azadirachtin inhibits secretion of trypsin in midgut of *Manduca sexta* caterpillars: reduced growth due to impaired protein digestion. *Entomol. Exp. Applic. 63*, 47–54.

Trisyono, A., Whalon, M.E., 1999. Toxicity of neem applied alone and in combinations with *Bacillus thuringiensis* to Colorado potato beetle (Coleoptera: Chrysomelidae). *J. Econ. Entomol. 92*, 1281–1288.

Trumm, P., Dorn, A., 2000. Effects of azadirachtin on the regulation of midgut peristalsis by the stomatogastric nervous system in *Locusta migratoria*. *Phytoparasitica 28*, 7–26.

Turner, C.J., Tempesta, M.S., Taylor, R.B., Zagorski, M.G., Termini, J.S., *et al.*, 1987. An NMR spectroscopic study of azadirachtin and its trimethyl ether. *Tetrahedron 43*, 2789–2803.

von Friesewinkel, D.C., Schmutterer, H., 1991. Kontaktwirkung von Niemol bei der afrikanischen Wanderheuschrecke *Locusta migratoria migratorioides* (R&F). *Z. Angew. Zool. 78*, 189–204.

Weintraub, P.G., Horowitz, A.R., 1997. Systemic effects of neem insecticide on *Lyriomyza huidobrensis* larvae. *Phytoparasitica 25*, 283–289.

Wong, W.L., Brostrom, M.A., Kuznetsov, G., Gmitter-Yellen, D., Brostrom, C.O., 1993. Inhibition of protein synthesis and early protein processing by thapsigargin in cultured cells. *Biochem. J. 289*, 71–79.

A5 Addendum: Azadirachtin, A Natural Product in Insect Control: An Update

A J Mordue (Luntz), University of Aberdeen, Aberdeen, UK
E D Morgan, Keele University, Keele, UK
A J Nisbet, Moredun Research Institute, Edinburgh, UK

© 2010 Elsevier B.V. All Rights Reserved

A5.1. Synthesis of Azadirachtin

A milestone in the chemistry of azadirachtin was reached in 2007. Forty one years after the isolation of azadirachtin and twenty one years after the full solution of its chemical structure, its synthesis was completed. This was by no means an indication that synthetic azadirachtin will be available in the future. It is the crowning glory of azadirachtin chemistry, and a remarkable feat for the art and science of synthetic organic chemistry (Jauch, 2008). Steven Ley and his group at Cambridge did it in 71 steps and the overall yield was 0.00015% (Veitch et al., 2007a). The synthesis was accomplished by a relay method (Veitch et al., 2007b). That is, azadirachtin was degraded to a simpler product A (**Figure A1**), which was then reconverted back to azadirachtin in the laboratory. Starting from a simple three-carbon compound, the "left-hand" decalin portion of the molecule was built up with an array of protecting groups that could be removed later. The "right-hand" portion was similarly constructed, starting from galactose. These two portions proved impossible to join by any known chemical method. A different "right-hand" molecule was constructed, and finally attached to the "left-hand" part, and in eight more steps the relay compound A was produced, completing the synthesis (**Figure A1**). The importance of azadirachtin chemically and biologically meant that several groups in other places were hard at work on the synthesis at the same time. The successful route enabled the Ley group to make five other compounds closely related to azadirachtin, all of which will be valuable on the way to understanding structure–activity relationships in this group.

A5.2. Emerging Technologies for Understanding Mode of Action

The mode of action of azadirachtin still requires resolution. A recent study showed azadirachtin promoted caspase-independent apoptosis in wing and eye imaginal discs when fed to larval *Drosophila melanogaster* and was associated with actin cytoskeletal depletion (Anuradha et al., 2007). These observations, accompanied by *in silico* modeling of azadirachtin binding to β actin suggest that azadirachtin exerts its effects via direct binding to cytoskeletal actin (Pravin Kumar et al., 2007). *In vitro* evidence from *Drosophila* cells suggests that azadirachtin binds directly to a protein complex, which contains a chaperone protein, HSP 60 (Robertson et al., 2007). Other studies point to tubulin polymerization as the biological target of the mode of action (reviewed in Mordue (Luntz) et al., 2005). From these studies it has therefore been impossible to attach a single unequivocal biochemical target for azadirachtin in insect cells and this may indeed point to promiscuity in regard to the biochemical target(s) of the molecule.

Current knowledge from these *in vivo*, *in vitro* and *in silico* models needs to be refined using a "systems biology approach" to clarify the mode of action. Using azadirachtin and the new analogues (see above) in a genetically-tractable model species should lead to a deeper understanding of its mode of action at the molecular level. Compound-centric chemical proteomics (reviewed by Rix and Superti-Furga, 2009) – a postgenomic variant on the classical drug-affinity studies previously performed with azadirachtin – will allow identification of proteins

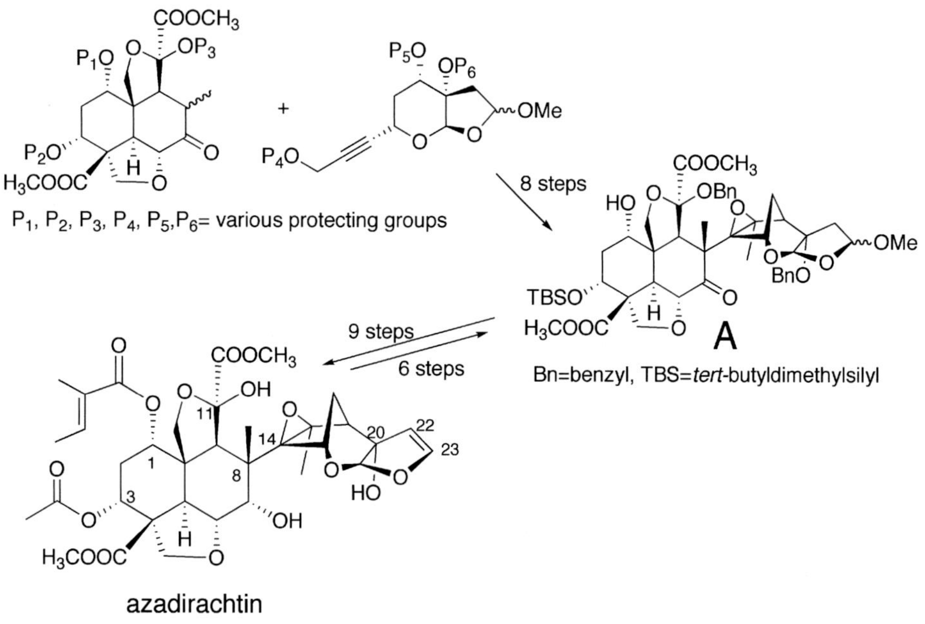

Figure A1 Summary of the key steps in the synthesis of intermediate A, which was also produced by degradation from azadirachtin, and has been converted back into azadirachtin. The sum of all the steps from simple intermediates to azadirachtin have thus been achieved.

which interact directly with azadirachtin at its cellular target. Two recent advances are central to this approach: (1) the total synthesis of azadirachtin (Veitch et al., 2007a) will allow the introduction of (for example) photoaffinity tags to permit covalent bonding between azadirachtin and its binding protein(s); (2) the introduction of linkers that do not affect azadirachtin:protein interaction but permit the purification of binding protein(s) via a ligand–linker–bead matrix; and (3) the availability of affordable "ultra-deep sequencing" allowing both genomic and transcriptomic resources to accurately identify target proteins arising from the proteomic analysis from model and target insects.

The availability of ultra-deep sequence data, particularly transcriptome data from insect tissues and cell lines, is also vital for gene expression analyses (Nielsen et al., 2008). This in conjunction with integrated bioinformatics will allow identification of individual biochemical and metabolic pathways affected by azadirachtin at timepoints immediately after exposure. Collectively such studies could finally provide links between target site binding and biological effects.

A5.3. Commercial Use of Neem Seed Insecticides

Papers continue to appear on various aspects of using azadirachtin as a pesticide. However, neem seed insecticides are still of minor importance in agriculture. Their unjustified reputation for low stability, though azadirachtin and its cogeners are more stable than pyrethrum (Morgan, 2009), high production costs of standardized neem products, regulatory issues as mixtures of natural products and the influx of a new generation of "reduced risk" neurotoxins (eg neonicotinoids), new chemical classes (diamides, ketoenols) and new microbial insecticides (spinosyns and semisynthetic spinosoids) has held the use of neem insecticides to niche markets. Neem insecticides are now established in organic agriculture, public health (mosquito, cockroach control), and home and garden use (Isman, 2006, 2008). However registration is limited in the western world with only the USA and some European countries allowing its use. Research is still ongoing to establish neem insecticides within integrated pest management (IPM) strategies for crop protection and to increase use against arthropods of medical and veterinary importance. Its role in crop protection, for resource poor farmers in developing countries where local products may be used, is still of paramount importance.

A5.4. Conclusions

With knowledge gained in the synthesis of azadirachtin and using a systems biology approach it should be possible to elucidate cellular and molecular mechanisms of azadirachtin action. Identification of this new target site would allow synthesis of simpler novel chemicals as insecticides. Given that more than 80% of global insecticide sales today utilize four different target sites all involving neurotransmission, there should be interest in pursuing the final goal of synthetic ligands at the azadirachtin target site. Only then will the true potential of this novel insecticide be realized as a safe, environmentally sound pesticide for crop protection in its widest sense.

References

Anuradha, A., Annadurai, R.S., Shashidhara, L.S., 2007. Actin cytoskeleton as a putative target of the neem limonoid Azadirachtin A. *Insect Biochem. Mol. Biol* 37, 627–634.

Jauch, J., 2008. Natural product synthesis – total synthesis of azadirachtin – finally completed after 22 years. *Angew. Chem. Int. Ed. 47*, 34–37.

Isman, M.B., 2006. Botanical insecticides, deterrents, and repellents in modern agriculture and an increasingly regulated world. *Annu. Rev. Entomol. 51*, 45–66.

Isman, M.B., 2008. Botanical insecticides: for richer, for poorer. *Pest Manage. Sci. 64*, 8–11.

Mordue (Luntz), A.J., Morgan, E.D., Nisbet, A.J., 2005. Azadirachtin, a natural product in insect control.

In: Gilbert, L.I, Iatrou, K., Gill, S.S. (Eds.), Comprehensive Molecular Insect Science, Vol. 6. Elsevier, Amsterdam, pp. 117–135.

Morgan, E.D., 2009. Azadirachtin, a scientific gold mine. *Bioorg. Med. Chem. 17*, 4096–4105.

Nielsen, K.L., Petersen, A.H., Emmersen, J., 2008. Deep-SAGE: Tag-based transcriptome analysis beyond microarrays. In: Janitz, M. (Ed.), Next Generation Genome Sequencing. Wiley-VCH, New York.

Pravin Kumar, R., Manoj, M.N., Kush, A., Annadurai, R.S., 2007. *In silico* approach of azadirachtin binding with actins. *Insect Biochem. Mol. Biol. 37*, 635–640.

Rix, U., Superti-Furga, G., 2009. Target profiling of small molecules by chemical proteomics. *Nat. Chem. Biol. 5*, 616–624.

Robertson, S.L., Ni, W., Dhadialla, T.S., Nisbet, A.J., McKusker, C., Ley, S.V., Mordue, W., Mordue (Luntz), A.J., 2007. Isolation of an azadirachtin-binding complex from *Drosophila* Kc167 Cells. *Arch. Insect Biochem. Physiol. 64*, 200–208.

Veitch, G.E., Beckmann, E., Burke, Boyer, A., Maslen, S.L., Ley, S.V., 2007a. Synthesis of azadirachtin: a long but successful journey. *Angew. Chem. Int. Ed. 46*, 7629–7632.

Veitch, G.E., Beckmann, E., Burke, Boyer, A., Ayats, C., Ley, S.V., 2007b. A relay route for the synthesis of azadirachtin. *Angew. Chem. Int. Ed. 46*, 7633–7635.

6 The Spinosyns: Chemistry, Biochemistry, Mode of Action, and Resistance

V L Salgado, Bayer CropScience AG, Monheim,
Germany
T C Sparks, Dow AgroSciences, Indianapolis,
IN, USA

© 2010, 2005 Elsevier B.V. All Rights Reserved

6.1. Introduction

The spinosyns and associated spinosoids (semisynthetic analogs of the spinosyns) constitute a new and unique class of insect control agents. The spinosyns are fermentation-derived natural products that are chemically unique (**Figure 1**). In addition to their unique chemical structure, the spinosyns possess a novel mode of action coupled with an insecticidal efficacy on par with many pyrethroid insecticides, plus a very favorable toxicological profile (Bret *et al.*, 1997; Sparks *et al.*, 1999). Thus, the spinosyns exhibit many highly desirable attributes for an insect control agent.

The first commercial product, spinosad, was launched in 1997. An extensive discovery effort followed the initial isolation and identification of the first spinosyns. This effort led to the identification of an array of naturally produced spinosyns, as well as a variety of semi-synthetic analogs; the spinosoids.

In this chapter attempts are made to bring together aspects of the discovery of the spinosyns and spinosoids, their chemistry, biology, mode of action, structure–activity relationships, and resistance management. The reader is also referred to several reviews (Crouse *et al.*, 1999; Sparks *et al.*, 1999; Anzeveno and Green, 2002; Kirst *et al.*, 2002b).

6.2. Discovery, Structure, and Biosynthesis of the Spinosyns

Among insecticidal insect control agents, natural products have in numerous instances become products or have been the model or inspiration for insecticidal products. At the very least, virtually all significant classes of insect control agents, including

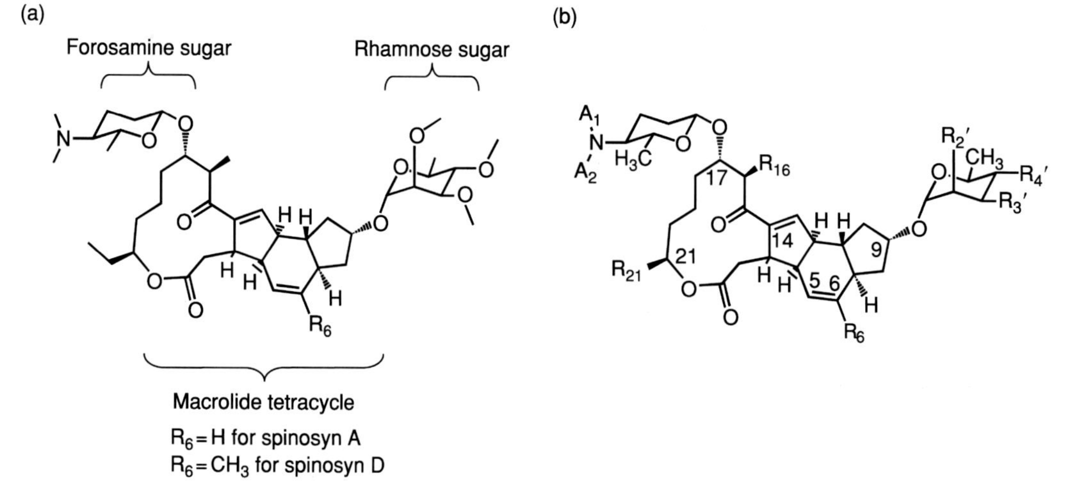

(a)

Forosamine sugar   Rhamnose sugar

Macrolide tetracycle

R₆ = H for spinosyn A
R₆ = CH₃ for spinosyn D

(b)

Figure 1   Structures of spinosad and spinosyns/spinosoids.

the organophosphates, carbamates, pyrethroids, neonicotinoids, and many others, can point to a model that exists among nature's chemical laboratory (Addor, 1995). As such, natural products have been and remain a very valuable source of inspiration for the discovery of new insect control agents, thereby providing a source of models, leads, and in some cases the insecticidal products.

There are two key elements to an effective natural products discovery program. The first is new, novel sources of natural products. The second is availability of new bioassays or approaches to screen or filter the natural products and their extracts. Many natural product discovery programs use the same source material coupled with new growth media, new extraction conditions, etc. Thus, new bioassays are central to the exploitation of these existing natural product libraries and hence, the discovery of new insecticidal natural products.

6.2.1. The Spinosyns

During the 1980s Lilly Research Laboratories had a program designed to look for rare and unique actinomycetes that were then fed into a whole series of biological screens. In late 1983 a new mosquito larvicide bioassay using early stadium *Aedes aegypti* larvae in 96-well microtiter plates was added to these biological screens. Because of the high throughput and small amount of material needed, a large number of broths could be screened at high sensitivity. An extract from a microorganism cultured from a soil sample collected in the Caribbean was found to be active in this assay (Thompson *et al.*, 1997). Refermentations of the sample (A83543) were active in subsequent mosquito bioassays, but it became of real interest when it showed contact and antifeedant

activity against larvae of the southern armyworm (*Spodoptera eridania*) in a plant-based assay, coupled with little activity on spider mites (*Tetranychus urticae*), aphids (*Aphis gossypii*), or corn rootworm (*Diabrotica undecimpunctata*), thus demonstrating some inherent selectivity – a rare occurrence (Thompson *et al.*, 1997). Characterization of the microorganism showed it to be a new species of actinomycete from an unusual genus, *Saccharopolyspora spinosa* (Mertz and Yao, 1990).

From extracts of *S. spinosa*, a series of related A83543 factors, later named "spinosyns," were identified. The spinosyns were found to consist of two sugars, forosamine and 2′,3′,4′-tri-O-methylrhamnose, attached to a unique tetracyclic ring system containing a 12-member macrocycle (**Figure 1**). Among macrocyclic lactones, the spinosyns are unique (Kirst *et al.*, 1992; Shomi and Omura, 2002), possessing a structure that is truly novel among insect control agents. Other naturally occurring insecticidal macrocyclic lactones are known (e.g., avermectins, milbemycins), but these are based on entirely different 14-member macrocycles, with modes of action and activity spectra distinct from those of the spinosyns.

As observed with many natural products, the parent (wild type, WT) strain of *S. spinosa* produces a family of closely related spinosyns, which vary in methyl substitution patterns on the forosamine nitrogen, the 2′- and 3′-positions of the rhamnose, and at the C6, C16, and C21 positions of the tetracycle (Kirst *et al.*, 1992) (**Table 1**). The spinosyns produced from the parent strain included spinosyns A–H and J (I was not assigned) (**Table 1**).

Within this mixture, the two major components produced are spinosyn A (primary) and spinosyn D (secondary). An extract of the fermentation broth

Table 1 Structures of the spinosyns, pseudoaglycones, aglycones, and spinosoids

Number	Compound	Strain	$A1^a$	A2	R21	R16	R6	R2'	R3'	R4'	5,6	13,14
Spinosyns												
1	A	WT	Me	Me	Et	Me	H	OMe	OMe	OMe	DB	DB
2	B	WT	H	b								
3	C	WT	H	H								
4	D	WT					Me					
5	E	WT				Me						
6	F	WT					H					
7	G	WT	c	c								
8	H	WT						OH				
9	J	WT							OH			
10	A C17-PSA	WT	na	na								
11	A C9-PSA	WT							na	na	na	
12	A aglycone	WT	na	na					na	na	na	
13	D C17-PSA	WT	na	na			Me					
14	D C9-PSA	WT					Me		na	na	na	
15	D aglycone	WT	na	na			Me		na	na	na	
7	H	H						OH				
16	Q	H					Me	OH				
17	R	H	H					OH				
18	S	H				Me		OH				
19	T	H						OH	OH			
8	J	J							OH			
20	L	J					Me		OH			
21	M	J	H						OH			
22	N	J	H				Me		OH			
23	K	K								OH		
24	O	K					Me			OH		
25	Y	K			Me					OH		
26	U	H + S						OH		OH		
27	V	H + S					Me	OH		OH		
28	P	J + S							OH	OH		
29	W	J + S					Me		OH	OH		
Novel spinosyns												
30	21-propyl A				nProp							
31	21-butenyl A	21B WT			Buten							
Spinosoids												
32	N-demethyl D	A	H				Me					
33	N-demethyl K	K	H							OH		
34	N,N-didemethyl K	K	H	H						OH		
35	2'-H A	H						H				
36	2'-H D	H					Me	H				
37	2'-O-ethyl A	H						OEt				
38	2'-O-vinyl	H						OVinyl				
39	2'-O-n-propyl A	H						OnProp				
40	2'-O-n-pentyl A	H						OnPent				
41	2'-O-benzyl A	H						OBenz				
42	2'-O-acetate A	H						OAcetate				
43	3'-H A	J							H			
44	3'-OH-epi A	J							OH-epi			
45	3'-O-methyl-epi A	J							OMe-epi			
46	3'-O-ethyl A	J							OEt			
47	3'O-vinyl A	J							OVinyl			
48	3'-O-n-propyl A	J							OnProp			
49	3'-O-isopropyl A	J							OIsoprop			
50	3'-O-isopropenyl A	J							OIsopropen			
51	3'-O-n-butyl A	J							OnBut			
52	3'-O-n-pentyl A	J							OnPent			
53	3'-O-allyl A	J							OAllyl			
54	3'-O-propargyl A	J							OProparg			
55	3'-O-CH_2CF_3 A	J							OCH_2CF_3			
56	3'-O-acetate A	J							OAcetate			
57	3'-O-methoxymethyl	J							OMeOMe			
58	3'-Methylene A	J							=Me			
59	4'-H A	K								H		
60	4'-O-ethyl A	K								OEt		
61	4'-O-acetate A	K								OAcetate		

Continued

Table 1   Continued

Number	Compound	Strain	A1[a]	A2	R21	R16	R6	R2'	R3'	R4'	5,6	13,14
62	2',3',4'-tri-O-ethyl A							OEt	OEt	OEt		
63	4''-N-ethyl B	C	Et	H								
64	4''-N-ethyl A	B	Et									
65	4''-N-n-propyl A	B	Prop									
66	4''-N-n-butyl A	B	Butyl									
67	4''-N,N-diethyl	C	Et	Et								
68	4''-OH	A	na	na								
69	5,6,13,14(α)-TH	A									SB	SB
70	5,6,13,14(β)-TH	A									SB	SB
71	13,14-α-DH	A										SB
72	13,14-β-DH	A										SB
73	7,11-Dehydro	A					Me					
74	5,6-α-epoxy	A									α-epoxy	
75	5,6-β-epoxy	A									β-epoxy	
76	5,6-DH	A									SB	
77	5,6-DH, 3'-O-ethyl	A							OEt		SB	
78	5,6-DH, 3'-O-n-propyl	A							OnProp		SB	
79	A C17-O-acetate	A	na	na								
80	A C17-O-DMAA	A	na	na								
81	A C17-O-DMAP	A	na	na								
82	A C17-O-DMAB	A	na	na								
83	A C17-O-NMPipz	A	na	na								
84	A C17-O-DMPipd	A	na	na								
85	A C9-O-L-lyxose[d]	A										
86	A C9-O-L-mannose[d]	A										

[a]Position on the spinosyn structure (see **Figure 1**).

[b]Dash indicates no change from substitution pattern for spinosyn A.

[c]The stereochemistry of 5''-methyl on the forosamine is axial.

[d]L-Lyxose is missing the 6-methyl of rhamnose; L-mannose is 6-hydroxy-rhamnose.

na, not applicable.; DB, double bond; DMAA, dimethylaminoacetate; DMAB, dimethylaminobutyrate; DMAP, dimethylaminoproprionate; DMPipd, dimethyl-N-piperidinyl acetate; NMPipz, N-methylpiperazinyl acetate; PSA, pseudoaglycone; SB, single bond; WT, wild type.

containing this naturally occurring mixture is known as spinosad, the active ingredient in a number of products (Tracer®, Success®, SpinTor®, Conserve®) marketed by Dow AgroSciences for insect control. Spinosad received registration from the US Environmental Protection Agency (EPA) for use in cotton insect control in February 1997. As discussed below (see Section 6.5.1.1.1), spinosyn A is the most active of the naturally occurring spinosyns from *S. spinosa*, followed closely by spinosyn D (**Table 2**). Thus, the most insecticidally active naturally occurring spinosyns produced by *S. spinosa* are also those that the microorganism naturally produces in the largest quantity.

Further research around the spinosyns required the availability of significant quantities of material. Because the wild-type parent strain only produced very small quantities of spinosyns, a strain improvement program was instituted to increase the titers of the spinosyns produced. During the process of strain improvement, a number of mutant strains were identified that had nonfunctional 2'- and/or 3'- and/or 4'-O-methyltransferases. Since these mutants were unable to methylate the hydroxyl groups on the 2'-, 3'-, or 4'-positions of the tri-O-methylrhamnose, a variety of new spinosyns were produced in addition to a few spinosyns seen only as minor factors in the wild-type strain (Sparks *et al.*, 1996, 1999).

The mutant strain possessing a nonfunctional 2'-O-methyltransferase produced the already existing spinosyn H (2'-O-demethyl analog of spinosyns A) as well as several new spinosyns: Q (2'-O-demethyl analog of spinosyn D), R (2'-O-demethyl analog of spinosyn B), S (2'-O-demethyl analog of spinosyn E), and T (2'-O-demethyl analog of spinosyn J). Likewise, a mutant with a nonfunctional 3'-O-methyltransferase produced spinosyn J (3'-O-demethyl analog of spinosyn A), a minor factor in the wild-type strain, along with the new spinosyns: M (3'-O-demethyl analog of spinosyn B), L (3'-O-demethyl analog of spinosyn D), and N (the 4'-N-demethyl analog of spinosyn L). The mutant possessing a nonfunctional 4'-O-methyltransferase produced three new spinosyns: K (4'-O-demethyl analog of spinosyn A), O (the 4'-O-demethyl analog of spinosyn D), and Y (the 4'-O-demethyl analog of spinosyn E) (Sparks *et al.*, 1996, 1999).

Table 2 Biological activity of spinosyns towards selected insect and mite species

Number	Compound	Heliothis virescens, neonate (LC$_{50}$, ppm)	Calliphora vicina, larvae (LC$_{50}$, ppm)	Stomoxys calcitrans, adult (LC$_{50}$, ppm)	Aphis gossypii (LC$_{50}$, ppm)	Macrosteles quadrilineatus (LC$_{50}$, ppm)	Tetranychus urticae (LC$_{50}$, ppm)
Spinosyns							
1	A	0.31	0.3–0.53	0.43	50 (42–88)	6.9	5.3
2	B	0.4	0.58	<0.5	11	5.1	0.9
3	C	0.8	0.63	0.45	33	14.5	8.4–29
4	D	0.8	0.44	0.55	50		3–19
5	E	4.6	1.95	0.16			100
6	F	4.5	2.50	0.80	>50		16
7	G	7.1					
8	H	5.7	1.1	0.31	>50		>50
9	J	>80	6.3	0.5	>50		~63
10	A C17-pseudoaglycone	>64		90			100
11	A C9-pseudoaglycone	>64					
12	A aglycone	>64					
13	D C17-pseudoaglycone	>64					100
14	D C9-pseudoaglycone	>64					
15	D aglycone	>64					
7	H	5.7	1.1	0.31	>50		95
16	Q	0.5	1.8	0.26	>50		14
17	R	14.5					
18	S	53				11.7	114
19	T	>64	>10	>10		>100	
8	J	>80	6.3	0.5	>50		~63
20	L	26	2.8	<5	>50	>100	48–67
21	M	22.6				>100	
22	N	40				61	>50
23	K	3.5	2.9	1.7	12	3.2	1.4
24	O	1.4		<5	11	6.9	0.8
25	Y	20					
26	U	22				8.2	>50
27	V	17				2.8	
28	P	>64	>10	6.1			
29	W	>64					
	Cypermethrin	0.18					
	Ethofenprox				0.3		
	Fenazaquin						1.0

Further variations in the methylation of the rhamnose sugar could also be obtained by the addition of sinefungin (Chen *et al.*, 1989), which in *S. spinosa* specifically blocked the 4'-O-methyltransferase during the fermentation process of the wild-type strain (Sparks *et al.*, 1996, 1999). When sinefungin was added to fermentations of the H (nonfunctional 2'-O-methyltranferase) and J (nonfunctional 3'-O-methyltranferase) mutant strains, several other new spinosyns (P, U, V, W) were produced that were missing methyl groups from the 4'-position in combination with missing methyl groups from either the 2'- or 3'-positions (Sparks *et al.*, 1996, 1999).

In general, the variations in all of the above-mentioned spinosyns center around (1) methylation of the forosamine amino group, (2) presence or absence of O-methyl groups on the rhamnose sugar,

and (3) presence or absence of alkyl group(s) at the C6, C16, and C21 positions of the tetracyclic ring system. Other factors related to spinosyn A include the C17 pseudoaglycone (PSA), which lacks the forosamine sugar, the corresponding C9 PSA, missing the rhamnose moiety, and the aglycone, where both sugars are absent. The corresponding PSAs and aglycone were also observed for spinosyn D (Table 1).

Through the above process of isolating major and minor factors from the wild-type and mutant strains of *S. spinosoa*, more than 20 different spinosyns were identified (Sparks *et al.*, 1996, 1999) (Table 1).

6.2.2. The 21-Butenyl Spinosyns

Like Eli Lilly and Company, Dow AgroSciences has also had a long-standing natural products screening program. During the mid-1990s a high-throughput screening (HTS) program was established that used a number of target insect pests including lepidopterous larvae, such as beet armyworm (*Spodoptera exigua*), in a 96-well format, as the primary natural products screening tool. This was accompanied by a number of secondary assays to further characterize extracts that were found to be active in the primary screen. One particular multiyear effort, to screen a very large number and variety of actinomycete broth extracts, relied on a novel bioassay as the secondary assay. From this screening effort an extract was found that was not only active in the primary diet-based screen, but that, on further testing, also exhibited interesting contact activity coupled with symptomology that was very reminiscent of the spinosyns. Further separation and isolation coupled to detailed physical characterization (nuclear magnetic resonance (NMR), liquid chromatography–mass spectroscopy (LC–MS), etc.) of this extract showed that the activity was indeed due to a spinosyn-like entity (Lewer *et al.*, 2001, 2003). However, there were some structural differences relative to the then known spinosyns. The dominant, distinctive feature was the presence of a 2-butenyl group at the C21 position of the macrocyclic ring system in place of the heretofore typical ethyl-group of the spinosyns (Lewer *et al.*, 2003) (**Figure 2**). The primary factor isolated from the mixture of 21-butenyl factors was the 21-butenyl homolog of spinosyn A (**Figures 1** and **2, Table 1**). To date more than 30 additional factors have been isolated, identified, and

characterized (Lewer *et al.*, 2003). These 21-butenyl spinosyns constitute a new family of spinosyns, for which research on the isolation, characterization, and chemistry is continuing (Crouse *et al.*, 2002). Characterization of the microorganism that produces the 21-butenyl spinosyns showed it to be a new species of *Saccharopolyspora* closely related to, but distinct from, *S. spinosa*, which has been named *Saccharopolysopra pogona* (Lewer *et al.*, 2003).

6.2.3. Spinosyn Biosynthesis

The spinosyns are 12-member macrocyclic lactones (macrolides). Based on incorporation studies using ^{13}C-labeled acetate, propionate, butyrate, and isobutyrate, the biosynthesis of the tetracycle of the spinosyns is consistent with a polyketide synthase-based (PKS) pathway (Kirst *et al.*, 1992). As a macrolide, the spinosyns differ from the norm in that the spinosyn PKS pathway ultimately leads to the introduction of three intramolecular carbon–carbon bonds to form the spinosyn tetracycle (Waldron *et al.*, 2001) (**Figure 3**). To this tetracycle a neutral sugar, rhamnose, is coupled at the C9 position, and then subsequently methylated via three O-methyltransferases. The final steps involve the addition of an amino sugar to the tetracycle at the C17 position. The amino sugar, forosamine, is also methylated prior to attachment to the spinosyn tetracycle – in this case at both positions of the nitrogen (Kirst *et al.*, 1992; Waldron *et al.*, 2000, 2001) (**Figure 3**).

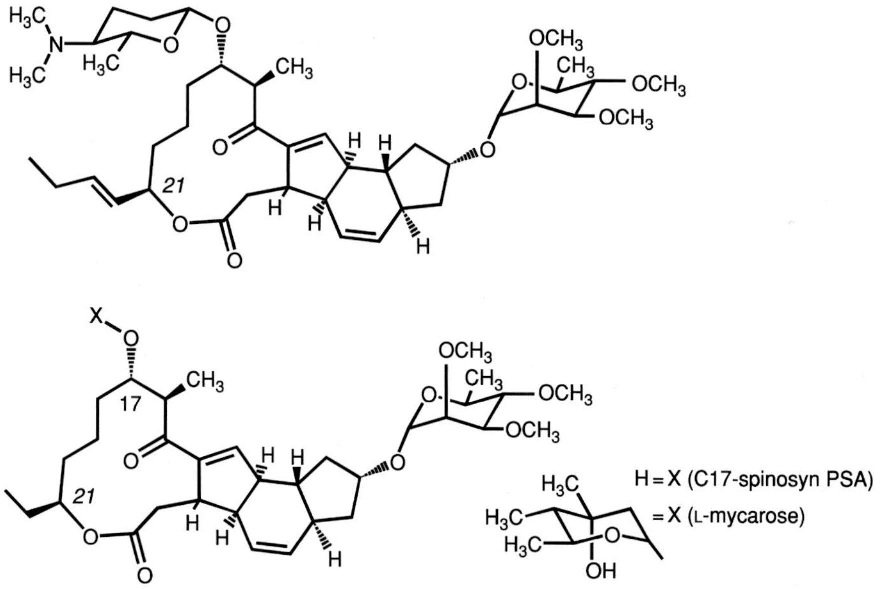

Figure 2   Structures for 21-butenyl spinosyn A (top), the spinosyn A C17-pseudoaglycone (PSA, bottom; X = H) and the C17 L-mycarose biotransformation product.

A large synthetic effort has gone into the modification of the forosamine and rhamnose moieties, as well as the core tetracycle (Crouse and Sparks, 1998; Crouse et al., 2001; Anzeveno and Green, 2002; Kirst et al., 2002b) (see Section 6.7). However, other approaches to macrolide modification are now possible. Sequencing the genes involved in spinosyn biosynthesis (Waldron et al., 2000, 2001; Madduri et al., 2001) has allowed the production of novel spinosyns through biotransformation and modification of the biosynthetic pathways. For example, a genetically engineered strain of Saccharopolyspora erythraea expressing the spnP glycosyltransferase gene was used to produce biotransformed spinosyns, wherein the β-D-forosamine moiety was replaced by α-L-mycarose at the C17 position (Gaisser et al., 2002). Alternatively, modifications to the basic tetracycle have been obtained via loading module swaps in the PKS, leading to a variety of novel substitutions at C21 and other positions (Burns et al., 2003; Martin, 2003). As one example, the ethyl group at C21 has been replaced with an n-propyl group (Martin, 2003), resulting in a unique spinosyn as active as spinosyn A (Tables 1 and 3).

6.2.4. Physical Properties of the Spinosyns

The spinosyns are high molecular weight (approximately 732 Da), relatively nonvolatile molecules (Table 4). As with the biological activity, small changes in structure can have far larger effects than might otherwise be expected. Although spinosyns A and D only differ by the presence of a methyl group at C6 (Table 1), this has a surprising effect on the physical properties, especially solvent solubilities. For example, spinosyn A is soluble in water at about 8.9 mg $100\,ml^{-1}$, while the water solubility of spinosyn D is only 0.05 mg $100\,ml^{-1}$. This pattern holds true for other solvents as well (Table 4).

6.3. Pharmacokinetics of Spinosad

6.3.1. Metabolism of the Spinosyns

6.3.1.1. Mammalian and avian spinosyn metabolism
Studies of ^{14}C-spinosyns A and D metabolism in rats identified fecal excretion as the major route of elimination for both spinosyns A and D. In the fecal material, between 52% (males) and 28% (females) of the administered dose of spinosyn A was excreted in the feces as parent or metabolites within 0–24 h post-dosing. Of this material only 6.4% (males) and 5.4% (females) was spinosyn A (Domoradzki et al., 1996). For spinosyn D, between 67.7% (males) and 73.3% (females) of the applied dose was present in the feces, of which

34.5% (males) and 35.2% (females) was the parent spinosyn D (Domoradzki et al., 1996). Thus, for both spinosyns A and D a substantial amount of metabolism had taken place in the first 24 h following administration. The primary pathways for the spinosyns in rats involved loss of a methyl group on the forosamine nitrogen (N-demethylation), and/or loss of one of the O-methyl groups on the rhamnose (O-demethylation) (Figure 4), both followed by conjugation. Conjugation of the parent was also noted as an important pathway in rats (Domoradzki et al., 1996).

In lactating goats, following 3 days of dosing with either spinosyn A or D, eight metabolites of spinosyn A and five metabolites of spinosyn D were detected. As with rats, N-demethylation was an important metabolic route. In addition, hydroxylation of the macrolide ring was noted as a primary metabolic pathway (Rainey et al., 1996) (Figure 4).

In studies with poultry, N-demethylation of the forosamine nitrogen and O-demethylation of the rhamnose moiety were also found to be the two primary metabolic pathways (Magnussen et al., 1996). In the latter case, the 2'/4'-O-demethyl metabolites (i.e., spinosyns H and K, respectively, for spinosyn A metabolism) predominated over the 3'-O-demethyl metabolites. Another pathway, involving removal of the forosamine sugar to form the C17 PSA, was identified, but appears to be secondary to the N- and O-demethylation (Magnussen et al., 1996) (Figure 4).

As demonstrated by these three studies, spinosyns A and D are readily metabolized in vertebrate and avian systems. Furthermore, N-demethylation of the forosamine nitrogen is a metabolic route that predominates in all three species. O-Demethylation of the rhamnose sugar is also important in rats and poultry. These particular pathways are both consistent with oxidative metabolism via monooxygenases and/or the action of glutathione transferases. In the rat studies, glutathione conjugates were noted for O-demethylated metabolites as well as for the parent spinosyns (A and D). The presence of hydroxylated macrolide metabolites in the goat provides direct support for involvement of monooxygenases in spinosyn metabolism.

6.3.1.2. Spinosyn metabolism in insects
Spinosyns A and D appear to have a number of potential sites for metabolism, including N-demethylation of the forosamine, O-demethylation of the rhamnose, epoxidation of the 5,6 or 13,14 double bonds, opening of the macrocyclic lactone, etc. As shown above (see Section 6.3.1.1), some of these potential metabolites have been observed. However, compared to more conventional insect control agents, the spinosyns are

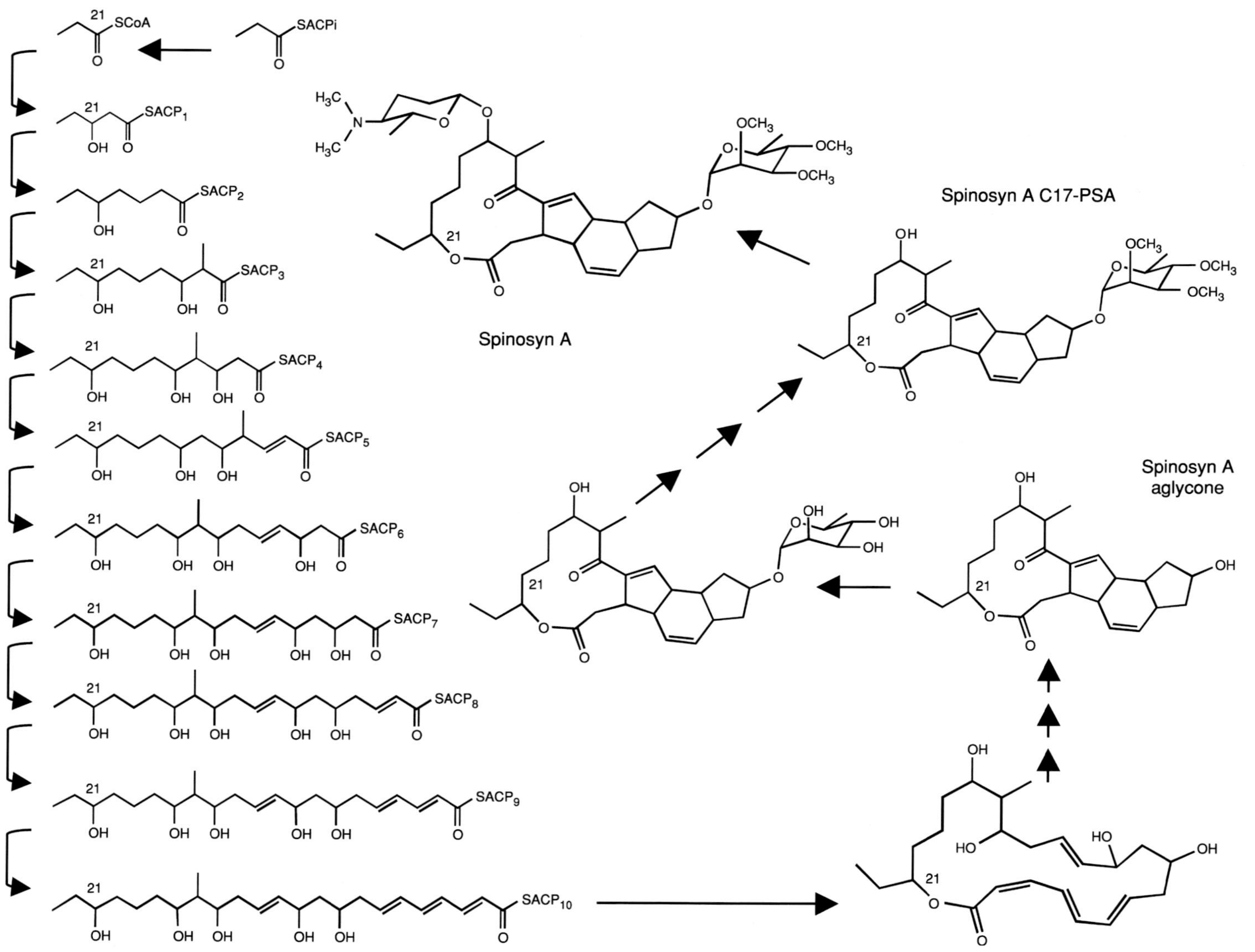
Spinosyn A
Spinosyn A C17-PSA
Spinosyn A aglycone

very large, unusual, complex molecules, and metabolic enzyme systems do exhibit distinct substrate specificities. Studies of spinosyn A metabolism in tobacco budworm (*H. virescens*) larvae show spinosyn A to penetrate slowly, and that the primary, and apparently only, component that was detected (within the limits of the experiment) was the parent, spinosyn A (Sparks *et al.*, 1997, 2001) (**Figure 5**). Thus, until 24 h post treatment there appears to be little or no metabolism of spinosyn A in pest insects such as larvae of *H. virescens*. In contrast, these same larvae readily metabolize the quinazoline acaricide, fenazaquin (Sparks *et al.*, 1997) (**Figure 5**). Subsequent studies with *H. virescens* larvae that were highly resistant (ca. 1000-fold) (Bailey *et al.*, 1999) to spinosad, also failed to show significant levels of metabolism (Young *et al.*, 2000). These observations are in clear distinction to those observed for mammals (see Section 6.3.1.1).

The apparent lack of spinosyn A metabolism by pest insects is also supported by synergist studies with the monooxygenase inhibitor, piperonyl butoxide (PBO). PBO is able to synergize the biological activity of the pyrethroid permethrin in house flies, but not that of spinosyn A (Sparks *et al.*, 1997). Likewise no PBO-based synergism was noted for spinosyn A or spinosad activity in studies involving *H. virescens* larvae (Sparks *et al.*, 1997) or in spinosad resistant larvae of the diamondback moth (*Plutella xylostella*) (Zhao *et al.*, 2002). Thus, within the limits of the available data, many pest insects appear to have a limited capacity to metabolize some spinosyns.

6.3.2. Spinosyn Penetration into Insects

The observed biological activity of any toxicant is a function of two primary parameters: (1) inherent activity at the target site, and (2) concentration of the toxicant at the target site as a function of time. Compared to more conventional insect control agents such as the organophosphates, carbamates, and pyrethroids, the spinosyns are large, complex, relatively high molecular weight molecules. Although the spinosyns, such as spinosyn A, exhibit good activity via topical or contact application, better activity is observed by oral application or by injection (Sparks *et al.*, 1995, 1997, 1998), suggesting that penetration through the insect cuticle may be comparatively slow. In contrast, insecticides such as cypermethrin are equitoxic to lepidopterous larvae by either topical application or injection (Sparks *et al.*, 1997), suggesting relatively rapid penetration through the cuticle of pest insect larvae. A time course of penetration of permethrin or spinosyn A into the hemolymph of larvae of the cabbage looper (*Trichoplusia ni*) (Sparks *et al.*, 1997) showed the rate of penetration for spinosyn A to be much slower than that of the pyrethroid permethrin. Likewise, studies of initial penetration rates (2–4 h) for a variety of insect control agents showed that spinosyn A possessed one of the slowest rates of penetration. Most organophosphate and pyrethroid insect control agents have initial rates of penetration that are several-fold higher; in some cases by more than an order of magnitude (**Table 5**). Interestingly, one compound that is reported to have a similarly slow rate of penetration is abamectin, another large macrolide (**Table 5**). Thus, in light of the very good biological activity possessed by many of the spinosyns, their comparatively slow rate of penetration may be well compensated for by a limited capacity of many pest insects to metabolize them.

6.3.3. Insecticidal Concentration in the Cockroach

A key parameter in understanding the mode of action of biologically active compounds is the aqueous concentration that needs to be in contact with the receptors in the body to cause the biological effect of interest. For pharmaceuticals, the therapeutic concentration is defined as the aqueous concentration of the substance in the patient's blood plasma needed to obtain a desired therapeutic effect. Likewise, the insecticidal concentration can be defined as the concentration of an insecticide in the aqueous phase of an insect's hemolymph needed to obtain a desired insecticidal effect. Hemolymph is a complex tissue, composed of cells suspended in a complex solution of salts and organic molecules, including high concentrations of proteins. Whole hemolymph concentration, as is usually measured for insecticides, is in itself not interesting for the mode of

---

Figure 3   Potential biosynthesis of spinosyn A. PSA, pseudoaglycone. (Adapted, in part, from information in Kirst, H.A., Michel, K.H., Mynderse, J.S., Chio, E.H., Yao, R.C., *et al.*, **1992**. Discovery, isolation, and structure elucidation of a family of structurally unique fermentation-derived tetracyclic macrolides. In: Baker, D.R., Fenyes, J.G., Steffens, J.J. (Eds.), Synthesis and Chemistry of Agrochemicals, vol. 3. American Chemical Society, Washington, DC, pp. 214–225; Waldron, C., Madduri, K., Crawford, K., Merlo, D.J., Treadway, P., *et al.*, **2000**. A cluster of genes for the biosynthesis of spinosyns, novel macrolide insect control agents produced by *Saccharopolyspora spinosa*. *Antonie van Leeuwenhoek 78*, 385–390; and Waldron, C., Matsushima, P., Rosteck, P.R., Jr., Broughton, M.C., Turner, J., *et al.*, **2001**. Cloning and analysis of the spinosad biosynthetic gene cluster of *Saccharopolyspora spinosa*. *Chem. Biol. 8*, 487–499.)

Table 3   Biological activity of novel spinosyns and spinosoids towards selected insect and mite species

Number	Compound	Heliothis virescens, neonate (LC$_{50}$, ppm)	Stomoxys calcitrans, adult (LC$_{50}$, ppm)	Aphis gossypii (LC$_{50}$, ppm)	Macrosteles quadrilineatus (LC$_{50}$, ppm)	Tetranychus urticae (LC$_{50}$, ppm)
Novel spinosyns						
30	21-Propyl A	0.16		27		
31	21-Butenyl A	0.29		5.2		1.4
Spinosoids						
32	*N*-demethyl D	5.6		>50		0.1
33	*N*-demethyl K	9.9			8.2	0.5
34	*N,N*-didemethyl K	7.4			11.1	3.4
35	2′-H A	0.23	0.24			134
36	2′-H D	0.23				25
37	2′-*O*-ethyl A	0.30				3.5
38	2′-*O*-vinyl	1.10				
39	2′-*O*-*n*-propyl A	0.30				
40	2′-*O*-*n*-pentyl A	2.00				
41	2′-*O*-benzyl A	1.80				
42	2′-*O*-acetate A	1.20	0.45		8.1	>100
43	3′-H A	0.36	0.47			5.2->50
44	3′-OH-*epi* A	13.2		>50	>50	>50
45	3′-*O*-methyl-*epi* A	1.9		>50	>50	~50
46	3′-*O*-ethyl A	0.03		12.7		2.0
47	3′-*O*-vinyl A	<0.06-0.3		>50		14
48	3′-*O*-*n*-propyl A	0.05		9.3	~50	2.6
49	3′-*O*-isopropyl A	0.27		>50		3.6
50	3′-*O*-isopropenyl A	2.85		>50		>50
51	3′-*O*-*n*-butyl A	0.38		2.5		1.5
52	3′-*O*-*n*-pentyl A					2.5
53	3′-*O*-allyl A	0.06		>50		11
54	3′-*O*-propargyl A	0.05		>50		10
55	3′-*O*-CH$_2$CF$_3$ A	0.33		25		>50
56	3′-*O*-acetate A	33.0	0.41	>50	>100	22
57	3′-*O*-Me-*O*-Me	0.17		19		1.8
58	3′-Methylene	0.95		2.8	>50	8.9
59	4′-H A	4.1				
60	4′-*O*-ethyl A	0.24		>50		4.1
61	4′-*O*-acetate A	1.30	0.90		3.5	1.8
62	2′,3′,4′-tri-*O*-ethyl A	0.02		6.8		1.3
63	4″-*N*-ethyl B			>50		3.8
64	4″-*N*-ethyl A	0.29		15.2	~50	0.3
65	4″-*N*-*n*-propyl A	0.27		17.3	~50	0.35
66	4″-*N*-*n*-butyl A	14.1		5.3		0.4
67	4″-*N,N*-diethyl	>64		4.8		0.3
68	4″-OH	3.7	>5	~0.4		
69	5,6,13,14(α)-TH	35	0.41			
70	5,6,13,14(β)-TH	>64	45			32
71	13,14-α-Dihydro	4.7				18
72	13,14-β-Dihydro	>80	45			>50
73	7,11-Dehydro	0.2		>50		47.4
74	5,6-α-Epoxy	8	0.96			
75	5,6-β-Epoxy	0.63	0.96			0.50
76	5,6-DH	0.46	2.4	6.4		0.44
77	5,6-DH, 3′-*O*-ethyl	0.05		15		0.7
78	5,6-DH, 3′-*O*-*n*-propyl	0.04		1.7		0.4
79	A C17-*O*-acetate	>64	47			47
80	A C17-*O*-DMAA	>64	22			21
81	A C17-*O*-DMAP	16	9.5			
82	A C17-*O*-DMAB	16.6	8.6			
83	A C17-*O*-NMPipz	2.5	4.7		4	

Continued

Table 3 Continued

Number	Compound	Heliothis virescens, neonate (LC_{50}, ppm)	Stomoxys calcitrans, adult (LC_{50}, ppm)	Aphis gossypii (LC_{50}, ppm)	Macrosteles quadrilineatus (LC_{50}, ppm)	Tetranychus urticae (LC_{50}, ppm)
84	A C17-O-DMPipd	4.1	4.5		24.8	
85	A C9-O-L-lyxose[a]	1.01				1.3
86	A C9-O-L-mannose[a]	0.04				
	Cypermethrin	0.18				
	Ethofenprox				0.3	
	Fenazaquin					1.0

[a]L-Lyxose is missing the 6-methyl of rhamnose; L-mannose is 6-hydroxy-rhamnose.
DMAA, dimethylaminoacetate; DMAB, dimethylaminobutyrate; DMAP, dimethylaminoproprionate; DMPipd, dimethyl-N-piperidinyl acetate; NMPipz, N-methylpiperazinyl acetate.

Table 4 Physical properties of spinosyns A and D and 21-butenyl spinosyn A

Property	Spinosyn A	Spinosyn D	21-Butenyl spinosyn A
Appearance	Colorless solid	Colorless solid	Colorless solid
Empirical formula	$C_{41}H_{65}NO_{10}$	$C_{42}H_{67}NO_{10}$	$C_{43}H_{67}NO_{10}$
Molecular weight	731.96	745.98	757.99
Melting point	120 °C	170 °C	110 °C
Solubility (20 °C)			
Water (distilled)	89.4 ppm	0.495 ppm	
Water (pH 5)	290 ppm	28.7 ppm	184 ppm
Methanol	190 000 ppm	2500 ppm	
Hexane	4 480 ppm	743 ppm	
Acetone	168 000 ppm	10 100 ppm	
Vapor pressure (25 °C)	3×10^{-11} kPa	2×10^{-11} kPa	
Octanol/water			
Partition coefficient (pH 7.0, log P)	4.01	4.53	4.37

Adapted, in part, from Anonymous, **2001**. Spinosad Technical Bulletin. Dow AgroSciences, Indianapolis, IN; Crouse, G.D., Sparks, T.C., **1998**. Naturally derived materials as products and leads for insect control: the spinosyns. *Rev. Toxicol. 2*, 133–146; and Thompson, G.D., Dutton, R., Sparks, T.C., **2000**. Spinosad: a case study – an example from a natural products discovery programme. *Pest Mgt Sci. 56*, 696–702.

action because it can be much higher than the true aqueous concentration, due to partitioning into cell membranes and binding to hemolymph proteins. For most insecticides, it is not possible to measure hemolymph aqueous concentration directly, because of the difficulty of removing cells and proteins from whole hemolymph without also removing a portion of the hydrophobic insecticide from the aqueous phase. This problem can be circumvented by using an indirect method to estimate the insecticidal concentration, whereby the insecticide content of nerve cords isolated from poisoned, symptomatic insects is determined and compared with the content of nerve cords incubated with the insecticide in saline until a steady state has been reached.

In adult male American cockroaches (*Periplaneta americana*) poisoned with a threshold dose of spinosyn A, the nerve cord content was found to be 4.2 pmol. On the other hand, incubation of isolated nerve cords with 100 nM spinosyn A in saline gave a steady-state tissue content of 19.4 pmol. Assuming passive distribution of spinosyn A between the nerve cord and the aqueous medium *in vivo* and *in vitro* allows us to calculate by simple proportionality that a saline concentration of 21 nM would give the nerve cord content of 4.2 pmol, as observed at the threshold dose (Salgado *et al.*, 1998). Thus, it appears that 20 nM spinosyn A in the hemolymph should have an effect on the target receptors that is strong enough to cause symptoms (Table 6).

6.4. Mode of Action of Spinosyns

Because spinosad causes excitatory neurotoxic symptoms, a systematic investigation of the symptoms using neurophysiology was the most direct way to identify the target site. Although spinosad is

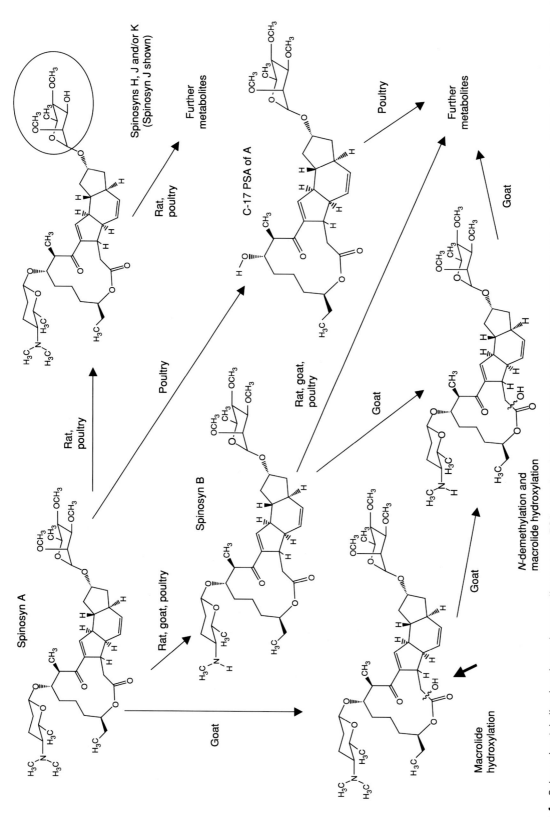

Figure 4 Spinosad metabolism in avian and mammalian systems. PSA, pseudoaglycone. (Adapted, in part, from information in Domoradzki, J.Y, Stewart, H.S., Mendrala, A.L., Gilbert, J.R., Markham, D.A., **1996**. Comparison of the metabolism and tissue distribution of ¹⁴C-labeled spinosyn A and ¹⁴C-labeled spinosyn D in Fischer 344 rats. Soc. Toxicol. Mtg. Anaheim, CA; Magnussen, J.D., Castetter, S.A., Rainey, D.P., **1996**. Characterization of spinosad related residues in poultry tissues and eggs following oral administration. In: 211th Natl. Mtg. American Chemical Society, New Orlean, LA, pp. AGRO 043; and Rainey, D.P., O'Neill, J.D., Castetter, S.A., **1996**. The tissue distribution and metabolism of spinosyn A and D in lactating goats. In: 211th Natl. Mtg. American Chemical Society, New Orlean, LA, pp. AGRO 045.)

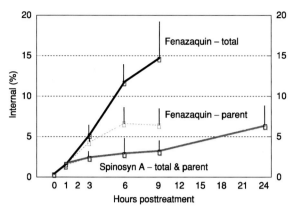

Figure 5 Metabolism of topically applied (1 µg per larva) fenazaquin and spinosyn A in last stadium *Heliothis virescens* larvae. Data points are means, bars give standard deviations. (Adapted from Sparks, T.C., Sheets, J.J., Skomp, J.R., Worden, T.V., Hertlein, M.B., *et al.*, **1997**. Penetration and metabolism of spinosyn A in lepidopterous larvae. In: Proc. 1997 Beltwide Cotton Conf., pp. 1259–1265 and previously unpublished data.)

targeted commercially at Lepidoptera and thrips, spinosyns and spinosoids are inherently broad spectrum, showing activity against insects in the orders Coleoptera, Diptera, Homoptera, Hymenoptera, Isoptera, Orthoptera, Lepidoptera, Siphonaptera, and Thysanoptera, as well as mites (Thompson *et al.*, 1995). American cockroaches were used as the test insects for mode of action studies because of their large size, well-described neuroanatomy and neurophysiology, and distinctive spinosad poisoning symptoms. Spinosyn A was used in all mode of action studies because it is the major component of spinosad, and is similar in biological activity to spinosyn D but much more soluble.

6.4.1. Poisoning Symptoms

Symptoms of spinosyn A poisoning in three insect species were described in detail by Salgado (1998) and can in all cases be divided into three phases.

Table 5 Initial rates of penetration and topical toxicities to *Heliothis virescens* larvae of various insecticides

Compound	Time (h)	(%) Penetration[a]	Ratio to spinosyn A	H. virescens third instar larvae (LD_{50}, $\mu g\, g^{-1}$)
Spinosyns				
Spinosyn A	3	2.3	1.0	1.12–2.4
Avermectins				
Abamectin	4	2.4	1.0	1.2
Pyrethroids and DDT				
DDT	4	10	4.3	52–152
λ-Cyhalothrin	3	8	3.5	0.929
Cypermethrin	3–4	35–36	15.2	0.25–1.61
Fenvalerate	3	3.3	1.4	0.870–1.89
Permethrin	2–3	9–18	3.9–7.8	1.33–2.79
Organophosphates				
Acephate	4	6	2.6	41.0–74.3
Chlorpyrifos	4	38.8	16.9	79.5
Chlorpyrifos-methyl	4	58	25.2	
Malathion	4	17	7.4	
Methyl parathion	4	28.9	12.6	8.3–20
Sulprofos	4	16	7.0	24
Carbamates				
Carbaryl	4	8	3.5	136–232
Cyclodienes				
Endrin	4	12	5.2	46.7
Formamidines				
Amitraz	4	12–15	5.2–6.5	
BTS-27271	4	9	3.9	
METI acaricide[b]				
Fenazaquin	3	5.3	2.3	~400

[a]Percent of material that was considered in the internal fraction.
[b]METI, mitochondrial electron transport inhibitor.
Adapted, in part, from Sparks, T.C., Sheets, J.J., Skomp, J.R., Worden, T.V., Hertlein, M.B., *et al.*, **1997**. Penetration and metabolism of spinosyn A in lepidopterous larvae. In: Proc. 1997 Beltwide Cotton Conf., pp. 1259–1265 and Crouse, G.D., Sparks, T.C., **1998**. Naturally derived materials as products and leads for insect control: the spinosyns. *Rev. Toxicol. 2*, 133–146.

Table 6   Dependence of various effects in *Periplaneta americana* on aqueous concentration of spinosyn A

Aqueous concentration (nM)	Symptoms[a]	Increased CNS efferent activity[a] (%)	Activation of nicotinic acetylcholine receptor (nAChN)[b] (%)	Inhibition of small-neuron γ-aminobutynic acid receptor (GABA-R)[c] (%)
1	None	0	4	>50
5	None	0	28	100
10	None	6	35	100
21	Difficulty righting	38	45	
30		54	52	
60	Prostration	68	65	
100		80	73	
1000		100	94	

[a]Symptoms and increased CNS afferent activity from Salgado *et al.* (1998).
[b]Activation of nAChN from Salgado and Saar (2004).
[c]Inhibition of small-neuron GABA-R from Watson (2001).

The earliest symptoms are due to prolonged involuntary muscle contractions that subtly alter the posture of the insect. This is evident in cockroaches and houseflies, where the first symptom is a lowering of the head and elevation of the tail, caused by involuntary straightening of the hindlegs. In the second phase of poisoning, the posture changes become so severe that the insects topple over and cannot right themselves, becoming prostrate. At this point, there are widespread fine tremors in all muscles, which sometimes can only be seen under a microscope. In hard-bodied insects, all appendages tremble constantly, and in soft-bodied insects such as caterpillars, the skin appears to crawl. In the last phase, the movements cease and the insects are paralyzed. The symptoms will be described in more detail for cockroaches, fruit flies, and tobacco budworm larvae.

The 24 h injection LD_{50} (median lethal dose) of spinosyn A was 0.74 (0.41–1.34) $\mu g\,g^{-1}$ for adult male *P. americana*. The first noticeable symptom was elevation of the body, due to depression of the coxae, extension of the legs, and flexion of the tarsae (**Figure 6b**). In this first poisoning phase, the cockroaches still walked around, and, if disturbed by touching, would resume a normal posture temporarily. These symptoms even persisted after the insect was decapitated (**Figure 6c**). As poisoning progressed to phase 2, the cockroaches became more uncoordinated and fell on their backs, as a result of asymmetric extension of the legs. The tarsae of the prostrate insects remained strongly flexed, and the insects showed tremors, which were not seen before prostration (**Figure 6d**). At the lowest effective dose, 0.625 $\mu g\,g^{-1}$, prostration occurred approximately 24 h after injection, whereas at 5 $\mu g\,g^{-1}$ it occurred within 4 h. The insects also

showed other symptoms of excitation, such as wing beating and abdominal bloating resulting from their swallowing large amounts of air. The prostrate cockroaches eventually ceased trembling as they entered the third or quiescent poisoning phase, during which movement could still be elicited by a disturbance, but grew gradually weaker. The legs gradually relaxed as the insects became paralyzed, and no further movement could be elicited. No recovery was observed.

Adult male fruit flies, *D. melanogaster,* exposed to spinosyn A in sugar water, exhibited a characteristic set of sublethal symptoms. The 24 h LC_{50} (median lethal concentration) was 8.0 (5.7–12.1) ppm Initially, flies were able to remain upright and even walk around, but most had difficulty maintaining normal posture. When standing still, the tibiae slowly flexed and the tarsae extended, causing the legs to pull together and the body to rise. Approximately every 5 s, the fly compensated by spreading its legs into a normal stance, after which the process was repeated, with alternate slow pulling together, followed by rapid spreading, of the legs. While this effect occurred in nearly all flies at this concentration, some flies also held the wings abnormally. Normally, the wings are folded over the back, but in many poisoned flies they were either open straight out or folded down.

Heliothis virescens larvae prostrated by spinosyn A typically curled up due to abdominal flexion and lay on their sides, exhibiting widespread fine tremors. The true legs were extended and trembling, as were the mouthparts and even the surface of the soft cuticle. These excitatory effects were not obvious to the naked eye, but were easily seen with a stereomicroscope. Diuresis was also a common symptom of spinosyn A poisoning in *H. virescens*

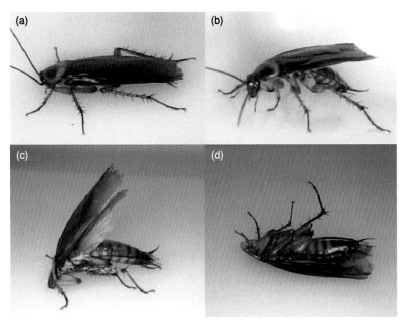

Figure 6 Typical symptoms shown by American cockroaches (*Periplaneta americana*) after injection with spinosyn A, in this case at 2 µg. (a) Control. (b) Depression of the legs elevates the body; note how all leg joints are extended. (c) The symptoms persist even after decapitation, and even worsen, as the whole body flexes. (d) Even after prostration, the tarsae remain strongly flexed, as seen in this decapitated insect. (Reprinted with permission from Salgado, V.L., **1998**. Studies on the mode of action of spinosad: insect symptoms and physiological correlates. *Pestic. Biochem. Physiol. 60*, 91–102; © Elsevier.)

larvae. The injection LD_{50} of spinosyn A against 50–70 mg *H. virescens* larvae was 0.23 (0.1–0.52) $\mu g \, g^{-1}$ (Salgado, 1998).

In summary, one of the first symptoms of spinosyn poisoning in all insects is leg extension, due to involuntary muscle contractions. Later in poisoning, all muscles appear to be affected.

6.4.2. Gross Electrophysiology

6.4.2.1. Spinosyns excite the central nervous system *in vivo* and *in vitro*
Although many muscles are activated during spinosyn poisoning, tarsal flexion was an excellent model because it is one of the first symptoms to appear and persists throughout poisoning, even when the poisoned insect is decapitated and pinned on its back, allowing its physiological basis to be easily studied. This symptom is caused by contraction of the tarsal flexor muscle, located in the tibia (Snodgrass, 1952). The onset of spinosyn-A-induced tarsal flexion was accompanied by continual firing of the tarsal flexor muscle (**Figure 7a**). In control insects, the muscle was relaxed except for activity during sporadic movements (not shown). Cutting all contralateral nerve roots did not diminish flexion, and when all ipsilateral nerves except nerve 5, which innervates the tarsal flexor muscle, were cut, the tarsus still remained flexed, and the firing of the flexor muscle continued (**Figure 7b**).

However, immediately upon severing nerve 5, the tarsus relaxed and the muscle recording became silent (not shown). Therefore, the smallest unit exhibiting the flexion was the leg and its ganglion attached through nerve 5, and it was concluded that spinosyn-induced tarsal flexion was caused by activation of the flexor motor neuron, driven by events occurring either within that neuron or elsewhere within that ganglion (Salgado, 1998).

Nerve recordings made from various nerves showed that central nervous system (CNS) and motor activity was generally increased in spinosyn A poisoned cockroaches, concomitant with appearance of excitatory symptoms and persisting even after paralysis. Alternatively, it could be demonstrated that the excitatory effects of spinosyn A were specifically on the CNS; there were no direct effects on axonal conduction, sensory receptor function, or neuromuscular function (Salgado, 1998).

The next step in analyzing the etiology of spinosyn-induced hyperactivity was to determine whether spinosyns could directly increase nerve activity when applied to isolated ganglia. **Figure 8** shows a rate-meter recording of the activity in a segmental nerve of an isolated larval housefly synganglion. In the control and before application of spinosyn A, the activity varied widely. Following treatment with 100 nM spinosyn A, the swings in activity became

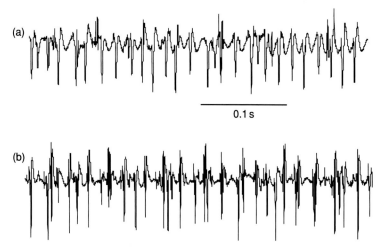

Figure 7　(a) Electromyogram showing continual firing of the metathoracic tarsal flexor muscle in a cockroach poisoned 4 h earlier by abdominal injection of 2.5 μg spinosyn A. (b) The activity persisted after all nerves to the metathoric ganglion, except ipsilateral nerve 5, were severed. (Reprinted with permission from Salgado, V.L., **1998**. Studies on the mode of action of spinosad: insect symptoms and physiological correlates. *Pestic. Biochem. Physiol. 60*, 91–102; © Elsevier.)

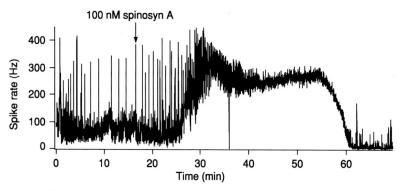

Figure 8　Spike rate recorded from a segmental nerve of an isolated housefly (*Musca domestica*) larval synganglion. Before spinosyn A, the activity varied mostly between 20 and 120 Hz, with occasional brief spikes to 400 Hz. After the addition of 100 nM spinosyn A, there was a delay of approximately 10 min, after which the activity increased greatly to a constant high level, with no dips below 200 Hz. After 25–30 min, the activity declined to below control levels. (Reprinted with permission from Salgado, V.L., **1998**. Studies on the mode of action of spinosad: insect symptoms and physiological correlates. *Pestic. Biochem. Physiol. 60*, 91–102; © Elsevier.)

much smaller and the activity stabilized at a high level, before eventually declining 20–30 min after the start of perfusion with spinosyn A. It was shown above that spinosyn A must attain an internal aqueous concentration near 20 nM in order to produce symptoms. This concentration applied in a saline bath was just sufficient to stimulate activity in isolated cockroach ganglia (**Table 6**). In other words, spinosyn A achieves a concentration in treated insects that is sufficient to cause the observed symptoms by direct actions on nerve ganglia. The ganglion from *Manduca sexta* is significantly more sensitive than the cockroach ganglion, giving a full response at 10 nM (Salgado *et al.*, 1998).

In summary, one of the first symptoms of spinosyn poisoning in all insects is leg extension, due to involuntary muscle contractions driven by activity in motor neurons, caused by a direct action of spinosyns within the ganglion. Later in poisoning, all muscles appear to be tonically activated. The elevated posture arises because, while opposing flexor and extensor muscles are both being activated, the stronger, weight-supporting, muscles overcome the weaker antagonists, so that each joint moves in the direction that elevates the body.

6.4.2.2. Mechanism of paralysis by spinosyns

Spinosyn-poisoned insects eventually become paralyzed, but only after prolonged excitation, suggesting that paralysis results secondarily, from prolonged overexcitation of the nervous system. Interestingly, neural activity continues at high levels in paralyzed insects, indicating that paralysis is due to fatigue-related failure of the neuromuscular system. This hypothesis was confirmed by showing that the maximal force exerted by the metathoracic femur in response to tetanic crural nerve stimulation faded more rapidly as poisoning progressed, and eventually disappeared concomitant with paralysis. The inhibitory neuromuscular synapses, which use γ-aminobutyric acid (GABA) as the neurotransmitter and are blocked after prostration with the GABA gated chloride channel blocker dieldrin, were shown to function normally in insects prostrated by spinosyn A (Salgado, 1998). In *in vitro* experiments on larval housefly muscle, spinosyn A at concentrations up to 1 µM had no effect on neuromuscular transmission, but at 5 µM and higher, the cell was slightly depolarized and the excitatory postsynaptic potential (EPSP) was enhanced. These concentrations are too high for these effects to be relevant during poisoning.

6.4.2.3. Spinosyns excite central neurons by inducing an inward, depolarizing, current

Because spinosyns potently stimulate CNS activity (see Section 6.4.2.1), effects on neurons of the cockroach 6th abdominal ganglion were investigated *in situ* with intracellular microelectrodes. Dorsal unpaired median (DUM) neurons were chosen because they are spontaneously active and are the best-characterized neurons in the insect CNS, and known to contain many different receptors important for insecticide action, including nicotinic acetylcholine receptors (nAChRs) (Buckingham *et al.*, 1997), GABA (Le Corronc *et al.*, 2002) and inhibitory glutamate (Raymond *et al.*, 2000) receptors, and a variety of sodium, calcium, and potassium channels (Grolleau and Lapied, 2000). As reported previously (Dubreil *et al.*, 1994), DUM neurons were spontaneously activate and had random spontaneous unitary inhibitory postsynaptic potentials (IPSPs). Furthermore, stimulation of the connective evoked an IPSP in the DUM cell (Pitman and Kerkut, 1970; Dubreil *et al.*, 1994) (Figure 9b). At 200 nM, but not at 50 nM, spinosyn A depolarized DUM cells *in situ* (10 and 13 mV in two cells; compare the resting levels of the traces in Figure 9b and c) and increased the rate of spontaneous IPSPs considerably, but did not block the spontaneous or evoked IPSPs, even at concentrations as high as 1 µM ($n = 3$) (Figure 9c). It also increased the spontaneous firing rate, as can be seen by the two action potentials in this trace (the evoked IPSP abolished an action potential that was just starting). Picrotoxinin at 10 µM abolished the spontaneous and evoked IPSPs (Figure 9d), leading to further depolarization and excitation (note the increased action potential frequency). Thus, while spinosyn A did not appear to affect the inhibitory postsynaptic receptors in DUM neurons,

or the ability to generate action potentials, it did have clear effects on the cells. While the increased spontaneous IPSP rates reflect changes occurring in presynaptic cells, the depolarization and increased firing rate appear to be due to a direct action of spinosyn A on the DUM cells themselves. It can be seen from the steady depolarization between action potentials (Figure 9d) that this firing is due to a continuous depolarizing stimulus flowing into the cell. This suggests that a spinosyn-induced inward current was responsible for the depolarization and increased firing rate. The depolarization was reversed by 10 µM methyllycaconitine (MLA), a selective nicotinic antagonist (Figure 10), indicating that it is mediated by nicotinic receptors. MLA also blocked the IPSP, consistent with the proposed multisynaptic pathway for this response, involving a nicotinic synapse (Dubreil *et al.*, 1994) (Figure 10).

6.4.2.4. Activation of nicotinic receptors by spinosyns generates a depolarizing inward current

The intracellular recordings (Figures 9 and 10) indicated that spinosyn-A-induced an inward current in central neurons, probably mediated by nAChRs. This current was then directly investigated with voltage clamp experiments on isolated central neurons. The cells studied were a mixed population of neuronal somata, with diameters between 35 and 80 µm, isolated from the thoracic ganglia of the cockroach *P. americana* and studied with the single-electrode voltage clamp method (SEVC) as detailed by Salgado and Saar (2004). In ten cells tested, spinosyn A did indeed induce an inward current at concentrations between 50 nM and 1 µM. Furthermore, this current was completely blocked by 100 nM α-bungarotoxin (α-BGTX), a highly specific antagonist of nAChRs. Because the interaction of both α-BGTX and spinosyn A with the receptors was relatively slow, it was difficult to measure the dose response for block quantitatively. In the example in Figure 11, 3 nM α-BGTX blocked the spinosyn-A-induced current by approximately 75% before washout was begun, while 10 nM α-BGTX blocked the spinosyn-A-induced current by 85%, and 1000 nM blocked it completely. This sensitivity to α-BGTX is comparable to that of the non-desensitizing nAChR, which is blocked $78 \pm 3\%$ by 3 nM and $93 \pm 1\%$ by 10 nM α-BGTX (Salgado and Saar, 2004).

6.4.3. Nicotinic Receptors as the Spinosyn Target Site

6.4.3.1. Two subtypes of nicotinic receptors

Two subtypes of α-BGTX-sensitive nicotinic receptors have been identified in insect neurons (van den

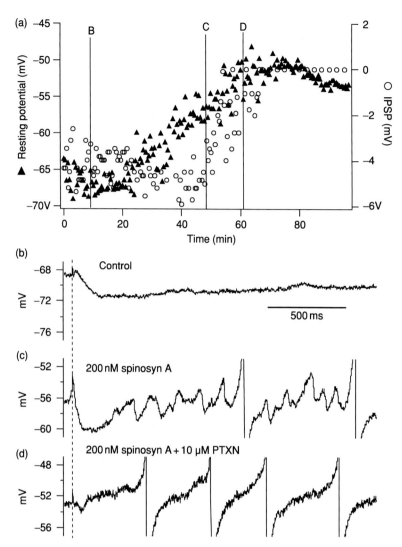

Figure 9   (a) Resting potential and inhibitory postsynaptic potential (IPSP) amplitude of a dorsal unpaired median (DUM) neuron in a cockroach terminal abdominal ganglion. (b–d) Sample recordings at the times indicated by the letters in panel A. The IPSP was evoked by connective stimulation at the vertical dashed line, 200 nM spinosyn A was added just after B, and 10 µM picrotoxinin (PTXN) was added just after point C. This example is representative of similar results obtained in three cells. The preparation for recording synaptic responses from DUM neurons is from Dubreil *et al.* (1994), with slight modifications. The terminal abdominal ganglion of adult male American cockroaches was removed with its associated peripheral nerves. The ganglion was desheathed, placed in an experimental chamber, and perfused with saline. Nerve XI and the anterior connectives were stimulated with suction electrodes, and compounds, dissolved in saline, were applied in the bath. Microelectrodes were filled with 3 M KOAc and had resistances between 110 and 150 MΩ (Salgado, unpublished data).

Beukel *et al.*, 1998; Nauen *et al.*, 2001). In American cockroaches, subtype nAChD receptors desensitize in the presence of agonists and are blocked by α-BGTX with a K_d of 21 nM, while subtype nAChN receptors are nondesensitizing and are blocked by α-BGTX with a K_d of 1.3 nM (Salgado and Saar, 2004). Both subtypes are blocked by MLA, with K_d values of 1 nM and 17 pM, respectively. Since nAChD receptors desensitize in the presence of agonists, they are antagonized during poisoning by nicotinic agonists such as the neonicotinoids (see **Chapter 3**), and this antagonism appears to

cause the subacute symptoms of these compounds, such as feeding inhibition in aphids and inhibition of cleaning behavior in termites (Salgado and Saar, 2004). On the other hand, activation of nicotinic receptors by nicotinic agonists has long been known to cause excitatory symptoms (Schroeder and Flattum, 1984; Nauen *et al.*, 2001), which appear to be mediated by nAChN receptors.

6.4.3.2. Selective action of spinosyns on the nAChN subtype of nicotinic receptor   Although almost all cockroach neurons studied contain both nAChD and nAChN receptors, the two can be separated and studied independently using subtype-selective antagonists. nAChN receptors are extremely sensitive to MLA, with a K_d of 17 pM. Even though nAChD receptors are also very sensitive to MLA, their K_d of 1 nM permits selective block of only nAChN receptors with the inclusion of 100 pM MLA in the bath. **Figure 12a** shows the total ACh-evoked current in an isolated cockroach neuron and the pure nAChD current remaining after specific block of nAChN receptors by 100 pM MLA.

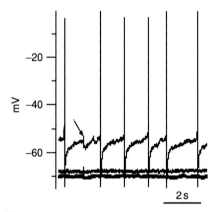

Figure 10 Reversal of spinosyn-A-induced depolarization by methyllycaconitine (MLA) in a DUM neuron. The lower trace is the control, with a resting potential near −70 mV. Spinosyn A at 1 μM depolarized the cell to −53 mV and induced firing. Addition of 10 μM MLA reversed the depolarization and stopped the firing (middle trace). Note that the small IPSP evoked by the stimulus (arrow) in the control became larger after the depolarization and was abolished by MLA. This example is representative of similar results obtained in two cells (Salgado, unpublished data).

On the other hand, selective block of nAChD receptors can be achieved by taking advantage of their property of desensitization. Desensitization is a property of the receptor itself, due to transition of activated receptors into a nonconducting state that binds agonists very tightly (Lena and Changeux, 1993). The desensitized state of the rat muscle nicotinic receptor, for example, has an affinity for agonists about 500-fold higher than that of the open state. The strong binding of agonists to the desensitized state can be used to specifically block the nAChD current (**Figure 12b**), where, in the presence of 100 nM imidacloprid (IMI), a potent nicotinic agonist (see **Chapter 3**), only the nAChN current remains functional, in a cell that has both receptor subtypes.

When nAChN receptors are blocked by the inclusion of 100 pM MLA in the bath, the effect of spinosyn A on nAChD receptors can be studied (**Figure 12c**). Spinosyn A blocked the nAChD receptors only very weakly, an average of only 23% at 10 μM (Salgado and Saar, 2004), and induced no inward current.

As mentioned above, the sensitivity of the spinosyn-A-induced nicotinic current to α-BGTX (IC$_{50}$ (50% inhibitory concentration) of 1.8 nM measured by Salgado et al. (1997)) is very similar to that of nAChN receptors (IC$_{50}$ of 1.2 nM measured by Salgado and Saar (2004), suggesting that spinosyn A activates nAChN receptors. **Figure 12d** shows the nAChN current, isolated with 100 nM IMI, in the presence of increasing concentrations of spinosyn A. As little as 1 nM spinosyn-A-induced a noticeable inward current and prolonged the action

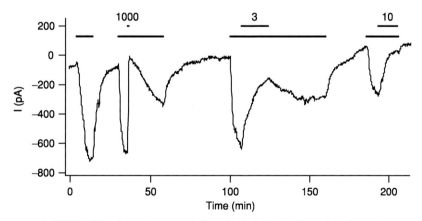

Figure 11 α-Bungarotoxin (BGTX) (thick bars, at concentrations indicated, in nM) block of the spinosyn A (thin bars, at 1 μM) induced inward current at −80 mV in an isolated voltage-clamped cockroach central neuron. The first application and washout of spinosyn A, between 0 and 35 min, reversibly induced a large inward (downward deflection) current. At the peak of the current evoked by the second application of spinosyn A, 1000 nM α-BGTX was applied, which rapidly blocked the current in the continued presence of spinosyn A; the inward current then slowly returned during washout of α-BGTX, but spinosyn A was also washed out, beginning at 60 min. The current evoked by the third application of spinosyn A, at 100 min, was blocked about 75% by 3 nM α-BGTX after about 15 min incubation, but the toxin was washed out before steady-state block was reached. The current evoked by the fourth spinosyn A application, though smaller, due to rundown, was blocked 85% by 10 nM α-BGTX (Salgado, unpublished data).

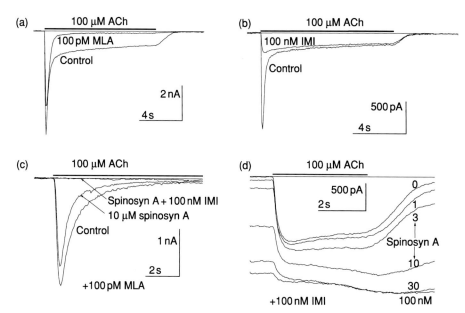

Figure 12   Pharmacological separation of cockroach nicotinic acetylcholine receptor (nAChR) subtypes and the action of spinosyn A on both subtypes. (a) Combined current in an isolated cockroach neuron and the pure nAChD current remaining after specific block of nAChN receptors by 100 pM MLA. (b) In the presence of 100 nM imidacloprid (IMI), only the nAChN current remains in a cell with both receptor subtypes. (c) When nAChN is blocked by the inclusion of 100 pM MLA in the bath, 10 μM spinosyn A slightly blocks the nAChD current. (d) nAChN current isolated with 100 nM IMI, alone and in the presence of increasing concentrations of spinosyn A (Salgado, unpublished data).

of ACh, and these effects increased dramatically with increasing concentrations of spinosyn A. Although the nAChN receptors do not desensitize, the control current drooped approximately 15% during the ACh pulse, possibly due to redistribution of ions near the membrane. At 3 nM spinosyn A, this droop was eliminated and at higher concentrations the current actually increased up to 10% during the ACh pulse, because of a synergistic interaction between spinosyn A and ACh on the nAChN receptor. This synergistic effect can also be seen from the fact that the current after the ACh pulses in the presence of spinosyn A decays much more slowly than in the control, an effect also noted by Salgado et al. (1997). The effect of spinosyn A on the nAChN receptors in **Figure 12d** was quantified by measuring the fraction of unmodified receptors remaining, taken as the amplitude of the rapid increase in current in the first second of ACh application as a fraction of the peak current in the control. For example, for the 10 nM spinosyn A treatment in **Figure 12d**, the result is 59% unmodified and for 100 nM it is 10%. The EC_{50} (concentration for 50% activation) for activation of nAChN receptors by spinosyn A, measured in the presence of 100 nM IMI to block nAChD receptors (**Figure 12d**) was 10 ± 1 nM ($n = 3$). Because of the synergistic interaction of spinosyns with competitive nicotinic agonists, the EC_{50} for activation of nAChN

receptors was also measured in the absence of IMI, giving a value of 27 ± 4 nM ($n = 4$) (Salgado and Saar, 2004).

The synergistic interaction between spinosyn A and classical, competitive agonists acting at the ACh binding site shows that spinosyns are in fact allosteric agonists of nAChN. This is further supported by the fact that α-BGTX blocks the spinosyn-induced current noncompetitively and that spinosyn A does not displace binding of [³H]α-BGTX (Salgado et al., 1997). Under certain circumstances the synergism between spinosad and classical nicotinic agonists, such as neonicotinoids, is demonstrable in practice (Andersch et al., 2003).

Note also in **Figure 12d** that the total current in the presence of 100 nM spinosyn A and 100 μM IMI is more than twice the total current evoked by ACh alone; spinosyn A is a more efficacious agonist than ACh and can therefore have very strong effects on cells that have a significant number of nAChN receptors. The sensitivity of any cell to the agonistic effects of spinosyns on nAChN receptors will be dependent on the number of those receptors and the input resistance of the cell, which together determine what proportion of the receptors need to be activated to significantly depolarize the cell and change its firing pattern. Symptoms arise in cockroaches at an internal aqueous concentration of 20 nM, at which 45% of nAChN receptors are activated (**Table 6**). This represents the sensitivity of the most sensitive cells. This value appears relatively large, suggesting that cells have homeostatic mechanisms to compensate for a relatively large amount of inward current. In fact, DUM neurons first altered their firing pattern when the spinosyn A concentration reached 200 nM (**Figure 9**), at which 80% of nAChN receptors are activated. This may be related to the fact that DUM neurons fire spontaneously and have a relatively low input resistance.

6.4.3.3. Correlation of spinosyn biological activity with potency on nAChN receptors

It has already been shown (**Table 6**) that symptoms appear when the internal aqueous concentration of spinosyn A reaches the range of 20 nM, which is also near the EC_{50} for activation of nAChN receptors, and is also in the range where significant nervous system stimulation is measurable. These results indicate that activation of nAChN receptors can account for the insecticidal activity of spinosyn A.

Potency in activating nAChN receptors was determined for 15 spinosyns and spinosoids, and the data are shown in **Table 7** and plotted in **Figure 13**. There is a clear relation between activity on cockroach nicotinic receptors and insecticidal activity against first instar *H. virescens* larvae. These results provide further support for the importance of nAChR activation in spinosyn poisoning.

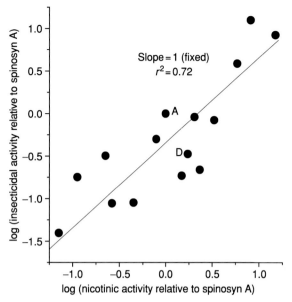

Figure 13 Relationship between relative insecticidal activity against first instar *Heliothis virescens* larvae in a drench assay and relative potency against nAChN receptors, for 15 spinosyns and spinosoids that were tested and had measurable activity in both assays. The regression line had a slope of 0.91, and $r^2 = 0.72$. The data and structural information for the spinosyns included in this figure are shown in **Table 7**. (Adapted from Salgado, V.L., Watson, G.B., Sheets, S.S., **1997**. Studies on the mode of action of spinosad, the active ingredient in Tracer® insect control. Proc. 1996 Beltwide Cotton Production Conf., pp. 1082–1086.)

Table 7 Spinosyns and spinosoids tested for nicotinic activity. Each cell was calibrated with 100 nM spinosyn A and the potency of the test compound was determined by increasing the concentration gradually to bracket the magnitude of the response to 100 nM spinosyn A. The potency of the test compound relative to spinosyn A was determined by dividing 100 nM by the interpolated equi-effective concentration of the test compound. At least two determinations were made for each compound. The data are plotted in **Figure 13**

Number	Spinosyn or spinosoid	Relative nicotinic potency	Relative Heliothis activity
	4″-N-Me Quat	0.071	0.039
	13,14-Epoxide	0.112	0.18
74	5,6-β-Epoxide	0.225	0.32
8	Spinosyn H	0.26	0.088
73	5,6-α-Epoxide	0.45	0.089
72	5,6-H2	0.79	0.50
1	Spinosyn A	1.0	1.0
	5,6-H2, 4″-N-deMe	1.49	0.18
4	Spinosyn D	1.73	0.33
2	Spinosyn B	2.0	0.91
23	Spinosyn K	2.3	0.22
	3′-O-Et, 4″-N-deMe	3.3	0.83
77	5,6-H2, 3′-O-ethyl	5.9	3.8
62	2′,3′,4′-tri-O-ethyl A	8.2	12.5
46	3′-O-ethyl A	15	8.3

6.4.4. Effects of Spinosyns on GABA Receptors in Small Diameter Neurons of Cockroach

Watson (2001), using the whole-cell patch clamp method, found that in cockroach neurons with a diameter less than 20 μm, the GABA-activated chloride current was completely and irreversibly antagonized by as little as 5 nM spinosyn A (**Figure 14**), whereas GABA-activated chloride currents in cells larger than 25 μm were not affected. Over the same concentration range, a picrotoxinin-sensitive chloride current was activated by spinosyn A. The chloride current activated by spinosyn A is small in relation to the GABA response, an average of 124 pA in cells with GABA-activated chloride currents generally exceeding 2 nA, but because of the occurrence of GABA receptor antagonism and chloride current activation in the same concentration range, both of these effects might be due to a low efficacy partial agonist action of spinosyn A on a subtype of GABA receptor that appears to be found specifically in small neurons. This would presumably be an allosteric effect, due to binding of spinosyns at a modulatory site analogous to that for spinosyns on nicotinic receptors or avermectins

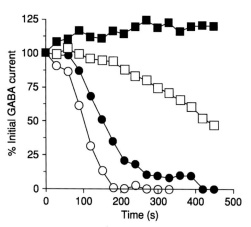

Figure 14 Representative data from individual neurons, showing loss of γ-aminobutyric acid (GABA) currents after the application of increasing concentrations of spinosyn A. With ATP in the recording electrode, responses to 20 mM GABA were fairly constant over time (filled squares). When spinosyn A was added to the perfusing medium at 1 nM (open squares), 5 nM (filled circles), and 100 nM (open circles), responses to GABA diminished over time. (Reprinted with permission from Watson, G.B., **2001**. Actions of insecticidal spinosyns on γ-aminobutyric acid responses from small-diameter cockroach neurons. *Pestic. Biochem. Physiol. 71*, 20–28; © Elsevier.)

on glutamate and GABA gated chloride channels. Spinosyns did not antagonize the binding of [³H]ivermectin to membranes prepared from several insect species, including cockroach (N. Orr, unpublished data). Although [³H]ivermectin appears to bind to both GABA and GluCl receptors (Ludmerer *et al.*, 2002), it is not known how much of this binding is to the spinosyn-sensitive GABA receptor subtype present in small neurons. That component may be so small as to be undetectable in binding studies, but could nevertheless be very important physiologically.

The observed effect on GABA receptors in small neurons would be expected to have significant effects on the function of affected inhibitory synapses in poisoned insects, but it is not yet clear what the functions of these synapses are or what effect their disruption by spinosyns has on the insect. Since the activated current is relatively small, the overriding effect may be loss of inhibition caused by the loss of GABA receptor function. However, it cannot be excluded that the induced chloride current induces inhibition in some neurons. The excitatory symptoms first become significant in cockroaches at internal aqueous concentrations in the range of 20 nM, whereas block of the GABA receptors in small neurons is strong, if not complete, at 1 nM, and quite rapid and complete at 5 nM (**Figure 14**). While the activation of nicotinic receptors requires a somewhat higher spinosyn A

concentration than the effects on GABA receptors in small neurons, its concentration dependence is such that it can account for the excitatory symptoms. Inhibition of GABA receptors in small neurons may act synergistically with nicotinic receptor activation to cause the excitatory symptoms.

In summary, spinosyns cause excitatory symptoms in insects by stimulating activity in the CNS. Activation of nondesensitizing nAChRs by an allosteric mechanism appears to be the primary mechanism by which spinosyns stimulate the neuronal activity, but it has also been shown that spinosyns potently antagonize GABA responses of small neurons, and this could also play a significant role in spinosyn poisoning. Both of these mechanisms are unique among insect control agents.

6.5. Biological Properties of the Spinosyns

6.5.1. Biological Activity and Spectrum of the Spinosyns

6.5.1.1. Pest insects The spinosyns are active against a broad range of insect pests, including many pest species among the Thysanoptera, Coleoptera, Isoptera, Hymenoptera, and Siphonaptera. However, for the spinosyns, the main focus for the activity is centered on the Lepidoptera and to a lesser degree the Diptera (Thompson *et al.*, 1995; Bret *et al.*, 1997; DeAmicis *et al.*, 1997; Sparks *et al.*, 1999; Kirst *et al.*, 2002a). The spinosyns are active against a wide variety of lepidopteran pest species including tobacco budworm (*Heliothis virescens*), cotton bollworm (*Helicoverpa zea*), diamondback moth (*Plutella xylostella*), fall armyworm (*Spodoptera frugiperda*), beet armyworm (*S. exigua*), cabbage looper (*Trichoplusia ni*), American bollworm (*Heliothis armigera*), rice stemborer (*Chilo suppressalis*), and many others (Thompson *et al.*, 1995; Bret *et al.*, 1997). This broad spectrum of pest insect activity is coupled with a high level of potency. Spinosyn A, the most abundant factor produced by *S. spinosa*, is very active against pest lepidopterans, with LC_{50} values in the same range as some pyrethroid insecticides, such as permethrin, cypermethrin, or λ-cyhalothrin (**Table 8**).

6.5.1.1.1. Structure–activity relationships of spinosyns – Lepidoptera Among the spinosyns produced by *S. spinosa*, spinosyn A is the most active against pest lepidopterans such as *H. virescens* (**Table 2**). Interestingly, spinosyn A is also the most abundant factor produced in the wild-type strains of *S. spinosa*. Other spinosyns, such as B and C (the 4″-N-demethyl and 4″-N,N-didemethyl derivatives of spinosyn A, respectively) (**Table 2**), are nearly as active as spinosyn A against *H. virescens*, as is the second most abundant factor, the 6-methyl analog, spinosyn D. Loss of a methyl group at the 2′-, 3′-, or 4′-position of the rhamnose moiety reduces lepidopteran efficacy by about an order of magnitude, as measured by activity against *H. virescens* larvae (spinosyns H, J, K, respectively) (**Table 2**). A comparable loss in activity is also associated with loss of a methyl group at C16 or a reduction in alkyl size (ethyl to methyl) at C21 (spinosyns F and E, respectively) (**Table 2**). Interestingly, an increase in alkyl size at C21 (ethyl to *n*-propyl) results in a slight improvement in activity towards *H. virescens* larvae (**Table 3**), while a further alkyl extension at C21 (2-*n*-butenyl) is about as active as spinosyn A (C21 = ethyl). Among the spinosyns arising from mutant strains (2′, 3′, or 4′-demethyl rhamnose derivatives) (**Table 2**), virtually all were far less active than spinosyn A. The one notable exception to this is spinosyn Q (LC$_{50}$ 0.5 ppm) which is close in activity to spinosyn A.

6.5.1.1.2. Structure–activity relationships of spinosyns – Diptera In general, the pattern of activity for the different spinosyns towards adult stable flies is similar to that observed for the Lepidoptera (Kirst *et al.*, 2002a). For example, spinosyns A–D and Q are nearly equitoxic to the adult stable fly (*Stomoxys calcitrans*), while the C16 demethyl analog (spinosyn F) is less active, as is spinosyn K (**Table 2**). However, unlike with *H. virescens* larvae, spinosyns E, H, and J are as active as spinosyn A (**Table 2**). The difference for spinosyn J is especially interesting since it is at least 100-fold less active than spinosyn A against larvae of *H. virescens*. Thus, there may be some significant differences in the spinosyn target site in adult stable flies, in the metabolism/conjugation of spinosyn J, or both, compared to larvae of *H. virescens*. Larval blowfly (*Calliphora vicina*) activity of the tested spinosyns was even more like lepidopteran activity than *S. calcitrans* activity (Kirst *et al.*, 2002a). As in *H. virescens* larvae, spinosyns A–D are as toxic as spinosyn A, while spinosyns E and F, and the 2′, 3′, or 4′-demethyl analogs (spinosyn H, J, and K, respectively) are less active than spinosyn A. However, unlike in larvae of *H. virescens*, spinosyn J still displayed a reasonable level of activity towards larvae of *C. vicina* (**Table 2**).

6.5.1.1.3. Structure–activity relationships of spinosyns – Homoptera Unlike members of the Lepidoptera and Diptera, homopterans such as

aphids are not very susceptible to the spinosyns (DeAmicis *et al.*, 1997; Sparks *et al.*, 1999; Crouse *et al.*, 2001). With the exception of spinosyns B, K, and O, the spinosyns of *S. spinosa* for which data are currently available are not very active against cotton aphid (*Aphis gossypii*) (**Table 2**). Spinosyn activity against the aster leafhopper (*Macrosteles quadrilineatus*) reflects a similar pattern in that spinosyn K is again the most active of the spinosyns tested, although comparatively, spinosyns A and B are also active. Interestingly, other 4′-O-demethyl spinosyns such as spinosyn O and the 2′,4′-di-O-demethyl spinosyns U and V are also quite active against the leafhoppers, relative to spinosyn A (**Table 2**).

6.5.1.1.4. Structure–activity relationships of spinosyns – Acarina The spinosyns are also active against tetranychid mite species such as the two-spotted spider mite (*Tetranychus urticae*) (Sparks *et al.*, 1999; Crouse *et al.*, 2001). As observed with some of the homopteran species, spinosyns K and O are among the most active of the natural spinosyns from *S. spinosa*, while spinosyns H and Q are comparatively weaker than spinosyn A (**Table 3**). Likewise, spinosyn B is also relatively active. As noted previously, there is only a weak correlation between activity towards neonate *H. virescens* larvae versus spinosyn activity towards *T. urticae* (Sparks *et al.*, 1999). While improvements in mite activity are possible, the physical properties affecting residuality and translaminar characteristics are such that the spinosyns are typically not well suited for mite control compared to available commercial standards (Crouse *et al.*, 2001).

6.5.1.2. Nontarget organisms
6.5.1.2.1. Beneficial insects The spinosyns, exemplified by spinosad, are highly efficacious against a variety of pest insect species, yet relatively weak against towards a variety of beneficial insect species, especially predatory insects. A detailed summary of studies involving the effects of spinosad on a wide variety of beneficial insects has recently been published (Williams *et al.*, 2003). Spinosyns such as spinosad typically have little effect on predatory insects and mites (Williams *et al.*, 2003). Where particular laboratory bioassays have shown some spinosad toxicity to predators (e.g., *Orius insidiosus*), corresponding greenhouse and field tests have generally shown spinosad to have little effect (Studebaker and Kring, 2003), principally due to the rapid environmental degradation of spinosad (Saunders and Bret, 1997; Crouse *et al.*, 2001; Williams *et al.*, 2003). Based on data in Williams *et al.* (2003), in more than 88% of the greenhouse/field assays

Table 8   Toxicity values for selected insect control agents

Compound	Rat/mouse/rabbit (LD_{50}, $mg\,kg^{-1}$) Oral	Dermal[a]	Avian; quail or mallard (LD_{50}, $mg\,kg^{-1}$)	Fish[b] (LC_{50}, ppm)	Daphnia (LC_{50}, ppm)	Heliothis virescens, third instar, topical treatment (LD_{50}, $\mu g\,g^{-1}$)	VSR, oral/Hv
DDT	87–500	1931	611	0.08t	0.0047	52–152	0.57–9.6
Cyclodienes							
Dieldrin	40–100	52–117	37	0.0012t	0.00024		
Endrin	3	12–19	14	0.00075t	0.0042	46.7	0.06
Endosulfan	18	74–130	805	0.0014t		73.3	0.25
Carbamates							
Carbaryl	307	>500 to >4000	>5000	1.95t	0.0056	136–232	1.32–2.25
Methomyl	17	>5000r	1100–3600	3.4t	0.0088	4.33–26.7	0.64–3.93
Aldicarb	0.9	2.5–5	594	0.6t	0.061	571	
Carbofuran	8	2550	190	0.38t			
Organophosphates							
Methyl parathion	9–42	63–72	90	3.7t	0.00014	11.6–65.0	0.14–3.62
EPN	7–65	22–262	349–437	0.21t	0.32	37	0.19–1.76
Chlorpyrifos	82–245	202–2000	940	0.0071t	0.0017	79.5	1.03–3.1
Acephate	866–945	>2000r		445t	>50	41.0–74.3	11.7–23
Methamidophos	13–30	110	8–11			85.7–150	0.09–0.35
Profenofos	400	472–1610		0.024t	0.0014	11	36.4
Pyrethroids							
Permethrin	2000 to >4000	>4000		0.0023t		1.33–2.79	717 to >3000
Fenvalerate	451	>5000	9932	0.0036t		0.40–1.89	239–1128
Cypermethrin	247	>2000	>10000	0.025t	0.0013	0.24–1.61	153–1029
λ-Cyhalothrin	56–79	632–696	>500–3950			0.93	60–85
Tralomethrin	1070–1250	>2000r				0.251	4263–4980
Insect growth regulators							
Pyriproxyfen	>5000	>2000	>2000	0.45–2.7	0.40		
Tebufenozide	>5000		>2150	3.0–5.7	3.8		
Methoxyfenozide	>5000	>2000	>5620	>4.3	3.7		
Neonicotinoids							
Imidacloprid	450	>5000	31–5000	211	32–85	350	na
Thiacloprid	444–836	>2000	2716q	30.5t	>85	12.6	na
Thiamethoxam	1563	>2000	576–1552	>114–125			
Acetamiprid	146–217	>2000		>100c	>100	~400	na
Dinetofuran	>2000	>2000	1000 to >2000	>40 to >1000			
Phenylpyrazoles							
Fipronil	100	>2000	31–2150	0.25–0.43	0.19		
Ethiprole	>2000	>2000					
Avermectins							
Abamectin	10.6–11.3		>2000	0.042	0.00032	1.16	9.1–9.7
Emamectin benzoate	70	>2000r	76–264	0.67		0.10	700
Pyrroles							
Chlorfenapyr	>626	>2000	10–34	7.4–11.6		4.5	>139
Oxadiazines							
Indoxacarb	>5000	>2000	>2250	0.5		0.93	>5376
Spinosyns							
Spinosad	3738 to >5000	>5000	>1333	6–30	93	1.28–2.25	1661 to >3906
Dihalopropenyl aryl ethers							
Pyridalyl	>5000	>5000	1133 to >5620	0.50t		0.82	>6097

[a]Dermal toxicity data, r = rabbit.

[b]Fish values are for carp unless otherwise indicated; t = trout.

VSR, vertebrate selectivity ratio: rat/mouse oral LD_{50}/topical LD_{50} for *H. virescens* larvae (see Section 6.5.1.2.2); na, not available. Data adapted from Ware, G.W., **1982**. Pesticides: Theory and Application. W.H. Freeman, San Francisco, CA; Larson, L.L., Kenaga, E.E., Morgan, R.W., **1985**. Commercial and Experimental Organic Insecticides. Entomological Society of America, College Park, MD, involving predatory insects, the effects of spinosad on the predators fall into the harmless to slightly harmful categories, with the bulk (79%) in the harmless category.

Initial studies with spinosad suggested that there was some activity against hymenopterous parasitoids (*Encarsia formosa*) and honeybees (*Apis mellifera*), but that these pests were much less sensitive to spinosad compared to more conventional insect control agents such as the pyrethroids (e.g., cypermethrin) (Thompson *et al.*, 1995, 2000). Subsequent studies, as recently reviewed by Williams *et al.* (2003), show hymenopterous parasitoids to have varying levels of susceptibility to spinosad, with many, unlike predaceous insects, tending to be relatively sensitive. While not all parasitoids are susceptible to spinosad (Elzen *et al.*, 2000; Papa *et al.*, 2002; Hill and Foster, 2003), studies with a wide variety of parasitoids show spinosad to be toxic to many of them, especially when applied directly to the parasite (Williams *et al.*, 2003). However, unlike most of the lepidopteran and dipteran pest insect larvae targeted for control by spinosad, parasitoids are typically highly mobile. As such if they are able to escape the initial, direct contact with wet residues, these same parasitoids may not be highly impacted by the dried residues, especially after a few days (e.g., Tillman and Mulrooney, 2000). As noted above, spinosad is rapidly degraded in the environment (Sanders and Bret, 1997; Crouse *et al.*, 2001). This rapid degradation contributes to observations in greenhouse and field studies that parasitoid populations, even those that are relatively sensitive to spinosad, can recover in a short period of time (1–2 weeks) (Williams *et al.*, 2003).

Thus, as with all insect control agents, the use and efficacy of spinosad should be based on a variety of factors, including the presence and role of any beneficial insects. Compared to many of the organophosphate and pyrethroid insect control agents available for pest insect control, spinosad, as well as other recently introduced insect control agents (Sparks, 2004), have a reduced overall potential to adversely impact beneficial insects in the field.

6.5.1.2.2. Mammalian toxicology and ecotoxicology

Among the critical attributes for any new insect control agent is the need to provide a high level of comparative safety to nontarget species and the environment.

The highly desirable trend with new insect control agents has been to increase efficacy towards target pest species while attempting to reduce the relative toxicity towards mammals and other nontarget species. Among the general populace, natural products are often viewed as inherently safer or "greener" than synthetic organic compounds, a view that ignores or is ignorant of the fact that many natural products are highly toxic to mammals and/or nontarget species (Thompson and Sparks, 2002). However, the spinosyns, in the form of spinosad, do indeed couple excellent levels of efficacy against pest insect species with a highly desirable environmental/toxicological profile.

Data for spinosad, as well as several new insect control agents, shows that toxicity to mammalian, avian, and aquatic species is relatively low when compared with many older and currently used insect control agents (**Table 8**). The relative selectivity of an insect control agent can be quantified (therapeutic index) by comparing the activity of a compound on a target insect pest with the activity towards a nontarget species. One such therapeutic index applied to insect versus mammalian activity is the vertebrate selectivity ratio (VSR) defined by Hollingworth (1976):

$$VSR = \frac{\text{Acute rat oral LD}_{50} \text{ in mg kg}^{-1}}{\text{Insect LD}_{50} \text{ in } \mu g\, g^{-1}} \quad [1]$$

The VSR for spinosad is among the most favorable in a set of now classic, current and newer insect control agents, when comparing *H. virescens* larvae topical toxicity and rat oral toxicity (Sparks *et al.*, 1999; Thompson and Sparks, 2002) (**Table 8**). Thus, the superb mammalian toxicity profile observed for spinosad is further enhanced by the high level of efficacy towards the target insect pests, thereby contributing to spinosad's very favorable toxicology

pp. 1–105; Sparks, T.C., Thompson, G.D., Kirst, H.A., Hertlein, M.B., Larson, L.L., *et al.*, **1998**. Biological activity of the spinosyns, new fermentation derived insect control agents, on tobacco budworm (Lepidoptera: Noctuidae) larvae. *J. Econ. Entomol.* 91, 1277–1283; Hollingworth, R.M., **2001**. New insecticides: mode of action and selective toxicity. In: Baker, D.R., Umetsu, N.K., (Eds.), Agrochemical Discovery: Insect, Weed and Fungal Control. American Chemical Society, Washington, DC, pp. 238–255; Smith, P. (Ed.), **2001**. AgroProjects: PestProjects – 2001. PJB Publications, Richmond, UK; Thompson, G.D., Sparks, T.C., **2002**. Spinosad: a green natural product for insect control. In: Lankey, R.L., Anastas, P.T. (Eds.), Advancing Sustainability through Green Chemistry and Engineering. American Chemical Society, Washington, DC, pp. 61–73; and references therein; Sparks, T.C., **1996**. Toxicology of insecticides and acaricides. In: King, E.G., Phillips, J.R., Coleman, R.J. (Eds.), Cotton Insects and Mites: Characterization and Management. The Cotton Foundation, Memphis, TN, pp. 283–322; Sparks, T.C., **2004**. New insect control agents: modes of action and selectivity. In: Endersby, N.M., Ridland, P.M. (Eds.), The Management of Diamond back Moth and Other Crucifers Pests: Proc. 4th Intl. Workshop, Melbourne, Australia, 26–29 November 2001, pp. 37–44; extoxnet.orst.edu; and unpublished data.

profile. Additionally, tests conducted for EPA registration showed spinosad to neither leach nor persist in the environment. These factors all contributed to the EPA's registration of spinosad as a reduced risk pesticide in early 1997 (Thompson et al., 2000; Thompson and Sparks, 2002), and the bestowing of the Presidential Green Chemistry Award in 1999.

6.6. Resistance Mechanisms and Resistance Management

6.6.1. Cross-Resistance

Many of the primary insect pests in major markets have a history of developing resistance to insect control agents. Thus, efficacy against insecticide resistant pest insects and the potential for the development of resistance are increasingly important considerations in the development decision for any new insect control agent. Early in the development process, spinosad was examined for efficacy against a variety of insecticide resistant strains of pest lepidopterans, such as larvae of *H. virescens* and *P. xylostella* (Sparks et al., 1995). *Heliothis virescens* larvae collected from a number of locations in the mid-southern USA were subjected to discriminating dose assays for a variety of insecticides. Relative to the susceptible strains, larvae collected from almost all of these locations were poorly controlled by cypermethrin (pyrethroid) or profenophos (organophosphate). In contrast, spinosad provided very good to excellent control of all of the strains tested (Sparks et al., 1995), demonstrating its utility against insecticide-resistant strains possessing what was likely a mixture of resistance mechanisms (based on the diverse nature of the other insect control agents). Likewise, studies with laboratory strains of *P. xylostella* selected for high levels of resistance to pyrethroid, avermectin, or acylurea insecticides and known to possess enhanced glutathione transferase and/or monooxygenase activity (Chih-Ning Sun, personal communication), showed no appreciable levels of cross-resistance to spinosad (Sparks et al., 1995) (**Table 9**). One subsequent study with field-selected strains of *P. xylostella* resistant to the pyrethroid permethrin, has shown a low level of, but broad, cross-resistance to several insect control agents including spinosad and the avermectin emamectin benzoate (Shelton et al., 2000) (**Table 9**). However, virtually all other studies involving a variety of insecticide resistant species and strains have shown no cross-resistance to spinosad (**Table 9**). For example, a multiresistant strain of housefly (*Musca domestica*) that was highly resistant (1800-fold and 4200-fold) to the pyrethroids permethrin, and cypermethrin, respectively, exhibited no cross-resistance to spinosad (Liu and Yue, 2000) (**Table 9**). Spinosad was highly effective against a number of strains that were highly resistant to other insect control agents including abamectin (decreased penetration and altered target site), cyclodienes (*rdl*), pyrethroids (monooxygenases, knockdown resistance, and reduced penetration), and organophosphates (altered acetylcholinesterase) (Scott, 1998). Only with the multiresistant LPR strain of housefly was any cross-resistance noted, and that was only at a very low level (Scott, 1998). Taken in total, these data are most consistent with the presence of a unique mode of action for the spinosyns as well as what may be an apparent limited susceptibility of the spinosyn structure to insect metabolic systems.

6.6.2. Resistance Mechanisms

While almost all insecticide-resistant strains of pest insects have exhibited little or no cross-resistance to the spinosyns/spinosad, it is nearly axiomatic that if sufficient pressure is put on an insect population, resistance will be developed to any insect control agent, and the spinosyns/spinosad are certainly no exception. This has indeed been borne out in that there are now a few examples of spinosad resistance being developed in the laboratory and in the field.

Resistance to spinosad has been selected for in the laboratory. Bailey et al. (1999) used a topical selection protocol to generate a high level of spinosad resistance (toxicity ratio, TR = 1068-fold, topical assay) in larvae of *H. virescens* after 14 generations of selection using methods that favored the development of resistance (NC spinosad-R) (Young et al., 2001, 2003). Likewise, a spinosad resistant strain (NYSPINR) of *M. domestica* (TR = >150) was selected in the laboratory starting from a composite of field strains (Shono and Scott, 2003). From this multiresistant NYSPINR strain, a further spinosad resistant strain (*rspin*) was developed that was specifically resistant to only spinosad and also incorporated a number of recessive mutant markers (Shono and Scott, 2003). For spinosad resistant strains of both *H. virescens* (NC-spinosad-R) and housefly (*rspin*), cross-resistance to other insect control agents such as emamectin benzoate, indoxacarb, acetamiprid, abamectin, etc. was negligible (Young et al., 2001; Shono and Scott, 2003) (**Table 9**). Both studies also showed spinosad resistance to be linked to a single, non sex-linked, recessive gene (Shono and Scott, 2003; Wyss et al., 2003). Studies with the NYSPINR housefly strain also showed no synergism of spinosad activity with piperonyl butoxide, *S,S,S*-tributyl-phosphorothioate (DEF), or diethyl maleate (DEM), suggesting that metabolism was not likely to be the basis for the observed resistance. In the case of the NC-spinosad-R *H. virescens* larvae, studies also showed that the resistance to spinosad did not appear to be associated with enhanced metabolism, decreased penetration, or excretion (Young *et al.*, 2000, 2001). At low concentrations, a decreased neural sensitivity was observed in the resistant strain (Young *et al.*, 2001). Thus, available data for both these spinosad resistant strains is most consistent with an altered target site resistance mechanism.

In addition to the laboratory selection studies, there are now a couple of studies indicating the presence of resistance to spinosad in field-collected insects. Surveys of beet armyworm (*Spodoptera exigua*) populations from the US and Thailand showed that there was a wide range in the biological response to spinosad when compared to a US Department of Agriculture (USDA) susceptible reference colony (Moulton *et al.*, 1999, 2000). Compared to the USDA larvae, toxicity ratios for an Arizona strain ranged from 14-fold to 20-fold, depending on assay method, while those of the Thailand strain ranged from 58-fold to 85-fold. Interestingly, the F_1 hybrids from the Thailand colony possessed toxicity ratios of 22-fold compared to the USDA strain, suggesting that the reduced susceptibility was incompletely dominant (Moulton *et al.*, 2000). Subsequent studies with the Thailand strain showed that susceptibility was returning after some time in laboratory culture (toxicity ratio declined to 15-fold) and that synergists (Piperonyl butoxide, DEF, and DEM) did not alter spinosad activity (Moulton *et al.*, 1999).

In mid-1998, spinosad was registered in Hawaii, for use on crucifers, for the control of *P. xylostella* larvae that had become resistant to all other then available insect control agents (Zhao *et al.*, 2002). Unfortunately, resistance management strategies incorporated into the spinosad label by the manufacturer did not foresee the particular cultural practices used in Hawaii. In many cases, the practices then in use resulted in nearly exclusive, almost continuous, year-round, weekly applications of spinosad to sequential crop plantings involving the same *P. xylostella* population, a situation that went on for more than 2 years. In 2000, field failures were observed with spinosad in Hawaii. Subsequent laboratory testing of the *P. xylostella* populations around the growing regions of the Hawaiian Islands demonstrated the presence of high levels of resistance to spinosad in several locations (Zhao *et al.*, 2002). Further laboratory selection, for nine generations, of one of the most resistant colonies (Pearl City, Oahu) with spinosad, increased resistance towards spinosad to an even larger degree (Zhao *et al.*, 2002). As noted above for the laboratory-selected spinosad-resistant

strains of *H. virescens* and *M. domestica*, resistance to spinosad in this spinosad-selected strain of *P. xylostella* was also an autosomal, incompletely recessive trait (Zhao *et al.*, 2002), which has led to the rapid dissipation of resistance. Likewise, the synergists piperonyl butoxide and DEF did not alter the level of resistance to spinosad, and there was no cross-resistance to either emamectin benzoate or indoxacarb. While these data are of a limited nature, like those above, they are most consistent with an altered target site as the resistance mechanism.

6.6.3. Resistance Management

The development of any new insect control agent is a very expensive proposition. For a company to derive full value, a new molecule must remain in the marketplace for many years. Thus, the development of resistance has become an increasingly important consideration in the development of any new insect control agent. In light of spinosad's uniqueness, as well as the expense and effort that led to its development, insecticide resistance management (IRM) was a consideration at an early stage in the development of spinosad, (Thompson *et al.*, 2000). The goal of this effort was to increase the likelihood of spinosad's long-term efficacy and utility in the marketplace. One aspect of the IRM program was the incorporation of use patterns on the product label that reduced the likelihood of selecting back-to-back generations of insect pests. The IRM recommendation from the manufacturer included limiting use to three or fewer applications within a 30-day period, followed by another 30-day period during which spinosad was not to be used (Zhao *et al.*, 2002). For most lepidopterous pests, this spinosad-free period would allow at least one complete generation that had not been selected. A total of only six applications was allowed per crop (Zhao *et al.*, 2002). Another aspect of the spinosad IRM plan was implementation of a program to monitor the susceptibility of key insect pests.

As demonstrated repeatedly over the past 50 years, given time and very heavy selection pressure, resistance can develop to any insect control agent (Sparks *et al.*, 1993; Clark and Yamaguchi, 2002). Likewise, as demonstrated by the development of spinosad resistance in *P. xylostella* in Hawaii, not every use pattern or scenario can be foreseen. The practice of having multiple "crops" at different stages in adjacent fields effectively negated IRM recommendations for spinosad. However, the monitoring program that was in place was able to quickly provide needed susceptibility data for a number of sites in the Hawaiian Islands. While the initial data showed most sites to be free of problems, several

Table 9 Resistance and cross-resistance to spinosad

Species	Strain	Select	Compound	Method	Toxicity ratio	Reference
Spinosad resistant strains						
Heliothis virescens	NC spinosad-R	Lab G17	Spinosad	Larval dip	>1000	Young *et al.* (2001)
	NC spinosad-R	Lab G17	Permethrin	Larval dip	0.65	
	NC spinosad-R	Lab G17	Profenofos	Larval dip	1.70	
	NC spinosad-R	Lab G17	Emamectin benzoate	Larval dip	1.91	
	NC spinosad-R	Lab G17	Indoxacarb	Larval dip	1.45	
	NC spinosad-R	Lab G17	Acetamiprid	Larval dip	0.41	
Plutella xylostella	Pearl City	Field	Spinosad	Leaf dip	1080	Zhao *et al.* (2002)
	Pearl City	Field	Emamectin benzoate	Leaf dip	4	
	Pearl City	Field	Indoxacarb	Leaf dip	1.28	
Musca domestica	rspin	Laboratory	Spinosad	Topical	>150	Shono and Scott (2002)
	rspin	Laboratory	Indoxacarb	Topical	0.11	
	rspin	Laboratory	Abamectin	Topical	1.2	
	rspin	Laboratory	Fipronil	Topical	0.34	
	rspin	Laboratory	Cyfluthrin	Topical	0.42	
	rspin	Laboratory	Dimethoate	Topical	0.31	
	rspin	Laboratory	Chlorfenapyr	Topical	0.77	
	rspin	Laboratory	Methomyl	Topical	0.61	
Spodoptera exigua	Arizona	Field	Spinosad	Leaf dip	14–20	Moulton *et al.* (2000)
	Thailand	Field	Spinosad	Leaf dip	58–85	Moulton *et al.* (2000)
Spinosad cross-resistance						
Plutella xylostella	Fenvalerate-R	Laboratory	Cypermethrin	Topical	1520	Sun (1991), cited in
	Fenvalerate-R	Laboratory	Spinosad	Topical	3.2	Sparks *et al.* (1995)
	Tebufenozide-R		Cypermethrin	Topical	27	Sun (1991), cited in
	Tebufenozide-R		Spinosad	Topical	0.9	Sparks *et al.* (1995)
	Abamectin-R	Laboratory	Cypermethrin	Topical	64	Sun (1991), cited in
	Abamectin-R	Laboratory	Spinosad	Topical	2.4	Sparks *et al.* (1995)
Plutella xylostella	Ocean Cliff	Field	Permethrin	Leaf dip	126	Shelton *et al.* (2000)
	Ocean Cliff	Field	Spinosad	Leaf dip	14.5	
	Ocean Cliff	Field	Emamectin benzoate	Leaf dip	13.0	
	Ocean Cliff	Field	Chlorfenapyr	Leaf dip	1.0	
	Ocean Cliff	Field	Methomyl	Leaf dip	7.1	
	Oxnard	Field	Permethrin	Leaf dip	206	Shelton *et al.* (2000)
	Oxnard	Field	Spinosad	Leaf dip	12.8	
	Oxnard	Field	Emamectin benzoate	Leaf dip	6.0	
	Oxnard	Field	Chlorfenapyr	Leaf dip	1.7	
	Oxnard	Field	Methomyl	Leaf dip	6.5	
Plutella xylostella	His-hu 2001	Field	Abamectin	Leaf dip	2497	Kao and Cheng (2001)
	His-hu 2001	Field	Emamectin benzoate	Leaf dip	305	
	His-hu 2001	Field	Fipronil	Leaf dip	65	
	His-hu 2001	Field	Chlorfeapyr	Leaf dip	9	
	His-hu 2001	Field	Spinosad	Leaf dip	65	
	His-hu 2001	Field	Azadirachtin	Leaf dip	>1000	
	Lu-chu 2001	Field	Abamectin	Leaf dip	4988	Kao and Cheng (2001)
	Lu-chu 2001	Field	Emamectin benzoate	Leaf dip	153	
	Lu-chu 2001	Field	Fipronil	Leaf dip	104	
	Lu-chu 2001	Field	Chlorfeapyr	Leaf dip	10	
	Lu-chu 2001	Field	Spinosad	Leaf dip	59	

Continued

Table 9 Continued

Species	Strain	Select	Compound	Method	Toxicity ratio	Reference
	Lu-chu 2001	Field	Azadirachtin	Leaf dip	>1000	
Blattella germanica	Apyr-R	Field	Permethrin	Topical	97	Wei et al. (2001)
	Apyr-R	Field	Deltamethrin	Topical	480	
	Apyr-R	Field	Spinosad	Topical	1.3	
Musca domestica	AVER	Field	Abamectin	Topical	>1000	Scott (1998)
	AVER	Field	Spinosad	Topical	1.9	
	LPR (Multi-R)	Field–Lab.	Spinosad	Topical	4.3	
	OCR (cyclodiene)		Cyclodiene	Topical	high	
	OCR (cyclodiene)		Spinosad	Topical	0.4	
	Cornell-R	Field-Lab.	Spinosad	Topical	0.9	
	R12 (CYP6D1)	Laboratory	Spinosad	Topical	1.8	
	R3 (kdr and pen)	Laboratory	Spinosad	Topical	1.4	
Musca domestica	ALHF (multi R)	Field	Permethrin	Topical	1800	Liu and Yue (2000)
	ALHF (multi R)	Field	Cypermethrin	Topical	4200	
	ALHF (multi R)	Field	Spinosad	Topical	1.4	
Musca domestica	Several	Field	Permethrin	Glass	high	Scott et al. (2000)
	Several	Field	Cyfluthrin	Glass	high	
	Several	Field	Spinosad	Feeding	low	
	Several	Field	Fipronil	Glass	very low	
	Several	Field	Dimethoate	Glass	low	
	Several	Field	Methomyl	Feeding	low-high	
Choristoneura rosaceana	Berrien	Field	Azinphosmethyl	Diet	27	Ahmad et al. (2002)
	Berrien	Field	Chlorpyrifos	Diet	25	
	Berrien	Field	Spinosad	Diet	0.83	
	Berrien	Field	Cypermethrin	Diet	8	
	Berrien	Field	Methoxyfenozide	Diet	3.1	
	Berrien	Field	Indoxacarb	Diet	705	
Choristoneura rosaceana	Site 2	Field	Azinphosmethyl	Leaf disc	15.5	Smirle et al. (2003)
	Site 2	Field	Spinosad	Leaf disc	1.8	
Choristoneura rosaceana	Brown 97 (OP-R)	Field	Tebufenozide	Leaf dip	12.8	Waldstein and Reissig (2000)
	Brown-96 (OP-R)	Field	Spinosad	Leaf dip	2.0	

sites were also shown to exhibit significant levels of resistance to spinosad that increased with time and were associated with failures to control *P. xylostella* larvae in the field (Zhao *et al.*, 2002). As a response to these developments, a regional IRM program was implemented (Zhao *et al.*, 2002; Mau and Gusukuma-Minuto, 2004). Spinosad was voluntarily withdrawn from use where field failures had occurred and two newly registered products, emamectin benzoate and indoxacarb, were put into a rotation scheme (Mau and Gusukuma-Minuto, 2004). Following its withdrawal from use, susceptibility to spinosad increased rapidly such that for some of the locations (Maui and Hawaii) the thresholds for spinosad reintroduction were met within 6-8 months, setting the stage for spinosad's reintroduction in early 2002 (Mau and Gusukuma-Minuto, 2004). By the fall of 2002, resistance to spinosad had also fallen below the threshold in the remaining region (Oahu) such that it was reintroduced in

early 2003 (Mau, personal communication). With its reintroduction, spinosad has become part of an IRM rotation scheme that also includes emamectin benzoate and indoxacarb (Mau, personal communication). Given the selection pressure that was placed on spinosad in several of the locations in Hawaii, it is not clear how long this rotation scheme will remain effective in the control of a pest that has developed resistance to virtually all insecticides previously used. Nevertheless, this experience with spinosad in Hawaii has demonstrated that IRM can indeed help extend, if not preserve, the life of an insect control agent.

6.7. Spinosyns and Spinosoid Structure–Activity Relationships

Exploration of the spinosyns included continued isolation and identification of new spinosyns (naturally occurring analogs of spinosyn A). This effort ultimately led to the isolation more than 20 new spinosyns from *S. spinosa* (see Section 6.2.1), and an expanding number from *S. pogona* (Hahn *et al.*, 2002; Lewer *et al.*, 2003). At the same time, running in parallel with the spinosyn isolation program, was a long-standing effort aimed at the preparation of semisynthetic derivatives/analogs of the spinosyns, termed spinosoids. In light of the excellent activity against lepidopterans and the desire for greater utility, the spinosoid synthesis program had two goals. The first was to increase activity against lepidopterans, typically using *H. virescens* larvae as an indicator species. The second goal was, if possible, to expand the spectrum of this unique chemistry. To date, the total number of spinosyns and spinosoids resulting from this long-standing and continuing effort easily exceeds 1000 molecules, the vast majority being spinosoids.

Chemistry around the spinosyns has been focused on modifications to various functionalities in the spinosyn structure, which has been both limited by and facilitated by the novel chemical nature of these molecules. These modifications can loosely be grouped into:

- modifications of the tetracycle,
- modification/replacement of the forosamine sugar, and
- modification/replacement of the tri-*O*-methyl rhamnose sugar.

Due to limited quantities of starting material, initial modifications to the spinosyn structure were limited to relatively simple or straightforward modifications, most often associated with a reduced level of activity. Later modifications did ultimately succeed in devising spinosoids that were more active than the naturally occurring spinosyns (Sparks *et al.*, 2000a, 2001; Crouse *et al.*, 2001).

6.7.1. Modifications of the Tetracycle

The tetracycle of spinosad is a rather rigid structure, the shape of which is influenced, in part, by the conjugated 13,14 double bond. Hydration of the 13,14 double bond (compounds 71 and 72) alters the three-dimensional shape of the tetracycle, which is associated with a reduction in activity (Crouse and Sparks, 1998; Crouse *et al.*, 1999) (**Table 3**). The 13,14-α-dihydro (compound 71) results in less of a conformational change and hence less of a distortion than the corresponding 13,14-β-dihydro (compound 72), which is reflected in the better activity of the α-analog (**Table 3**). Not surprisingly, reduction of both the 5,6 and 13,14 double bonds (compounds 69 and 70) reduces activity against *H. virescens* larvae (Crouse *et al.*, 1999) and is

neutral to negative for stable flies (Kirst *et al.*, 2002a) and mites (**Table 3**). In contrast, a reduction of only the 5,6 double bond (compound 70) has little effect on the three-dimensional shape or on activity towards *H. virescens* larvae, but reduces activity to stable flies, and is associated with an increase in activity against aphids and mites (Crouse *et al.*, 2001) (**Table 3**). Likewise, the 5,6-β-epoxy analog of spinosyn A is about as active against *H. virescens* larvae as spinosyn A, while the 5,6-α-epoxy analog (compound 74) is much less active (**Table 3**). Both 5,6-epoxides are slightly less effective against stable fly adults compared to spinosyn A, while the 5,6-β-epoxy analog is, just like the 5,6-dihydro derivative (compound 74), more active against mites (**Table 3**). Thus, for larvae of *H. virescens* and stable fly adults, alteration of the 5,6 double bond provides no improvement in activity, while some interesting increases in activity are noted for *T. urticae* (**Table 3**). In general, for *H. virescens* larvae, any modification to the internal structure of the tetracycle reduces activity (Crouse *et al.*, 1999), with the exception of the addition of extra unsaturation at the 7–11-position of spinosyn D (compound 73), which is about as active as spinosyn A (**Table 3**). Many other synthetic modifications have been made to the tetracycle (Crouse and Sparks, 1998; Crouse *et al.*, 1999) but none has provided a significant overall improvement in biological activity.

Where chemical synthesis has been unable to thus far succeed, mother nature, and the genetic manipulation thereof, have been able, in part, to fill the gap. Extension of the alkyl group at C21 to *n*-propyl (compound 30), produced through alteration of polyketidesyntase (PKS) modules (Burns *et al.*, 2003), provides a slight improvement in activity over spinosyn A against *H. virescens* larvae and aphids (**Table 3**). Further extension of the C21 position to 21-butenyl (compound 31), the primary factor produced by *S. pogona* (Lewer *et al.*, 2003), provides activity against *H. virescens* equivalent to spinosyn A. However, for *Aphis gossypii*, activity is further improved over C21-*n*-propyl (compound 30) (**Table 3**). Unfortuantely, as with the other modifications that have improved intrinsic aphid and mite activity, the physical properties of the spinosyns thus far discovered generally preclude the effective plant mobility and residuality necessary for effective control of aphids or mites in the field (Crouse *et al.*, 2001).

6.7.2. Modification or Replacement of the Forosamine Sugar

As described above (see Section 6.7.1) for the spinosyns, removal of a methyl group from the forosamine nitrogen tends to be a fairly neutral modification, resulting in a slight decrease (spinosyns A to B, spinosyns H to R, spinosyn L to N) or a modest increase in activity (spinosyn J to M). An increase in alkyl size for one of the positions on the forosamine nitrogen (i.e., ethyl and *n*-propyl; compounds 64 and 65) has little effect on *H. virescens* activity, while further increases in bulk greatly reduce activity (compounds 66 and 67). In contrast, *T. urticae* activity is only increased by these same modifications as is, to a lesser degree, activity against *A. gossypii* (Table 3). Replacement of the forosamine *N,N*-dimethylamine moiety with a hydroxyl (compound 68), reduces activity against *H. virescens* larvae while, apparently (response can be variable), increasing aphid activity (Table 3). Loss of the entire forosamine moiety results in major loss of activity (Sparks *et al.*, 1999) (Table 2) with only a small improvement noted by replacement with an acetate (C17-acetate, compound 79) group. Substitution of the forosamine with *N,N*-dimethylaminoacyl esters (compounds 80–82) only partially restores activity (Kirst *et al.*, 2002b) (Table 3). *N*-methylamino piperazinylacetate (compound 83) or *N,N*-dimethylaminopiperidinylacetate esters (compound 84) at C17 go further towards restoring activity against *H. virescens* and *Stomoxys calcitrans*, but still fall far short of the activity observed with forosamine (spinosyn, compound 1) (Kirst *et al.*, 2002b) (Table 3). A similar trend is observed with *S. calcitrans* (stable fly). Thus, for the lepidopterans, the above modifications to the forosamine do not lead to improvements in activity. In contrast, both aphid and mite activity were improved by replacement of the forosamine *N*-methyl groups with larger alkyl groups (compounds 64–67).

6.7.3. Modification or Replacement of the Tri-*O*-Methyl-Rhamnose Sugar

While many of the early synthetic modifications to the spinosyn structure were insecticidal, all of the early spinosoids were typified by a somewhat reduced or most often an almost total lack of *H. virescens* activity compared to spinosyn A. However, an exception was found in the form of the desmethoxy (2′-H or 3′-H) analogs of spinosyns H/Q and J, respectively (compounds 35, 36 and 43) (Table 3). These "desmethoxy" analogs were found to be as active against *H. virescens* larvae or slightly more so, than spinosyn A (Creemer *et al.*, 2000; Kirst *et al.*, 2002b) (Table 2). These modifications constituted the first clear indication that spinosoids at least matching the activity of the most active of the natural spinosyns (i.e., spinosyn A) were possible.

Concurrent with the synthetic efforts, several computer-aided modeling and design (CAMD)

approaches were examined in an attempt to identify potential synthetic directions for improved analogs. Included among the computer modeling approaches initially used were comparative molecular field analysis (CoMFA) and Hansch-style quantitative structure–activity relationships (QSAR) (Crouse and Sparks, 1998). However, the large molecular size and complex structure rendered the available computing power insufficient to the task for these more conventional methodologies and generally made QSAR difficult. However, where conventional CAMD techniques were not successful, the application of artificial neural networks to spinosyn QSAR (Sparks *et al.*, 2000a) identified new directions that led to, as one example, new modified-rhamnose spinosoids. These latter compounds were in many cases far more active than spinosyn A against pest lepidopterans (Crouse and Sparks, 1998; Sparks *et al.*, 2000a, 2001; Anzeveno and Green, 2002).

As shown by the activity of the 2′, 3′, 4′-demethyl analogs from the mutant strains, *O*-demethylation generally leads to less active compounds compared to spinosyn A (Table 2). Loss of the rhamnose also leads to a large reduction in activity (Table 2). The first spinosoids possessing any kind of improvement in lepidopteran activity over spinosyn A were (as noted above) the des-methoxy (2′-H (compounds 35 and 36), or 3′-H (compound 43) analogs of spinosyns H/Q and J, respectively (Table 3). For the 2′-substituted spinosoids, the 2′-des-methoxy analog (compound 35) along with the 2′-*O*-ethyl (compound 37) and 2′-*O*-*n*-propyl (compound 39) represent the most active of the 2′-substitution-based analogs, which are as active or very slightly more so than spinosyn A. Likewise, modifications of the 4′-position of the rhamnose (e.g., compound 60) results in no real activity improvement over spinosyn A, with the exception of the 4′-des-methyoxy analog (compound 59), which generally remains equivalent or nearly so, to spinosyn K (4′-*O*-demethyl, compound 23) (Table 3). Thus, in general, modifications to the 2′- or 4′-positions do little to improve activity compared to spinosyn A.

Unlike the 2′- and 4′-positions (e.g., compounds 37 and 60), a genuine improvement in *H. virescens* activity was found to be associated with an increased size of the alkoxy groups on the rhamnose at the 3′-position (compounds 46–52) (Sparks *et al.*, 2000a). At the 3′-position, the activity optimum is between two to three carbons. Further increases in bulk or length resulted in a loss of activity (Sparks *et al.*, 2000a, 2001; Crouse *et al.*, 2001). The introduction of unsaturation into these short alkyl chains (e.g., compounds 47, 53, 54) was generally neutral in effect, while branching was detrimental (compounds

49 and 50) (**Table 3**). The 3'-O-vinyl analog (compound 47) could represent a very active configuration, but stability issues resulted in erratic results (Crouse *et al.*, 2001). Other modifications to the alkyl chains (halogenation, introduction of another oxygen, etc.) were generally not as effective as the simple alkyl chains (compounds 55 and 57) (Crouse *et al.*, 2001) (**Table 3**). While the optimum in 3'-O-alkyl chain length for lepidopterous larvae such as *H. virescens* (Crouse *et al.*, 2001), beet armyworm (*Spodoptera exigua*) and cabbage looper (*Trichoplusia ni*) (data not shown) is between two and three carbons (e.g., compounds 46, 48, 53 and 54), the optimum for aphids such as *A. gossypii* extends to four carbons (compound 51) (Crouse *et al.*, 2001) (**Table 3**). Unlike either aphids or lepidopterans, the longer chain 3'-O-alkyl groups (C2–C4) (compounds 46, 48, and 51, respectively) were essentially equivalent in activity for *T. urticae* (**Table 3**) and were only modestly more active than spinosyn A.

Interestingly, the 2',3',4'-tri-O-ethyl analog (compound 62) is essentially as potent against *H. virescens* larvae, *A. gossypii*, and *T. urticae* as the 3'-O-ethyl derivative (compound 46) (**Table 3**), further supporting the importance of the 3'-position of the rhamnose to insecticidal activity.

In addition to modifications to the tri-O-methylrhamnose moiety, the effects of rhamnose replacement with other sugar and nonsugar moieties were also investigated (Anzeveno and Green, 2002). These studies demonstrated that very little change around the rhamnose moiety is tolerated. Replacement sugars such as D-rhamnose, the β-anomeric rhamnose, furanose, and forosaminyl analogs (not shown) were much less active or inactive against *H. virescens* larvae and *T. urticae* (Anzeveno and Green, 2002). In contrast, removal of the methyl group at the 6'-position (L-lyxose analog, compound 85) had little effect on activity relative to spinosyn A, while addition of hydroxy to the 6'-methyl group (L-mannose, compound 86) provided a very significant improvement in activity over spinosyn A (Anzeveno and Green, 2002), nearly equaling that of the 3'-O-ethyl analog (compound 46). Replacement of the tri-O-methylrhamnose with nonsugar moieties such as simple alkyl groups or methoxy-substituted benzoyl groups (compounds not shown) all were found to be inactive (Anzeveno and Green, 2002).

6.7.4. Quantitative Structure–Activity Relationships and the Spinosyns

The above examples (see Sections 6.7.1–6.7.3) clearly demonstrate that very significant improvements in activity of the spinosyns are possible, both in terms of potency towards a particular pest or group of insect pests and greater spectrum. These improvements, especially those involving rhamnose ring modifications, were facilitated through the application of both classical and less conventional CAMD techniques early in the project. The QSAR analysis of the spinosyns conducted using artificial neural networks directly led to the identification and synthesis of far more potent spinosoids than had been seen up to that time (Sparks *et al.*, 2000a, 2001; Anzeveno and Green, 2002). The analysis identified several potential modifications to improve activity, including the lengthened alkyl groups on the rhamnose, and also pinpointed the 3'-position as the most important for activity (Sparks *et al.*, 2000a, 2001). While the neural net-based QSAR was able to identify useful targets for synthesis, understanding the physicochemical basis for the improved activity is not a question easily addressed with neural nets (Sparks *et al.*, 2001). Thus, more classical approaches have also been employed. Hansch-style multiple regression analysis (Kubinyi, 1993) of the spinosyns and spinosoids was able to establish some relationships between insecticidal activity against *H. virescens* larvae and several physicochemical parameters.

Among the spinosyns, the biological response of *H. virescens* larvae was best described by the three whole-molecule parameters, CLogP, Mopac dipole (dipole moment of the whole spinosyn molecule), and HOMO (highest occupied molecular orbital) (Sparks *et al.*, 2000b):

$$\text{Log } Hv \, LC_{50} = -2.18 \, CLogP + 0.61 \, \text{Mopac dipole} \\ + 2.89 \, HOMO + 33.74$$
$$r^2 = 0.824, \quad s = 0.372, \quad F = <0.0001,$$
$$q^2 = 0.724, \quad n = 18 \tag{2}$$

For a larger set of spinosyns and spinosoids, these same parameters (eqn [3]) were also able to explain much of the observed insecticidal activity towards *H. virescens* larvae (Sparks *et al.*, 2001). The addition of the squared term for CLogP (eqn [3]) also reflects the presence of an optimum of between two and three carbons at the rhamnose 3'-position, that is observed for *H. virescens* larval activity (Sparks *et al.*, 2001) (**Table 3**):

$$\text{Log } Hv \, LC_{50} = -6.62 \, CLogP + 0.67 \, CLogP^2 \\ + 0.59 \, \text{Mopac dipole} \\ + 3.09 \, HOMO + 43.05$$
$$r^2 = 0.816, \quad s = 0.475, \quad F = <0.0001,$$
$$q^2 = 0.706, \quad n = 34 \tag{3}$$

For the spinosyns and spinosoids, HOMO appears to be directly influenced by substitution on the forosamine nitrogen (Sparks *et al.*, 2000b) suggesting that the primary drivers for activity are CLogP and Mopac dipole moment. In turn, Mopac dipole moment appears rather well correlated with the substitution pattern on the rhamnose moiety, and secondarily by substitution on the forosamine nitrogen (Sparks *et al.*, 2001). As implied by the above equations, the most active spinosyns and spinosoids tend to have smaller values for Mopac dipole moment and larger values (up to a point) for CLogP (Sparks *et al.*, 2001). Thus, these relationships provide a useful basis for understanding spinosyn and spinosoid activity not only against lepidopteran insect pests but also for other pest insects, including aphids and leafhoppers (Dintenfass *et al.*, 2001). CAMD and QSAR continue to play an important role in the investigation and exploitation of the spinosyn family of chemistry.

6.8. Conclusion

The natural spinosyns produced by select members of the genus *Saccharopolyspora*, and their semisynthetic derivatives, known as spinosoids, constitute a new and unique class of insect control agents. The spinosyns possess a novel mode of action, a very favorable toxicological profile coupled with high insecticidal efficacy against a broad range of pest insects, with commercial use focused on the Lepidoptera and, to a lesser degree, the Diptera. The commercial fermentation-derived product, spinosad, is a naturally occurring mixture of spinosyns A and D, which also happen to be the most insecticidally active of the natural spinosyns produced by *S. spinosa*. While possessing good contact activity, the spinosyns are most often more active orally, in part due to the slow cuticular penetration of the spinosyns compared to many insect control agents. Although spinosyns A and D are readily metabolized in vertebrate and avian systems, metabolism in insects is limited, compensating in part for slow penetration into insects. The spinosyns cause excitatory symptoms in insects by stimulating activity in the CNS. Activation of nondesensitizing nicotinic acetylcholine receptors by an allosteric mechanism appears to be the primary mechanism by which spinosyns stimulate neuronal activity, but antagonism of an as-yet-unidentified subtype of GABA receptor may also play a significant role in spinosyn poisoning. Through an extensive program of strain selection, genetic manipulation of the biosynthetic pathways and chemical synthesis, the effects of many modifications to the forosamine and rhamnose

moieties as well as to the core tetracycle have been investigated. This experience, with guidance from classical modeling approaches as well as the application of QSAR methodologies utilizing artificial intelligence, has led to the synthesis of spinosoids with biological activity far exceeding that of spinosyn A. Because of its unique mode of action and limited susceptibility to metabolism in insects, pest insects resistant to other insecticides would typically not show cross-resistance to spinosad. However, as expected for any insect control agent put under heavy selection pressure, resistance to spinosad has been developed in the laboratory and has also recently appeared in the field. Available information suggests that resistance to spinosad is most likely target-site based, reinforcing the importance of the IRM recommendations developed by the manufacturer at the time of its launch that spinosad be used in rotation with products with other modes of action to ensure the long-term utility of this novel chemistry.

Acknowledgments

The authors would like to thank their many colleagues from Eli Lilly and Co., and Dow AgroSciences (DAS) who have contributed to our knowledge of the spinosyns. Special thanks are due to Mr. Jerry Watson (DAS), Dr Nailah Orr (DAS), Dr Gary Thompson (DAS), Dr Mark Hertlein (DAS), Dr Gary Crouse (DAS), Dr Paul Lewer (DAS), and Dr Ron Mau (University of Hawaii) for their many helpful discussions during the writing of this chapter.

References

Addor, R.W., 1995. Insecticides. In: Godfrey, C.R.A. (Ed.), Agrochemicals from Natural Products. Dekker, New York, pp. 1–62.

Ahmad, M., Hollingworth, R.M., Wise, J.C., 2002. Broad-spectrum insecticide resistance in obliquebanded leafroller *Choristoneura rosaceana* (Lepidoptera: Tortricidae) from Michigan. *Pest Mgt Sci. 58*, 834–838.

Anzeveno, P.B., Green, F.R., III, 2002. Rhamnose replacement analogs of spinosyn A. In: Baker, D.R., Fenyes, J.G., Lahm, G.P., Selby, T.P., Stevenson, T.M. (Eds.), Synthesis and Chemistry of Agrochemicals, vol. VI. American Chemical Society, Washington, DC, pp. 262–276.

Anonymous, 2001. Spinosad Technical Bulletin. Dow AgroSciences, Indianapolis, IN.

Andersch, W., Schnorbach, H.-J., Wollweber, D., 2003. Synergistic insecticidal mixtures, Bayer AG. US Patent No. 6,686,387.

Bailey, W.D., Young, H.P., Roe, R.M., 1999. Laboratory selection of a Tracer® resistant strain of tobacco budworm and comparisons with field selected strains from the southeasetern US. Proc. 1999 Beltwide Cotton Production Conf., pp. 1221–1224.

Brett, B.L., Larson, L.L., Schoonover, J.R., Sparks, T.C., Thompson, G.D., 1997. Biological properties of spinosad. *Down to Earth 52*, 6–13.

Buckingham, S.D., Lapied, B., Le Corronc, H., Satelle, D.B., 1997. Imidacloprid actions on insect neuronal acetylcholine receptors. *J. Exp. Biol. 200*, 2685–2692.

Burns, L.S., Graupner, P.R., Lewer, P., Martin, C.J., Vousden, W.A., et al., 2003. Spinosyn Polyketide Synthase Fusion Products Synthesizing Novel Spinosyns and their Preparation and Use. Dow AgroSciences LLC, Indianapolis, IN.

Chen, T.S., Hensens, O.D., Schulman, M.D., 1989. Biosynthesis. In: Campbell, W.C. (Ed.), Ivermectin and Abamectin. Springer, New York, pp. 55–72.

Clark, J.M., Yamaguchi, I., 2002. Scope and status of pesticide resistance. In: Clark, J.M., Yamaguchi, I. (Eds.), Agrochemical Resistance: Extent, Mechanism and Detection. American Chemical Society, Washington, DC, pp. 1–22.

Creemer, L.C., Kirst, H.A., Paschal, J.W., Worden, T.V., 2000. Synthesis and insecticidal activity of spinosyn analogs functionally altered at the 2'-, 3'- and 4'-positions of the rhamnose moiety. *J. Antibiotics 53*, 171–178.

Crouse, G.D., Sparks, T.C., 1998. Naturally derived materials as products and leads for insect control: the spinosyns. *Rev. Toxicol. 2*, 133–146.

Crouse, G.D., Sparks, T.C., DeAmicis, C.V., Kirst, H.A., Martynow, J.G., et al., 1999. Chemistry and insecticidal activity of the spinosyns. In: Brooks, G.T., Roberts, T.R. (Eds.), Pesticide Chemistry and Bioscience: The Food–Environment Challenge. Royal Society of Chemistry, Cambridge, pp. 155–166.

Crouse, G.D., Sparks, T.C., Schoonover, J., Gifford, J., Dripps, J., et al., 2001. Recent advances in the chemistry of spinosyns. *Pest Mgt Sci. 57*, 177–185.

Crouse, G.D., Hahn, D.R., Graupner, P.R., Gilbert, J.R., Lewer, P., et al., 2002. Synthetic Macrolides of 21-Butenyl and Related Spinosyn Glycosides as Insecticides and Acaricides. PCT Int. Appl., Indianapolis, IN.

DeAmicis, C.V., Dripps, J.E., Hatton, C.J., Karr, L.L., 1997. Physical and biological properties of the spinosyns: novel macrolide pest-control agents from fermentation. In: Hedin, P.A., Hollingworth, R.M., Masler, E.P., Miyamoto, J., Thompson, D.G. (Eds.), Phytochemicals for Pest Control. American Chemical Society, Washington, DC, pp. 144–154.

Dintenfass, L., DeAmicis, C.V., Dripps, J.D., Crouse, G.D., Kempe, M., et al., 2001. Structure–activity relationships of spinosyns and spinosoids. In: 221st Natl Mtg. American Chemical Society, San Diego, CA, Book of Abstracts, pp. AGRO 077.

Domoradzki, J.Y, Stewart, H.S., Mendrala, A.L., Gilbert, J.R., Markham, D.A., 1996. Comparison of the metabolism and tissue distribution of ^{14}C-labeled spinosyn A and ^{14}C-labeled spinosyn D in Fischer 344 rats. Soc. Toxicol. Mtg. Anaheim, CA.

Dubreil, V., Sinakevitch, I.G., Hue, B., Geffard, M., 1994. Neuritic GABAergic synapses in insect neurosecretory cells. *Neurosci. Res. 19*, 235–240.

Elzen, G.W., Maldonado, S.N., Rojas, M.G., 2000. Lethal and sublethal effects of insecticides and an insect growth regulator on the boll weevil (Coleoptera: Curculionidae) ectoparasitoid *Catolaccus grandis* (Hymenoptera: Pteromalidae). *J. Econ. Entomol. 93*, 300–303.

Gaisser, S., Martin, C.J., Wilkinson, B., Sheridan, R.M., Lill, R.E., et al., 2002. Engineered biosynthesis of novel spinosyns bearing altered deoxyhexose substituents. *Chem. Commun. 2002*, 618–619.

Grolleau, F., Lapied, B., 2000. Dorsal unpaired median neurons in the insect central nervous system: towards a better understanding of the ionic mechanisms underlying spontaneous electrical activity. *J. Exp. Biol. 203*, 1633–1648.

Hahn, D.R., Balcer, J.L., Lewer, P., Gilbert, J.R., Graupner, P.R., 2002. Pesticidal Spinosyn Derivatives Produced by Culturing Mutant *Saccharopolyspora* sp. Strains. PCT Int. Appl., Indianapolis, IN.

Hill, T.A., Foster, R.E., 2003. Influence of selected insecticides on the population dynamics of diamondback moth (Lepidoptera: Plutellidae) and its parasitoid *Diadegma insulare* (Hymenoptera: Ichneumonidae), in cabbage. *J. Entomol. Sci. 38*, 59–71.

Hollingworth, R.M., 1976. The biochemical and physiological basis of selective toxicity. In: Wilkinson, C.F. (Ed.), Insecticide Biochemistry and Physiology. Plenum, New York, pp. 431–506.

Hollingworth, R.M., 2001. New insecticides: mode of action and selective toxicity. In: Baker, D.R., Umetsu, N.K. (Eds.), Agrochemical Discovery: Insect, Weed and Fungal Control. American Chemical Society, Washington, DC, pp. 238–255.

Kao, C-.H., Cheng, E.Y., 2001. Insecticide resistance in *Plutella xylostella* L. 11. Resistance to new introduced insecticides in Taiwan (1990–2001). *J. Agric. Res. China 50*, 80–89.

Kirst, H.A., Michel, K.H., Mynderse, J.S., Chio, E.H., Yao, R.C., et al., 1992. Discovery, isolation, and structure elucidation of a family of structurally unique fermentation-derived tetracyclic macrolides. In: Baker, D.R., Fenyes, J.G., Steffens, J.J. (Eds.), Synthesis and Chemistry of Agrochemicals, vol. 3. American Chemical Society, Washington, DC, pp. 214–225.

Kirst, H.A., Creemer, L.C., Naylor, S.A., Pugh, P.T., Snyder, D.E., et al., 2002a. Evaluation and development of spinosyns to control ectoparasites on cattle and sheep. *Curr. Topics Medic. Chem. 2*, 675–699.

Kirst, H.A., Creemer, L.C., Broughton, M.C., Huber, M.B.L., Turner, J.R., 2002b. Chemical and microbiological modifications of spinosyns: exploring synergies between fermentation microbiology and organic chemistry. In: Baker, D.R., Fenyes, J.G., Lahm, G.P., Selby, T.P., Stevenson, T.M. (Eds.), Synthesis and Chemistry of Agrochemicals, vol. 6. American Chemical Society, Washington, DC, pp. 251–261.

Kubinyi, H., 1993. QSAR: Hansch Analysis and Related Approaches. VCH Publishers, New York.

Larson, L.L., Kenaga, E.E., Morgan, R.W., 1985. Commercial and Experimental Organic Insecticides. Entomological Society of America, College Park, MD, pp. 1–105.

Le Corronc, H., Alix, P., Hue, B., 2002. Differential sensitivity of two insect GABA-gated chloride channels to dieldrin, fipronil and picrotoxinin. J. Insect Physiol. 48, 419–431.

Lena, C., Changeux, J.-P., 1993. Allosteric modulations of the nicotinic acetylcholine receptor. Trends Neurosci. 16, 181–186.

Lewer, P., Hahn, D.R., Karr, L.L., Graupner, P.R., Gilbert, J.R., et al., 2001. Pesticidal Macrolides. PCT Int. Appl., Indianapolis, IN.

Lewer, P., Hahn, D.R., Karr, L.L., Graupner, P.R., Gilbert, J.R., et al., 2003. A new family of insecticidal spinosyns from a novel Saccharopolyspora strain. In: Proc. 44th Natl Mtg. Am. Soc. Pharmacognosy, Book of Abstracts, pp. 0:37.

Liu, N., Yue, X., 2000. Insecticide resistance and cross-resistance in the house fly (Diptera: Muscidae). J. Econ. Entomol. 93, 1269–1275.

Ludmerer, S.W., Warren, V.A., Williams, B.S., Zheng, Y., Hunt, D.C., et al., 2002. Ivermectin and nodulisporic acid receptors in Drosophila melanogaster contain both γ-aminobutyric acid-gated Rdl and glutamate-gated GluCl α chloride channel subunits. Biochemistry 41, 6548–6560.

Magnussen, J.D., Castetter, S.A., Rainey, D.P., 1996. Characterization of spinosad related residues in poultry tissues and eggs following oral administration. In: 211th Natl Mtg. American Chemical Society, New Orlean, LA, pp. AGRO 043.

Madduri, K., Waldron, C., Merlo, D.J., 2001. Rhamnose biosynthesis pathway supplies precursors for primary and secondary metabolism in Saccharopolyspora spinosa. J. Bacteriol. 183, 5632–5638.

Martin, C., 2003. Genetic engineering of Saccharopolyspora spinosa to generate a library of spinosyn analogues. In: 13th Int. Soc. Biol. Actinomycetes, Book of Abstracts.

Mau, R.F.L., Gusukuma-Minuto, L., 2004. Diamondback moth, Plutella xylostella (L.), resistance management in Hawaii. In: Endersby, N.M., Ridland, P.M. (Eds.), The Management of Diamondback Moth and Other Crucifer Pests: Proc. 4th Intl workshop, Melbourne, Australia, 26–29 November 2001, pp. 307–311.

Mertz, F.P., Yao, R.C., 1990. Saccharopolyspora spinosa sp. nor. isolated from soil collected in a sugar rum still. Int. J. Syst. Bacteriol. 40, 34–39.

Moultron, J.K., Pepper, D.A., Dennehy, T.J., 1999. Studies of resistance of beet armyworm (Spodoptera exigua) to spinosad in field populations from the southern USA and southeast Asia. In: Proc. 1999 Beltwide Cotton Production Conf., pp. 884–889.

Moulton, J.K., Pepper, D.A., Dennehy, T.J., 2000. Beet armyworm (Spodoptera exigua) resistance to spinosad. Pest Mgt Sci. 56, 842–848.

Nauen, R., Ebbinghaus-Kintscher, U., Elbert, A., Jeschke, P., Tietjen, K., 2001. Acetylcholine receptors as sites for developing neonicotinoid insecticides. In: Ishaya, I. (Ed.), Biochemical Sites of Insecticide Action and Resistance. Springer, Berlin, pp. 288–321.

Papa, G., de Almeida, F.J., Rotundo, M., 2002. Effect of insecticides and methods of sampling on the population of beneficial arthropods, in cotton in Brazil. In: Proc. 2003 Beltwide Cotton Conf., pp. 1197–1201.

Pitman, R.M., Kerkut, G.A., 1970. Comparison of the actions of iontophoretically applied acetylcholine and gamma aminobutyric acid with the EPSP and IPSP in cockroach central neurons. Comp. Gen. Pharmacol. 1, 221–230.

Rainey, D.P., O'Neill, J.D., Castetter, S.A., 1996. The tissue distribution and metabolism of spinosyn A and D in lactating goats. In: 211th Natl Mtg. American Chemical Society, New Orlean, LA, pp. AGRO 045.

Raymond, V., Sattelle, D.B., Lapied, B., 2000. Coexistence in DUM neurones of two GluCl channels that differ in their picrotoxin sensitivity. Neuroreport 11, 2695–2701.

Salgado, V.L., Watson, G.B., Sheets, S.S., 1997. Studies on the mode of action of spinosad, the active ingredient in Tracer® insect control. Proc. 1996 Beltwide Cotton Production Conf., pp. 1082–1086.

Salgado, V.L., 1998. Studies on the mode of action of spinosad: insect symptoms and physiological correlates. Pestic. Biochem. Physiol. 60, 91–102.

Salgado, V.L., Sheets, J.J., Watson, G.B., Schmidt, A.L., 1998. Studies on the mode of action of spinosad: the internal effective concentration and the concentration dependence of neural excitation. Pestic. Biochem. Physiol. 60, 103–110.

Salgado, V.L., Saar, R., 2004. Desensitizing and non-desensitizing subtypes of α-bungarotoxin-sensitive nicotinic acetylcholine receptors in cockroach neurons. J. Insect Physiol. (in press).

Saunders, D.G., Bret, B.L., 1997. Fate of spinosad in the environment. Down to Earth 52, 14–20.

Schroeder, M.E., Flattum, R.F., 1984. The mode of action and neurotoxic properties of the nitromethylene heterocycle insecticides. Pestic. Biochem. Physiol. 22, 148–160.

Scott, J.G., 1998. Toxicity of spinosad to susceptible and resistant strains of house flies, Musca domestica. Pestic. Sci. 54, 131–133.

Scott, J.G., Alefantis, T.G., Kaufman, P.E., Rutz, D.A., 2000. Insecticide resistance in house flies from caged-layer poultry facilities. Pest Mgt Sci. 56, 147–153.

Shelton, A.M., Sances, F.V., Hawley, J., Tang, J.D., Boune, M., et al., 2000. Assessment of insecticide resistance after the outbreak of diamondback moth (Lepidoptera: Plutellidae) in California in 1997. J. Econ. Entomol. 93, 931–936.

Shomi, K., Omura, S., 2002. Discovery of new macrolides. In: Omura, S. (Ed.), Macrolide Antibiotics: Chemistry, Biology and Practice. Academic Press, New York, pp. 1–56.

Shono, T., Scott, J.G., 2003. Spinosad resistance in the housefly, *Musca domestica*, is due to a recessive factor on autosome 1. *Pestic. Biochem. Physiol.* 75, 1–7.

Smirle, M.J., Lowery, D.T., Zurowski, C.L., 2003. Susceptibility of leafrollers (Lepidoptera: Tortricidae) from organic and conventional orchards to azinphosmethyl, spinosad and *Bacillus thuringiensis*. *J. Econ. Entomol.* 96, 879–884.

Smith, P. (Ed.) 2001. AgroProjects: PestProjects – 2001. PJB Publications, Richmond, UK.

Snodgrass, R.E., 1952. A Textbook of Arthropod Anatomy. Comstock, Ithaca, NY.

Sparks, T.C., 1996. Toxicology of insecticides and acaricides. In: King, E.G., Phillips, J.R., Coleman, R.J. (Eds.), Cotton Insects and Mites: Characterization and Management. The Cotton Foundation, Memphis, TN, pp. 283–322.

Sparks, T.C., 2004. New insect control agents: modes of action and selectivity. In: Endersby, N.M., Ridland, P.M. (Eds.), The Management of Diamond back Moth and Other Crucifers Pests: Proc. 4th Intl. Workshop, Melbourne, Australia, 26–29 November 2001, pp. 37–44.

Sparks, T.C., Graves, J.B., Leonard, B.R., 1993. Insecticide resistance and the tobacco budworm: past, present and future. *Rev. Pestic. Toxicol.* 2, 149–183.

Sparks, T.C., Thompson, G.D., Larson, L.L., Kirst, H.A., Jantz, O.K., et al., 1995. Biological characteristics of the spinosyns: a new class of naturally derived insect control agents. In: Proc. 1995 Beltwide Cotton Production Conf., pp. 903–907.

Sparks, T.C., Kirst, H.A., Mynderse, J.S., Thompson, G.D., Turner, J.R., et al., 1996. Chemistry and biology of the spinosyns: components of spinosad (Tracer®), the first entry into DowElanco's naturalyte class of insect control products. In: Proc. 1996 Beltwide Cotton Production Conf., pp. 692–696.

Sparks, T.C., Sheets, J.J., Skomp, J.R., Worden, T.V., Hertlein, M.B., et al., 1997. Penetration and metabolism of spinosyn A in lepidopterous larvae. In: Proc. 1997 Beltwide Cotton Conf., pp. 1259–1265.

Sparks, T.C., Thompson, G.D., Kirst, H.A., Hertlein, M.B., Larson, L.L., et al., 1998. Biological activity of the spinosyns, new fermentation derived insect control agents, on tobacco budworm (Lepidoptera: Noctuidae) larvae. *J. Econ. Entomol.* 91, 1277–1283.

Sparks, T.C., Thompson, G.D., Kirst, H.A., Hertlein, M.B., Mynderse, J.S., et al., 1999. Fermentation-derived insect control agents: the spinosyns. In: Hall, F.R., Menn, J.J. (Eds.), Biopesticides: Use and Delivery. Humana Press, Totowa, NJ, pp. 171–188.

Sparks, T.C., Anzeveno, P.B., Martynow, J.G., Gifford, J.M., Hertlein, M.B., et al., 2000a. The application of artificial neural networks to the identification of new spinosoids with improved biological activity toward larvae of *Heliothis virescens*. *Pestic. Biochem. Physiol.* 67, 187–197.

Sparks, T.C., Durst, G., Worden, T.V., 2000b. Structure–activity relationships of the spinosyns. In: Proc. 2000 Beltwide Cotton Conf., pp. 1225–1229.

Sparks, T.C., Crouse, G.D., Durst, G., 2001. Natural products as insecticides: the biology, biochemistry and quantitative structure–activity relationships of spinosyns and spinosoids. *Pest Mgt Sci.* 57, 896–905.

Studebaker, G.E., Kring, T.J., 2003. Effects of insecticides on *Orius insidiosus* (Hemiptera: Anthocoridae) measured by field, greenhouse and Petri dish bioassays. *Florida Entomol.* 86, 178–185.

Thompson, G.D., Busacca, J.D., Jantz, O.K., Kirst, H.A., Larson, L.L., et al., 1995. Spinosyns: an overview of new natural insect management systems. In: Proc. 1995 Beltwide Cotton Conf., pp. 1039–1043.

Thompson, G.D., Michel, K.H., Yao, R.C., Mynderse, J.S., Mosburg, C.T., et al., 1997. The discovery of *Saccharopolyspora spinosa* and a new class of insect control products. *Down to Earth* 52, 1–5.

Thompson, G.D., Dutton, R., Sparks, T.C., 2000. Spinosad: a case study – an example from a natural products discovery programme. *Pest Mgt Sci.* 56, 696–702.

Thompson, G.D., Sparks, T.C., 2002. Spinosad: a green natural product for insect control. In: Lankey, R.L., Anastas, P.T. (Eds.), Advancing Sustainability through Green Chemistry and Engineering. American Chemical Society, Washington, DC, pp. 61–73.

Tillman, P.G., Mulrooney, J.E., 2000. Effect of selected insecticides on the natural enemies *Coleomegilla maculata* and *Hippodamia convergens* (Coleoptera: Coccinellidae), *Geocoris punctipes* (Hemiptera: Lygaeidae), and *Bracon mellitor*, *Cardiochiles nigriceps*, and *Cotesia marginiventris* (Hymenoptera: Braconidae) in cotton. *J. Econ. Entomol.* 93, 1638–1643.

van den Beukel, I., van Kleef, R.G.D.M., Zwart, R., Oortgiesen, M., 1998. Physostigmine and acetylcholine differentially activate nicotinic receptor subpopulatons in *Locusta migratoria* neurons. *Brain Res.* 789, 263–273.

Waldron, C., Madduri, K., Crawford, K., Merlo, D.J., Treadway, P., et al., 2000. A cluster of genes for the biosynthesis of spinosyns, novel macrolide insect control agents produced by *Saccharopolyspora spinosa*. *Antonie van Leeuwenhoek* 78, 385–390.

Waldron, C., Matsushima, P., Rosteck, P.R., Jr., Broughton, M.C., Turner, J., et al., 2001. Cloning and analysis of the spinosad biosynthetic gene cluster of *Saccharopolyspora spinosa*. *Chem. Biol.* 8, 487–499.

Waldstein, D.E., Reissig, W.H., 2000. Synergism of tebufenozide in resistant and susceptible strains of obliquebanded leafroller (Lepidoptera: Tortricidae) and resistance to new insecticides. *J. Econ. Entomol.* 93, 1768–1772.

Ware, G.W., 1982. Pesticides: Theory and Application. W.H. Freeman, San Francisco, CA.

Watson, G.B., 2001. Actions of insecticidal spinosyns on γ-aminobutyric acid responses from small-diameter cockroach neurons. *Pestic. Biochem. Physiol. 71*, 20–28.

Wei, Y., Appel, A.G., Moar, W.J., Liu, N., 2001. Pyrethroid resistance and cross-resistance in the German cockroach, *Blattella germainica* (L). *Pest Mgt Sci. 57*, 1055–1059.

Williams, T., Valle, J., Viñuela, E., 2003. Is the naturally derived insecticide spinosad compatible with insect natural enemies? *Biocont. Sci. Tech. 13*, 459–475.

Wyss, C.F., Young, H.P., Shukla, J., Roe, R.M., 2003. Biology and genetics of a laboratory strain of the tobacco budworm, *Heliothis virescens* (Lepidoptera: Noctuidae), highly resistant to spinosad. *Crop Protect. 22*, 307–314.

Young, H.P., Bailey, W.D., Wyss, C.F., Roe, R.M., Sheets, J.J., *et al.*, 2000. Studies on the mechanism(s) of tobacco budworm resistance to spinosad (Tracer®). In: Proc. 2000 Beltwide Cotton Conf., pp. 1197–1201.

Young, H.P., Bailey, W.B., Roe, R.M., Iwasa, T., Sparks, T.C., *et al.*, 2001. Mechanism of resistance and cross-resistance in a laboratory, spinosad-selected strain of the tobacco budworm and resistance in laboratory-selected cotton bollworms. In: Proc. 2001 Beltwide Cotton Conf., pp. 1167–1171.

Young, H.P., Bailey, W.D., Roe, R.M., 2003. Spinosad selection of a laboratory strain of the tobacco budworm, *Heliothis virescens* (Lepidoptera: Noctuidae), and characterization of resistance. *Crop Protect. 22*, 265–273.

Zhao, J.-Z., Li, Y.-X., Collins, H.L., Gusukuma-Minuto, L., Mau, R.F.L., *et al.*, 2002. Monitoring and characterization of diamondback moth (Lepidoptera: Plutellidae) resistance to spinosad. *J. Econ. Entomol. 95*, 430–436.

Relevant Websites

http://www.extoxnet.orst.edu – Extension Toxicology Network, Oregon State University.

A6   Addendum: The Spinosyns

T C Sparks, G B Watson, and J E Dripps,
Dow AgroSciences, Indianapolis, IN, USA

© 2010 Elsevier B.V. All Rights Reserved

A6.1. Spinosyn Chemistry and Biosynthesis

Modifications to the spinosyn rhamnose ring result in significant improvements in its insecticidal activity (Salgado and Sparks, 2005). This discovery led to the development of spinetoram (**Figure A1**), a new semi-synthetic spinosyn insecticide launched in 2007 by Dow AgroSciences, and recipient of a 2008 Presidential Green Chemistry Challenge Award (Crouse et al., 2007; Sparks et al., 2008; Dripps et al., 2008). Spinetoram is produced by a two-step post-fermentation synthetic modification of a mixture of spinosyns J and L. These spinosyns lack a 3'-O-methyl group but are otherwise identical to spinosyns A and D. Ethylation of the free hydroxyl group, followed by a reduction of the 5,6-double bond of the J component, produces spinetoram. Compared to spinosad, spinetoram controls an expanded spectrum of insect pests, has greater potency, and has longer duration of control, all coupled with a very favorable toxicological and environmental profile (Dripps et al., 2008).

More recently, modifications to the C21 position of the tetracycle have received particular interest (Crouse et al., 2007; Huang et al., 2009). The discovery of a new spinosyn producing organism, *Saccharopolyspora pogona* (Salgado and Sparks, 2005; Hahn et al., 2006), led to identification of >30 new spinosyns, typically characterized by a but-1-enyl moiety at C21 (**Figure A2**, Crouse et al., 2007; Lewer et al., 2009). Other new structural motifs include a hydroxyl at C8, and other sugars in place of forosamine (Lewer et al., 2009). Although many of these analogs are similar to spinosyn A in their activity against lepidopteran pests, they are often more potent against sap-feeding insects (e.g., aphids, whiteflies; **Figure A2**, Lewer et al., 2009).

Other C21 modifications have been made through replacement of the polyketide synthesis (PKS) loading module from *S. spinosa* with the module from avermectin PKS. This modification allowed incorporation of longer chain alkyl and cyclic alkyl groups at the C21 position by introducing exogenous novel carboxylic acids into the fermentation medium (Sheehan et al., 2006). Many of these C21 modifications, especially longer chain and cyclobutyl moieties, were more active than spinosyn A against lepidopteran pests and sap-feeding insects such as aphids (**Figure A2**, Sheehan et al., 2006; Crouse et al., 2007).

Another approach to exploring the C21 position involved synthetic modification of the 21-but-1-enyl group via olefin cross-metathesis (Daeuble et al., 2009). This procedure has generated a range of novel substitutions such as vinyl, hex-1-enyl, and styryl at the C21 position. Most of the analogs examined were similar in activity to corresponding 3'-O-ethyl spinosyn J analogs (**Figure A2**, Daeuble et al., 2009). Thus, many modifications to the C21 position are well tolerated and, in some cases, have improved the spectrum of insecticidal activity.

Other novel spinosyn derivatives have also been reported including analogs that possess a structurally simplified tetracycle (Tietze et al., 2007). In addition, research continues to explore the biosynthesis of spinosyns, including pathways involved in the formation of spinosyn tetracycle (e.g., Hahn et al., 2006; Kim et al., 2007; Huang et al., 2009) and sugar moieties (e.g., Hong et al., 2008; Chen et al., 2009).

A6.2. Mode of Action

A molecular target site for the spinosyns has recently been described. Mutant strains of *Drosophila melanogaster* with high levels of resistance to spinosyns were created by chemical mutagenesis (Watson et al., 2010). This resistance was shown to be due to a mutation in the nicotinic acetylcholine receptor (nAChR) subunit Dα6. Dα6 was cloned and expressed in *Xenopus* oocytes as a heteromeric protein which is co-expressed with the nAChR subunit Dα5. The resulting receptor was found to be sensitive to nAChR agonists (e.g., acetylcholine and nicotine) and insecticidal spinosyns such as spinosad and spinetoram. A related study by Perry et al. (2007) found that a Dα6 knockout strain of

Drosophila also shows high levels of spinosad resistance. Together, these two studies strongly suggest that the Dα6 nAChR is an important target site for

spinosyns in *Drosophila*. Earlier work with the American cockroach also showed spinosyns interact with certain nAChRs (Salgado and Saar, 2004). In total, these findings suggest that nAChRs are important target sites for spinosyns in all sensitive insect species. However, houseflies that have target site resistance to spinosad do not appear to have similar mutations in nAChR subunits homologous to Dα6 (Scott, 2008). Therefore, the spinosyn target site in housefly could be a novel nAChR subunit, or another, as yet undetermined target, indicating a need for further research in the housefly, and other insect species.

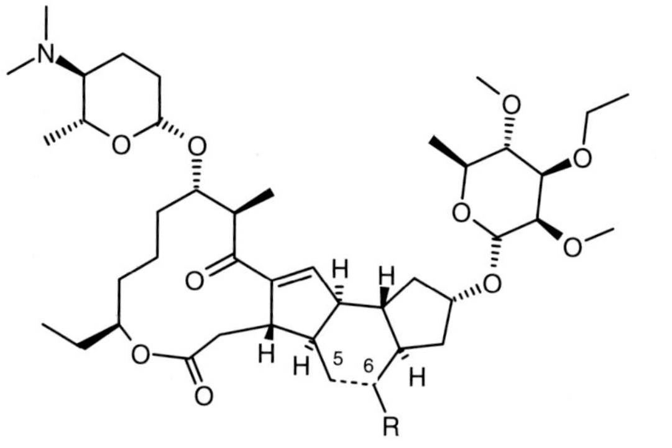

A6.3. Spinosad Resistance and Cross Resistance

As the use of spinosyns has expanded, more studies have examined resistance to spinosad in the laboratory (e.g., Hsu and Feng, 2006; Wang *et al.*, 2006;

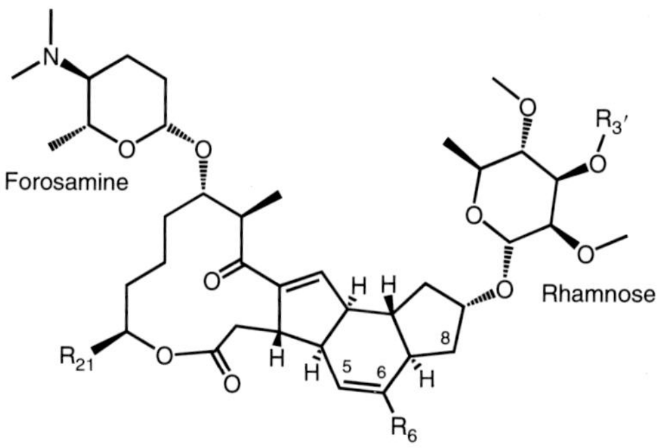

Figure A1 Structure of spinetoram: Mixture of 3′-*O*-ethyl-5,6-dihydro-spinosyn J (major component; single bond at C5–C6, R = H) and 3′-*O*-ethyl-spinosyn L (minor component; double bond at C5–C6, R = CH$_3$).

R21	R6	R3′	TBWn	BAWo	BAWt	CA	WF
1. ethyl	H	Me	0.31	0.079	0.63	18-55	5- 23
2. ethyl	Me	Me	0.8	--		65	1.7
3. ethyl	H	Et	0.03	0.012	0.039	8.1	5.3
4 but-1-enyl	H	Me	0.29	--	0.035	5-11	2
5. but-1-enyl	Me	Me	0.30	--	--	2	--
6. *n*-propyl	H	Me	0.16	--	0.031	27	--
7. cyclopropyl	H	Me	--	--	~1	~46	>50
8. cyclobutyl	H	Me	--	--	>1	~3	55
							13.5
10. cyclobutyl*	H	Me	--	--	0.13	1.7	0.7
11. vinyl	H	Et	0.5	0.018	--	--	--
12. hex-1-enyl	H	Et	--	<0.012	--	--	--
13. oct-1-enyl	H	Et	--	0.022	--	--	--
14. styryl	H	Et	--	0.013	--	--	--
15. MeO$_2$CC=CH-	H	Et	--	0.20	--	--	--

TBWn = tobacco budworm neonate drench bioassay (LC$_{50}$ ppm)
BAWo = beet armyworm oral (diet) bioassay (LC$_{50}$ μg/cm^2)
BAWt = beet armyworm topical bioassay (LD$_{50}$ μg/larva)
CA = cotton aphid on plant bioassay (LC$_{50}$ ppm)
WF = whitefly on plant bioassay (LC$_{50}$ ppm)
*5,6-dihydro

Figure A2 Structures and activity of selected modifications at the C21 position of the spinosyn tetracycle. Data adapted from Sheehan *et al.* (2006), Lewer *et al.* (2009), Daeuble *et al.* (2009).

Scott, 2008) and in the field (e.g., Zhao *et al.*, 2006; Bielza, 2008). With few exceptions, target site resistance continues to be the primary mechanism with little cross-resistance to other insecticides in spinosad-resistant insect strains. As a continuance of earlier trends (Salgado and Sparks, 2005), there is little spinosad cross-resistance in insect strains resistant to other classes of insecticides, including the avermectins and neonicotinoids.

References

Bielza, P., 2008. Insecticide resistance management strategies against the western flower thrips, *Frankliniella occidentalis*. *Pest. Manage. Sci.* 64, 1131–1138.

Chen, Y.-L., Chen, Y.-H., Lin, Y.-C., Tsai, K.-C., Chiu, H.-T., 2009. Functional characterization and substrate specificity of spinosyn rhamnosyltransferase by *in vitro* reconstitution of spinosyn biosynthetic enzymes. *J. Biol. Chem.* 284, 7352–7363.

Crouse, G.D., Dripps, J.E., Orr, N.T., Sparks, C., Waldron, C., 2007. DE-175 (Spinetoram), a new semisynthetic spinosyn in development. In: Kramer, W., Schirmer, U. (Eds.), Modern Crop Protection Compounds, Vol. 3. Wiley, New York, pp. 1013–1031.

Daeuble, J., Sparks, T.C., Johnson, P., Graupner, P.R., 2009. Modification of the butenyl-spinosyns utilizing cross-metathesis. *Bioorg. Med. Chem.* 17, 4197–4205.

Dripps, J., Olsen, B., Sparks, T., Crouse, G., 2008. Spinetoram: how artificial intelligence combined natural fermentation with synthetic chemistry to produce a new spinosyn insecticide Online. *Plant Health Progress* doi:10.1094/PHP-2008-0822-01-PS.

Hahn, D.R., Gustafson, G., Waldron, C., Bullard, R., Jackson, J.D., Mitchell, J., 2006. Butenyl-spinosyns, a natural example of genetic engineering of antibiotic biosynthetic genes. *J. Ind. Microbiol. Biotechnol.* 33, 94–104.

Hong, L., Zhao, Z., Melanon, C.E., Zhang, H., Liu, H.-W., 2008. *In vitro* characterization of the enzymes involved in TDP-d-forosamine biosynthesis in the spinosyn pathway of *Saccharopolyspora spinosa*. *J. Am. Chem. Soc.* 130, 4954–4967.

Hsu, J.-C., Feng, H.-T., 2006. Development of resistance to spinosad in Oriental fruit fly (Diptera: Tephritidae) in laboratory selection and cross-resistance. *J. Econ. Entomol.* 99, 931–936.

Huang, K.-X., Xia, L., Zhang, Y., Ding, X., Zahn, J.A., 2009. Recent advances in the biochemistry of the spinosyns. *Appl. Microbiol. Biotechnol.* 82, 13–23.

Kim, H.J., Pongdee, R., Wu, Q., Hong, L., Liu, H.-W., 2007. The biosynthesis of spinosyn in *Saccharopolyspora spinosa*: synthesis of the cross-bridging precursor and identification of the function of SpnJ. *J. Am. Chem. Soc.* 129, 14582–14584.

Lewer, P., Hahn, D.R., Karr, L.L., Duebelbeis, D.O., Gilbert, J.R., Crouse, G.D., Worden, T., Sparks, T.C., McKamey, P., Edwards, R., Graupner, P.R., 2009. Discovery of the butenyl-spinosyn insecticides: novel macrolides from the new bacterial strain, *Saccharopolyspora pogona*. *Bioorg. Med. Chem.* 17, 4185–4196.

Orr, N., Chouinard, S.W., Cook, K.R., Geng, C., Gifford, J.M., Gustafson, G.D., Hasler, J.M., Larrinua, I.M., Letherer, T.J., Mitchell, J.C., Pak, W.L., Salgado, V.L., Sparks, T.C., Watson, G.B., 2009. Heterologous expression of a spinosyn-sensitive *Drosophila melanogaster* nicotinic acetylcholine receptor identified through chemically induced target site resistance and resistance gene identification. *Insect Biochem. Mol. Biol.* (Accepted).

Perry, T., McKenzie, J.A., Batterham, P., 2007. A Dα6 knockout strain of *Drosophila melanogaster* confers a high level of resistance to spinosad. *Insect. Biochem. Mol. Biol.* 37, 184–188.

Scott, J.G., 2008. Unraveling the mystery of spinosad resistance in insects. *J. Pestic. Sci.* 33, 221–227.

Salgado, V.L., Saar, R., 2004. Desensitizing and nondesensitizing subtypes of alpha-bungarotoxin-sensitive nicotinic acetylcholine receptors in cockroach neurons. *J. Insect Physiol.* 50, 867–879.

Salgado, V.L., Sparks, T.C., 2005. The Spinosyns: chemistry, biochemistry, mode of action and resistance. In: Gilbert, L.I., Iatrou, K., Gill, S.S. (Eds.), Comprehensive Insect Molecular Science, Vol. 6 *Control*. Elsevier, Amsterdam, pp. 137–173.

Sheehan, L.S., Lill, R.E., Wilkinson, B., Sheridan, R.M., Vousden, W.A., Kaja, A.L., Crouse, G.D., Gifford, J., Graupner, P.R., Karr, L., Lewer, P., Sparks, T.C., Leadlay, P.F., Waldron, C., Martin, C.J., 2006. Engineering of the spinosyn PKS: directing starter unit incorporation. *J. Nat. Prod.* 69, 1702–1710.

Sparks, T.C., Crouse, G.D., Dripps, J.E., Anzeveno, P., Martynow, J., DeAmicis, C.V., Gifford, J., 2008. Neural network-based QSAR and insecticide discovery: spinetoram. *J. Comput. Aided Mol. Des.* 22, 393–401.

Tietze, L.F., Brasche, G., Grube, A., Bohnke, N., Stadler, C., 2007. Synthesis of novel spinosyn A analogs by Pd-mediated transformations. *Chem. Eur. J.* 13, 8543–8563.

Wang, W., Mo, J., Cheng, J., Zhuang, P., Tang, Z., 2006. Selection and characterization of spinosad resistance in *Spodoptera exigua* (Hubner) (Lepidoptera: Noctuidae). *Pestic. Biochem. Physiol.* 84, 180–187.

Watson, G.B., Chouinard, S.W., Cook, K.R., Geng, C., Gifford, J.M., Gustafson, G.D., Hasler, J.M., Larrinua, I.M., Letherer, T.J., Mitchell, J.C., Pak, W.L., Salgado, V.L., Sparks, T.C., 2010. Heterologous expression of a spinosyn-sensitive *Drosophila melanogaster* nicotinic acetylcholine receptor identified through chemically induced target site resistance and resistance gene identification. *Insect Biochem. Molec. Biol. In Press, (This article is still in press, SSGill)*.

Zhao, J.-Z., Collins, H.L., Li, Y.-X., Mau, R.F.L., Thompson, G.D., Hertlein, M., Andaloro, J.T., Boykin, R., Shelton, A.M., 2006. Monitoring of diamondback moth (Lepidoptera: Plutellidae) resistance to spinosad, indoxacarb and emamectin benzoate. *J. Econ. Entomol.* 99, 176–181.

7 *Bacillus thuringiensis*: Mechanisms and Use

A Bravo and M Soberón, Instituto de
Biotechnología, Cuernavaca Morelos, Mexico
S S Gill, University of California, Riverside, CA,
USA

© 2010, 2005 Elsevier B.V. All Rights Reserved

7.1. General Characteristics

Bacillus thuringiensis is a member of the *Bacillus cereus* group that also includes *B. cereus*, *B. anthracis*, and *B. mycoides* (Helgason *et al.*, 2000). The feature that distinguishes *B. thuringiensis* from the other members of the *B. cereus* group is its entomopathogenic properties. This bacterial species produces insecticidal proteins (δ-endotoxins) during sporulation phase as parasporal inclusions, which predominantly comprise one or more proteins, called Cry and Cyt toxins. These protein toxins are highly selective to their target insect, are innocuous to humans, vertebrates, and plants, and are completely biodegradable. Therefore, *B. thuringiensis* is a viable alternative for the control of insect pests in agriculture and disease vectors of importance in human public health.

Numerous *B. thuringiensis* strains have been isolated that show activity towards lepidopteran, dipteran, or coleopteran insects (Schnepf *et al.*, 1998). In recent years *B. thuringiensis* strains active against Hymenoptera, Homoptera, Orthoptera, and Mallophaga insect orders and to other noninsect organisms like nematodes, mites, and protozoa have been isolated (Crickmore *et al.*, 1998; Wei *et al.*, 2003). The entomopathogenic activity of *B. thuringiensis* is mainly due to Cry toxins. One feature that distinguishes these Cry proteins is their high selectivity for their target insect. More than 200 different *cry* genes have been isolated, and this constitutes an important arsenal for the control of a wide variety of insect pests. In this chapter we will summarize the present knowledge of the pathogenic properties of *B. thuringiensis*, the mode of action of Cry toxins, and their application in agriculture and disease-vector control.

7.2. Virulence Factors and the PlcR Regulon

The *B. cereus* group is characterized by its pathogenic capabilities to a diverse group of organism, such as mammals in the case of *B. anthracis* and *B. cereus*, and insects in the case of *B. thuringiensis*. The pathological characteristics of the different members of the *B. cereus* group are mainly due to the presence of specific toxin genes that for the most part encoded by large, self-transmissible extrachromosomal plasmids (Schnepf *et al.*, 1998). However, other chromosomal encoded virulence factors contribute to the pathogenic effects of different members of the *B. cereus* group (Dubois and Dean, 1995; Lereclus *et al.*, 1996; Agaisse *et al.*, 1999; Guttman and Ellar, 2000). In the case of *B. thuringiensis*, it has been known for some time that the spores synergize the effects of Cry proteins (Dubois and Dean, 1995). This synergistic effect was shown to be due to the production of several virulence factors, such as α-exotoxins, β-exotoxins, chitinases, phospholipases, and enterotoxins (Lereclus *et al.*, 1996; Agaisse *et al.*, 1999; Guttman and Ellar, 2000).

Mutation of genes encoding some of these virulence factors showed that they contribute to the pathogenicity of *B. thuringiensis* to insects (Lereclus *et al.*, 1996; Fedhila *et al.*, 2002). In the case of phosphatidylinositol specific phospholipase C (PI-PLC), disruption of the structural gene *plcA* diminished the synergistic effects of spores indicating that at least this virulence factor was necessary for an efficient virulence of the bacterium (Lereclus *et al.*, 1996). This is also the case for the *inhA2* gene that codes for a zinc-requiring metalloprotease (Fedhila *et al.*, 2002). Characterization of the mechanism controlling *plcA* gene expression led to the discovery of a pleiotropic transcriptional regulator, PlcR. Mutations in the *plcR* gene abolished *plcA* gene expression and the virulence of the bacterium (Lereclus *et al.*, 1996; Agaisse *et al.*, 1999). A genetic screening of PlcR regulated genes identified several extra cellular virulence factors including a secreted RNase, an S-layer protein, phospholipase C, and enterotoxins, indicating that the PlcR regulon includes a diverse set of virulence factors (Agaisse *et al.*, 1999). Also, the *inhA2* gene was shown to be under the control of PlcR (Fedhila *et al.*, 2003). A proteome analysis of secreted proteins from *B. cereus* strain ATCC 14579 showed that disruption of the *plcR* gene reduced the levels at least 56 exported proteins (Gohar *et al.*, 2002). Alignment of the promoter regions of the

PlcR regulated genes identified a 20-nucleotide palindromic sequence ("PlcR box") that is required for PlcR activation (Agaisse *et al.*, 1999). Although, several PlcR regulated genes are also present in *B. cereus* and *B. anthracis* (Guttman and Ellar, 2000), only the *B. cereus* PlcR protein is functionally equivalent to that of *B. thuringiensis*, since the *B. anthracis plcR* gene is disrupted by a transposon-like sequence indicating that production of virulence factors in *B. anthracis* is blocked or may be different from that of *B. thuringiensis* and *B. cereus* (Agaisse *et al.*, 1999). In fact, expression of a functional PlcR in *B. anthracis* leads to the activation of a large set of genes encoding enzymes and toxins that have protease, phospholipase, and hemolysis activities (Mignot *et al.*, 2001).

PlcR regulated genes are expressed at the end of the vegetative growth phase during the stationary phase of growth (Lereclus *et al.*, 1996). PlcR expression is negatively regulated by the sporulation key factor Spo0A, indicating that this gene is only temporally expressed in vegetative cells before the onset of sporulation (Lereclus *et al.*, 1996). The PlcR protein senses cell density and is, therefore, part of a quorum sensing mechanism (Slamti and Lereclus, 2002). The *papR* gene, which encodes a 48 amino acid peptide and is located downstream of the *plcR* gene, is activated by PlcR and the gene product is secreted. Outside the cell, PapR is degraded by extra cellular proteases and the C-terminus end pentapeptide is internalized to the bacterium cytoplasm by the oligopeptide permease Opp (Slamti and Lereclus, 2002). Inside the cell, the pentapeptide binds PlcR activating the capacity of PlcR to bind to the PlcR box sequence. This binding then induces the activation of PlcR specific gene expression (Slamti and Lereclus, 2002). It was also shown that the PlcR activating peptide is strain specific for the activation of the PlcR regulon. Single amino acid changes on the activating PapR peptide were enough for strain-specific induction of the PlcR regulon (Slamti and Lereclus, 2002). These results suggest that quorum sensing of related *B. thuringiensis* strains is restricted by the sequence homology of the C-terminal pentapeptide of PapR. **Figure 1** shows the PlcR regulon network including the gene targets and the regulation of PlcR activation by PapR protein.

7.3. Insecticidal Toxins

7.3.1. Classification and Nomenclature

The major determinants of *B. thuringiensis* insecticidal properties are the δ-endotoxins (Schnepf *et al.*, 1998). These endotoxins form two multigenic families, *cry* and *cyt*. Cry proteins are specifically toxic to orders of insects, Lepidoptera, Coleoptera,

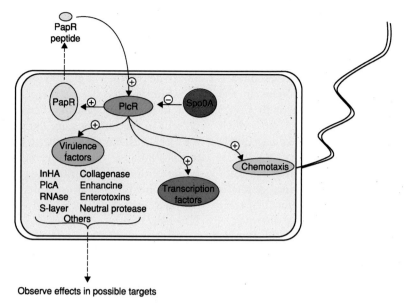

Figure 1 The PlcR regulon in *Bacillus thuringiensis*. Positive (+) and negative (−) regulation in the PlcR regulon are shown.

Hymenoptera, and Diptera. In contrast, Cyt toxins are mostly found in *B. thuringiensis* strains active against Diptera, although a few exceptions of coleopteran active strains containing Cyt proteins have been documented (Guerchicoff *et al.*, 2001). Besides these crystals proteins, some *B. thuringiensis* and *B. cereus* strains produce a third group of insecticidal proteins called vegetative insecticidal proteins (Vip proteins) (Estruch *et al.*, 1996). In contrast to crystal proteins, the Vip proteins are synthesized during the vegetative growth phase of the bacterium and secreted into the medium without forming crystal inclusions.

The Cry proteins comprise 40 subgroups with more than 200 members. The definition of Cry proteins is rather broad: a parasporal inclusion protein from *B. thuringiensis* that exhibits toxic effect to a target organism, or any protein that has obvious sequence similarity to a known Cry protein (Crickmore *et al.*, 1998). This definition includes the Cyt toxins although the mnemonic Cyt is given for Cry proteins that are structurally related to Cyt toxins (Crickmore *et al.*, 1998, 2002). The nomenclature of Cry and Cyt proteins is based on primary sequence identity among the different protein sequences. Crystal proteins receive the mnemonic Cry or Cyt and four hierarchical ranks consisting of numbers, capital letters, lowercase letters, and numbers (e.g., Cry1Ab1). The ranks are given depending on the sequence identity shared with other Cry or Cyt proteins. A different number (first rank) is given to a protein if it shares less than 45% identity with all the others Cry proteins (Cry1,

Cry2, etc.). The capital letter (second rank) is given if the protein shares less than 78% but more than 45% identity with a particular group of Cry proteins (e.g., Cry1A, Cry1B, etc.). The third rank, a lowercase letter, is given to distinguish proteins that share more than 78% but less than 95% identity with other Cry proteins (e.g., Cry1Aa, Cry1Ab, etc.). Finally a number is given to distinguish proteins that share more than 95% identity, but which are not identical and should be considered variants of the same protein (Crickmore *et al.*, 1998, 2002). Although this system does not take into account the insect selectivity of Cry toxins, some subgroups of Cry toxins show toxicity against a particular genus of insects. Thus Cry1 proteins are active principally against lepidopteran insects, while the Cry3 proteins are toxic towards coleopterans. In the case of dipteran-active toxins, a surprisingly high number of different subgroups are active. For example, Cry proteins that share low sequence similarity (e.g., Cry1C, Cry2Aa, Cry4, Cry11, etc.) are active against mosquitoes.

Bacillus thuringiensis strains produce different proteins that are not related phylogenetically and all these subfamilies comprise the Cry family. Among these are: the three-domain Cry family; the larger group of Cry proteins (formed by 30 different Cry subgroups); the binary-like toxins (Cry35 and Cry36); and Mtx-like toxins (Cry15, Cry23, Cry33, Cry38, and Cry40), which are related to toxins produced by *B. sphaericus* (Crickmore *et al.*, 1998, 2002). The Cry35 requires the Cry34 toxin to induce the lethal effects against *Diabrotica virgifera*

(Ellis *et al.*, 2002). The Cry34 is not related to other toxins but as mentioned above is the binary partner of Cry35. Similarly the coleopteran specific Cry23A toxin functions effectively only in the presence of the Cry37 protein (Donovan *et al.*, 2000). The sequence of Cry37 is not related to other proteins in the database but showed some similarities with Cry22 (Baum and Light Mettus, 2000) and Cry6, which are coleopteran specific (Thompson *et al.*, 2001) and also nematode specific toxins (Wei *et al.*, 2003).

The members of the three-domain family are globular molecules containing three distinct domains connected by single linkers. The alignment of their protein sequences revealed the presence of five conserved sequence blocks, although, some blocks are absent in some subgroups of the three-domain family. One particular feature of the members of this family is the presence of protoxins with two different lengths. One large group of protoxins is approximately twice as long as the majority of the toxins (130 or 70 kDa). The C-terminal extension found in the longer protoxins is dispensable for toxicity but is believed to play a role in the formation of the crystal inclusion bodies within the bacterium (de Maagd *et al.*, 2001).

Cyt toxins comprise two highly related gene families (*cyt1* and *cyt2*) (Crickmore *et al.*, 1998, 2002). Analysis of amino acid sequences shows that the different Cyt versions show a high degree of conservation in predicted α-helices and β-sheets (Guerchicoff *et al.*, 2001). Cyt toxins are also synthesized as protoxins and small portions of the N-terminus and C-terminus are removed to activate the toxin (Li *et al.*, 1996). Cyt proteins are almost exclusively found in dipteran active strains, although a few exceptions have been found as the presence of Cyt in coleopteran specific strains (Guerchicoff *et al.*, 2001). Cyt toxins synergize the toxic effect of some Cry proteins against different targets and also that of the *B. sphaericus* binary toxin (Wu *et al.*, 1994; Sayyed *et al.*, 2001; Wirth *et al.*, 2001). In the case of the three-domain Cry and Cyt proteins it is widely accepted that the primary action of these toxins is the formation of lytic pores in the midgut cells of susceptible insects. **Figure 2** shows the phylogenetic dendogram of Cry and Cyt toxins (Crickmore *et al.*, 2002).

Finally, Vip proteins comprise three families with seven members. The *vip1* and *vip2* genes are located in a single operon, and the proteins encoded by these genes constitute a binary toxin since both are necessary for toxicity against coleopteran larvae. The Vip1 and Vip2 proteins are 100 and 52 kDa in size, respectively (Warren, 1997). The mechanism of action of these proteins has not been clearly described. Nevertheless, the three-dimensional structure of Vip2 protein indicates that it may have ADP-ribosylating activity (Craig *et al.*, 1999). The Vip3 protein (88 kDa) is toxic to several lepidopteran larvae (Estruch *et al.*, 1996). A preliminary mode of action of Vip3 has been recently described. This protein is proteolytically activated by proteases present in the midgut juice extract or by trypsin to a 62 kDa resistant fragment, which binds two proteins of 80 and 100 kDa located in the apical membrane of the lepidopteran midgut cells. Vip3A also forms ionic pores that are voltage independent and cation selective characterized by long open times (Lee *et al.*, 2003).

7.3.2. Structure of Toxins

The crystal structures of the trypsin activated toxin Cry1Aa (lepidopteran specific), Cry3A and Cry3B (coleopteran specific), and Cry2Aa (dipteran–lepidopteran specific) protoxin have been solved (Li *et al.*, 1991; Grouchulski *et al.*, 1995; Galistki *et al.*, 2001; Morse *et al.*, 2001). Although the sequence similarity between these toxins is very low (20% and 17% identity of Cry2Aa with Cry3Aa and Cry1Aa, respectively) the overall structural topology of these proteins is very similar. **Figure 3** shows the three-dimensional structure of Cry1A, Cry3Aa toxins, and Cry2Aa protoxin (Morse *et al.*, 2001). The structure is composed of three structural domains. Domain I is a seven α-helix bundle in which a central helix α-5 is surrounded by six outer helices. This domain has been implicated in the membrane ion channel formation. The six α-helices are amphipathic and are long enough to span the 30-Å thick hydrophobic region of a membrane bilayer. Cry2Aa protoxin has an extra 49 amino acid region forming two extra α-helices in the N-terminal end that are cleaved out after proteolytic activation of the toxin (Morse *et al.*, 2001) (**Figure 3**). Domain II consist of three antiparallel β-sheets packed around a hydrophobic core in a "β-prism." This domain represents the most divergent part in structure among Cry toxin molecules (de Maagd *et al.*, 2001) and has been described as the selectivity-determining domain. Finally, domain III is a β-sandwich of two antiparallel β-sheets. Several lines of evidence indicate that domain III is implicated in insect selectivity and therefore involved in receptor binding. However, additional roles for this domain, such as in pore formation, have been suggested (Schwartz *et al.*, 1997c; de Maagd *et al.*, 2001).

Figure 4 shows the analysis of structural similarities of the three domains of Cry toxins with other proteins. This analysis revealed interesting features

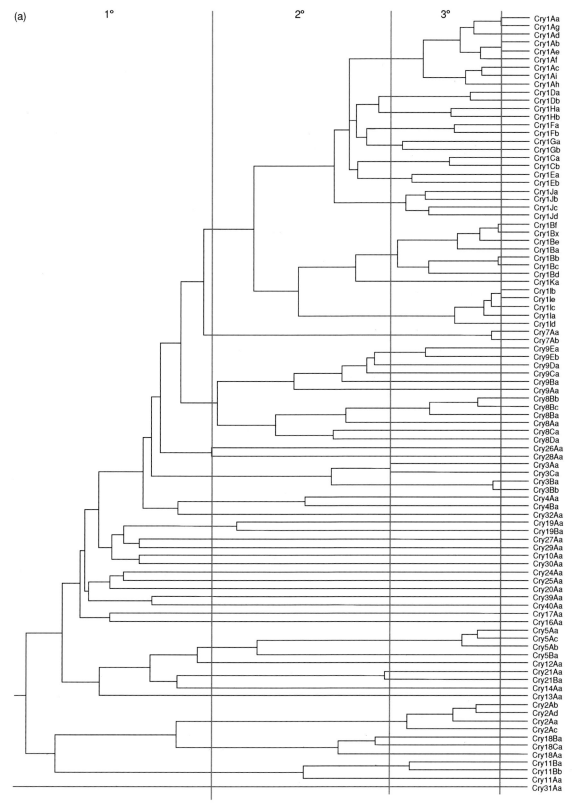

Figure 2   Continued

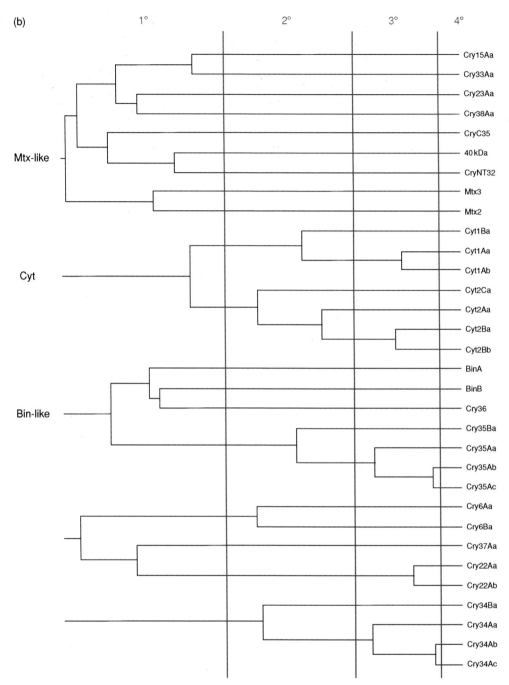

Figure 2 Dendograms of *cry* gene sequences. (a) The three domain sequences; (b) other *cry* gene sequences. Update and dendogram analysis can be found at http://www.biols.susx.ac.uk.

that could give hints on the possible role of these domains in the mode of action of Cry proteins. Domain I shares structural similarities with other pore forming toxins like colicin Ia and N (Protein Data Bank (PDB) codes: 1cii, 1a87), hemolysin E (1qoy), and diphtheria toxin (1ddt). This similarity suggests the role of this domain is in pore formation. In the case of domain II, structural similarities with several carbohydrate binding proteins

like vitelline (1vmo), lectin jacalin (1jac), and lectin Mpa (1jot) supports the conclusion that this domain is involved in binding. Also for domain III, carbohydrate binding proteins with similar structures were identified as the cellulose binding domain of 1,4-β-glucanase CenC (1ulo), galactose oxidase (1gof), sialidase (1eut), β-glucuronidase (1bgh), the carbohydrate binding domain of xylanase U (1gmm), and β-galactosidase (1bgl). These results suggest

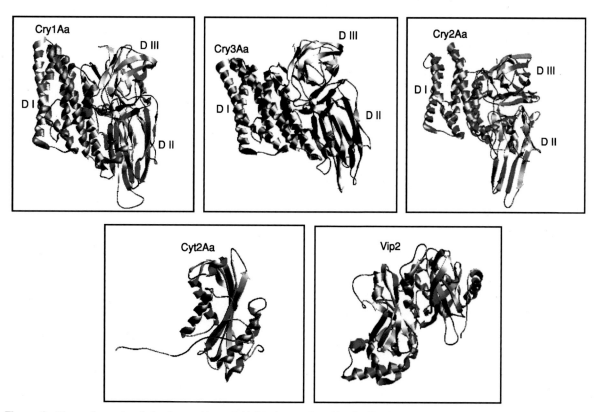

Figure 3   Three-dimensional structures of insecticidal toxins produced by *Bacillus thuringiensis*. D I, II, III, domains I, II, and III.

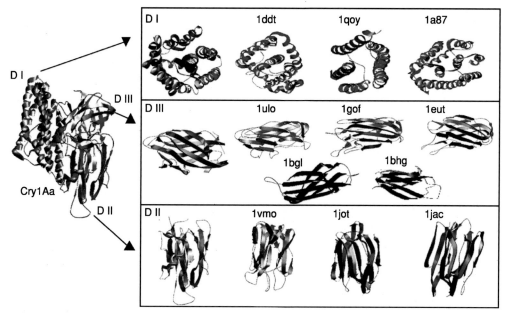

Figure 4   Proteins with structural similarity to the three domains of Cry toxins.

that carbohydrate moieties could have an important role in the mode of action of these three-domain Cry toxins, specifically in the recognition interaction of the toxin and its membrane receptors. As will be discussed later, the involvement of carbohydrate moieties in the mode of action of some Cry toxins has been demonstrated since different glycosylated proteins (aminopeptidase-N, a cadherin-like protein, and a glycoconjugate) have been identified as Cry toxin receptors in lepidopteran larvae.

Figure 3 also shows the ribbon representation of the three dimensional structure of Cyt2Aa toxin (Li *et al.*, 1996). Cyt2Aa has single domain of α/β architecture comprising two outer layers of α-helical hairpins wrapped around a β-sheet. The Cyt2Aa protoxin is organized as a dimer, and proteolytic activation releases the active Cyt2Aa toxin monomer (Li *et al.*, 1996; Gazit *et al.*, 1997). Based on the length of the secondary structures it was proposed that the inner core of β-sheets (β-5, β-6, β-7, and β-3) could span the cell membrane (Gazit *et al.*, 1997). However, analysis of membrane insertion capabilities of synthetic peptides corresponding to the different secondary structural elements suggested that α-helices A and C, rather than the β-sheets, could be the elements responsible for membrane interaction (Gazit *et al.*, 1997).

Vip2A and Vip1A were isolated and characterized from a *B. cereus* strain (Warren, 1997). As mentioned above the structural genes *vip2* and *vip1* form an operon, and both Vip1 and Vip2 proteins constitute a binary toxin since both proteins are required for toxicity. **Figure 3** shows the three-dimensional structure of Vip2 protein. Vip2 is a mixed α/β protein and is divided into two domains. The domains are structurally homologous although they share low amino acid identity. The overall fold of each domain resembles the catalytic domains of classical A-B toxins that have two components, a binding component (Vip1 presumably) and a catalytic ADP-ribosylating component (Vip2) (Warren, 1997). The cellular target of Vip2 is still unknown.

7.3.3. Evolution of Three-Domain Toxins

The increasing number of Cry toxins and the wide variety of target organisms that these proteins have puts forward the question of how evolution of this protein family created such a diverse arsenal of toxins. It has been proposed that toxins coevolved with their target insects (de Maagd *et al.*, 2001). However, there are no studies that correlate the geographical distribution and Cry protein content of *B. thuringiensis* strains and the distribution of their target insect species in nature. In one particular study, a *B. thuringiensis* strain collection obtained from different climate regions of Mexico revealed that putative novel *cry* genes were more frequently found in *B. thuringiensis* strains isolated from tropical regions where the insect diversity is high (Bravo *et al.*, 1998). Also the dipteran specific *cry11* and *cyt* genes were more frequently found in the rainy tropical regions than in the semiarid regions, which correlates with the distribution of dipteran insects (Bravo *et al.*, 1998). A similar correlation between the frequencies of active strains with the geographical origin of the samples was presented by Bernhard *et al.* (1997) who reported that high number of *B. thuringiensis* strains active against *Heliothis virescens* in samples collected from North America where this species is a major agricultural pest. These results suggest that Cry toxins and insects may coevolve, although more comprehensive studies are needed to support this assumption.

Phylogenetic analysis of the complete Cry toxin or protoxin protein showed a broad correlation with toxicity (Bravo, 1997). However, analyses of the phylogenetic relationships of the isolated domains revealed interesting features regarding the creation of diversity in this family of proteins (Bravo, 1997; de Maagd *et al.*, 2001). The analyses of phylogenetic relationships of isolated domain I and domain II sequences revealed that these domains coevolved. In the case of domain II, phylogenetic relationships showed a topology clearly related to the specificity of the toxin proteins (Bravo, 1997). This result is not surprising in spite of the fact that domain II is involved in receptor recognition and is therefore a determinant of insect selectivity (Bravo, 1997; de Maagd *et al.*, 2001). Surprisingly, phylogenetic relationships of domain I sequences, that is involved in the pore formation of the toxin, showed a topology also clearly related to the selectivity of the toxin proteins (Bravo, 1997; de Maagd *et al.*, 2001), suggesting that different types of domain I have been selected for acting in particular membrane conditions of their target insect (Bravo, 1997; de Maagd *et al.*, 2001). The analysis of domain III sequences revealed a different topology due to the fact that several examples of domain III swapping among toxins occurred in nature (Bravo, 1997; de Maagd *et al.*, 2001). Some toxins with dual specificity (coleopteran, lepidopteran) are clear examples of domain III swapping among coleopteran and lepidopteran specific toxins (Bravo, 1997; de Maagd *et al.*, 2001). Cry1B toxins are good examples that illustrate domain III swapping in natural toxins. Five different Cry1B toxins have been described. These toxins share almost identical domain I and domain II sequences. However, Cry1B domain III sequences cluster with several Cry genes suggesting a very active process of domain III exchanges. The Cry1Ba domain III sequence is closely related to domain III sequences of Cry1Jb, Cry8Aa, and Cry9Da (**Figure 5**) but, in contrast, Cry1Bb and Cry1Bc share similar domain III sequences between them. In the case of Cry1Be, domain III clusters with domain III from Cry1Cb and Cry1Eb. Finally Cry1Bd domain III is similar to domain III of Cry1Ac toxin. Indeed Cry1B toxins

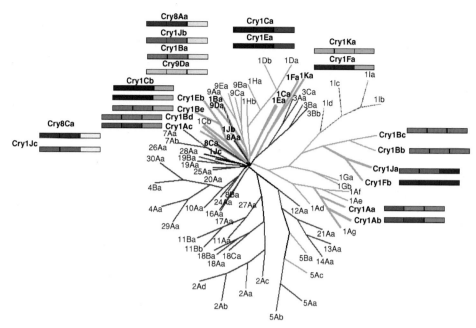

Figure 5 Phylogenetic relationships of Cry domain III sequences. Genes with similar domain III sequences are shown in the same color.

bioassays show that some of these toxins have different insect selectivities. For example, Cry1Ba is active against the coleopteran Colorado potato beetle, *Leptinotarsa decemlineata*, while Cry1Bd has higher activity against the lepidopteran diamondback moth, *Plutella xylostella,* than Cry1Ba and Cry1Bb toxins. A more comprehensive analysis of toxicity of Cry1B toxins could establish the relevance of domain III swapping in toxicity. However, this result suggests that domain III swapping could create novel specificities. Indeed, *in vitro* domain III swapping of certain Cry1 toxins results in changes in insect specificity as demonstrated for the hybrid protein containing domain I and II of Cry1Ab toxin and domain III of Cry1C which resulted in a toxin protein with increased toxicity towards *Spodoptera* sp. (Bosch *et al.*, 1994; de Maagd *et al.*, 2000, 2001). **Figure 5** shows the topology of the phylogenetic relationships of domain III sequences and the representation of Cry toxins in which domain III swapping occurred in nature during evolution of these proteins.

Phylogenetic analyses of the Cry toxin family shows that the great variability in the biocidal activity of this family has resulted from two fundamental evolutionary processes: (1) independent evolution of the three structural domains, and (2) domain III swapping among different toxins. These two processes have generated proteins with similar modes of action but with very different insect selectivities (Bravo, 1997; de Maagd *et al.*, 2001).

7.4. Mode of Action of Three-Domain Cry Toxins

7.4.1. Intoxication Syndrome of Cry Toxins

As mentioned above (see Section 7.3.2), it is widely accepted that the primary action of Cry toxins is to lyse midgut epithelial cells in the target insect by forming lytic pores in the apical microvilli membrane of the cells (Gill *et al.*, 1992; Schnepf *et al.*, 1998; Aronson and Shai, 2001; de Maagd *et al.*, 2001). In order to exert its toxic effect, Cry protoxins in the crystalline inclusions are modified to membrane-inserted oligomers that cause ion leakage and cell lysis. The crystal inclusions ingested by susceptible larvae dissolve in the environment of the gut, and the solubilized inactive protoxins are cleaved by midgut proteases yielding 60–70 kDa protease resistant proteins (Choma *et al.*, 1990). Toxin activation involves the proteolytic removal of an N-terminal peptide (25–30 amino acids for Cry1 toxins, 58 for Cry3A, and 49 for Cry2Aa) and approximately half of the remaining protein from the C-terminus in the case of the long Cry protoxins (130 kDa protoxins). The activated toxin then binds to specific receptors on the brush border membrane of the midgut epithelium columnar cells of susceptible insects (Schnepf *et al.*, 1998; de Maagd *et al.*, 2001) before inserting into the membrane (Schnepf *et al.*, 1998; Aronson and Shai, 2001). Toxin insertion leads to the formation of lytic pores in microvilli apical membranes (Schnepf *et al.*, 1998; Aronson and Shai, 2001). Cell lysis and disruption of the midgut epithelium releases the cell contents providing spores a rich medium that is suitable for spore germination leading to a severe septicemia and insect death (Schnepf *et al.*, 1998; de Maagd *et al.*, 2001).

7.4.2. Solubilization and Proteolytic Activation

Solubilization of long protoxins (130 kDa) depends on the highly alkaline pH that is present in guts of lepidopteran and dipteran insects, in contrast to coleopteran insect guts that have a neutral to slightly acidic pH (Dow, 1986). In a few cases, protoxin solubilization has been shown to be a determinant for insect toxicity. Cry1Ba is toxic to the coleopteran *L. decemlineata* only if the protoxin is previously solubilized *in vitro*, suggesting insolubility of the toxin at the neutral–acidic pH of coleopteran insects (Bradley *et al.*, 1995). The C-terminal portion of protoxins contains many cysteine residues that form disulfide bonds in the crystal inclusions and, therefore, reducing the disulfide bonds is a necessary step for the solubilization of long Cry protoxins (Du *et al.*, 1994). Differences in the midgut pH between lepidopteran and coleopteran midguts may be a reason for the bias in the utilization of arginine as basic amino acid over lysine in the lepidopteran specific toxins Cry1, Cry2, and Cry9 with exception of the Cry1I toxin, which is also active against coleopteran insects (Grochulski *et al.*, 1995; de Maagd *et al.*, 2001). The higher pK_a of arginine, compared with that of lysine, might be required for maintaining a positive charge even at the high pH of lepidopteran guts (up to pH 11) resulting in soluble toxins at alkaline pH.

Proteolytic processing of Cry toxins is a critical step involved not only on toxin activation but also on specificity (Haider and Ellar, 1989; Haider *et al.*, 1989) and insect resistance (Oppert *et al.*, 1997; Shao *et al.*, 1998). Besides pH, lepidopteran and coleopteran insects differ in the type of proteases present in the insect gut; serine proteases are the main digestive proteases of Lepidoptera and Diptera, whereas cysteine and aspartic proteases are abundant in the midguts of Coleoptera (Terra and Ferreira, 1994). It has been reported that enhanced degradation of Cry toxins is associated with the loss of sensitivity of fifth instar *Spodoptera litoralis* larvae to Cry1C (Keller *et al.*, 1996) and that serine protease inhibitors enhanced the insecticidal activity of some *B. thuringiensis* toxins up to 20-fold (MacIntosh *et al.*, 1990). More recently, it was found that the low toxicity of Cry1Ab toxin to

S. frugiperda could be explained in part by rapid degradation of the toxin on the insect midgut (Miranda *et al.*, 2001). For several Cry proteins inactivation within the insect gut involves intramolecular processing of the toxin (Choma *et al.*, 1990; Lambert *et al.*, 1996; Audtho *et al.*, 1999; Pang *et al.*, 1999; Miranda *et al.*, 2001). However, for several other Cry toxins, intramolecular processing is not always related to loss of toxicity and sometimes is required for proper activation of the toxin (Dai and Gill, 1993; Zalunin *et al.*, 1998; Yamagiwa *et al.*, 1999). Therefore, in some cases, differential proteolytic processing of Cry toxins in different insects could be a limiting step in the toxicity of Cry proteins (Miranda *et al.*, 2001).

One interesting feature of Cry toxin activation is the processing of the N-terminal end of the toxins. The three-dimensional structure of Cry2Aa protoxin showed that two α-helices of the N-terminal region occlude a region of the toxin involved in the interaction with the receptor (Morse *et al.*, 2001) (**Figure 3**). Several lines of evidences suggest that the processed N-terminal peptide of Cry protoxins might prevent binding to nontarget membranes (Martens *et al.*, 1995; Kouskoura *et al.*, 2001; Bravo *et al.*, 2002b). *Escherichia coli* cells producing Cry1Ab or Cry1Ca toxins lacking the N-terminal peptide were severely affected in growth (Martens *et al.*, 1995; Kouskoura *et al.*, 2001). It was speculated that the first 28 amino acids prevented the Cry1A toxin from inserting into the membrane (Martens *et al.*, 1995; Kouskoura *et al.*, 2001). Recently, it was found that a Cry1Ac mutant that retains the N-terminus end after trypsin treatment binds nonspecifically to *Manduca sexta* membranes and was unable to form pores on *M. sexta* brush border membrane vesicles (Bravo *et al.*, 2002b). Therefore, processing of the N-terminal end of Cry protoxins may unmask a hydrophobic patch of the toxin involved in toxin–receptor or toxin–membrane interaction (Morse *et al.*, 2001; Bravo *et al.*, 2002b).

7.4.3. Receptor Identification

The major determinant of Cry toxin selectivity is the interaction with specific receptors on the insect gut of susceptible insects (Jenkins and Dean, 2000). Therefore, receptor identification is fundamental for determining the molecular basis of Cry toxin action and also in insect resistance management that in many cases has been shown to correlate with defects in receptor binding (Ferré and Van Rie, 2002). A number of putative receptor molecules for the lepidopteran specific Cry1 toxins have been identified. In *M. sexta*, the Cry1Aa, Cry1Ab, and Cry1Ac proteins bind to a 120 kDa aminopep-tidase-N (APN) (Knight *et al.*, 1994; Garczynski and Adang, 1995; Denolf *et al.*, 1997) and to a 210 kDa cadherin-like protein (Bt-R$_1$) (Belfiore *et al.*, 1994; Vadlamudi *et al.*, 1995). Cadherins represent a large family of glycoproteins that are classically responsible for intercellular contacts. These proteins are transmembrane proteins with a cytoplasmic domain and an extracellular ectodo-main with several cadherin repeats (12 in the case of Bt-R$_1$). The ectodomain contain calcium binding sites, integrin interaction sequences, and cadherin binding sequences. In *Bombyx mori*, Cry1Aa binds to a 175 kDa cadherin-like protein (Bt-R$_{175}$) (Nagamatsu *et al.*, 1998, 1999) and to a 120 kDa APN (Yaoi *et al.*, 1997). In *H. virescens* Cry1Ac binds to two proteins of 120 kDa and 170 kDa both identified as APN (Gill *et al.*, 1995; Oltean *et al.*, 1999). Also, a cadherin-like protein is involved in the mode of action of Cry1Ac toxin in *H. virescens* (Gahan *et al.*, 2001). In *P. xylostella* and *Lymantria dispar* APNs were identified as Cry1Ac receptors (Valaitis *et al.*, 1995; Lee *et al.*, 1996; Denolf *et al.*, 1997; Luo *et al.*, 1997). In *L. dispar*, besides APN and cadherin-like receptors, a high molecular weight anionic protein (Bt-R$_{270}$) that binds Cry1A toxins with high affinity was identified (Valaitis *et al.*, 2001). For Cry1C toxin an APN receptor molecule was identified in *Spodoptera litura* (Agrawal *et al.*, 2002). Sequence analysis of various APN from lepidopteran insects suggests that APN's group into at least four classes (Oltean *et al.*, 1999). *Plutella xylostella* and *B. mori* produce the four classes of APN and it is anticipated that several other lepidopteran insects may also produce the four isoforms (Nakanishi *et al.*, 2002). Cry1A binding to the different *B. mori* APN isoforms revealed that this toxin only binds the 115 kDa isoform in ligand blot binding analysis (Nakanishi *et al.*, 2002).

Surface plasmon resonance experiments showed that the binding affinity of Cry1A toxins to the *M. sexta* APN is in the range of 100 nM (Jenkins and Dean, 2000), while that of cadherin-like recep-tors (Bt-R$_1$) is in the range of 1 nM (Vadlamudi *et al.*, 1995). This difference in the binding affinities be-tween APN and Bt-R$_1$ suggest that binding to Bt-R$_1$ might be the first event on the interaction of Cry1A toxins with microvilli membranes and, therefore, the primary determinant of insect specificity.

Different experimental evidence supports the in-volvement of both Cry1A toxins receptors (APN and cadherin-like) in toxicity. Expression of the *M. sexta* and *B. mori* cadherin-like proteins, Bt-R$_1$ and Bt-R$_{175}$ respectively, on the surface of different cell lines render these cells sensitive to Cry1A toxins,

although the toxicity levels were low (Nagamatsu *et al.*, 1999; Dorsch *et al.*, 2002; Tsuda *et al.*, 2003). Cry1Aa toxin was shown to lyse isolated midgut epithelial cells; this toxic effect was inhibited if the cells were preincubated with anti-Bt-R$_{175}$ antisera in contrast with the treatment with anti-APN antisera, suggesting that cadherin-like protein Bt-R$_{175}$ is a functional receptor of Cry1Aa toxin (Hara *et al.*, 2003). Also, a single-chain antibody (scFv73) that inhibits binding of Cry1A toxins to cadherin-like receptor, but not to APN, reduced the toxicity of Cry1Ab to *M. sexta* larvae (Gómez *et al.*, 2001). Moreover, disruption of a cadherin gene by a retro-transposon-mediated insertion and its linkage to high resistance to Cry1Ac toxin in *H. virescens* YHD2 (a laboratory selected line) larvae supports a role for cadherin as a functional receptor (Gahan *et al.*, 2001). Overall, these results suggest that binding to the cadherin-like receptor is an important step in the mode of action Cry1A toxins (**Figure 6**).

Regarding the APN receptor, several reports argue against the involvement of this protein in toxin activity. Mutants of Cry1Ac toxin affected on APN binding retained similar toxicity levels to *M. sexta* as the wild-type toxin (Burton *et al.*, 1999). However, there are several reports describing point mutations of Cry toxins located in domain II that affected both APN binding and toxicity (Jenkins and Dean, 2000). Expression of APN in heterologous systems did not result in Cry1A sensitivity (Denolf *et al.*, 1997) and, as mentioned above, an APN antibody did not protect *B. mori* midgut cells from Cry1Aa toxic effect in contrast to a cadherin anti-body (Hara *et al.*, 2003). However, two recent reports clearly demonstrate the importance of this molecule on the mode of action of Cry1 toxins. Inhibition of APN production in *S. litura* larvae by dsRNA interference showed that insects with low APN levels became resistant to Cry1C toxin (Rajagopal *et al.*, 2002). Also, heterologous expres-sion of *M. sexta* APN in midguts and mesoder-mal tissues of transgenic *Drosophila melanogaster* caused sensitivity to Cry1Ac toxin (Gill and Ellar, 2002). Additionally, previous reports demonstrated that incorporation of *M. sexta* APN into black lipid bilayers lowers the concentration of toxin needed for pore formation activity of Cry1Aa toxin (Schwartz *et al.*, 1997b). Finally, all Cry1A APN receptors are anchored to the membrane by a glyco-syl phosphatidylinositol (GPI) (Knight *et al.*, 1994; Garczynski and Adang, 1995; Denolf *et al.*, 1997; Oltean *et al.*, 1999; Agrawal *et al.*, 2002; Nakanishi *et al.*, 2002). And treatment of *Trichoplusia ni* brush border membrane vesicles with PI-PLC, which cleaves GPI-anchored proteins from the membrane,

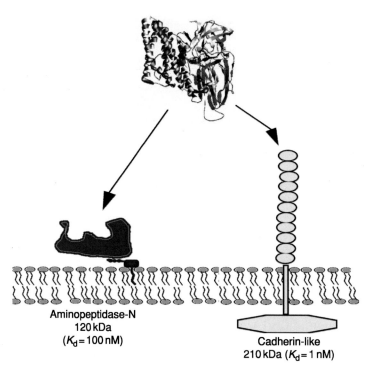

Figure 6   Receptor molecules of Cry1A proteins. K_d values are average affinity values of Cry1A toxins to aminopeptidase-N and cadherin receptors.

reduced Cry1Ac pore formation activity (Lorence *et al.*, 1997). These reports suggest that APN binding is also an important step in the mode of action of Cry1 toxins.

In the case of the dipteran specific Cry11A and Cry4B toxins, two binding proteins of 62 and 65 kDa were identified on brush border membrane vesicles from *Aedes aegypti* larvae (Buzdin *et al.*, 2002). The identity of these binding molecules and their role as receptors of Cry11A and Cry4B toxins still remains to be analyzed.

One interesting feature of the Cry1 receptor molecules identified so far is that all these proteins are glycosylated. As mentioned before, it is interesting to note that domain II and domain III of Cry toxins have structural homology to several protein domains that interact with carbohydrates. In the case of Cry1Ac, Cry5, and Cry14 toxins different experimental evidence suggest a role of carbohydrate recognition in the mode of action of these toxins. The domain III of Cry1Ac interacts with a sugar N-acetylgalactosamine on the aminopeptidase receptor (Masson *et al.*, 1995). In *Caenorhabditis elegans* resistance to the Cry5B and Cry14 toxins arises because of changes in the expression of enzymes involved in glycosylation (Griffits *et al.*, 2001). However, the structural similarities of domain II and III with carbohydrate binding proteins does not exclude the possibility that these

domains may have protein–protein interactions with the receptor molecules. In fact some loop regions of domain II of Cry1Ab and the Bt-R$_1$ receptor do have protein–protein interaction (Gómez *et al.*, 2002a). The interaction between the toxin and its receptors can be complex involving multiple interactions with different toxin–receptor epitopes or carbohydrate molecules. Interestingly, the binding of Cry1Ac to two sites on the purified APN from *M. sexta* has been reported, and N-acetylgalactosamine could inhibit 90% of Cry1Ac binding, which does not inhibit the binding of Cry1Aa and Cry1Ab to the same APN receptor (Masson *et al.*, 1995). Only one of the Cry1Ac binding sites on APN is shared and recognized by Cry1Aa and Cry1Ab toxins (Masson *et al.*, 1995).

7.4.4. Toxin Binding Epitopes

The identification of epitopes involved in Cry toxin–receptor interaction will provide insights into the molecular basis of insect specificity and could help in the characterization of insect resistant populations in nature. Such studies would also aid in developing strategies to design toxins that could overcome receptor point mutations leading to Cry toxin resistance.

The toxin binding epitopes have been mapped for several Cry toxins. Domains II and III are the most variable regions of Cry toxins and had, therefore, been subject to mutagenesis studies to determine their role in receptor recognition. Domain II was first recognized as an insect toxicity determinant domain based on hybrid toxin construction using *cry* genes with different selectivities (Ge *et al.*, 1989). Site-directed mutagenesis studies of Cry1A toxins showed that some exposed loop regions, loop α-8, loop 2, and loop 3, of domain II are involved in receptor recognition (Rajamohan *et al.*, 1996a, 1996b; Jenkins *et al.*, 2000; Lee *et al.*, 2000, 2001). Specifically, the loop 2 region of Cry1Ab toxin was shown to be important in the interaction with *M. sexta* brush border membrane vesicles. Characterization of mutants affected at Arg368-Arg369 in loop 2 indicated that these residues have an important role in the reversible binding of the toxin to brush border membrane vesicles, while Phe371 is involved in the irreversible binding of the toxin (Rajamohan *et al.*, 1995, 1996b; Jenkins and Dean, 2000). Reversible binding is related to the initial interaction of the toxin to the receptor while irreversible binding is related to the insertion of the toxin into the membrane (Rajamohan *et al.*, 1995). This result was interpreted as suggesting that Phe371 might be involved in membrane insertion (Rajamohan *et al.*, 1995, 1996b). Also, evidence has been provided showing that Arg368-Arg369 in loop 2 of Cry1Ab are involved in APN binding since this mutant showed no binding to purified APN in surface plasmon resonance experiments (Jenkins and Dean, 2000; Lee *et al.*, 2000). In this regard it is interesting to note that a truncated derivative of Cry1Ab toxin containing only domains II and III was still capable of receptor interaction but was affected in irreversible binding (Flores *et al.*, 1997). Therefore, it is likely that mutations in Phe371 affect membrane insertion probably by interfering with conformational change, necessary for membrane insertion, after initial recognition of the receptor or as originally proposed this residue could be directly involved in membrane interaction (Rajamohan *et al.*, 1995). Phylogenetic relationship studies of different Cry1 loop 2 sequences show that there is a correlation between the loop 2 amino acid sequences and cross-resistance with several Cry1 toxins in resistant insect populations, implying that this loop region is an important determinant of receptor recognition (Tabashnik *et al.*, 1994; Jurat-Fuentes and Adang, 2001). Overall these results suggest that loop 2 of Cry1A toxins plays an important role in receptor recognition. Regarding loop α-8 and loop 3, mutations of some residues in these regions of Cry1A toxins affected reversible binding to brush border membrane vesicles or the purified APN receptor (Rajamohan *et al.*, 1996a; Lee *et al.*,

2001). The closely related toxins Cry1Ab and Cry1Aa have different loop 3 amino acid sequences even though they interact with the same receptor molecules (Rajamohan *et al.*, 1996a). Analysis of Cry1Ab and Cry1Aa binding to the cadherin-like Bt-R$_1$ receptor in ligand blot experiments and competition with synthetic peptides corresponding to the toxin exposed loop regions, showed that, besides loop 2, loop 3 of Cry1Aa toxin was important for Bt-R$_1$ recognition (Gómez *et al.*, 2002a). Although the Cry1A loop 1 region seems not to play an important role in receptor recognition this is certainly not the case for other Cry toxins. Mutagenesis of Cry3A loop 1 residues showed that this region, besides loop 3, is important for receptor interaction in coleopteran insects (Wu and Dean, 1996). Loop 1 residues of two Cry toxins with dual insecticidal activity (Cry1Ca and Cry2Aa, active against dipteran and lepidopteran insects) are important for toxicity against mosquitoes but not against lepidopterous insects (Widner and Whiteley, 1990; Smith and Ellar, 1994; Morse *et al.*, 2001). Besides loop1 of Cry1C toxin, loops 2 and 3 are important for toxicity to dipteran and lepidopteran insects (Abdul-Rauf and Ellar, 1999). In the case of Cry2Aa toxin, an amino acid region involved in lepidopteran activity was not located in loops 1, 2, or 3 regions but was located to a different part of domain II in loops formed by β-5 and β-6, and by β-7 and β-8 (Morse *et al.*, 2001). These results show that the loop regions of domain II are important determinants for receptor interaction, although other regions of this domain, in certain toxins, could also participate in receptor recognition.

As mentioned previously (see Section 7.3.3), domain III swapping indicated that this domain is involved in receptor recognition (Bosch *et al.*, 1994; de Maagd *et al.*, 2000), and it has been proposed that domain swap has been used as an evolutionary mechanism of these toxins (Bravo, 1997; de Maagd *et al.*, 2001). The swapping of domain III between Cry1Ac and Cry1Ab toxins showed that the Cry1Ac domain III was involved in APN recognition (Lee *et al.*, 1995; de Maagd *et al.*, 1999a). The interaction of Cry1Ac domain III and APN was dependent on N-acetylgalactosamine (Gal-Nac) residues (Burton *et al.*, 1999; Lee *et al.*, 1999). Mutagenesis studies of Cry1Ac domain III identified Gln509-Asn510-Arg511 Asn506, and Tyr513 as the epitope for sugar recognition (Burton *et al.*, 1999; Lee *et al.*, 1999). The three-dimensional structure of Cry1Ac in the presence of the sugar confirmed that these residues are important for Gal-Nac recognition and that sugar binding has a conformational effect on the pore forming domain I (Li *et al.*, 2001).

In the case of Cry1Ac toxin, a sequential binding mechanism to the APN receptor has been proposed (Jenkins *et al.*, 2000). The interaction of domain III to an *N*-acetylgalactosamine moiety in the receptor precedes the binding of loop regions of domain II (Jenkins *et al.*, 2000).

7.4.5. Receptor Binding Epitopes

In regard to the receptor binding epitopes, a region of 63 residues (Ile135–Pro198) involved in Cry1Aa binding was identified by analysis of truncated derivatives of *B. mori* APN. This site was specific for Cry1Aa toxin since it was not involved in Cry1Ac binding (Yaoi *et al.*, 1999; Nakanishi *et al.*, 2002). Nevertheless, this binding region is present in other APN molecules that do not bind Cry1Aa toxin when assayed by toxin overlay assays (Nakanishi *et al.*, 2002). This result can be explained if the epitope mapped was not accessible in native conditions. In fact it has been shown that denaturation of *M. sexta* APN exposes binding epitopes hidden under nondenaturating conditions (Daniel *et al.*, 2002). Therefore, the role of the mapped binding epitope in *B. mori* APN in toxicity remains to be analyzed.

Regarding cadherin-like receptors, a Cry1A binding epitope was mapped in Bt-R_1 and Bt-R_{175} receptor molecules by the analysis of truncated derivatives of these receptors in toxin overlay assays (Nagamatsu *et al.*, 1999; Dorsch *et al.*, 2002). In the case of Bt-R_1 and Bt-R_{175}, a toxin binding region of 70 amino acid residues was mapped in the cadherin repeat number 11 which is close to the membrane spanning region (Nagamatsu *et al.*, 1999; Dorsch *et al.*, 2002). The binding epitope was narrowed to 12 amino acids ([1331]IPLPASILTVTV[1342]) by using synthetic peptides as competitors. Binding of Cry1Ab toxin to the 70 residue toxin binding peptide was inhibited by synthetic peptides corresponding to loop α-8 and loop 2, suggesting that these loop regions are involved in the interaction with this receptor epitope (Gómez *et al.*, 2003). Using a library of single-chain antibodies displayed in M13 phage, a second Cry1A toxin binding region was mapped in the Bt-R_1 receptor (Gómez *et al.*, 2001). An scFv antibody (scFv73) that inhibited binding of Cry1A toxins to the cadherin-like receptor Bt-R_1, but not to APN, and reduced the toxicity of Cry1Ab to *M. sexta* larvae was identified (Gómez *et al.*, 2001). Sequence analysis of CDR3 region of the scFv73 molecule led to the identification of an eight amino acid epitope of *M. sexta* cadherin-like receptor, Bt-R_1 ([869]HITDTNNK[876]) involved in binding of Cry1A toxins. This amino acid region maps in the cadherin repeat 7 (Gómez *et al.*, 2001).

Using synthetic peptides of the exposed loop regions of domain II of Cry1A toxins, loop 2 was identified as the cognate binding epitope of the *M. sexta* receptor Bt-R_1 [869]HITDTNNK[876] site (Gómez *et al.*, 2002a). This finding highlights the importance of the [869]HITDTNNK[876] binding epitope since extensive mutagenesis of loop 2 of Cry1A toxins has shown that this loop region is important for receptor interaction and toxicity (Rajamohan *et al*, 1995, 1996b; Jenkins and Dean, 2000; Jenkins *et al.*, 2000). Nevertheless, binding to cadherin repeat 7 was only observed in small truncated derivatives of Bt-R_1 (Gómez *et al.*, 2003) in contrast with larger truncated derivatives (Nagamatsu *et al.*, 1999; Dorsch *et al.*, 2002). Analysis of the dissociation constants of Cry1Ab binding to similar 70 amino acid peptides containing both toxin binding regions revealed that the toxin binds the epitope located in cadherin repeat 7 with sixfold higher affinity than cadherin repeat 11. Based on these results a sequential binding mechanism was proposed where binding of toxin to cadherin repeat 11 facilitates the binding of toxin loop 2 to the epitope in cadherin repeat 7 (Gómez *et al.*, 2003). Accumulating evidence indicate that proteins can interact with amino acid sequences displaying inverted hydropathic profiles (Blalock, 1995). The interactions of loop 2 with Bt-R_1 [865]NITI-HITDTNN[875] region and of loops α-8 and 2 with [133]1IPLPASILTVTV[1342] region were shown to be determined by hydropathic complementarity (Gómez *et al.*, 2002a, 2003).

As mentioned previously, it is generally accepted that the toxic effect of Cry proteins is exerted by the formation of a lytic pore. However, the fact that Cry1A toxins interact with protein molecules involved in cell–cell interactions (cadherin) within susceptible hosts could be relevant for the intoxication process as has been described for several other pathogens (Dorsch *et al.*, 2002). Targeting cell junction molecules seems to be representative of those bacteria that disrupt or evade epithelial barriers in their hosts. In this regard, it is remarkable that Cry1A toxins interact with at least two structural regions that are not close together in the primary sequence of Bt-R_1 (cadherin repeats 7 and 11). Although we cannot exclude the possibility that both sites could be located close together in the three-dimensional structure of Bt-R_1, we speculate that binding of Cry1A toxins could cause a conformational change in cadherin molecule that could interact with other cell-adhesion proteins, and consequently disrupt the epithelial cell layer (**Figure 7**).

As mentioned previously, mutagenesis studies have shown that besides domain II loop 2 and α-8,

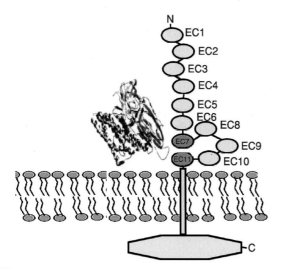

Figure 7 Structural regions on *Manduca sexta* cadherin receptor (Bt-R₁) involved in Cry1A binding. Ectodomains (EC) 7 and 11 contain Cry1A binding epitopes.

loop 3 of Cry1A toxins is important for receptor interaction and toxicity (Rajamohan *et al.*, 1996a; Lee *et al.*, 2001). The Bt-R₁ and the APN epitope involved in binding loop 3 regions still remains to be identified.

7.4.6. Cry Toxin–Receptor Binding Function in Toxicity

It is proposed that, following binding, at least part of the toxin inserts into the membrane resulting in pore formation (Schnepf *et al.*, 1998). However, it is still largely unknown what is the role of receptor binding in promoting the insertion and oligomerization of the toxin. In the case of Cry1A toxins the differences of binding affinities between APN and Bt-R₁ (100 nM versus 1 nM K_d respectively) suggest that binding to Bt-R₁ is the first event in the interaction of Cry1A toxins with microvilli membranes.

In other noninsecticidal pore forming toxins, receptor binding facilitates a complete proteolytic activation of the toxin resulting in the formation of functional oligomers that are membrane insertion competent (Abrami and Van der Goot, 1999). In the case of Cry1A toxins, Gómez *et al.* (2002b, 2003) provided evidence that interaction of Cry toxin with its cadherin-like receptor is a necessary step for a complete proteolytic of the toxin. Incubation of Cry1Ab protoxin with the single-chain antibody scFv73 that mimics the cadherin-like receptor or with the toxin-binding peptides of Bt-R₁, and treatment with *M. sexta* midgut juice, resulted in toxin preparations with high pore formation activity *in vitro*. In contrast toxin preparations

activated with same midgut juice proteases in the absence of the antibody showed very low pore formation activity (Gómez *et al.*, 2002b, 2003). The high pore formation activity correlated with the formation of a 250 kDa oligomer composed of four Cry toxins that lacked the α-1 helix of domain I. The oligomer, in contrast to the 60 kDa monomer, has higher hydrophobicity as judged by 8-anilino-1-naphthalenesulfonate binding (Gómez *et al.*, 2002b). The oligomer was membrane insertion competent in contrast with the monomer, as judged by measuring toxin membrane insertion using the intrinsic fluorescence of tryptophan residues (Raussel and Bravo, unpublished data). The 250 kDa oligomer could also be obtained by incubation of the Cry1Ab protoxin with brush border membrane vesicles isolated in the absence of protease inhibitors, presumably by the action of a membrane associated protease (Gómez *et al.*, 2002b). Therefore toxin binding to cadherin facilitates the complete proteolytic activation of the toxin and the formation of a prepore structure that consists of four monomers. Characterization of the membrane insertion capabilities of the prepore, the kinetic properties of the ionic pore formed by this structure, and solving its three-dimensional structure could be important steps towards understanding the role of this insertion-intermediate structure in the mode of action of Cry toxins.

As mentioned previously APN receptors are also key molecules involved in the toxicity of Cry1 toxins. Cry1A APN receptors are anchored to the membrane by a GPI anchor. An indication of its possible role came from the characterization of membrane microdomains (lipid rafts) from microvilli membranes of *M. sexta* and *H. virescens* (Zhuang *et al.*, 2002). Like their mammalian counterparts, *H. virescens* and *M. sexta* lipid rafts are enriched in cholesterol, sphingolipids, and glycosylphosphatidylinositol anchored proteins (Zhuang *et al.*, 2002). Lipid rafts have been implicated in membrane and protein sorting and in signal transduction (Simons *et al.*, 2000). They have also been described as portals for different viruses, bacteria and toxins. The interaction of different bacterial toxins with their receptors located in lipid rafts is a crucial step in the oligomerization and insertion of toxins into the membrane (Cabiaux *et al.*, 1997; Abrami and van der Goot, 1999; Rosenberg *et al.*, 2000). Several Cry1A receptors, including the GPI anchored proteins, 120 and 170 kDa APNs from *H. virescens*, and the 120 kDa APN from *M. sexta* were preferentially partitioned into lipid rafts. After toxin exposure, Cry1A toxins were associated with lipid rafts and the integrity of these microdomains was essential for *in vitro* Cry1Ab pore forming activity (Zhuang *et al.*, 2002). Additionally, PI-PLC treatment of *T. ni* brush border membrane vesicles, resulting in cleavage of GPI-anchored proteins from membrane, reduced drastically Cry1Ac pore formation (Lorence *et al.*, 1997). Therefore, the possible role of APN binding could be to drive Cry1A toxins, probably the prepore, to lipid rafts microdomains where the toxin inserts and forms pores. The participation of lipid rafts in the mode of action of Cry toxins could suggest a possible role of signal transduction events and/or the internalization of Cry toxins, since lipid rafts have an active role in these cellular processes. In the case of Cry5 toxin evidence was provided showing that toxin is internalized into the epithelial midgut cells after exposure of *C. elegans* nematodes to the toxin (Griffits *et al.*, 2001).

Binding of Cry1A toxins to cadherin and APN receptors are key steps in the mode of action of these toxins. However, the participation of other uncharacterized molecules in the process of membrane insertion and pore formation of these toxins cannot be excluded.

7.4.7. Toxin Insertion

Following binding to their receptors, Cry toxins insert into membrane of midgut epithelial cells to form lytic pores. Insertion of Cry toxins into membranes requires a major conformational change in the toxin to expose a hydrophobic surface that can interact with the membrane bilayer. Domain I has been recognized as the pore forming domain based on mutagenesis studies (Wu and Aronson, 1992; Cooper *et al.*, 1998), and on the similarity of the structure of this domain with other pore forming domains of bacterial toxins, such as colicin Ia, diphtheria translocation domain, and hemolysin (Aronson and Shai, 2001). The α-helices of domain I of Cry1 toxins are long enough to span the membrane and have an amphipathic character (Aronson and Shai, 2001). Helix α-5 is highly hydrophobic and conserved (conserved block 1) among all three-domain Cry toxin families (de Maagd *et al.*, 2001). Cross-linking experiments done in Cry1Ac, that was genetically engineered to create disulfide bridges between some α-helices of domain I, showed that domain I swings away from the rest of the toxin exposing the α-7 helix for the initial interaction with the membrane, resulting in membrane insertion of the hairpin formed by helices α-4 and α-5 (Schwartz *et al.*, 1997b). Analyses of the insertion capabilities of synthetic peptides corresponding to the seven α-helices of domain I from Cry3A showed that α-1 is the only helix that does not interact with the

membrane, in contrast to the other helices (Gazit *et al.*, 1998). Helices α-4 and α-5 were the only helices capable of adopting a transmembrane orientation (Aronson and Shai, 2001; Gazit *et al.*, 1998). These results suggest that helices α-4 and α-5 insert into the membrane, while the rest of the helices remain on the membrane surface. These data agree with the proposed "umbrella model" of toxin insertion in which helices α-4 and α-5 insert into the membrane, leaving the rest of the α-helices in the interface of the membrane (Schwartz *et al.*, 1997b). In view of the proposed role of the prepore structure in toxin insertion (Gómez *et al.*, 2002b), it is tempting to speculate that in the three-dimensional structure of the prepore, four helices α-7 are exposed and form an hydrophobic surface that participates in the initial interaction of the tetramer with the membrane. Subsequently, the four hairpins formed by helices α-4 and α-5 insert into the membrane. The location of domains II and II in the membrane inserted state is unknown. However, with exception of helix α-1, membrane inserted Cry1Ac toxin resist proteinase K treatment suggesting that domains II and III might also, somehow, be inserted into the membrane (Aronson, 2000).

A different model of the inserted toxin, based on calorimetric determinations, proposed that the three-dimensional structure of the toxin does not change dramatically compared to the structure of the soluble monomer (Loseva *et al.*, 2001). This model proposed that helices α-1 to α-3 are cleaved, and that the pore lumen is provided by residues of domain II and III, while the domain I hydrophobic surface (without helices α-1 to α-3) faces the lipid bilayer in a oligomeric structure (Loseva *et al.*, 2001). In this regard it is interesting to note that mutagenesis of conserved arginine residues in Cry1Aa domain III affected the pore formation activity of Cry1Aa toxin, suggesting that domain III participates in channel activity (Schwartz *et al.*, 1997b). Analysis of the pore formation activity of several Cry1 chimeric proteins established that the characteristics of the pore are influenced by domain II and domain III, because combination of domain III from Cry1Ab with domains I and II of Cry1C gave a protein that was more active than Cry1C in spodoptera frugiperda (Sf9) cells (Rang *et al.*, 1999). Solving the structure of Cry toxins in the membrane inserted state will be important to determine the possible roles of domains II and III in pore formation.

7.4.8. Pore Formation

Based on the observation of large conductance states formed by several Cry toxins in synthetic planar lipid bilayers (Lorence *et al.*, 1995; Peyronnet *et al.*, 2002) and the estimation of pore size at 10–20 Å (Von-Tersch *et al.*, 1994), it has been proposed that the pore could be formed by an oligomer of Cry toxins containing four to six toxin monomers. Moreover, intermolecular interaction between Cry1Ab toxin monomers is a necessary step for pore formation and toxicity (Soberón *et al.*, 2000). This conclusion was derived from studies that used two Cry1Ab mutant proteins that affected different steps in toxicity (binding and pore formation). Individually these mutant proteins had decreased toxicity to *M. sexta*; however, when assayed as a mixture of the two toxins, pore formation activity and toxicity against *M. sexta* larvae was recovered. These results show that monomers affected in different steps of their mode of action can form functional heterooligomers, and that oligomerization is a necessary step for toxicity (Soberón *et al.*, 2000). Recently, the structure of the pore formed by Cry1Aa toxin was analyzed by atomic force microscopy showing that the pore is a tetramer (Vie *et al.*, 2001). These data are in agreement with the proposition of a prepore structure composed of four monomers (Gómez *et al.*, 2002b). The regions of the toxin involved in oligomerization have not been determined; however, based on mutagenesis studies and analysis of toxin aggregation it has been suggested that some residues of helix α-5 may be implicated in this process (Aronson *et al.*, 1999; Vie *et al.*, 2001).

The pore activity of Cry toxins has been studied by a variety of electrophysiological techniques (Schwartz and Laprade, 2000), for example using synthetic membranes without receptor or in isolated brush border membrane vesicles containing natural receptors (Lorence *et al.*, 1995; Peyronnet *et al.*, 2001, 2002; Bravo *et al.*, 2002a). Also ion channels induced by various activated Cry1 toxins in its monomeric form – Cry1Aa (Grochulski *et al.*, 1995; Schwartz *et al.*, 1997a), Cry1Ac (Slatin *et al.*, 1990; Schwartz *et al.*, 1997a; Smedley *et al.*, 1997), and Cry1C (Schwartz *et al.*, 1993; Peyronnet *et al.*, 2002) – have been analyzed in black lipid bilayers. The channel formation of these toxins was extremely inefficient and in some studies was only achieved mechanically (Peyronnet *et al.*, 2001, 2002). The toxin concentrations needed to achieve channel formation in these conditions were two to three orders of magnitude higher than their *in vivo* insecticidal concentration. Conductance varied from 11 to 450 pS and multiple subconducting states are frequently observed showing unstable traces with current jumps of intermediate levels that are difficult to resolve. These high conductances are probably related to clusters of various numbers of identical size pores operating

synchronously rather than pore oligomer structures of different sizes (Peyronnet *et al.*, 2002). Under non-symmetrical ionic conditions, the shift in the reversal potential (zero current voltage E_{rev}) towards the K^+ equilibrium potential (E_K) indicated that channels of Cry1 toxins are slightly cation selective. In fact, several reports indicated that Cry toxins form pores that are poorly selective to cationic ions including divalent cations (Lorence *et al.*, 1995; Kirouac *et al.*, 2002). As mentioned previously, the presence of receptor (APN) diminished the concentration, more than 100-fold, of Cry1Aa toxin required for pore formation activity in synthetic planar bilayers (Schwartz *et al.*, 1997a). Studies performed in lipid bilayers containing fused brush border membrane vesicles isolated from the target insect suggested that the channels formed by Cry1 toxins in the presence of their receptors have higher conductance than those formed in receptor free bilayers. The conductance of monomeric Cry1C induced channels ranged from 50 pS to 1.9 nS in bilayers containing brush border membrane vesicles from *S. frugiperda* (Lorence *et al.*, 1995). Similarly, the conductance of channels induced by the monomeric form of Cry1Aa toxin in bilayers containing membranes from *L. dispar* were about eightfold larger than the channels formed in the absence of receptor (Peyronnet *et al.*, 2001). However, the presence of multiple conductances is still observed and the instability of the currents induced in these studies suggested that even in the presence of receptors the insertion of monomers into the membrane does not involve a single conformation. In contrast, preliminary analysis of the currents induced by pure oligomer preparations in the absence of receptor showed highly stable conductance, suggesting a stable insertion of a single conformation of the toxin into the membrane (Muñoz-Garay and Bravo, unpublished data). Finally it is important to mention that, in contrast to other pore forming toxins, the pore formation activity of Cry1 proteins is not regulated by low pH, suggesting that Cry toxins are not internalized into acidic vesicles for insertion as other pore forming toxins (Tran *et al.*, 2001).

Overall, the mode of action of Cry toxins can be visualized as follows (**Figure 8**):

1. Solubilization of the crystal and activation of the protoxin by midgut proteases resulting in the monomer toxin production.
2. Binding of the monomer to the cadherin receptor located in the apical membrane of midgut cells, probably accompanied by a mild denaturation of the monomer that allows proteolytic cleavage of helix α-1. This cleavage might result in a conformational change and the formation of a

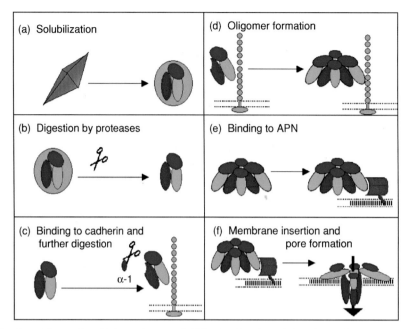

Figure 8 Mode of action of Cry toxins. (a) Crystals are solubilized and activated give rise to the monomeric toxin. (b) The toxin monomer binds the cadherin receptor, followed by proteolytic cleavage of helix α-1. (c) The tetramer is formed by intermonomeric contacts. (d) The toxin oligomer binds to the APN receptor. (e) The APN receptor and oligomeric Cry toxin localize to lipid rafts. (f) Following a conformational change the oligomer inserts into membrane forming a tetrameric pore.

molten globule state of the monomer exposing hydrophobic regions.

3. Formation of a tetramer by intermonomeric contacts.

4. Binding of the toxin oligomer to the APN receptor.

5. Mobilization of the APN receptor and Cry toxin to membrane microdomains.

6. A second conformational change of the oligomer, resulting in the insertion of the toxin into the membrane and formation of the membrane active tetrameric pore.

7.5. Synergism of Mosquitocidal Toxins

As previously mentioned (see Section 7.3.1) Cyt toxins synergizes the insecticidal effect of some Cry toxins. *Bacillus thuringiensis* subsp. *israelensis* is highly toxic to different mosquito species like *Aedes* spp., *Culex* spp., and blackfly, and also to *Anopheles* spp. but with lower toxicity (Margalith and Ben-Dov, 2000). This bacterium produces a crystal inclusion composed of at least four toxins: Cry4A, Cry4B, Cry11A, and Cyt1A (Huber and Luthy, 1981; Guerchicoff *et al.*, 1997). The toxicity of the crystal inclusion is greater, by far, than the toxicity of the isolated Cry and Cyt components. These Cry toxins are more toxic against mosquito larvae than Cyt1A with a difference in toxicity of

at least 25-fold (Wirth *et al.*, 1997). However, the Cyt1A toxin synergizes the effect of the other Cry toxins. Toxicity of different combinations of Cry toxins with CytA1 was higher than the addition of expected toxicities of the isolated components (Crickmore *et al.*, 1995; Khasdan *et al.*, 2001). Also, the Cyt toxins synergize the toxicity of the binary toxins produced by *Bacillus sphaericus*. A ratio of 10:1 of *B. sphaericus* toxins to Cyt1A was 3600-fold more toxic to *A. aegypti* than *B. sphaericus* alone (Wirth *et al.*, 2001), and the presence of Cyt toxin can overcome the resistance of *Culex quinquefasciatus* to the binary toxins (Wirth *et al.*, 2001).

The mode of action of Cyt toxins includes the following steps: solubilization of the toxin under the alkaline pH and reducing conditions of the midgut, proteolytic processing of the protoxin, binding of the toxin to the epithelium surface of midgut cells, and pore formation leading to cell lysis (Li *et al.*, 1996, 2001). An important difference between Cyt and Cry toxins is the lack of a protein receptor for the Cyt toxin. Cyt toxins bind phospholipids and are capable of forming pores in cell lines of different origin (Gill *et al.*, 1987; Chow *et al.*, 1989; Li *et al.*, 1996, 2001). There are currently two proposed mechanisms of insertion and membrane perturbation by Cyt toxins. The first mechanism proposes that multimers of four Cyt toxin subunits form a β-barrel structured pore within the membrane, where the transmembrane region involves the long β-sheets of the toxin (Promdonkoy and Ellar, 2003), whereas the second proposed mechanism suggests that the Cyt toxins exert their effect through a less specific detergent action (Butko *et al.*, 1997; Butko, 2003).

Besides the synergistic effect of Cyt toxins, these proteins overcome or suppress resistance of mosquitoes to the Cry toxins (Wirth *et al.*, 1997). *Culex quinquefasciatus* populations resistant to Cry4A, Cry4B, or Cry11A recovered sensitivity when assayed in the presence of Cyt toxin (Wirth *et al.*, 1997). Furthermore, it has not been possible to select resistant mosquitoes against Cry toxins when selection is performed in the presence of Cyt toxin (Wirth *et al.*, 1997). As mentioned previously (see Section 7.3.1), Cyt toxins are principally found in dipteran-active strains. However, a coleopteran-active strain containing a Cyt toxin has been described, although the possible synergistic effect of this Cyt toxin to other Cry proteins has not been analyzed (Guerchicoff *et al.*, 2001). The molecular mechanism of the synergistic effect of Cyt toxins is still unknown. Cyt toxins could enhance any of the steps in the mode of action of Cry toxins, such as proteolytic activation, receptor binding, or toxin insertion. One plausible mechanism could be that Cyt toxins bind to microvilli membranes and then act as receptors of Cry toxins thus providing a binding site even in resistant insect populations where the Cry receptor molecules are affected. It would be interesting to investigate whether mosquitocidal Cry proteins are able of interacting with Cyt toxins and, if so, to map the binding epitopes. Engineering Cyt toxins for synergizing coleopteran- and lepidopteran-active toxins could be very useful in preventing the development of Cry resistant insect populations in nature.

7.6. Genomics

The genome size of different *B. thuringiensis* strains is in the range of 2.4–5.7 Mb (Carlson *et al.*, 1994). Comparison of physical maps with that of *B. cereus* strains suggests that all *B. cereus* and *B. thuringiensis* have chromosomes with very similar organizations (Carlson *et al.*, 1996). In May 2003 the complete genome sequence of type strains of the close *B. thuringiensis* relatives, *B. anthracis* (GenBank Accession no. AE016879) and *B. cereus* (GenBank Accession no. AE016877), were completed (Ivanova *et al.*, 2003; Read *et al.*, 2003). The *B. cereus* strain ATCC 14579 could be considered a *B. thuringiensis* strain based on its ribosomal 16S sequence, but does not produce crystals inclusions. The complete sequenced genome of *B. cereus*

strain ATCC 1479 has 5.427 Mb (Ivanova *et al.*, 2003). Analysis of putative coding regions and comparison to the sequence of the *B. anthracis* type strain showed that this group of bacteria lacks the capacity of metabolizing diverse types of carbohydrates that characterize typical soil bacteria like *Bacillus subtilis*. Carbohydrate catabolism in the *B. cereus* group is limited to glycogen, starch, chitin, and chitosan (Ivanova *et al.*, 2003). In contrast, *B. cereus*, *B. anthracis*, and presumably *B. thuringiensis*, have a wide variety of peptide, amino acid transporters, and amino acid degradation pathways indicating that peptides and amino acids may be the most preferred nitrogen and carbon sources for these bacteria. These data led to the speculation that probably the habitat of the common ancestor of the *B. cereus* group was the insect intestine (Ivanova *et al.*, 2003). The presence in the genome of several hydrolytic activities support this speculation, chitinolytic activities could participate in the degradation of chitin present in the peritrophic membrane, while the zinc metalloprotease could enhance degradation of the major component of the peritrophic membrane mucin. Also, several genes related to possible invasion, establishment, and propagation in the host were identified in both genome sequences (Ivanova *et al.*, 2003; Read *et al.*, 2003). Finally, the analysis of putative PlcR regulated genes gave a figure of 55 PlcR boxes controlling over 100 genes including four other transcriptional activators implying that the PlcR regulon is more complex than expected (Ivanova *et al.*, 2003).

7.6.1. Sequence of Plasmid pBtoxis

Bacillus thuringiensis isolates usually harbor large plasmids, which are self-transmisible, containing δ-endotoxin genes that distinguish them from *B. cereus* strains (Schnepft *et al.*, 1998). Therefore, the sequencing of plasmids containing *cry* genes could help to understand the evolution of *B. thuringiensis* and the spread of *cry* genes in nature. In 2002, the complete 128 kb sequence of a toxin encoded plasmid from *B. thuringiensis* subsp. *israelensis* (pBtoxis) was released (Berry *et al.*, 2002). This plasmid contains 125 putative coding sequences with an average gene length of 725 bp (Berry *et al.*, 2002). Sequence analysis showed that besides the known Cry toxins genes (*cry4Aa*, *cry4Ba*, *cry11A*, *cry10Aa*, *cyt2Ba*, and *cyt1Aa*) a putative novel *cry* gene was identified. Based on phylogenetic relationships with other Cyt toxins, the novel gene was named *cyt1Ca* (Berry *et al.*, 2002). In contrast to the Cyt toxins, Cyt1Ca contains an additional domain located on the C-terminus (280 residues) which shares similarity with domains present in other bacterial toxins that are involved on carbohydrate recognition (Berry *et al.*, 2002). Therefore, it was proposed that Cyt1Ca could act by recognizing carbohydrate moieties in the cell surface (Berry *et al.*, 2002). Based on sequence analysis, eight pseudogenes were identified. Among these, several gene fragments that shared significative sequence identity to described *cry* genes were identified, suggesting that the ancestors of plasmid pBtoxis contained other *cry* genes that were lost subsequently (Berry *et al.*, 2002). Interestingly the *cry* gene fragments are located in nearby transposon related sequences indicating that gene transposition may have caused these genetic rearrangements and could be an important mechanism for toxin gene spread and evolution (Berry *et al.*, 2002). Beside *cry* genes, the plasmid pBtoxis contains several putative genes coding for virulence factors (Berry *et al.*, 2002). A second *plcA* gene, coding for Pl-PLC, was identified. However, this gene is probably not translated and no PlcR box was found in the promoter region of this gene. Interestingly, a PlcR box was identified in front of a peptide antibiotic production and export system, and this peptide antibiotic may be a novel virulence factor (Berry *et al.*, 2002). Other genes related to sporulation and germination were identified, suggesting that plasmids could influence the sporulation and germination of the host strain (Berry *et al.*, 2002). An interesting feature of the plasmid pBtoxis is that it shares similarity with a limited number of genes of plasmid pXO1, the virulence plasmid of *B. anthracis*; 29 of 125 open reading frames show amino acid sequence similarity to open reading frames present in pXO1 (Berry *et al.*, 2002).

7.7. Mechanism of Insect Resistance

One of the major concerns regarding the use of *B. thuringiensis* is the generation of insect populations resistant to Cry toxins. Resistance to these toxins could be obtained by interfering with any of the steps involved in the mode of action of Cry toxins. The analysis of the molecular mechanism of insect resistance to *B. thuringiensis* toxins has been, for the most part, based on the study of insect resistant populations selected under laboratory conditions. These studies could provide important knowledge that allows for the rational design of procedures needed for resistance management in the field. However, resistance mechanisms of natural insect populations are likely not to compromise significantly the fitness of the insects to guarantee the fixation of resistant mutant alleles in the population. Therefore, the mutant alleles found in

laboratory selected insect populations may not necessarily be found in nature although alleles found in nature are likely to be present in laboratory selected lines.

Selection of insects resistant to *B. thuringiensis* toxins in the laboratory has been performed for a variety of insect species. Lepidopteran insects resistant to *B. thuringiensis* toxins include *Plodia interpunctella*, *P. xylostella*, *H. virescens*, and *Spodoptera exigua* (Ferré and Van Rie, 2002), while coleopteran insect species include *L. decemlineata* and *Chrysomela scripta* (Ferré and Van Rie, 2002). As mentioned above (see Section 7.5), selection of Cry toxin resistant dipteran insects has been performed for *C. quinquefasciatus* (Wirth *et al.*, 1997). Finally resistance to Cry toxins has also been achieved for the (noninsect) nematode *C. elegans* (Griffits *et al.*, 2001). Until now only one insect species, *P. xylostella*, has developed resistance to Cry toxins in the field (Ferré and Van Rie, 2002). Only the resistance mechanisms in lepidopteran insects and in *C. elegans* nematode, have been characterized biochemically. Below we will summarize recent findings on the mechanisms involved in resistance to *B. thuringiensis* toxins.

7.7.1. Proteolytic Activation

As mentioned previously (see Section 7.4.2) Cry protoxins are activated by the action of midgut proteases. A line of *P. interpunctella* resistant to *B. thuringiensis* toxin had lower protease and protoxin activating activities. These differences were shown to be due to the lack of a major trypsin-like proteinase in the gut. Genetic analysis showed a cosegregation for the lack of the major protease and insect resistance to Cry1Ab toxin (Oppert *et al.*, 1997). The consequence of reduced gut protease was a lower concentration of the activated toxin in the gut (Oppert *et al.*, 1997). The resistant strain showed altered growth and morphology suggesting that this allele is unlikely to be selected in natural populations.

7.7.2. Receptor Binding

The most frequent mechanism of resistance to Cry toxins is altered receptor binding (Ferré and Van Rie, 2002). Different resistant populations from different lepidopteran species show an effect on toxin binding (Ferré and Van Rie, 2002). In a *P. interpunctella* strain selected for resistance to Cry1Ab, toxin binding affinity (K_d) was reduced 50-fold. Interestingly, this strain showed an increased susceptibility to Cry1C toxin which correlated with an increase number of binding sites to this toxin (Ferré and Van Rie, 2002). Several receptor molecules for Cry1 toxins in *H. virescens* and *P. xylostella* have been hypothesized based on cross-resistance and susceptibility of different Cry1 resistant populations (Ferré and Van Rie, 2002). For *H. virescens*, three different receptors have been proposed: receptor A which binds Cry1Aa, Cry1Ab, and Cry1Ac toxins; receptor B which binds Cry1Ab and Cry1Ac toxins; and receptor C which recognizes Cry1Ac toxin. Based on the susceptibility of resistant insect populations, it was concluded that receptors B and C are unlikely to be involved in toxicity (Ferré and Van Rie, 2002). In the case of *P. xylostella*, four binding sites have been proposed (Ferré and Van Rie, 2002): receptor 1 is only recognized by Cry1Aa toxin; receptor 2 binds Cry1Aa, Cry1Ab, Cry1Ac, Cry1F, and Cry1J; receptor 3 binds Cry1B; and receptor 4 binds Cry1C (Ferré and Van Rie, 2002). Cross-resistance of Cry1A toxins with Cry1F and Cry1J correlates with sequence similarity of loop regions in domain II, in particular loop 2, of these toxins (Tabashnik *et al.*, 1994; Jurat-Fuentes and Adang, 2001) (see Section 7.3.3). In the case of four *P. xylostella* strains that developed resistance in the field, reduced Cry1A binding correlated with resistance (Schnepf *et al.*, 1998; Ferré and Van Rie, 2002).

In the majority of the selected resistant insect lines described above there is no information regarding the characterization at the molecular level of mutations that leads to Cry resistance. In the case of the laboratory selected *H. virescens* Cry1Ac resistant line, YHD2, it was shown that a single mutation was responsible for 40–80% of Cry1Ac resistance levels. The mutation was mapped and shown to be linked to a retrotransposon insertion in the cadherin-like gene (Gahan *et al.*, 2001). As mentioned previously (see Section 7.4.5), this result shows that binding of Cry1A toxins to cadherin-like receptors is an important event in the mode of action of these toxins. Based on these results, it has been suggested that if mutations of cadherin genes are the primary basis of resistance in the field, the characterization of cadherin alleles could be helpful in monitoring field resistance (Morin *et al.*, 2003). The characterization of cadherin alleles in field derived and laboratory selected strains of the cotton pest pink bollworm (*Pectinophora gossypiella*) revealed three mutated cadherin alleles that were associated with resistance in this lepidopteran insect (Morin *et al.*, 2003). **Figure 9** shows the structural features of the four mutated alleles of cadherin genes associated with resistance in *H. virescens* and *P. gossypiella*.

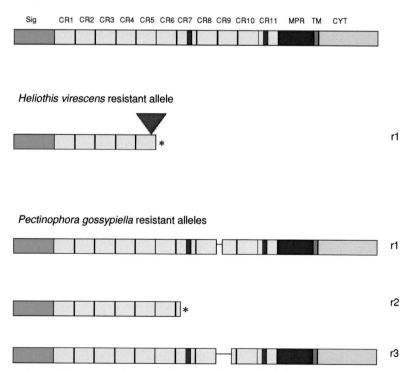

Figure 9 Gene structure of resistant cadherin alleles of *Heliothis virescens* (Gahan *et al.*, 2001) and *Pectinophora gossypiella* (Morin *et al.*, 2003). Sig corresponds to signal sequence for protein export; CR, cadherin repeat or repeated ectodomains; MPR, membrane proximal region; TM, transmembrane region; CYT, cytosolic domain; *, stop codon; red triangle, retrotransposon insertion; solid lines, deletions. Red sections on CR7 and CR11 are Cry1A binding epitopes mapped on *M. sexta* Bt-R$_1$ cadherin receptor.

7.7.3. Oligosaccharide Synthesis

Domain II and domain III of Cry toxins show structural similarity to carbohydrate binding domains present in other proteins, and in the case of Cry1Ac toxin, binding to the APN receptor involves the carbohydrate moiety *N*-acetylgalactosamine (de Maagd *et al.*, 1999a; Burton *et al.*, 1999) (see Section 7.3.3). Nevertheless, the most compelling evidence of the involvement of carbohydrate moieties on the mode of action of Cry toxins has come from the selection and characterization of nematode *C. elegans* Cry5B resistant lines (Griffits *et al.*, 2001). Five independent *C. elegans* resistant strains were selected by feeding with *E. coli* cells expressing Cry5B toxin. Cloning and sequence of one gene responsible for the resistant phenotype (*bre-5*) showed that the mutated gene shares high sequence similarity to β-1,3-galactosyltransferase enzyme involved in carbohydrate synthesis (Griffits *et al.*, 2001). The *bre-5* mutation confers cross-resistance to the Cry14A toxin which also has been shown to be toxic to insects (Griffits *et al.*, 2001). The *bre-5* mutation had no effect on the fitness of the resistant nematodes (Griffits *et al.*, 2001). These results suggest that carbohydrate moieties could play an important role in the mode of action (probably binding) of some Cry toxins.

In the case of the *H. virescens* YHD2 Cry1Ac resistant line, an altered pattern of glycosylation of two microvillar proteins of 63 and 68 kDa was shown to correlate with resistance (Jurat-Fuentes *et al.*, 2002). As mentioned above (see Section 7.4.5), YHD2 strain contains several mutations responsible for its Cry1Ac resistant phenotype. Although the most important resistant allele is the disruption of the cadherin-like gene, altered glycosylation of microvilli membrane proteins increases its resistant phenotype (Jurat-Fuentes *et al.*, 2002).

7.8. Applications of Cry Toxins

7.8.1. Forestry

One of the most successful applications of *B. thuringiensis* has been the control of lepidopteran defoliators that are pests of coniferous forests mainly in Canada and the United States. The strain HD-1, producing Cry1Aa, Cy1Ab, Cry1Ac, and Cry2Aa toxins, has been the choice for controlling these forests defoliators. Use of *B. thuringiensis* for the control spruce worm (*Choristoneura fumiferana*) started in the mid-1980s (van Frankenhuyzen, 2000). Afterwards *B. thuringiensis* has been used for the control of jackpine budworm (*C. pinus pinus*), Western spruce worm (*C. occidentalis*),

Eastern hemlock looper (*Lambdina fiscellaria fiscellaria*) gypsy moth (*L. dispar*), and white-marked tussock moth (*Orgyia leucostigma*) (van Frankenhuyzen, 2000). In Canada, for the control of spruce worm, up to 6 million ha have been sprayed with *B. thuringiensis*, while up to 2.5 million ha have been sprayed for the control of other defoliators. In the case of United States up to 2.5 million ha are known to have been sprayed with *B. thuringiensis*. In these countries, 80–100% of the control of forests defoliators relies on the use of *B. thuringiensis*. In Europe, *B. thuringiensis* use for the control of defoliators greatly increased in the 1990s, reaching up to 1.8 million ha of forests treated by the end of the decade (van Frankenhuyzen, 2000). Successful application for the control of defoliators is highly dependent on proper timing, weather conditions, and high dosage of spray applications. These factors combine to determine the probability of larvae ingesting a lethal dose of toxin (van Frankenhuyzen, 2000). The use of *B. thuringiensis* in the control of defoliators has resulted in a significant reduction in the use of chemical insecticides for pest control in forests.

7.8.2. Control of Mosquitoes and Blackflies

As mentioned previously (see Section 7.5), *B. thuringiensis* subsp. *israelensis* is highly active against disease vector mosquitoes like *Aedes aegypti* (the vector of dengue fever), *Simulium damnosum* (the vector of onchocerciasis), and certain *Anopheles* species (vectors of malaria). The high insecticidal activity, the lack of resistance to this *B. thuringiensis* subspecies, the lack of toxicity to other organisms, and the appearance of insect resistant populations to chemical insecticides have resulted in the rapid use of *B. thuringiensis* subsp. *israelensis* as an alternative method for control of mosquitoes and blackfly populations.

In 1983, very soon after the discovery of *B. thuringiensis* subsp. *israelensis*, a control program for the eradication of onchocerciasis was launched in 11 countries in West Africa. This program was needed since the *S. damnosum* populations had developed resistance to organophosphates used for larval control (Guillet *et al.*, 1990). At the present time, more than 80% of the region is protected by applications of this bacterial formulation and 20% with temephos (Guillet *et al.*, 1990). As a consequence the disease has been controlled, protecting over 15 million children without the appearance of resistance this bacteria (Guillet *et al.*, 1990).

In the Upper Rhine Valley in Germany, *B. thuringiensis* subsp. *israelensis* along with *B. sphaericus* has been used to control *Aedes vexans* and *Culex pipiens pipiens* mosquitoes. Each year, approximately 300 km of river and 600 km^2 inundation areas are treated. Since 1981, more than 170 000 ha of mosquito breeding sites have been treated with different *B. thuringiensis* subsp. *israelensis* formulations. This program (KABS) has resulted in the reduction of mosquito populations by over 90% each year (Becker, 2000). These results are comparable to the use of *B. thuringiensis* subsp. *israelensis* for the control of *A. vexans* in the United States and in southern Switzerland (Becker, 2000). Again, no resistance to has been observed even after more than 20 years of the launch of the program (Becker, 2000).

High incidences of malaria in Hubei Province in China were reduced by more than 90% by the application of *B. thuringiensis* subsp. *israelensis* and *B. sphaericus* into the Yangtze River. The vector mosquitoes *Anopheles sinensis* and *A. anthropophagus* populations were reduced 90% by weekly applications of fluid formulations of two *B. thuringiensis* subsp. *israelensis* and *B. sphaericus* strains (Becker, 2000). In Brazil the outbreak of *A. aegypti* resistant populations to chemical insecticides in Rio de Janeiro and São Paulo led to the application of *B. thuringiensis* subsp. *israelensis* formulations for mosquito control (Regis *et al.*, 2000).

The success of vector control using *B. thuringiensis* subsp. *israelensis* will certainly spread the use of this *B. thuringiensis* subspecies around the world. However, the low activity of *B. thuringiensis* subsp. *israelensis* against certain vector mosquitoes, particular *Anopheles* spp., will require the isolation of strains with novel *cry* genes that are effective against different disease vectors.

7.8.3. Transgenic Crops

Bacillus thuringiensis is the leading biopesticide used in agriculture. However, only 2% of the insecticidal market consists of *B. thuringiensis* sprayable products; the rest consists mostly of chemical insecticides (Navon, 2000). Since the discovery of the HD-1 strain, formulations for lepidopteran pests control were released and products for the control of coleopteran insects are also available. However, the use of sprayable *B. thuringiensis* for insect control has been limited. Many reasons have contributed to this low acceptance of such commercial *B. thuringiensis* products. These include: the higher costs of commercial *B. thuringiensis* products; the lower efficacy of *B. thuringiensis* as compared to chemical insecticides; the limited persistence of *B. thuringiensis* products; the narrow insect spectrum of *B. thuringiensis*; and the fact that sucking and borer insects are poorly controlled (Navon, 2000).

However, the creation of transgenic crops that produce crystal *B. thuringiensis* proteins has been a major breakthrough, and substitution of chemical insecticides provides environmental friendly alternatives. In transgenic plants Cry toxin is produced continuously protecting the toxin from degradation and making it available to chewing and boring insects. Cry protein production has been improved by engineering *cry* genes with a codon usage compatible with that of plants, by removal of putative splicing signal sequences and by deletion of the C-terminal region of the protoxin (Schuler *et al.*, 1998). The first *B. thuringiensis* crop (Bt-potato) was commercialized in 1995. These transgenic plants expressed the Cry3A protein for protection against Colorado potato beetles resulting in 40% reduction in insecticidal use in 1997 (Schuler *et al.*, 1998). In 1996 Bt-cotton, producing the Cry1Ac toxin, was released to protect cotton from tobacco budworm, cotton bollworm, and pink bollworm. Also in 1996 Bt-maize, producing Cry1Ab toxin, was released for the control of European corn borer (*Ostrimia nubilalis*) larvae (Schuler *et al.*, 1998). By 2000, transgenic crops were grown on 44.3 million ha globally. Of this, 23% were insect resistant Bt-maize and 74% were herbicide resistant crops (James, 2001). Yield effects of insect resistant crops in the United States are less than 10% on average, but nevertheless use of insect resistant crops has considerably reduced the use of chemical pesticides (Qaim and Zilberman, 2003). However, the use of Bt-cotton in countries like China, Mexico, and India has showed that the use of this Bt-crop has a significant effect on yields and also in the use of chemical pesticides, since in underdeveloped countries yield loss is mainly due to technical and economical constraints that are overcome in part by the use of insect resistant crops (Qaim and Zilberman, 2003; Toenniessen *et al.*, 2003).

7.9. Public Concerns on the Use of *B. thuringiensis* Crops

Three major concerns have been raised on the use of Bt-crops: insect resistance, a direct effect on nontarget insects, and transgene flow to native landraces.

Widescale introduction of Bt-crops endangers the durability of *B. thuringiensis* as an insecticide in crops and in sprays since continuous exposure to Bt-plants will lead to selection for Cry toxin resistant insect populations. In response to this threat, resistance management strategies associated with the release of Bt-crops have been developed. The predominant approach involves the combination of high production levels of the Cry protein combined with the establishment of refuge zones (de Maagd *et al.*, 1999b; Conner *et al.*, 2003). Refuges are areas of nontransgenic crops that are within close proximity of fields with Bt-plants. This procedure aims to maintain a population of susceptible insects that contain wild-type alleles. In most cases studied, resistance to Cry toxins is conferred by recessive mutations (de Maagd *et al.*, 1999b; Ferré and Van Rie, 2002; Conner *et al.*, 2003). Thus mating of homozygous individuals having susceptible alleles with heterozygous individuals having a resistant allele will give rise to susceptible heterozygote progeny, thereby delaying the development of resistance. Also, second generation transgenic crops are now being developed that produce two different Cry proteins that bind to different receptor molecules in the same insect. Thus, the possibilities of selecting mutants affected in receptor–toxin interaction are diminished exponentially (de Maagd *et al.*, 1999b).

The massive use of *B. thuringiensis* sprays for forest defoliators raised concern on the effect of *B. thuringiensis* toxins on nontarget insects (Scriber *et al.*, 2001). More recently, public concerns over the impact of Bt-crops on nontarget insects has increased based on results obtained on the toxicity of Bt-maize pollen to the Monarch butterfly (*Danaus plexippus*) (Losey *et al.*, 1999). In this study, pollen, from a commercial variety of Bt-maize (N4640) expressing the *cry1Ab* gene, when spread onto milkweed (*Asclepias syriaca*) leaves was lethal to Monarch butterfly caterpillars in the laboratory (Losey *et al.*, 1999). However, follow-up studies showed that low expression of *B. thuringiensis* toxin genes in pollen of most commercial transgenic hybrids, the lack of toxicity at the expected field rates, the absence of overlap between the pollen shed and larval activity, and the limited overlap in distribution of Bt-maize and milkweed, made field risks to the Monarch populations negligible (Gatehouse *et al.*, 2002; Hellmich *et al.*, 2001; Oberhauser *et al.*, 2001; Pleasants *et al.*, 2001; Sears *et al.*, 2001; Stanley-Horn *et al.*, 2001; Zangerl *et al.*, 2001).

Concerns have also been raised about the effects of transgene introduction, or gene flow, on the genetic diversity of crop landraces and wild relatives in areas from which the crop originated. In 2001 the presence of transgenic DNA (from Bt-maize) constructs was reported in native maize landraces in Oaxaca, Mexico (Quist and Chapela, 2001). Oaxaca is considered part of the Mesoamerican center of origin and diversification of maize. Transgenic DNA was identified by polymerase chain reaction (PCR) based methodology, and the 35S promoter sequence was found in five of seven different native Criollo maize samples analyzed (Quist and Chapela, 2001). Moreover, inverse PCR reactions of genomic DNA of native landraces showed a high frequency of transgene insertion into a range of genomic contexts (Quist and Chapela, 2001). Nevertheless, based on criticisms on the interpretation of these data and possible false priming in the PCR reactions (Kaplinsky *et al.*, 2002; Metz and Futterer, 2002), the journal *Nature* reconsidered its previous judgement on the acceptability of the results reported by Quist and Chapela, and concluded that the evidence provided was not sufficient to justify the publication (Editorial, 2002). However, more recent independent studies performed by the Mexican government (National Institute of Ecology (INE) and National Commission of Biodiversity (Conabio) claim to have evidence on the presence of transgenic DNA in native maize landraces, although publication of these results is still pending. In any case, gene flow is not restricted to transgenic plants, and naturally selected hybrids could also impact the populations and gene content of native landraces. Even if gene flow is confirmed, further studies on the possible impact of the gene flow from commercial hybrids (transgenic or nontransgenic) to native landraces should include long-term studies on the integrity of the transgenic constructs (or genes from nontransgenic hybrids) and the maintenance of these genes in the native landraces. These studies should provide a rational framework for taking decisions regarding the introduction and commercialization of transgenic crops in center of origin and diversification of crops. The final goal should be to assure the use of this environmental friendly technology without affecting the genetic diversity of natural landraces.

Acknowledgments

Research reported in this review was supported in part by DGAPA/UNAM IN216300, CONACyT 36505-N, and USDA 2002-35302-12539.

References

Abdul-Rauf, M., Ellar, D.J., **1999**. Mutations of loop2 and loop3 in domain II of *Bacillus thuringiensis* Cry1C delta-endotoxin affect insecticidal specificity and initial binding to *Spodoptera littoralis* and *Aedes aegypti* midgut membranes. *Curr. Microbiol.* 39, 94–98.

Abrami, L., van der Goot, F.G., **1999**. Plasma membrane microdomains act as concentration platforms to facilitate intoxication by aerolysin. *J. Cell Biol.* 147, 175–184.

Agaisse, H., Gominet, M., Okstad, O.A., Kolsto, A.-B., Lereclus, D., **1999**. PlcR is a pleiotropic regulator of extracellular virulence factor gene expression in *Bacillus thuringiensis*. *Mol. Microbiol.* 32, 1043–1053.

Agrawal, N., Malhotra, P., Bhatnagar, R.K., 2002. Interaction of gene-cloned and insect cell-expressed aminopeptidase N of *Spodoptera litura* with insecticidal crystal protein Cry1C. *Appl. Environ. Microbiol. 68,* 4583–4592.

Aronson, A.I., Geng, C., Wu, I., 1999. Aggregation of *Bacillus thuringiensis* Cry1A toxins upon binding to target insect larval midgut vesicles. *Appl. Environ. Microbiol. 65,* 2503–2507.

Aronson, A.I., 2000. Incorporation of protease K in insect larval membrane vesicles does not result in the disruption of the integrity or function of the pore-forming *Bacillus thuringiensis* δ-endotoxin. *Appl. Environ. Microbiol. 66,* 4568–4570.

Aronson, A.I., Shai, Y., 2001. Why *Bacillus thuringiensis* insecticidal toxins are so effective: unique features of their mode of action. *FEMS Microbiol. Lett. 195,* 1–8.

Audtho, M., Valaitis, A.P., Alzate, O., Dean, D.H., 1999. Production of chymotrypsin-resistant *Bacillus thuringiensis* Cry2Aa1 δ-endotoxin by protein engineering. *Appl. Environ. Microbiol. 65,* 4601–4605.

Baum, J.A., Light Mettus, A.-M., 2000. Polypeptide compositions toxic to Diabrotica insects, obtained from *Bacillus thuringiensis*; CryET70, and methods of use. Journal Patent PCT WO/00/26378-B.

Becker, N., 2000. Bacterial control of vector-mosquitoes and black flies. In: Charles, J.F., Delécluse, A., Nielsen-LeRoux, C. (Eds.), Entomopathogenic Bacteria: From Laboratory to Field Application. Kluwer, London, pp. 383–398.

Belfiore, C.J., Vadlamudi, R.K., Osman, Y.A., Bulla, L.A. Jr., 1994. A specific binding protein from *Tenebrio molitor* for the insecticidal toxin of *Bacillus thuringiensis* subsp. *thuringiensis*. *Biochem. Biophys. Res. Commun. 200,* 359–364.

Bernhard, K., Jarret, P., Meadows, M., Butt, J., Ellis, D.J., et al., 1997. Natural isolates of *Bacillus thuringiensis*: worldwide distribution, characterization and activity against insect pests. *J. Invertebr. Pathol. 70,* 59–68.

Berry, C., O'Neil, S., Ben-Dov, E., Jones, A.F., Murphy, L., et al., 2002. Complete sequence and organization of pBtoxis, the toxin-coding plasmid of *Bacillus thuringiensis* subsp. *israeliensis*. *Appl. Environ. Microbiol. 68,* 5082–5095.

Blalock, J.E., 1995. Genetic origins of protein shape and interaction rules. *Nature Med. 1,* 876–878.

Bosch, D., Schipper, B., van der Kleij, H., de Maagd, R.A., Stiekema, J., 1994. Recombinant *Bacillus thuringiensis* insecticidal proteins with new properties for resistance management. *Biotechnology 12,* 915–918.

Bradley, D., Harkey, M.A., Kim, M.-K., Biever, K.D., Bauer, L.S., 1995. The insecticidal Cry1B crystal protein of *Bacillus thuringiensis* subsp. *thuringiensis* has dual specificity to coleopteran and lepidopteran larvae. *J. Invertebr. Pathol. 65,* 162–173.

Bravo, A., 1997. Phylogenetic relationships of *Bacillus thuringiensis* δ-endotoxin family proteins and their functional domains. *J. Bacteriol. 179,* 2793–2801.

Bravo, A., Sarabia, S., Lopez, L., Ontiveros, H., Abarca, C., et al., 1998. Characterization of *cry* genes in a Mexican *Bacillus thuringiensis* strain collection. *Appl. Environ. Microbiol. 164,* 4965–4972.

Bravo, A., Miranda, R., Gómez, I., Soberón, M., 2002a. Pore formation activity of Cry1Ab toxin from *Bacillus thuringiensis* in an improved membrane vesicle preparation from *Manduca sexta* midgut cell microvilli. *Biochem. Biophys. Acta 1562,* 63–69.

Bravo, A., Sánchez, J., Kouskoura, T., Crickmore, N., 2002b. N-terminal activation is an essential early step in the mechanism of action of the *B. thuringiensis* Cry1Ac insecticidal toxin. *J. Biol. Chem. 277,* 23985–23987.

Burton, S.L., Ellar, D.J., Li, J., Derbyshire, D.J., 1999. N-Acetylgalactosamine on the putative insect receptor aminopeptidase N is recognized by a site on the domain III lectin-like fold of a *Bacillus thuringiensis* insecticidal toxin. *J. Mol. Biol. 287,* 1011–1022.

Butko, P., Huang, F., Pusztai-Carey, M., Surewicz, W.K., 1997. Interaction of the delta-endotoxin CytA from *Bacillus thuringiensis* var. *israelensis* with lipid membranes. *Biochemistry 36,* 12862–12868.

Butko, P., 2003. Cytolytic toxin Cyt1A and its mechanism of membrane damage: data and hypotheses. *Appl. Environ. Microbiol. 69,* 2415–2422.

Buzdin, A.A., Revina, L.P., Kostina, L.I., Zalunin, I.A., Chestukhina, G.G., 2002. Interaction of 65- and 62-kDa proteins from the apical membranes of the *Aedes aegypti* larvae midgut epithelium with Cry4B and Cry11A endotoxins of *Bacillus thuringiensis*. *Biochemistry* (Moscow) *67,* 540–546.

Cabiaux, V., Wolff, C., Ruysschaert, J.M., 1997. Interaction with a lipid membrane: a key step in bacterial toxins virulence. *Int. J. Biol. Macromol. 21,* 285–298.

Carlson, C.R., Caugant, D.A., Kolsto, A.B., 1994. Genotypic diversity among *Bacillus cereus* and *Bacillus thuringiensis* strains. *Appl. Environ. Microbiol. 60,* 1719–1725.

Carlson, C.R., Johansen, T., Lecadet, M.-M., Kolsto, A.-B., 1996. Genomic organization of the entomopathogenic bacterium *Bacillus thuringiensis* subsp. *berliner* 1715. *Microbiology 142,* 1625–1634.

Choma, C.T., Surewicz, W.K., Carey, P.R., Pozsgay, M., Raynor, T., et al., 1990. Unusual proteolysis of the protoxin and toxin from *Bacillus thuringiensis*: structural implications. *Eur. J. Biochem. 189,* 523–527.

Chow, E., Singh, G.J.P., Gill, S.S., 1989. Binding and aggregation of the 25 kDa toxin of *Bacillus thuringiensis israelensis* to cell membranes: alteration by monoclonal antibodies and amino acid modifiers. *Appl. Environ. Microbiol. 55,* 2779–2788.

Conner, A.J., Glare, T.R., Nap, J.-P., 2003. The release of genetically modified crops into the environment. 2. Overview of ecological risk assessment. *Plant J. 33,* 19–46.

Cooper, M.A., Carroll, J., Travis, E., Williams, D.H., Ellar, D.J., 1998. *Bacillus thuringiensis* Cry1Ac toxin interaction with *Manduca sexta* aminopeptidase N in a model membrane envionment. *Biochem. J. 333*, 677–683.

Craig, H.S., Putnam, C., Carozi, N., Tainer, J., **1999**. Evolution and mechanism from structures of an ADP-ribosylating toxin and NAD complex. *Nature Struct. Biol. 6*, 932–936.

Crickmore, N., Bone, E.J., Wiliams, J.A., Ellar, D., **1995**. Contribution of individual components of the δ-endotoxins crystal to the mosquitocidal activity of *Bacillus thuringiensis* subsp. *israeliensis. FEMS Microbiol. Lett. 131*, 249–254.

Crickmore, N., Zeigler, D.R., Feitelson, J., Schnepf, E., Van Rie, J., et al., **1998**. Revision of the nomenclture for the *Bacillus thuringiensis* pesticidal crystal proteins. *Microbiol. Mol. Biol. Rev. 62*, 807–813.

Crickmore, N., Zeigler, D.R., Schnepf, E., Van Rie, J., Lereclus, D., et al., **2002**. *Bacillus thuringiensis* toxin nomenclature. http:// www.biols.susx.ac.uk/Home/Neil_Crickmore/Bt/index.html.

Dai, S.M., Gill, S.S., **1993**. *In vitro* and *in vivo* proteolysis of the *Bacillus thuringiensis* subsp. *israelensis* CryIVD protein by *Culex quinquefasciatus* larval midgut proteases. *Insect Biochem. Mol. Biol. 23*, 273–283.

Daniel, A., Sangadala, S., Dean, D.H., Adang, M.J., **2002**. Denaturation of either *Manduca sexta* aminopeptidase N or *Bacillus thuringiensis* Cry1A toxins exposes binding epitopes hidden under nondenaturating conditions. *Appl. Environmen. Microbiol. 68*, 2106–2112.

de Maagd, R.A., Bakker, P., Masson, L., Adang, M.J., Sangadala, S., et al., **1999a**. Domain III of the *Bacillus thuringiensis* delta-endotoxin Cry1Ac is involved in binding to *Manduca sexta* brush border membranes and to its purified aminopeptidase N. *Mol. Microbiol. 31*, 463–471.

de Maagd, R.A., Bosch, D., Stiekema, W., **1999b**. *Bacillus thuringiensis* toxin-mediated insect resistance in plants. *Trends Plant Sci. 4*, 9–13.

de Maagd, R.A., Weemen-Hendriks, M., Stiekema, W., Bosch, D., **2000**. Domain III substitution in *Bacillus thuringiensis* delta-endotoxin Cry1C domain III can function as a specific determinant for *Spodoptera exigua* in different, but not all, Cry1-Cry1C hybrids. *Appl. Environ. Microbiol. 66*, 1559–1563.

de Maagd, R.A., Bravo, A., Crickmore, N., **2001**. How *Bacillus thuringiensis* has evolved specific toxins to colonize the insect world. *Trends Genet. 17*, 193–199.

Denolf, P., Hendrickx, K., VanDamme, J., Jansens, S., Peferoen, M., et al., **1997**. Cloning and characterization of *Manduca sexta* and *Plutella xylostella* midgut aminopeptidase N enzymes related to *Bacillus thuringiensis* toxin-binding proteins. *Eur. J. Biochem. 248*, 748–761.

Donovan, W.P., Donovan, J.C., Slaney, A.C., **2000**. *Bacillus thuringiensis* cryET33 and cryET34 compositions and uses therefore. US Patent No. 6 063 756.

Dorsch, J.A., Candas, M., Griko, N.B., Maaty, W.S.A., Midbo, E.G., et al., **2002**. Cry1A toxins of *Bacillus thuringiensis* bind specifically to a region adjacent to the membrane-proximal extracelllar domain of Bt-R₁ in *Manduca sexta*: involvement of a cadherin in the entomopathogenicity of *Bacillus thuringiensis*. *Insect Biochem. Mol. Biol. 32*, 1025–1036.

Dow, J.A.T., **1986**. Insect midgut function. *Adv. Insect Physiol. 19*, 187–238.

Dubois, N.R., Dean, D.H., **1995**. Synergism between Cry1A insecticidal crystal proteins and spores of *Bacillus thuringiensis*, other bacterial spores and vegetative cells against *Lymantria dispar* (Lepidoptera: Lymantriidae) larvae. *Environ. Entomol. 24*, 1741–1747.

Du, C., Martin, A.W., Nickerson, K.W., **1994**. Comparison of disulfide contents and solubility at alkaline pH of insecticidal and noninsecticidal *Bacillus thuringiensis* protein crystals. *Appl. Environ. Microbiol. 60*, 3847–3853.

Editorial, **2002**. Biodiversity (communication arising): editorial note. *Nature 416*, 600.

Ellis, R.T., Stockhoff, B.A., Stamp, L., Schnepf, H.E., Schwab, G.E., et al., **2002**. Novel *Bacillus thuringiensis* binary insecticidal crystal proteins active on western corn rootworm, *Diabrotica virgifera virgifera* LeConte. *Appl. Environ. Microbiol. 68*, 1137–1145.

Estruch, J.J., Warren, G.W., Mullins, M.A., Nye, G.J., Craig, J.A., et al., **1996**. Vip3A, a novel *Bacillus thuringiensis* vegetative insecticidal protein with a wide spectrum of activities against lepidopteran insects. *Proc. Natl Acad. Sci. USA 93*, 5389–5394.

Fedhila, S., Nel, P., Lereclus, D., **2002**. The InhA2 metalloprotease of *Bacillus thuringiensis* strain 407 is required for pathogenicity in insects infected via the oral route. *J. Bacteriol. 184*, 3296–3304.

Fedhila, S., Gohar, M., Slamti, L., Nel, P., Lereclus, D., **2003**. The *Bacillus thuringiensis* PlcR-regulated gene inhA2 is necessary, but not sufficient, for virulence. *J. Bacteriol. 185*, 2820–2825.

Ferré, J., Van Rie, J., **2002**. Biochemestry and genetics of insect resistance to *Bacillus thuringiensis*. *Annu. Rev. Entomol. 47*, 501–533.

Flores, H., Soberón, X., Sánchez, J., Bravo, A., **1997**. Isolated domain II and III from *Bacillus thuringiensis* Cry1Ab δ-endotoxin binds to lepidopteran midgut membranes. *FEBS Lett. 414*, 313–318.

Gahan, L.J., Gould, F., Heckel, D.G., **2001**. Identification of a gene associated with Bt resistance in *Heliothis virescens*. *Science 293*, 857–860.

Galitsky, N., Cody, V., Wojtczak, A., Debashis, G., Luft, J.R., et al., **2001**. Structure of insecticidal bacterial δ-endotoxin Cry3Bb1 of *Bacillus thuringiensis*. *Acta Crystallogr. D57*, 1101–1109.

Garczynski, S.F., Adang, M.J., **1995**. *Bacillus thuringiensis* CryIA(c) δ-endotoxin binding aminopeptidase in the *Manduca sexta* midgut has a glycosyl-phosphatidylinositol anchor. *Insec. Biochem. Mol. Biol. 25*, 409–415.

Gatehouse, A.M.R., Ferry, N., Raemaekers, R.J.M., **2002**. The case of the monarch butterfly: a verdict is returned. *Trends Genet. 18*, 249–251.

Gazit, E., Burshtein, N., Ellar, D.J., Sawyer, T., Shai, Y., **1997**. *Bacillus thuringiensis* cytolytic toxin associates specifically with its synthetic helices A and C in the membrane state: implications for the assembly of oligomeric transmembrane pores. *Biochemistry 36,* 15546–15556.

Gazit, E., La Rocca, P., Sansom, M.S.P., Shai, Y., 1998. The structure and organization within the membrane of the helices composing the pore-forming domain of *Bacillus thuringiensis* δ-endotoxin are consistent with an umbrella-like structure of the pore. *Proc. Natl Acad. Sci. USA 95,* 12289–12294.

Ge, A.Z., Shivarova, N.I., Dean, D.H., 1989. Location of the *Bombyx mori* specificity domain on a *Bacillus thuringiensis* delta-endotoxin. *Proc. Natl Acad. Sci. USA 79,* 6951–6955.

Gill, M., Ellar, D., 2002. Transgenic *Drosophila* reveals a functional *in vivo* receptor for the *Bacillus thuringiensis* toxin Cry1Ac1. *Insect Mol. Biol. 11,* 619–625.

Gill, S.S., Cowles, E.A., Pietrantonio, P., 1992. The mode of action of *Bacillus thuringiensis* endotoxins. *Annu. Rev. Entomol. 37,* 615–636.

Gill, S.S., Singh, G.J.P., Hornung, J.M., 1987. Cell membrane interaction of *Bacillus thuringiensis* subsp. *israelensis* cytolytic toxins. *Infect. Immun. 55,* 1300–1308.

Gill, S.S., Cowles, E.A., Francis, V., 1995. Identification, isolation, and cloning of a *Bacillus thuringiensis* Cry1Ac toxin-binding protein from the midgut of the lepidopteran insect *Heliothis virescens. J. Biol. Chem. 270,* 27277–27282.

Gohar, M., Okstad, O.A., Gilois, N., Sanchis, V., Kolsto, A.-B., et al., 2002. Two-dimensional electrophoresis analysis of the extrcellular proteome of *Bacillus cereus* reveals the importance of the PlcR regulon. *Proteomics 2,* 784–791.

Gómez, I., Oltean, D.I., Sanchez, J., Gill, S.S., Bravo, A., et al., 2001. Mapping the epitope in cadherin-like receptors involved in *Bacillus thuringiensis* Cry1A toxin interaction using phage display. *J. Biol. Chem. 276,* 28906–28912.

Gómez, I., Miranda-Rios, J., Rudiño-Piñera, E., Oltean, D.I., Gill, S.S., et al., 2002a. Hydropathic complementarity determines interaction of epitope ^{869}HITDTNNK^{876} in *Manduca sexta* Bt-R$_1$ receptor with loop 2 of domain II of *Bacillus thuringiensis* Cry1A toxins. *J. Biol. Chem. 277,* 30137–30143.

Gómez, I., Sánchez, J., Miranda, R., Bravo, A., Soberón, M., 2002b. Cadherin-like receptor binding facilitates proteolytic cleavage of helix α-1 in domain I and oligomer pre-pore formation of *Bacillus thuringiensis* Cry1Ab toxin. *FEBS Lett. 513,* 242–246.

Gómez, I., Dean, D.H., Bravo, A., Soberón, M., 2003. Molecular basis for *Bacillus thuringiensis* Cry1Ab toxin specificity: two structural determinants in the *Manduca sexta* Bt-R$_1$ receptor interact with loops α-8 and 2 in domain II of Cy1Ab toxin. *Biochemistry 42,* 10482–10489.

Griffits, J.S., Whitacre, J.L., Stevens, D.E., Aroian, R.V., 2001. *Bacillus thuringiensis* toxin resistance from loss of a putative carbohydrate-modifying enzyme. *Science 293,* 860–864.

Grochulski, P., Masson, L., Borisova, S., Pusztai-Carey, M., Schwartz, J.L., et al., 1995. *Bacillus thuringiensis* Cry1A(a) insecticidal toxin: crystal structure and channel formation. *J. Mol. Biol. 254,* 447–464.

Guerchicoff, A., Ugalde, R.A., Rubinstein, C.P., 1997. Identification and characterization of a previously undescribed *cyt* gene in *Bacillus thuringiensis* subsp. *israeliensis. Appl. Environ. Microbiol. 62,* 2716–2721.

Guerchicoff, A., Delécluse, A., Rubinstein, C.P., 2001. The *Bacillus thuringiensis cyt* genes for hemolytic endotoxin constitute a gene family. *Appl. Envion. Microbiol. 67,* 1090–1096.

Guillet, P., Kurstak, D.C., Philippon, B., Meyer, R., 1990. Use of *Bacillus thuringiensis israelensis* for onchocerciasis control in West Africa. In: de Barjac, H., Sutherland, D.J. (Eds.), Bacterial Control of Mosquitoes and Blackflies. Rutgers University Press, New Brunswick, NJ, pp. 187–201.

Guttmann, D.M., Ellar, D.J., 2000. Phenotypic and genotypic comparisons of 23 strains from the *Bacillus cereus* complex for a selection of known and putative *B. thuringiensis* virulence factors. *FEMS Microbiol. Lett. 188,* 7–13.

Haider, M.Z., Ellar, D.J., 1989. Functional mapping of an entomocidal δ-endotoxin: single amino acid changes produce by site directed mutagenesis influence toxicity and specificity of the protein. *J. Mol. Biol. 208,* 183–194.

Haider, M.Z., Smith, G.P., Ellar, D.J., 1989. Delineation of the toxin coding fragments and an insect-specificity region of a dual toxicity *Bacillus thuringiensis* crystal protein gene. *FEMS Microbiol. Lett. 49,* 157–163.

Hara, H., Atsumi, S., Yaoi, K., Higurashi, S., Miura, N., et al., 2003. A cadherin-like protein functions as a receptor for *Bacillus thuringiensis* Cry1Aa and Cry1Ac toxins to epithelial cells of *Bombyx mori* larvae. *FEBS Lett. 538,* 29–34.

Helgason, E., Okstad, O.A., Caugant, D.A., Johansen, H.A., Fouet, A., et al., 2000. *Bacillus anthracis, Bacillus cereus,* and *Bacillus thuringiensis*: one species on the basis of genetic evidence. *Appl. Environ. Microbiol. 66,* 2627–2630.

Hellmich, R.L., Siegfried, B.D., Sears, M.K., Stanley-Horn, D.E., Daniels, M.J., et al., 2001. Monarch larvae sensitivity to *Bacillus thuringiensis*-purified proteins and pollen. *Proc. Natl Acad. Sci. USA 98,* 11925–11930.

Huber, H.E., Luthy, P., 1981. *Bacillus thuringiensis* δ-endotoxin; composition and activation. In: Davidson, E.W. (Ed.), Pathogenesis of Microbiol Diseases. Allanheld Osmum, Totowa, NJ, pp. 209–234.

Ivanova, N., Sorokin, A., Anderson, I., Galleron, N., Candelon, B., et al., 2003. Genome sequence of *Bacillus cereus* and comparative analysis with *B. anthracis. Nature 423,* 87–91.

James, C., 2001. Global review of commercialized transgenic crops: 2001. In: International Service for the Acquisition of AgricBiotech Applications Briefs No. 24. ISAAA, Ithaca, New York.

Jenkins, J.L., Dean, D.H., 2000. Exploring the mechanism of action of insecticidal proteins by genetic engineering methods. In: Setlow, K. (Ed.), Genetic Engineering: Principles and Methods, vol. 22. Plenum, New York, pp. 33–54.

Jenkins, J.L., Lee, M.K., Valaitis, A.P., Curtiss, A., Dean, D.H., 2000. Bivalent sequential binding model of a *Bacillus thuringiensis* toxin to gypsy moth aminopeptidase N receptor. *J. Biol. Chem.* 275, 14423–14431.

Jurat-Fuentes, J.L., Adang, M.J., 2001. Importance of Cry1 δ-endotoxin domain II loops for binding specificity in *Heliothis virescens* (L). *Appl. Environ. Microbiol.* 67, 323–329.

Jurat-Fuentes, J.L., Gould, F.L., Adang, M.J., 2002. Altered glycosylation of 63- and 68-kilodalton microvillar proteins in *Heliothis virescens* correlated with reduced Cry1 toxin binding, decreased pore formation, and increased resistance to *Bacillus thuringiensis* Cry1 toxins. *Appl. Environ. Microbiol.* 68, 5711–5717.

Kaplinsky, N., Braun, D., Lisch, D., Hay, A., Hake, S., et al., 2002. Biodiversity (communication arising): maize transgene results in Mexico are artifacts. *Nature* 416, 601.

Keller, M., Sneh, B., Strizhov, N., Prudovsky, E., Regev, A., et al., 1996. Digestion of δ-endotoxin by gut proteases may explain reduced sensitivity of advanced instar larvae of *Spodoptera littoralis* to CryIC. *Insect Biochem. Mol. Biol.* 26, 365–373.

Khasdan, V., Ben-Dov, E., Manasherob, R., Boussiba, S., Zaritsky, A., 2001. Toxicity and synergism in transgenic *Escherichia coli* expressing four genes from *Bacillus thuringiensis* subsp. *israelensis*. *Environ. Microbiol.* 3, 798–806.

Kirouac, M., Vachon, V., Noel, J.F., Girard, F., Schwartz, J.L., et al., 2002. Amino acid and divalent ion permeability of the pores formed by the *Bacillus thuringiensis* toxins Cry1Aa and Cry1Ac in insect midgut brush boreder membrane vesicles. *Biochim. Biophys. Acta* 1561, 171–179.

Knight, P., Crickmore, N., Ellar, D.J., 1994. The receptor for *Bacillus thuringiensis* CryIA(c) delta-endotoxin in the brush border membrane of the lepidopteran *Manduca sexta* is aminopeptidase N. *Mol. Microbiol.* 11, 429–436.

Kouskoura, T., Tickner, C., Crickmore, N., 2001. Expression and cristalization of an N-terminally activated form of the *Bacillus thuringiensis* Cry1Ca toxin. *Curr. Microbiol.* 43, 371–373.

Lambert, B., Buysse, L., Decock, C., Jansens, S., Piens, C., et al., 1996. A *Bacillus thuringiensis* insecticidal crystal protein with a high activity against members of the family Noctuidae. *Appl. Environ. Microbiol.* 62, 80–86.

Lee, M.K., Young, B.A., Dean, D.H., 1995. Domain III exchanges of *Bacillus thuringiensis* CryIA toxins affect binding to different gypsy moth midgut receptors. *Biochem. Biophys. Res. Comm.* 216, 306–312.

Lee, M.K., You, T.H., Young, B.A., Cotrill, J.A., Valaitis, A.P., et al., 1996. Aminopeptidase N purified from gypsy moth brush border membrane vesicles is a specific receptor for *Bacillus thuringiensis* CryIAc toxin. *Appl. Environ. Microbiol.* 62, 2845–2849.

Lee, M.K., You, T.H., Gould, F.L., Dean, D.H., 1999. Identification of residues in domain III of *Bacillus thuringiensis* Cry1Ac toxin that affect binding and toxicity. *Appl. Environ. Microbiol.* 65, 4513–4520.

Lee, M.K., Rajamohan, F., Jenkins, J.L., Curtiss, A., Dean, D.H., 2000. Role of two arginine residues in domain II, loop 2 of Cry1Ab and Cry1Ac *Bacillus thuringiensis* δ-endotoxin in toxicity and binding to *Manduca sexta* and *Lymantria dispar* aminopetidase N. *Mol. Microbiol.* 38, 289–298.

Lee, M.K., Jenkins, J.L., You, T.H., Curtiss, A., Son, J.J., et al., 2001. Mutations at the arginine residues in alpha 8 loop of *Bacillus thuringiensis* delta-endotoxin Cry1Ac affect toxicity and binding to *Manduca sexta* and *Lymantria dispar* aminopeptidase N. *FEBS Lett.* 497, 108–112.

Lereclus, D., Agaisse, H., Gominet, S., Salamitou, S., Sanchis, V., 1996. Identification of a gene that positively regulates transcription of the phosphatidylinositol-specific phospholipase C gene at the onset of the stationary phase. *J. Bacteriol.* 178, 2749–2756.

Li, J., Carroll, J., Ellar, D.J., 1991. Crystal structure of insecticidal δ-endotoxin from *Bacillus thuringiensis* at 2.5 A resolution. *Nature* 353, 815–821.

Li, J., Pandelakis, A.K., Ellar, D., 1996. Structure of the mosquitocidal δ-endotoxin CytB from *Bacillus thuringiensis* ssp. *kyushuensis* and implications for membrane pore formation. *J. Mol. Biol.* 257, 129–152.

Li, J., Derbyshire, D.J., Promdonkoy, B., Ellar, D.J., 2001. Structural implications for the transformation of the *Bacillus thuringiensis* δ-endotoxins from water-soluble to membrane-inserted forms. *Trans. Biochem. Soc.* 29, 571–577.

Lorence, A., Darszon, A., Díaz, C., Liévano, A., Quintero, R., et al., 1995. δ-Endotoxins induce cation channels in *Spodoptera frugiperda* brush border membranes in suspension and in planar lipid bilayers. *FEBS Lett.* 360, 217–222.

Lorence, A., Darszon, A., Bravo, A., 1997. The pore formation activity of *Bacillus thuringiensis* Cry1Ac toxin on *Trichoplusia ni* membranes depends on the presence of aminopeptidase N. *FEBS Lett.* 414, 303–307.

Loseva, O.I., Toktopulo, E.I., Vasilev, V.D., Nikulin, A.D., Dobritsa, A.P., et al., 2001. Structure of Cry3A δ-endotoxin within phopholipid membranes. *Biochemistry* 40, 14143–14151.

Losey, J.E., Raylor, L.S., Carter, M.E., 1999. Transgenic pollen harms monarch larvae. *Nature* 399, 214.

Luo, K., Tabashnik, B.E., Adang, M.J., 1997. Binding of *Bacillus thuringiensis* Cry1Ac toxin to aminopeptidase in susceptible and resistant diamondback moths (*Plutella xylostella*). *Appl. Environ. Microbiol.* 63, 1024–1027.

MacIntosh, S.C., Kishore, G.M., Perlak, F.J., Marrone, P.G., Stone, T.B., et al., 1990. Potentiation of *Bacillus thuringiensis* insecticidal activity by serine protease inhibitor. *J. Agric. Food Chem.* 38, 1145–1152.

Margalith, Y., Ben-Dov, E., **2000**. Biological control by *Bacillus thuringiensis* subsp. *israeliensis*. In: Rechcigl, J.E., Rechcigl, N.A. (Eds.), Insect Pest Management: Techniques for Environmental Protection. CRC Press, Boca Raton, FL, pp. 243–301.

Martens, J.W.M., Visser, B., Vlak, J.M., Bosch, D., **1995**. Mapping and characterization of the entomocidal domain of the *Bacillus thuringiensis* CryIA(b) protoxin. *Mol. Gen. Genet. 247*, 482–487.

Masson, L., Lu, Y.-J., Mazza, A., Brosseau, R., Adang, M.J., **1995**. The Cry1A(c) receptor purified from *Manduca sexta* displays multiple specificities. *J. Biol. Chem. 270*, 20309–20315.

Metz, M., Futterer, J., **2002**. Biodiversity (communication arising): suspect evidence of transgenic contamination. *Nature 416*, 600–601.

Mignot, T., Mock, M., Robichon, D., Landier, A., Lereclus, D., *et al.*, **2001**. The incompatibility between the PlcR- and AtxA-controlled regulons may have selected a nonsense mutation in *Bacillus anthracis*. *Mol. Microbiol. 42*, 1189–1198.

Miranda, R., Zamudio, F., Bravo, A., **2001**. Processing of Cry1Ab δ-Endotoxin from *Bacillus thuringiensis* by midgut proteases: role in toxin activation and inactivation. *Insect Biochem. Mol. Biol. 31*, 1155–1163.

Morin, S., Biggs, R.W., Shriver, L., Ellers-Kirk, C., Higginson, D., *et al.*, **2003**. Three cadherin alleles associated with resistance to *Bacillus thuringiensis* in pink bollworm. *Proc. Natl Acad. Sci. USA 100*, 5004–5009.

Morse, R.J., Yamamoto, T., Stroud, R.M., **2001**. Structure of Cry2Aa suggests an unexpected receptor binding epitope. *Structure 9*, 409–417.

Nakanishi, K., Yaoi, K., Nagino, Y., Hara, H., Kitami, M., *et al.*, **2002**. Aminopeptidase N isoforms from the midgut of *Bombyx mori* and *Plutella xylostella*: their classification and the factors that determine their binding specificity to *Bacillus thuringiensis* Cry1A toxin. *FEBS Lett. 519*, 215–220.

Nagamatsu, Y., Toda, S., Yagamuchi, F., Ogo, M., Kogure, M., *et al.*, **1998**. Identification of *Bombyx mori* midgut receptor for *Bacillus thuringiensis* insecticidal CryIA(a) toxin. *Biosci. Biotechnol. Biochem. 62*, 718–726.

Nagamatsu, Y., Koike, T., Sasaki, K., Yoshimoto, A., Furukawa, Y., **1999**. The cadherin-like protein is essential to specificity determination and cytotoxic action of the *Bacillus thuringiensis* insecticidal CryIAa toxin. *FEBS Lett. 460*, 385–390.

Navon, A., **2000**. *Bacillus thuringiensis* application in agriculture. In: Charles, J.F., Delécluse, A., Nielsen-LeRoux, C. (Eds.), Entomopathogenic Bacteria: From Laboratory to Field Application. Kluwer, London, pp. 355–369.

Oberhauser, K.S., Prysby, M.D., Mattila, H.R., Stanley-Horn, D.E., Sears, M.K., *et al.*, **2001**. Temporal and spatial overlap between monarch larvae and corn pollen. *Proc. Natl Acad. Sci. USA 98*, 11913–11918.

Oltean, D.I., Pullikuth, A.K., Lee, H.-K., Gill, S.S., **1999**. Partial purification and characterization of *Bacillus thuringiensis* Cry1A toxin receptor A from *Heliothis*

virescens and cloning of the corresponding cDNA. *Appl. Environ. Microbiol. 65*, 4760–4766.

Oppert, B., Kramer, K.J., Beeman, R.W., Johnson, D., McGaughey, W.H., **1997**. Proteinase-mediated insect resistance to *Bacillus thuringiensis* toxins. *J. Biol. Chem. 272*, 23473–23476.

Pang, A.S., Gringorten, J.L., Bai, C., **1999**. Activation and fragmentation of *Bacillus thuringiensis* δ-endotoxin by high concentrations of proteolytic enzymes. *Can. J. Microbiol. 45*, 816–825.

Peyronnet, O., Vachon, V., Schwartz, J.L., Laprade, R., **2001**. Ion channels induced in planar lipid bilayers by the *Bacillus thuringiensis* toxin Cry1Aa in the presence of gypsy moth (*Lymantria dispar*) brush border membranes. *J. Membr. Biol. 184*, 45–54.

Peyronnet, O., Nieman, B., Généreux, F., Vachon, V., Laprade, R., **2002**. Estimation of the radius of the pores formed by the *Bacillus thuringiensis* Cry1C δ-endotoxin in planar lipid bilayers. *Biochim. Biophys. Acta 1567*, 113–122.

Pleasants, J.M., Hellmich, R.L., Diveley, G.P., Sears, M.K., Stanley-Horn, D.E., *et al.*, **2001**. Corn pollen deposition on milkweeds in and near cornfields. *Proc. Natl Acad. Sci. USA 98*, 11919–11924.

Promdonkoy, B., Ellar, D.J., **2003**. Investigation of the pore forming mechanism of a cytolitic δ-endotoxin from *Bacillus thuringiensis*. *Biochem. J. 374*, 255–259.

Qaim, M., Zilberman, D., **2003**. Yield effects of genetically modified crops in developing countries. *Science 299*, 900–902.

Quist, D., Chapela, I.H., **2001**. Transgenic DNA introgressed into traditional maize landraces in Oaxaca, Mexico. *Nature 414*, 541–543.

Rajagopal, R., Sivakumar, S., Agrawal, N., Malhotra, P., Bhatnagar, R.K., **2002**. Silencing of midgut aminopeptidase N of *Spodoptera litura* by double-stranded RNA establishes its role as *Bacillus thuringiensis* toxin receptor. *J. Biol. Chem. 277*, 46849–46851.

Rajamohan, F., Alcantara, E., Lee, M.K., Chen, X.J., Curtiss, A., *et al.*, **1995**. Single amino acid changes in domain II of *Bacillus thuringiensis* Cry1Ab δ-endotoxin affect irreversible binding to *Manduca sexta* midgut membrane vesicles. *J. Bacteriol. 177*, 2276–2282.

Rajamohan, F., Hussain, S.-R.A., Cotrill, J.A., Gould, F., Dean, D.H., **1996a**. Mutations at domain II, loop 3, of *Bacillus thuringiensis* Cry1Aa and Cry1Ab δ-endotoxins suggest loop 3 is involved in initial binding to lepidopteran insects. *J. Biol. Chem. 271*, 25220–25226.

Rajamohan, F., Cotrill, J.A., Gould, F., Dean, D.H., **1996b**. Role of domain II, loop 2 residues of *Bacillus thuringiensis* Cry1Ab δ-endotoxin in reversible and irreversible binding to *Manduca sexta* and *Heliothis virescens*. *J. Biol. Chem. 271*, 2390–2396.

Rang, C., Vachon, V., de Maagd, R.A., Villalon, M., Schwartz, J.-L., *et al.*, **1999**. Interaction between functional domains of *Bacillus thuringiensis* insecticidal proteins. *Appl. Environ. Microbiol. 65*, 2918–2925.

Read, T.D., Peterson, A.N., Tourasse, N., Baillie, L.W., Paulsen, I.T., *et al.*, **2003**. The genome sequence of

Bacillus anthracis Ames and comparison to closely related bacteria. *Nature 423*, 81–86.

Regis, L., da Silva, S.B., Melo-Santos, M.A.V., 2000. The use of bacterial larvicides in mosquito and black fly control programmes in Brazil. *Mem. Inst. Oswaldo Cruz, Rio de Janeiro 95*(Suppl. I), 207–210.

Rosenberger, C.M., Brumel, J.H., Finlay, B.B., 2000. Microbial pathogenesis: lipid rafts as pathogen portals. *Current Biol. 10*, 823–825.

Sayyed, A.H., Crickmore, N., Wright, D.J., 2001. Cyt1Aa from *Bacillus thuringiensis* subsp. *israelensis* is toxic to the diamondback moth, *Plutella xylostella*, and synergizes the activity of Cry1Ac towards a resistant strain. *Appl. Environ. Microbiol. 67*, 5859–5861.

Schnepf, E., Crickmore, N., Van Rie, J., Lereclus, D., Baum, J.R., *et al.*, 1998. *Bacillus thuringiensis* and its pesticidal crystal proteins. *Microbiol. Mol. Biol. Rev. 62*, 705–806.

Schuler, T.H., Poppy, G.M., Kerry, B.R., Denholm, I., 1998. Insect-resistant transgenic plants. *Trends Biotechnol. 16*, 168–175.

Schwartz, J.L., Garneau, L., Savaria, D., Masson, L., Brousseau, R., *et al.*, 1993. Lepidopteran-specific crystal toxins from *Bacillus thuringiensis* form cation- and anion-selective channels in planar lipid bilayers. *J. Membr. Biol. 132*, 53–62.

Schwartz, J.L., Lu, Y.J., Sohnlein, P., Brousseau, R., Laprade, R., *et al.*, 1997a. Ion channels formed in planar lipid bilayers by *Bacillus thuringiensis* toxins in the presence of *Manduca sexta* midgut receptors. *FEBS Lett. 412*, 270–276.

Schwartz, J.L., Juteau, M., Grochulski, P., Cygler, M., Préfontaine, G., *et al.*, 1997b. Restriction of intramolecular movements within the Cry1Aa toxin molecule of *Bacillus thuringiensis* through disulfide bond engineering. *FEBS Lett. 410*, 397–402.

Schwartz, J.L., Potvin, L., Chen, X.J., Brousseau, R., Laprade, R., *et al.*, 1997c. Single-site mutations in the conserved alternating arginine region affect ionic channels formed by CryIAa, a *Bacillus thuringiensis* toxin. *Appl. Environ. Microbiol. 63*, 3978–3984.

Schwartz, J.L., Laprade, R., 2000. Membrane permeabilization by *Bacillus thuringiensis* toxins: protein insertion and pore formation. In: Charles, J.F., Delécluse, A., Nielsen-LeRoux, C. (Eds.), Entomopathogenic Bacteria: From Laboratory to Field Application. Kluwer, London, pp. 199–217.

Scriber, J.M., 2001. *Bt* or not *Bt*: Is that the question? *Proc. Natl Acad. Sci. USA 98*, 12328–12330.

Sears, M.K., Hellmich, R.L., Stanley-Horn, D.E., Oberhauser, K.S., Pleasants, J.M., *et al.*, 2001. Impact of *Bt* corn pollen on monarch butterfly populations: a risk assessment. *Proc. Natl Acad. Sci. USA 98*, 11937–11942.

Shao, Z., Cui, Y., Liu, X., Yi, H., Ji, J., *et al.*, 1998. Processing of δ-endotoxin of *Bacillus thuringiensis* subsp. *kurstaki* HD-1 in *Heliothis armigera* midgut juice and the effects of protease inhibitors. *J. Invertebr. Pathol. 72*, 73–81.

Simons, K., Toomre, D., 2000. Lipid rafts and signal transduction. *Nature Rev. Mol. Cell Biol. 1*, 31–39.

Slamti, L., Lereclus, D., 2002. A cell–cell signaling peptide activates the PlcR virulence regulon in bacteria of the *Bacillus cereus* group. *EMBO J. 17*, 4550–4559.

Slatin, S.L., Abrams, C.K., English, L.H., 1990. Delta-endotoxins form cation-selective channels in planar lipid bilayers. *Biochem. Biophys. Res. Commun. 169*, 765–772.

Smedley, D.P., Armstrong, G., Ellar, D.J., 1997. Channel activity caused by a *Bacillus thuringiensis* delta-endotoxin preparation depends on the method of activation. *Mol. Membr. Biol. 14*, 13–18.

Smith, D.J., Ellar, D.J., 1994. Mutagenesis of two surface-exposed loops of the *Bacillus thuringiensis* CryIC δ-endotoxin affects insecticidal specificity. *Biochem. J. /title>302*, 611–616.

Soberón, M., Perez, R.V., Nuñez-Valdéz, M.E., Lorence, A., Gómez, I., *et al.*, 2000. Evidences for inter-molecular interaction as a necessary step for pore-formation activity and toxicity of *Bacillus thuringiensis* Cry1Ab toxin. *FEMS Microbiol. Lett. 191*, 221–225.

Stanley-Horn, D.E., Dively, G.P., Hellmich, R.L., Mattila, H.R., Sears, M.K., *et al.*, 2001. Assesing the impact of Cry1Ab-expressing corn pollen on monarch butterfly in field studies. *Proc. Natl Acad. Sci. USA 98*, 11931–11936.

Tabashnik, B.E., Finson, N., Johnson, M.W., Heckel, D.G., 1994. Cross-resistance to *Bacillus thuringiensis* toxin Cry1F in the diamondback moth (*Plutella xylostella*). *Appl Environ. Microbiol. 60*, 4627–4629.

Terra, W., Ferreira, C., 1994. Insect digestive enzymes: properties, compartamentalization and function. *Comp. Biochem. Physiol. B 109*, 1–62.

Toenniessen, G.H., O'Toole, J.C., DeVries, J., 2003. Advances in plant biotechnology and its adoption in developing countries. *Curr. Opin. Plant Biol. 6*, 191–198.

Thompson, M., Knuth, M., Cardineau, G., 2001. *Bacillus thuringiensis* toxins with improved activity. US Patent No. 6 303 364.

Tran, L.B., Vachon, V., Schwartz, J.L., Laprade, R., 2001. Differential effects of pH on the pore-forming properties of *Bacillus thuringiensis* insecticidal crystal toxins. *Appl. Environ. Microbiol. 67*, 4488–4494.

Tsuda, Y., Nakatani, F., Hashimoto, K., Ikawa, S., Matsuura, C., *et al.*, 2003. Cytotoxic activity of *Bacillus thuringiensis* Cry proteins on mammalian cells transfected with cadherin-like Cry receptor gene of *Bombyx mori* (silkworm). *Biochem. J. 369*, 697–703.

Vadlamudi, R.K., Weber, E., Ji, I., Ji, T.H., Bulla, L.A., Jr., 1995. Cloning and expression of a receptor for an insecticidal toxin of *Bacillus thuringiensis*. *J. Biol. Chem. 270*, 5490–5494.

Valaitis, A.P., Lee, M.K., Rajamohan, F., Dean, D.H., 1995. Brush border membrane Aminopeptidase-N in the midgut of the Gypsy Moth serves as the receptor for the CryIA(c) δ-endotoxin of *Bacillus thuringiensis*. *Insect Biochem. Mol. Biol. 25*, 1143–1151.

Valaitis, A.P., Jenkins, J.L., Lee, M.K., Dean, D.H., Garner, K.J., 2001. Isolation and partial characterization of Gypsy moth BTR-270 an anionic brush border membrane glycoconjugate that binds *Bacillus thuringiensis* Cry1A toxins with high affinity. *Arch. Insect Biochem. Physiol.* 46, 186–200.

van Frankenhuyzen, K., 2000. Application of *Bacillus thuringiensis* in forestry. In: Charles, J.F., Delécluse, A., Nielsen-LeRoux, C. (Eds.), Entomopathogenic Bacteria: From Laboratory to Field Application. Kluwer, London, pp. 371–382.

Vie, V., Van Mau, N., Pomarde, P., Dance, C., Schwartz, J.L., et al., 2001. Lipid-induced pore formation of the *Bacillus thuringiensis* Cry1Aa insecticidal toxin. *J. Membr. Biol.* 180, 195–203.

Von-Tersch, M.A., Slatin, S.L., Kulesza, C.A., English, L.H., 1994. Membrane-permeabilizing activities of *Bacillus thuringiensis* coleopteran-active toxin CryIIIB2 and CryIIIB2 domain I peptide. *Appl. Environ. Microbiol.* 60, 3711–3717.

Warren, G., 1997. Vegetative insecticidal proteins: novel proteins for control of corn pests. In: Carozzi, N., Koziel, M. (Eds.), Advances in Insect Control: The Role of Transgenic Plants. Taylor and Francis, London, pp. 109–121.

Wei, J.-Z., Hale, K., Carta, L., Platzer, E., Wong, C., et al., 2003. *Bacillus thuringiensis* crystal proteins that target nematodes. *Proc. Natl Acad. Sci. USA* 100, 2760–2765.

Widner, W.R., Whiteley, H.R., 1990. Location of the dipteran specific region in a lepidopteran–dipteran crystal protein from *Bacillus thuringiensis*. *J. Bacteriol.* 172, 2826–2832.

Wirth, M.C., Georghiou, G.P., Federeci, B.A., 1997. CytA enables CryIV endotoxins of *Bacillus thuringiensis* to overcome high levels of CryIV resistance in the mosquito, *Culex quinquefasciatus*. *Proc. Natl Acad. Sci. USA* 94, 10536–10540.

Wirth, M.C., Delécuse, A., Walton, W.E., 2001. CytAb1 and Cyt2Ba1 from *Bacillus thuringiensis* subsp. *medellin* and *B. thuringiensis* subsp. *israelensis* synergize *Bacillus sphaericus* against *Aedes aegypti* and resistant *Culex quinquefasciatus* (Diptera: Culicidae). *Appl. Environ. Microbiol.* 67, 3280–3284.

Wu, D., Aronson, A.I., 1992. Localized mutagenesis defines regions of the *Bacillus thuringiensis* δ-endotoxin involved in toxicity and specificity. *J. Biol. Chem.* 267, 2311–2317.

Wu, D., Johnson, J.J., Federeci, B.A., 1994. Synergism of mosquitocidal toxicity between CytA and CryIVD proteins using inclusions produced from cloned genes of *Bacillus thuringiensis*. *Mol. Microbiol.* 13, 965–972.

Wu, S.-J., Dean, D.H., 1996. Functional significance of loops in the receptor binding domain of *Bacillus thuringiensis* CryIIIA δ-endotoxin. *J. Mol. Biol.* 255, 628–640.

Yamagiwa, M., Esaki, M., Otake, K., Inagaki, M., Komano, T., et al., 1999. Activation process of dipteran-specific insecticidal protein produced by *Bacillus thuringiensis* subsp. *israelensis*. *Appl. Environ. Microbiol.* 65, 3464–3469.

Yaoi, K., Kadotani, T., Kuwana, H., Shinkawa, A., Takahashi, T., et al., 1997. Aminopeptidase N from *Bombyx mori* as a candidate for the receptor of *Bacillus thuringiensis* Cry1Aa toxin. *Eur. J. Biochem.* 246, 652–657.

Yaoi, K., Nakanishi, K., Kadotani, T., Imamura, M., Koizumi, N., et al., 1999. *Bacillus thuringiensis* Cry1Aa toxin-binding region of *Bombyx mori* aminopeptidase N. *FEBS Lett.* 463, 221–224.

Zangerl, A.R., McKenna, D., Wraight, C.L., Carroll, M., Ficarello, P., et al., 2001. Effects of exposure to event 176 *Bacillus thuringiensis* corn pollen on monarch and black swallowtail caterpillars under field conditions. *Proc. Natl Acad. Sci. USA* 98, 11908–11912.

Zalunin, I.A., Revina, L.P., Kostina, L.I., Chestukhina, G.G., Stepanov, V.M., 1998. Limited proteolysis of *Bacillus thuringiensis* CryIG and CryIVB δ-endotoxins leads to formation of active fragments that do not coincide with the structural domains. *J. Protein Chem.* 17, 463–471.

Zhuang, M., Oltean, D.I., Gómez, I., Pullikuth, A.K., Soberón, M., et al., 2002. *Heliothis virescens* and *Manduca sexta* lipid rafts are involved in Cry1A toxin binding to the midgut epithelium and subsequent pore formation. *J. Biol. Chem.* 277, 13863–13872.

A7 Addendum: *Bacillus thuringiensis*, with Resistance Mechanisms

A Bravo and M Soberón, Universidad Nacional Autónoma de México, Cuernavaca Morelos, Mexico
S S Gill, University of California, Riverside, USA

© 2010 Elsevier B.V. All Rights Reserved

Since the publication of the chapter in 2005, there have been important developments in understanding the mode of action of *Bacillus thuringiensis* (Bt) toxins in different insect orders. Here we highlight some of these important developments.

A7.1. Mode of Action of Cry Toxins

The proposed model of the mode of action of Cry1A toxins involves sequential interaction of these toxins with at least two different receptor molecules, a transmembrane cadherin protein and GPI-anchored aminopeptidase-N (APN) (Bravo *et al.*, 2005). In recent years, GPI-anchored alkaline phosphatase (ALP) was identified as a Cry toxin-receptor in different insect species (Jurat-Fuentes and Adang, 2004; Fernández *et al.*, 2006). Monomeric Cry1A toxin binds to the cadherin receptor, facilitating additional cleavage of the N-terminal helix α-1 inducing oligomerization of the toxin (Bravo *et al.*, 2005). Cry1A toxin oligomers then bind APN, due to its higher affinity, resulting in membrane insertion and pore formation (**Figure A1**). However, a second model proposed that insect cell death is triggered by the binding of the monomeric Cry1A toxin to cadherin, which in turn activates adenylyl cyclase resulting in increased cell levels of cAMP, that then activates protein kinase-A resulting in oncotic cell death (Zhang *et al.*, 2006). In the signal transduction model, GPI-anchored receptors are involved neither in Cry toxicity nor in oligomer formation (Zhang *et al.*, 2006).

Oligomerization of Cry1Ab toxin was subsequently shown to be an indispensable step in its toxicity to *Manduca sexta*. Site directed mutagenesis of specific amino acids in domain I helix α-3 resulted in the isolation of Cry1Ab mutants that could not form oligomers, still bound cadherin receptor and lost toxicity to *M. sexta* (Jiménez-Juarez *et al.*, 2007). It was also demonstrated that helix α-4 mutants affected in pore-formation have a dominant negative phenotype since they inhibited Cry1Ab toxicity and pore-formation when mixed in low ratios with wild-type Cry1Ab toxin, indicating that toxin hetero-oligomers were not functional (Rodriíguez-Almazan *et al.*, 2009). Overall, these data indicate that Cry toxin oligomerization is an essential step in the mode of action of Cry toxins.

In addition, it was shown in different insect species that GPI-anchored proteins are involved in Cry toxin action. In *M. sexta*, it was shown that an scFv antibody that bound Cry1Ab domain III and inhibited binding to APN but not to cadherin reduced Cry1Ab toxicity in bioassays (Gómez *et al.*, 2006). Furthermore, populations of *Spodoptera exigua* and *Helicoverpa armigera* resistant to Cry1C and Cry1Ac, respectively, showed reduced levels of APN transcripts (Herrero *et al.*, 2005; Zhang *et al.*, 2009). These results demonstrate that binding to GPI-anchored proteins is an important step in the mode of action of Cry toxins.

Finally, it was shown that modified Cry1Ab and Cry1Ac toxins (Cry1AbMod or Cry1AcMod) could skip cadherin binding and retain toxicity (Soberón *et al.*, 2007). It was shown that these Cry1AbMod or Cry1AcMod toxins engineered to lack helix α-1 form oligomers in the absence of cadherin binding (**Figure A1**) and were toxic to a *Pectinophora gossypiella* population that is resistant to these toxins because of a mutation in the cadherin gene. These results show that cadherin binding alone is not sufficient for insect death. Cry1AMod

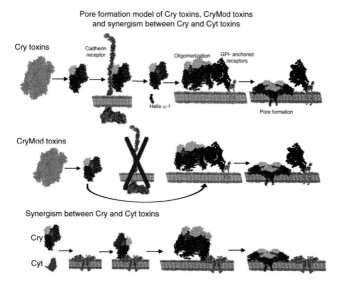

Pore formation model of Cry toxins, CryMod toxins
and synergism between Cry and Cyt toxins

Figure A1 Mode of action of Cry1A toxin. Upper panel, Mode of action of Cry1A toxins. Middle panel, Mode of action of Cry1AMod toxins that skip cadherin binding. Lower panel, Cyt1A functions as Cry11Aa receptor explaining synergism between these toxins. The proteins depicted in the figure are hypothetical structures except for Cry and Cyt1Aa toxins.

toxins will be useful in countering the potential problem of insect resistance in the field (Bravo and Soberón, 2008).

Overall, the results discussed here clearly indicate that toxicity of Cry toxins involves formation of oligomers and binding to GPI-anchored proteins. It remains to be determined whether signal transduction is involved in toxin action but if so, it should be triggered by pore-formation and not by binding to cellular receptors.

A7.2. Conserved Cry Toxin-binding Proteins Among Different Insect Orders

In the past few years, significant progress has been made in the identification of Cry toxin receptors in coleopteran and dipteran species. Cadherin proteins in the coleopterans *Tenebrio molitor* and *Diabrotica virgifera* were identified as Cry3A receptors (Fabrick *et al.*, 2009; Park *et al.*, 2009). In the case of *T. molitor*, it was shown that binding of Cry3Aa to cadherin induced oligomerization of the toxin, while in *D. virgifera* a cadherin fragment containing a Cry3Aa binding region enhanced Cry3Aa toxicity (Fabrick *et al.*, 2009; Park *et al.*, 2009). Regarding GPI-anchored receptors, an ADAM-metalloprotease was identified as a Cry3Aa receptor in *Leptinotarsa decemlineata* (Ochoa-Campuzano *et al.*, 2007).

In the case of mosquitocidal Cry toxins, it is proposed that the mechanism of action is also conserved, since similar Cry-binding molecules have been identified in their midgut cells, such as cadherin of *Anopheles gambiae* and *Aedes aegypti*, that bind Cry4Ba or Cry11Aa, respectively (Hua *et al.*, 2008; Chen *et al.*,

2009a), or GPI-anchored APN in *A. quadrimaculatus*, *A. gambiae*, and *Ae. aegypti* that binds Cry11Ba or Cry11Aa (Abdulla *et al.*, 1996; Zhang *et al.*, 2008; Chen *et al.*, 2009b), and ALP in *Ae. aegypti* or *A. gambiae* that binds Cry11Aa toxin or Cry11Ba, respectively (Fernández *et al.*, 2006; Hua *et al.*, 2009).

A7.3. Other Cry Proteins Used in Insect Control

Besides the three-domain Cry toxins, several Bt strains produce other Cry toxins that are active against different insect pests. In particular, the binary toxin Cry34/Cry35 that has activity against the western corn rootworm (*D. virgifera virgifera*) has been successfully produced in transgenic maize and shown to protect corn roots from insect attack (Moellenbeck *et al.*, 2001).

A7.4. Cyt1Aa Functions as a Surrogate Receptor for Cry11Aa

Bacillus thuringiensis subs *israelensis* (Bti) is used worldwide in the control of mosquitoes that are vectors of important human diseases. Despite the large-scale use of Bti in mosquito control programs, there are no reports of resistance to Bti under field conditions.

The lack of insect resistance development to Bti is mainly due to the fact that this strain produces at least four different crystal proteins that belong to two unrelated family of toxins, namely Cry (Cry4Aa, Cry4Ba, Cry11Aa) and Cyt (Cyt1Aa). Interestingly, the insecticidal activity of the isolated Bti toxins is magnitudes of order less toxic than the crystal inclusion containing all toxins (Bravo *et al.*, 2005). This synergistic effect and the lack of resistance is mainly due to Cyt1Aa that enhances the activity of Cry4Aa, Cry4Ba, or Cry11Aa (Bravo *et al.*, 2005). The mechanism of synergism of Cyt1Aa and Cry11Aa depends on specific interaction between these toxins since single point mutations on Cyt1Aa affected Cry11Aa binding and synergism (Pérez *et al.*, 2005). Also, it was shown that Cry11Aa domain II loop regions involved in receptor binding were also involved in binding and synergism with Cyt1Aa (Pérez *et al.*, 2005). These data led to the hypothesis that Cyt1Aa functions as a membrane-bound receptor for Cry11Aa (Perez *et al.* 2005). Furthermore, binding of Cry11Aa to Cyt1Aa facilitates formation of Cry11Aa oligomers, which were efficient in pore-formation indicating Cyt1Aa fulfils the role of the primary cadherin receptor facilitating oligomer formation (Pérez *et al.*, 2007). This is the first example of an insect pathogenic bacterium that carries a toxin and also its functional receptor, promoting toxin binding to target membranes and toxicity (**Figure A1**).

References

Abdullah, M.A., Valaitis, A.P., Dean, D.H., **1996**. Identification of a *Bacillus thuringiensis* Cry11Ba toxin-binding aminopeptidase from the mosquito. *Anopheles quadrimaculatus. BMC Biochem 22*, 7–16.

Bravo, A., Soberón, M., **2008**. How to cope with resistance to Bt toxins? *Trends Biotechnol. 26*, 573–579.

Bravo, A., Gill, S.S., Soberón, M., **2005**. *Bacillus thuringiensis* mechanisms and use. In: Gilbert, L.I., Iatrou, K., Gill, S.S. (Eds.), Comprehensive Molecluar Insect Science. Elsevier, pp. 175–206 ©2005 Elsevier BV ISBN (Set): 0–44–451516-X.

Chen, J., Aimanova, K.G., Pan, S., Gill, S.S., **2009a**. Identification and characterization of *Aedes aegypti* aminopeptidase N as a putative receptor of *Bacillus thuringiensis* Cry11A toxin. *Insect Biochem Mol Biol. 39*, 688–696.

Chen, J., Aimanova, K.G., Fernandez, L.E., Bravo, A., Soberón, M., Gill, S.S., **2009b**. *Aedes aegypti* cadherin serves as a putative receptor of the Cry11Aa toxin from *Bacillus thuringiensis* subsp. *israelensis. Biochem. J.* doi:10.1042/BJ20090730.

Fabrick, J., Oppert, C., Lorenzen, M.D., Morris, K., Oppert, B., Jurat-Fuentes, J.L., **2009**. A novel Tenebrio molitor cadherin is a functional receptor for Bacillus thuringiensis Cry3Aa toxin. *J. Biol. Chem. 284*, 18401–18410.

Fernández, L.E., Aimaniova, K.G., Gill, S.S., Bravo, A., Soberón, M., **2006**. A GPI-anchored alkaline phosphatase is a functional midgut receptor of Cry11Aa toxin in *Aedes aegypti* larvae. *Biochem. J. 394*, 77–84.

Fernández, L.E., Martinez-Anaya, C., Lira, E., Chen, J., Evans, J., Hernández-Martínez, S., Lanz-Mendoza, H., Bravo, A., Gill, S.S., Soberón, M., **2009**. Cloning and epitope mapping of Cry11Aa-binding sites in the Cry11Aa-receptor alkaline phosphatase from *Aedes aegypti. Biochemistry 48*, 8899–8907.

Gómez, I., Arenas, I., Benitez, I., Miranda-Rííos, J., Becerril, B., Grande, G., Almagro, J.C., Bravo, A., Soberón, M., **2006**. Specific epitopes of Domains II and III of *Bacillus thuringiensis* Cry1Ab toxin involved in the sequential interaction with cadherin and aminopeptidase-N receptors in *Manduca sexta. J. Biol. Chem. 281*, 34032–34039.

Herrero, S., Gechev, T., Bakker, P.L., Moar, W.J., de Maagd, R.A., **2005**. *Bacillus thuringiensis* Cry1Ca-resistant *Spodoptera exigua* lacks expression of one of four Aminopeptidase N genes. *BMC Genomics 24*, 6–96.

Hua, G., Zhang, R., Abdullah, M.A., Adang, M.J., **2008**. *Anopheles gambiae* cadherin AgCad1 binds the Cry4Ba toxin of *Bacillus thuringiensis israelensis* and a fragment of AgCad1 synergizes toxicity. *Biochemistry 47*, 5101–5110.

Hua, G., Zhang, R., Bayyareddy, K., Adang, M.J., **2009**. *Anopheles gambiae* alkaline phosphatase is a functional receptor of *Bacillus thuringiensis jegathesan* Cry11Ba toxin. *Biochemistry 48*, 9785–9793.

Jiménez-Juárez, A., Muñoz-Garay, C., Gómez, I., Saab-Rincon, G., Damian-Alamazo, J.Y., Gill, S.S., Soberón, M., Bravo, A., **2007**. *Bacillus thuringiensis* Cry1Ab mutants affecting oligomer formation are non-toxic to *Manduca sexta* larvae. *J. Biol. Chem. 282*, 21222–21229.

Jurat-Fuentes, J.L., Adang, M.J., **2004**. Characterization of a Cry1Ac-receptor alkaline phosphatase in susceptible and resistant Heliothis virescens larvae. *Eur. J. Biochem. 271*, 3127–3135.

Ochoa-Campuzano, C., Real, M.D., Martiínez-Ramiírez, A.C., Bravo, A., Rausell, C., **2007**. An ADAM metalloprotease is a Cry3Aa *Bacillus thuringiensis* toxin receptor. *Biochem. Biophys. Res. Commun. 362*, 437–442.

Moellenbeck, D.J., Peters, M.L., Bing, J.W., Rouse, J.R., Higgins, L.S., Sims, L., Neveshmal, T., Marshall, L., Ellis, R.T., Bystrak, P.G., Lang, B.A., Stewart, J.L., Kouba, K., Sondag, V., Gustafson, V., Nour, K., Xu, D., Swenson, J., Zhang, J., Czpala, T., Schwab, G., Jayne, S., Dtorckhoff, B.A., Narva, K., Schnepf, H.E., Stelman, S.J., Poutre, C., Koziel, M., Duck, N., **2001**. Insecticidal proteins from *Bacillus thuringiensis* protect corn from corn rootworms. *Nat. Biotechnol. 19*, 668–672.

Park, Y., Abdullah, M.A., Taylor, M.D., Rahman, K., Adang, M.J., **2009**. Enhancement of *Bacillus thuringiensis* Cry3Aa and Cry3Bb toxicities to coleopteran larvae by a toxin-binding fragment of an insect cadherin. *Appl. Environ. Microbiol. 75*, 3086–3092.

Pérez, C., Fernandez, L.E., Sun, J., Folch, J.L., Gill, S.S., Soberón, M., Bravo, A., **2005**. *Bacillus thuringiensis* subsp. *israeliensis* Cyt1Aa synergizes Cry11Aa toxin by functioning as a membrane-bound receptor. *Proc.-Natl. Acad. Sci. 102*, 18303–18308.

Pérez, C., Muñoz-Garay, C.C., Portugal, L., Sánchez, J., Gill, S.S., Soberón, M., Bravo, A., 2007. *Bacillus thuringiensis* subsp. *israelensis* Cyt1Aa enhances activity of Cry11Aa toxin by facilitating the formation of a pre-pore oligomeric structure. *Cell. Microbiol. 9*, 2931–2937.

Rodríguez-Almázan, C.R., Zavala, L.E., Muñoz-Garay, C., Jaménez-Juárez, N., Pacheco, S., Masson, L., Soberón, M., Bravo, A., 2009. Dominant negative mutants of *Bacillus thuringiensis* Cry1Ab toxin function as anti-toxins: demonstration of the role of oligomerization in toxicity. *PLoS ONE 4*(5), e5545. doi:10.1371/journal.pone.0005545.

Soberón, M., Pardo-López, L., López, I., Gómez, I., Tabashnik, B., Bravo, A., 2007. Engineering modified Bt toxins to counter insect resistance. *Science 318*, 1640–1642.

Zhang, R., Hua, G., Andacht, T.M., Adang, M.J., 2008. A 106-kDa aminopeptidase is a putative receptor for *Bacillus thuringiensis* Cry11Ba toxin in the mosquito *Anopheles gambiae*. *Biochemistry 47*, 11263–11272.

Zhang, S., Cheng, H., Gao, Y., Wang, G., Liang, G., Wu, K., 2009. Mutation of an aminopeptidase N gene is associated with *Helicoverpa armigera* resistance to *Bacillus thuringiensis* Cry1Ac toxin. *Insect Biochem. Mol. Biol. 39*, 421–429.

Zhang, X., Candas, M., Griko, N.B., Taussig, R., Bulla, L.A., 2006. A mechanism of cell death involving an adenylyl cyclase/PKA signaling pathway is induced by the Cry1Ab toxin of *Bacillus thuringiensis*. *Proc. Natl. Acad. Sci. 103*, 9897–9902.

8 Mosquitocidal *Bacillus sphaericus*: Toxins, Genetics, Mode of Action, Use, and Resistance Mechanisms

J-F Charles, Institut Pasteur, Paris, France
I Darboux and D Pauron, INRA, Sophia Antipolis, France
C Nielsen-Leroux, Institut Pasteur, Paris, France

© 2010, 2005 Elsevier B.V. All Rights Reserved

8.1. Introduction

8.1.1. Generalities

Bacillus thuringiensis was the first bacterium described to have insecticidal activity. The discovery of a particular strain of *B. thuringiensis* serovar *israelensis* in the late seventies (Goldberg and Margalit, 1977; de Barjac, 1978) opened up the possibility of using microorganisms to control Diptera Nematocera vectors of disease agents, such as mosquitoes (Culicidae) and black-flies (Simuliidae).

The first strains of *B. sphaericus* to be active against mosquito larvae were isolated from moribund mosquito larvae (Kellen *et al.*, 1965). The larvicidal activity of these isolates was very low, so they could not be used to control mosquitoes (**Table 1**). The identification of the SSII-1 strain in India (Singer, 1973) renewed interest in the use of this species as a biological control agent. However, it was only after the isolation of strain 1593 from dead mosquito larvae in Indonesia (Singer, 1974) that the potential of *B. sphaericus* as a real biological control agent for mosquito populations was taken seriously (Singer, 1977). This occurred because of the high level of mosquitocidal activity of this strain.

8.1.2. Comparison of the Properties of Mosquitocidal Strains of *Bacillus sphaericus*

B. sphaericus Meyer and Neide (Neide, 1904) is an aerobic bacterium that produces terminal spherical spores. It lacks a number of biochemical pathways, meaning that it is unable to metabolize sugars. This species is generally considered to be heterogeneous, based on the findings obtained through a variety of genetic and biochemical analyses. Given the large number of strains isolated, it became difficult to distinguish between toxic and atoxic strains, or between the toxic strains themselves.

The relationships between *B. sphaericus* strains have been examined by DNA homology (Krych *et al.*, 1980). Five groups were initially identified but group II has since been further subdivided into IIA (over 79% identity, containing all toxic strains), and IIB. Two further systems were then developed: bacteriophage typing (Yousten *et al.*, 1980; Yousten *et al.*, 1984) and serotyping using flagellar antigens as for *B. thuringiensis* (de Barjac *et al.*, 1985; de Barjac, 1990). Both techniques gave very similar classifications, with the mosquitocidal strains in group IIA. Currently, nine serotypes containing active strains have been identified (**Table 1**). Other

Table 1 Comparative properties of some mosquitocidal strains of *B. sphaericus*[a]

Strain	Origin	Source	Flagellar serotype	Phage type	DNA group	RNA group	Fatty acid group	Mosquitocidal activity[b]	Crystal (Bin) genes[c]	Mtx genes 1	Mtx genes 2	Mtx genes 3
Kellen K, Q	USA	*Culiseta incidens*	1a	1	IIA	?	AIII	Low	–	+	+	+
BS-197	India	?	1a	?	?	?	?	High	+	+	+	+
SSII-1	India	*Culex fatigans*	2a,2b	2	IIA	RIIA	AII	Moderate	–	+	+	+
IAB 881	Ghana	*Culex* sp. breeding site	3	–[d]	?	?	AV	High	1	±[e]	?	?
LP1-G	Singapore	Mosquito breeding site	3	8	IIA	?	?	Moderate	4	+	?	?
1593	Indonesia	*Culex fatigans*	5a,5b	3	IIA	RIIA	AIII	High	2	+	+	?
2362	Nigeria	*Simulium damnosum*	5a,5b	3	IIA	RIIA	AIII	High	2	+	+	+
1691	El Salvador	*Anopheles albimanus*	5a,5b	3	IIA	?	?	High	2	+	+	+
IAB 59	Ghana	*Anopheles gambiae*	6	3	?	?	?	High	1	+	+	+
31	Turkey	Mosquito breeding site	9a,9c	8	IIA	?	?	Low	–	+	+	+
2314-2	Thailand	?	9a,9b	–[d]	?	?	?	Low		+	+	+
2297	Sri Lanka	*Culex pipiens*	25	4	IIA	RIIA	AV	High	3	+	+	+
2173 (ISPC5)	India	*Culex fatigans*	26a,26b	–[d]	IIA	?	AIII	Moderate	–	–	–	–
IAB 872	Ghana	*Culex* sp. breeding site	48	3	?	?	?	High	1	+	+	+

[a]Data compiled from Krych *et al.* (1980), Yousten *et al.* (1980), de Barjac *et al.* (1985), Alexander and Priest (1990), Frachon *et al.* (1991), Guerineau *et al.* (1991), Thanabalu *et al.* (1991), Liu *et al.* (1996a,1996b), Thanabalu and Porter (1995, 1996), Priest *et al.* (1997), Humphreys and Berry (1998), and from data of the IEBC Collection (Pasteur Institute). +: present; –: absent; ?: no information available.

[b]Based on the lethal concentration 50% (LC$_{50}$) on *Culex pipiens* fourth instars.

[c]Based on the deduced amino-acid sequences, when known (see also **Table 2**).

[d]Strains not responding to any of the bacteriophages tested.

[e]– following the original publication (Humphreys and Berry, 1998); + following Z. Yuan, personal communication.

techniques such as numerical classification based on the taxonomy of phenotypic features (Alexander and Priest, 1990; Guerineau *et al.*, 1991), cellular fatty acid composition analysis (Frachon *et al.*, 1991), ribotyping (Aquino de Muro *et al.*, 1992; Miteva *et al.*, 1998), and random amplified polymorphic DNA analysis (Woodburn *et al.*, 1995) showed that most of the pathogenic strains are grouped in a few clusters. None of these methods can be used to predict the level of toxicity of a given strain. Strains can be divided into three groups according to their toxicity, but these groups do not correspond to the groups defined by any other classification method (**Table 1**). The genes encoding a number of toxins have been cloned (see below), and it has been suggested that hybridization experiments could be used to predict activity (Woodburn *et al.*, 1995). However, some strains that harbor toxin genes but display only weak mosquitocidal activity have been reported. Therefore, the only way to identify potentially valuable strains and to increase our knowledge of the real nature of the mosquito-larvicidal activity is to determine toxicity experimentally using mosquito larvae.

B. sphaericus strains are generally highly active against *Anopheles* and *Culex* larvae but are poorly, or are not toxic to *Aedes* larvae. However, susceptibility appears to depend on the species of mosquito considered, and thus can vary within a given genus (Berry *et al.*, 1993). Toxicity levels vary greatly between the larvicidal serotypes and even within the same serotype (Thiéry and de Barjac, 1989). Therefore, the relative potency of each strain is expressed as the specific titer on the different mosquito species and as the activity ratios derived from such titers (Thiéry and de Barjac, 1989).

Singer was the first to suggest that *B. sphaericus* produces toxins, rather than producing septicemia, like *B. thuringiensis* on Lepidoptera (Singer, 1973). A few years later, Davidson and Myers (Davidson and Myers, 1981) observed parasporal inclusions that they believed to participate in the toxic action of *B. sphaericus*. Indeed, all the most toxic strains produce such inclusions during sporulation (**Figure 1**). These inclusions have a crystalline ultrastructure and are released into the medium along with the spore after the completion of sporulation. The relationship between sporulation, crystal formation, and mosquitocidal activity, was first clearly established for strains belonging to serotype H5a,5b (Myers *et al.*, 1979; Broadwell and Baumann, 1986), H25 (Yousten and Davidson, 1982; de Barjac and Charles, 1983; Kalfon *et al.*, 1984), and H6 (de Barjac *et al.*, 1988). Later on, the partial purification of crystals toxic to mosquito larvae (Payne

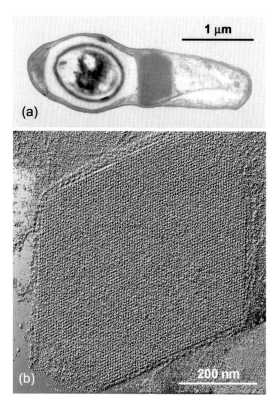

Figure 1 (a) *B. sphaericus* strain 2297 spore and crystal during the lysis of the sporangium, at the end of the sporulation process. (b) Detail of the crystal lattice.

and Davidson, 1984) and studies showing mutants that failed to make crystals were not mosquitocidal (Charles *et al.*, 1988), confirmed the toxic nature and protein composition of the crystals. More recently, toxins different to the crystal toxins (Mtx toxins) were identified, thus renewing interest in *B. sphaericus* as a mosquitocidal agent (Thanabalu *et al.*, 1991; Thanabalu and Porter, 1995).

There are three reviews summarizing *B. sphaericus* genetics, toxins and mode of action (Baumann *et al.*, 1991; Porter *et al.*, 1993; Charles *et al.*, 1996). The increasing field use of *B. sphaericus* has led to some cases of resistance. It is important to understand the nature and mode of action of *B. sphaericus* toxins and the mechanisms of resistance, if we are to restrict the development of resistance. These issues have been extensively investigated and an overview is given below.

8.2. Biochemistry and Genetics of *B. sphaericus* Toxins

At least two kinds of toxin (crystal toxins, or Bin toxins, and Mtx toxins) seem to account for the larvicidal activity of *B. sphaericus*. They differ both in their composition and time of synthesis. The crystal toxins are present in all highly active strains (**Table 1**) and are produced during sporulation. The Mtx toxins are responsible for the activity of most of the less active strains, and seem to be synthesized only during the vegetative phase.

8.2.1. Crystal Toxins

The crystal toxin is composed of a 42-kDa protein and a 51-kDa protein, hereafter designated BinA and BinB, respectively (Baumann _et al._, 1987; Berry and Hindley, 1987; Hindley and Berry, 1987; Arapinis _et al._, 1988; Baumann _et al._, 1988; Hindley and Berry, 1988; Berry _et al._, 1989). These two proteins are synthesized in approximately equimolar amounts, assembled in crystal structures and visible at about stage III of sporulation (Yousten and Davidson, 1982; Kalfon _et al._, 1984; Baumann _et al._, 1985). The genes encoding both proteins have been cloned from several highly toxic strains. They encode proteins with predicted molecular masses of 51.4 and 41.9 kDa (Baumann _et al._, 1987; Berry and Hindley, 1987; Hindley and Berry, 1987; Arapinis

_et al._, 1988; Baumann _et al._, 1988; Hindley and Berry, 1988; Berry _et al._, 1989). They appear to be organized in an operon, with a 174- to 176-bp intergenic region. A stem-loop structure (characteristic of transcription terminators) has been identified downstream from the _binA_ gene (**Figure 2**, top; Hindley and Berry, 1987; Baumann _et al._, 1988). No sequences similar to _B. subtilis_ sporulation promoters have been found upstream from the _binB_ gene. However, both genes are expressed only during sporulation in _B. subtilis_ (Baumann and Baumann, 1989). Moreover, _lacZ_ fusion experiments have shown that transcription begins immediately before the end of exponential growth and continues into stationary phase in _B. sphaericus_ (Ahmed _et al._, 1995). This probably allows sufficient amounts of the proteins for crystal formation to accumulate by stage III.

In mutants of _B. sphaericus_ 2362 strain in which _spo0A_ or _spoIIAC_ sporulation genes have been disrupted by insertional mutations, toxicity on mosquito larvae is greatly reduced and no crystal

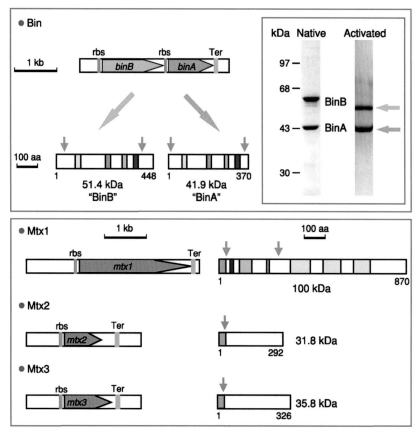

Figure 2   _B. sphaericus_ toxin genes and schematic representation of corresponding polypeptides. Top, crystal (Bin) toxin: regions shared by BinB and BinA are represented by identical colors. Inset: SDS-PAGE (PolyAcrylamide Gel Electrophoresis) of native toxin and of toxin after activation by trypsin or by mosquito midgut juice. Bottom, Mtx toxins: the orange boxes indicate potential signal sequences; the colors in Mtx1 represent the potential transmembrane sequence (red), the two regions shared with ADP-ribosyltransferase toxins (green), and internal repeated sequences (yellow). Tryptic cleavage sites for each toxin are represented by gray arrows. Potential ribosome binding sites (rbs) and terminator sequences (Ter) are also indicated.

proteins are detected by SDS-PAGE or Western blotting (El-Bendary *et al.*, 2004). These results indicate that crystal protein synthesis is dependent on early and late mother cell sigma factors, as is the synthesis of most crystal proteins in *B. thuringiensis*.

In all the studied strains, toxin genes are chromosomal, although these strains contain plasmids (Aquino de Muro *et al.*, 1992). Both *binA* and *binB* have been identified in a wide variety of strains, and DNA hybridization has shown that all highly toxic strains tested contain very similar sequences (Aquino de Muro and Priest, 1994). The amino acid sequences of BinB and BinA are not similar to those of dipteran active toxins produced by *B. thuringiensis* serovar *israelensis* (*Bti*), or to any other toxins. However, BinB and BinA share four segments of sequence similarity (**Figure 2**, top; Baumann *et al.*, 1988), the significance of which remains unclear. Nevertheless, this indicates that BinB and BinA of *B. sphaericus* constitute a separate family of insecticidal toxins (Baumann *et al.*, 1988).

The toxin coding and flanking regions of strains 1593, 2362 and 2317.3 are identical over a span of 3479 nucleotides, whereas the sequences in strains IAB-59 and 2297 differ by 7 and 25 nucleotides, respectively (Berry *et al.*, 1989). These differences result in 5 and 3 amino acid differences in BinB, respectively, and in 1 and 5 amino acid differences in BinA, respectively. Partial or total sequencing of toxin genes led to the establishment of a classification system including four types of Bin toxin (types 1 to 4) (**Table 2**) (Priest *et al.*, 1997; Humphreys and Berry, 1998). The amino acid differences at positions 99, 135 (hydrophobic) and 267 (basic) of BinA are conservative. Type 1 and 2 BinA have a His at position 125, whereas types 3 and 4 contain an Asn. Type I BinA contains an acidic amino acid at position 104, whereas the other three types do not. Interestingly, BinA from type 4 (from strain LP1-G, only weakly active on mosquito larvae) has a serine (polar) at position 93 whereas types 1 to 3 contain a hydrophobic residue (Leu) (**Table 2**). Replacement of the serine at position 93 of BinA from LP1-G with a leucine restored the toxicity of the mutated type 4 toxin, whereas the replacement of the leucine with a serine at position 93 of the type 2 BinA from strain 1593 led to a significant loss in toxicity (Yuan *et al.*, 2001). This confirms that the amino acid at position 93 plays a key role in the toxicity of the BinA part of Bin toxins.

Amino acid residue substitutions at selected sites in the N- and C-terminal regions of type 2 BinA revealed the importance of other amino acids at the C-terminal end of the BinA peptide. For example, replacement of the Arg residue at position 312 by lysine or histidine (positively charged amino acids) completely destroys the biological activity of the binary toxin on *C. quinquefasciatus*, even though the binding of these mutants to mosquito brush border membranes is not altered (Elangovan *et al.*, 2000).

Table 2 Nomenclature and comparison of the four known Bin amino acid sequences from various *B. sphaericus* strains

Polypeptide	Amino acid position	Type 1 IAB-59[a], IAB-872[b] IAB-881[b], 9002[b]	Type 2 1593[a], 2362[a], 2317.3[a] BSE18[c], C3-41[d]	Type 3 2297[a]	Type 4 LP1-G[b]
BinA (41.9 kDa)		BinA1	BinA2	BinA3	BinA4
	93	Leu	Leu	Leu	Ser
	99	Val	Val	Phe	Val
	104	Glu	Ala	Ser	Ser
	125	His	His	Asn	Asn
	135	Tyr	Tyr	Phe	Phe
	267	Arg	Arg	Lys	Lys
BinB (51.4 kDa)		BinB1	BinB2	BinB3	BinB4
	69	Ala	Ser	Ser	Ser
	70	Lys	Asn	Asn	Asn
	110	Ile	Thr	Thr	Ile
	314	His	Leu	Tyr	His
	317	Leu	Phe	Leu	Leu
	389	Leu	Leu	Met	Met

[a]Berry *et al.* (1989).
[b]Humphreys and Berry (1998), Priest *et al.* (1997).
[c]Berry (personal communication).
[d]Yuan *et al.* (1998).

The replacement of alanine at some sites in the N- and C-terminal regions of both the BinA and BinB peptides results in the total loss of mosquitocidal activity. Surprisingly, toxicity is restored by mixing two nontoxic derivatives of the same peptide; *i.e.*, one mutated at the N-terminal end and the other mutated at the C-terminal end of either BinA or BinB (Shanmugavelu *et al.*, 1998). Thus, the altered binary toxins can functionally complement each other by forming oligomers.

The aggregation of both BinB and BinA has been analyzed by expressing crystal toxin components separately or together in homologous or heterologous *Bacillus* hosts. The proteins form amorphous inclusions when expressed independently in *B. subtilis* and in *B. sphaericus* or *B. thuringiensis* crystal-negative hosts (Charles *et al.*, 1993; Nicolas *et al.*, 1993). In contrast, crystals similar to those produced by natural *B. sphaericus* strains are found when the two genes are simultaneously expressed in either *B. sphaericus* or *B. thuringiensis* (Charles *et al.*, 1993; Nicolas *et al.*, 1993). No crystals can be detected in *B. subtilis* when both genes are present, unless they are fused, eliminating the intergenic region (Charles *et al.*, 1993). These results suggest that *B. sphaericus* and *B. thuringiensis*, but not *B. subtilis*, contain factors that help stabilize their protein and subsequent crystallization.

In vivo, BinA is slowly converted into a stable ~39-kDa protein, whereas BinB is rapidly converted into a stable ~43-kDa fragment (Baumann *et al.*, 1985; Broadwell and Baumann, 1987). *In vitro* deletion analysis was used to delineate the minimal active fragments of both proteins, which indeed correspond to the activated fragments (Broadwell *et al.*, 1990b; Clark and Baumann, 1990; Oei *et al.*, 1990; Sebo *et al.*, 1990). Thirty-two and 53 amino acids, at the N- and C-termini of BinB, respectively, can be eliminated without loss of toxicity (Clark and Baumann, 1990); deletions of 10 and 17 amino acids, at the N- and C-terminus of BinA, respectively, result in a protein similar to the 39-kDa activated fragment (**Figure 2**, top; Broadwell *et al.*, 1990b).

Sub-cloning experiments have shown that BinA alone is toxic for mosquito larvae (*C. pipiens*), although the activity is weaker than that of crystals containing both proteins (Nicolas *et al.*, 1993). In contrast, BinB alone is not toxic, although its presence enhances the larvicidal activity of BinA, suggesting synergy between the two polypeptides (de la Torre *et al.*, 1989; Broadwell *et al.*, 1990b; Oei *et al.*, 1990; Nicolas *et al.*, 1993). *In vitro* assays confirmed that the activated form of BinA alone is toxic for *C. quinquefasciatus* cells, whereas BinB appears to be inactive; however, no synergy

between the components was observed *in vitro* (Baumann and Baumann, 1991). Although the simultaneous presence of both proteins appears necessary for full toxicity, the differing activities of the different *B. sphaericus* strains towards various mosquito species depends on the origin of BinA, as shown by analysis of *in vitro* mutated toxins (Berry *et al.*, 1993). Indeed, when amino acids were substituted in a region centered around position 100 of BinA, rendering BinA from strains IAB-59 and 2297 similar to that from strain 2362, the activity and specificity of these mutant toxins towards *Culex* and *Aedes* larvae was comparable, unlike the wild-type, indicating that this region is involved in specificity. Taken together, these observations suggest that BinA is the most important determinant of specificity and activity.

8.2.2. Mtx Toxins

Three types of Mtx toxin have been described to date: Mtx1, Mtx2 and Mtx3, with molecular weights of 97, 31.8 and 35.8 kDa, respectively (**Figure 2**, bottom). The genes encoding these proteins were initially cloned from the SSII-1 strain (Thanabalu *et al.*, 1991; Thanabalu and Porter, 1995; Liu *et al.*, 1996a), which has a moderate mosquitocidal activity (**Table 1**). In contrast to the crystal toxin genes, *mtx* genes are expressed during the vegetative growth phase, and sequences resembling vegetative promoters from *B. subtilis* have been found upstream from each gene (Thanabalu *et al.*, 1991; Thanabalu and Porter, 1995; Liu *et al.*, 1996a). *lacZ* fusion experiments confirmed these findings (Ahmed *et al.*, 1995). These proteins possess short N-terminal leader sequences characteristic of Gram-positive bacterial signal peptides (Thanabalu *et al.*, 1991; Thanabalu and Porter, 1995; Thanabalu and Porter, 1996). However, these toxins have been found associated with the cell membrane of *B. sphaericus*, indicating little or no cleavage of the signal sequence. The mature Mtx1 toxin can be further processed by gut proteases, leading to two fragments of 27 and 70 kDa, corresponding to the N- and C-terminal regions, respectively (Thanabalu *et al.*, 1992). The 70-kDa fragment possesses three repeated regions of about 90 amino acids (**Figure 2**, bottom), the function of which is unknown. The 27-kDa fragment contains a short putative transmembrane domain. These toxins do not display any similarity with each other or with crystal proteins or any other insecticidal proteins. In contrast, the 27-kDa fragment shares weak sequence similarities with the catalytic domains of various bacterial ADP-ribosyltransferases (**Figure 2**, bottom; Thanabalu *et al.*, 1991; Thanabalu *et al.*, 1992). Mtx2 and Mtx3 show similarities to two toxins active against mammalian cells; namely the epsilon-toxin from *Clostridium perfringens* and the cytotoxin from *Pseudomonas aeruginosa*, respectively (Thanabalu and Porter, 1995; Liu *et al.*, 1996a). The genes encoding Mtx toxins are found in numerous *B. sphaericus* strains (Liu *et al.*, 1993; Thanabalu and Porter, 1995; Liu *et al.*, 1996a; Thanabalu and Porter, 1996), including low, moderate and high toxicity strains (see **Table 1**).

8.3. Mode of Action of *B. sphaericus* Toxins

8.3.1. The Crystal Toxins

The mode of action of *B. sphaericus* crystal toxins has only been studied in mosquito larvae. However,

there is one report of activity in adult *C. quinquefasciatus* after introduction by enema into the midgut, whereas no effect was recorded for *A. aegypti* adults (Stray *et al.*, 1988). Following ingestion of the spore/crystal complex by mosquito larvae, the protein crystal matrix is quickly dissolved in the lumen of the anterior stomach (Yousten and Davidson, 1982; Charles and Nicolas, 1986; Charles, 1987) by the combined action of midgut proteinases and of the high pH (Dadd, 1975; Charles and de Barjac, 1981). The toxin is released from *B. sphaericus* crystals in all species, even in the midgut lumen of nonsusceptible species such as *A. aegypti* (**Figure 3a**). Indeed, a number of studies have shown that the differences in susceptibility to *B. sphaericus* between mosquito species are not due to differences in the solubility and/or activation of the binary toxin (Baumann *et al.*, 1985; Davidson *et al.*, 1987; Nicolas *et al.*, 1990).

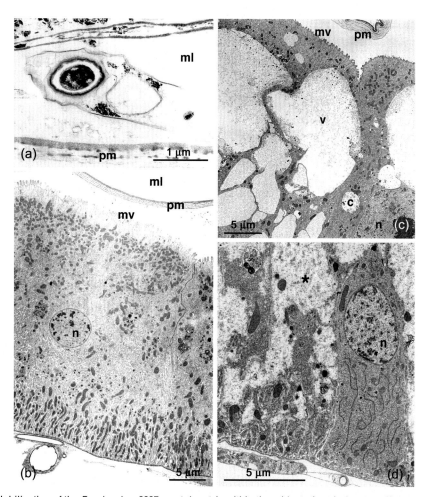

Figure 3 (a) Solubilization of the *B. sphaericus* 2297 crystal matrix within the midgut of an *Aedes aegypti* larva, under the combined action of proteinases and alkaline pH: only the envelopes initially surrounding the crystal are still visible. (b) Control midgut cell of an untreated *A. stephensi* larva. (c) Midgut cell of *C. pipiens* after feeding on *B. sphaericus*. (d) Midgut cell of *A. stephensi* after feeding on *B. sphaericus*. c, cytolysosome; ml, midgut lumen; mv, micovilli; n, nucleus; pm, peritrophic membrane; v, vacuoles; arrows indicate cytolysosomes; the star indicate areas of low electron density.

8.3.1.1.   Cytopathology   and   physiological effects   *Bti* toxins completely breakdown the larval midgut epithelium, whereas *B. sphaericus* does not. Nevertheless, midgut alterations start within 15 min of feeding on *B. sphaericus* spore/crystal complex (Davidson, 1981; Karch and Coz, 1983; Charles, 1987; Singh and Gill, 1988). There is no difference between midgut damage in *C. pipiens* after ingestion of spore/crystals from strains 1593 or 2297 (Davidson, 1981; Charles, 1987; Singh and Gill, 1988), even though these two strains belong to different Bin types. In contrast, symptoms of intoxication are not identical in different mosquito species. Large vacuoles (and/or cytolysosomes) appear in *C. pipiens* midgut cells, whereas large areas of low electron density appear in *Anopheles stephensi* midgut cells (compare **Figure 3c** and **d** with the control, **Figure 3b**). Mitochondrial swelling is a general feature described for *C. pipiens* and *A. stephensi*, as well as for *A. aegypti*, when intoxicated with a very high dose of spore/crystals (Charles, 1987). The midgut cells, especially cells of the posterior stomach and the gastric caecae, are the most severely damaged by the toxin, and Singh and Gill (Singh and Gill, 1988) also reported late damage to neural tissue and skeletal muscles.

Ultrastructural effects have also been observed in cultured cells of *C. quinquefasciatus* within few minutes of treatment with soluble and activated *B. sphaericus* toxin. These alterations consisted mainly in the swelling of mitochondria cristae and endoplasmic reticula, followed by the enlargement of vacuoles and the condensation of the mitochondrial matrix (Davidson *et al.*, 1987).

The physiological effects of *B. sphaericus* crystal toxin have been very poorly documented. The oxygen uptake of mitochondria isolated from *B. sphaericus*-treated *C. quinquefasciatus* larvae is inhibited, as is the activity of larval choline acetyl transferase (Narasu and Gopinathan, 1988). In addition, oxygen uptake by mitochondria isolated from rat liver is reduced by the *B. sphaericus* toxin (Narasu and Gopinathan, 1988).

8.3.1.2. Binding to a specific receptor in the brush border membrane   As the differences in susceptibility between mosquito species do not seem to be due to the ability to solubilize and/or to activate the binary toxin, the variation in susceptibility is presumably due to cellular differences. Indeed, fluorescently-labeled toxin binds to the gastric caecae and the posterior stomach only in very susceptible *Culex* species. BinB does not bind to the gut of *Aedes aegypti*, whereas BinA weakly and nonspecifically binds in this species, and no regionalized

binding occurs for *Anopheles* (Davidson, 1988; Davidson, 1989; Davidson and Yousten, 1990; Oei *et al.*, 1992). Furthermore, in *C. quinquefasciatus*, only BinB binds to the caecae and posterior stomach, and the binding of BinA appears to be conditioned by the binding of the BinB (**Figure 4b**), whereas BinA alone binds nonspecifically throughout the midgut (**Figure 4a**). This indicates that only one of the two components of the crystal toxin (putatively BinB) specifically binds to the midgut of susceptible larvae. The hypothesis that the N-terminal region of BinB is involved in regional binding of this protein in the larval midgut, and that the C-terminal region of BinB, as well as both the N- and C-terminal regions of BinA are involved in the interaction of the two components leading to the binding of BinA in the same regions as BinB, was supported by early *in vivo* binding studies using nontoxic deletion mutants of the crystal toxin (Oei *et al.*, 1992). The same authors also showed that the Bin toxin is only internalized when both components are present. However, further intracellular investigations are required to determine whether one or both components are internalized.

In vitro binding assays, using ^{125}I-labeled activated crystal toxin and brush border membrane fractions (BBMFs) isolated from susceptible or nonsusceptible mosquito larvae, confirmed this hypothesis. Indeed, direct binding experiments with *C. pipiens* BBMFs indicated that the toxin binds to a single class of specific receptor (**Figure 5**, left; Nielsen-LeRoux and Charles, 1992). No significant specific binding was detected with BBMFs from *A. aegypti* (**Figure 5**, right), which is consistent

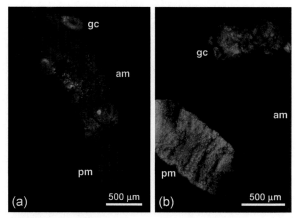

Figure 4   (a) Midgut of *C. quinquefasciatus* larvae only fed with fluorescently-labeled BinA. BinA weakly binds over the entire midgut. (b) Larvae fed with fluorescently-labeled BinA and unlabeled BinB; BinA binding is restricted to the gastric caecae and posterior stomach. am, anterior stomach; gc, gastric caeca; pm, posterior stomach. (Micrographs kindly provided by Dr C. Berry, Department of Biochemistry, Cardiff.)

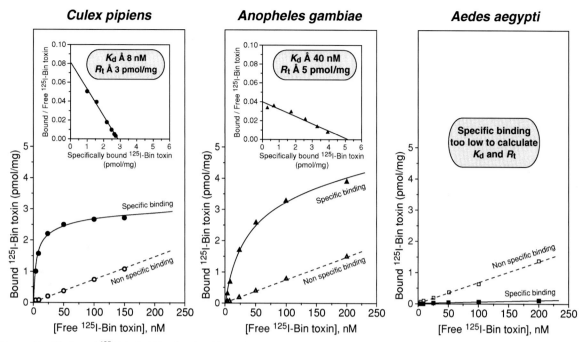

Figure 5 Binding of ^{125}I-labeled *B. sphaericus* Bin toxin to midgut brush border membranes of *C. pipiens*, *A. gambiae* and *A. aegypti*. Insets, specific binding in Scatchard plots. Values of dissociation constant (K_d) and receptor concentrations (R_t) are given when known.

with the lack of specific binding in fluorescence-labeling studies (Davidson, 1989; Davidson *et al.*, 1990). Similar results were obtained with other Culicidae species such as *Anopheles gambiae*, with minor differences in the values of the toxin/receptor binding characteristics: dissociation constants (K_d) were ~8 nM for *C. pipiens* and ~40 nM for *A. gambiae*, and receptor concentrations (R_t) were about 3 and 5 pmol mg^{-1} of BBMF protein respectively (**Figure 5**, center; Silva-Filha *et al.*, 1997). Both crystal toxin components bound to the membranes of all susceptible species, but the linearity of the Scatchard plots clearly indicated that only one of the components specifically bound to the receptor (**Figure 5**, insets).

Binding studies in which radiolabeled BinA or BinB were exposed separately to BBMFs confirmed that only BinB was able to bind specifically to *C. pipiens* midgut membranes, but that BinA alone was also able to bind midgut membranes from *A. gambiae* (Charles *et al.*, 1997).

8.3.1.3. Permeabilization of artificial lipid membranes Complementary studies using artificial lipid vesicles have been carried out to investigate further the respective roles of BinA and BinB in the toxic effect of the crystal toxin. This revealed that the whole Bin toxin and its individual components (BinA and BinB) are able to permeabilize

receptor-free, large unilamellar phospholipid vesicles (LUVs) and planar lipid bilayers (PLBs) by forming pores (Schwartz *et al.*, 2001). Calcein-release experiments showed that LUV permeabilization is optimal at alkaline pH (which corresponds to the pH inside the mosquito larval midgut *in vivo*) and in the presence of acidic lipids. BinA is more efficient than BinB, and BinB facilitates the BinA effect. Stoichiometric (1:1) mixtures are more effective than the full Bin toxin, consistent with previous *in vivo* bioassay observations (Broadwell *et al.*, 1990a; Oei *et al.*, 1990; Baumann and Baumann, 1991; Nicolas *et al.*, 1993). In PLBs, BinA forms voltage-dependent channels of 100–200 pS with long open times and a high open probability. Larger channels (~400 pS) have also been observed. BinB, which inserts less easily, forms smaller channels (~100 pS) with shorter mean open times. The channels formed after the sequential addition of the two components, or when they are mixed in equal amounts (w/w), display BinA-like activity, supporting the hypothesis that BinA is principally responsible for pore formation in lipid membranes whereas BinB, the binding component of the toxin, promotes channel activity (Schwartz *et al.*, 2001).

8.3.1.4. Identification and cloning of the Bin toxin receptor Two important biochemical features have made it possible to purify the Bin toxin receptor: (1) the availability of an iodinated derivative of the toxin that strongly binds to its receptor both in intact and solubilized membranes prepared from susceptible *Culex* mosquitoes (Nielsen-LeRoux and Charles, 1992); (2) the availability of midgut brush border membrane fractions (BBMFs) prepared from 4th instar larvae, constituting a rich source of receptors (Silva-Filha *et al.*, 1999). The Bin toxin receptor can be purified from BBMFs prepared from the midguts of larvae of a susceptible laboratory strain of *C. pipiens* (IP) by means of a classical protocol including solubilization with CHAPS plus anion exchange chromatography (Silva-Filha *et al.*, 1999). This receptor is a protein of approximately 60 kDa that is thought to be linked to the plasma membrane by a glycosylphosphatidylinositol (GPI) moiety, given the effect of phosphatidylinositol-specific phospholipase C (PI-PLC). This protein displays α-amylase-like activities, i.e., it has α-glucosidase (or maltase) and α-amylase *stricto sensu* activities. Moreover, partial microsequencing of the 60 kDa protein revealed significant similarities with identified insect maltases. Hence, this protein was named Cpm1 for *Culex pipiens* maltase 1 (Silva-Filha *et al.*, 1999).

The full-length *cpm1* cDNA was first cloned from the IP strain (Darboux *et al.*, 2001). The coding sequence is 1740 base pair-long and corresponds to a 580-amino acid protein (with a theoretical molecular weight of 66 kDa). It contains the tryptic peptides previously identified (Silva-Filha *et al.*, 1999) and peculiar structural features on both ends of the molecule. A hydrophobic signal peptide located at the N-terminal end of the protein targets the protein to the cell membrane. The C-terminal end contains a domain composed of three small amino acids (Ser^{558}-Ser^{559}-Ala^{560}) followed by hydrophilic residues and then a hydrophobic tail. These sequences fulfill the criteria for posttranslational maturation by linkage of a GPI moiety (Udenfriend and Kodukula, 1995; Eisenhaber *et al.*, 1998). Multiple alignments showed that this sequence shares a high level of similarity with maltase-like proteins from insects, especially over the first two-thirds of the molecule that are thought to bear the enzymatic activity responsible for hydrolyzing the terminal (α-1,4-linked glucose residues of various carbohydrates. Northern blotting experiments identified a single transcript of about 2 kb in the midgut of *Culex* larvae (Darboux *et al.*, 2001).

Cpm1 is not fully processed when expressed in *Escherichia coli*, but it is able to bind the Bin toxin specifically and has weak but significant α-glucosidase activity by hydrolyzing 4-nitrophenyl α-D-glucopyranoside. Much more conclusive data

were obtained when *cpm1* cDNA was cloned in the insect Sf9 cell line. These cells are derived from pupal ovarian tissue originating from the Lepidoptera *Spodoptera frugiperda*. They have the enzymatic equipment required to mature Cpm1 by adding the GPI moiety (Darboux *et al.*, 2002). Cpm1 can easily be detected at the surface of Sf9-IP cells, which express the receptor. Western blotting on the expression product of the *cpm1*/Sf9 construct, using a polyclonal antibody directed against the C-terminal end of the molecule (Darboux *et al.*, 2002), detects only one protein of about 66 kDa in the membrane fraction of transfected cells. Moreover, when transfected cells are treated with PI-PLC, no signal is observed in the membrane compartment whereas a strong signal is observed in the soluble fraction, indicating that the Cpm1 produced in Sf9 cells is indeed attached to the cell membranes by a GPI (Darboux *et al.*, 2002). The number of GPI-anchored proteins found in vertebrates and in invertebrates is increasing rapidly. They have numerous physiological roles (enzymatic activities, cell adhesion, transport and cell–cell signaling), but seem to be present in specialized micro-domains of the plasma membranes known as lipid rafts. Many studies have concentrated on the role of these rafts in cell trafficking and signal transduction (Simons and Toomre, 2000). When cells are transfected with a truncated *cpm1* sequence lacking its 3′ end (IP-Mut), no Cpm1 is found in the membrane fraction. Instead, Cpm1 is secreted and recovered in the culture medium. This secreted form exhibits strong α-glucosidase activity but little α-amylase activity compared to untransfected Sf9 cells (Darboux *et al.*, 2002). Binding experiments on *cpm1*/Sf9 cell membrane preparations showed that the native receptor is indeed expressed as the affinity of Bin for its site is very similar to that reported for *in vitro* experiments done on *Culex* BBMFs (about 10 nM). An anti-Cpm1 antibody was used to localize receptor expression sites in mosquito tissues. In the midguts of 4th instar larvae, a strong signal was detected both in the gastric caecum and in the posterior midgut, with a clear cutoff between the intermediate and posterior midguts (Darboux *et al.*, 2002). These data confirmed the localization of the Bin toxin binding site, which had been previously reported by use of fluorescently labeled derivatives (Davidson, 1988; Oei *et al.*, 1992).

8.3.2. The Mtx1 Toxin

Almost nothing is known about the mode of action of Mtx toxins other than Mtx1, because few studies have been carried out. The mosquitocidal activity of

Mtx1 was initially measured using fusion proteins synthesized in *E. coli*: Mtx1 is highly active against *C. quinquefasciatus* larvae (Thanabalu *et al.*, 1992), as the LC_{50} value is comparable to that of the crystal Bin proteins (15 ng/ml). Deletion analysis suggested that the 27 kDa fragment of Mtx1 is able to self ADP-ribosylate, whereas the 70 kDa fragment is assumed to play a role in the toxicity to cultured *C. quinquefasciatus* cells; however, both regions are necessary for toxicity to mosquito larvae (Thanabalu *et al.*, 1993).

Due to its cytotoxicity towards bacterial cells, the 27 kDa enzyme fragment cannot be produced alone, in the activated form, in *E. coli* expression systems. However, a nontoxic 32 kDa N-terminal truncated version of Mtx1 has been successfully expressed in *E. coli* and subsequently cleaved to an active 27 kDa enzyme fragment (Schirmer *et al.*, 2002b). *In vitro*, this 27 kDa fragment of Mtx1 is able to ADP-ribosylate numerous proteins in *E. coli* lysates, especially a 45 kDa protein identified as the *E. coli* elongation factor, Tu (EF-Tu). The inactivation of EF-Tu by Mtx1-mediated ADP-ribosylation and the resulting inhibition of bacterial protein synthesis probably play important roles in the cytotoxicity of the 27 kDa enzyme fragment of Mtx1 towards *E. coli* (Schirmer *et al.*, 2002b). Glu^{197} of the 27 kDa enzyme component was identified as the "catalytic" glutamate that is conserved in all ADP-ribosyltransferases (Schirmer *et al.*, 2002a). Transfection of mammalian HeLa cells with a vector encoding the 27 kDa fragment fused with a green fluorescent protein leads to cytotoxic effects characterized by cell rounding and formation of filopodia-like protrusions (Schirmer *et al.*, 2002a).

In vitro binding assays using isolated BBMFs from *C. pipiens* midguts indicate the presence of a membrane receptor to the 70-kDa fragment. The affinity of this receptor/toxin interaction is low (K_d ~1.4 µM), indicating that the 70-kDa component has a very low affinity for its binding site (**Figure 6a**; Charles and Berry, unpublished data). In addition, saturation of this binding site was not possible. Complementary competition experiments, using either the labeled P70 or the labeled Bin toxin as the labeled component, and unlabeled P70 or Bin toxin as the unlabeled competitors, showed that the Bin toxin can compete with the labeled P70 component, whereas the opposite is not true (**Figure 6b** and **c**). This suggests that the P70-Mtx1 component shares a common binding site with the Bin toxin, and that the Bin toxin can also bind to this binding site with a very low affinity.

After cleavage of Mtx1 by chymotrypsin-like proteases, the P70 proteolytic fragment of Mtx1 remains noncovalently bound to the N-terminal 27-kDa fragment (Schirmer *et al.*, 2002a). Thus, from a functional point of view, the binding of the P70 component of the toxin to a membrane receptor may play a key role in carrying the 27-kDa enzymatic fragment to its target site in mosquito midgut cells.

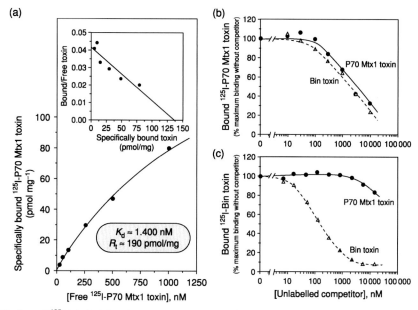

Figure 6 Direct binding of ^{125}I-labeled *B. sphaericus* P70-Mtx1 toxin to midgut brush border membranes of *C. pipiens* (a). Inset, specific binding in Scatchard plots. Competition experiments between ^{125}I-P70 and unlabeled P70 or Bin toxin (b), or between ^{125}I-Bin toxin and unlabeled P70 or Bin toxin (c).

8.4. Field Use of *B. sphaericus*

Due to the special properties of mosquitocidal *Bti* and *B. sphaericus* (e.g., environmental safety, high specificity of *B. sphaericus* and *Bti* toxins, relative ease of mass production, formulation and application, suitability for integrated control programs, and relatively low cost for development and registration), these bacilli were rapidly developed and used in many mosquito and/or black fly control programs. *Bti* has a broader application because unlike *B. sphaericus* it also has larvicidal activity against important vectors belonging to the *Aedes* and *Simulium* genera. *B. sphaericus* has been used to control *C. pipiens* and *C. quinquefasciatus* larvae since the late 1980s, and it is also used to control *Anopheles* spp. in some areas (**Table 3**; Karch *et al.*, 1992; Hougard *et al.*, 1993; Kumar *et al.*, 1994). Even though the larvicidal activity of *B. sphaericus* is mainly restricted to *Culex* and *Anopheles* species, it is stable, can recycle in organically polluted water (major breeding sites of *Culex* species), and is highly persistent (Lacey *et al.*, 1987; Correa and Yousten, 1995), giving it an advantage over *Bti*. Both *B. sphaericus* and *Bti* overcome the resistance of *Culex* and *Anopheles* to conventional insecticides.

In temperate regions, *B. sphaericus* is mainly used to reduce nuisance due to mosquito bites. It has been used on a large scale in Germany, especially in the Rhine Valley (Becker, 2000), and in France, where it was used to control *C. pipiens* (resistant to the organophosphate insecticide temephos) on the Mediterranean coast for more than 7 years before any problems of resistance arose (Sinègre *et al.*, 1993). However, the main interest (amount of product used and treated areas) is in tropical regions where *B. sphaericus* is used to control *Culex* and *Anopheles* populations, particularly to reduce the incidence of vector-born diseases like filariasis and malaria. In Goa (India), the use of *B. sphaericus* against *A. stephensi* larvae resulted in a notable reduction in the number of cases of malaria (Kumar *et al.*, 1994). This was also the case in China against *A. sinensis* (Becker, 2000; Yuan *et al.*, 2000). In the Democratic Republic of Congo, successful experimental field trials were run against *A. gambiae* in rice fields (Karch *et al.*, 1992). *C. quinquefasciatus* populations were reduced in the town of Maroua in Cameroon (Barbazan *et al.*, 1997). In the city of Recife, in Brazil, the experimental application of *B. sphaericus* successfully reduced the number of cases of filariasis by reducing the adult vector population of *C. quinquefasciatus* (Regis *et al.*, 1995).

Table 3 Examples of large-scale field use of *Bacillus sphaericus* against mosquito populations and the development of resistance

Target species	*B. sphaericus* strain	Treatment frequency	Treatment period	Treated area	Locality/country	Resistance RR [a]	References
A. stephensi	B-101	Weekly	9 months	2 km²	Panaji city (India)	–	Kumar *et al.* (1996)
C. pipiens	2362	Every 21 days to 6 months	7 years	210 villages	Mediterranean Coast (France)	>20 000 [b]	Sinègre *et al.* (1993, 1994)
C. quinquefasciatus	2362	Monthly/fortnightly	26 months	1.2 km²	Recife city (Brazil)	10	Silva–Filha *et al.* (1995)
C. quinquefasciatus	1593M	Fortnightly	2 years	8 km²	Kochi (India)	146	Rao *et al.* (1995)
C. quinquefasciatus	C3-41	3 times a month	8 years	8 km²	Dongguan city (China)	22 000	Yuan *et al.* (2000)
C. quinquefasciatus	2362	Every 3 months	4 years	200 ha	Yaoundé (Cameroon)	–	Hougard *et al.* (1993)
C. quinquefasciatus	2362	Twice a year	2 years	2000 ha	Maroua (Cameroon)	–	Barbazan *et al.* (1997)

[a]The level of resistance is calculated as the ratio of the LC$_{50}$ values of the resistant population to that of susceptible laboratory colonies.
[b]At isolated breeding sites.
–, resistance ratio value not reported.

Only a few of the highly larvicidal *B. sphaericus* strains are commercially sold: strain 2362 (Vecto-Lex® and Spherimos®, Valent–Bioscience Corporation, USA) in the USA and Europe, strains 1593 (Biocide-S®) and 101-B (Sphericide®, Biotech International Ltd) in India, and strain C3-41 in China. These products are more or less adapted to various applications and larval breeding sites. As the success of a given strain is strongly dependent on the performance of the formulation, they have been improved over recent years, so that new products are now much more efficient (Skovmand and Bauduin, 1997; Fillinger *et al.*, 2003). There have been several attempts to produce these products locally in both Asia (Wuhan, China) and South America (Rio de Janeiro, Brazil), but only a few large international companies remain on the market.

8.5. Resistance To *B. sphaericus*

8.5.1. Introduction

Unlike *Bti*, towards which no cases of field resistance have appeared so far, several cases of larval resistance to *B. sphaericus* crystal toxin have been reported in *C. pipiens* and *C. quinquefasciatus* (**Table 3**). However, no resistance has yet been found in *B. sphaericus*-treated *Anopheles* populations. Indications for resistance to *B. sphaericus* were first reported in laboratory-selected populations (Georghiou *et al.*, 1992; Rodcharoen and Mulla, 1994), then by Pei *et al.* (2002), and a bit later from field-treated populations: from France (Sinègre *et al.*, 1994), India (Rao *et al.*, 1995), and Brazil (Silva-Filha *et al.*, 1995), and more recently from China (Yuan *et al.*, 2000) and Thailand (Mulla *et al.*, 2003) (**Table 4**).

The emergence of resistance was not that surprising given: (1) the apparent "one site" mode of action of the Bin toxin; (2) the ability of *B. sphaericus* to recycle and to persist in larval breeding zones, thus leading to increasing selection pressure (Lacey *et al.*, 1987; Correa and Yousten, 1995); and (3) the apparent high natural variation in susceptibility of *Culex* to *B. sphaericus* (Georghiou and Lagunes-Tejeda, 1991). However, resistance to *B. sphaericus* is also characterized by a large variation in resistance level, speed of appearance, and stability, indicating that factors such as selection pressure and the initial frequencies of resistance alleles in the mosquito population are very important.

We will summarize what is presently known about the genetics and mechanism of resistance to *B. sphaericus* with examples from both *C. pipiens* and *C. quinquefasciatus*. We will also discuss the mode of action and mechanism of resistance of the

Bin toxin, as well as issues concerning the preservation of *B. sphaericus* as a mosquito control agent by searching for either natural or recombinant strains with low cross-resistance. Finally, we will discuss implications for application strategies.

Table 4 reports several examples of both laboratory-selected and initially field-selected resistance, from tropical and temperate areas, from places with high and low treatment frequencies and from various types of breeding sites. All populations reach very high levels of resistance (resistance ratio >10 000), either directly in the field or following further laboratory selection (Rao *et al.*, 1995; Nielsen-LeRoux *et al.*, 1997; Yuan *et al.*, 2000; Mulla *et al.*, 2003).

8.5.2. Genetics and Mechanisms of Resistance

Investigations into mosquito resistance to the Bin toxin under laboratory conditions were pioneered by Georghiou and co-workers. This group treated large numbers of early-4th instar larvae of a Californian population of susceptible *C. quinquefasciatus* (CpqS) with high concentrations of spore-crystal mixtures (Georghiou *et al.*, 1992; Wirth *et al.*, 2000b). Selection was very effective and resistance arose so rapidly in this colony (named CpqR or GEO) that the larvae were virtually insensitive to the *B. sphaericus* toxins by generation 12 (resistance ratio >100 000). Moreover, the F_1 offspring resulting from crosses between resistant and susceptible mosquitoes were almost totally susceptible, indicating that the resistance is recessive and borne by a single locus (Nielsen-LeRoux *et al.*, 1995; Wirth *et al.*, 2000b). Interestingly, a high level of resistance was also detected in first instar larvae, which suggests that the genetic factor of resistance is constitutive and not related to a particular developmental stage. Similar investigation with field-selected *B. sphaericus* resistant *C. pipiens* also showed monofactorial and recessive resistance, as F_1 progenies were as susceptible as the parental susceptible colony. In addition, dose-mortality Probit lines from back-cross progenies indicated that resistance of these colonies were likewise due to one major gene (Nielsen-LeRoux *et al.*, 1997; Oliveira *et al.*, 2003a). Until now, all studied cases of resistance were due to one major recessive gene that is sex linked in Mediterranean populations of *C. pipiens* (Nielsen-LeRoux *et al.*, 1997; Chevillon *et al.*, 2001; Nielsen-LeRoux *et al.*, 2002) but autosomally inherited in *C. quinquefasciatus* from Brazil and China (Oliveira *et al.*, 2003a) and possibly also in the GEO (CpqR) colony (Wirth *et al.*, 2000b). In general, no signs of a fitness cost have been reported in the laboratory (Oliveira *et al.*, 2003b), except for

Table 4 Characteristics of various *Culex pipiens* or *C. quinquefasciatus* colonies resistant to *B. sphaericus* strains

Country/year	Field (F) or Laboratory (L) selection	Colony	B. sphaericus strain used for selection	Level of resistance[a]	Genetics of resistance			Mechanisms of resistance[b] Binding characteristics		Reference
					Nb. genes	Dom = D Res = R	Linkage	K_d (nM)	R_t (pmol mg⁻¹)	
USA/1992	L	GEO	2362	>100 000	1	R	Autosomal	No binding	–	1, 2
USA/1994	L	JRMM-R	2362	37	?[c]	?	?	Binding[d]	–	3
Brazil/1995	F	Brazil-R	2362	10	?	?	?	11 ± 1	7 ± 1	4
India/1995	F/L	Kochi	1593M	146 (F)/>10 000 (L)	?	"D"[e]	?	No binding[e]	–	5
France/1994	F/L	SPHAE	2362	16 000 (F)/>50 000 (L)	1	R	Sex	3 ± 1	16 ± 1	6
Tunisia/1994	F/L	TUNIS	2362	>10 000 (L)	1	R	Sex	4 ± 1	3 ± 1	7
France/2000	F/L	BP	2362	>10 000 (F)/>50 000 (L)	1	R	Sex	No binding	–	7
China/2000	F/L	RLCq2(C3-41)	C3-41	22 000 (F)/>144 000 (L)	1	R	Autosomal	No binding	–	8
Brazil/2002	L	CqSF	2362	>162 000	1	R	Autosomal	No binding	–	8

[a]Level of resistance is calculated as the ratio of the LC₅₀ values of the resistant colony to that of a susceptible colony.

[b]Binding experiments are done with (BBMF) brush border membrane fraction from homozygous laboratory selected resistant larvae.

[c]? No information available.

[d]Nielsen-LeRoux et al. (Unpublished data); not enough data to calculate K_d value.

[e]Poopathi and Nielsen-LeRoux (Unpublished data).

[1]Nielsen-LeRoux et al. (1995), [2]Wirth et al. (2000a), [3]Rodcharoen & Mulla (1994), [4]Silva-Filha et al. (1995), [5]Rao et al. (1995), [6]Nielsen-LeRoux et al. (1997), [7]Nielsen-LeRoux et al. (2002), [8]Oliviera et al. (2003a).

one, the stable low level resistant JRMM-R (Rodcharoen and Mulla, 1994; Rodcharoen and Mulla, 1997). In the field, resistance appears to be unstable in the absence of selection pressure because, as soon after *B. sphaericus* treatment is interrupted, the exposed *C. quinquefasciatus* populations become susceptible to *B. sphaericus* again (Silva-Filha and Regis, 1997; Yuan *et al.*, 2000).

8.5.2.1. Molecular basis of laboratory-selected resistance

As described above, the binding of BinB to the α-glucosidase (Cpm1) receptor at the apical membrane of midgut cells is essential for larvicidal activity. Thus, when resistance to *B. sphaericus* was first reported, the mechanisms of resistance were analyzed by comparing the kinetics of Bin receptor binding in the midguts of GEO and susceptible *C. quinquefasciatus* larvae (CpqS). As expected, the binding characteristics of ^{125}I-Bin to CpqS BBMFs were very similar to those previously reported for *C. pipiens* membranes (Nielsen-LeRoux and Charles, 1992). In contrast, Bin cannot specifically bind BBMFs from GEO mosquitoes (Nielsen-LeRoux *et al.*, 1995). In addition, when assayed on BBMFs prepared from F_1 (CpqS X GEO) larvae, ^{125}I-Bin bound to a single class of receptor, which is consistent with the suspected recessive nature of the resistance trait. Analysis of binding data with BBMFs from the backcross (BC) progeny showed that the binding sites are not saturated in the range of toxin concentrations used, and that the total amount of bound toxin is much lower than for the susceptible parental strain. LIGAND analysis showed that the experimental data obtained with the BC progeny fit a two-site model better than a one-site model. This possible existence of two classes of binding sites in the BC population is suggestive of genetic heterogeneity.

The next step was to determine whether this receptor was absent from the GEO mosquito midgut or whether it was present in a form that could no longer recognize the Bin toxin. In Northern blot experiments, total RNA extracted from GEO midguts was probed with a *cpm1* DNA fragment. The 2-kb transcript previously identified in IP mosquitoes was present in similar amounts in GEO mosquitoes (Darboux and Pauron, unpublished data). *In situ* hybridization experiments confirmed that *cpm1* transcripts are equally distributed both quantitatively and qualitatively in CpqS and GEO midguts. These transcripts were found in the regions that had been previously identified as Cpm1 reservoirs, i.e., cardia cells, the gastric caecae and the posterior midgut (Darboux *et al.*, 2002).

Nevertheless, the amount of Cpm1 protein differed completely in the two populations. Firstly, immunolocalization performed on the same type of cryosections as the *in situ* hybridizations failed to detect Cpm1 in any of the above-mentioned structures or anywhere else in the larval midgut. Secondly, Cpm1 was not detected in BBMFs from GEO mosquitoes by Western blotting (Darboux *et al.*, 2002). Taken together, these results suggest that the sequence encoding Cpm1 is altered in GEO. Analysis of the full cDNA sequence of $cpm1_{GEO}$ confirmed this hypothesis. The $cpm1_{GEO}$ sequence contains seven nonsilent mutations compared to $cpm1_{IP}$. Six of these mutations result in amino acid substitutions in the protein itself (Ala95Asp; Lys115Met; Glu178Thr; Asp230His; Asn265Asp; Leu486Met), and the seventh introduces a termination signal in the hydrophobic tail of the protein (Leu569Stop). To determine which of these mutations were involved in the resistance mechanism, Sf9 cells were transfected with various constructs corresponding to natural or chimeric forms of Cpm1. Western blotting, binding experiments and enzymatic assays were performed on the expression products of each construct. The $Cpm1_{GEO}$ form was unable to link to the plasma membranes of Sf9 cells (**Figure 7**), but accumulated in the extracellular medium, which explains why no signal could be detected both in BBMF Western blotting and by immunolocalization in midgut sections. Concomitantly, no binding activity was detected on membranes prepared from $Sf9/Cpm1_{GEO}$ cells, which confirms the previous results of *in vitro* binding of ^{125}I-Bin toxin to $BBMF_{GEO}$ (Darboux *et al.*, 2002).

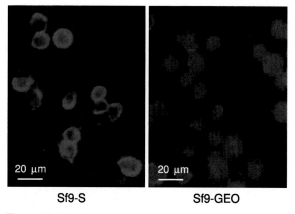

Figure 7 Ectopic expression of Cpm1. Sf9 cells were transfected with the $cpm1_{IP}$ sequence (Sf9-S) or the $cpm1_{GEO}$ one (Sf9-GEO). The anti-Cpm1 antibody gives a signal only on the plasma membrane of the Sf9-S cells (Castella and Pauron, unpublished data).

A chimeric form of Cpm1 was also constructed, where the GEO sequence was modified by replacing the early TAG codon by TTG (Stop569Leu) by site-directed mutagenesis in order to recover a "susceptible-like" hydrophobic tail. When transfected into Sf9 cells, this construct, named Sf9-GEOMut, was able to mature fully, and Cpm1$_{GEOMut}$ was detected in the membrane fraction of the cells to which it was linked by a GPI, as shown by its sensitivity to PI-PLC (Darboux *et al.*, 2002). Moreover, competition as well as saturation binding assays showed that ^{125}I-Bin binds membranes prepared from Sf9-GEOMut cells with the same affinity as those prepared from Sf9-IP cells. Hence, the six mutations present in the protein itself have no effect on the capacity of the receptor to recognize the toxin and are not involved in the resistance mechanism. They have no effect on the enzymatic activity of Cpm1, as shown by measurements done using the secreted form of Cpm1$_{GEO}$ (**Figure 8**).

Cpm1$_{GEO}$ hydrolyzes 4-nitrophenyl α-D-gluco-pyranoside as well as Cpm1$_{IPMut}$, which suggests that Cpm1$_{GEO}$ retains its physiological activity even when it is no longer bound to the membrane *in vivo*. If this is the case, this resistance mechanism may cause less pleiotropic effects than other mechanisms that result in the severe dysfunction in the physiology of the insect. In that respect, resistance to other *Bacillus* toxins may impair important functions such as cell–cell signaling in the case of the cadherin-like protein (Gahan *et al.*, 2001) or carbohydrate modifications in the case

of the β-1,3-galactosyltransferase (Griffitts *et al.*, 2001). Resistance in GEO also contrasts with identified mechanisms of resistance to chemical insecticides by target modification, which is usually caused by a small number of point mutations in the receptor sequence that decrease its affinity for the pesticide (Pauron *et al.*, 1989; ffrench-Constant *et al.*, 1993; Mutero *et al.*, 1994).

A single point mutation in the *cpm1$_{GEO}$* sequence results in Cpm1 molecules with a shortened hydrophobic tail (**Figure 9**). This hydrophobic stretch becomes too small to be recognized by the transamidase complex, thus preventing Cpm1 from being matured by the anchoring of a GPI moiety. Nevertheless, the unprocessed form of the receptor is translocated towards the apical membrane of the cell, where it is secreted. In the extracellular space, Cpm1$_{GEO}$ may still encounter and bind the Bin toxin but the complex formed can no longer promote the cytopathological effects of the toxin and the larva is not affected. This mechanism is the first example of a putative combination of two types of resistance mechanism to insecticides, i.e., target modification and sequestration of the toxic ligand.

8.5.2.2. Mechanisms of field-selected resistance

In vitro binding studies have also been conducted with other field-resistant mosquito colonies (see **Table 4**). In a number of cases, the mechanism of resistance was due to lack of measurable Bin toxin–receptor interaction. This was recently found for *C. quinquefasciatus* colonies from Brazil and China (Oliveira *et al.*, 2003a), from India (KOCHI colony, unpublished data), and for the *C. pipiens* BP colony from France (Nielsen-LeRoux *et al.*, 2002). Interestingly, the initial Bin toxin–receptor interactions of the highly *B. sphaericus*-resistant *C. pipiens* colonies SPHAE (Nielsen-LeRoux *et al.*, 1997) and TUNIS (Nielsen-LeRoux *et al.*, 2002), are similar to those of susceptible *C. pipiens* (*sensu lato*) colonies (reviewed in Nielsen-LeRoux *et al.*, 2002). Functional receptors have also been found in a low-level and stably resistant colony (Rodcharoen and Mulla, 1994; Nielsen-LeRoux, unpublished data). Another low-level resistant colony has been shown to have functional receptors, although the larvae collected in the field may not all have been homozygous resistant (Silva-Filha *et al.*, 1995).

There is evidence for more than one resistance mechanism to the Bin toxin in the *C. pipiens* complex worldwide and even within small areas like the West Mediterranean countries. Indeed, mutations in three West Mediterranean *C. pipiens* populations give rise to three different mutants: *sp-1^R* (SPHAE strain), *sp-2^R* (BP strain), and *sp-T^R* (TUNIS strain).

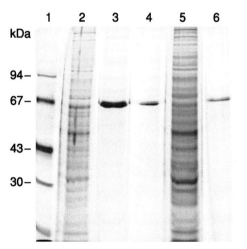

Figure 8 Silver staining of secreted forms of Cpm1. Lane 1, molecular weight markers. Lanes 2 and 5, crude extracellular medium of Sf9-IPMut and Sf9-GEO cells, respectively. Lanes 3 and 4, Cpm1$_{IPMut}$ purified by affinity chromatography. Lane 6 Cpm1$_{GEO}$ purified by affinity chromatography (Pauchet and Pauron, unpublished data).

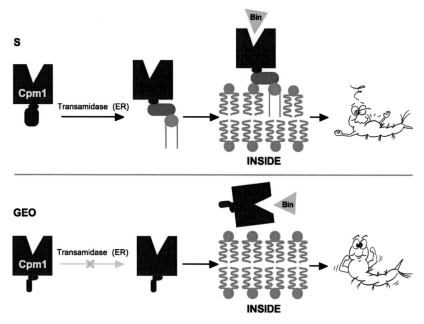

Figure 9 Schematic representation of the interaction between the Bin toxin and its receptor in a susceptible (S, top) and in a resistant (GEO, bottom) mosquito larva.

Mutations are all sex-linked and separately confer the same totally recessive resistance to *B. sphaericus*. However, binding experiments showed clear differences between the BP strain mutant and the other two mutants (Nielsen-LeRoux *et al.*, 2002). It is not clear whether mutations of TUNIS and SPHAE are identical or not, even though toxin receptor affinities are high in both cases and no gene recombination was observed when TUNIS and SPHAE were crossed (Chevillon *et al.*, 2001; Nielsen-LeRoux *et al.*, 2002). Moreover, the *cpm1* coding sequences of TUNIS and SPHAE are strictly identical (Darboux and Pauchet, unpublished data). *B. sphaericus* resistance in Californian *C. quinquefasciatus* populations may also be based on different mutations, as the laboratory-selected resistance of two Californian *C. quinquefasciatus* collected in the field was reported to be stable, one at a very high level (>100 000 fold, GEO strain) (Georghiou *et al.*, 1992; Wirth *et al.*, 2000b) and the other at a low level (35 fold, JRMM-R strain) (Rodcharoen and Mulla, 1994). A second study supports this hypothesis, because of the numerous *B. sphaericus*-resistant colonies tested for cross-resistance to other *B. sphaericus* strains, the only one that was resistant to all tested *B. sphaericus* strains was the JRMM-R colony (Nielsen-LeRoux *et al.*, 2001).

Cpm1 from the SPHAE strain appears to be totally functional, and its distribution in the larval midgut is indistinguishable from that in susceptible individuals (Castella and Pauron, unpublished data).

In addition, the transcription pattern of $cpm1_{SPHAE}$ is identical to that of $cpm1_{IP}$ and the coding sequence of $cpm1_{SPHAE}$ displays no nonsilent mutations compared to the susceptible allele (Darboux and Pauchet, unpublished data). Hence, as no apparent differences in the activation process of the Bin toxin have been reported in SPHAE larvae (Nielsen-LeRoux *et al.*, 1997), the step involved in resistance in SPHAE presumably occurs after the Bin toxin interacts with its receptor. Determining whether this is dependent on the endocytosis of BinA, BinB, and/or the receptor, and whether they persist with $sp-1^{RR}$ larvae is likely to improve our understanding of how $sp-1^{R}$ interferes with the intoxication process.

When considering the mechanism underlying the resistance of the BP strain, it is worth noting the possibility of a tight physiological interaction between *sp-2* (BP) and *sp-1* (SPHAE). For example, the resistance of $sp-1^{RS} sp-2^{RS}$ double heterozygotes is increased by over 100-fold, even though each mutation was totally recessive alone (Chevillon *et al.*, 2001). However, the BP colony may be different from SPHAE, as immunological characterization of Cpm1 in BP confirmed the binding data: no signal was ever detected either by Western blotting analysis of $BBMF_{BP}$ proteins or in cryosections of BP larva midgut (Castella and Pauron, unpublished data). In this strain, the resistance mechanism appears to be rather complex, as preliminary results indicate that both the coding sequence and the pattern of transcription of *cpm1* are altered in BP (Darboux and Pauron, unpublished data).

8.6. Conclusions and Perspectives: Management of *B. sphaericus*-Resistance

The genes encoding BinA and BinB have been sequenced, and the regions important for the specificity and activity of these components identified. It is clear that both BinA and BinB are needed for full activity of the *B. sphaericus* crystal toxin: BinA alone is slightly toxic and might form pores in the cell membranes, and BinB appears to be responsible for the regionalized binding of the toxin to a specific midgut membrane receptor. Even if the nature of the Bin receptor, an α-glucosidase, is known in *Culex* spp., the possible intracellular actions of the Bin toxin(s) remain unclear. Indeed, immunolocalization investigations on the midguts of intoxicated *C. quinquefasciatus* larvae revealed that both BinA and BinB are present within cells a few hours after intoxication (Silva-Filha and Peixoto, 2003). These results show that the Bin toxin is internalized during the intoxication process and confirms previous results using fluorescently labeled Bin toxin (Davidson, 1988; Oei *et al.*, 1992). The importance of internalization of the Bin toxins into midgut cells is demonstrated by the fact that in the presence of the pore-forming or "detergent-like" Cyt1A toxin from *Bti*, the Bin toxin is internalized into *B. sphaericus*-resistant larval midgut cells (Federici *et al.*, personal communication). Further studies are required to elucidate the mode of action fully, as the final target molecules are unknown.

Further studies are also needed to determine the mode of action of the Bin toxin towards *Anopheles*, because there is some evidence suggesting that it differs from that towards *Culex*. As previously mentioned, no cases of resistance have yet been reported in *Anopheles* and *in vitro* binding experiments indicated that both BinA and BinB can bind to a receptor-binding site (Charles *et al.*, 1997).

The Mtx toxins have been studied less extensively than the Bin toxins, so further studies are needed, especially to determine how the enzymatic fragment from the Mtx1 toxin enters midgut cells to ADP-ribosylate target proteins.

As indicated above, there is a real risk that resistance to *B. sphaericus* will appear when it is used in large-scale *Culex* spp. control programs. Thus, it is important to investigate possible variations in toxic activity and specificity of different natural *B. sphaericus* strains, in order to evaluate the level of possible cross-resistance among different

strains and the Bin toxins they produce. All *B. sphaericus*-resistant *Culex* spp. populations were selected using strains 2362, 1593 or C3-41 (Rodcharoen and Mulla, 1994; Nielsen-LeRoux *et al.*, 1995; Rao *et al.*, 1995; Yuan *et al.*, 2000), all of which belong to the same serotype and have identical Bin toxin genes (Bin2 type). However, there are small differences in the amino acid sequences of Bin toxins (**Table 2**), which may be important in the structure/function of the toxin/receptor complex, and therefore for larvicidal activity. Several groups have investigated the level of cross-resistance towards active *B. sphaericus* strains: 2297, IAB-881, IAB-872, LP1-G and IAB-59, in *Culex* spp. colonies resistant to strains 2362, 1593, or C3-41. The 2362-derived resistant JRMM-R colony is resistant to strains 2297 and 1593 (Rodcharoen and Mulla, 1994), and to IAB-59, IAB-881 and IAB-872 (Nielsen-LeRoux *et al.*, 2001). The GEO, SPHAE and the KOCHI colonies are resistant to all strains except IAB-59 (Poncet *et al.*, 1997; Poopathi *et al.*, 1999; Nielsen-LeRoux *et al.*, 2001). In addition, strain LP1-G is able to overcome resistance in *C. quinquefasciatus* larvae resistant to strain C3-41 from China (Yuan *et al.*, 2000).

Investigations on both strains IAB-59 and LP1-G suggested the presence of toxic factors other than the Bin and Mtx toxins. First, the crystals from LP1-G contain a natural nontoxic variant of the Bin toxin (Bin4 type) (Yuan *et al.*, 2001). Thus, although the whole strain is larvicidal, the toxic activity may come from other compounds produced by this strain. Second, crystals from the IAB-59 or IAB-872 Bin toxin (Bin1 type) are also nontoxic to the *B. sphaericus*-resistant *Culex* spp. colonies, which are also susceptible to the whole IAB-59 strain (Nielsen-LeRoux, unpublished data; Shi *et al.*, 2001; Pei *et al.*, 2002). In addition, the low level of cross-resistance to IAB-59 has recently been validated by selecting for resistance to this strain in laboratory conditions, in two *C. quinquefasciatus* populations collected in the field in China and Brazil (**Table 4**). The speed of appearance was much slower and the level of resistance much lower with the IAB-59 strain than with strains 2362 or C3-41. Finally, colonies that carry only a low level of resistance to IAB-59 are highly resistant to strains 2362 and C3-41, suggesting that resistance is directed against the Bin toxins regardless of whether it belongs to the Bin1 or Bin2 type. This is in agreement with *in vitro* toxin–receptor binding studies, which showed high competition among Bin toxins (Bin1, Bin2 and Bin4) and provided evidence for a common receptor-binding site (Nielsen-LeRoux *et al.*, unpublished data). Further studies are required to evaluate whether strain IAB-59 is of commercial value as an alternative to *B. sphaericus* strains 1593, 2362, 101-B, and C3-41 (the only strains currently produced for use in the field). This depends on its potency and field performance. Additionally, several groups are currently trying to identify the genes encoding the additional toxic compounds from strains IAB-59 and LP1-G.

The fact that the Bin toxin behaves as a unique moiety during the binding step indicates that the probability of selecting resistance to *B. sphaericus* is much higher than for *Bti*, which contains at least four proteins that potentially act on different target molecules. Despite reports of laboratory and field resistance, the use of *B. sphaericus* for the control of mosquito larvae is still of interest. Nevertheless, it is important to collect information about baseline susceptibility before starting treatment. However, as resistance to *B. sphaericus* is recessive, it is only possible to detect resistance when it has already become homozygous and therefore difficult to handle. Thus, it may be possible to monitor the risk of resistance development in a given area by use of molecular nucleotide probes based on the mutation of resistance genes. However, resistance to *B. sphaericus* seems to be very complex and different probes would be needed to cover all possible mutations, only one of which has been identified so far (Darboux *et al.*, 2002).

Both from a practical and an ecological point of view, the best and most direct way to overcome and to manage *B. sphaericus* resistance is of course *Bti*, which is already commercialised. Fortunately, all *Culex* populations reported to display field resistance to *B. sphaericus* have been tested for susceptibility to *Bti*. No cross-resistance has been observed. There is even some evidence for increased susceptibility in some colonies (Rao *et al.*, 1995; Silva-Filha *et al.*, 1995; Yuan *et al.*, 2000). This was expected, because the multi-toxin complex of this bacterium does not share any receptor-binding sites with the *B. sphaericus* binary toxin (Nielsen-LeRoux and Charles, 1992). In several situations, it is recommended that *Bti* be used in rotation with *B. sphaericus* (Regis and Nielsen-LeRoux, 2000; Yuan *et al.*, 2000). Recent results obtained in Thailand have suggested that a mixture of *Bti* and *B. sphaericus* may help reduce the risk of appearance of *B. sphaericus* resistance while preserving the superior field performance of *B. sphaericus* compared to *Bti*, particularly in breeding sites rich in organic matter (Mulla *et al.*, 2003).

Alternatives to natural *Bti* and *B. sphaericus* strains include genetically modified organisms, as combining different toxins binding to different receptors in the same organism is also a good way of preventing resistance. In addition to the Bin toxins, other toxins from either *B. sphaericus* or other insecticidal microorganisms could be produced. For example, the expression of *Mtx1* genes in *B. sphaericus* during sporulation, i.e., under the control of a sporulation promoter, could allow the diversification of toxins. Similarly, *Bti* mosquitocidal toxin genes could be combined with *B. sphaericus* genes; some groups have already attempted this, for example by introducing the *Bti* genes encoding Cry4B, Cry4D or Cry11A into toxic *B. sphaericus* strains (Trisrisook *et al.*, 1990; Bar *et al.*, 1991; Poncet *et al.*, 1994; Poncet *et al.*, 1997). Conversely, *B. sphaericus* crystal toxin genes have been successfully introduced into toxic *Bti* (Bourgouin *et al.*, 1990). Although these studies did not demonstrate any increase in toxicity against *Anopheles* and *Culex*, such recombinants may delay the emergence of resistant insects.

Several attempts have been made to introduce or to combine toxin genes from *Bti*, *B. thuringiensis* serovar *jegathesan* (*Bt jeg*) and serovar *medellin* (*Bt med*), and to evaluate their activity towards both susceptible and *B. sphaericus*-resistant colonies. The introduction of Cry11A from *Bti* and Cry11Ba from *Bt jeg* partially restored the susceptibility of resistant colonies (Poncet *et al.*, 1997; Servant *et al.*, 1999). Nonspecific cytolytic toxins (Cyt1A and Cyt1B) also partially restored its toxicity towards two resistant colonies (Thiéry *et al.*, 1998; Wirth *et al.*, 2000a). A *Bti* strain harboring genes encoding the binary toxin as well as Cyt1 and Cry11B was more toxic than unmodified *Bti* IPS-82 (Park *et al.*, 2003). Several groups have also attempted to introduce and to overproduce entomopathogenic toxins into organisms living in the environment and serving as natural foods for mosquitoes, such as cyanobacteria (Liu *et al.*, 1996b), or to stably integrate the toxins encoding these genes into the chromosomes of other bacteria such as *Enterobacter amnigenus* (Tanapongpipat *et al.*, 2003). However, given our current knowledge of *B. sphaericus* resistance, the expression of Bin toxin as the only active compound should be avoided. No recombinant strains have yet proved to have a better field performance than the natural strains, and no products based on recombinant strains have yet been registered for use in mosquito control programs. Therefore there is a continuing need to identify new strains with novel modes of action against the principal mosquito vectors of human diseases.

References

Ahmed, H.K., Mitchell, W.J., Priest, F.G., **1995**. Regulation of mosquitocidal toxin synthesis in *Bacillus sphaericus*. *Appl. Microbiol. Biotechnol. 43*, 310–314.

Alexander, B., Priest, F.G., **1990**. Numerical classification and identification of *Bacillus sphaericus* including some strains pathogenic for mosquito larvae. *J. Gen. Microbiol. 136*, 367–376.

Aquino de Muro, M., Priest, F.G., **1994**. A colony hybridization procedure for the identification of mosquitocidal strains of *Bacillus sphaericus* on isolation plates. *J. Invertebr. Pathol. 63*, 310–313.

Aquino de Muro, M., Mitchell, W.J., Priest, F.G., **1992**. Differentiation of mosquito-pathogenic strains of *Bacillus sphaericus* from non-toxic varieties by ribosomal RNA gene restriction patterns. *J. Gen. Microbiol. 138*, 1159–1166.

Arapinis, C., de la Torre, F., Szulmajster, J., **1988**. Nucleotide and deduced amino acid sequence of the *Bacillus sphaericus* 1593M gene encoding a 51.4 kD polypeptide which acts synergistically with the 42 kD protein for expression of the larvicidal toxin. *Nucleic Acids Res. 16*, 7731.

Bar, E., Lieman-Hurwitz, J., Rahamim, E., Keynan, A., Sandler, N., **1991**. Cloning and expression of *Bacillus thuringiensis israelensis* δ-endotoxin DNA in *B. sphaericus*. *J. Invertebr. Pathol. 57*, 149–158.

Barbazan, P., Baldet, T., Darriet, F., Escaffre, H., Haman, D.D., *et al.*, **1997**. Control of *Culex quinquefasciatus* (Diptera: Culicidae) with *Bacillus sphaericus* in Maroua, Cameroon. *J. Am. Mosq. Control Assoc. 13*, 263–269.

Baumann, L., Baumann, P., **1989**. Expression in *Bacillus subtilis* of the 51- and 42-kilodalton mosquitocidal toxin genes of *Bacillus sphaericus*. *Appl. Environ. Microbiol. 55*, 252–253.

Baumann, L., Baumann, P., **1991**. Effects of components of the *Bacillus sphaericus* toxin on mosquito larvae and mosquito-derived tissue culture-grown cells. *Curr. Microbiol. 23*, 51–57.

Baumann, P., Baumann, L., Bowditch, R.D., Broadwell, A.H., **1987**. Cloning of the gene of the larvicidal toxin of *Bacillus sphaericus* 2362: evidence for a family of related sequences. *J. Bacteriol. 169*, 4061–4067.

Baumann, L., Broadwell, A.H., Baumann, P., **1988**. Sequence analysis of the mosquitocidal toxin genes encoding 51.4- and 41.9-kilodalton proteins from *Bacillus sphaericus* 2362 and 2297. *J. Bacteriol. 170*, 2045–2050.

Baumann, P., Clark, M.A., Baumann, L., Broadwell, A.H., **1991**. *Bacillus sphaericus* as a mosquito pathogen: properties of the organism and its toxins. *Microbiol. Rev. 55*, 425–436.

Baumann, P., Unterman, B.M., Baumann, L., Broadwell, A.H., Abbene, S.J., *et al.*, **1985**. Purification of the larvicidal toxin of *Bacillus sphaericus* and evidence for high-molecular-weight precursors. *J. Bacteriol. 163*, 738–747.

Becker, N., **2000**. Bacterial control of vector-mosquitoes and black flies. In: Charles, J.-F., Delécluse, A., Nielsen-LeRoux, C. (Eds.), Entomopathogenic Bacteria: from Laboratory to Field Application. Kluwer Academic Publisher, Dordrecht, Boston, London, pp. 383–398.

Berry, C., Hindley, J., **1987**. *Bacillus sphaericus* strain 2362: identification and nucleotide sequence of the 41.9 kDa toxin gene. *Nucleic Acids Res. 15*, 5591.

Berry, C., Hindley, J., Ehrhardt, A.F., Grounds, T., de Souza, I., *et al.*, **1993**. Genetic determinants of host ranges of *Bacillus sphaericus* mosquito larvicidal toxins. *J. Bacteriol. 175*, 510–518.

Berry, C., Jackson-Yap, J., Oei, C., Hindley, J., **1989**. Nucleotide sequence of two toxin genes from *Bacillus sphaericus* IAB59: sequence comparisons between five highly toxinogenic strains. *Nucleic Acids Res. 17*, 7516.

Bourgouin, C., Delécluse, A., de la Torre, F., Szulmajster, J., **1990**. Transfer of the toxin protein genes of *Bacillus sphaericus* into *Bacillus thuringiensis* subsp. *israelensis* and their expression. *Appl. Environ. Microbiol. 56*, 340–344.

Broadwell, A.H., Baumann, P., **1986**. Sporulation-associated activation of *Bacillus sphaericus* larvicide. *Appl. Environ. Microbiol. 52*, 758–764.

Broadwell, A.H., Baumann, P., **1987**. Proteolysis in the gut of mosquito larvae results in further activation of the *Bacillus sphaericus* toxin. *Appl. Environ. Microbiol. 53*, 1333–1337.

Broadwell, A.H., Baumann, L., Baumann, P., **1990a**. Larvicidal properties of the 42 and 51 kilodalton *Bacillus sphaericus* proteins expressed in different bacterial hosts: evidence for a binary toxin. *Curr. Microbiol. 21*, 361–366.

Broadwell, A.H., Clark, M.A., Baumann, L., Baumann, P., **1990b**. Construction by site-directed mutagenesis of a 39-kilodalton mosquitocidal protein similar to the larva-processed toxin of *Bacillus sphaericus* 2362. *J. Bacteriol. 172*, 4032–4036.

Charles, J.-F., **1987**. Ultrastructural midgut events in Culicidae larvae fed with *Bacillus sphaericus* 2297 spore/crystal complex. *Ann. Inst. Pasteur/Microbiol. 138*, 471–484.

Charles, J.-F., de Barjac, H., **1981**. Variation du pH de l'intestin moyen d'*Aedes aegypti* en relation avec l'intoxication par les cristaux de *Bacillus thuringiensis* var. *israelensis* (sérotype H 14). *Bull. Soc. Path. Exot. 74*, 91–95.

Charles, J.-F., Nicolas, L., **1986**. Recycling of *Bacillus sphaericus* 2362 in mosquito larvae: a laboratory study. *Ann. Inst. Pasteur/Microbiol. 137B*, 101–111.

Charles, J.-F., Hamon, S., Baumann, P., **1993**. Inclusion bodies and crystals of *Bacillus sphaericus* mosquitocidal proteins expressed in various bacterial hosts. *Res. Microbiol. 144*, 411–416.

Charles, J.-F., Nielsen-LeRoux, C., Delécluse, A., **1996**. *Bacillus sphaericus* toxins: molecular biology and mode of action. *Annu. Rev. Entomol. 41*, 389–410.

Charles, J.-F., Kalfon, A., Bourgouin, C., de Barjac, H., **1988**. *Bacillus sphaericus* asporogenous mutants:

ultrastructure, mosquito larvicidal activity and protein analysis. *Ann. Inst. Pasteur/Microbiol. 139*, 243–259.

Charles, J.-F., Silva-Filha, M.H., Nielsen-LeRoux, C., Humphreys, M., Berry, C., 1997. Binding of the 51- and 42-kDa individual components from the *Bacillus sphaericus* crystal toxin on mosquito larval midgut membranes from *Culex* and *Anopheles* sp. *(Diptera: Culicidae)*. *FEMS Microbiol. Lett. 156*, 153–159.

Chevillon, C., Bernard, C., Marquine, M., Pasteur, N., 2001. Resistance to *Bacillus sphaericus* in *Culex pipiens* (Diptera: Culicidae): interaction between recessive mutants and evolution in southern France. *J. Med. Entomol. 38*, 657–664.

Clark, M.A., Baumann, P., 1990. Modification of the *Bacillus sphaericus* 51- and 42-kilodalton mosquitocidal proteins: effects of internal deletions, duplications, and formation of hybrid proteins. *Appl. Environ. Microbiol. 57*, 267–271.

Correa, M., Yousten, A., 1995. *Bacillus sphaericus* spore germination and recycling in mosquito larval cadavers. *J. Invertebr. Pathol. 66*, 76–81.

Dadd, R.H., 1975. Alkalinity within the midgut of mosquito larvae with alkaline-active digestive enzymes. *J. Insect Physiol. 21*, 1847–1853.

Darboux, I., Nielsen-LeRoux, C., Charles, J.-F., Pauron, D., 2001. The receptor of *Bacillus sphaericus* binary toxin in *Culex pipiens* (Diptera: Culicidae) midgut: molecular cloning and expression. *Insect Biochem. Mol. Biol. 31*, 981–990.

Darboux, I., Pauchet, Y., Castella, C., Silva-Filha, M.H., Nielsen-LeRoux, C., *et al.*, 2002. Loss of the membrane anchor of the target receptor is a mechanism of bioinsecticide resistance. *Proc. Natl Acad. Sci. USA 99*, 5830–5835.

Davidson, E.W., 1981. A review of the pathology of bacilli infecting mosquitoes, including an ultrastructural study of larvae fed *Bacillus sphaericus* 1593 spores. *Dev. Industr. Microbiol. 22*, 69–81.

Davidson, E.W., 1988. Binding of the *Bacillus sphaericus* (Eubacteriales: Bacillaceae) toxin to midgut cells of mosquito (Diptera: Culicidae) larvae: relationship to host range. *J. Med. Entomol. 25*, 151–157.

Davidson, E.W., 1989. Variation in binding of *Bacillus sphaericus* toxin and wheat germ agglutinin to larval midgut cells of six species of mosquitoes. *J. Invertebr. Pathol. 53*, 251–259.

Davidson, E.W., Myers, P., 1981. Parasporal inclusions in *Bacillus sphaericus*. *FEMS Microbiol. Lett. 10*, 261–265.

Davidson, E.W., Bieger, A.L., Meyer, M., Shellabarge, R.C., 1987. Enzymatic activation of the *Bacillus sphaericus* mosquito larvicidal toxin. *J. Invertebr. Pathol. 50*, 40–44.

Davidson, E.W., Oei, C., Meyer, M., Bieber, A.L., Hindley, J., *et al.*, 1990. Interaction of the *Bacillus sphaericus* mosquito larvicidal proteins. *Can. J. Microbiol. 36*, 870–878.

Davidson, E.W., Yousten, A.A., 1990. The mosquito larval toxin of *Bacillus sphaericus*. In: de Barjac, H.,

Sutherland, D.J. (Eds.), Bacterial Control of Mosquitoes and Blackflies. Rutgers University Press, New Brunswick, pp. 237–255.

de Barjac, H., 1978. Une nouvelle variété de *Bacillus thuringiensis* très toxique pour les moustiques: *Bacillus thuringiensis* var. *israelensis*, sérotype 14. *C. R. Acad. Sci. Paris, sér. D 286*, 797–800.

de Barjac, H., 1990. Classification of *Bacillus sphaericus* strains and comparative toxicity to mosquito larvae. In: de Barjac, H., Sutherland, D.J. (Eds.), Bacterial Control of Mosquitoes and Blackflies. Rutgers University Press, New Brunswick, pp. 228–236.

de Barjac, H., Charles, J.-F., 1983. Une nouvelle toxine active sur les moustiques, présente dans des inclusions cristallines produites par *Bacillus sphaericus*. *C. R. Acad. Sci. Paris (série III) 296*, 905–910.

de Barjac, H., Larget-Thiéry, I., Cosmao Dumanoir, V., Ripouteau, H., 1985. Serological classification of *Bacillus sphaericus* strains in relation with toxicity to mosquito larvae. *Appl. Microbiol. Biotechnol. 21*, 85–90.

de Barjac, H., Thiéry, I., Cosmao Dumanoir, V., Frachon, E., Laurent, P., *et al.*, 1988. Another *Bacillus sphaericus* serotype harbouring strains very toxic to mosquito larvae: serotype H6. *Ann. Inst. Pasteur/Microbiol. 139*, 363–377.

de la Torre, F., Bennardo, T., Sebo, P., Szulmajster, J., 1989. On the respective roles of the two proteins encoded by the *Bacillus sphaericus* 1593M toxin genes expressed in *Escherichia coli* and *Bacillus subtilis*. *Biochem. Biophys. Res. Comm. 164*, 1417–1422.

Eisenhaber, B., Bork, P., Eisenhaber, F., 1998. Sequence properties of GPI-anchored proteins near the omega-site: constraints for the polypeptide binding site of the putative transamidase. *Protein Eng. 11*, 1155–1161.

El-Bendary, M., Priest, F., Charles, J.-F., Mitchell, W., 2004. Crystal protein synthesis is dependent on early sporulation gene expression in *Bacillus sphaericus*. *Appl. Environ. Microbiol.* (in press).

Elangovan, G., Shanmugavelu, M., Rajamohan, F., Dean, D.H., Jayaraman, K., 2000. Identification of the functional site in the mosquito larvicidal binary toxin of *Bacillus sphaericus* 1593M by site-directed mutagenesis. *Biochem. Biophys. Res. Commun. 276*, 1048–1055.

ffrench-Constant, R.H., Steichen, J.C., Ode, P.J., 1993. Cyclodiene insecticide resistance in *Drosophila melanogaster* (Meigen) is associated with a temperature-sensitive phenotype. *Pest. Biochem. Physiol. 46*, 73–77.

Fillinger, U., Knols, B.G., Becker, N., 2003. Efficacy and efficiency of new *Bacillus thuringiensis* var. *israelensis* and *Bacillus sphaericus* formulations against Afrotropical anophelines in Western Kenya. *Trop. Int. Health 8*, 37–47.

Frachon, E., Hamon, S., Nicolas, L., de Barjac, H., 1991. Cellular fatty acid analysis as a potential tool for predicting mosquitocidal activity of *Bacillus sphaericus* strains. *Appl. Environ. Microbiol. 57*, 3394–3398.

Gahan, L.J., Gould, F., Heckel, D.G., 2001. Identification of a gene associated with Bt resistance in *Heliothis virescens*. *Science* 293, 857–860.

Georghiou, G.P., Lagunes-Tejeda, A., 1991. The occurrence of resistance to pesticides in arthropods. Food & Agriculture Organization of the United Nations, Rome.

Georghiou, G.P., Malik, J.I., Wirth, M., Sainato, K., 1992. Characterization of resistance of *Culex quinquefasciatus* to the insecticidal toxins of *Bacillus sphaericus* (strain 2362). In: Cost, J., Chase, L. (Eds.), Mosquito Control Research, Annual Report. University of California Press, Berkeley, pp. 34–35.

Goldberg, L.J., Margalit, J., 1977. A bacterial spore demonstrating rapid larvicidal activity against *Anopheles sergentii, Uranotaenia unguiculata, Culex univitattus, Aedes aegypti* and *Culex pipiens*. *Mosq. News.* 37, 355–358.

Griffitts, J.S., Whitacre, J.L., Stevens, D.E., Aroian, R.V., 2001. Bt toxin resistance from loss of a putative carbohydrate-modifying enzyme. *Science* 293, 860–864.

Guerineau, M., Alexander, B., Priest, F.G., 1991. Isolation and identification of *Bacillus sphaericus* strains pathogenic for mosquito larvae. *J. Invertebr. Pathol.* 57, 325–333.

Hindley, J., Berry, C., 1987. Identification, cloning and sequence analysis of the *Bacillus sphaericus* 1593 41.9 kD larvicidal toxin gene. *Mol. Microbiol.* 1, 187–194.

Hindley, J., Berry, C., 1988. *Bacillus sphaericus* strain 2297: nucleotidic sequence of a 41.9 kDa toxin gene. *Nucleic Acids Res.* 16, 4168.

Hougard, J.M., Mbentengam, R., Lochouarn, L., Escaffre, H., Darriet, F., et al., 1993. Control of *Culex quinquefasciatus* by *Bacillus sphaericus*: results of a pilot campaign in a large urban area in equatorial Africa. *Bull. WHO* 71, 367–375.

Humphreys, M.J., Berry, C., 1998. Variants of the *Bacillus sphaericus* binary toxins: implications for differential toxicity of strains. *J. Invertebr. Pathol.* 71, 184–185.

Kalfon, A., Charles, J.-F., Bourgouin, C., de Barjac, H., 1984. Sporulation of *Bacillus sphaericus* 2297: an electron microscope study of crystal-like inclusions biogenesis and toxicity to mosquito larvae. *J. Gen. Microbiol.* 130, 893–900.

Karch, S., Coz, J., 1983. Histopathologie de *Culex pipiens* Linné (Diptera, Culicidae) soumis à l'activité larvicide de *Bacillus sphaericus* 1593-4. *Cah. ORSTOM, sér. Ent. méd. Parasitol. XXI*, 225–230.

Karch, S., Asidi, N., Manzambi, M., Salaun, J.J., 1992. Efficacy of *Bacillus sphaericus* against the malaria vector *Anopheles gambiae* and other mosquitoes in swamps and rice fields in Zaire. *J. Am. Mosq. Control Assoc.* 8, 376–380.

Kellen, W., Clark, T., Lindergren, J., Ho, B., Rogoff, M., et al., 1965. *Bacillus sphaericus* Neide as a pathogen of mosquitoes. *J. Invertebr. Pathol.* 7, 442–448.

Krych, V.K., Johnson, J.L., Yousten, A.A., 1980. Deoxyribonucleic acid homologies among strains of *Bacillus sphaericus*. *Int. J. Syst. Bact.* 30, 476–482.

Kumar, A., Sharma, V., Sumodan, P., Thavaseluam, D., Kamat, R., 1994. Malaria control utilising *Bacillus sphaericus* against *Anopheles stephensi* in Pinaji, Goa. *J. Am. Mosq. Control Assoc.* 10, 534–539.

Kumar, A., Sharma, V.P., Thavaselvam, D., Sumodan, P.K., Kamat, R.H., et al., 1996. Control of *Culex quinquefasciatus* with *Bacillus sphaericus* in Vasco City, Goa. *J. Am. Mosq. Control Assoc.* 12, 409–413.

Lacey, L.A., Day, J., Heitzman, C.M., 1987. Long-term effects of *Bacillus sphaericus* on *Culex quinquefasciatus*. *J. Invertebr. Pathol.* 49, 116–123.

Liu, J.W., Hindley, J., Porter, A.G., Priest, F.G., 1993. New high-toxicity mosquitocidal strains of *Bacillus sphaericus* lacking a 100-kilodalton-toxin gene. *Appl. Environ. Microbiol.* 59, 3470–3473.

Liu, J.W., Porter, A.G., Wee, B.Y., Thanabalu, T., 1996a. New gene from nine *Bacillus sphaericus* strains encoding highly conserved 35.8-Kilodalton mosquitocidal toxins. *Appl. Environ. Microbiol.* 62, 2174–2176.

Liu, J.W., Yap, W.H., Thanabalu, T., Porter, A.G., 1996b. Efficient synthesis of mosquitocidal toxins in *Asticcacaulis excentricus* demonstrates potential of Gram-negative bacteria in mosquito control. *Nature Biotechnol.* 14, 343–347.

Miteva, V., Gancheva, A., Mitev, V., Ljubenov, M., 1998. Comparative genome analysis of *Bacillus sphaericus* by ribotyping, M13 hybridization, and M13 polymerase chain reaction fingerprinting. *Can. J. Microbiol.* 44, 175–180.

Mulla, M.S., Thavara, U., Tawatsin, A., Chomposri, J., Su, T., 2003. Emergence of resistance and resistance management in field populations of tropical *Culex quinquefasciatus* to the microbial control agent *Bacillus sphaericus*. *J. Am. Mosq. Control Assoc.* 19, 39–46.

Mutero, A., Pralavorio, M., Bride, J.M., Fournier, D., 1994. Resistance-associated point mutations in insecticide-insensitive acetylcholinesterase. *Proc. Natl Acad. Sci. USA* 91, 5922–5926.

Myers, P., Yousten, A.A., Davidson, E.W., 1979. Comparative studies of the mosquito-larval toxin of *Bacillus sphaericus* SSII-1 and 1593. *Can. J. Microbiol.* 25, 1227–1231.

Narasu, L.M., Gopinathan, K.P., 1988. Effect of *Bacillus sphaericus* 1593 toxin on choline acetyl transferase and mitochondrial oxidative activities of the mosquito larvae. *Ind. J. Biochem. Biophys.* 25, 253–256.

Neide, E., 1904. Botanische Beschneibung einiger sporenbildenden Bakterien. *Zentralbl. Bakteriol. Parasitenk. Infektionskr. Hyg. Abt.* 12, 1–32.

Nicolas, L., Lecroisey, A., Charles, J.-F., 1990. Role of the gut proteinases from mosquito larvae in the mechanism of action and the specificity of the *Bacillus sphaericus* toxin. *Can. J. Microbiol.* 36, 804–807.

Nicolas, L., Nielsen-LeRoux, C., Charles, J.-F., Delécluse, A., 1993. Respective role of the 42- and 51-kDa component of the *Bacillus sphaericus* toxin overexpressed in *Bacillus thuringiensis*. *FEMS Lett.* 106, 275–280.

Nielsen-LeRoux, C., Charles, J.-F., 1992. Binding of *Bacillus sphaericus* binary toxin to a specific receptor on midgut brush-border membranes from mosquito larvae. *Eur. J. Biochem. 210*, 585–590.

Nielsen-LeRoux, C., Charles, J.-F., Thiéry, I., Georghiou, G.P., 1995. Resistance in a laboratory population of *Culex quinquefasciatus* (Diptera: Culicidae) to *Bacillus sphaericus* binary toxin is due to a change in the receptor on midgut brush-border membranes. *Eur. J. Biochem. 228*, 206–210.

Nielsen-LeRoux, C., Pasquier, F., Charles, J.-F., Sinègre, G., Gaven, B., et al., 1997. Resistance to *Bacillus sphaericus* involves different mechanisms in *Culex pipiens* mosquito larvae (Diptera: Culicidae). *J. Med. Entomol. 34*, 321–327.

Nielsen-LeRoux, C., Pasteur, N., Prêtre, J., Charles, J.-F., Ben Cheick, H., et al., 2002. High resistance to *Bacillus sphaericus* binary toxin in *Culex pipiens* (Diptera: Culicidae): the complex situation of West-Mediterranean countries. *J. Med. Entomol. 39*, 729–735.

Nielsen-LeRoux, C., Rao, D.R., Rodcharoen-Murhy, J., Carron, A., Mani, T.R., et al., 2001. Various levels of cross-resistance to *Bacillus sphaericus* strains in *Culex pipiens* (Diptera: Culicidae) colonies resistant to *B. sphaericus* strain 2362. *Appl. Environ. Microbiol. 67*, 5049–5054.

Oei, C., Hindley, J., Berry, C., 1990. An analysis of the genes encoding the 51.4- and 41.9-kDa toxins of *Bacillus sphaericus* 2297 by deletion mutagenesis: the construction of fusion proteins. *FEMS Microbiol. Lett. 72*, 265–274.

Oei, C., Hindley, J., Berry, C., 1992. Binding of purified *Bacillus sphaericus* binary toxin and its deletion derivates to *Culex quinquefasciatus* gut: elucidation of functional binding domains. *J. Gen. Microbiol. 138*, 1515–1526.

Oliveira, C.M.O., Silva-Filha, M.H., Nielsen-LeRoux, C., Pei, G., Yuan, Z., et al., 2003a. Inheritance and mechanism of resistance to *Bacillus sphaericus* in *Culex quinquefasciatus* from China and Brazil (Diptera: Culicidae). *J. Med. Entomol. 41*, 58–64.

Oliveira, C.M.O., Silva-Filha, M.H., Beltrán, J., Nielsen-LeRoux, C., Pei, C.G., et al., 2003b. Biological fitness of *Culex quinquefasciatus* (Diptera: Culicidae) larvae resistant to *Bacillus sphaericus*. *J. Am. Mosq. Control Assoc. 19*, 125–129.

Park, H.W., Bideshi, D.K., Federici, B.A., 2003. Recombinant strain of *Bacillus thuringiensis* producing Cyt1A, Cry11B and the *Bacillus sphaericus* binary toxin. *Appl. Environ. Microbiol. 69*, 1331–1334.

Pauron, D., Barhanin, J., Amichot, M., Pralavorio, M., Bergé, J.-B., et al., 1989. Pyrethroid receptor in the insect Na$^+$ channel: alteration of its properties in pyrethroid-resistant flies. *Biochemistry 28*, 1673–1677.

Payne, J.M., Davidson, E.W., 1984. Insecticidal activity of crystalline parasporal inclusions and other components of the *Bacillus sphaericus* 1593 spore complex. *J. Invertebr. Pathol. 43*, 383–388.

Pei, G.F., Oliveira, C.M.F., Yuan, Z.M., Nielsen-LeRoux, C., Silva-Filha, M.H., et al., 2002. A strain of *Bacillus*

sphaericus causes slower development of resistance in *Culex quinquefasciatus*. *Appl. Environ. Microbiol. 68*, 3003–3009.

Poncet, S., Delécluse, A., Guido, A., Klier, A., Rapoport, G., 1994. Transfer and expression of the *cryIVB* and *cryIVD* genes of *Bacillus thuringiensis* subsp. *israelensis* in *Bacillus sphaericus* 2297. *FEMS Microbiol. Lett. 117*, 91–96.

Poncet, S., Bernard, C., Dervyn, E., Cayley, J., Klier, A., et al., 1997. Improvement of *Bacillus sphaericus* toxicity against dipteran larvae by integration, via homologous recombination, of the *cry11A* toxin gene from *Bacillus thuringiensis* subsp. israelensis. *Appl. Environ. Microbiol. 63*, 4413–4420.

Poopathi, S., Mani, T., Rao, R.D., Baskaran, G., Kabilan, L., 1999. Cross-resistance to *Bacillus sphaericus* strains in *Culex quinquefasciatus* resistant to *B. sphaericus* 1593M. *Southeast Asian J. Trop. Med. Public Health 30*, 477–481.

Porter, A.G., Davidson, E.W., Liu, J.W., 1993. Mosquitocidal toxins of bacilli and their genetic manipulation for effective biological control of mosquitoes. *Microbiol. Rev. 57*, 838–861.

Priest, F.G., Ebrup, L.V.Z., Carter, P.E., 1997. Distribution and characterization of mosquitocidal toxin genes in some strains of *Bacillus sphaericus*. *Appl. Environ. Microbiol. 63*, 1195–1198.

Rao, D.R., Mani, T.R., Rajendran, R., Joseph, A.S.J., Gajanana, A., et al., 1995. Development of a high level of resistance to *Bacillus sphaericus* in a field population of *Culex quinquefasciatus* from Kochi, India. *J. Am. Mosq. Control Assoc. 11*, 1–5.

Regis, L., Nielsen-LeRoux, C., 2000. Management of resistance to bacterial vector control. In: Charles, J.-F., Delécluse, A., Nielsen-LeRoux, C. (Eds.), Entomopathogenic Bacteria: from Laboratory to Field Application. Kluwer Academic Publisher, Dordrecht, Boston, London, pp. 419–433.

Regis, L., Silva-Filha, M.H.N.L., de Oliveira, C.M.F., Rios, E.M., da Silva, S.B., et al., 1995. Integrated control measures against *Culex quinquefasciatus*, the vector of filariasis in Recife. *Mem. Inst. Oswaldo Cruz. 90*, 115–119.

Rodcharoen, J., Mulla, M.S., 1994. Resistance development in *Culex quinquefasciatus* (Diptera: Culicidae) to the microbial agent *Bacillus sphaericus*. *J. Econ. Entomol. 87*, 1133–1140.

Rodcharoen, J., Mulla, M.S., 1997. Biological fitness of *Culex quinquefasciatus* (Diptera: Culicidae) susceptible and resistant to *Bacillus sphaericus*. *J. Med. Entomol. 34*, 5–10.

Schirmer, J., Wieden, H.-J., Rodnina Marina, V., Aktories, K., 2002a. Inactivation of the elongation factor Tu by mosquitocidal toxin-catalyzed mono-ADP-ribosylation. *Appl. Environ. Microbiol. 68*, 4894–4899.

Schirmer, J., Just, I., Aktories, K., Schirmer, J., Wieden, H.-J., et al., 2002b. The ADP-ribosylating mosquitocidal toxin from *Bacillus sphaericus*: proteolytic activation, enzyme activity, and cytotoxic effects. *J. Biol. Chem.* 277, 11941–11948.

Schwartz, J.-L., Potvin, L., Coux, F., Charles, J.-F., Berry, C., *et al.*, **2001**. Permeabilization of model lipid membranes by *Bacillus sphaericus* mosquitocidal binary toxin and its individual components. *J. Membr. Biol.* 184, 171–183.

Sebo, P., Bennardo, T., de la Torre, F., Szulmajster, J., **1990**. Delineation of the minimal portion of the *Bacillus sphaericus* 1593M toxin required for the expression of larvicidal activity. *Eur. J. Biochem.* 194, 161–165.

Servant, P., Rosso, M.L., Hamon, S., Poncet, S., Delécluse, A., *et al.*, **1999**. Production of Cry11A and Cry11Ba toxins in *Bacillus sphaericus* confers toxicity towards *Aedes aegypti* and resistant *Culex* populations. *Appl. Environ. Microbiol.* 65, 3021–3026.

Shanmugavelu, M., Rajamohan, F., Kathirvel, M., Elangovan, G., Dean, D.H., *et al.*, **1998**. Functional complementation of nontoxic mutant binary toxins of *Bacillus sphaericus* 1593M generated by site-directed mutagenesis. *Appl. Environ. Microbiol.* 64, 756–759.

Shi, Y., Yuan, Z., Cai, Q., Yu, J., Yan, J., *et al.*, **2001**. Cloning and expression of the binary toxin gene from *Bacillus sphaericus* IAB872 in a crystal-minus *Bacillus thuringiensis* subsp. *israelensis*. *Curr. Microbiol.* 43, 21–25.

Silva-Filha, M.H., Nielsen-LeRoux, C., Charles, J.-F., **1997**. Binding kinetics of *Bacillus sphaericus* binary toxin to midgut brush border membranes of *Anopheles* and *Culex* sp. *larvae*. *Eur. J. Biochem.* 247, 754–761.

Silva-Filha, M.H., Nielsen-LeRoux, C., Charles, J.F., **1999**. Identification of the receptor for *Bacillus sphaericus* crystal toxin in the brush border membrane of the mosquito *Culex pipiens* (Diptera: Culicidae). *Insect Biochem. Mol. Biol.* 29, 711–721.

Silva-Filha, M.H., Peixoto, C.A., **2003**. Immunocytochemical localization of the *Bacillus sphaericus* binary toxin components in *Culex quinquefasciatus* (Diptera: Culicidae) larvae midgut. *Pest. Biochem. Physiol.* 77, 138–146.

Silva-Filha, M.-H., Regis, L., **1997**. Reversal of a low-level resistance to *Bacillus sphaericus* in a field population of *Culex quinquefasciatus* (Diptera: Culicidae) from an urban area of Recife, Brazil. *J. Econ. Entomol.* 90, 299–303.

Silva-Filha, M.-H., Regis, L., Nielsen-LeRoux, C., Charles, J.-F., **1995**. Low-level resistance to *Bacillus sphaericus* in a field-treated population of *Culex quinquefasciatus* (Diptera: Culicidae). *J. Econ. Entomol.* 88, 525–530.

Simons, K., Toomre, D., **2000**. Lipid rafts and signal transduction. *Nat. Rev. Mol. Cell Biol.* 1, 31–39.

Sinègre, G., Babinot, M., Vigo, G., Jullien, J.-L., **1993**. *Bacillus sphaericus* et démoustication urbaine. In: Bilan de cinq années d'utilisation expérimentale de la spécialité Spherimos® dans le sud de la France. Document E.I.D.L.M. N°62. Entente Interdépartementale pour la Démoustication du Littoral Méditerranéen, Montpellier. 21pp.

Sinègre, G., Babinot, M., Quermel, J.-M., Gavon, B., **1994**. First field occurrence of *Culex pipiens* resistance to *Bacillus sphaericus* in southern France. Abstr. VIIIth Eur. Meet. Soc. Vector Ecol., 5–8 September, Barcelona, p. 17.

Singer, S., **1973**. Insecticidal activity of recent bacterial isolates and their toxins against mosquito larvae. *Nature* 244, 110–111.

Singer, S., **1974**. Entomogenous bacilli against mosquito larvae. *Dev. Industr. Microbiol.* 15, 187–194.

Singer, S., **1977**. Isolation and development of bacterial pathogens in vectors. In: "Biological regulation of vectors," DHEW Publication No. 77-1180. National Institutes of Health, Bethesda, MD, pp. 3–18.

Singh, G.J.P., Gill, S.S., **1988**. An electron microscope study of the toxic action of *Bacillus sphaericus* in *Culex quinquefasciatus* larvae. *J. Invertebr. Pathol.* 52, 237–247.

Skovmand, O., Bauduin, S., **1997**. Efficacy of granular formulation of *Bacillus sphaericus* against *Culex quinquefasciatus* and *Anopheles gambiae* in West African countries. *J. Vector Ecol.* 22, 43–51.

Stray, J.E., Klowden, M.J., Hurlbert, R.E., **1988**. Toxicity of *Bacillus sphaericus* crystal toxin to adult mosquitoes. *Appl. Environ. Microbiol.* 54, 2320–2321.

Tanapongpipat, S., Nantapong, N., Cole, J., Panyim, S., **2003**. Stable integration and expression of mosquito-larvicidal genes from *Bacillus thuringiensis* subsp. *israelensis* and *Bacillus sphaericus* into the chromosome of *Enterobacter amnigenus*: a potential breakthrough in mosquito biocontrol. *FEMS Microbiol. Lett.* 221, 343–248.

Thanabalu, T., Berry, C., Hindley, J., **1993**. Cytotoxicity and ADP-ribosylating activity of the mosquitocidal toxin from *Bacillus sphaericus* SSII-1: possible roles of the 27-kilodalton and 70-kilodalton peptides. *J. Bacteriol.* 175, 2314–2320.

Thanabalu, T., Hindley, J., Berry, C., **1992**. Proteolytic processing of the mosquitocidal toxin from *Bacillus sphaericus* SSII-1. *J. Bacteriol.* 174, 5051–5056.

Thanabalu, T., Hindley, J., Jackson-Yap, J., Berry, C., **1991**. Cloning, sequencing, and expression of a gene encoding a 100-kilodalton mosquitocidal toxin from *Bacillus sphaericus* SSII-1. *J. Bacteriol.* 173, 2776–2785.

Thanabalu, T., Porter, A.G., **1995**. *Bacillus sphaericus* gene encoding a novel class of mosquitocidal toxin with homology to *Clostridium* and *Pseudomonas* toxins. *Gene* 61, 4031–4036.

Thanabalu, T., Porter, A.G., **1996**. A *Bacillus sphaericus* gene encoding a novel type of mosquitocidal toxin of 31.8 kDa. *Gene* 170, 85–89.

Thiéry, I., de Barjac, H., **1989**. Selection of the most potent *Bacillus sphaericus* strains, based on activity ratios determined on three mosquito species. *Appl. Microbiol. Biotechnol.* 31, 577–581.

Thiéry, I., Hamon, S., Delécluse, A., Orduz, S., **1998**. The introduction into *Bacillus sphaericus* of the *Bacillus thuringiensis* subsp. *medellin* cyt1Ab1 gene results in higher susceptibility of resistant mosquito larva populations to _B. sphaericus_. Appl. Environ. Microbiol. _64_, 3910–3916.

Trisrisook, M., Pantuwatana, S., Bhumiratana, A., Panbangred, W., **1990**. Molecular cloning of the 130-kilodalton mosquitocidal δ-endotoxin gene of _Bacillus thuringiensis_ subsp. _israelensis_ in _Bacillus sphaericus_. Appl. Environ. Microbiol. _56_, 1710–1716.

Udenfriend, S., Kodukula, K., **1995**. How glycosylphosphatidylinositol-anchored membrane proteins are made. _Annu. Rev. Biochem. 64_, 563–591.

Wirth, M.C., Walton, W.E., Federici, B.A., **2000a**. Cyt1A from _Bacillus thuringiensis_ restores toxicity of _Bacillus sphaericus_ against resistant _Culex quinquefasciatus_ (Diptera: Culicidae). _J. Med. Entomol. 37_, 401–407.

Wirth, M.C., Georghiou, G.P., Malik, J.I., Abro, G.H., **2000b**. Laboratory selection for resistance to _Bacillus sphaericus_ in _Culex quinquefasciatus_ (Diptera: Culicidae) from California, USA. _J. Med. Entomol. 37_, 534–540.

Woodburn, M.A., Yousten, A.A., Hilu, K.H., **1995**. Random amplified polymorphic DNA fingerprinting of mosquito-pathogenic and nonpathogenic strains of _Bacillus sphaericus_. _Int. J. Syst. Bacteriol. 45_, 212–217.

Yousten, A.A., de Barjac, H., Hedrick, J., Cosmao Dumanoir, V., Myers, P., **1980**. Comparison between bacteriophage typing and serotyping for the differentiation of _Bacillus sphaericus_ strains. _Ann. Microbiol. (Inst. Pasteur) 131B_, 297–308.

Yousten, A.A., Davidson, E.W., **1982**. Ultrastructural analysis of spores and parasporal crystals formed by _Bacillus sphaericus_ 2297. _Appl. Environ. Microbiol. 44_, 1449–1455.

Yousten, A.A., Madhekar, N., Wallis, D.A., **1984**. Fermentation conditions affecting growth, sporulation, and mosquito larval toxin formation by _Bacillus sphaericus_. _Dev. Industr. Microbiol. 25_, 757–762.

Yuan, Z., Nielsen-Le Roux, C., Pasteur, N., Delécluse, A., Charles, J.-F., _et al._, **1998**. Detection of the binary toxin genes and proteins of several _Bacillus sphaericus_ strains and their toxicity against susceptible and resistant _Culex pipiens_. _Acta Entomol. Sinica. 41_, 337–342.

Yuan, Z., Rang, C., Maroun, R.C., Juarez-Perez, V., Frutos, R., _et al._, **2001**. Identification and molecular structural prediction analysis of a toxicity determinant in the _Bacillus sphaericus_ crystal larvicidal toxin. _Eur. J. Biochem. 268_, 2751–2760.

Yuan, Z., Zhang, Y., Cai, Q., Liu, E.Y., **2000**. High-level field resistance to _Bacillus sphaericus_ C3-41 in _Culex quinquefasciatus_ from southern China. _Biocontrol. Sci. Technol. 10_, 41–49.

A8 Addendum: *Bacillus sphaericus* Taxonomy and Genetics

C Berry, Cardiff University Park Place, Wales, UK
M H N L Silva Filha, Centro de Pesquisas Aggeu Magalhães-Fundação Oswaldo Cruz, Recife-PE, Brazil

© 2010 Elsevier B.V. All Rights Reserved

Although still widely referred to as *Bacillus sphaericus*, a taxonomic revision has redesignated this bacterium as *Lysinibacillus sphaericus* (Ahmed *et al.*, 2007). Our understanding of this bacterium has been enhanced by genome sequencing of the mosquitocidal serotype H5a,5b strain C3-41 (Hu *et al.*, 2008). The genome revealed a lack of important enzymes and transport systems that explain the inability of this species to metabolize sugars. A new gene encoding an Mtx2/Mtx3-like protein (Mtx4) was also identified along with a further pseudogene in this family that is located upstream of the *mtx3* gene. This implies that recombination may be a feature of this protein family and all the genes in this family appear to be associated with mobile genetic elements (Hu *et al.*, 2008). The *mtx2* gene in strain C3-41 lies close to the *mtx1* gene and the *mtx4* gene lies close upstream of the *bin* genes, but other toxin genes are distributed around the genome. It is of interest to note that in this strain, *mtx1* contains a stop codon and appears, therefore, to be a pseudogene. The *bin* genes encoding BinA and BinB toxins are present along with the *mtx4* gene on an ~35 kb region that is duplicated in the genome and on a large plasmid pBsph. The existence of such duplication had been suggested earlier for strain 2297 (Poncet *et al.*, 1997).

A8.1. Mtx1 Toxin Structure

The structure and activity of the Mtx1 toxin has been studied in some detail (reviewed in Carpusca *et al.*, 2006). The 100 kDa protein is processed in the insect gut to a 27 kDa fragment with ADP-ribosylating activity and a 70 kDa fragment thought to mediate entry of the 27 kDa portion into target cells (Thanabalu *et al.*, 1992, 1993; Schirmer *et al.*, 2002). ADP-ribosylation of target proteins in mosquito cells and in other eukaryotic and bacterial cells (Schirmer *et al.*, 2002) has been reported and appears to occur by modification of arginine residues (Schirmer *et al.*, 2002). Enzymatic activity of the activated 27 kDa protein can be inhibited by the binding of the 70 kDa protein and a 20 amino acid region responsible for this effect by competition with NAD$^+$ (Carpusca *et al.*, 2004). The C-terminal 70 kDa fragment contains four QxW3 repeats, characteristic of ricin toxin lectin-like sequence motifs (Hazes and Read, 1995). A further, noncompetitive inhibition of ADP-ribosyl transferase activity by these lectin domains has also been reported (Carpusca *et al.*, 2004). Crystal structures for the N-terminal region (Reinert *et al.*, 2006) and the holotoxin (Treiber *et al.*, 2008) illustrate many of the features of the toxin including auto-inhibition and activation sites.

A8.2. New Toxins

In addition to identification of the *mtx4* gene mentioned earlier, other toxins have been identified recently in *B. sphaericus* strains. Sphaericolysin is a 53 kDa-secreted protein that, on injection, shows insecticidal activity against the German cockroach, *Blattela germanica* and to a lesser degree, against *Spodoptera litura* (Nishiwaki *et al.*, 2007). Originally isolated from strain A3-2, it is also encoded by the genome of strain C3-41 (Hu *et al.*, 2008). Strain A3-2 was isolated from the ant lion *Myrmeleon bore*, which may use the toxins to kill its prey (Nishiwaki *et al.*, 2007). Sphaericolysin appears to be a member of the cholesterol-dependent cytolysins and causes neuronal damage in intoxicated insects.

As detailed in section 8.6 of the main article, strains such as LP1-G and IAB59 were found to be able to overcome resistance in mosquitoes that had been selected to be insensitive to other Bin toxin-producing strains. A new crystal toxin pair responsible for the additional mosquitocidal activity has now been identified in IAB59, LP1-G, and some other strains (Jones et al., 2007). One component of the new toxin, the Cry49 protein, is related to both BinA and BinB. This extends the family of Bin-like toxins, which now also includes the *Bacillus thuringiensis* Cry35 and Cry36 proteins. The second component, Cry48, is related to the family of three-domain Cry toxins of *B. thuringiensis*. Neither Cry48 nor Cry49 shows activity alone and both components are required for activity, which is optimal at approximately a molar ratio of 1:1 and under these conditions, the Cry48/Cry49 pair has comparable toxicity to the Bin toxins (Jones *et al.*, 2007). However, in *B. sphaericus* strains that produce Cry48/Cry49 toxicity appears to be suboptimal because of the production of Cry48 at relatively low levels. Insecticidal activity of Cry48/Cry49 appears to be restricted to mosquitoes in the genus *Culex* (Jones et al., 2008) and ultrastructural effects on intoxicated cells include disruption of the endoplasmic reticulum and production of cytoplasmic vacuoles (de Melo *et al.*, 2009) as observed previously with Bin-intoxicated cells. Mitochondrial vacuolation was also seen with Cry48/Cry49 but this is not a feature of Bin activity although it has been reported previously for 3-domain Cry toxins from *B. thuringiensis* (de Melo *et al.*, 2009).

A8.3. Characterization of Other Bin Toxin Receptors in Culicide Larvae

Bin toxin receptors in *C. quinquefasciatus* and *An. gambiae*, two target species of *Bacillus sphaericus*, have been identified and named Cqm1 and Agm3, respectively (Romão *et al.*, 2006; Opota *et al.*, 2008). An ORF 1743 base pair long from *C. quinquefasciatus cqm1* cDNA showed 84 nucleotide differences compared to *cpm1* sequence from *C. pipiens*, previously described, resulting in 16 amino acid substitutions and 97% identity in the deduced protein (Romão *et al.*, 2006). The ortholog Agm3 cloned from a 1767 bp *An. gambiae* cDNA has 66% amino acid identity to Cpm1 (Opota *et al.*, 2008). Both Cqm1 and Agm3 show features similar to those of Cpm1 and are GPI-anchored α-glucosidases located on the gastric caeca and posterior midgut of larvae with capacity to bind specifically to the Bin toxin. Sequence alignments showed a strong conservation of α-glucosidases among dipterans,

and *Ae. aegypti*, despite being a Bin toxin-refractory species, shows an ortholog sharing 74% identity to Cpm1 (Opota *et al.*, 2008). The conservation of this group indicates that slight differences in the amino acid sequence or post-translational modifications might be critical to the capacity to bind the Bin toxin, and characterization of these epitopes opens perspectives for the improvement of its spectrum of action.

A8.4. Characterization of Resistance Alleles and Resistance Management

The molecular basis of *B. sphaericus* resistance in *Culex* larvae has advanced particularly because of the identification of alleles of the receptor genes that confer larval refractoriness to the Bin toxin. The first allele, $cpm1_{GEO}$, from the laboratory-selected *C. pipiens* GEO colony has a nonsense mutation, which resulted in the production of a truncated Cpm1 protein that has no GPI-anchor and hence not present in the cell membrane, as described in the main article (Darboux *et al.*, 2002). A second independent event was detected in a highly resistant laboratory colony of *C. quinquefasciatus* from Brazil that displayed a failure of Bin toxin binding to midgut microvilli of larvae (Oliveira *et al.*, 2004). Resistance in this colony is conferred by the allele $cqm1_{REC}$ which contains a 19-nucleotide deletion that introduces a stop codon and also prevents the expression of a full-length GPI-anchored protein in midgut epithelium (Romão *et al.*, 2006). Complexity of field-selected *C. pipiens* BP population has recently been clarified, and two alleles, $cpm1_{BP}$ and $cpm1_{BP}$-del, were found to coexist (Darboux *et al.*, 2007). The $cpm1_{BP}$ allele displays a nonsense mutation and encodes a truncated protein missing GPI-anchoring signals and produces similar functional effects to those reported for $cpm1_{GEO}$ and $cqm1_{REC}$ alleles. The $cpm1_{BP}$-del allele displays a novel mechanism in which a transposable element has been inserted in the coding sequence thereby inducing a new mRNA splicing event. Functionally, the protein coded by $cpm1_{BP}$-del lacks 66 amino acids and does not bind to the Bin toxin, despite being a GPI-anchored protein. Mechanisms underlying the resistance of *C. pipiens* field populations SPHAE and TUNIS, which have functional Cpm1 receptors, remain unsolved (Nielsen-LeRoux *et al.*, 2002).

Resistance in laboratory- and field-selected strains already studied was shown to be a trait inherited recessively, which is an obstacle to resistance detection during the early stages of selection (Nielsen-LeRoux *et al.*, 2002; Oliveira *et al.*, 2004; Amorim *et al.*, 2007). Characterization of alleles is essential to develop tools for monitoring the risk of resistance, which is far the most important issue concerning the large-scale utilization of *B. sphaericus*. Detection of the $cqm1_{REC}$ allele in field populations of *C. quinquefasciatus* from Recife urban area, Brazil, was performed by diagnostic polymerase chain reaction and revealed the presence of $cqm1_{REC}$ at detectable frequencies in the order of 10^{-2} and 10^{-3} in exposed and nonexposed populations, respectively (Chalegre *et al.*, 2009). Considering that multiple events are involved in resistance, further investigation is needed to characterize the repertoire of mutations to develop diagnostic tools. Management strategies are needed to optimize the use of *B. sphaericus* and besides monitoring resistance, the development of biolarvicides combining insectidal proteins with different sites of action can dramatically reduce the potential of resistance development (Federici *et al.*, 2007). Associations among the Bin toxin with *B. thuringiensis* dipteran-active toxins have proved to be effective and other components, such as Mtx and Cry48Aa/Cry49Aa toxins from *B. sphaericus* strains, are also candidates for overcoming resistance (Wirth *et al.*, 2004, 2007; Jones *et al.*, 2008). Bin toxin, along with other insecticidal proteins, is to date the most potent and environmentally safe molecule and can be effectively employed in mosquito-control programs.

References

Ahmed, I., Yokota, A., Yamazoe, A., Fujiwara, T., 2007. Proposal of *Lysinibacillus boronitolerans* gen. nov. sp. nov., and transfer of *Bacillus fusiformis* to *Lysinibacillus fusiformis* comb. nov. and *Bacillus sphaericus* to *Lysinibacillus sphaericus* comb. nov. *Int. J. Syst. Evol. Microbiol.* 57, 1117–1125.

Amorim, L.B., Oliveira, C.M.F., Rios, E.M., Regis, L., Silva-Filha, M.H.N.L., 2007. Development of *Culex quinquefasciatus* resistance to *Bacillus sphaericus* strain IAB59 needs long term selection pressure. *Biol. Control* 42, 155–160.

Carpusca, I., Schirmer, J., Aktories, K., 2004. Two-site autoinhibition of the ADP-ribosylating mosquitocidal toxin (MTX) from *Bacillus sphaericus* by its 70-kDa ricin-like binding domain. *Biochemistry* 43, 12009–12019.

Carpusca, I., Jank, T., Aktories, K., 2006. *Bacillus sphaericus* mosquitocidal toxin (MTX) and pierisin: the enigmatic offspring from the family of ADP-ribosyltransferases. *Mol. Microbiol.* 62, 621–630.

Chalegre, K.D., Romao, T.P., Amorim, L.B., Anastacio, D.B., de Barros, R.A., de Oliveira, C.M., Regis, L., de-Melo-Neto, O.P., Silva-Filha, M.H., 2009. Detection of an allele conferring resistance to *Bacillus sphaericus* binary toxin in *Culex quinquefasciatus* populations by molecular screening. *Appl. Environ. Microbiol.* 75, 1044–1049.

Darboux, I., Pauchet, Y., Castella, C., Silva-Filha, M.H., Nielsen-LeRoux, C., Charles, J.F., Pauron, D., 2002. Loss of the membrane anchor of the target receptor is a mechanism of bioinsecticide resistance. *Proc. Natl. Acad. Sci. USA* 99, 5830–5835.

Darboux, I., Charles, J.F., Pauchet, Y., Warot, S., Pauron, D., 2007. Transposon-mediated resistance to *Bacillus sphaericus* in a field-evolved population of *Culex pipiens* (Diptera: Culicidae). *Cell. Microbiol.* 9, 2022–2029.

de Melo, J.V., Jones, G.W., Berry, C., Vasconcelos, R.H., de Oliveira, C.M., Furtado, A.F., Peixoto, C.A., Silva-Filha, M.H., 2009. Cytopathological effects of *Bacillus sphaericus* Cry48Aa/Cry49Aa toxin on binary toxin-susceptible and -resistant *Culex quinquefasciatus* larvae. *Appl. Environ. Microbiol.* 75, 4782–4789.

Federici, B.A., Park, H.W., Bideshi, D.K., Wirth, M.C., Johnson, J.J., Sakano, Y., Tang, M., 2007. Developing recombinant bacteria for control of mosquito larvae. *J. Am. Mosq. Control Assoc.* 23, 164–175.

Hazes, B., Read, R.J., 1995. A mosquitocidal toxin with a ricin-like cell-binding domain. *Struct. Biol* 2, 358–359.

Hu, X., Fan, W., Han, B., Liu, H., Zheng, D., Li, Q., Dong, W., Yan, J., Gao, M., Berry, C., Yuan, Z., 2008. Complete genome sequence of the mosquitocidal bacterium *Bacillus sphaericus* C3–41 and comparison with those of closely related *Bacillus* species. *J. Bacteriol.* 190, 2892–2902.

Jones, G.W., Nielsen-Leroux, C., Yang, Y., Yuan, Z., Dumas, V.F., Monnerat, R.G., Berry, C., 2007. A new Cry toxin with a unique two-component dependency from *Bacillus sphaericus*. *FASEB J.* 21, 4112–4120.

Jones, G.W., Wirth, M.C., Monnerat, R.G., Berry, C., 2008. The Cry48Aa-Cry49Aa binary toxin from *Bacillus sphaericus* exhibits highly restricted target specificity. *Environ. Microbiol.* 10, 2418–2424.

Nielsen-LeRoux, C., Pasteur, N., Pretre, J., Charles, J.F., Sheikh, H.B., Chevillon, C., 2002. High resistance to *Bacillus sphaericus* binary toxin in *Culex pipiens* (Diptera: Culicidae): the complex situation of west Mediterranean countries. *J. Med. Entomol.* 39, 729–735.

Nishiwaki, H., Nakashima, K., Ishida, C., Kawamura, T., Matsuda, K., 2007. Cloning, functional characterization, and mode of action of a novel insecticidal pore-forming toxin, sphaericolysin, produced by *Bacillus sphaericus*. *Appl. Environ. Microbiol.* 73, 3404–3411.

Oliveira, C.M., Silva-Filha, M.H., Nielsen-Leroux, C., Pei, G., Yuan, Z., Regis, L., 2004. Inheritance and mechanism of resistance to *Bacillus sphaericus* in *Culex quinquefasciatus* (Diptera: Culicidae) from China and Brazil. *J. Med. Entomol.* 41, 58–64.

Opota, O., Charles, J.F., Warot, S., Pauron, D., Darboux, I., 2008. Identification and characterization of the receptor for the *Bacillus sphaericus* binary toxin in the malaria vector mosquito, *Anopheles gambiae*. *Comp. Biochem. Physiol. B Biochem. Mol. Biol.* 149, 419–427.

Poncet, S., Bernard, C., Dervyn, E., Cayley, J., Klier, A., Rapoport, G., 1997. Improvement of *Bacillus sphaericus* toxicity against dipteran larvae by integration, via homologous recombination, of the Cry11A toxin gene from *Bacillus thuringiensis* subsp. *israelensis*. *Appl. Environ. Microbiol. 63*, 4413–4420.

Reinert, D.J., Carpusca, I., Aktories, K., Schulz, G.E., 2006. Structure of the mosquitocidal toxin from *Bacillus sphaericus*. *J. Mol. Biol. 357*, 1226–1236.

Romão, T.P., de Melo Chalegre, K.D., Key, S., Ayres, C.F., Fontes de Oliveira, C.M., de-Melo-Neto, O.P., Silva-Filha, M.H., 2006. A second independent resistance mechanism to *Bacillus sphaericus* binary toxin targets its alpha-glucosidase receptor in *Culex quinquefasciatus*. *FEBS J. 273*, 1556–1568.

Schirmer, J., Just, I., Aktories, K., 2002. The ADP-ribosylating mosquitocidal toxin from *Bacillus sphaericus*: proteolytic activation, enzyme activity, and cytotoxic effects. *J. Biol. Chem. 277*, 11941–11948.

Thanabalu, T., Berry, C., Hindley, J., 1993. Cytotoxicity and ADP-ribosylating activity of the mosquitocidal toxin from *Bacillus sphaericus* SSII-1: possible roles of the 27- and 70-kilodalton peptides. *J. Bacteriol. 175*, 2314–2320.

Thanabalu, T., Hindley, J., Berry, C., 1992. Proteolytic processing of the mosquitocidal toxin from *Bacillus sphaericus* SSII-1. *J. Bacteriol. 174*, 5051–5056.

Treiber, N., Reinert, D.J., Carpusca, I., Aktories, K., Schulz, G.E., 2008. Structure and mode of action of a mosquitocidal holotoxin. *J. Mol. Biol. 381*, 150–159.

Wirth, M.C., Jiannino, J.A., Federici, B.A., Walton, W.E., 2004. Synergy between toxins of *Bacillus thuringiensis* subsp. *israelensis* and *Bacillus sphaericus*. *J. Med. Entomol. 41*, 935–941.

Wirth, M.C., Yang, Y., Walton, W.E., Federici, B.A., Berry, C., 2007. Mtx toxins synergize *Bacillus sphaericus* and Cry11Aa against susceptible and insecticide-resistant *Culex quinquefasciatus* larvae. *Appl. Environ. Microbiol. 73*, 6066–6071.

9 Insecticidal Toxins from *Photorhabdus* and *Xenorhabdus*

**R H ffrench-Constant, N Waterfield, and
P Daborn**, University of Bath, Bath, UK

© 2010, 2005 Elsevier B.V. All Rights Reserved

9.1. Introduction

9.1.1. The Biology of *Photorhabdus* and *Xenorhabdus*

9.1.1.1. Bacteria, nematodes, and insects
Photorhabdus and *Xenorhabdus* bacteria live in association with nematodes from the families Heterorhabditidae and Steinernematidae, respectively (Forst *et al.*, 1997). Both genera bacteria are members of the Enterobacteriaceae, and are found as symbionts within the guts of the infective juvenile nematodes (ffrench-Constant *et al.*, 2003). These infective juvenile nematodes seek out and invade insects, whereupon the bacteria are released from the mouth of the nematode directly into the insect blood system or hemocoel (Forst *et al.*, 1997). The bacteria then multiply rapidly within the insect, apparently unaffected by the insect immune system (Silva *et al.*, 2002). In the case of *Photorhabdus*, the bacteria first colonize the gut and grow rapidly in the hemocoel only, later colonizing all the other insect tissues (Silva *et al.*, 2002). The nematodes also multiply within the insect cadaver, feeding off both the bacteria and the degraded insect tissues (Forst *et al.*, 1997). Finally, a new generation of infective juvenile nematodes reacquires the bacteria and leaves the insect in search of new hosts.

During the course of this infection process the insect dies and the bacteria are inferred to secrete a range of toxins capable of killing the insect or attacking specific tissues (ffrench-Constant *et al.*, 2003). For the purposes of this chapter, it is important to note that during this process the bacteria lie within the hemocoel of the insect or within the insect tissues. Therefore, any toxin action on the gut *in vivo* is likely to be from the hemocoel side of the gut. The existence of toxins acting from the lumen side of the gut is therefore, unexpected from the biology of the organism (Bowen *et al.*, 1998). In this respect, this chapter is divided into sections on the orally active Toxin complexes (Tc's) and on the injectably active Makes *c*aterpillars *f*loppy (Mcf) toxins.

9.1.1.2. Bacterial nomenclature
As this chapter involves the description of toxins from a wide range of *Photorhabdus* and *Xenorhabdus* strains, some description of current bacterial taxonomy is necessary to clarify subsequent discussion. The genus *Xenorhabdus* contains a number of species including *X. nematophilus*, *X. beddingii*, *X. bovienii*, and *X. poinarii* (Forst *et al.*, 1997). The taxonomy of the genus *Photorhabdus*, however, is more complicated and has recently been revised. In this revision (Fischer-Le Saux *et al.*, 1999), the genus was recently split into three species, *P. luminescens*, *P. temperata*, and *P. asymbiotica*. Two of these genera have also been subdivided into new subspecies. Thus, the *P. luminescens* group has three subspecies, *luminescens*, *akhurstii*, and *laumondii*, and the *P. temperata* group has one subspecies, *temperata*. The third species group, *P. asymbiotica*, which consists of isolates from human wounds recovered in the apparent absence of a nematode vector (Gerrard *et al.*, 2003), has not yet been subdivided. Most of the genetic analyses have been performed on a limited number of strains: *P. luminescens* subsp. *akhurstii* strain

W14, *P. luminescens* subsp. *laumondii* strain TT01, and *P. temperata* strains K122 and NC1. For brevity these strains will be referred to as *Photorhabdus* strains W14, TT01, K122, or NC1, or simply W14, TT01, K122, and NC1, except where reference to their different specific designation is relevant to the discussion.

9.1.2. The Need for Alternatives to Bt

To date, the majority of transgenes deployed in insect resistant crops are *Cry* genes from *Bacillus thuringiensis* or "Bt." Despite the diversity of *Cry* genes cloned from *B. thuringiensis*, only a few Cry proteins are toxic towards specific pests. This observation has led to concerns about the development of resistance to specific Cry toxins and over the subsequent management of cross-resistance to other toxins occupying the same or similar binding sites (Ives, 1996; McGaughey *et al.*, 1998). These concerns are increasing as specific cases of laboratory-developed Bt resistance are documented in pest insects (McGaughey, 1985; Gahan *et al.*, 2001). *Bacillus thuringiensis* has also proved to be a useful source of other non-*Cry* genes such as the vegetative insecticidal proteins or "Vips" (Estruch *et al.*, 1996; Yu *et al.*, 1997); however, this organism clearly has a limited capacity to produce further toxins. The work described here on the isolation and characterization of insecticidal toxins from *Photorhabdus* and *Xenorhabdus* has been performed partly in the search for alternative novel insecticidal proteins for insect control.

9.2. The Toxin Complexes

9.2.1. Discovery of the Toxin Complexes

9.2.1.1. Purification and cloning of *Photorhabdus* toxins
Despite the fact that *Photorhabdus* is released directly into the insect hemocoel by its nematode host, the culture supernatant of *Photorhabdus* strain W14 shows unexpected oral toxicity to the lepidopteran model *Manduca sexta* (Bowen and Ensign, 1998). A toxic high molecular weight-protein fraction was purified from the W14 supernatant by sequential ultrafiltration, dimethyl aminoethyl (DEAE) anion-exchange chromatography, and gel filtration (Bowen and Ensign, 1998). As a final purification step, high-performance liquid chromatography (HPLC) anion-exchange chromatography was used to separate four peaks or "Toxin complexes" A, B, C, and D (Bowen *et al.*, 1998). Purified Toxin complex A (Tca) has a median lethal dose of 875 ng cm^{-2} of diet against *M. sexta*, and is therefore as active as some Bt Cry proteins (Bowen

et al., 1998). The histopathology of orally ingested Tca shows the primary site of action to be the midgut, where cells of the midgut epithelium produce blebs as the epithelium itself disintegrates (Blackburn *et al.*, 1998). Interestingly, injection of purified Tca also results in destruction of the midgut with a similar histopathology, suggesting that Tca can act on the gut from either the lumen or the hemocoel (Blackburn *et al.*, 1998).

Each of the HPLC purified complexes migrates as a single or double species on a native gel but resolves into numerous different polypeptides on a denaturing sodium dodecylsulfate (SDS) gel (Bowen *et al.*, 1998). The *toxin complex* (*tc*) encoding genes were cloned by raising both monoclonal and polyclonal antisera against the purified complexes and using these to screen an expression library. Each of the four complexes Tca, Tcb, Tcc, and Tcd is encoded by four independent loci *tca*, *tcb*, *tcc*, and *tcd* (Bowen *et al.*, 1998). Each of these loci consists of an operon with successive open reading frames, for example, *tcaA*, *tcaB*, *tcaC*, and *tcaZ* (**Figure 1**). These open reading frames encode the different polypeptides that can be resolved from the native complexes as confirmed by N-terminal sequencing of the individual polypeptides resolved on an SDS gel (Bowen *et al.*, 1998). Genetic knockout of each of the *tca*, *tcb*, *tcc*, and *tcd* loci in turn showed that both *tca* and *tcd* contribute to oral toxicity to *M. sexta* and that removal of both these loci in the same strain renders the double mutant nontoxic (Bowen *et al.*, 1998).

Similar independent purification approaches to the supernatant of strain W14 identified two high molecular weight complexes, confusingly termed "toxin A" and "toxin B," with oral activity to the coleopteran *Diabrotica undecimpunctata howardi* (Guo *et al.*, 1999). These two toxins correspond to the species TcdA and TcbA, respectively, therefore confirming that both Tcd and Tcb also have activity against Coleoptera. The native molecular weights of TcdA and TcbA were estimated at 860 kDa, leading to the suggestion that they are tetramers of the 208 kDa species observed on an SDS gel (Guo *et al.*, 1999). These species can also be proteolytically cleaved by proteases found in the culture supernatant and an increase in insecticidal activity associated with the cleaved form of the toxin was reported (Guo *et al.*, 1999). However, the nature of the protease responsible and the relevance of the cleavage in the biological activity of the toxins remain obscure.

9.2.1.2. Cloning of *Xenorhabdus* toxins
Supernatants of some *Xenorhabdus* strains also show oral

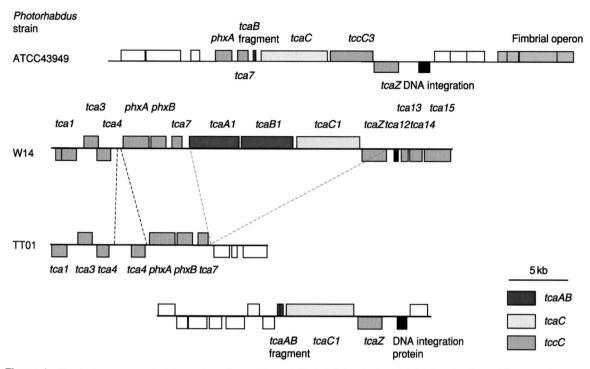

Figure 1   The *toxin complex a* (*tca*) locus from three different *Photorhabdus* strains highlighting the three different color-coded elements *tcaAB*, *tcaC*, and *tccC* (see key). Note that this locus is only complete in strain W14 (in which it confers oral toxicity) and that in TT01 the locus is deleted from the equivalent location in the genome and found as a partially complete copy elsewhere (see text for discussion).

toxicity to insects. Oral insecticidal activity against another lepidopteran, *Pieris brassicae*, was used to identify *tc* gene homologs from *X. nematophilus* PMF1296 (Morgan *et al.*, 2001). In this study, a cosmid library was made in *Escherichia coli* and individual cosmids were screened for oral activity. Two overlapping cosmids with oral activity were recovered and one of these was completely sequenced. Transposon mutagenesis of the cosmids was then carried out to discover which open reading frames were involved in oral toxicity (Morgan *et al.*, 2001). The genes encoding insecticidal activity were termed *Xenorhabdus protein toxins* (*xpt*) and insertions within the large gene *xptA1* were associated with a loss of oral activity (Morgan *et al.*, 2001). The predicted amino acid sequences of the *xpt* genes show high similarity to the *tc* genes of *Photorhabdus* and their closest homologs have been indicated in a revised map of the open reading frames found in the cosmid (**Figure 2**). The orally toxic *Xenorhabdus* cosmid therefore contains two *tcdA* homologs transcribed in opposite directions (*xptA1* and *xpta2*) with *tcaC* (*xptC1*) and *tccC* (*xptB1*) homologs between them. Expression of one *tcdA*-like gene (*xptA1*) alone in *E. coli* was sufficient to reproduce oral activity in *E. coli* (Morgan *et al.*, 2001). This data suggests that the two different genera of

nematode symbionts use similar *toxin complex*-like genes.

9.2.1.3. Genomic organization of *Photorhabdus tc* genes
Extended sequencing of the *tcd* locus in strain W14 revealed that clusters of *tcdA*-like genes were present in the *Photorhabdus* W14 genome (**Figure 3**). Further, these multiple *tcdA* and *tcdB*-like genes are inserted next to an AspV tRNA in a potential pathogenicity island (Waterfield *et al.*, 2002). This island is a region of DNA inserted in putative "core" sequence, i.e., sequence similar to core sequences in *E. coli* and *Yersinia pestis*. The island contains multiple copies of *tccC*-like sequences, which may promote recombination, and also enteric repetitive intergenic consensus (ERIC) sequences (Waterfield *et al.*, 2002). The region also contains numerous directly duplicated open reading frames and a fragment of a *tcdA*-like gene, suggesting that it may be prone to rearrangement. Finally, a comparison of pathogenicity islands or phage inserted at the Asp tRNA in other related bacteria shows similar genomic islands carrying other toxin genes and *rhs* elements in place of *tccC*-like elements (**Figure 4**). All these factors combine to support the concept that the *tcd*-island is a pathogenicity island. Moreover it is interesting to note that this island is adjacent to a

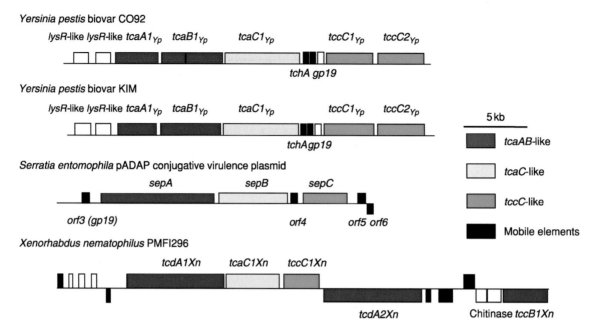

Figure 2   Comparison of *tc*-like loci in entomopathogenic bacteria. *Yersinia pestis* is vectored by the flea and carries a *tca*-like locus, complete in biovar KIM but with *tcaB1* disrupted in biovar CO92. *Serratia entomophila* is a free-living bacteria that infects grass grubs. In this species the *tc*-homologous *sepABC* is carried on a conjugative plasmid required for insect pathogenicity. In *Xenorhabdus nematophilus*, vectored by a nematode, *tc*-homologous genes lie on a cosmid clone recovered on the basis of oral toxicity to caterpillars. Note again that the color coding indicates homology with the three central *tcaAB*, *tcaC*, and *tccC*-like elements (see key).

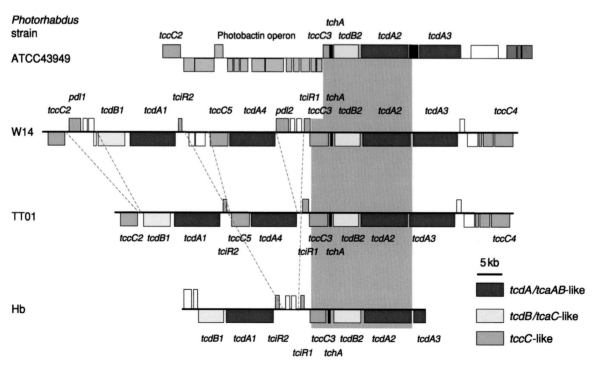

Figure 3   The *toxin complex d* (*tcd*) locus from four different *Photorhabdus* strains highlighting the three different color-coded *tcaAB/tcdA*, *tcaC/tcdB*, and *tccC*-like elements (see key). Comparison of the four loci shows a conserved core (boxed in gray) and then localized deletions within and between the different *tcd* homologs (see text). Note that the *tcd* locus from W14 can confer oral toxicity on recombinant *E. coli*.

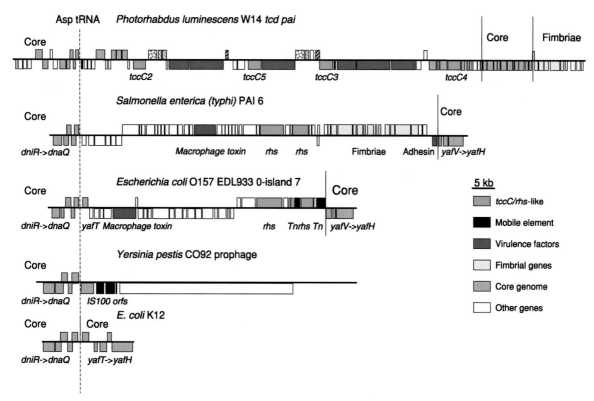

Figure 4   Comparison of *Photorhabdus* W14 *tcd* pathogenicity island with pathogenicity islands also inserted at the Asp tRNA in other related bacteria. The "core" sequence flanking the inserted island is shown. Note the common elements of the pathogenicity islands sharing insertion near a tRNA, toxin genes and mobile elements (see key).

gene *ngrA* implicated in nematode symbiosis (Ciche *et al.*, 2001). Thus regions containing insect pathogenicity genes and nematode symbiosis factors may be linked in the W14 genome.

9.2.2. Homologs in Other Bacteria

Homologs of *tc* genes have now been found in other bacteria, both those pathogenic to insects and also bacteria with no obvious insect association (Waterfield *et al.*, 2001b). Gene homologs are found in *Serratia entomophila*, a free-living bacterium causing "amber disease" in the New Zealand grass grub, *Costelytra zealandica* (Hurst *et al.*, 2000). Larval disease symptoms after oral ingestion of *S. entomophila* include cessation of feeding, clearance of the gut, amber coloration, and eventual death. In *S. entomophila*, the disease determinants are encoded on a 115 kb plasmid designated pADAP, for amber disease-associated plasmid (Glare *et al.*, 1993). Introduction of pADAP into *E. coli* is sufficient to confer oral toxicity on the whole bacteria (Hurst *et al.*, 2000). The genes encoding the amber disease toxins were mapped to a 23 kb fragment of pADAP and then identified by insertional mutagenesis. The three disease-associated genes were termed *Serratia entomophila pathogenicity* (*sep*)

genes (Hurst *et al.*, 2000). These predicted products of these genes show high similarity to those of the *tc* genes and are again clearly homologs. Thus *sepA* is *tcd*-like, *sepB* is *tcaC*-like, and *sepC* is a *tccC*-like gene (**Figure 2**). The finding of *tc*-like genes in a bacterium with no known association with a nematode shows that this class of genes is not specific to nematode symbionts. Further, the presence of the *sep* genes on a plasmid supports the concept, derived from the *tcd* island of *Photorhabdus* W14 (Waterfield *et al.*, 2002; ffrench-Constant *et al.*, 2003), that the *tc* genes can be found on unstable or mobile stretches of DNA.

More recently, *tca*-like genes have also been discovered in both the sequenced genomes of *Y. pestis* (Parkhill *et al.*, 2001; Deng *et al.*, 2002), the causative agent of bubonic plague (Perry and Fetherston, 1997). Although highly pathogenic to man, *Y. pestis* is vectored by fleas and is therefore insect associated for part of its life cycle. Following uptake of an infected blood meal the midgut of the flea becomes "blocked," a process facilitating transmission of the infected blood to the next mammalian host (Hinnebusch *et al.*, 1996). The hypothesis that the *tca* homolog in *Y. pestis* may be involved in midgut blockage was investigated by making a knockout mutant of the gene and examining its effect on blood feeding in the flea. However, no direct effect of the *tca* mutant on midgut blockage was seen (ffrench-Constant, Perry, and Hinnebusch, unpublished data). The functional role of this gene in *Y. pestis* therefore remains unknown. The observation that one of the *tca* open reading frames is apparently disrupted in strain CO92 led Parkhill *et al.* (2001) to suggest the alternative hypothesis, that loss of *tca* function may actually be required for persistence of *Y. pestis* in the flea gut. Further, the presence of an intact *tca* homolog in the recent putative ancestor of *Y. pestis*, *Y. pseudotuberculosis* (S. Hinchcliffe, personal communication), supports the concept that *tca* function may have recently been lost.

Gene homologs can also be detected in the genomes of bacteria with no currently known associations with insects. In *Pseudomona syringae* pv tomato, there is a locus with *tcaC*, *tccB*, and *tccC* homologs (**Figure 5**). Although this bacterium has no known association with insects, as a plant pathogen it may have an undescribed insect vector. Finally, and most recently, *tc* homologs have been found in the genome of *Fibrobacter succinogenes*, a commensal of ruminants with no obvious insect association. This growing range of noninsect associated bacteria with *tc* gene homologs may suggest that this class of genes is widely found in Gram-negative bacteria and that insect toxicity has been acquired in only a limited subset of *tc* gene homologs. The inferred ancestral function of *tc* homologs, however, remains obscure.

9.2.3. Molecular Biology of the *toxin complex* Genes

9.2.3.1. *toxin complex* gene organization The previous survey of *tc* gene homologs found in a range of bacteria leaves us with a confusing array of gene names and homologies. However, several aspects of the genomic organization of these genes in different species begin to indicate which may be the minimal functional units of insect toxicity. First, a color-coded comparison of *tc* loci in different bacteria shows that the different loci are mainly variations on a simple theme of three basic gene types: (1) *tcdA*-like genes, (2) *tcaC*-like genes, and (3) *tccC*-like genes (**Figures 1–3**). Second, where isolated pieces of DNA can confer oral toxicity, such as the pADAP plasmid of *S. entomophila* (Hurst *et al.*, 2000) and the cosmids isolated from *X. nematophilus* (Morgan *et al.*, 2001), only these same three gene types are required for full toxicity. This effectively gives us a working hypothesis for attempts to recombinantly express *Photorhabdus* toxins, i.e., that combinations of these three types of genes are required for the expression of full oral toxicity.

9.2.3.2. Expression and structure of *Photorhabdus* toxins Following the demonstration that knock-out of either *tca* or *tcd* in *Photorhabdus* W14 decreases oral toxicity and in the absence of orally toxic cosmids carrying these loci, it was required to define the minimal components necessary for full oral toxicity in *E. coli*. In these experiments plasmids were inserted into *E. coli* carrying either (1) *tcdA* alone, (2) *tcdA* and *tcdB*, or (3) *tcdA*, *tcdB*, and *tccC* together. Only the plasmid carrying all three genes confers oral toxicity when expressed in *E. coli* and toxicity is increased by induction of the culture with mitomycin, suggesting that Tc production may be related to stress induced by this phage-inducing agent. Electron microscopy of high molecular weight particle preparations of the recombinant protein showed the Tc complexes to be "ball

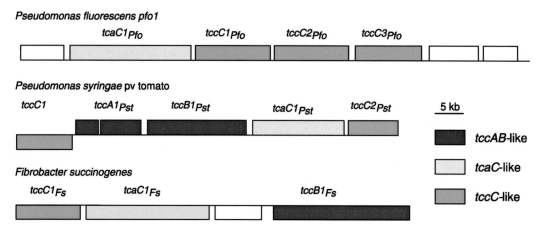

Figure 5 Comparison of *tc*-like loci in bacteria with no known insect association showing that *tc*-like genes can also be found in plant pathogens with no known insect vector like *Pseudomonas syringae* and even ruminal commensals like *Fibrobacter succinogenes* (see text for discussion).

and stick" shaped particles of 25 nm in length (Waterfield *et al.*, 2001a). Further, despite the fact that the presence of *tccC* is necessary for full oral toxicity, the particles themselves can be made by *tcdA* and *tcdB* alone. This suggests that TccC plays a critical role in the activation or export of the complex but does not change the physical structure of the *tcdAB*-derived structure.

Finally, despite the unexpected oral activity of the Tc toxins, how does *Photorhabdus* itself employ these toxins in a normal infection? Immunocytochemistry using an anti-TcaC antibody in insects infected with *Photorhabdus* W14 shows that Tca is produced during infection of the insect midgut (Silva *et al.*, 2002). During infection, the bacteria occupy specific grooves in the midgut, penetrating under the sheathing extracellular matrix to lie directly underneath the midgut epithelium. Within this sheltered niche the bacteria produce Tca, much of which appears to be associated with the bacterial outer membrane (Silva *et al.*, 2002). In this fashion the gut active toxin can be delivered in extreme proximity to the midgut epithelium itself.

9.3. The Makes Caterpillars Floppy Toxins

9.3.1. Discovery of Makes Caterpillars Floppy

9.3.1.1. Cloning of *mcf1*
Although the Tc's account for the oral toxicity of some *Photorhabdus* supernatants, the nature of the toxins used by the bacteria growing within the hemocoel to actually kill the insect host remain unclear. To investigate

potentially injectably active toxins individual cosmids were employed but this time injecting *E. coli* cultures carrying each cosmid individually into *M. sexta* caterpillars (Daborn *et al.*, 2002). By injecting 2×10^7 cells per culture, 300 clones were injected and a single cosmid that resulted in death of the insect was recovered. As well as killing the insect, the cosmid also induced a rapid loss of body turgor, termed the "floppy" phenotype (Daborn *et al.*, 2002). The cosmid was mutagenized with 126 single transposons to look for insertions showing the loss of either phenotype. The same insertions were also used as entry points to sequence the complete cosmid and identify open reading frames. Out of 126 insertions, 39 negated both phenotypes and all of these lay within a single 8.8 kb gene. This gene also caused both death and loss of body turgor when *E. coli* carrying this gene alone were injected into carterpillars. Therefore, this open reading frame was termed the *makes caterpillars floppy* gene or *mcf*. This open reading frame predicts a high molecular weight protein of 324 kDa. However, the predicted protein shows little homology to other proteins in the databases. Three regions of limited homology are present (**Figure 6**). First, the C-terminus shows similarity to a repeated sequence found in RTX-like toxins (Schaller *et al.*, 1999) potentially involved in toxin secretion. Second, part of the central domain is similar to the translocation domain from toxin B of *Clostridium difficile* (Hofmann *et al.*, 1997). Finally, a region in the N-terminus matches the consensus sequence for a Bcl-2 homology domain 3 (BH3) domain. Vertebrate proteins involved in apoptosis that contain only this domain

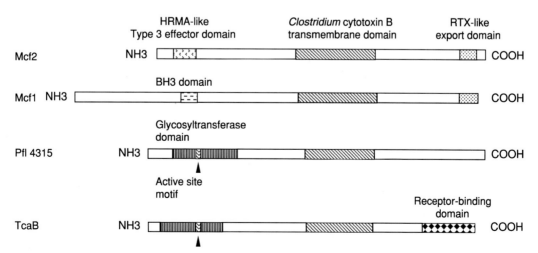

Figure 6 The Makes caterpillars floppy (Mcf) toxins 1 and 2 compared to toxins from *Pseudomonas fluorescens* (Pfl 4315) and *Clostridium* (confusingly termed TcaB but not a Tc toxin). Both Mcf1 and Mcf2 have domains in their N-termini putatively associated with apoptosis, namely a BH3 and HRMA-like domain, respectively. Two regions of similarity within this class of toxins are highlighted. First a transmembrane domain found in the *Clostridium* toxin and second an RTX-like export domain found in the C-terminus of both Mcf proteins.

are proapoptotic (Budd, 2001), leading to the working hypothesis that Mcf itself may also promote apoptosis. Interestingly, if proved, this would be the first BH3 domain found outside of eukaryotes.

The observation that *E. coli* expressing Mcf alone can survive in the presence of the insect immune system is unexpected. Normally, *E. coli* carrying a pUC18 plasmid are cleared from the insect, presumably via the action of both antibacterial peptides and the insect phagocytes or hemocytes. In contrast, *E. coli* carrying pUC18-*mcf* and a second green fluorescent protein (GFP) expressing plasmid persist longer in the hemolymph. Dissection of infected animals shows that the Mcf and GFP expressing bacteria can colonize tissues in the insect and that the hemocytes fail to encapsulate the bacteria in melanin. Therefore, it was speculated that this failure to encapsulate Mcf expressing bacteria was due to hemocytes, dying via Mcf induced apoptosis. To investigate whether Mcf can promote apoptosis in hemocytes, the hemocyte monolayers were exposed to recombinant Mcf. The results show that exposed hemocytes, disintegrate rapidly, within 6 h, by producing numerous characteristic blebs. The residual actin cytoskeleton also shows a characteristic punctate pattern of staining. This proapoptotic effect on hemocytes helps to explain how recombinant *E. coli* can persist in the insect but fails to explain how the insects lose body turgor and die.

To examine the cause of the floppy phenotype, the histopathology of insects infected with *E. coli* expressing Mcf was investigated. Given that both the Malpighian tubules and the midgut are responsible for osmoregulation, destruction of either of these organs could result in a rapid loss of body turgor. Sections of infected caterpillars showed that the Malpighian tubules were intact at the time of death but cells of the midgut epithelium were severely affected as early as 12 h post injection. Both goblet and columnar cells shed circular blebs, often including the nucleus, into the midgut lumen. Nuclei within the dying cells appear pycnotic and stain terminal deoxynucleotidyl transferase-mediated duTP end labeling (TUNEL) positive, suggesting that they are apoptotic. These observations are consistent with Mcf having a proapoptotic function and with its primary site of action being the insect midgut. However, some caution is necessary in interpreting these results as insects ingesting *B. thuringiensis* Cry of Vip toxins also show a similar histopathology of the midgut. More specific experiments to show that Mcf is proapoptotic are therefore needed.

Comparison of the genomic location of *mcf1*-like genes between *Photorhabdus* strains confirms two important points (**Figure 7**). First, *mcf1*-like genes appear to be present in all strains of *Photorhabdus*, as would be expected if Mcf1 is a dominant toxin. Second, like the *tc* toxins, the location of *mcf1* differs between genomes and is often associated with either a Phe tRNA or integrase genes. In fact in strain TT01, *mcf1* is adjacent to a Phe tRNA on one side and the downstream of the *tccC* locus, suggesting that in this case both a *tc* and *mcf* locus may be found together on the same pathogenicity

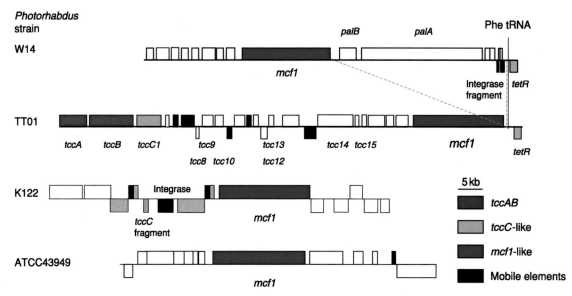

Figure 7 Comparison of *mcf1* containing loci from four different *Photorhabdus* strains. Note that *mcf1* lies alongside a Phe tRNA in both W14 and TT01; however, W14 also carries *palBA*. In the other two strains sequenced, K122 and ATCC43949, *mcf1* appears to occupy different genomic locations suggesting that the toxin is mobile within the *Photorhabdus* genome.

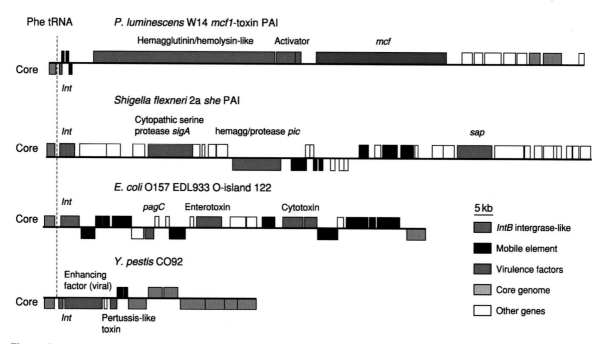

Figure 8 Comparison of *Photorhabdus* W14 *mcf1* pathogenicity island with pathogenicity islands also inserted at the Phe tRNA in related bacteria. The "core" sequence flanking the inserted island is shown. Note the common elements of the pathogenicity islands sharing insertion near a tRNA, toxin genes and mobile elements (see key). Note also the similarities in genomic organization with the *tcd* island in **Figure 4**.

island. Finally, comparison of the relative genomic location of the island containing *mcf1* in *Photorhabdus* strain W14 with other islands inserted at the Phe tRNA in related bacteria again show a similar pattern with other toxins being inserted alongside integrase genes or integrase gene fragments (**Figure 8**). This confirms that *Photorhabdus*, like related enteric bacteria, carries a range of toxin encoding pathogenicity islands inserted at tRNA genes.

9.3.1.2. Cloning of *mcf2* homologs Sample sequencing of the *P. luminescens* W14 genome revealed the presence of a second *mcf*-like open reading frame termed *mcf2*. The predicted Mcf2 protein is similar to Mcf1 (**Figure 6**) but is significantly shorter at the N-terminus. Further, rather than carrying BH3-like domain, part of the N-terminus shows similarity to HrmA from *Pseudomonas syringae* pv. *syringae* (**Figure 6**). The *hrmA* gene encodes a protein that has an avirulence (*avr*) phenotype in plants, travels a type III (Hrp) secretion system, and also elicits cell death when transiently expressed directly in tobacco cells. This suggests that HrmA is a type III delivered effector that can cause apoptosis in plant cells. The similarity of HrmA with part of Mcf2 suggests that Mcf2 may also be involved in inducing apoptosis but via a different mechanism than Mcf1. In this case, it is

interesting to note how a type III effector protein such as HrmA, which is normally delivered directly into the cell, has become fused within the N-terminus of Mcf2 to be delivered within the high molecular weight protein via an undescribed mechanism. Comparisons of available sequence data from different *Photorhabdus* strains suggests that copies of *mcf1* are always present, suggesting it may be the dominant insect killing toxin (**Figure 7**). However, we remain uncertain about the distribution of *mcf2* and although this toxin is present in W14 and TT01, this gene is not present in currently available sequence from the *P. asymbiotica* isolate being sequenced at the Sanger Center. It is therefore not clear if *mcf2* is a required toxin present in all *Photorhabdus* strains like *mcf1* or if it is only present in a subset of strains.

9.4. Toxin Genomics

9.4.1. Microarray Analysis

The widespread distribution of the *tc* genes between different bacteria raises interesting questions as to their origins and transmission. To examine these questions in *Photorhabdus* strains the simple question should be answered: is oral toxicity monophyletic? In other words, do all *Photorhabdus* strains with orally toxic supernatants belong to the same taxonomic group? To address this question, two parallel approaches were used. First, phylogenetic analysis of orally toxic and nonorally toxic strains using 16S DNA sequences was carried out. Second, hybridized genomic DNA from each strain to a microarray carrying 96 putative virulence factor genes from the orally toxic *Photorhabdus* strain W14 was done to investigate the minimal subset of *tc* genes required for oral toxicity.

The phylogenetic analysis showed that the six orally toxic strains examined fell into two distinct taxonomic groups. The first (IS5, EG2, IND, and W14 itself) lies within the previously recognized *P. luminescens* subsp. *akhurstii* subspecies. The second (Hb1 and Hm1) represents a different subgroup, *P. luminescens* subsp. *luminescens* (**Figure 9**). The microarray analysis also revealed groups of

genes either conserved or variable between the strains examined (**Figure 10**). Conserved genes included genes such as *lux*, which produces light, a phenotype common to all *Photorhabdus* (Forst and Nealson, 1996). A series of putative virulence factors also appear conserved in most, or all, strains, including the toxin encoding gene *mcf1* (Daborn *et al.*, 2002), *rtxA1* and *rtxA2*-like genes, the operon encoding the *prtA* protease, and an *attachment and invasion locus* (*ail*) homolog (ffrench-Constant *et al.*, 2003). The type III secretion system (Waterfield *et al.*, 2002), often associated with virulence in other Gram-negative bacteria (Galan and Collmer, 1999), is also present in all strains. Other conserved genes include those encoding catalase, chitinase, ferrochetalese, and flagellae. Variable genes include bacteriocins, toxins/hemolysins, insertion elements, pili,

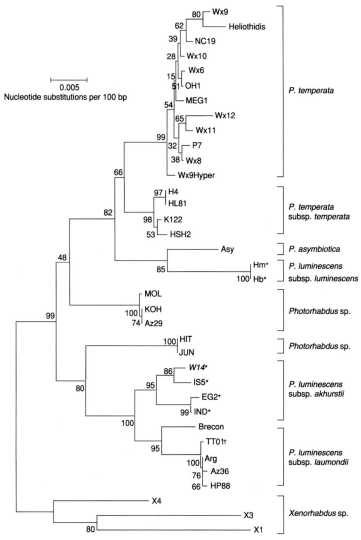

Figure 9 Phylogenetic analysis of orally and nonorally toxic strains used in the microarray analysis based on 16s DNA sequencing. Note that orally toxic strains (∗) fall into two well-supported phylogenetic groups. Strain TT01 (†) has recently been completely sequenced (see text).

and genes involved in iron acquisition. Some toxin and hemolysin/hemagglutinin genes are also variable between strains. These include *pnf*, a *Photorhabdus necrotizing factor* that is found within a recently acquired region of the W14 genome (Waterfield et al., 2002) which is absent from TT01, and the *palA* and *palB* genes, encoding a hemolysin/hemagglutinin and its export activator (Waterfield et al., 2002). In W14, *palBA* is immediately downstream of the *mcf1* toxin gene and adjacent to a Phe tRNA (Waterfield et al., 2002) whereas in TT01 (**Figure 7**), *palBA* is deleted from the equivalent location, as predicted. In contrast, a second two-component hemolysin locus *phlAB* (Brillard et al., 2002), is predicted to be similar in all strains, and indeed potentially duplicated in strain Hm.

The *tc* genes correspond to variable genes, whose distribution is strikingly split between orally toxic and nontoxic strains (**Figure 11**). Orally toxic strains carry all three genes in the *tca* operon (*tcaA*, *tcaB*, and *tcaC*) whereas those lacking toxicity lack *tcaA* and *tcaB*. The genes of the *tca* operon are the only genes to show a perfect correlation with oral toxicity, and genes from *tcb*, *tcc*, and *tcd* are all variable across orally toxic strains. For the *tca* locus (**Figure 1**), a comparison of W14 and TT01 shows two important findings. First, the W14 *tcaABC* operon is absent from the equivalent location in TT01. Second, a *tca*-like locus is present elsewhere in TT01 but lacks most of *tcaA* and *tcaB*, which have been deleted, and retains only a *tcaC1*-like gene. These observations confirm that the presence of *tcaAB* is necessary for oral toxicity of the bacterial supernatant. Further, the fact that *tca*-like loci can be found at different locations in different genomes supports the concept that the *tca* locus is mobile, as suggested by the presence of either a transposase, in W14, or an integration protein, in TT01, adjacent to both *tca*-like loci (**Figure 1**). For the *tcb* locus, the suggestion that *tcbA* is lacking from TT01 is again confirmed by analysis of the genomic sequence, as *tcbA* is deleted from the equivalent location in TT01 and a transposase left in its place. Again comparing W14 and TT01, the array suggests that the *tcc* locus should be completely conserved, as supported by an examination of the genomic sequence. Finally for *tcd*, the genomic sequence (**Figure 3**) confirms that all the *tc* genes of the island are present in both W14 and TT01, including both *lysR*-like regulators. Further, the predicted loss of *tcdA4* in group 2 orally toxic strains is confirmed by an examination of this locus in the Hb strain. The array even predicts the loss of *pdl1* and *pdl2* from within the *tcd* island of TT01, again confirmed by the sequence

(**Figure 3**). The localized deletion of *tcdA4*, and the apparent rearrangements mediated by *tccC*-like genes, again support the hypothesis that the *tcd* genes are encompassed in an unstable pathogenicity island. However, the consistent maintenance of four genetic elements within this island (*tcdA2*, *tcdB2*, a *gp13*-like holin, and the conserved region of *tccC3*), and the conserved organization of similar *tcd*-like genes in other bacteria (such as *sepA*, *sepB*, *orf4*, and *sepC* in *Serratia*), suggest that these genes form an invariant, but not orally toxic, core.

Previous analysis of strain W14 showed that either *tca* or *tcd* can independently contribute to oral toxicity. Thus plasmid clones of either *tca* or *tcd* from W14 confer recombinant oral toxicity when expressed in other nonorally toxic *Photorhabdus* strains, such as K122. This data is consistent with the current observation that the presence or absence of *tcaAB* is perfectly correlated with toxicity of the supernatant. However, all the *tc* genes within the *tcd* island are also present in some strains, such as TT01 (**Figure 3**), which lack orally toxic supernatants. To investigate the apparent lack of oral toxicity associated with *tcd* in TT01, the toxicity of the bacterial cells was examined independently of their supernatant and it was found that toxicity is associated with the cells rather than the supernatant (**Figure 12**). Confirmation that this novel cell-associated toxicity is *tcd* related now requires knockout or heterologous expression of this locus. In conclusion, even a very limited microarray can provide a powerful and predictive tool for correlating observed phenotypes with bacterial genotypes. Similar microarrays may therefore be useful in investigating other *Photorhabdus* phenotypes involved either in insect pathogenicity or symbiosis with their nematode hosts.

9.5. Conclusions

With the publication of the full genome sequence of strain TT01 and the ongoing sequencing of a *P. asymbiotica* strain at the Sanger Center, we are entering an era of comparative *Photorhabdus* genomics. As this review has indicated this now allows us to begin to determine where given toxin genes might have come from, and how they are passed between different strains. In the case of the *tc* genes, the appearance of orally toxic *tca* genes in divergent strains and the finding that these genes can occur at different locations in the genome support the hypothesis that they are mobile between strains. Microarray analysis of more toxin genes in a wider array of strains may also cast light on the origins of

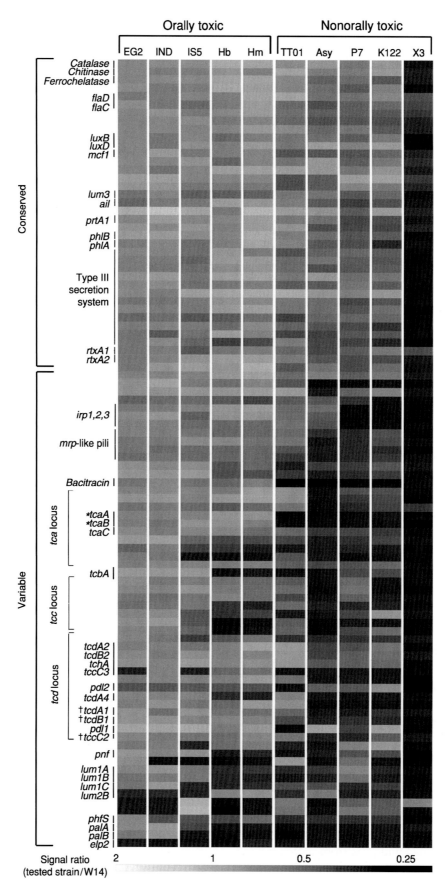

Figure 10   (Continued)

other toxin genes such as *mcf1* and *mcf2*. Despite this recent explosion in sequence data, perhaps the most important biological questions remain largely unanswered. Thus, what is the biological role of the array of toxins that *Photorhabdus* encodes? Although it is easy to infer that these toxins may be necessary to kill the insect host, assessing the role of individual toxins via genetic knockout may be hampered by their functional redundancy. Thus, *tc* gene knockouts may show no oral activity but

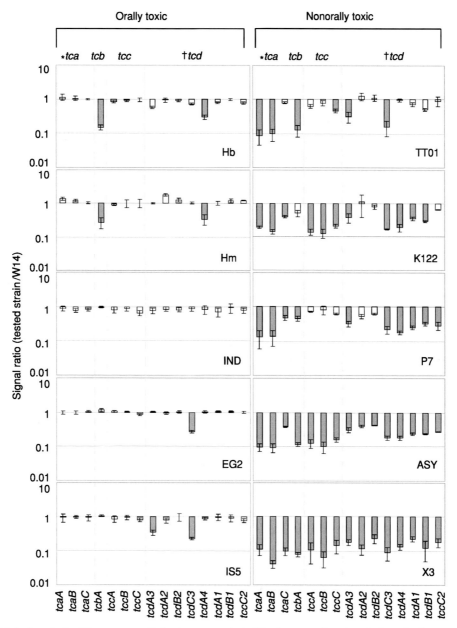

Figure 11 Detailed analysis of the array results for *tc* genes only. In this diagram ratios close to 1 predict the presence of a gene and ratios greater than 0.5 (shaded histograms) predict it to be absent. Note that only the presence of *tcaAB* (∗) is perfectly correlated with the presence of oral toxicity in the different strains (see text) and that *tcd* (†) is variable between strains.

Figure 10 Diagrammatic representation of microarray results where DNA from orally and nonorally toxic strains was hybridized to 96 W14 virulence factor genes. Genes have been divided into conserved (showing a consistent hybridization signal in all strains) and variable (those showing a variable signal between strains). Note for example that the type III secretion system appears present in all *Photorhabdus* strains, whereas members of the *tc* genes are variable (see text). The two loci implicated in oral toxicity are indicated: *tcaAB* (∗) and *tcdAB* (†).

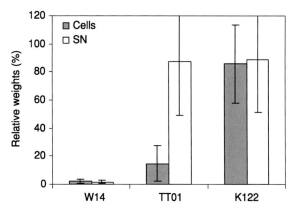

Figure 12 Comparison of supernatant (SN) versus cell-associated oral toxicity for three different *Photorhabdus* strains. Note that strain TT01 containing a complete copy of the *tcd* locus has no oral toxicity associated with the bacterial supernatant but does show oral toxicity associated with the bacterial cells (see text for discussion).

they can still kill the insect host. Further, mutants lacking *mcf1* take a longer time to kill insects but are still lethal. The current challenge is therefore to dissect out the individual roles of each toxin gene, even when they are members of large and rapidly growing gene families.

References

Blackburn, M., Golubeva, E., Bowen, D., ffrench-Constant, R.H., 1998. A novel insecticidal toxin from *Photorhabdus luminescens*: histopathological effects of Toxin complex A (Tca) on the midgut of *Manduca sexta*. *Appl. Environ. Microbiol. 64*, 3036–3041.

Bowen, D., Rocheleau, T.A., Blackburn, M., Andreev, O., Golubeva, E., et al., 1998. Insecticidal toxins from the bacterium *Photorhabdus luminescens*. *Science 280*, 2129–2132.

Bowen, D.J., Ensign, J.C., 1998. Purification and characterization of a high molecular weight insecticidal protein complex produced by the entomopathogenic bacterium *Photorhabdus luminescens*. *Appl. Environ. Microbiol. 64*, 3029–3035.

Brillard, J., Duchaud, E., Boemare, N., Kunst, F., Givaudan, A., 2002. The PhlA hemolysin from the entomopathogenic bacterium *Photorhabdus luminescens* belongs to the two-partner secretion family of hemolysins. *J. Bacteriol. 184*, 3871–3878.

Budd, R.C., 2001. Activation-induced cell death. *Curr. Opin. Immunol. 13*, 356–362.

Ciche, T.A., Bintrim, S.B., Horswill, A.R., Ensign, J.C., 2001. A phosphopantetheinyl transferase homolog is essential for *Photorhabdus luminescens* to support growth and reproduction of the entomopathogenic nematode *Heterorhabditis bacteriophora*. *J. Bacteriol. 183*, 3117–3126.

Daborn, P.J., Waterfield, N., Silva, C.P., Au, C.P.Y., Sharma, S., et al., 2002. A single *Photorhabdus* gene makes caterpillars floppy (*mcf*) allows *Escherichia coli* to persist within and kill insects. *Proc. Natl Acad. Sci. USA 99*, 10742–10747.

Deng, W., Burland, V., Plunkett, G., 3rd, Boutin, A., Mayhew, G.F., et al., 2002. Genome sequence of *Yersinia pestis* KIM. *J. Bacteriol. 184*, 4601–4611.

Estruch, J.J., Warren, G.W., Mullins, M.A., Nye, G.J., Craig, J.A., 1996. Vip3A, a novel *Bacillus thuringiensis* vegetative insecticidal protein with a wide spectrum of activities against lepidopteran insects. *Proc. Natl Acad. Sci. USA 93*, 5389–5394.

ffrench-Constant, R., Waterfield, N., Daborn, P., Joyce, S., Bennett, H., et al., 2003. *Photorhabdus*: towards a functional genomic analysis of a symbiont and pathogen. *FEMS Microbiol. Rev. 26*, 433–456.

Fischer-Le Saux, M., Viallard, V., Brunel, B., Normand, P., Boemare, N.E., 1999. Polyphasic classification of the genus *Photorhabdus* and proposal of new taxa: *P. luminescens* subsp. *luminescens* subsp. nov., *P. luminescens* subsp. *akhurstii* subsp. nov., *P. luminescens* subsp. *laumondii* subsp. nov., *P. temperata* sp. nov., *P. temperata* subsp. *temperata* subsp. nov. and *P. asymbiotica* sp. nov. *Int. J. Syst. Bacteriol. 49*, 1645–1656.

Forst, S., Dowds, B., Boemare, N., Stackebrandt, E., 1997. *Xenorhabdus* and *Photorhabdus* spp.: bugs that kill bugs. *Annu. Rev. Microbiol. 51*, 47–72.

Forst, S., Nealson, K., 1996. Molecular biology of the symbiotic–pathogenic bacteria *Xenorhabdus* spp. and *Photorhabdus* spp. *Microbiol. Rev. 60*, 21–43.

Gahan, L.J., Gould, F., Heckel, D.G., 2001. Identification of a gene associated with Bt resistance in *Heliothis virescens*. *Science 293*, 857–860.

Galan, J.E., Collmer, A., 1999. Type III secretion machines: bacterial devices for protein delivery into host cells. *Science 284*, 1322–1328.

Gerrard, J.G., McNevin, S., Alfredson, D., Forgan-Smith, R., Fraser, N., 2003. *Photorhabdus* species: bioluminescent bacteria as emerging human pathogens? *Emerg. Infect. Dis. 9*, 251–254.

Glare, T.R., Corbett, G.E., Sadler, A.J., 1993. Association of a large plasmid with amber disease of the New Zealand grass grub, *Costelytra zealandica*, caused by *Serratia entomophila* and *Serratia proteamaculans*. *J. Invertebr. Pathol. 62*, 165–170.

Guo, L., Fatig, R.O., 3rd, Orr, G.L., Schafer, B.W., Strickland, J.A., et al., 1999. *Photorhabdus luminescens* W-14 insecticidal activity consists of at least two similar but distinct proteins. *J. Biol. Chem. 274*, 9836–9842.

Hinnebusch, B.J., Perry, R.D., Schwan, T.G., 1996. Role of the *Yersinia pestis* hemin storage (*hms*) locus in the transmission of plague by fleas. *Science 273*, 367–370.

Hofmann, F., Busch, C., Prepens, U., Just, I., Aktories, K., 1997. Localization of the glucosyltransferase activity of *Clostridium difficile* toxin B to the N-terminal part of the holotoxin. *J. Biol. Chem. 272*, 11074–11078.

Hurst, M.R., Glare, T.R., Jackson, T.A., Ronson, C.W., 2000. Plasmid-located pathogenicity determinants of *Serratia entomophila*, the causal agent of amber disease of grass grub, show similarity to the insecticidal toxins of *Photorhabdus luminescens*. *J. Bacteriol. 182*, 5127–5138.

Ives, A.R., **1996**. Evolution of insect resistance to Bt-transformed plants. *Science 273*, 1412–1413.

McGaughey, W.H., **1985**. Insect resistance to biological insecticide *Bacillus thuringiensis*. *Science 229*, 193–195.

McGaughey, W.H., Gould, F., Gelernter, W., **1998**. Bt resistance management. *Nat. Biotechnol. 16*, 144–146.

Morgan, J.A., Sergeant, M., Ellis, D., Ousley, M., Jarrett, P., **2001**. Sequence analysis of insecticidal genes from *Xenorhabdus nematophilus* PMFI296. *Appl. Environ. Microbiol. 67*, 2062–2069.

Parkhill, J., Wren, B.W., Thomson, N.R., Titball, R.W., Holden, M.T., *et al.*, **2001**. Genome sequence of *Yersinia pestis*, the causative agent of plague. *Nature 413*, 523–527.

Perry, R.D., Fetherston, J.D., **1997**. *Yersinia pestis*: etiologic agent of plague. *Clin. Microbiol. Rev. 10*, 35–66.

Schaller, A., Kuhn, R., Kuhnert, P., Nicolet, J., Anderson, T.J., *et al.*, **1999**. Characterization of apxIVA, a new RTX determinant of *Actinobacillus pleuropneumoniae*. *Microbiology 145*, 2105–2116.

Silva, C.P., Waterfield, N.R., Daborn, P.J., Dean, P., Chilver, T., *et al.*, **2002**. Bacterial infection of a model insect: *Photorhabdus luminescens* and *Manduca sexta*. *Cell. Microbiol. 6*, 329–339.

Waterfield, N., Daborn, P.J., ffrench-Constant, R.H., **2002**. Genomic islands in *Photorhabdus*. *Trends Microbiol. 10*, 541–545.

Waterfield, N., Dowling, A., Sharma, S., Daborn, P.J., Potter, U., *et al.*, **2001a**. Oral toxicity of *Photorhabdus luminescens* W14 toxin complexes in *Escherichia coli*. *Appl. Environ. Microbiol. 67*, 5017–5024.

Waterfield, N.R., Bowen, D.J., Fetherston, J.D., Perry, R.D., ffrench-Constant, R.H., **2001b**. The toxin complex genes of *Photorhabdus*: a growing gene family. *Trends Microbiol. 9*, 185–191.

Yu, C.-G., Mullins, M.A., Warren, G.W., Koziel, M.G., Estruch, J.J., **1997**. The *Bacillus thuringiensis* vegetative insecticidal protein Vip3A lyses midgut epithelium cells of susceptible insects. *Appl. Environ. Microbiol. 63*, 532–536.

A9 Addendum: Recent Advances in *Photorhabdus* Toxins

A J Dowling, P A Wilkinson, and R H ffrench-Constant, University of Exeter in Cornwall, Tremough Campus, Penryn, UK

© 2010 Elsevier B.V. All Rights Reserved

Here we briefly review recent work on toxins from *Photorhabdus* bacteria and put this work into the context of recent developments in novel screening techniques for bacterial virulence.

A9.1. Discovery of New Toxins

Recently documented toxins are the "*Photorhabdus* insect related A and B" (PirAB) toxins and the *Photorhabdus* Virulence Cassettes or PVCs. PirAB are binary toxins, found as gene pairs at two loci in *Photorhabdus luminescens* TT01, with oral activity against both lepidoptera and diptera (Duchaud *et al.*, 2003; Waterfield *et al.*, 2005; ffrench-Constant *et al.*, 2007; Ahantarig *et al.*, 2009). Sequence similarity to a developmentally regulated beetle neurotoxin gene, β-leptinosarin-h, and the pore-forming domain of *Bacillus thuringiensis* δ-endotoxins has led to the hypothesis that PirAB is a pore/channel-former. Notably, PirAB is effective against larval *Aedes* mosquitoes, which are dengue vectors, and importantly does not affect the predatory *Mesocyclops thermocyclopoides* (Ahantarig *et al.*, 2009).

PVCs consist of phage-like open-reading frames (ORFs) and effector components, which act as delivery systems carrying toxic effectors to target cells (Yang *et al.*, 2006). *P. luminescens* TT01 and *P. asymbiotica* strain ATCC43949 contain numerous PVC loci with different putative effector proteins, these contain homology to regions of: Mcf, Toxin A (*Clostridium difficile*), YopT (*Yersinia enterocolitica*), the active domain of Cytotoxic Necrosis Factor 1 (CNF1) from *Escherichia coli* and others with no obvious similarities, possibly representing novel effectors (ffrench-Constant *et al.*, 2007). PVCs are injectably toxic to the wax moth *Galleria mellonella*, destroying hemocytes; expression of effector domains inside eukaryotic cells causes dramatic cytoskeletal rearrangement. The variety of PVCs may equip *Photorhabdus* for activity against a range of different hosts.

A9.2. A New Genome Sequence for *P. asymbiotica*

The genome of *P. asymbiotica* strain ATCC43949, isolated from a human infection in North America, has recently been completed and presented as a comparative study with the insect pathogenic *P. luminescens* TT01 (Wilkinson *et al.*, 2009). *P. asymbiotica* ATCC43949 has a smaller genome and has acquired a plasmid, pAU1 (related to pMT1 from *Yersinia pestis*). Notably, the genome has a reduced diversity of insecticidal toxin genes. This needs to be altered to "The Tc pathogenicity islands encode fewer orthologs of Tca and Tcd-like toxins. All of *tcaA* and the majority of *tcaB* are deleted and a new *tccC* homolog has been acquired." Further, there is deletion of most of *tcbA* and also lost or gained *tccC* genes. Five PVCs are present in *P. asymbiotica* as apposed to six in TT01, an *mcf1*-like gene is present but not *mcf2*, only a single locus encoding PirAB, and four fewer hemolysins are observed. Despite these absences, *P. asymbiotica* ATCC43949 is, in fact, more lethal to insect hosts than *P. luminescens* TT01 or *Photorhabdus temperata* K122 (Eleftherianos *et al.*, 2006).

A9.3. New Systems for the Dissection of Bacterial Virulence

Rapid Virulence Annotation (RVA) is a technique recently developed for functionally annotating bacterial genomes. A genomic DNA library of *P. asymbiotica* ATCC43949 was screened, in parallel, for Gain Of Toxicity (GOT) against invertebrate and vertebrate hosts (insects, amoeba, nematodes, and mammalian macrophages) (Waterfield *et al.*, 2008). Numerous possible virulence factors, genomic islands and biological functions have been documented. Interestingly, RVA has identified biological activities from regions encoding secondary metabolite gene clusters including polyketide synthesis and nonribosomal peptide synthesis-like regions notoriously difficult to ascribe functionality to. Hemolysis and hemagglutinin-like genes, pili/fimbrial operons, putative specialized secretion systems, and toxin-like genes were also identified.

New functions of *Photorhabdus* bacteria and toxins have also been discovered using a novel technique employing the powerful genetics of *Drosophila* mutants and live-imaging (Vlisidou *et al.*, 2009). *Drosophila* embryonic hemocytes were challenged with *P. asymbiotica*, the bacteria were recognized, but not phagocytosed, caused rapid "freezing" and significant cytoskeletal alteration of these motile cells, and effects were phenocopied by both recombinant *E. coli* expressing Mcf1 and pure Mcf1. *Drosophila* mutants revealed a requirement for Mcf1 internalization and early activity on the cytoskeleton, demonstrating applicability of this technique for ascribing toxin mode of action.

A9.4. Advances in Toxin Mode of Action

Each mature Toxin complex, or Tc, consists of three components A, B, and C, and for each toxin gene, the name is combined with the relevant component, for example, for Tcd, TcdA is component A, TcdB component B, and TccC component C (ffrench-Constant and Waterfield, 2006; ffrench-Constant *et al.*, 2007). Oral toxicity to *Manduca sexta* has been achieved by cloning A (*tcdA*) from *P. luminescens* strain W14 into *Arabidopsis* (Liu *et al.*, 2003). High expression levels of A (*tcdA*) alone in recombinant *E. coli* reproduce oral toxicity; however, for full toxicity, B and C (*tcdB* and *tccC*) are needed (Waterfield *et al.*, 2005). Components B and C on their own are not orally toxic leading to the hypothesis that they may potentiate A. Cross-potentiation between *Xenorhabdus*, *Paenibacillus*, and *Photorhabdus* Tc components demonstrates the potential of "mixing and matching" to insect targets or to evade resistance (Hey *et al.*, 2004). Recently *tcdA1* and *tcdB1* have been cloned into recombinant termite gut bacteria and shown effective for termite control (Zhao *et al.*, 2008).

Current work has revealed activity of *Yersinia* and *P. luminescens* Tcs against mammalian tissue culture cells (Hares *et al.*, 2008). *Yersinia* Tcs show no oral activity against *M. sexta* and they are not toxic to lepidopteran *Sf*9 cells, unlike *P. luminescens* Tcs; leading to speculation of toxicity evolution toward their respective vertebrate and invertebrate target

hosts. Finally, structural and biophysical studies of *Xenorhabdus nematophila* PMF1296 Tc component XptA1 suggests a mechanism of action where it binds to the cell membrane, forming a structure with a central cavity, and complexes with XptB1 and XptC1 (Lee *et al.*, 2007). XptA1 has a cage-like structure that survives at alkaline pH, indicating persistence in host lepidopteran midguts.

The "Makes Caterpillars Floppy" toxin 1, or Mcf1, has been further characterized, and shown to require endocytosis to induce apoptosis which it does via the mitochondria; further, a double mutation within the putative BH3-like domain attenuates apoptosis (Dowling *et al.*, 2004, 2007). Antiphagocytic effects of Mcf1 occurring earlier than apoptotic events have been observed implying a possible dual mode of action for Mcf1 (Vlisidou *et al.*, 2009). Mcf1 may therefore facilitate *Photorhabdus* infection by early freezing of the insect phagocytes and subsequent destruction of the midgut epithelium via apoptosis.

References

Ahantarig, A., Chantawat, N., *et al.*, 2009. PirAB toxin from *Photorhabdus asymbiotica* as a larvicide against dengue vectors. *Appl. Environ. Microbiol.* 75(13), 4627–4629.

Dowling, A.J., Daborn, P.J., *et al.*, 2004. The insecticidal toxin makes caterpillars floppy (Mcf) promotes apoptosis in mammalian cells. *Cell. Microbiol.* 6(4), 345–353.

Dowling, A.J., Waterfield, N.R., *et al.*, 2007. The Mcf1 toxin induces apoptosis via the mitochondrial pathway and apoptosis is attenuated by mutation of the BH3-like domain. *Cell. Microbiol.* 9(10), 2470–2484.

Duchaud, E., Rusniok, C., *et al.*, 2003. The genome sequence of the entomopathogenic bacterium *Photorhabdus luminescens*. *Nat. Biotechnol.* 21(11), 1307–1313.

Eleftherianos, I., Marokhazi, J., *et al.*, 2006. Prior infection of *Manduca sexta* with non-pathogenic *Escherichia coli* elicits immunity to pathogenic *Photorhabdus luminescens*: roles of immune-related proteins shown by RNA interference. *Insect Biochem. Mol. Biol.* 36(6), 517–525.

ffrench-Constant, R., Waterfield, N., 2006. An ABC guide to the bacterial toxin complexes. *Adv. Appl. Microbiol.* 58, 169–183.

ffrench-Constant, R.H., Dowling, A., *et al.*, 2007. Insecticidal toxins from Photorhabdus bacteria and their potential use in agriculture. *Toxicon* 49(4), 436–451.

Hares, M.C., Hinchliffe, S.J., *et al.*, 2008. The *Yersinia pseudotuberculosis* and *Yersinia pestis* toxin complex is active against cultured mammalian cells. *Microbiology* 154(Pt 11), 3503–3517.

Hey, T.D., Schleper, A.D., *et al.*, 2004. Mixing and matching Tc proteins for pest control. World Intellectual Property Patent No. WO 2,004,067,727.

Lee, S.C., Stoilova-McPhie, S., *et al.*, 2007. Structural characterisation of the insecticidal toxin XptA1, reveals a 1.15 MDa tetramer with a cage-like structure. *J. Mol. Biol. 366*(5), 1558–1568.

Liu, D., Burton, S., *et al.*, 2003. Insect resistance conferred by 283-kDa *Photorhabdus luminescens* protein TcdA in *Arabidopsis thaliana*. *Nat. Biotechnol. 21*(10), 1222–1228.

Vlisidou, I., Dowling, A.J., *et al.*, 2009. Drosophila embryos as model systems for monitoring bacterial infection in real time. *PLoS Pathog. 5*(7), e1000518.

Waterfield, N., Hares, M., *et al.*, 2005. Potentiation and cellular phenotypes of the insecticidal toxin complexes of Photorhabdus bacteria. *Cell. Microbiol. 7*(3), 373–382.

Waterfield, N., Kamita, S.G., *et al.*, 2005. The Photorhabdus Pir toxins are similar to a developmentally regulated insect protein but show no juvenile hormone esterase activity. *FEMS Microbiol. Lett. 245*(1), 47–52.

Waterfield, N.R., Sanchez-Contreras, M., *et al.*, 2008. Rapid virulence annotation (RVA): identification of virulence factors using a bacterial genome library and multiple invertebrate hosts. *Proc. Natl Acad. Sci. USA 105*(41), 15967–15972.

Wilkinson, P., Waterfield, N.R., *et al.*, 2009. Comparative genomics of the emerging human pathogen *Photorhabdus asymbiotica* with the insect pathogen *Photorhabdus luminescens*. *BMC Genomics 10*, 302.

Yang, G., Dowling, A.J., *et al.*, 2006. Photorhabdus virulence cassettes confer injectable insecticidal activity against the wax moth. *J. Bacteriol. 188*(6), 2254–2261.

Zhao, R., Han, R., *et al.*, 2008. Cloning and heterologous expression of insecticidal-protein-encoding genes from *Photorhabdus luminescens* TT01 in *Enterobacter cloacae* for termite control. *Appl. Environ. Microbiol. 74*(23), 7219–7226.

10 Genetically Modified Baculoviruses for Pest Insect Control

S G Kamita, K-D Kang, and B D Hammock,
University of California, Davis, CA, USA
A B Inceoglu, Ankara University, Ankara, Turkey

© 2010, 2005 Elsevier B.V. All Rights Reserved

10.1. Introduction

There are presently more than 20 known groups of insect pathogenic viruses, which are classified into 12 viral families (Tanada and Kaya, 1993; Blissard *et al.*, 2000). Among insect pathogenic viruses, members of the family Baculoviridae are the most commonly found and most widely studied. Baculoviruses are rod shaped, enveloped viruses with large, covalently closed, double-stranded DNA (dsDNA) genomes. Baculoviruses are made up of two genera: nucleopolyhedrovirus (NPV) and granulovirus (GV). The NPVs can be further divided into two groups (I and II) on the basis of the phylogenic relationships of 20 distinguishing genes (Herniou *et al.*, 2001). Baculoviruses produce two phenotypes, the budded virus (BV) and the occluded virus (OV), during their life cycles (Granados and Federici, 1986; Miller, 1997). BVs are predominantly produced during an early phase of infection and acquire their envelopes as they bud through the plasma membrane. BVs are responsible for the systemic or cell-to-cell spread of the virus within an infected insect. Continuous cell lines that support high-level production of BVs are available for NPVs, but not for GVs. The availability of these cell lines has been critical for the development of genetically modified NPVs. OVs are produced during a late phase of infection and are involved in the horizontal or larva-to-larva transmission of the virus. The OV of the NPV is known as a polyhedron (plural polyhedra) or polyhedral inclusion body, and that of the GV is known as a granule. Within the proteinaceous crystal matrix of the OV are embedded virions termed the occlusion derived virions (ODVs). Each ODV is formed by single or multiple nucleocapsids surrounded by an envelope. In the field, the polyhedron or granule structure protects the embedded ODV(s) from the environment.

Owing to their inherent insecticidal activities, natural baculoviruses have been used as safe and effective biopesticides for the protection of field and orchard crops, and forest in the Americas, Europe, and Asia (Black *et al.*, 1997; Hunter-Fujita *et al.*, 1998; Vail *et al.*, 1999; Moscardi, 1999; Copping and Menn, 2000; Lacey *et al.*, 2001). Both NPVs and GVs have been successfully registered for use as microbial pesticides by commercial companies and government agencies. NPVs of the velvetbean caterpillar *Anticarsia gemmatalis* (AgMNPV) and *Helicoverpa armigera* (HaSNPV) are being used with particular success for the protection of soybean in Brazil (Moscardi, 1999) and cotton in China (Sun *et al.*, 2002), respectively. Natural baculoviruses, however, are slower acting and more target specific (i.e., their host specificity is narrow) compared to synthetic chemical pesticides such as the pyrethroids. The general use of natural baculoviruses in developed countries has been limited except against forest pests primarily due to their slow speed of insect killing compared to chemical insecticides, and partially due to their relatively narrow host specificity, low field stability, and cost of production. As a natural control agent the "slow kill" characteristic allows the virus to replicate to tremendous numbers while allowing its host to feed for several days. Although an attribute in a natural control strategy, this trait is a severe limitation in modern agriculture. This trait and others such as narrow host specificity can and have been addressed by genetically modifying the baculovirus using recombinant DNA technology.

During the 1980s, the birth of genetically modified (GM) baculoviruses came along with exciting new research in the laboratories of Summers (Summers and Smith, 1987) and Miller (1988). They simultaneously exploited a combination of unique characteristics of NPVs to establish the baculovirus expression vector systems (BEVS) that are now in common use for basic research and commercial applications. These characteristics include: (1) the availability of the exceptionally strong polyhedrin gene (*polh*) promoter to drive foreign gene expression; (2) a selection system based upon the visualization of the nonessential (in cultured cells) gene product of the polyhedrin gene; (3) a dsDNA genome that can be easily modified; (4) a rod-shaped capsid that can extend to package additional DNA; and (5) a eukaryotic cell line that supports virus replication at high levels. *Autographa californica* multicapsid NPV (AcMNPV), originally isolated from the alfalfa looper *A. californica* (Vail *et al.*, 1973, 1999), is the baculovirus type species. AcMNPV was used by the Summers and Miller laboratories as the parental baculovirus for BEVS. Another baculovirus, *Bombyx mori* NPV (BmNPV), isolated from the silk moth *B. mori*, was used by Maeda (1989a) as the parental

baculovirus in an alternative BEVS that used larvae of *B. mori* for *in vivo* expression. The methodologies for the construction and use of recombinant AcMNPVs and BmNPVs for the expression of heterologous genes have been thoroughly described (Summers and Smith, 1987; O'Reilly *et al.*, 1992; Richardson, 1995; Merrington *et al.*, 1999). These methodologies, with slight modifications, have also been used for the construction of GM baculovirus pesticides.

The studies to date indicate that GM baculoviruses can easily become an integral part of pest insect control, especially in developing countries and for the control of insects that have become resistant to synthetic chemical pesticides (reviews: Hammock *et al.*, 1993; McCutchen and Hammock, 1994; Miller, 1995; Bonning and Hammock, 1996; Wood, 1996; Harrison and Bonning, 2000a; Inceoglu *et al.*, 2001a; Bonning *et al.*, 2002). Several innovative and successful approaches have been taken to improve the speed of kill of a baculovirus by genetic modification. These approaches include: (1) insertion of a foreign gene into the baculovirus genome whose product alters the physiology of the target host insect or is toxic towards the target host; (2) deletion of an endogenous gene from the baculovirus genome; and (3) incorporation of active toxin into the OV. Combinations of these approaches have also been successful in terms of decreasing the time required to kill the host or more importantly the time required to stop host feeding. Safe and effective protection of the crop from feeding damage should be the goal of a GM baculovirus pesticide. Baculoviruses have been transformed from natural disease agents to efficient pesticides through the above-mentioned discoveries and innovative approaches. Here, numerous innovations that have been used to improve the efficacy of baculoviruses for crop protection are discussed. Additionally, studies that have addressed the safety of natural and GM baculoviruses, especially in terms of risk to humans, the environment, and nontarget beneficial insects are also covered. The individual sections of this chapter provide a detailed and comprehensive summary of the field, especially in terms of improving the insecticidal activity of GM baculoviruses. In Section 10.7, the reader will also find a discussion of the reasons why GM baculoviruses do not receive the interest that they deserve. The authors hope that after reading this review, the reader will be convinced that the currently available GM baculovirus pesticides are effective and safe, and should be used as biological pesticides.

10.2. Insertion of Hormone and Enzyme Genes

10.2.1. Hormones

Keeley and Hayes (1987) were two of the first to suggest the use of an insect neurohormone gene to increase the insecticidal activity of the baculovirus. They wrote:

> use of an insect baculovirus as an expression vector for neurohormone genes has several advantages as a pest control strategy. (1) Insect viruses are genera- or species-specific so that the virus has a limited host range and would not affect nontarget insects. (2) The natural hormone would be produced at continuous, high levels by the viral expression vector. (3) The combination of natural insect hormones with insect-specific viruses constitutes an ideal insect pest control agent from the environmental standpoint.

Menn and Bořkovec (1989) further suggested "a neuropeptide gene placed behind a strong nonessential viral promoter is capable of turning an infected cell into a neuropeptide factory within the insect..." Maeda (1989b) pioneered the field by being the first to put these concepts into practice by generating a recombinant BmNPV expressing a diuretic hormone gene that disrupted the normal physiology of larvae of the silkworm *B. mori*. Subsequently, at least four biologically active peptide hormones have been expressed using recombinant baculoviruses: eclosion hormone (Eldridge *et al.*, 1991), prothoracicotropic hormone (O'Reilly *et al.*, 1995), pheromone biosynthesis activating neuropeptide (Vakharia *et al.*, 1995; Ma *et al.*, 1998), and neuroparsin (Girardie *et al.*, 2001). Unfortunately, expression of a biologically active hormone by the recombinant baculovirus has only been modestly successful or unsuccessful (**Figure 1**) in terms of improving the speed of kill of the baculovirus. In hindsight, this lack of success is not completely unexpected since critical events in the insect's physiology and life cycle are often protected by sequestration (by physical means or through time) or by overlapping systems. As we learn more about the regulatory mechanisms and timing of the hormonal control systems of insects, we should be able to develop more refined approaches to the use of hormone genes to improve the baculovirus as a pesticide.

10.2.1.1. Diuretic hormone

Diuretic and antidiuretic hormones play critical roles in the excretion and retention of water by insects in response to changes in their environment (Holman *et al.*, 1990; Coast *et al.*, 2002; Gade, 2004). The tobacco hornworm *Manduca sexta* encodes a neuropeptide hormone consisting of 41 amino acid residues that stimulates diuresis (Kataoka *et al.*, 1989). Because of the relatively short length of this peptide hormone, Maeda (1989b) was able to generate a synthetic gene encoding the peptide and attempted to express this gene in fifth instar larvae of *B. mori* using BmNPV. The synthetic diuretic hormone (DH) gene was designed on the basis of the codon usage of the polyhedrin gene of BmNPV and the gene structure of an amidated peptide from the giant silk moth *Hyalophora cecropia*. The DH gene construct also included a signal sequence for secretion from a cuticle protein (CPII) of *Drosophila melanogaster* (Meigan) (Snyder *et al.*, 1982) and a glycine residue at the C-terminus for amidation. The recombinant BmNPV (BmDH5) carrying the synthetic DH gene killed fifth instar larvae about 1 day faster than the wild-type BmNPV (Maeda, 1989b). This corresponded to a roughly 20% improvement in speed of kill in comparison to the wild-type BmNPV. BmDH5 also caused a 30% reduction in hemolymph volume, which was hypothesized to be due to the excretion of water into the Malpighian tubules, and subsequently hindgut and outside (Maeda, 1989b). Biologically active DH, however, was not detected in the circulating hemolymph of these animals (D.A. Schooley, unpublished data). The 20% improvement in speed of kill is modest in comparison to more recent GM baculovirus constructs, and the bioassays were based on injection of BV rather than oral infection with polyhedra (the normal mode of baculovirus infection in the field). Nevertheless, these experiments generated a lot of excitement in the study of genetically modified baculoviruses as pesticides and established some of the groundwork for subsequent GM baculovirus pesticide constructs.

10.2.1.2. Eclosion hormone

Eclosion hormones (EH) are neuropeptides that influence several aspects of pupal–adult ecdysis (i.e., eclosion) as well as larval–larval ecdyses (Holman *et al.*, 1990; Nijhout, 1994). The release of EH from the brain is controlled by a circadian clock within the brain and declining ecdysteroid titers. The overexpression of EH by a recombinant baculovirus has been hypothesized to induce the premature onset of eclosion behavior and molting. A cDNA encoding EH of *M. sexta* has been inserted into recombinant AcMNPVs and shown to express high levels of biologically active and secreted EH (Eldridge *et al.*, 1991, 1992b). Time-mortality bioassays in fifth or sixth instar larvae of *Spodoptera frugiperda* injected with vEHEGTD (a recombinant AcMNPV carrying the EH encoding cDNA at the ecdysteroid

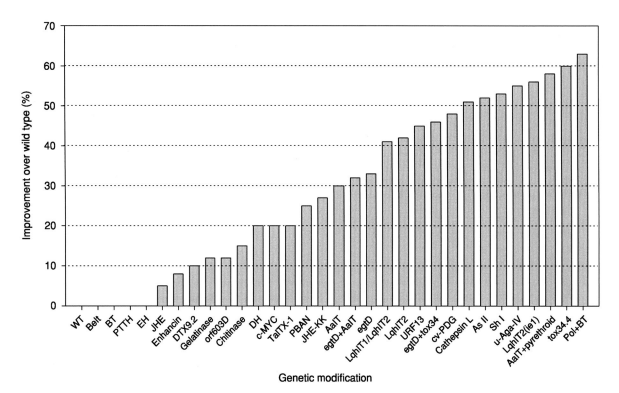

Figure 1   The speed of kill of the wild-type baculovirus can be dramatically improved by genetic modification. The genetic modification (insertion of a foreign gene or deletion of an endogenous gene) and percent improvement in speed of kill (or paralysis) relative to the wild-type virus or control virus is given. Because of differences in the parent virus, promoter, secretion signal, host strain and age, virus dose, and inoculation methods that were used, comparison between the different virus constructs is not possible. Abbreviations and reference(s): WT, wild type; Belt, insectotoxin-1 (Carbonell *et al.*, 1988); BT, *Bacillus thuringiensis* endotoxin (Merryweather *et al.*, 1990); PTTH, prothoracicotropic hormone (O'Reilly, D.R., **1995**. Baculovirus-encoded ecdysteroid UDP-glucosyltransferases. *Insect Biochem. Mol. Biol. 25*, 541–550); EH, eclosion hormone (Eldridge *et al.*, 1992b); JHE, juvenile hormone esterase (Hammock *et al.*, 1990a); enhancin, MacoNPV enhancin (Li *et al.*, 2003); DTX9.2, spider toxin (Hughes *et al.*, 1997); gelatinase, human gelatinase A (Harrison and Bonning, 2001); orf603D, deletion of AcMNPV *orf603* (Popham *et al.*, 1998a); chitinase (Gopalakrishnan *et al.*, 1995); DH, diuretic hormone (Maeda, 1989b); c-MYC, transcription factor (Lee *et al.*, 1997); TalTX-1, spider toxin (Hughes *et al.*, 1997); PBAN, pheromone biosynthesis activating neuropeptide (Ma *et al.*, 1998); JHE-KK, stabilized JHE (Bonning *et al.*, 1997b); AaIT, scorpion *Androctonus australis* insect toxin (McCutchen *et al.*, 1991; Stewart *et al.*, 1991); egtD + AaIT, insertion of *aait* at the *egt* gene locus (Chen *et al.*, 2000); egtD, deletion of ecdysteroid UDP-glucosyltransferase (*egt*) gene (O'Reilly and Miller, 1991); LqhIT1/LqhIT2, simultaneous expression of LqhIT1 and LqhIT2 (Regev *et al.*, 2003); LqhIT2, scorpion *Leiurus quinquestriatus* insect toxin 2 (Froy *et al.*, 2000); URF13, maize pore forming protein (Korth and Levings, 1993); egtD + tox34, insertion of *tox34* under the early *DA26* promoter at the *egt* gene locus (Popham *et al.*, 1997); cv-PDG, glycosylase (Petrik *et al.*, 2003); cathepsin L (Harrison and Bonning, 2001); As II, sea anemone toxin (Prikhod'ko *et al.*, 1996); Sh I, sea anemone toxin (Prikhod'ko *et al.*, 1996); μ-Aga-IV, spider toxin (Prikhod'ko *et al.*, 1996); LqhIT2(ie1), LqhIT2 under the early *ie1* promoter (Harrison and Bonning, 2000b); AaIT + pyrethroid, co-application of low doses of AcAaIT and pyrethroid (McCutchen *et al.*, 1997); tox34.4, mite toxin (Tomalski *et al.*, 1988; Burden *et al.*, 2000); Pol + BT, double expression of authentic polyhedrin and polyhedrin-BT-GFP fusion proteins (Chang *et al.*, 2003).

UDP-glucosyltransferase (*egt*) gene locus) showed that the median survival times (ST_{50}s) were reduced by approximately 29% (83 versus 117 h) and 17% (100 versus 120 h), respectively, in comparison to control larvae injected with wild-type AcMNPV. However, in comparison to control larvae injected with vEGTDEL (a control virus in which the *egt* gene was deleted), the ST_{50}s were not significantly different. Similar results were generated by time-mortality bioassays in neonate *S. frugiperda* that were orally inoculated with vEHEGTD, vEGTDEL,

or AcMNPV. These findings indicated that deletion of the *egt* gene (rather than expression of EH) was responsible for the reduction in the ST_{50}. Eldridge *et al.* (1992b) thus concluded that EH expression does not provide enhanced biological control properties to AcMNPV. The role of the *egt* gene in improving speed of kill is discussed later in this chapter.

10.2.1.3.   Prothoracicotropic hormone   Prothoracicotropic hormones (PTTHs) or "brain hormones" are neurosecretory polypeptides that stimulate the secretion of ecdysteroids from the prothoracic glands (Holman *et al.*, 1990; Nijhout, 1994). The overexpression of PTTH by a recombinant baculovirus has been hypothesized to artificially elevate ecdysteroid levels resulting in the premature induction of molting, which in turn could disrupt insect feeding behavior (Menn and Borkovec, 1989; O'Reilly *et al.*, 1995; Black *et al.*, 1997). O'Reilly *et al.* (1995) have constructed a recombinant AcMNPV expressing the mature PTTH of *B. mori* (fused to a secretion signal sequence of sarcotoxin 1A of the flesh fly). This virus expressed PTTH in larvae of *S. frugiperda* resulting in the induction of higher than normal levels of hemolymph ecdysteroids. However, the expressed PTTH produced no observable effects on the development of *S. frugiperda* and, furthermore, inhibited the pathogenicity of the virus. In the same study, O'Reilly *et al.* (1995) expressed the mature PTTH-signal sequence construct using a mutant AcMNPV in which the AcMNPV encoded *egt* gene was knocked out. This virus apparently generated an even higher level of peak hemolymph ecdysteroid titer (1162 ± 476 versus $724 \pm 194 \, \text{pg} \, \mu\text{l}^{-1}$) in comparison to a control virus that expressed PTTH but retained a normal *egt* gene. This higher ecdysteroid peak, however, was delayed by about 48 h (i.e., the peak occurred at 96 versus 48 h postinfection (p.i.)). These findings suggest that higher levels of hemolymph ecdysteroids are detrimental to the pathogenicity of the virus. In contrast, other studies (as will be discussed later) show that inactivation of the *egt* gene (putatively resulting in higher ecdysteroid levels) results in improved pathogenicity.

10.2.1.4. Pheromone biosynthesis activating neuropeptide

Pheromone biosynthesis activating neuropeptide (PBAN) is a neurosecretory peptide (33–34 amino acid residues) of the brain that appears to be essential for pheromone biosynthesis and release in some lepidopterans (Holman *et al.*, 1990; Nijhout, 1994). PBAN stimulates pheromone biosynthesis in conjunction with nervous stimuli during the dark phase of the photoperiod. Ma *et al.* (1998) expressed a PBAN of the corn earworm *Helicoverpa zea* (Hez-PBAN) fused to the bombyxin (an insect neurohormone) signal sequence for secretion using a recombinant AcMNPV (AcBX-PBAN). AcBX-PBAN expressed biologically active PBAN in cultured cells and larvae of *Trichoplusia ni*. In droplet feeding assays using preoccluded virus, neonate and third instar larvae of *T. ni* infected with AcBX-PBAN showed ST_{50}s that were reduced by 26% (58.3 versus 79.1 h) and 19% (70.2 versus 87.0 h), respectively, in comparison to larvae

infected with a control virus. Although the mechanism for the improved speed of kill is unknown, Ma *et al.* (1998) did not observe any morphological differences between larvae that were infected with AcBX-PBAN or the control virus.

10.2.2. Juvenile Hormone Esterase

Wigglesworth (1935, 1936) was the first to identify a "juvenile factor" produced by the corpora allata that keeps larval insects in the juvenile state. Subsequently, Röller *et al.* (1967) and Meyer *et al.* (1970) showed the chemical structure of "juvenile hormone (JH)." Six JHs (JH-0, JH-I, JH-II, JH-III, 4-methyl JH-1, and JHB_3) have been identified to date, all of which are terpenoids derived from farnesenic acid (or its homologs) with an epoxide group at the 10, 11 position of one end and a conjugated methyl ester at the other end (Nijhout, 1994; Cusson and Palli, 2000). JH-III appears to be the most common of the JHs being found in all of the insect orders examined to date. In addition to their function as a "juvenile factor," JHs and/or their metabolites are involved in a diverse array of other functionalities including roles in development, metamorphosis, reproduction, diapause, migration, polyphenism, and metabolism (Riddiford, 1994; Gilbert *et al.*, 2000; Truman and Riddiford, 2002). These diverse functionalities clearly indicate that the biosynthesis, transport, sequestration, and degradation of JH and/or its metabolites must be carefully regulated. Conversely, this careful regulation of JH titer opens a window of attack where a recombinant baculovirus expressing an appropriate protein(s) could disrupt the fine balance in JH titer and consequently the insect life cycle. Two pathways for the degradation of JH have been intensively studied in insects (Hammock, 1985; Roe and Venkatesh, 1990; de Kort and Granger, 1996). One involves a soluble esterase, JH esterase (JHE), that hydrolyzes the methyl ester moiety at one end of the JH molecule resulting in a carboxylic acid moiety (Kamita *et al.*, 2003a). The other involves a microsomal epoxide hydrolase, JH epoxide hydrolase (JHEH) that hydrolyzes an epoxide moiety at the other end of the JH molecule to produce a diol.

Hammock *et al.* (1990a) first hypothesized that the natural insecticidal activity of the baculovirus AcMNPV could be improved by expression of a gene encoding JHE. The rationale behind this was that the recombinant AcMNPV would be ingested by early larval instars, and subsequently JHE would be produced at a point in development that is inappropriate for the insect. The observed result was that infected larvae reduced feeding and weight gain, and subsequently died slightly more quickly in comparison to the wild-type AcMNPV (Hammock *et al.*, 1990a, 1990b; Eldridge *et al.*, 1992a). Two approaches to improve this technology have included: (1) improving the *in vivo* stability (i.e., reducing removal from the hemolymph and/or degradation) of the JHE enzyme by genetic modification of the JHE gene; and (2) increasing or altering the timing of gene expression by using alternative promoters to drive JHE expression. Recombinant baculoviruses expressing JHE have also been used as tools for hypothesis testing with regard to the biological activity of JHE within the insect host (van Meer *et al.*, 2000). JHE proteins from at least five different insect species including *Heliothis virescens* (Hammock *et al.*, 1990a; Bonning *et al.*, 1992), *Choristoneura fumiferana* (Feng *et al.*, 1999), *B. mori* (Hirai *et al.*, 2002), *M. sexta* (Hinton and Hammock, 2003a), and *T. molitor* (Hinton and Hammock, 2003b) have been expressed using recombinant AcMNPVs.

10.2.2.1. Increase in *in vivo* stability
Studies have shown that recombinant JHE from *H. virescens*, when injected into larvae of *M. sexta*, is rapidly recognized and taken up into the pericardial cells from the hemolymph (Booth *et al.*, 1992; Ichinose *et al.*, 1992b, 1992a). This removal of JHE from the hemolymph occurs by a receptor mediated, endocytotic, saturable mechanism that does not involve passive filtration (Ichinose *et al.*, 1992a, 1992b; Bonning *et al.*, 1997a). The JHE is presumed to be degraded in lysosomes (Booth *et al.*, 1992). At least two putative JHE binding proteins may be involved in transport and/or degradation of JHE in the pericardial cells, including a putative heatshock cognate protein (HSP) (Bonning *et al.*, 1997a) and P29 (Shanmugavelu *et al.*, 2000, 2001). Receptor mediated endocytosis of JHE has been demonstrated in early and late larval instars of *M. sexta* (Ichinose *et al.*, 1992a). Although JHE is normally stable in hemolymph, the half-life of JHE injected into the hemolymph can be as little as 20 min under conditions where endogenous and exogenous proteins including bovine serum albumin, ovalbumin, and hemolymph JH binding protein have half-lives of days (Ichinose *et al.*, 1992a). Thus, the mechanism by which JHE is specifically recognized and removed by the pericardial cells could be a very important target for increasing the half-life of JHE in the hemolymph. There are three ways in which the uptake and degradation of JHE can be disrupted: (1) prevention of receptor mediated uptake by the pericardial cells; (2) disruption of transport to the lysosomes; and (3) prevention of lysosomal degradation.

Bonning *et al.* (1997b) identified two lysine residues that are likely to be on the surface of the JHE protein of *H. virescens* and potentially involved in uptake or degradation of JHE. These lysine residues are Lys-29 near the N-terminus, which is potentially involved in ubiquitin conjugation, and Lys-524, which is located within a putative lysosome targeting sequence. A mutated JHE protein (JHE-KK) in which both of these lysine residues were mutated to arginine residues showed decreased efficiency of lysosomal targeting (Bonning *et al.*, 1997b), and binds to the putative JHE binding protein (P29) with significantly less affinity than authentic JHE (Shanmugavelu *et al.*, 2000). However, JHE-KK as well as mutant JHEs in which only one of the lysine residues was mutated to an arginine residue showed similar catalytic activities and removal rates of JH from the hemolymph as the authentic JHE. A high-resolution crystal structure should assist with determining the basis of the interaction of JHE and the putative JHE binding protein P29 described by Shanmugavelu *et al.* (2000). These researchers hypothesize that P29 interacts with JHE of *H. virescens* and facilitates targeting of JHE to lysosomes within pericardial cells. Other putative JHE binding proteins have been identified in *M. sexta*, and their possible roles in the degradation of JHE remain to be elucidated (Shanmugavelu *et al.*, 2001). First instar larvae of *H. virescens* that were infected with a recombinant AcMNPV expressing JHE-KK (AcJHE-KK) under a very late viral promoter died approximately 22% faster (ST_{50} of 90 versus 116 h) than control larvae infected with a recombinant AcMNPV expressing the authentic JHE (AcJHE) (Bonning *et al.*, 1999). In similar experiments, first instar *T. ni* that were infected with AcJHE-KK died approximately 27% faster (ST_{50} of 83 versus 113 h) compared to control larvae infected with AcJHE. Kunimi *et al.* (1996) have analyzed the survival times of third to fifth instar *T. ni* that were infected with AcJHE-KK. In these older insects, they found that the ST_{50}s of AcJHE-KK infected third, fourth, and fifth larvae were reduced by roughly 4% (124.8 versus 130.2 h), 9% (126.6 versus 138.6 h), and 8% (135.0 versus 147.0 h), respectively, in comparison to control AcMNPV infected larvae. In similar bioassays of second to fifth instars of the soybean looper *Pseudoplusia includens*, AcJHE-KK infected larvae did not show a reduced survival time in comparison to AcMNPV infected control larvae (Kunimi *et al.*, 1997). The reasons for these differences in the survival times between species and instars are unknown, however, Bonning *et al.* (1999) speculated that host and/or age specific effects may be involved.

10.2.2.2. Gene silencing and RNA interference

AcPR1 is a recombinant AcMNPV that produces biologically active JHE of *H. virescens* (Hanzlik *et al.*, 1989) under the very late baculoviral *p10* gene promoter (Roelvink *et al.*, 1992). AcPR2 is a recombinant AcMNPV in which the same JHE gene is placed under the *p10* promoter in the antisense direction. Transcription of this gene generates mRNAs that are antisense to the JHE mRNAs generated by AcPR1. These antisense JHE mRNAs are able to reduce the JHE activity produced by AcPR1 when AcPR1 and AcPR2 are coinfected into *Spodoptera frugiperda* (Roelvink *et al.*, 1992). By injection of AcPR2 into fifth instar *H. virescens*, Hajos *et al.* (1999) showed that AcPR2 downregulates JHE activity in more than 95% of the injected larvae. This effect was putatively due to direct RNA–RNA interaction between sense and antisense JHE RNAs. Although 95% of the infected larvae showed a reduction in JHE activity, only 25% of these larvae showed morphogenic alterations (Hajos *et al.*, 1999). These alterations were similar to those induced by the exogenous application of JH (Cymborowski and Zimowska, 1984) or JHE inhibitors (Hammock *et al.*, 1984) to the larvae. On the basis of the effectiveness of this baculovirus mediated gene silencing approach, Hajos *et al.* (1999) suggest that the identification of other host gene targets could greatly reduce lethal times or feeding damage of a GM baculovirus pesticide.

RNA interference (RNAi) is a sequence specific mechanism by which a targeted gene is silenced by the introduction of double-stranded RNA (dsRNA) that is homologous to the targeted gene (Hutvagner and Zamore, 2002; Denli and Hannon, 2003). Mechanistically, it is likely that the antisense RNA

mediated gene silencing described by Hajos *et al.* (1999) involves the RNAi pathway. RNAi has been demonstrated in a diverse range of organisms including plants, fungi, arthropods, and mammals. One biological function of RNAi appears to be as a defense mechanism against RNA viruses in plants and other organisms. Recently, RNAi has been shown to be effective in lepidopteran cell lines (Means *et al.*, 2003; Valdes *et al.*, 2003) and larvae such as the giant silk moth, *H. cecropia* (Bettencourt *et al.*, 2002) and cabbage looper *T. ni* (Kramer and Bentley, 2003). Our laboratory has explored the use of RNAi to block the activity and effects of JHE in cultured *S. frugiperda* cells and larvae of *H. virescens*. In the larval experiments, a 0.5 kb dsRNA fragment that corresponds to the 5′-coding sequence of the JHE gene of *H. virescens* (Hanzlik *et al.*, 1989) was generated. Neonate *H. virescens* that ingested these dsRNAs within 6 h of hatching showed a statistically significant (two-sided Mann-Whitney test with an error type I of 99%) increase in weight (in comparison to control larvae) by the second and third days of the fifth larval instar. Additionally, injection of the same 0.5 kb dsRNAs into the larvae of *H. virescens* on the first day of the fifth instar resulted in a 16 h delay in the median time to pupation in comparison to water injected or untreated controls (**Figure 2**). Treatments were compared with Wilcoxon (Gehan) statistics on a 99% error type I. These comparisons indicated that RNAi against the JHE of *H. virescens* is effective at prolonging the juvenile state. The authors are currently in the process of constructing recombinant AcMNPVs carrying JHE gene derived sequences in a "head-to-head" manner such that the expressed RNAs will form a hairpin loop structure. Such a

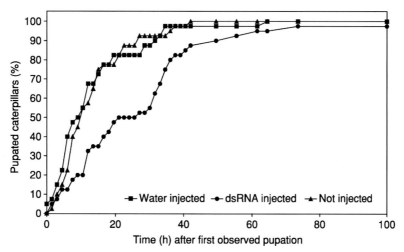

Figure 2 Percentage of larvae that pupate over time following treatment with double-stranded RNA (dsRNA), double-distilled water, or untreated. Pupation was scored on the basis of head capsule slippage at 1.5 h or longer intervals.

recombinant AcMNPV may be able to induce the RNAi response more efficiently. Of course, a recombinant virus that is effective at prolonging the juvenile state would be a disaster as a biological insecticide. However, these experiments are important in that they demonstrate the effectiveness of an RNAi approach in the regulation of insect neuroendocrinology and the results should be easily transferred to other significant physiological effectors.

The use of an authentic, insect-derived protein to combat the insect is conceptually elegant and potentially safer because one is only trying to express an endogenous protein at an inappropriate time or at higher levels in order to alter the physiology of the insect. Additionally, there is a perception that expressing the insect's own protein at an inappropriate time in development offers public relations advantages over the expression of toxins or proteases as will be discussed below. In practice, however, overexpression of an insect hormone or protein involved in hormone metabolism has not been dramatically effective. For example, the limited number of JHEs tested to date not enhance the speed of kill as well as scorpion toxins or proteases. However, with JHE there is a clear mechanism of action and an obvious path toward generating improved viruses by engineering the protein for greater stability in the insect. The recent homology model of Thomas et al. (1999) and crystal structure of the protein will greatly enhance these efforts.

10.2.3. Proteases

In order to better understand the rationale behind the use of protease encoding genes for improving speed of kill, a brief review of the midgut barriers faced by the baculovirus is given. The first barrier that the ODV encounters in the midgut is the fluid within the lumen. The high pH (8–11) and endogenous proteases found in the lumen are initially required to release ODVs from the polyhedra or granules. However, digestive enzymes or other factors within the lumen could also lead to the inactivation of the released ODV, if the ODVs do not quickly initiate infection. One such factor is a lipase, Bmlipase-1, isolated from the digestive juice of *B. mori* that shows antiviral activity against BmNPV (Ponnuvel et al., 2003). The peritrophic membrane (PM; or peritrophic envelope) is the first physical barrier that the ODV faces (Richards and Richards, 1977; Brandt et al., 1978; Tellam, 1996). The PM is composed of chitin fibrils in a protein–carbohydrate matrix. The specific composition and layout of the PM differs in different lepidopteran

species and at different larval instars. The manner in which the ODVs pass this barrier is still not well understood. The second physical barrier that the virus faces is the midgut epithelium. The midgut epithelium is primarily composed of a single layer of columnar cells. The columnar cells show directionality (polarity) in which the luminal border is formed by microvilli (brush border membrane) and the hemocoelic border shows characteristic infolding with large numbers of mitochondria associated with the infolds. Regenerative and endocrine cells are also part of the midgut epithelium. Several pathways have been proposed for how the nucleocapsids pass through the midgut epithelium barrier following the fusion of ODVs with the microvillar membrane (Federici, 1997). One pathway involves entry of the nucleocapsid into the columnar cells, translocation through the cytoplasm, and budding through the hemocoelic border of the same cell without any viral replication. The other pathways involve the generation of progeny nucleocapsids within the columnar cells that can bud (1) through the hemocoelic border, (2) into the tracheal matrix via tracheoblasts, or (3) into regenerative cells. Once in the regenerative cells, the nucleocapsids may translocate through the cell without replication and through the hemocoelic border or virus replication may occur in these cells as well. Washburn et al. (2003) have proposed a "hybrid" pathway in which some nucleocapsids of MNPVs uncoat and start replication whereas others pass directly through the columnar cells. The third midgut associated barrier to systemic infection is the basement membrane (or basal lamina). The basement membrane (BM) is a fibrous matrix composed primarily of glycoproteins, type IV collagen, and laminin, which are secreted by the epithelial cells (Ryerse, 1998). The BM functions in several roles including structural support of the epithelium or surrounding tissues, filtration, and differentiation (Yurchenco and O'Rear, 1993).

10.2.3.1. Enhancins
Enhancins are baculovirus encoded proteins that can enhance the oral infectivity of a heterologous or homologous baculovirus. Enhancin, also known as synergistic factor (SF) or virus enhancing factor (VEF), was first described by Tanada as a lipoprotein of 93–126 kDa found within the capsule of the GV of the armyworm *Pseudaletia unipuncta* (Yamamoto and Tanada, 1978a, 1978b) that enhances the infection of baculoviruses in lepidopteran larvae (Tanada, 1959). The enhancin of the GV of *T. ni* (TnGV) has been shown to possess metalloprotease activity (Lepore et al., 1996; Wang and Granados, 1997). Tanada first suggested

that this factor is virus encoded (Tanada and Hukuhara, 1968) and Hashimoto *et al.* (1991) confirmed this by cloning and sequencing the *vef* gene from TnGV. Subsequently, enhancin gene homologs have been sequenced from the GVs of *H. armigera* and *P. unipuncta* (Roelvink *et al.*, 1995) and *Xestia c-nigrum* (four homologs; Hayakawa *et al.*, 1999). Enhancin gene homologs have also been identified in two NPVs, *Lymantria dispar* NPV (two homologs, E1 (Bischoff and Slavicek, 1997) and E2 (Popham *et al.*, 2001)) and *Mamestra configurata* NPV (Li *et al.*, 2003). Disruption of the E1 enhancin gene homolog of LdMNPV reduces viral potency from 1.4- to 4-fold in comparison to the wild-type LdMNPV but does not affect the killing speed of the virus (Bischoff and Slavicek, 1997). Enhancins appear to enhance virus infection by: (1) degrading high molecular weight proteins such as mucin (Wang and Granados, 1997), which are found in the PM (Derksen and Granados, 1988; Gallo *et al.*, 1991; Lepore *et al.*, 1996; Wang and Granados, 1997); and/or (2) enhancing fusion of the virus to the midgut epithelium cells (Ohba and Tanada, 1984; Kozuma and Hukuhara, 1994; Wang *et al.*, 1994).

Enhancin genes have been expressed by recombinant AcMNPVs in order to improve the ability of the virus to gain access to the midgut epithelium cells. A recombinant AcMNPV, AcMNPV-enMP2, carrying the enhancin gene of *M. configurata* NPV under its authentic promoter has been constructed (Popham *et al.*, 2001; Li *et al.*, 2003). In dose-mortality studies, AcMNPV-enMP2 shows a 4.4-fold lower median lethal dose ($LD_{50} = 2.8$ versus 12.4 polyhedra per larva) in second instar larvae of *T. ni* in comparison to the wild-type virus (Li *et al.*, 2003). In time-mortality studies, the ST_{50} of AcMNPV-enMP2 infected larvae was not significantly different from that of wild-type AcMNPV infected larvae when the larvae were inoculated at biologically equivalent doses (i.e., at an LD_{90} dose). However, when the larvae were inoculated with the same number (i.e., 32 polyhedra per larva) of AcMNPV-enMP2 or wild-type AcMNPV, the AcMNPV-enMP2 inoculated larvae showed an approximately 8% faster ST_{50} (130 versus 141 h). Li *et al.* (2003) suggest that this enhancement in speed of kill may be the result of increasing the number of initial infection foci in the host insect midgut rather than increasing the speed of the spread of the virus infection.

Hayakawa *et al.* (2000b) have generated a recombinant AcMNPV (AcEnh26) expressing the enhancin gene from *T. ni* granulovirus. In bioassays in which third instar larvae of *S. exigua* were fed

a mixture of AcEnh26 infected Sf-9 cells and AcMNPV polyhedra, the LD_{50} of AcMNPV was 21-fold lower in comparison to feeding with a mixture of control virus infected Sf-9 cells and AcMNPV (2.69×10^3 versus 5.75×10^4 polyhedra per larva). In similar experiments, the LD_{50} of another baculovirus, *Spodoptera exigua* NPV (SeNPV), was ninefold lower (19.1 versus 162 polyhedra per larva) when coinfected with AcEnh26 infected Sf-9 cells. In an interesting twist to these experiments, Hayakawa *et al.* (2000b) engineered the enhancin gene from *T. ni* into tobacco plants and used a mixture of AcMNPV and lyophilized transgenic tobacco leaves for bioassay. They found a tenfold reduction in the LD_{50} of AcMNPV when the enhancin expressing tobacco was added to the bioassay. These findings suggest that expression of the enhancin gene in plants (Granados *et al.*, 2001; Cao *et al.*, 2002) or mixing enhancin expressing plant material in baculovirus pesticide formulations may be an effective method for improving the speed of kill of natural and recombinant baculoviruses.

10.2.3.2. Basement membrane degrading proteases

As mentioned above, the basement membrane is the last midgut-associated barrier that the baculovirus virion must cross prior to the initiation of systemic infection. Harrison and Bonning (2001) have constructed recombinant AcMNPVs expressing three different proteases (rat stromelysin-1, human gelatinase A, and cathepsin L from the flesh fly *Sarcophaga peregrina*) that are known to digest BM proteins. Of these recombinant AcMNPVs, AcMLF9.ScathL expressing cathepsin L under the late baculovirus *p6.9* gene promoter generated a 51% faster speed of kill (ST_{50} of 48 versus 98 h) in comparison to wild-type AcMNPV in neonate larvae of *H. virescens*. Expression of gelatinase under the *p6.9* promoter also improved speed of kill by approximately 12% (ST_{50} of 86.5 versus 98 h) in comparison to wild-type AcMNPV. On the basis of lethal concentration bioassays, there were no statistically significant differences in the virulence (i.e., LD_{50}) of these viruses in comparison to the wild-type AcMNPV. In bioassays to determine feeding damage, AcMLF9.ScathL infected larvae consumed approximately 27- and 5-fold less lettuce (29.27 and 5.21 cm^2, respectively) in comparison to mock or AcMNPV infected (1.1 cm^2) second instars of *H. virescens*, respectively. Although the mechanism of this improvement in speed of kill has yet to be determined, Harrison and Bonning (2001) speculated that digestion of BM proteins by the cathepsin L may: (1) hasten the spread of infection through the BM; (2) inappropriately activate prophenoloxidase, a key enzyme involved in the formation of melanin during wound healing and the immune response in insects; and/or (3) cause BM damage that results in deregulation of ionic and molecular traffic between the hemocoel and the tissues. Harrison and Bonning (2001) also expressed cathepsin L under the baculovirus early *ie1* promoter using the recombinant baculovirus AcIE1TV3.ScathL. This virus, however, did not show any improvement in speed of kill. They speculated that expression under the early promoter did not produce sufficient cathepsin L to have an effect on the infected host. The effectiveness of protein expression using early, late, and/or constitutively expressed promoters will be further discussed later.

10.2.4. Other Enzymes and Factors

Chitin is an insoluble structural polysaccharide that is found in the exoskeleton and gut linings of insects (Cohen, 1987). Chitinases are enzymes that can degrade chitin into low molecular weight, soluble, and insoluble oligosaccharides (Kramer and Muthukrishnan, 1997). Genes encoding chitinases have been identified in several lepidopteran insects including *M. sexta* (Gopalakrishnan *et al.*, 1993), *B. mori* (Kim *et al.*, 1998; Daimon *et al.*, 2003), *Hyphantria cunea* (Kim *et al.*, 1998), *S. litura* (Shinoda *et al.*, 2001), and *C. fumiferana* (Zheng *et al.*, 2002). The genome of *B. mori* appears to encode multiple, potentially five different, chitinase genes (Daimon *et al.*, 2003). Chitinase and cysteine protease genes are also found in the genomes of NPVs and GVs (Ohkawa *et al.*, 1994; Hawtin *et al.*, 1995; Slack *et al.*, 1995; Kang *et al.*, 1998; Gong *et al.*, 2000). Baculovirus encoded chitinases and proteases (V-CATH) are thought to be involved in the degradation of the chitinous and proteinaceous components, respectively, of the host cadaver in order to induce liquefaction (O'Reilly, 1997; Hom and Volkman, 2000). Hom and Volkman (2000) have suggested another role of the baculovirus encoded chitinase, namely, proper folding of the nascent V-CATH polypeptides in the endoplasmic reticulum (ER). A recombinant AcMNPV, vAcMNPV·chi, has been generated that expresses the chitinase gene of *M. sexta* (Gopalakrishnan *et al.*, 1995). When fourth instar *S. frugiperda* were injected with vAcMNPV·chi, larval death occurred nearly 1 day earlier than those infected with the wild-type virus AcMNPV.

Carbohydrates are the major source of energy that powers insect metabolism and form the material basis of chitin in the cuticle of insects (Cohen,

1987). Thus, the availability of sugars is essential for normal growth and development. Trehalose, a disaccharide composed of two glucose molecules, is the major carbohydrate in the hemolymph of most insects (Friedman, 1978). Sato *et al.* (1997) have generated two recombinant AcMNPVs, vTREVL, and vERTVL, expressing a trehalase cDNA from the mealworm beetle *T. molitor* in the sense and antisense directions, respectively. In Sf-9 cell cultures and larvae of the cabbage armyworm *Mamestra brassicae* that were infected with vTREVL, biologically active trehalase was secreted into the supernatant and found in the hemolymph, respectively. The survival time of fifth instar larvae of *M. brassicae* that were infected with vTREVL was longer than control larvae infected with the wild-type AcMNPV. Interestingly, the vTREVL infected larvae showed considerably reduced melting in comparison to the AcMNPV infected larvae. The survival time of larvae infected with vERTVL was also slightly longer than that of control larvae infected with AcMNPV, but less than that of vTREVL.

URF13 is a mitochondrially encoded protein of maize that forms pores in the inner mitochondrial membrane (Korth *et al.*, 1991). Korth and Levings (1993) have generated recombinant, occlusion-negative AcMNPVs expressing authentic (BV13T) or mutated (BV13.3940) URF13 encoding genes under the *polh* promoter. When third to fourth instar larvae of *T. ni* were injected with BV13T or BV13.3940, 100% died by 60 h post injection. Control larvae that were injected with wild-type AcMNPV or a control β-galactosidase expressing AcMNPV lived up to 106 and 100 h p.i., respectively. The median lethal time (LT_{50}) of BV13T or BV13.3940 appeared reduced by about 45% (50 versus 94 h) in comparison to control virus infected larvae (estimate based on mortality graph). Although the mechanism of this toxicity was not determined, Korth and Levings (1993) showed that baculovirus expressed URF13 is localized in the membranes of cells of Sf-9 cultures and larvae of *T. ni*. They suggested that this membrane localization interferes with normal cellular functions. The URF13 induced cytotoxicity, however, appeared to be unrelated to pore formation, because the mutated URF13 expressed by BV13.3940 is unable to form membrane pores (Braun *et al.*, 1989), but the toxicity of BV13.3940 was the same as that of BV13T.

c-MYC is a transcription factor that is involved in various physiological processes including cell growth, proliferation, loss of differentiation, and apoptosis in a variety of organisms (Pelengaris and Khan, 2003). Lee *et al.* (1997) have generated a recombinant AcMNPV carrying a 754 bp-long fragment of human *c-myc* exon II in the antisense direction under the *polh* promoter. When third or fourth instar *S. frugiperda* are injected with this virus at an average dose of 3500 plaque forming units (pfu) per mg of body weight, 75% of the larvae stop feeding by 3 days postinfection. This is roughly a twofold reduction in feeding in comparison to wild-type AcMNPV injected larvae, in which about 35% of the larvae stop feeding. This feeding cessation occurs roughly 2 days before death of the larva. Furthermore, the average survival time of the larvae is reduced by approximately 1 day. Lee *et al.* (1997) indicate two important advantages to this antisense strategy. First, this strategy does not require protein synthesis and also avoids any potential problems associated with posttranslational modifications or secretion. Second, since no foreign protein is produced the likelihood of insect resistance will be reduced.

Sunlight (i.e., UV radiation) is a major factor in the inactivation of baculoviruses in the field (Ignoffo and Garcia, 1992; Dougherty *et al.*, 1996; Black *et al.*, 1997; Ignoffo *et al.*, 1997). In an attempt to reduce UV inactivation, Petrik *et al.* (2003) have generated a recombinant AcMNPV that expresses an algal virus pyrimidine dimer specific glycosylase, cv-PDG (Furuta *et al.*, 1997), which is involved in the first steps of the repair of UV damaged DNA. The recombinant AcMNPV (vHSA50L) expresses the cv-PDG encoding the *A50L* gene under the *hsp70* promoter of *D. melanogaster*. Polyhedra of vHSA50L were not more resistant to UV inactivation than the wild-type AcMNPV, however, BV of vHSA50L were threefold more resistant to UV inactivation. In dose-mortality bioassays in neonate *T. ni*, the median lethal concentrations (LC_{50}s) of vHSA50L and AcMNPV were not significantly different. In neonates of *S. frugiperda*, however, the LC_{50} of vHSA50L was reduced by approximately 16-fold (7.1×10^5 versus 1.15×10^7 polyhedra per ml). Surprisingly, time-mortality bioassays (at an LC_{90} dose) in vHSA50L infected neonates of *S. frugiperda* indicated that the LT_{50} of vHSA50L was reduced by approximately 41% (73.6 versus 124.4 h). This reduction in LT_{50} was even greater (48%, 65.0 versus 124.4 h) for a construct (vHSA50-LORF) that carried *A50L* excluding the 5′-untrans-untranslated region under the *hsp70* promoter. In contrast to the neonate *S. frugiperda*, significant differences were not found in the LT_{50}s of neonate *T. ni* that were infected with vHSA50L or vHSA50LORF.

10.3. Insertion of Insect Selective Toxin Genes

Arthropods that feed on or parasitize other arthropods possess potent venoms that rapidly immobilize their prey. The venoms of arthropods such as scorpions, spiders, and parasitic wasps are composed of a mixture of salts, small molecules, proteins, and peptides (Zlotkin *et al.*, 1978, 1985; Zlotkin, 1991; Loret and Hammock, 1993; Gordon *et al.*, 1998; Possani *et al.*, 1999). Peptide toxins that are specifically active against invertebrates, vertebrates, or both invertebrates and vertebrates have been identified in these venoms. These peptide toxins generally target major ion channels such as the Na^+, K^+, Ca^{2+}, and Cl^- channels although there are several examples of peptides with unusual targets such as the intracellular calcium activated ryanodine channel (Fajloun *et al.*, 2000). Kinin-like peptides that interfere with physiological events have also been identified in venom (Inceoglu and Hammock, unpublished data). Arthropod venoms have been utilized as highly useful sources for the mining of highly potent and selective peptide toxins that selectively paralyze insects. Several of these peptides have been expressed by recombinant baculoviruses for the purpose of improving their speed of kill and insecticidal activity.

10.3.1. Scorpion Toxins

On average scorpion venom contains from 200 to 500 different peptides. It is conservatively assumed that about 10% of these peptides are insect selective toxins. Considering that there are roughly 1250 identified species of scorpion, this represents an enormous diversity (25 000 to 62 500) of potent and selective insecticidal peptides for pest control using baculovirus expression. Scorpion toxins are classified on the basis of size and pharmacological target site into two main groups: long and short chain neurotoxins (Zlotkin *et al.*, 1978; Loret and Hammock, 1993). Toxins with selective activity against insects are found in both the long and short chain groups. Long chain neurotoxins are polypeptides of 58–70 amino acid residues, cross-linked by four disulfide bonds, which mainly target voltage gated sodium channels and to some degree calcium channels (Zlotkin, 1991; Loret and Hammock, 1993; Gordon *et al.*, 1998). Short chain neurotoxins are peptides of 31–37 amino acid residues, cross-linked by three or four disulfide bonds, which mainly target potassium and chloride channels (Loret and Hammock, 1993). The long chain neurotoxins, the better studied of the two main groups, are further classified into mammal or insect selective toxins, with the insect selective toxins being further classified on the basis of their mode of action into three subgroups: alpha insect toxins, excitatory toxins, and depressant toxins (Zlotkin *et al.*, 1995; Gordon *et al.*, 1998; Inceoglu *et al.*, 2001b). Each of these subgroups target different molecular sites and produce distinct effects upon the voltage gated sodium channel of the insect (Cestele and Catterall, 2000). In addition, when injected into larvae of the blowfly *S. falculata*, toxins in each of these subgroups display unique symptoms. Alpha insect toxins cause contractive paralysis that is delayed and sustained (Eitan *et al.*, 1990). Excitatory toxins, such as AaIT from the North African scorpion *Androctonus australis*, cause paralysis that is immediate and contractive (Zlotkin *et al.*, 1971, 1985; Zlotkin, 1991). In contrast, depressant toxins, such as LqhIT2 from the yellow Israeli scorpion *Leiurus quinquestriatus hebraeus*, cause transient (until 5 min post injection) contractive paralysis, followed by sustained, flaccid paralysis (Zlotkin *et al.*, 1985; Zlotkin, 1991).

Carbonell *et al.* (1988) were the first to attempt to express biologically active scorpion toxin, insectotoxin-1 of *Buthus eupeus* (*BeIt*), under control of the very late polyhedrin promoter using recombinant AcMNPV constructs (vBeIt-1, vBeIt-2, and vBeIt-3). The vBeIt-1 construct carried the 112 nucleotide-long *BeIt* gene 6 nucleotides downstream of the last nucleotide of the polyhedrin leader. The vBeIt-2 construct carried the *BeIt* gene fused to a 21 amino acid-long signal sequence for secretion of human β-interferon (Ohno and Taniguchi, 1981) 18 nucleotides downstream of the polyhedrin leader. The vBeIt-3 construct expressed *BeIt* as a fusion with the N-terminal 58 codons of the polyhedrin gene. All three constructs produced similar, high levels of *BeIt* specific transcripts from the polyhedrin promoter. The vBeIt-1 and vBeIt-2 constructs produced exceptionally low levels of the 4 kDa BeIt peptide, whereas the vBeIt-3 construct produced significant amounts of a 13–14 kDa polyhedrin-BeIt fusion peptide. Toxin specific biological activity, however, was not observed by bioassay in larvae of *T. ni*, *Galleria mellonella*, and *Sarcophaga*. Carbonell *et al.* (1988) speculated that this was the result of: (1) toxin instability; (2) a low sensitivity threshold of the toxin bioassay; and/or (3) inability of the polyhedrin-BeIt fusion to properly fold.

10.3.1.1. AaIT

The insect-selective neurotoxin AaIT (*Androctonus australis* insect toxin 1), found in the venom of the scorpion *A. australis*, was the first scorpion toxin to be expressed by recombinant

baculoviruses that showed biological activity (Maeda *et al.*, 1991; McCutchen *et al.*, 1991; Stewart *et al.*, 1991). AaIT is composed of a single polypeptide chain of 70 amino acid residues cross-linked by four disulfide bonds and is highly specific for the voltage gated sodium channel of insects (Zlotkin *et al.*, 2000). The toxin induces a neurological response similar to that evoked by the pyrethroid insecticides, but apparently acts at a different site within the sodium channel (see discussion below regarding toxin interactions with chemical pesticides). Maeda *et al.* (1991) constructed a recombinant BmNPV carrying a synthetic AaIT gene (Darbon *et al.*, 1982) that was linked to a signal sequence for secretion of bombyxin and driven by the very late polyhedrin (*polh*) gene promoter. The recombinant virus, BmAaIT, expressed biologically active AaIT that was secreted into the hemolymph of BmAaIT infected silkworm larvae. The BmAaIT infected larvae (second to fifth instar *B. mori*) displayed symptoms that were consistent with sodium channel binding by AaIT including body tremors, dorsal arching, feeding cessation, and paralysis beginning at 40 h p.i. Death occurred by 60 h p.i. This corresponded to an improvement in speed of kill of approximately 40% in comparison to control larvae infected with the wild-type BmNPV.

In related experiments, McCutchen *et al.* (1991) and Stewart *et al.* (1991) independently generated recombinant AcMNPVs expressing AaIT under the very late baculoviral *p10* promoter. The Stewart *et al.* construct, AcST-3, expressed AaIT as a fusion protein with the signal sequence of the baculoviral GP67 protein. The McCutchen *et al.* construct, AcAaIT, expressed AaIT as a fusion protein with the bombyxin signal sequence. Bioassays using orally inoculated, second instar larvae of *T. ni* indicated that the median lethal dose (LD_{50}) of AcST-3 was reduced by about 30% (44 versus 31 polyhedra per larva) in comparison to the wild-type AcMNPV. The median survival time (ST_{50}) of AcST-3 infected larvae (infected with approximately 17.4 polyhedra per larva) was reduced by about 25% (85.8 versus 113.1 h) in comparison to control larvae infected with AcMNPV. Third instar *T. ni* that are infected with a dose of AcST-3 that resulted in 100% mortality showed a 50% reduction in feeding damage in comparison to AcMNPV infected control larvae. The McCutchen *et al.* construct, AcAaIT, showed similar results by bioassay using second instar larvae of *H. virescens*. The LD_{50} of AcAaIT was reduced by about 39% (13.3 versus 21.9 polyhedra per larva) in comparison to AcMNPV. The ST_{50} of AcAaIT infected larvae (infected with 250 polyhedra per larva) was reduced by about 30% (88.0 versus 125 h) in comparison to control larvae infected with AcMNPV. By droplet feeding assays of neonate *H. virescens*, the ST_{50} of AcAaIT infected larvae was reduced by up to 46% (Inceoglu and Hammock, unpublished data). McCutchen *et al.* conducted additional experiments in which third instar larvae of *M. sexta* (an unnatural host of AcMNPV) were injected with AcAaIT (5 μl containing 1×10^6 pfu). They detected AaIT specific symptoms as early as 72 h post injection and found a 29% decrease (120 versus 168 h) in the speed of kill. Additional observations made during this study showed that larvae infected with AcAaIT typically were paralyzed and stopped feeding several hours prior to death. Furthermore, the yields of progeny viruses (polyhedra per cadaver) in AcAaIT infected larvae (third, fourth, or fifth instar *T. ni*) were only 20–32% of those of control larvae infected with AcMNPV (Kunimi *et al.*, 1996; Fuxa *et al.*, 1998). This suggested that the recombinant virus will be quickly outcompeted in the field by the wild-type virus (see below).

Hoover *et al.* (1995) have further characterized the paralytic effect of AaIT by bioassay using third instar *H. virescens* that were inoculated with either AcAaIT or AcMNPV. When the AcAaIT or AcMNPV infected larvae were placed on greenhouse cultivated cotton plants, it was found that the AcAaIT infected larvae fell off the plants approximately 5–11 h before death. This "knock-off"

effect occurred before the induction of feeding cessation. As a consequence, the amount of leaf area consumed by the AcAaIT infected larvae was up to 62% and 72% less than that consumed by the AcMNPV and mock infected larvae, respectively (**Figure 3**). Knock-off effects have also been observed in field trials (on cotton) to assess the efficacy of recombinant AcMNPV (Cory *et al.*, 1994) or HaSNPV (Sun *et al.*, 2004) expressing AaIT. On the basis of this knock-off effect, Hoover *et al.* (1995) suggested that median survival time is not necessarily predictive of the reduction in the amount of feeding damage that results from the application of a recombinant baculovirus. Cory *et al.* (1994) and Hoover *et al.* (1995) also emphasized that the knock-off effect should reduce foliage contamination because unlike AcAaIT infected larvae that fall off the plant, wild-type virus infected larvae tend to die on the plant. Consequently, they suggested that the reduction in virus inoculum on the foliage should decrease the spread and recycling of the recombinant virus in comparison to the wild-type virus. A discussion of the fitness of a recombinant virus in comparison to the wild-type parent is given later in this chapter.

The AaIT gene has been expressed using other baculovirus vectors, for example, the NPVs of the mint looper *Rachiplusia ou* (RoMNPV; Harrison and Bonning, 2000b) and cotton bollworms *Helicoverpa zea* (HzNPV; Treacy *et al.*, 2000) and

Figure 3 Damage to cotton plants by third instar larvae of *Heliothis virescens* that were inoculated with an LD_{99} dose of AcMNPV or AcAaIT or mock infected. On average, the leaf damage caused by AcAaIT infected larvae was reduced by 62% in comparison to AcMNPV infected larvae and 71.7% in comparison to mock infected larvae (Hoover *et al.*, 1995). (Reprinted with permission from Hoover, K., Schultz, C.M., Lane, S.S., Bonning, B.C., Duffey, S.S., *et al.*, **1995**. Reduction in damage to cotton plants by a recombinant baculovirus that knocks moribund larvae of *Heliothis virescens* off the plant. *Biol. Control 5*, 419–426; © Elsevier.)

H. armigera (HaSNPV; Chen *et al.*, 2000; Sun *et al.*, 2002, 2004). RoMNPV is also known as *Anagrapha falcifera* Kirby MNPV (Harrison and Bonning, 1999). Harrison and Bonning (2000b) have generated recombinant RoMNPVs (Ro6.9AaIT) expressing AaIT fused to the bombyxin signal sequence under the late *p6.9* promoter of AcMNPV. The *p6.9* promoter, like the very late promoters, requires the products of viral early genes and the initiation of viral DNA replication for its activation (Lu and Miller, 1997). However, it is activated earlier and drives higher levels of expression than the *p10* or *polh* promoters in tissue culture (Hill-Perkins and Possee, 1990; Bonning *et al.*, 1994). In dose-mortality bioassays (droplet feeding), there were no significant differences between the LC$_{50}$s of Ro6.9AaIT and the wild-type RoMNPV in neonate European corn borer *Ostrinia nubilalis* (8.1 × 10^6 and 1.0 × 10^7 polyhedra per ml, respectively) or *H. zea* (9.5 × 10^4 and 5.4 × 10^4 polyhedra per ml, respectively). The LC$_{50}$s, however, were significantly different in neonate *H. virescens* (1.67 × 10^5 and 8.4 × 10^4 polyhedra per ml, respectively). Survival-time bioassays (at an LC$_{99}$ dose) of neonates of *O. nubilalis*, *H. zea*, and *H. virescens* infected with Ro6.9AaIT generated ST$_{50}$s that were reduced by approximately 34% (120 versus 181 h), 37% (81 versus 128 h), and 19% (72.5 versus 89.5 h), respectively, in comparison to control larvae infected with RoMNPV. Later in this chapter, recombinant HzNPV and HaSNPV constructs in which the AaIT gene was placed under a very late promoter or other alternative promoter (e.g., early, chimeric, or constitutive) and inserted at the *egt* gene locus of the virus are discussed.

10.3.1.2. Lqh and Lqq   A series of highly potent insecticidal toxins have been identified and characterized from the venom of yellow Israeli scorpions, *L. quinquestriatus hebraeus* and *L. quinquestriatus quinquestriatus*. Both excitatory (e.g., LqqIT1, LqhIT1, and LqhIT5) and depressant (e.g., LqhIT2 and LqqIT2) insect selective toxins have been isolated from these scorpions (Zlotkin *et al.*, 1985, 1993; Kopeyan *et al.*, 1990; Zlotkin, 1991; Moskowitz *et al.*, 1998). Gershburg *et al.* (1998) have generated recombinant AcMNPVs expressing the excitatory LqhIT1 toxin under the very late *p10* (AcLIT1.p10) and early *p35* (AcLIT1.p35) gene promoters, and the depressant LqhIT2 toxin under the very late *polh* gene promoter (AcLIT2.pol). These constructs expressed LqhIT1 and LqhIT2 that were secreted to the outside of the cell under endogenous secretion signals. On the basis of time-mortality bioassays using neonate *H. armigera*, they found

that the median effective times (ET$_{50}$s) for paralysis and/or death of AcLIT1.p10 and AcLIT2.pol were 66 and 59 h. This was an improvement of roughly 24% and 32%, respectively, in comparison to the wild-type AcMNPV (ET$_{50}$ of 87 h). The ET$_{50}$ was also reduced by about 16% (to 73 h) when LqhIT1 was expressed under the early promoter by AcLIT1.p35, although not as dramatically in comparison to AcLIT1.p10. Imai *et al.* (2000) have generated a recombinant BmNPV (BmLqhIT2) expressing LqhIT2 fused to a bombyxin signal sequence under the *polh* gene promoter. Fourth instar larvae of *B. mori* that were injected with BmLqhIT2 (2 × 10^5 pfu per larva) initially showed LqhIT2 specific symptoms at 48 h post injection and stopped feeding at roughly 82 h post injection. This was an improvement of roughly 35% in the ET$_{50}$ in comparison to the wild-type BmNPV.

Harrison and Bonning (2000b) have generated recombinant AcMNPVs expressing LqhIT2 fused to a bombyxin signal sequence under the late *p6.9* (AcMLF9.LqhIT2) or very late *p10* (AcUW21.LqhIT2) gene promoters. Survival-time bioassays (at an LC$_{99}$ dose) of neonate *H. virescens* infected with AcMLF9.LqhIT2 or AcUW21.LqhIT2 showed that the ST$_{50}$s (63.5 h) of larvae infected with these viruses were the same (Harrison and Bonning, 2000b). However, log-rank tests of the Kaplan-Meier survival curves of larvae infected by these viruses were significantly different ($P < 0.05$). In comparison to AcUW21.LqhIT2 infected neonate *H. virescens*, more of the AcMLF9.LqhIT2 infected neonates died at earlier times postinfection (Harrison and Bonning, personal communication). In comparison to control larvae infected with wild-type AcMNPV, AcMLF9.AaIT (recombinant AcMNPV expressing AaIT under the *p6.9* gene promoter), or AcAaIT, neonate *H. virescens* infected with AcMLF9.LqhIT2 or AcUW21.LqhIT2 showed reductions in median survival times of approximately 34% (63.5 versus 96.0 h), 10% (63.5 versus 71.0 h), and 10% (63.5 versus 70.5 h), respectively. Harrison and Bonning (2000b) have also generated recombinant RoMNPVs expressing LqhIT2 under the *p6.9* (Ro6.9LqhIT2) and *p10* (Ro10LqhIT2) promoters of AcMNPV. Ro10LqhIT2 produced visibly fewer polyhedra that potentially occluded fewer virions in Sf-9 cell cultures and thus were not used in bioassays. Survival-time bioassays (at an LC$_{99}$ dose) in neonate larvae of the European corn borer *O. nubilalis* Hübner, *H. zea*, and *H. virescens* infected with Ro6.9LqhIT2 generated ST$_{50}$s that were approximately 41% (107 versus 181 h), 42% (74.5 versus 128 h), and 27% (65.5 versus 89.5 h) faster, respectively, than those generated in control neonates infected with wild-type RoMNPV. The ST_{50}s of Ro6.9LqhIT2 infected neonate *H. zea* and *H. virescens* were also significantly lower than control neonates infected with a recombinant RoMNPV expressing AaIT under the *p6.9* promoter.

Chejanovsky *et al.* (1995) have generated a recombinant AcMNPV (AcLα22) that expresses the *L. quinquestriatus hebraeus* derived alpha toxin, LqhαIT (Eitan *et al.*, 1990). The LT_{50} of AcLα22 was roughly 35% faster (78 versus 120 h) than that of the wild-type AcMNPV in larvae of *H. armigera*. Since the LqhαIT toxin binds at a different site on the insect sodium channel from that of the excitatory toxins (Zlotkin *et al.*, 1978; Cestele and Catterall, 2000), Chejanovsky *et al.* (1995) suggested that a baculovirus expressing both alpha and excitatory toxins may yield a synergistic interaction between the toxins. However, they cautioned that LqhαIT toxin lacks absolute selectivity for insects (Eitan *et al.*, 1990); thus, a recombinant baculovirus expressing LqhαIT is not appropriate as a biological pesticide. The LqhαIT toxin, however, should be an effective tool to study the targeting of different types of toxins on the voltage gated sodium channel. The effectiveness of the use of multiple synergistic toxins is discussed below. Other examples of the expression of insect selective toxins from *L. quinquestriatus hebraeus* using alternative promoters, signal sequences for secretion, viral vectors, and/or insertion of the toxin gene at the *egt* gene locus are given later in this chapter.

10.3.2. Mite Toxins

The insect-predatory straw itch mite *Pyemotes tritici* encodes an insect paralytic neurotoxin TxP-I that induces rapid, muscle contracting paralysis in larvae of the greater wax moth *G. mellonella* (Tomalski *et al.*, 1988, 1989). TxP-I is encoded by the *tox34* gene as a precursor protein of 291 amino acid residues. The mature protein is secreted from the insect cell following cleavage of a 39 amino acid long signal sequence for secretion (Tomalski and Miller, 1991). The mode of action of TxP-I is unknown, however, it is highly toxic (even at a dose of $500\,\mu g\,kg^{-1}$) to lepidopteran larvae but not toxic to mice at a dose of $50\,mg\,kg^{-1}$. A recombinant, occlusion-negative AcMNPV (vEV-Tox34) expressing the *tox34* gene under a modified polyhedrin promoter (P_{LSXIV}; Ooi *et al.*, 1989) was shown to paralyze or kill fifth instar larvae of *T. ni* by 2 days post injection with 400 000 pfu of BV. In contrast, control larvae injected with BV of wild-type AcMNPV never showed symptoms of paralysis (Tomalski and Miller, 1991). Tomalski and Miller

(1991, 1992) have constructed two other occlusion-negative AcMNPVs that express the *tox34* gene under early (vETL-Tox34) or hybrid late/very late (vCappolh-Tox34) gene promoters as well as an occlusion-positive AcMNPV (vSp-Tox34) expressing *tox34* under a different hybrid late/very late promoter. The activities of these constructs will be discussed later in this chapter. Lu *et al.* (1996) have constructed an occlusion-positive AcMNPV (vp6.9tox34) that utilizes the late *p6.9* promoter to drive expression of *tox34*. Use of this late promoter resulted in earlier, by at least 24 h, and higher level of TxP-I expression in comparison to TxP-I expression under a very late promoter. As discussed above, the *p6.9* gene promoter is not an early promoter, but is activated earlier and can drive higher levels of expression than very late promoters in tissue culture (Hill-Perkins and Possee, 1990; Bonning *et al.*, 1994). In time-mortality bioassays (at an LC_{95} dose), the median time to effect (ET_{50}) of vp6.9tox34 in neonate larvae of *S. frugiperda* and *T. ni* was reduced by approximately 56% (44.7 versus 101.3 h) and 58% (41.7 versus 99.0 h), respectively, in comparison to wild-type AcMNPV. In neonate *S. frugiperda* and *T. ni*, the earlier and higher level of TxP-I expression under the late promoter resulted in a 20–30% faster induction of effective paralysis or death in comparison to TxP-I expression under the very late gene promoter.

Burden *et al.* (2000) have constructed a slightly different TxP-I encoding gene (*tox34.4*) by RT-PCR of mRNAs purified from total RNA extracted from *P. tritici* using primers designed to amplify the *tox34* open reading of Tomalski and Miller (1991). A recombinant AcMNPV (AcTOX34.4) expressing TxP-I under the *p10* promoter has been generated. The LD_{50}s of AcTOX34.4 were not significantly different than those of the wild-type AcMNPV in dose-mortality bioassays of second (9.3 polyhedra per larva) and fourth (13.1 polyhedra per larva) instar larvae of *T. ni*. In time-mortality bioassays, second and fourth instar larvae of *T. ni* infected with AcTOX34.4 showed a 50–60% reduction (depending on virus dose and instar) in the mean time to death in comparison to control larvae infected with wild-type AcMNPV. There was also a dramatic reduction in the yield of progeny virus (number of polyhedra per μg of cadaver) at the time of death. Second and fourth instar larvae of *T. ni* that were infected with AcTOX34.4 produced roughly 85% and 95% lower yields of polyhedra per unit weight, respectively, in comparison to control larvae infected with AcMNPV. On the basis of pathogen–host model systems that describe how insect viruses may regulate host population density (as will be discussed later in this chapter), Burden *et al.* (2000) suggested that the dramatic reduction in yield of AcTOX34.4 will cause it to be outcompeted by the wild-type AcMNPV.

10.3.3. Other Toxins

In addition to peptide toxins of scorpion and mite origin, insect-selective and highly potent toxins have been identified from other organisms including spiders, anemones, and bacteria. The spider *Agelenopsis aperta* and sea anemones *Anemonia sulcata* and *Stichadactyla helianthus* possess the insect selective toxins μ-Aga-IV, As II, and Sh I, respectively (Prikhod'ko *et al.*, 1996). Recombinant AcMNPVs carrying synthetic genes that express μ-Aga-IV (vMAg4p+), As II (vSAt2p+), or Sh I (vSSh1p+) under the very late P_{SynXIV} promoter have been constructed (Prikhod'ko *et al.*, 1996). The As II and Sh I toxins were each expressed as fusion peptides with a signal sequence derived from the flesh fly sarcotoxin IA, whereas μ-Aga-IV was expressed as a fusion peptide with the mellitin signal sequence of the honeybee, *Apis mellifera*. In neonate larvae of *T. ni*, the LC_{50} of vSSh1p+ was slightly lower than that of AcMNPV, but those of vMAg4p+ and vSAt2p+ were essentially identical. In contrast, in neonate larvae of *S. frugiperda* the LC_{50}s of all three viruses were increased by 3.75- to 5.25-fold in comparison to AcMNPV. In time-mortality bioassays, neonate larvae of *T. ni* and *S. frugiperda* infected (at an LC_{95} dose) with any of the recombinant viruses died more quickly than control larvae infected with AcMNPV. In neonate *T. ni* infected with vMAg4p+, vSAt2p+, and vSSh1p+, the ET_{50}s (median times to effectively paralyze or kill) were reduced by 17% (84.5 versus 102.0 h), 38% (62.8 versus 102.0 h), and 36% (65.3 versus 102.0 h), respectively. In neonate *S. frugiperda*, the ET_{50}s were reduced by 37% (65.7 versus 104.3 h), 31% (71.8 versus 104.3 h), and 35% (67.5 versus 104.3 h), respectively. Related constructs in which these toxins were expressed under a constitutive promoter or in tandem will be discussed below.

Hughes *et al.* (1997) have generated recombinant AcMNPVs expressing insect specific toxins DTX9.2 and TalTX-1 from the spiders *Diguetia canities* and *Tegenaria agrestis*, respectively. DTX9.2 was expressed by vAcDTX9.2 as a fusion protein with a signal sequence from the AJSP-1 gene of *T. ni* under the *polh* promoter. TalTX-1 (authentic signal sequence and mature protein) was expressed by vAcTalTX-1 under the *polh* promoter. In time-mortality bioassays, neonate larvae of *T. ni*, *S. exigua*, and *H. virescens* infected (at a less than

LC_{50} dose) with vAcDTX9.2 generated ST_{50}s that were reduced by approximately 9% (62.6 versus 68.8 h), 9% (71.3 versus 78.5 h), and 10% (74.1 versus 81.9 h), respectively, in comparison to control larvae infected with a control virus (Ac-Bb1). Neonate *T. ni*, *S. exigua*, and *H. virescens* infected (at a less than LC_{50} dose) with vAcTalTX-1 generated ST_{50}s that were reduced by approximately 20% (55.5 versus 68.8 h), 18% (64.4 versus 78.5 h), and 19% (66.1 versus 81.9 h), respectively, in comparison to control larvae infected with Ac-Bb1. In all cases, larvae infected with the toxin expressing viruses stopped feeding prior to larvae infected with Ac-Bb1 or wild-type AcMNPV. Neonate *T. ni*, *S. exigua*, and *H. virescens* infected with vAcDTX9.2 showed FT_{50}s (median times to cessation of feeding) that were approximately 33, 28, and 42%, respectively, faster than control larvae infected with Ac-Bb1. Similarly, neonate *T. ni*, *S. exigua*, and *H. virescens* infected with vAcTalTX-1 showed FT_{50}s that were approximately 26, 17, and 37%, respectively, faster than control larvae. Interestingly, although the speed of kill of vAcTalTx-1 was faster, vAcDTX9.2 was able to stop feeding more quickly. Thus, Hughes *et al.* (1997) emphasized that enhanced speed of kill is not necessarily a reliable indicator of the enhanced speed with which feeding is stopped.

The genus *Bacillus* is composed of Gram-positive, endospore forming bacteria. During endospore formation, *Bacillus thuringiensis* (*Bt*) produces parasporal, proteinaceous, crystalline inclusion bodies that possess insecticidal properties. The biology and genetics of *Bt* and *Bt* toxins have been previously reviewed (Gill *et al.*, 1992; Schnepf *et al.*, 1998; Aronson and Shai, 2001). *Bt* encodes two major classes of lepidopteran-active toxins, cytolytic toxins and δ-endotoxins that are encoded by the *cyt* and *cry* genes, respectively. The δ-endotoxins of *Bt* are composed of large (>120 kDa) protoxin proteins that are solubilized in the alkaline conditions of the insect midgut and processed into the active toxin or toxins by proteases. The active toxins bind to specific receptors found on the insect's midgut epithelial cells and aggregate to form ion channels. Channel formation leads to disruption of the cell's transmembrane potential, which leads to osmotic cell lysis. Lethality is believed to be caused by (1) disruption of normal midgut function, which results in feeding cessation and starvation of young larvae, or (2) *Bt* septicemia.

Several studies have looked at the effectiveness of the expression of *Bt* toxin in terms of improving the insecticidal activity of the baculovirus. These studies have investigated the expression of full-length or truncated forms of *Bt* toxin genes (e.g., *cryIAb*, *cryIAc*, etc.) that are placed under a very late promoter and expressed in AcMNPV (Martens *et al.*, 1990, 1995; Merryweather *et al.*, 1990; Ribeiro and Crook, 1993, 1998) or *Hyphantria cunea* NPV (Woo *et al.*, 1998). These studies showed that the *Bt* toxin is highly expressed by the baculovirus and subsequently processed into the biologically active form. However, *Bt* toxin expression did not improve the virulence of the virus (i.e., the LD_{50} of the recombinant virus is similar to or higher than the LD_{50} of the wild-type virus) or decrease the ST_{50} of larvae infected with the recombinant virus. These findings are not completely unexpected considering that the site of action of *Bt* toxins is the midgut epithelial cell, whereas the baculovirus expressed *Bt* protoxin is produced within the cytoplasm of cells within the insect body. Furthermore, the expressed protoxin (1) may not be digested to the active form because of the lack of appropriate proteases within the cytoplasm, (2) may be poorly secreted, or (3) the mature toxin may be cytotoxic to the cell (Martens *et al.*, 1995). In order to improve toxin secretion, several *Bt* toxin gene constructs have been generated in which a signal sequence of JHE of *H. virescens* was fused to the N-terminus of the toxin (Martens *et al.*, 1995). These toxins were translocated across the ER membrane, but they appeared to be retained within one of the ER or Golgi compartments. These findings suggest that the expression of the *Bt* toxin gene will have little or no effect in improving insecticidal activity once the virus crosses the midgut.

Chang *et al.* (2003) have taken a novel and apparently highly successful approach for the delivery of *Bt* toxin directly to the insect midgut epithelial cells using a recombinant baculovirus that occludes the toxin within its polyhedra. Their strategy was based upon the coexpression of (1) native polyhedrin and (2) a polyhedrin-foreign protein fusion from the same baculovirus (Je *et al.*, 2003). Chang *et al.* (2003) used this system to coexpress native polyhedrin and a polyhedrin-Cry1Ac-green fluorescent protein (GFP) fusion from the recombinant baculovirus ColorBtrus. ColorBtrus is a recombinant AcMNPV that produces apparently normal polyhedra that occlude *Bt* toxin and GFP. Although the Cry1Ac toxin was fused at both the N- and C-termini, this fusion protein could be digested by trypsin resulting in an immunoreactive protein that behaved identically to authentic Cry1Ac toxin. Chang *et al.* estimated that approximately 10 ng of Cry1Ac was contained in 1.5×10^6 ColorBtrus polyhedra. Furthermore, because of GFP incorporation,

the ColorBtrus polyhedra fluoresced under UV light. This characteristic should allow the rapid monitoring and detection of ColorBtrus infected insects in the field (Chao *et al.*, 1996; Chang *et al.*, 2003). In dose-mortality bioassays using second or third instar larvae of the diamondback moth, *Plutella xylostella*, Chang *et al.* (2003) showed that the LD_{50} of ColorBtrus was reduced by roughly 100-fold in comparison to wild-type AcMNPV (28.3 versus 2798.3 polyhedra per larva, respectively). By time-mortality bioassays, the ST_{50} of larvae of *P. xylostella* infected with ColorBtrus was reduced by 63% (33.9 versus 92.8 h) in comparison to control larvae infected with wild-type AcMNPV when the larvae were inoculated at LD_{90} doses (i.e., 80 ColorBtrus or 10 240 AcMNPV polyhedra per larva). Chang *et al.* (2003) indicated that the ColorBtrus inoculated larvae displayed symptoms, such as feeding cessation, that were consistent with exposure to Cry1Ac. This suggested that feeding damage to the plant should cease very quickly after ingestion of ColorBtrus. Chang *et al.* (2003) further suggested that the "stacking" of multiple effectors (i.e., Cry1Ac toxicity and viral pathogenesis) by ColorBtrus should reduce the occurrence of resistance because the few individuals within the population that evolve resistance to the *Bt* toxin would also have to simultaneously evolve resistance to the baculovirus. Additionally, although the Cry1Ac toxin is highly toxic to a wide range of lepidopteran insects, it is not effective against *Spodoptera* species such as the beet armyworm *S. exigua* (Bai *et al.*, 1993). Since the beet armyworm is susceptible to AcMNPV, they suggested that this insect can be used for the efficient *in vivo* production of ColorBtrus or similar virus.

Besides the toxins mentioned here, new and interesting toxins are frequently being discovered. These novel insect specific toxins can potentially be expressed by baculoviruses. There is a great diversity of available toxins that can be expressed by GM baculoviruses either alone or in combination. We may have not yet reached the limits of improvement of the speed of kill of GM baculoviruses.

10.3.4. Improvement of Toxin Efficacy by Genetic Modifications

Thus far, the discussion on improving baculovirus pesticides has primarily focused on the expression of an insect selective toxin gene under late or very late baculoviral promoters. Here, the focus is on studies that have attempted to improve the efficacy of the toxin expressing constructs by (1) altering the timing and level of toxin expression, (2) improving folding and secretion efficiency of the toxin, and/or

(3) expressing multiple toxins that target different sites within the insect sodium channel. Related to this last point, studies on synergistic interactions between the expressed toxin and chemical pesticides are also discussed.

10.3.4.1. Alteration of the timing and level of expression

In general, the first generation of recombinant baculovirus constructs used a single baculovirus derived very late (e.g., *polh* or *p10*) or late (e.g., *p6.9*) promoter to drive expression of the toxin gene (or other foreign gene). These promoters generally drive exceptionally high levels of protein expression (i.e., they are very strong promoters), but they require the products of viral early genes for activation. Thus, the very late/late promoters are not activated until a late stage of the baculovirus replication (Lu and Milller, 1997). At this late stage of infection, the translational machinery and processing pathways of the cell may be compromised (Jarvis *et al.*, 1990). Thus, a large amount of protein may be produced but not all of the protein will be functionally active due to poor folding or poor posttranslational modifications and the protein may also be retained within the cell due to poor secretion. Baculovirus early gene promoters, alternatively, are recognized by host transcription factors and are activated very soon after the viral DNA is uncoated in the nucleus. At this early stage of infection, the processing and secretory pathways of the cell should be in a less compromised state (in comparison to the late stage of infection). Thus, it is likely that a higher percentage of the expressed proteins will be properly folded, modified, and secreted. Earlier expression may be especially important for the proper folding of neurotoxins because they possess a relatively large number of disulfide bonds (e.g., the 70 amino acid-long AaIT peptide contains four disulfide bonds). Early gene promoters, however, generally drive very low levels of protein expression (i.e., they are weak promoters).

Several groups have attempted to circumvent the potential problems associated with protein expression under very late promoters using: (1) early gene promoters; (2) dual or chimeric promoters; and/or (3) constitutively expressed promoters to drive the expression of a foreign gene. Jarvis *et al.* (1996) have constructed a recombinant AcMNPV (*ie1*-AaIT) expressing the AaIT gene under the control of an immediate early gene (*ie1*) promoter of AcMNPV. In cultured Sf-9 cells infected with the *ie1*-AaIT construct, AaIT was detected as early as 4 h p.i. and continued to accumulate until at least 24 h p.i. At 24 h p.i., the amount of the expressed AaIT peaked and was about equal to that expressed by AcAaIT

(AaIT expressed under the very late *p10* promoter). *In vitro* expression assays indicated that there was early and abundant expression of AaIT under the *ie1* promoter. These findings suggested that the *ie1*-AaIT construct should kill infected larvae more quickly than the AcAaIT construct. The results of the insect bioassays using neonate *H. virescens*, however, did not support this. Time-mortality bioassays showed that the *ie1*-AaIT virus killed neonate *H. virescens* only about 10% faster than the wild-type AcMNPV (LT$_{50}$ of 87.0 versus 96.5 h). In comparison, AcAaIT killed the neonate *H. virescens* about 22% faster than AcMNPV (LT$_{50}$ of 75.0 versus 96.5 h) in parallel experiments. Although Jarvis *et al.* (1996) were unclear as to the reason(s) for the slower speed of kill of *ie1*-AaIT in comparison to AcAaIT, they speculated that: (1) *ie1*-AaIT does not produce a threshold dose of AaIT that must be reached before kill rates can be enhanced; (2) duplication of the *ie1* promoter usurps limited cellular transcription factors needed by the endogenous *ie1* promoter for expression of the IE1 protein; (3) early expression of AaIT in midgut epithelial cells induces these cells to be shed thereby reducing the primary infection in these cells and/or dissemination of the virus into the hemocoel; and (4) the *ie1* promoter of AcMNPV has a narrower host range and/or tissue specificity in comparison to the virus as a whole. Interestingly, although *ie1*-AaIT infected larvae did not die as quickly as AcAaIT infected larvae, the relative growth rate of the *ie1*-AaIT infected larvae was about 13% lower (0.8 versus 0.92 mg per larva per day) than that of AcAaIT infected larvae. This suggested that even though they do not die as quickly, the *ie1*-AaIT infected larvae may consume less food.

van Beek *et al.* (2003) have compared the efficacy of recombinant AcMNPVs expressing either AaIT or LqhIT2 under various baculoviral early (*ie1* or *lef3*), early/late (*39K*), or very late (*p10*) gene promoters in larvae of *H. virescens*, *T. ni*, and *S. exigua*. The larvae were inoculated at both low (approximately LD$_{50}$) and high (approximately 100 times the LD$_{50}$) doses of each recombinant virus. By time-to-response bioassays using neonate and third instar *H. virescens* or *T. ni*, and second instar *S. exigua*, they found that the ET$_{50}$ of the recombinant AcMNPV expressing AaIT under the early *hr5/ie1* promoter led in the majority of cases to a faster response, but a generalized pattern of response in terms of the effect of viral dose and instar was not found. The "*hr5*" designation indicates that the baculovirus *hr5* enhancer sequence was placed immediately upstream of the promoter sequence. By bioassays using third instar larvae of *H. virescens*, van Beek *et al.* found that there were no significant differences in the ET$_{50}$ of the recombinant AcMNPVs expressing LqhIT2 under an early (*hr5/ie1* or *hr5/lef3*) or early/late (*hr5/39K*) gene promoter. Furthermore, regardless of the larval instar or dose applied, when the early *hr5/ie1* promoter was used to drive the expression of LqhIT2 or AaIT, the ET$_{50}$s of the recombinant AcMNPV expressing LqhIT2 were consistently lower (by 8–32 h) in comparison to those of the recombinant AcMNPV expressing AaIT. This was putatively due to differences in the effectiveness of the two toxins.

As described above, Gershburg *et al.* (1998) have constructed recombinant AcMNPVs expressing the excitatory toxin LqhIT1 or depressant toxin LqhIT2 under either the early *p35* gene or very late *p10* gene promoters. They found that the recombinant AcMNPV expressing LqhIT1 under the *p35* promoter had a slightly slower ET$_{50}$ (73 versus 66 h) in comparison to the recombinant AcMNPV expressing LqhIT1 under the *p10* promoter. However, they detected a clear paralytic effect when the *p35* promoter was used to express LqhIT1 even when they were unable to detect the LqhIT1 by immunochemical analysis. Thus, they suggested that baculovirus expression of toxin in cells that are adjacent to the target sites of the motor neural tissues overcomes the pharmacokinetic barriers that the toxin may face when, for example, purified toxin is injected to the insect. Consistent with this hypothesis, Elazar *et al.* (2001) have shown that the concentration of AaIT in the hemolymph of paralyzed *B. mori* larvae is approximately 50-fold lower when the paralyzing dose is delivered by a baculovirus (BmAaIT) rather than by the direct injection of purified AaIT. They hypothesized that baculovirus expression of AaIT provides a continuous, local supply of freshly produced toxin (via the tracheal system) to the neuronal receptors, thus providing access to critical target sites that are inaccessible to toxin that is injected. Therefore, a lower level of continuous expression under an early promoter may be sufficient to elicit a paralytic response.

As discussed above, Tomalski and Miller (1992) have constructed an occlusion-negative recombinant AcMNPV (vETL-Tox34) that expresses the *tox34* gene under the baculoviral early P$_{ETL}$ promoter. In early fifth instar larvae of *T. ni*, this virus induced much slower paralysis (occurring after 48 h in 95% of larvae) in comparison to control larvae infected with recombinant AcMNPVs (vEV-Tox34, vCappolh-Tox34, or vSp-Tox34) expressing *tox34* under a very late or chimeric promoter (all larvae were paralyzed or dead by 48 h). Additionally, the average cumulative weight gain of vETL-Tox34

infected larvae at 24 h post injection was not significantly different to that of control larvae injected with AcMNPV or cell culture medium (mock infection). Lu *et al.* (1996) have also constructed a recombinant AcMNPV (vDA26tox34) expressing the *tox34* gene under the early *Da26* gene promoter. In comparison to vp6.9tox34 (*tox34* under a late promoter) or vSp-tox34 (*tox34* under a very late promoter) infected Sf-21 or TN-368 cells (derived from *S. frugiperda* and *T. ni*, respectively), TxP-I expression was detected at least 24 h earlier (but at dramatically lower levels) in Sf-21 and TN-368 cells infected with vDA26tox34. In time-mortality bioassays (at an LC$_{95}$ dose), the ET$_{50}$ of vDA26tox34 in neonate larvae of *S. frugiperda* and *T. ni* was reduced by approximately 39% (61.8 versus 101.3 h) and 28% (71.2 versus 99.0 h), respectively, in comparison to wild-type AcMNPV. However, this was 17.1 and 29.5 h slower, respectively, in comparison to neonate larvae of *S. frugiperda* and *T. ni* that were infected with vp6.9tox34 (ET$_{50}$s of 44.7 and 41.7 h, respectively). Even though the TxP-I is expressed significantly earlier under the early promoter, it is apparently not expressed in sufficient quantity, at least initially, to paralyze larvae of *S. frugiperda* or *T. ni*. Popham *et al.* (1997) have also expressed the *tox34* gene under the early *DA26* and late *p6.9* promoters (of AcMNPV) by recombinant HzSNPVs (HzEGTDA26tox34 and HzEGTp6.9tox34, respectively). They found that the ET$_{50}$s of HzEGTDA26tox34 and HzEGTp6. 9tox34 in neonate larvae of *H. zea* were reduced by approximately 47% (35.4 versus 67.3 h) and 44% (38.0 versus 67.3 h), respectively, in comparison to a control virus. In contrast to the results of Lu *et al.* (1996), the early promoter was more effective in the case of HzSNPV and neonate *H. zea* suggesting that the effectiveness of a promoter will be virus and insect specific. With toxins of current potency, the use of early (and weak) promoters such as *ie1*, *lef3*, *p35*, *etl*, or *DA26* may or may not provide any benefit in terms of improved crop protection. However, if toxins that act at lower concentration (i.e., with much greater potency) are identified, the use of early promoters may offer a great advantage.

Tomalski and Miller (1992) have used a chimeric promoter (P$_{cap/polh}$) comprised of both the late capsid and very late polyhedrin promoter elements to drive the expression of the *tox34* gene in an occlusion-negative recombinant AcMNPV (vCappolh-Tox34). In Sf-21 cell cultures infected with vCappolh-Tox34, TxP-I is detected in the culture medium at 12 h p.i. at levels that are similar to those found in vEV-Tox34 (*tox34* under a very late promoter) infected Sf-21 cells at 24 h p.i. (i.e., high-level expression occurred approximately 12 h earlier). When early fifth instar larvae of *T. ni* were injected with vCappolh-Tox34 approximately 50% of the insects were paralyzed by 24 h post injection. In comparison, only 10% or less of control virus (e.g., vEV-Tox34) infected larvae were paralyzed. Larvae infected with vCappolh-Tox34 also showed a significantly lower cumulative weight gain in comparison to control virus infected larvae. An occlusion-positive construct (vSp-Tox34) in which the *tox34* gene was placed under another hybrid late/very late promoter (P_{SynXIV}; Wang *et al.*, 1991) has also been constructed by Tomalski and Miller (1991, 1992). The induction of paralysis by vSp-Tox34 in fifth instar *T. ni* was slightly delayed, but all of the larvae were paralyzed or dead by 48 h post injection (as were vCappolh-Tox34 injected larvae; Tomalski and Miller, 1991, 1992). In time-mortality bioassays (at an LC_{95} dose), the ET_{50} of vSp-Tox34 in neonate larvae of *S. frugiperda* and *T. ni* was reduced by approximately 45% (55.4 versus 101.3 h) and 41% (58.5 versus 99.0 h), respectively, in comparison to control larvae infected with AcMNPV (Tomalski and Miller, 1992; Lu *et al.*, 1996). Furthermore, the yield (2.1×10^9 versus 3.5×10^9 polyhedra per larva) of vSP-Tox34 polyhedra from infected last instar larvae of *S. frugiperda* was reduced by approximately 40% in comparison to AcMNPV infected control larvae. Tomalski and Miller (1992) suggested that this reduction in yield will cripple the virus in terms of its ability to compete effectively with the wild-type virus in the environment.

Sun *et al.* (2004) have generated a recombinant HaSNPV (HaSNPV-AaIT) that expresses the AaIT gene under a chimeric promoter (ph-p69p) consisting of the late *p6.9* gene promoter inserted immediately downstream of the very late *polh* gene promoter in the same direction. The AaIT gene cassette was inserted at the *egt* gene locus of HaSNPV. Detailed dose-mortality bioassays indicated that the LD_{50}s of HaSNPV-AaIT was unchanged compared to wild-type (HaSNPV-WT) or *egt* gene-deleted (HaSNPV-EGTD) viruses in first to fifth instar larvae of the cotton bollworm, *H. armigera*. In time-mortality bioassays, first to fifth instar larvae of *H. armigera* infected with HaSNPV-AaIT showed reductions in the ST_{50}s of 10% (68.5 versus 76.5 h), of 20% (60.5 versus 75.5 h), 8% (71.0 versus 77.5 h), 4% (104.6 versus 109.0 h), and 10% (108.5 versus 121.0 h), respectively, in comparison to control larvae infected with HaSNPV-EGTD. In comparison to control (first to fifth instar) larvae infected with HaSNPV-WT, the corresponding

reductions were 19, 28, 25, 21, and 18%, respectively. In third to fifth instar of *H. armigera* infected with HaSNPV-AaIT, the median times to feeding cessation (FT_{50}s) were reduced by 25% (51.5 versus 68.5 h), 22% (66.5 versus 85.0 h), and 12% (84.5 versus 96.5 h), respectively, in comparison to control larvae infected with HaSNPV-EGTD. In comparison to control larvae infected with HaSNPV-WT, the corresponding reductions in the FT_{50}s were 39%, 43%, and 30%, respectively. The role of the *egt* deletion in improving speed of kill is discussed further below. Sun *et al.* (2004) did not test constructs in which only the *polh* or *p6.9* gene promoter was used to drive expression of the AaIT gene, thus it was not possible to predict if expression under the dual promoter is more effective than expression under a single late or very late promoter.

The *hsp70* promoter of *D. melanogaster* (Snyder *et al.*, 1982) is constitutively active in insect cells (Vlak *et al.*, 1990; Zuidema *et al.*, 1990). Unlike very late or late gene promoters of the baculovirus, the *hsp70* promoter does not require the expression of baculovirus early gene products or the initiation of viral DNA synthesis for its activity. The *hsp70* promoter shows higher activity than early gene promoters, but it is not as active as baculoviral very late or late gene promoters (Morris and Miller, 1992, 1993). McNitt *et al.* (1995) have constructed occlusion-positive, recombinant AcMNPVs expressing AaIT (vHSP70AaIT) and TxP-I (vHSP70tox34) under the *hsp70* promoter. TxP-I was expressed at a significantly higher level under the *hsp70* promoter than under the early *DA26* promoter in both vHSP70tox34 infected Sf-21 and TN-368 cell cultures (McNitt *et al.*, 1995; Lu *et al.*, 1996). Additionally, TxP-I expression was detected as early as 6 and 12 h p.i., respectively, in the culture medium of vHSP70tox34 infected Sf-21 and TN-368 cells. Dose-mortality bioassays using neonate larvae of *S. frugiperda* or *T. ni* indicated that the LC_{50}s of vHSP70tox34 were unchanged from that of the AcMNPV. Time-mortality bioassays (at an LC_{95} dose) in neonate *S. frugiperda* and *T. ni* showed that the ET_{50}s of vHSP70tox34 were reduced by approximately 59% (41.8 versus 101.3 h) and 46% (53.8 versus 99.0 h), respectively, in comparison to control larvae infected with AcMNPV. In comparison to occlusion-positive, recombinant AcMNPVs expressing the *tox34* gene under very late (vSp-tox34) or late (vp6.9tox34) promoters, the ET_{50}s of vHSP70tox34 were faster than both vp6.9tox34 (44.7 h) and vSp-tox34 (55.4 h) in neonate *S. frugiperda*; and faster than vp6.9tox34 (41.7 h) and slower than vSp-tox34 (58.5 h) in neonate *T. ni*. This is surprising considering that the overall level of TxP-I secretion under the *hsp70* promoter is substantially lower than that under the *p6.9* promoter (Lu *et al.*, 1996). On the basis of these findings Lu *et al.* (1996) suggested that expression from a relatively strong constitutive promoter such as *hsp70* is more effective, in some cases, than expression from a very strong late baculoviral promoter. Popham *et al.* (1997) have expressed the *tox34* gene under the *hsp70* promoter (HzEGT hsptox34) at the *egt* gene locus of a recombinant HzSNPV. Time-mortality bioassays (at an LC_{95} dose) in neonate *H. zea* showed that the ET_{50} of HzEGThsptox34 was reduced by 34% (44.1 versus 67.3 h) in comparison to control larvae infected with an *egt* deletion mutant of HzSNPV (HzEGT-del). In contrast to the results of Lu *et al.* (1996), however, expression of the *tox34* gene under the *hsp70* promoter was not as effective as expression under the early *Da26* (ET50 of 35.4 h) or late *p6.9* (ET_{50} of 38.0 h) promoters (both of these promoters were of AcMNPV origin).

The *hsp70* promoter has also been used to drive the expression of spider and sea anemone toxins. Prikhod'ko *et al.* (1998) have constructed recombinant AcMNPVs vhsMAg4p+, vhsSAt2p+, and vhsSSh1p+ that express the insect selective toxins μ-Aga-IV, As II, and Sh I, respectively, under the *hsp70* promoter. All of these constructs showed significantly reduced ET_{50}s (at an LD_{95} dose) in comparison to AcMNPV in neonate larvae of *S. frugiperda* and *T. ni*. In neonate *S. frugiperda* infected with vhsMAg4p+, vhsSAt2p+, or vhsSSh1p+, the ET_{50}s were reduced by approximately 55% (49.0 versus 110 h), 52% (52.5 versus 110 h), and 53% (51.9 versus 110 h), respectively. In neonate *T. ni*, the ET_{50}s were reduced by 42% (53.6 versus 99.7 h), 50% (49.0 versus 99.7 h), and 46% (52.9 versus 99.7 h), respectively. In both larval species, expression of μ-Aga-IV (*mag4* gene), As II (*sat2* gene), and Sh I (*ssh1* gene) under the *hsp70* promoter was more effective than expression under the very late P_{SynXIV} promoter (ET_{50}s of 60.0, 63.6, and 56.8 h, respectively, in *S. frugiperda*, and 84.4, 62.8, and 60.6 h, respectively, in *T. ni*).

Numerous studies (including several that are not included in our discussion above) have investigated the effectiveness of constitutive and baculoviral promoters (early, late, and very late) used either alone or in combination for the expression of the current generation of toxin genes. These studies, however, do not form a consensus opinion as to which promoter system is the most optimal in terms of improving insecticidal effectiveness of the toxin expressing baculovirus. The toxin gene, parental virus, and host-specific factors all appear to play important roles in determining the effectiveness of a particular promoter system. Given that an early or constitutive promoter does not provide a clear advantage (at least within the context of toxin genes of current potency), the use of a baculoviral late or very late promoter may provide a slight reduction in the risk that a toxin gene could be expressed in a nontarget insect. The argument here is that very late and late gene promoters are inactive in beneficial insects and other organisms (Heinz *et al.*, 1995; McNitt *et al.*, 1995) because the activation of these promoters requires (1) viral DNA replication and (2) late expression factors (LEF; Lu and Miller, 1997). In AcMNPV there are 18 *lef* genes that are required for optimal transactivation of expression from late and very late gene promoters. At least seven of the 18 *lef* genes appear to be involved in DNA replication and at least three *lef* gene products are part of a virus induced or virus encoded RNA polymerase. This RNA polymerase is required for transcription from late and very late genes. Thus, without viral DNA replication and the expression of essential *lef* genes, transcription from a late or very late gene promoter will not occur. In contrast, early and constitutive gene promoters can potentially be activated in nontarget insect cells (Morris and Miller, 1992, 1993; McNitt *et al.*, 1995). A discussion of the potential risks associated with gene transfer to nontarget insects and other organisms (e.g., mammals) is given below.

10.3.4.2. Better secretion and folding
In order for a peptide or protein to be secreted to the outside of a eukaryotic cell, it must possess a signal sequence consisting of 15–30 hydrophobic amino acid residues at its N-terminal. Following the initiation of protein synthesis by ribosomes within the cytoplasm, the newly synthesized signal sequence is bound by a signal recognition particle (SRP). Subsequently, the SRP signal sequence–ribosome complex binds to a SRP receptor on the cytosolic surface of the rough ER membrane. The signal sequence peptide is then able to cross the ER membrane (and is generally cleaved off within the lumen of the ER) and the remainder of the protein is produced and secreted into the lumen of the ER. Once inside the ER lumen the peptide can undergo appropriate disulfide bond formation and folding. The mature protein then continues to the Golgi apparatus for subsequent release via secretory vesicles to the outside of the cell. In general, once inside the ER lumen, proteins are automatically transported through the Golgi apparatus and secreted (i.e., this is thought to be the default pathway) unless they possess a

specific signal that directs them to another location. Thus, the ability of the insect cell SRP to recognize the signal sequence on the baculovirus expressed protein is a key determinant for its eventual secretion to the outside of the cell. Signal sequences that originate from a variety of organisms (e.g., mammal, plant, yeast, insect, virus, etc.) are recognized by lepidopteran SRPs and can help direct baculovirus expressed proteins to the ER for subsequent processing and secretion from the cell (O'Reilly et al., 1992). Signal sequences derived from proteins of insect origin, e.g., bombyxin from the silk moth B. mori (Adachi et al., 1989), mellitin from the honeybee A. mellifera (Tessier et al., 1991), and cuticle protein II (CPII) from D. melanogaster (Snyder et al., 1982), have been shown to be functional when attached to a wide variety of proteins.

In order to improve processing, folding, and/or secretion of baculovirus expressed toxin, van Beek et al. (2003) have generated recombinant AcMNPVs expressing LqhIT2 (under the early hr5/ie1 promoter) fused to various secretion signals of insect and noninsect origin. Signal sequences originating from bombyxin, CPII, adipokinetic hormone from M. sexta L. (Jaffe et al., 1986), chymotrypsin of Lucilia cuprina (Weidemann) (Casu et al., 1994), AcMNPV gp67 (Whitford et al., 1989), and scorpion neurotoxins BjIT from Hottentota judaicus (Simon) (Zlotkin et al., 1993), AaIT (Bougis et al., 1989), and LqhIT2 (Zlotkin et al., 1993) were tested. In time-mortality bioassays in third instar larvae of H. virescens (at a dose of polyhedra that was estimated to elicit a response in less than 50% of the larvae), they found that a recombinant AcMNPV expressing LqhIT2 fused to the bombyxin signal sequence elicited the fastest response time (ET_{50} of 62 h, estimate based on the graph). Recombinant AcMNPVs expressing LqhIT2 fused to the signal sequences of adipokinetic hormone (ET_{50} of 78 h), chymotrypsin (ET_{50} of 78 h), LqhIT2 (ET_{50} of 85 h), and gp67 (ET_{50} of 72 h) elicited intermediate response times. All of these viruses putatively expressed and secreted active toxin as evidenced by reduced survival times and the occurrence of paralysis. In contrast, recombinant AcMNPV expressing LqhIT2 fused to the signal sequences of CPII, BjIT, and AaIT appeared not to secrete active toxin because insects infected with these constructs exhibited symptomology that is typical of AcMNPV infection. In general, they found that the baculoviral gp67 or insect derived signal sequences (excluding CPII) worked better than the scorpion toxin derived signal sequences for the secretion of biologically active LqhIT2.

Lu et al. (1996) have constructed three recombinant AcMNPVs (vSp-BSigtox34, vSp-DCtox34, and vSp-tox21A/tox34) that express TxP-I (under the very late P_{SynXIV} promoter) fused to signal sequences originating from sarcotoxin IA of the flesh fly (Matsumoto et al., 1986), CPII, and tox21A (a homolog of tox34) (Tomalski et al., 1993), respectively. The levels of TxP-I secreted into the culture medium of Sf-21 cells infected with vSp-DCtox34 or vSp-tox21/tox34 were similar to control cells infected with vSp-tox34 (TxP-I fused to its own signal sequence). TxP-I secretion was undetectable on the basis of Western blotting in Sf-21 cell cultures infected with vSp-BSigtox34. By comparing the levels of TxP-I that were found within the cell (i.e., cell lysate) and in the culture medium (i.e., secreted), it appeared that the CPII secretion signal (vSp-DCtox34) was better than the tox21A signal sequence (vSp-tox21/tox34) in terms of being able to direct a higher percentage of the expressed TxP-I to the outside of the cell. TxP-I was only detected at very low levels within the cell lysates of Sf-21 cells infected with vSp-BSigtox34, thus it is difficult to determine whether the lack of TxP-I in the cell culture medium was the result of poor secretion or poor translation. In dose-mortality bioassays using neonate larvae of T. ni, the LC_{50}s of vSp-BSigtox34 and vSp-DCtox34 were similar to that of AcMNPV, whereas the LC_{50} of vSp-tox21A/tox34 was roughly double (5.3×10^4 versus 2.2×10^4 polyhedra per ml) that of AcMNPV. In time-mortality bioassays (at an LC_{95} dose) in neonate T. ni, the ET_{50}s of vSp-BSigtox34, vSp-DCtox34, and vSp-tox21A/tox34 were reduced by 26% (70.0 versus 94.6 h), 47% (49.9 versus 94.6 h), and 36% (60.8 versus 94.6 h), respectively, in comparison to control larvae infected with AcMNPV. However, in comparison to neonate T. ni infected with vSp-tox34 (a control virus expressing TxP-I fused to its authentic secretion signal, ET_{50} of 51.1 h), neonate T. ni infected with viruses expressing TxP-I fused to an alternative signal sequence did not show any improvements in ET_{50}. Additionally, although TxP-I secretion was undetected in Sf-21 cells infected with vSP-BSigtox34, neonate T. ni infected with this virus displayed paralysis indicating that the threshold level of toxin required for paralysis is low or that some cells within the larvae may more efficiently produce toxin. On the basis of these studies, Lu et al. (1996) suggested that the level of toxin secreted into the supernatant is generally diagnostic of how the virus will perform in vivo.

Chaperones found in the cytosol and ER (such as BiP, murine immunoglobulin heavy chain binding protein; Bole *et al.*, 1986) play key roles in protein folding, transport, and quality control in a wide range of eukaryotic cells including insect cells (Ruddon and Bedows, 1997; Ailor and Betenbaugh, 1999; Trombetta and Parodi, 2003). The coexpression of BiP by a recombinant baculovirus has been shown to reduce aggregation and increase secretion of baculovirus expressed murine IgG in *T. ni* derived cells (Hsu *et al.*, 1994). Taniai *et al.* (2002) have investigated whether coexpression of BiP can improve the secretion of AcAaIT expressed AaIT (rAaIT) in Sf-21 cells. They found that coexpression of BiP increased the amount of soluble rAaIT, however, the amount of active rAaIT that was secreted into the culture medium was not improved. Protein disulfide isomerase (PDI) is another factor within the ER that is known to help protein folding and secretion in insect cells by catalyzing the oxidization, reduction, and isomerization of disulfide bonds *in vitro* (Hsu *et al.*, 1996; Ruddon and Bedows, 1997). Coexpression of a PDI expressing baculovirus with an AaIT expressing baculovirus was also ineffective in increasing the amount of active AaIT that was secreted into the culture medium (Taniai, Inceoglu, and Hammock, unpublished data).

10.3.4.3. Expression of multiple synergistic toxins

At least six distinct receptor sites are found on the voltage gated sodium channels of mammals and insects that are the molecular targets of a broad range of neurotoxins. Two additional receptor sites are found on the insect sodium channel that are the molecular targets of insect selective excitatory and depressant scorpion toxins (Cestele and Catterall, 2000). On the basis of binding experiments using radiolabeled toxins, the depressant toxins bind to two noninteracting binding sites (one showing high affinity and the other low affinity) on the insect sodium channel (Gordon *et al.*, 1992; Zlotkin *et al.*, 1995; Cestele and Catterall, 2000). The excitatory toxins bind only to the high-affinity receptor site (Gordon *et al.*, 1992). Herrmann *et al.* (1995) have shown that when excitatory and depressant toxins are simultaneously coinjected into larvae of the blow-fly *S. falculata* or *H. virescens*, the amount of toxin required to give the same paralytic response is reduced five- to tenfold in comparison to the amount required when only one of the toxins is injected. On the basis of this synergism, they suggested that speed of kill of recombinant baculovirus(es) could be further reduced by: (1) the coinfection of two or more recombinant baculoviruses that each express toxin genes with synergistic properties; or (2) genetic modification such that two or more synergistic toxin genes are simultaneously expressed.

Regev *et al.* (2003) have generated a recombinant AcMNPV (vAcLqIT1-IT2) that expresses both the excitatory LqhIT1 and the depressant LqhIT2 toxins under the very late *p10* and *polh* promoters, respectively. Time-response bioassays (at an LC_{95} dose) using neonate *H. virescens* showed that the ET_{50} of vAcLqIT1-IT2 was reduced by 41% (46.9 versus 79.1 h) in comparison to AcMNPV. In comparison, the ET_{50}s of recombinant AcMNPVs expressing only LqhIT1 (vAcLqIT1) or only LqhIT2 (vAcLqIT2) were 57.3 and 60.0 h, respectively. Thus, expression of both LqhIT1 and LqhIT2 by vAcLqIT1-IT2 resulted in a synergistic effect that reduced the ET_{50} by 10.4 h (18%) and 13.1 h (22%), respectively. A synergistic effect (15–19%) was also observed following coinfection of neonate *H. virescens* with a mixture of both vAcLqIT1 and vAcLqIT2 (ET_{50} of 48.8 h). Time-response bioassays (at an LC_{95} dose) using neonate *H. armigera* showed that the ET_{50} of vAcLqIT1-IT2 was reduced by 24% (69.1 versus 90.8 h) in comparison to AcMNPV. However, a synergistic effect was not elicited in comparison to neonate *H. armigera* infected with vAcLqIT1 or vAcLqIT2 alone. Since *H. armigera* is only a semi-permissive host of AcMNPV, Regev *et al.* (2003) hypothesized that inefficient oral infection may have diminished any synergistic effect. By intrahemocoelic injection (at an LD_{95} dose) of third instar larvae of *H. armigera*, they found that the ET_{50} of vAcLqIT1-IT2 (72 h) was reduced by 13 h (15%) and 70 h (49%) in comparison to larvae injected with only vAcLqIT1 or vAcLqIT2, respectively. In similar intrahemocoelic injection assays of third instar larvae of *S. littoralis* (nonpermissive by oral infection for AcMNPV), they found that the ET_{50} of vAcLqIT1-IT2 (54 h) was reduced by 43 h (44%) and 20 h (27%) in comparison to control larvae injected with only vAcLqIT1 or vAcLqIT2, respectively. Similar or decreased levels of synergism were found with a recombinant AcMNPV (vAcLqαIT-IT2) expressing both alpha and depressant scorpion toxins in orally and intrahemocoelically infected larvae of *H. virescens*, *H. armigera*, and *S. littoralis*.

Using thoracic neurons of the grasshopper *Schistocerca americana* Drury, Prikhod'ko *et al.* (1998) have shown that the spider and sea anemone toxins, μ-Aga-IV and As II, respectively, act at distinct sites on the insect sodium channel and synergistically promote channel opening. These toxins also showed synergism when injected into third instar blowfly *Lucilia sericata* and fourth instar fall armyworm *S. frugiperda*. Prikhod'ko *et al.* (1998) have generated recombinant AcMNPVs (vMAg4Sat2 and vhsMAg4SAt2) that coexpress both μ-Aga-IV and

As II toxins under P_{SynXIV} or *hsp70* promoters, respectively. Time-response bioassay (at an LC_{95} dose) using neonate *T. ni* or *S. frugiperda* showed that the ET_{50} of vMAg4Sat2 was reduced by 39% (60.6 versus 99.7 h) and 45% (56.8 versus 104.0 h), respectively, in comparison to AcMNPV. A synergistic effect, however, was not found in comparison to larvae infected with recombinant AcMNPVs expressing only µ-Aga-IV (vMAg4p+) or As II (vDASAt2). In similar bioassays, the ET_{50} of vhsMAg4SAt2 was reduced by 47% (52.3 versus 99.2 h) and 57% (48.3 versus 111.1 h), respectively, in comparison to AcMNPV. A synergistic effect, however, was again not found in comparison to larvae infected with vhsMAg4p+ or vDAhsSAt2. They hypothesized that synergism was not detected in the case of coexpression of these toxins under the very late P_{SynXIV} promoter (by vMAg4Sat2) because it takes time to activate this promoter during infection and, once activated, high enough levels of expression of either one of the toxins is sufficient to effectively debilitate the host insect. In the case of coexpression under the *hsp70* promoter by vhsMAg4SAt2, they hypothesized that lower ET_{50}s (i.e., a synergistic effect) may not be possible since it takes a minimum period of time (i.e., approximately 50 h) for the virus to initiate both primary and secondary infections within the host. Although synergism was not observed in terms of the ET_{50} (mean time at which 50% of the test larvae cease to respond to stimulus), synergism was observed in terms of the FT_{50} (mean time at which 50% of the test larvae cease feeding) of neonate *T. ni* infected with vhsMAg4SAt2 (at an LC_{95} dose by droplet feeding). The FT_{50} of vhsMAg4SAt2 infected neonates was lower than that of vMAg4p+, vDASAt2, and AcMNPV by approximately 25% (24.3 versus 32.3 h), 29% (24.3 versus 34.3 h), and 63% (24.3 versus 65.8 h). Prikhod'ko *et al.* (1998) suggested that 24 h may be the fastest time at which a recombinant baculovirus could stop feeding because of the time (approximately 12 h) that it takes to complete the primary infection cycle within the midgut. Furthermore, they suggested that an FT_{50} of 24 h is sufficiently rapid to make recombinant baculovirus pesticides competitive with chemical pesticides.

10.3.5. Toxin Interactions with Chemical Pesticides

Pyrethroids, carbamates, and organophosphates are common synthetic chemical pesticides that are used for protection against pest insects worldwide. Pyrethroids currently account for about 20% of the total market for insecticides (Khambay, 2002). Pyrethroids are the most commonly used chemical insecticides for the control of the tobacco budworm *H. virescens* and cotton bollworm *H. zea* in North America. Pyrethroids are divided into two groups, type I and type II, on the basis of the absence or presence, respectively, of a cyano (-CN) group in the alpha position of the carboxyl moiety. Both type I and type II pyrethroids target receptors in the sodium channel causing them to remain open and resulting in prolonged sodium influx that causes hyper-excitation of the nervous system of both insects and mammals. Insects, however, are more sensitive to pyrethroid action than mammals due to several factors including the relative concentration of pyrethroid metabolizing enzymes, binding kinetics of the pyrethroid to the receptor, and differential sensitivity of the insect and mammalian sodium channels to pyrethroid action (Soderlund *et al.*, 2002). Although the precise location of the pyrethroid receptor within the insect sodium channel is unknown, the binding site is apparently different from that of peptide toxins produced by spiders, sea anemones, and scorpions (Ghiasuddin and Soderlund, 1985; Herrmann *et al.*, 1995). The differential binding of pyrethroid and toxin molecules indicate that when they are both present within the sodium channel they can act independently, synergistically, or antagonistically.

McCutchen *et al.* (1997) have characterized the effect of low concentrations of six chemical pesticides (allethrin, cypermethrin, DDT, endosulfan, methomyl, and profenofos) on the efficacy of wild-type and recombinant baculoviruses. They found synergistic interactions when neonate *H. virescens* were exposed to AcAaIT and low concentrations (LC_{10}–LC_{20} at 24 h) of the type II pyrethroid (cypermethrin) and a carbamate (methomyl). The interactions of AcAaIT with allethrin, DDT, endosulfan, or profenofos were not synergistic but additive. Specifically, neonate *H. virescens* that were inoculated with 2000 polyhedra per larva (a greater than LC_{99} dose) of AcAaIT and immediately exposed to an LC_{10}–LC_{20} dose of cypermethrin or methomyl showed an LT_{50} of 30 h, a reduction of roughly 58% compared to control larvae infected with the virus alone. Neither synergistic nor antagonistic effects were observed in control experiments in which AcJHE.KK (a recombinant AcMNPV expressing a modified juvenile hormone esterase) was used in place of AcAaIT. Since the mode and site of action of JHE is putatively different from that of AaIT, this was not surprising. In contrast, the interactions of wild-type AcMNPV and the chemical pesticides (except for methomyl) were slightly antagonistic. McCutchen *et al.* (1997) have also examined the efficacy of AcAaIT in combination with a low dose of pyrethroid in a pyrethroid resistant strain (PEG strain) of *H. virescens*. Pyrethroid resistant neonate *H. virescens* that were infected with AcAaIT (at a greater than LC_{99} dose) died approximately 11% faster (LT_{50} of 63 versus 71 h) than pyrethroid sensitive neonate *H. virescens* infected with AcAaIT. Furthermore, this decrease in the LT_{50} was not found in pyrethroid resistant neonates that were infected with wild-type AcMNPV (LT_{50}s of 94 and 92 h in the resistant and sensitive strains, respectively). McCutchen *et al.* (1997) thus suggested that the pyrethroid resistant larvae are more sensitive to the neurotoxin AaIT. They further suggested that the use of AcAaIT or other recombinant baculovirus might provide the means to deter the onset of insecticide resistance in the field and might be used to drive resistant populations towards susceptibility.

In related experiments, Popham *et al.* (1998b) tested the effectiveness of the coapplication of the type II pyrethroid deltamethrin and recombinant AcMNPVs expressing μ-Aga-IV, AS II, or Sh I (genes *mag4*, *sat2*, and *ssh1*, respectively) in neonate larvae of *T. ni* and *H. virescens*. The genes encoding these toxins were placed under the very late P_{SynXIV} (vMAg4p+, vSAt2p+, and sVVh1p+, respectively) or constitutively expressed *hsp70* (vhsMAg4p+, vhsSAt2p+, and vhsSSh1p+, respectively) promoters. By diet incorporation bioassays in neonate *T. ni* treated with a low dose (LC_{20}) of deltamethrin and a low dose (LC_{20}) of vMAg4p+, vSAt2p+, or vSSh1p+, they found additive effects in terms of larval mortality, but no synergistic effects. By droplet feeding bioassays in neonate *T. ni* treated with a low dose of deltamethrin and a high (LC_{75} or greater) dose of vMAg4p+, vSAt2p+, or vSSh1p+, they found that only the larvae exposed to both deltamethrin and vMAg4p+ showed a synergistic response in terms of FT_{50}, ET_{50}, and LT_{50}. In similar droplet feeding bioassays in neonate *T. ni* treated with a low dose of deltamethrin and a high dose of vhsMAg4p+, vhsSAt2p+, and vhsSSh1p+, they found no synergistic effects. Synergistic effects were observed in at least one time response parameter (FT_{50}, ET_{50}, or LT_{50}) in droplet feeding bioassays in neonate *H. virescens* treated with a low dose of deltamethrin and high dose of a toxin expressing AcMNPV. In all of the cases where a synergistic effect was observed, the coapplication of a high viral dose was also required. Thus, Popham *et al.* (1998b) suggested that (in contrast to the McCutchen *et al.* (1997) study discussed above) the coapplication of a pyrethroid and a recombinant baculovirus would confer little or no advantage in the field because low doses of the virus would most likely be applied. Furthermore,

they suggested that the different outcomes of the McCutchen *et al.* (1997) study and their study might result from differences in the method of application of the pyrethroid.

A wide variety of venomous and nonvenomous animals, bacteria, and even plants possess peptide toxins that are highly potent and selective for lepidopteran insects and which have the potential to improve the insecticidal activity of recombinant baculoviruses. Rapid methods for the separation, purification, and identification of toxins from venom and other biological matrices have been developed (Nakagawa *et al.*, 1997, 1998). These methods have been successfully used for the identification of a number of insecticidal peptide toxins, for example, AaIT5 from *A. australis* (Nakagawa *et al.*, 1997) and ButaIT from the South Indian red scorpion *B. tamulus* (Wudayagiri *et al.*, 2001). ButaIT is a short chain neurotoxin that targets the insect potassium channel, but is particularly interesting because it is highly selective towards the *Heliothine* subfamily but nontoxic against blowfly larvae and other insects. This high level of specificity may provide some unique advantages in comparison to other scorpion toxins. The concept of delivering individual toxins or combinations of toxins via baculoviruses remains a major route to practical application of insect selective peptide toxins. Whether used alone or in combination, the ever-increasing number of characterized insect selective toxins, and their compatibility and complementary nature to existing pest control strategies including chemical insecticides, provide us with extraordinary tools for "green" pest control.

10.4. Modification of the Baculovirus Genome

10.4.1. Deletion of the Ecdysteroid UDP-Glucosyltransferase Gene

Ecdysteroids are key molecules in the regulation of physiological events such as molting in insects. More than 60 different ecdysteroids have been isolated from insects and other arthropods, and about 200 phytoecdysteroids from plants (Nijhout, 1994; Henrich *et al.*, 1999; Lafont, 2000; Dinan, 2001). Ecdysone was the first ecdysteroid to be isolated and is the best characterized. Following its production and secretion from the prothoracic glands, ecdysone is converted into 20-hydroxyecdysone in the hemolymph, epidermis, and fat body. 20-Hydroxyecdysone is the primary active form of the molting hormone in most insects. Ecdysteroids are often bound to specific carrier proteins that putatively help the ecdysteroid to penetrate cell membranes or increase the capacity of the hemolymph for ecdysteroids. Like juvenile hormone, the titer of ecdysteroids in the hemolymph is determined by both their synthesis and their metabolism. During normal insect development, ecdysteroids are transported to the nucleus where they initiate a cellular response. The prevention of ecdysteroid transport results in the interruption of insect growth or abnormal development (or death).

Ecdysteroid UDP-glucosyltransferase is a baculovirus encoded enzyme that catalyzes the conjugation of sugar molecules to ecdysteroids (O'Reilly and Miller, 1989; O'Reilly, 1995). The conjugation of a hydrophilic sugar molecule to the ecdysteroid prevents it from crossing cellular membranes. Thus, conjugation effectively inactivates ecdysteroid function resulting in the inhibition of molting and pupation. O'Reilly and Miller (1989) were the first to identify a gene, *egt*, that encodes ecdysteroid UDP-glucosyltransferase in a baculovirus (AcMNPV). Homologs of the AcMNPV *egt* gene have been identified in approximately 90% (20 NPVs and 8 GVs, **Table 1**) of the baculovirus genomes that have been searched (Clarke *et al.*, 1996; Tumilasci *et al.*, 2003). Although the *egt* gene is commonly found in both NPVs and GVs, it is not essential for *in vitro* (in cultured insect cells) or *in vivo* (in larval hosts) replication of AcMNPV (O'Reilly and Miller, 1989, 1991). The *egt* gene has also been shown to be nonessential in the NPVs of *M. brassicae* (Clarke *et al.*, 1996), *L. dispar* (Slavicek *et al.*, 1999), *H. armigera* (Chen *et al.*, 2000), *H. zea* (Treacy *et al.*, 2000), *B. mori* (Kang *et al.*, 2000), and *A. gemmatalis* (Rodrigues *et al.*, 2001) (**Table 2**). Larvae of *S. frugiperda* or *T. ni* infected with vEGTDEL, an *egt* deletion mutant of AcMNPV, show earlier mortality and reduced feeding (by about 40%) in comparison to control larvae infected with wild-type AcMNPV (O'Reilly and Miller, 1991; Eldridge *et al.*, 1992a; Wilson *et al.*, 2000). Specifically, in time-mortality bioassays (at an LC_{90} or LC_{97} dose), the ST_{50} of vEGTDEL infected neonate *S. frugiperda* is reduced by approximately 22% (99.7 versus 127.2 h) in comparison to control larvae infected with AcMNPV. Furthermore, vEGTDEL infected larvae yield 23% fewer progeny viruses (polyhedra) in comparison to AcMNPV infected larvae. In second or fourth instar larvae of *T. ni*, vEGTDEL infection reduces the time to death by about 11% in comparison to AcMNPV infection. Virus yield is also lower in fourth instar

Table 1   Baculoviruses carrying an ecdysteroid UDP-glucosyltransferase (*egt*) gene homolog

Virus	Virus designation	Reference/accession No.
Nucleopolyhedrovirus	*Autographa californica* MNPV (AcMNPV)	O'Reilly and Miller (1989)
	Anticarsia gemmatalis MNPV (AgMNPV)	Rodrigues *et al.* (2001)
	Amsacta albistriga NPV (AmalNPV)	AF204881
	Bombyx mori NPV (BmNPV)	Gomi *et al.* (1999)
	Buzura suppressaria NPV (BusuNPV)	Hu *et al.* (1997)
	Choristoneura fumiferana defective MNPV (CfDEFMNPV)	Barrett *et al.* (1995)
	Choristoneura fumiferana MNPV (CfMNPV)	Barrett *et al.* (1995)
	Ecotropis oblique NPV (EcobNPV)	AF107100
	Epiphyas postvittana MNPV (EppoMNPV)	Caradoc-Davies *et al.* (2001)
	Heliocoverpa armigera NPV (HearNPV)	Chen *et al.* (1997)
	Heliocoverpa zea NPV (HzNPV)	Popham *et al.* (1997)
	Lymantria dispar MNPV (LdMNPV)	Riegel *et al.* (1994)
	Mamestra configurata NPV (MacoNPV)	Li *et al.* (2002)
	Mamestra brassicae MNPV (MbMNPV)	Clarke *et al.* (1996)
	Orgyia pseudotsugata MNPV (OpMNPV)	Pearson *et al.* (1993)
	Rachiplusia ou MNPV (RoMNPV)	Harrison and Bonning (2003)
	Spodoptera exigua MNPV (SeMNPV)	Ijkel *et al.* (1999)
	Spodoptera frugiperda MNPV (SfMNPV)	Tumilasci *et al.* (2003)
	Spodoptera littoralis NPV (SpliNPV)	Kikhno *et al.* (2002)
	Spodoptera litura NPV (SpltNPV)	Pang *et al.* (2001)
Granulovirus	*Adoxophyes honmai* GV (AdhoGV)	Nakai *et al.* (2002)
	Adoxophyes orana GV (AdorGV)	Wormleaton and Winstanley (2001)
	Choristoneura fumiferana GV (CfGV)	AF058690
	Cydia pomonella GV (CpGV)	Luque *et al.* (2001)
	Epinotia aporema GV (EpapGV)	Manzan *et al.* (2002)
	Lacanobia oleracea GV (LoGV)	Smith and Goodale (1998)
	Phthorimaea operculella (PhopGV)	Taha *et al.* (2000)
	Plutella xylostella GV (PxGV)	Hashimoto *et al.* (2000)

Table 2   Genetic modifications at the *egt* gene locus

Parental virus	Virus designation	Reference
AcMNPV	VEGTZ	O'Reilly and Miller (1989)
	vEGTDEL	O'Reilly and Miller (1991), Treacy *et al.* (1997)
	vJHEEGTD	Eldridge *et al.* (1992a)
	vEGHEGTD	Eldridge *et al.* (1992b)
	VEGT⁻PTTHM	O'Reilly *et al.* (1995)
	AcMNPV-Δegt	Bianchi *et al.* (2000)
	vV8EGTdel, vp6.9*tox34*, vV8EE6.9*tox34*, vHSP70*tox34*, vV8EEHSP*tox34*	Popham *et al.* (1998a)
AgMNPV	VEGTDEL	Rodrigues *et al.* (2001)
	vAgEGTΔ-*lacZ*	Pinedo *et al.* (2003)
BmNPV	BmEGTZ	Kang *et al.* (2000)
HaSNPV (HearNPV)	HaLM2, HaCXW1, HaCXW2	Chen *et al.* (2000), Sun *et al.* (2002)
HzSNPV	HzEGTdel, HzEGTp6.9tox34, HzEGThsptox34, HzEGTDA26tox34	Popham *et al.* (1997)
LdMNPV	vEGT-	Slavicek *et al.* (1999)
MbMNPV	vEGTDEL	Clarke *et al.* (1996)

larvae but not in the second instars. The *egt* gene thus provides a selective advantage to the virus in terms of the production of progeny. Conversely, this suggests that removal of the *egt* gene generates a virus that is less fit in comparison to the wild-type virus. In general, deletion of the *egt* genes of other NPVs has resulted in improvements in the speed of kill of 15–33% in comparison to the wild-type virus (e.g., Treacy *et al.*, 1997; Popham *et al.*, 1998a; Slavicek *et al.*, 1999; Treacy *et al.*, 2000; Chen *et al.*, 2000; Pinedo *et al.*, 2003). However, examples in which deletion of the *egt* gene does not reduce speed of kill are also found. Popham *et al.* (1997) found that deletion of the *egt* gene of HzSNPV had no effect in terms of reducing the ET_{50} (67.3 versus 65.4 h) in neonate larvae of *H. zea* in comparison to the wild-type virus. Bianchi *et al.* (2000) found that LT_{50} of AcMNPV-Δ*egt*, an *egt* gene deletion mutant of AcMNPV, was the same as that of wild-type AcMNPV in second and fourth instar larvae of *S. exigua*. Sun *et al.* (2004) found that the ST_{50} of neonate larvae of *H. armigera* infected with HaSNPV-EGTD (*egt* gene deletion mutant of HaSNPV) was not significantly improved over HaSNPV infected larvae, whereas second–fifth instar larvae were killed more rapidly. Thus, they suggested that *egt* gene deletion affects the speed of

kill of the virus more strongly in later instars than in earlier instars.

Mechanistically, it is still unclear how deletion of the *egt* gene improves the speed of kill. In second instar larvae of *S. exigua*, Flipsen *et al.* (1995) found that an *egt* gene deletion mutant of AcMNPV caused earlier degradation of the Malpighian tubules in comparison to the wild-type AcMNPV. They speculated that this earlier degradation results in the faster speed of kill. Another possibility is that the general level of protein synthesis is elevated in larvae infected with the *egt* deletion virus (due to relatively higher levels of active ecdysteroids). Consistent with this hypothesis, Kang *et al.* (2000) found that virus encoded and virus induced proteins are expressed earlier in the infection cycle and at higher levels in larvae of *B. mori* that are infected with BmEGTZ, an *egt* gene deletion mutant of BmNPV, in comparison to the larvae infected with the wild-type BmNPV. Thus, mature progeny virions may be made and released more quickly in larvae infected with BmEGTZ than in larvae infected with BmNPV resulting in a faster systemic infection and death. Kang *et al.* (2000) have also shown that virus specific and certain host specific proteins are induced in a concentration dependent manner by the injection of purified ecdysteroids into BmEGTZ infected larvae (24 h prior to injection of ecdysteroids), but not in larvae infected with BmNPV. Furthermore, using a histochemical assay they showed that virus transmission is positively correlated with the amount of ecdysteroids that are injected into a larva (i.e., the higher the amount of ecdysteroids injected, the faster the transmission of the virus). In contrast, O'Reilly *et al.* (1995) found that the pathogenicity of a recombinant *egt*-gene-deleted AcMNPV that expresses biologically active PTTH (vEGT⁻PTTHM) is dramatically reduced (LC_{50} of 280×10^5 versus 3.8×10^5 polyhedra per ml) in comparison to vEGTDEL. This effect was not found in a PTTH expressing recombinant AcMNPV (vWTPTTHM) in which the *egt* gene was functional. In vEGT⁻PTTHM infected larvae of *S. frugiperda*, the titer of ecdysteroids spiked at around 96 h p.i. ($1162 \pm 476 \, \text{pg} \, \mu l^{-1}$) and quickly dropped at 120 h p.i. ($45 \pm 12 \, \text{pg} \, \mu l^{-1}$). In contrast, in vEGTDEL infected larvae, the titer of ecdysteroids increased at 72 h p.i. ($519 \pm 199 \, \text{pg} \, \mu l^{-1}$) and remained high at 96 h p.i. ($282 \pm 55 \, \text{pg} \, \mu l^{-1}$) and 120 h p.i. ($334 \pm 80 \, \text{pg} \, \mu l^{-1}$). Although the peak ecdysteroid titer in vEGT⁻PTTHM infected larvae appeared to be significantly higher than that found in vEGTDEL infected larvae, this may not be the case because a higher percentage ($54.5 \pm 21.0\%$ versus $32.3 \pm 5.1\%$) of the ecdysteroids in the vEGT⁻PTTHM infected larvae were in the conjugated form in comparison to vEGTDEL infected larvae.

An elegant approach to further improve the *egt* gene deletion construct has been to insert an insect selective neurotoxin gene cassette into the *egt* gene locus, thus generating a toxin expressing, *egt* gene inactivated construct. For example, Chen *et al.* (2000) inserted an AaIT-GFP gene cassette (*aait* and green fluorescent protein (*gfp*) genes driven by the *polh* gene promoters of HaSNPV and AcMNPV, respectively) into the *egt* gene locus of HaSNPV (also known as HearNPV) in order to generate HaCXW2. In time-mortality bioassays, they found that the ST$_{50}$ of second instar *H. armigera* infected with HaCXW2 was reduced by 32% (63.9 versus 94.4 h) and 8% (63.9 versus 69.5 h) in comparison to control larvae infected with wild-type HaSNPV or HaCXW1 (a recombinant HaSNPV carrying *gfp* at its *egt* gene locus). Sun *et al.* (2002, 2004) also showed that HaCXW1 and HaCXW2 were effective pest control agents under field conditions. In their field experiments, second instar *H. armigera* were allowed to feed on HaCXW1 or HaCXW2 treated plots (1.8 × 10^{10} polyhedra per 0.015 hectare plot containing 800 cotton plants) for 12 h, then the larvae were transferred to artificial diet and reared at ambient temperature. The ST$_{50}$ values of the larvae infected with HaCXW1 or HaCXW2 in the field were reduced by 15% (101 versus 119 h) and 26% (88 versus 119 h), respectively, in comparison to control larvae that were field infected with HaSNPV. Treatment of cotton plants in the field with HaCXW1 or HaCXW2 reduced leaf consumption by approximately 50% and 63%, respectively. The results of additional field trials using HaSNPV-AaIT (a recombinant HaSNPV expressing AaIT under a chimeric ph-p69p promoter at the *egt* gene locus) are discussed below.

From these results it appears that deletion of the *egt* gene from the baculovirus genome is a useful strategy to improve its speed of kill and consequently its insecticidal activity, but by itself the reduction in speed of kill is too small to be of commercial value. Deletion of the *egt* gene, however, reduces the number of progeny viruses that are produced in comparison to the wild-type virus. Thus, the fitness of the *egt* gene deleted virus appears to be reduced in comparison to the wild-type virus. Conceptually, removal of an endogenous gene from its genome is safer than the insertion of a heterologous gene because a "new" gene will not be introduced into the environment. In practice, however, it is believed that there will be no significant difference in the risk of a recombinant baculovirus pesticide in which

an endogenous gene is deleted from the genome or in which a heterologous gene is inserted into the genome. Safety issues and fitness of the virus are further discussed below.

10.4.2. Removal of Other Nonessential Genes

In a comparative study of seven completely sequenced lepidopteran baculovirus genomes, Hayakawa *et al.* (2000a) identified 67 putative genes that were common to all seven. When Herniou *et al.* (2001, 2003) added four additional lepidopteran NPVs and a GV (i.e., a total of 12 completely sequenced lepidopteran baculoviruses) 62 common genes were identified. Considering that the genomes of most baculoviruses encode well over 100 putative genes, these findings suggest that a large number of genes in the genome of lepidopteran baculoviruses may serve auxiliary or host specific functions or help to improve the fitness of the virus. This hypothesis is consistent with the finding that 61 out of the 136 putative genes of BmNPV appear not to be essential for viral replication in BmN cells, i.e., some level of viral replication was detected even when a gene was deleted by replacement with a *lacZ* gene cassette (Kamita *et al.*, 2003b; Gomi, Kamita, and Maeda, unpublished data). These findings suggest that there are a number of target genes that can be removed from the baculovirus genome in order to decrease fitness, alter host specificity, or simply to obtain a virus that may replicate faster or more efficiently.

Dai *et al.* (2000) have identified a mutant of *S. exigua* MNPV (SeXD1) that is lacking 10.6 kb of sequence (encoding 14 complete or partial genes including an *egt* gene homolog) that corresponds to the region between 13.7 and 21.6 map units in the wild-type SeMNPV genome. Dose-mortality bioassays in third instar larvae of *S. exigua* showed that the LD$_{50}$s of SeXD1 and SeMNPV (403 and 124 polyhedra per larva, respectively) were not significantly different. Time-mortality bioassays indicated that the ST$_{50}$ of third instar *S. exigua* infected with SeXD1 was reduced by 25% (70.2 versus 93.1 h) in comparison to control larvae infected with SeMNPV. Dai *et al.* (2000) did not further examine which gene or genes in this 10.6 kb region are involved in decreasing the virulence of SeXD1 in comparison to SeMNPV. Considering that deletion of the *egt* gene from other baculoviruses has been shown to reduce the survival times of larvae infected with these viruses, the *egt* gene homolog of SeMNPV appears to be a prime candidate as the gene involved in reducing the virulence. However, several of the other genes that were deleted in this region are commonly found in other baculoviruses and may play roles in enhancing infection (e.g., *gp37*) or dissemination (e.g., *v-cath*, *chiA*, and *ptp-2*) of the virus. Genes that help to enhance virulence or improve dissemination, but are nonessential, are potential targets for deletion in order to reduce the fitness of the virus.

As discussed above, the ST_{50} of larvae that are infected with an *egt* gene-deleted mutant baculovirus is often reduced in comparison to larvae that are infected with the wild-type virus. This reduction in survival time often results in the production of fewer progeny viruses (i.e., fewer polyhedra per larva) suggesting that the fitness of the *egt* gene deleted virus will also be reduced in comparison to the wild-type virus. Two other baculovirus genes that are involved in altering the virulence of the baculovirus are *orf603* (Popham *et al.*, 1998a) and *Ac23* (Lung *et al.*, 2003) both of AcMNPV origin. The LC_{50} of an AcMNPV variant (V8) in which the *orf603* gene is truncated is not significantly different from that of AcMNPV. The AcMNPV V8 variant (as well as other mutants in which the *orf603* gene was specifically inactivated) shows an ET_{50} that is reduced by approximately 12% (e.g., 88.5 versus 100.8 h) in comparison to a revertant virus (i.e., a virus in which the *orf603* mutation was repaired). In an attempt to further decrease the speed of kill of the V8 variant of AcMNPV, Popham *et al.* (1998a) inserted the *tox34* gene under the control of either the *p6.9* or *hsp70* promoter into the *egt* gene locus of V8 generating V8EEp6.9*tox34* and V8EEHSP*tox34*, respectively. In time-mortality bioassays in neonate larvae of *S. frugiperda* (at an LC_{95} dose), the ET_{50}s of V8EEp6.9*tox34* and V8EEHSP*tox34* were reduced by approximately 34% (58.2 versus 88.2 h) and 42% (51.2 versus 88.2 h), respectively, in comparison to control larvae infected with V8EGTdel, an *egt* gene deletion mutant of V8. However, in comparison to larvae infected with viruses in which the *orf603* gene was not deleted (but expressing the *tox34* gene under the *p6.9* or *hsp70* promoter at the *egt* gene locus), no differences were found in the ET_{50}s. They speculated that the relatively small decrease in speed of kill elicited by deletion of the *orf603* gene was masked by expression of the toxin gene.

A low-pH-activated fusion protein called GP64 is found in the envelopes of all group I NPVs, but lacking in group II NPVs. A different low-pH-activated envelope fusion protein called the F protein (Westenberg *et al.*, 2002) has been identified in group II NPVs. F proteins were first identified in LdMNPV (Pearson *et al.*, 2000) and SeMNPV (Ijkel *et al.*, 2000); however, F protein homologs are found in the genomes of all of the sequenced baculoviruses (i.e., both group I and group II NPVs and GVs) (Hayakawa *et al.*, 2000a; Herniou *et al.*, 2001). The F proteins of group II NPVs can functionally substitute for GP64 (Lung *et al.*, 2002), whereas the F proteins of group I NPVs appear not to function in membrane fusion (Pearson *et al.*, 2001). Lung *et al.* (2003) found that Ac23, the F protein homolog of AcMNPV, is not essential for AcMNPV replication in Sf-9 or High 5 cells or larvae of *T. ni*. The growth curve of *Ac23*null, an *Ac23* deletion mutant of AcMNPV, is indistinguishable to those of a repair virus of *Ac23*null (*Ac23*null-repair) or wild-type AcMNPV. *Ac23*null and *Ac23*null-repair also show indistinguishable lethal doses in larvae of *T. ni*. However, in survival-time bioassays, the survival times of neonate or fourth instar *T. ni* that are infected with *Ac23*null are at least 28% (26 h) longer than those of control larvae infected with *Ac23*null-repair or AcMNPV (i.e., viruses that carry the *Ac23* gene). Interestingly, although not quantified, Sf-9 cells infected with *Ac23*null appear to produce significantly lower yields of polyhedra in comparison to Sf-9 cells infected with the control viruses (G.W. Blissard and O.Y. Lung, personal communication).

Wandering is a behavior that normally occurs only at the end of the larval stage in lepidopteran larvae (Nijhout, 1994; Goulson, 1997). At this stage, holometabolous insects stop feeding, void their gut, and look for a suitable location to pupate. Baculoviruses can artificially induce wandering in infected larvae to increase virus spread (Evans, 1986; Vasconcelos *et al.*, 1996; Goulson, 1997). It is recently found that baculovirus induced wandering requires a virus encoded protein tyrosine phosphatase (*ptp*) gene using a *ptp* gene deleted mutant of BmNPV (BmPTPD; S.G. Kamita, S. Maeda, and B.D. Hammock, unpublished data). Time-mortality bioassays (at an LC_{99} dose) showed that the ST_{50}s of BmPTPD or BmNPV infected neonate larvae of *B. mori* were not significantly different (88.1 and 81.7 h, respectively). BmPTPD infected neonates, however, did not display any symptoms of virus induced wandering as did the BmNPV infected neonates. Thus, the *ptp* gene is directly involved in increasing the dissemination of the virus. Removal of this gene may reduce the dispersal (and consequently fitness) of the virus in the field.

10.4.3. Alteration of Host Range

Although baculoviruses have been isolated from more than 400 insect species (Tanada and Kaya, 1993), most baculovirus isolates are permissive only in the original insect host from which they were isolated. Thus, the host range of individual baculoviruses is considered to be narrow. Among baculoviruses, AcMNPV has a relatively wide host range being infectious against more than 30 insect species (Gröner, 1986) including a large number of pest insect species. The baculovirus must overcome a number of obstacles both at the organismal and cellular levels in order to replicate and eventually kill the insect host. At the organismal level, the polyhedra must first dissociate and release ODVs that can establish a primary infection. As discussed above, the dissociation of the polyhedra requires high pH and the presence of proteases, whereas the released ODVs must pass through at least three midgut barriers prior to establishing a systemic infection. At the cellular level, the BV must recognize and attach to a receptor on the cell surface, and enter the cell via endocytosis. Subsequently, the nucleocapsid must translocate to a nuclear pore, enter the nucleus, uncoat to release viral DNA, express early genes, initiate viral DNA synthesis, express late genes, and assemble and release mature virions. The baculovirus is able to express some early genes in a wide variety of insect cells (Miller and Lu, 1997; Thiem, 1997). Significantly, reduced levels of viral DNA synthesis and late gene promoter activity are also observed in some nonpermissive insect cells lines. Thus, the host range of a baculovirus may be limited, at least in terms of insect hosts, to its ability to: (1) express the entire complement of early genes; (2) synthesize genomic DNA at normal levels; (3) express late/very late genes at normal levels; and/or (4) assemble and release mature virions. Other key factors in the ability of a baculovirus to productively infect a host cell include its ability to inhibit: (1) the premature or global cessation of protein synthesis (Thiem, 1997) and (2) virus induced apoptosis (Miller et al., 1998; Clem, 2001).

The gypsy moth L. dispar and the L. dispar derived Ld652Y cell line are nonpermissive hosts of AcMNPV (Gröner, 1986; Du and Thiem, 1997a; Chen et al., 1998). AcMNPV infection of Ld652Y cells induces cytotoxicity and a premature cessation of host and viral protein synthesis (Du and Thiem, 1997b; Mazzacano et al., 1999). An L. dispar multicapsid NPV (LdMNPV) gene, host range factor 1 (hrf-1), has been identified that promotes AcMNPV replication in Ld652Y cells (Thiem et al., 1996; Du and Thiem, 1997a) and gypsy moth larvae (Chen et al., 1998). LdMNPV hrf-1 does not function as an apoptotic suppressor (Du and Thiem, 1997b), however, in vitro translation assays indicate that the mechanism of translation arrest involves defective or depleted tRNA species (Mazzacano et al., 1999). In neonate larvae of L. dispar, the LC_{50} of a

recombinant AcMNPV carrying the hrf-1 gene of LdMNPV (vAcLdPD) is reduced by greater than 1800-fold (1.2×10^5 versus 2.2×10^8 polyhedra per ml) in comparison to AcMNPV. The LC_{50} of vAcLdPD, however, was still tenfold higher (1.2×10^5 versus 1.1×10^4 polyhedra per ml) than that of LdMNPV in neonate L. dispar. In second instar larvae of H. zea, the LC_{50} of vAcLdPD is 5.8-fold lower (5.49×10^5 versus 3.17×10^6 polyhedra per ml) than that of AcMNPV. However, the LC_{50} of AcLdPD was not significantly different to that of AcMNPV in second instar larvae of P. xylostella (AcMNPV resistant insect) and S. exigua (AcMNPV sensitive insect).

Two different types of antiapoptotic genes, p35 and iap, have been identified in baculoviruses (Clem et al., 1996; Clem, 2001; Bortner and Cidlowski, 2002). The p35 gene of AcMNPV was first shown by Clem et al. (1991) to function as an inhibitor of apoptosis in baculovirus infected Sf-21 cells. The P35 protein is an inhibitor of caspases (Bump et al., 1995; Xue and Horvitz, 1995) in a very wide variety of organisms (Clem, 2001; Bortner and Cidlowski, 2002). The first iap gene was identified by Crook et al. (1993) in Cydia pomonella GV. Subsequently, iap gene homologs have been identified in at least 10 other baculoviruses and a wide range of other organisms (Clem, 2001). The wild-type AcMNPV induces apoptosis in SL2 cells, a cell line derived from the Egyptian cotton worm S. littoralis (Chejanovsky and Gershburg, 1995; Du et al., 1999). The yield of AcMNPV BV in SL2 cells infected with AcMNPV is approximately 2700-fold lower than that of Sf-9 cells infected with AcMNPV. A recombinant AcMNPV (vHSP-P35) that overexpresses the p35 gene of AcMNPV under the hsp70 promoter does not induce apoptosis in SL2 cells; however, vHSP-P35 still produces low yields of BV (Gershburg et al., 1997). Lu et al. (2003) have found that AcMNPV is able to efficiently replicate in SL2 cells and larvae of S. littoralis, if it expresses higher levels of IE1 (an essential, multifunction protein that was first identified by its ability to transactivate baculovirus early genes) relative to IE0 (another protein that is capable of transactivating baculovirus early genes). The IE0 product is a longer form (by 54 N-terminal amino acids) of the IE1 product that is the result of transcriptional initiation from the ie0 transcription initiation site and removal of the ie0 intron (Friesen, 1997). Lu et al. used two approaches to increase the level of IE1 relative to IE0. First, they generated a mutant virus that expressed lower levels of IE0 by disrupting the ie0 promoter by insertion of a 519 bp-long DNA fragment or a chloramphenicol acetyltransferase gene. Second, they generated a recombinant AcMNPV (vHsp-1) that expressed a second copy of the *ie1* gene under the constitutively expressed *hsp70* promoter. Both of these approaches generated genetically modified AcMNPVs with host ranges that extended to *S. littoralis*.

At present, overexpression of AcMNPV *ie1* and insertion of LdMNPV *hrf-1* into the AcMNPV genome are the only ways in which the host range of AcMNPV can be extended to include pest insect species. Other AcMNPV genes that are involved in extending host range or dramatically increasing virulence include *p35* (Clem *et al.*, 1991; Hershberger *et al.*, 1992; Clem and Miller, 1993; Griffiths *et al.*, 1999), the DNA helicase gene *p143* (Maeda *et al.*, 1993; Croizier *et al.*, 1994), host cell specific factor 1 (*hcf-1*; Lu and Miller, 1995), late expression factor 7 (*lef-7*; Lu and Miller, 1995), and immediate early gene 2 (*ie2*; Prikhod'ko *et al.*, 1999). The roles of these genes in host range expansion and increase of virulence is discussed in the previous chapter.

10.5. Safety of GM Baculoviruses

The safety of a synthetic chemical or biologically based pesticide can never be assured with absolute confidence. Thus, when society considers whether or not a new pesticide is "safe," a better question to ask might be: What level of risk are we as a society willing to accept for a given benefit? In the case of chemical pesticides, the general public and a wide array of researchers (scientists, medical doctors, governmental regulators, statisticians, pathologists, ecologists, etc.) have analyzed the actual, potential, and perceived risks that they pose to human health and the environment. In general, all of these stakeholders agree that the benefits of chemical pesticides sufficiently outweigh any short- or long-term risks that they may pose. Society is thus willing (and should be willing) to accept some level of risk for a given benefit (e.g., increased yields of food, fiber, and feed from a given amount of natural resource). There are of course instances in which society has determined that the costs and risks associated with a particular pesticide (e.g., DDT) outweigh the benefits. In terms of the use of a GM baculovirus pesticide, this leads us to the following questions: (1) what are the risks associated with the field release of a GM baculovirus? (2) are these risks worth the benefits? and (3) are these risks any greater than the risks associated with the use of a chemical insecticide? A number of research studies and essays have addressed the first question. Miller

(1995) and Hails (2001) suggest that the primary focus of assessing the risks of the field release of a GM baculovirus should include:

1. How the GM baculovirus affects nontarget species.
2. Whether the introduced gene provides a selective advantage to viral replication, survival, and/or host range (i.e., is fitness improved).
3. Should the introduced gene transfer to another organism, will this event provide a selective advantage to that organism.

10.5.1. Potential Effects of a GM Baculovirus on Nontarget Species

The potential of natural and recombinant baculoviruses to induce a deleterious effect upon vertebrates has been extensively studied (Burges, 1981; Black *et al.*, 1997). More than two dozen different baculoviruses have been tested in a wide range of vertebrates including 10 different mammalian species, birds, and fish. The baculoviruses in these tests were administered by various routes including inhalation, topical application, injection (intravenous, intracerebral, intramuscular, intradermal), and orally. Baculovirus induced deleterious effects were not observed in any of these tests. Direct and indirect tests on human subjects have also been conducted including oral administration of high (greater than 10^9 polyhedra) doses of polyhedra of HzSNPV over a period of 5 days (Heimpel and Buchanan, 1967). The blood of workers who were involved in the production of HzSNPV during a 26 month long period was analyzed for infectious baculovirus, baculoviral antigens, or baculoviral antibodies – none were detected (Black *et al.*, 1997). Furthermore, the ubiquitous baculovirus load from foods in our diet may be quite high (Heimpel *et al.*, 1973; Thomas *et al.*, 1974). For example, Heimpel *et al.* (1973) found that a square inch (approximately 6.5 cm^2) of cabbage taken from a store shelf may contain up to 2×10^6 polyhedra suggesting that consumption of a single serving of coleslaw (16 in^2 of cabbage) exposes us to up to 3.2×10^7 polyhedra. These studies indicate that we live under constant exposure to baculoviruses with no apparent deleterious effects. The lack of deleterious effects is not completely unexpected since polyhedra are extremely insensitive to the neutral or acidic pH conditions of the vertebrate digestive system. Thus, any polyhedra that are normally ingested while eating coleslaw, for example, pass through the digestive tract and are either excreted intact or should any ODVs be released they are quickly inactivated first by the action of pepsin at low pH and subsequently by the pancreatic proteases at neutral pH (Black *et al.*, 1997; Miller and Lu, 1997).

Beginning in 1995, two studies have shown that baculoviruses can serve as highly efficient vehicles for the transfer of foreign genes into the nuclei of primary cultures of human, mouse, and rat hepatocytes and human hepatoma cell lines (Hofmann *et al.*, 1995; Boyce and Bucher, 1996). Since then, baculoviruses have been shown to be able to transduce a number of mammalian cell types including human neural cells (Sarkis *et al.*, 2000), human fibroblasts (Dwarakanath *et al.*, 2001), human keratinocytes (Condreay *et al.*, 1999), and several established mammalian cell lines (Boyce and Bucher, 1996; Shoji *et al.*, 1997). Nonmammalian, vertebrate cell lines (e.g., fish cells) have also been transduced with a recombinant baculovirus (Leisy *et al.*, 2003; Wagle and Jesuthasan, 2003). At first glance, these studies appear to raise concerns about the risks of baculoviruses that are used for pest control. However, upon careful consideration it is believed that they provide further evidence as to the safety of GM baculoviruses. First, these studies show that although the baculovirus can enter the mammalian cell, productive baculovirus infection does not occur even following inoculation at exceptionally high multiplicities of infection (MOI) of up to 1500 pfu per cell using a virus inoculum that was concentrated by ultracentrifugation (Hofmann *et al.*, 1995; Barsoum *et al.*, 1997; Shoji *et al.*, 1997; Kost and Condreay, 2002; Hu *et al.*, 2003). Furthermore, the inoculated cells did not show any visible cytopathic effects, grew normally, and showed normal plating efficiencies. Second, in all of these examples, the foreign gene is expressed only when placed under a promoter that is specific for mammalian cells (e.g., cytomegalovirus immediate early, Rous sarcoma virus long terminal repeat, simian virus 40, etc.). Foreign gene expression was never observed under a very late baculoviral promoter. Third, although transduction was successful in established cell lines and primary cultures, the direct application of baculoviruses for gene delivery to the liver *in vivo* is strongly reduced or inhibited because the baculovirus activates the complement (C) system (Sandig *et al.*, 1996; Hofmann and Strauss, 1998). The C system represents a first line of host defense of the innate immune system for the elimination of foreign elements (Liszewski and Atkinson, 1993). Baculoviruses possess a number of characteristics including their inability to replicate in mammalian cells and induce any deleterious effects that make them ideal gene delivery vectors for human gene therapy and as a screening systems for proteins and

other molecules of human health interest (Lotze and Kost, 2002).

All of the studies to date indicate that there are no known risks associated with human (or other vertebrate) exposure to baculoviruses. Thus, a second focus in the discussion of the safety of recombinant baculoviruses should focus on nontarget lepidopterans and other invertebrates such as predatory or beneficial insects. As discussed above and in the previous chapter, the normal host range of a baculovirus is generally limited to the lepidopteran insect species from which it was originally isolated (Gröner, 1986). This assessment of "host range," however, is potentially problematic because it is often based upon the ability of a baculovirus to elicit virus specific symptomology or death. Huang *et al.* (1997) have tested seven different recombinant NPVs (based on AcMNPV, BmNPV, LdMNPV, or OpMNPV) carrying a reporter gene encoding β-galactosidase, secreted alkaline phosphatase (SEAP), or luciferase under the control of an early (ETL) or very late (*polh*) promoters. These reporter viruses were tested in 23 different insect species from eight insect orders (Blattodea, Coleoptera, Diptera, Hemiptera, Homoptera, Lepidoptera, Neuroptera, and Orthoptera) and 17 families. The reporter viruses were initially injected into the test insects by hemocoelic injection. Insects that supported virus replication (i.e., on the basis of expression of the reporter protein) by injection were subsequently tested by *per os* (oral) inoculation with the preoccluded form of the virus. The use of these reporter viruses allowed the detection of both symptomless and pathogenic infections. By the injection experiments, β-galactosidase, SEAP, and luciferase activities were found only in the lepidopteran larvae. Consistent with previous reports, AcMNPV had the widest host range in the 10 lepidopteran hosts that were tested, whereas LdMNPV was the most host specific. Host range following *per os* inoculation was more limited in comparison to inoculation by injection. In fact two species, the monarch butterfly *Danaus plexippus* and gypsy moth *L. dispar*, that appeared to be susceptible (to AcMNPV, OpMNPV, and BmNPV) by injection were not susceptible by *per os* inoculation. Additionally, as pointed out by Cory and Myers (2003) the host range of a baculovirus in the field may be considerably narrower due to spatial or temporal differences that are not found in the laboratory.

Other studies have looked at whether natural enemies of lepidopterans such as parasitoids, scavengers, and predators (McNitt *et al.*, 1995; McCutchen *et al.*, 1996; Heinz *et al.*, 1995; Li *et al.*, 1999; Smith *et al.*, 2000b; Boughton *et al.*,

2003) are adversely affected when they prey upon larvae that are infected with recombinant or wild-type viruses. The social wasp *Polistes metricus* is a beneficial insect that preys on lepidopteran larvae (Gould and Jeanne, 1984). McNitt *et al.* (1995) analyzed five developmental parameters (i.e., larval development time, pupal development time, larval mortality, mean nest size, and mean number of adults per nest) in order to assess colony health and individual vigor of *P. metricus* that fed on lepidopteran larvae infected with recombinant AcMNPVs expressing AaIT or TxP-I. These toxins were expressed under either the constitutive *hsp70* or very late *polh* promoter. There were no significant differences in any of the developmental parameters tested when the *P. metricus* fed on uninfected or virus infected (at 3 days before exposure to *P. metricus*) fourth or fifth instar *S. frugiperda*. In related experiments, McCutchen *et al.* (1996) allowed the parasitic wasp *Microplitis croceipes* to parasitize second instar *H. virescens* that were subsequently infected with AcAaIT, AcJHE.KK, or AcMNPV (at a greater than LC$_{99}$ dose, 5×10^4 polyhedra per larva) at various times (0, 48, 72, 96, or 120 h) postparasitization. The survival of the parasitoid was less than 4% at 0 and 48 h postparasitization but increased gradually until 120 h postparasitization, when there was no significant difference in the virus or mock treated *H. virescens*. There were no significant differences in terms of time taken for emergence of the adult wasp, ability of the F$_1$ wasps to oviposit, or sex ratios of the F$_1$ wasps. However, there were significant differences in the time (205–216 versus 228 h) taken for emergence of the larval parasitoid from virus infected *H. virescens* in comparison to mock infected *H. virescens*. This was particularly evident in the AcAaIT infected larvae. McCutchen *et al.* (1996) suggested that the parasitoid larva responds to the poor condition of its host by prematurely emerging. Additionally, wasps that developed from AcAaIT or AcJHE.KK infected larvae were significantly smaller than wasps that developed from mock infected larvae.

The green lacewing, *Chrysoperla carnea* Stephens, insidious flower bug, *Orius insidiousus* (Say), red imported fire ant, *Solenopsis invicta* Buren, big-eyed bug, *Geocoris punctipes* (Say), convergent lady beetle, *Hippodamia convergens* Guerin-Meneville, and twelve-spotted lady beetle, *Coleomegilla maculata* DeGeer are generalist predators that attack larval lepidopterans such as *H. virescens*. Heinz *et al.* (1995) have looked at the development times of green lacewings and flower bugs that fed upon near-dead second instar larvae

of *H. virescens* (three per day) that were infected with a greater than LC$_{99}$ dose (5×10^4 polyhedra per larva) of AcAaIT or AcMNPV. They found no significant differences in the survival percentages of insects that fed upon AcAaIT-, AcMNPV-, or mock-infected *H. virescens*. There was a significant but short (19.1 versus 20.6 days) decrease in the larva-to-adult development time in green lacewings that fed upon AcAaIT or AcMNPV infected *H. virescens* in comparison to lacewings that fed upon mock infected *H. virescens*. One may speculate that the protein quality in the prey might be reduced due to overexpression and agglomeration of the toxin proteins resulting in these subtle differences. Li *et al.* (1999) have investigated the life history traits (rate of food consumption, travel speed, fecundity, and adult survival) of fire ants, big-eyed bugs, and convergent lady beetles fed on second instar *H. virescens* that were infected with one of seven viruses (AcMNPV, HzSNPV, HzSNPV expressing LqhIT2 under the *ie1* gene promoter, or AcMNPV expressing LqhIT2 or AaIT under the *ie1* or *p10* gene promoters) at 24 or 60 h prior to predation by the predator. No significant shifts in the life history characteristics were detected in predators that fed on any of the virus infected larvae in comparison to predators that fed on healthy larvae.

Boughton *et al.* (2003) have analyzed the adult survival, developmental time, and oviposition rates of green lacewings and twelve-spotted lady beetles fed upon first or second instar *H. virescens* that were infected with $100 \times$ LC$_{50}$ doses of AcMLF9.ScathL (recombinant AcMNPV expressing a basement membrane degrading protease under the *p6.9* gene promoter) or AcMNPV. Unexpectedly, lacewings that fed on the AcMLF9.ScathL infected *H. virescens* showed significantly elevated survival rates in comparison to AcMNPV infected or uninfected *H. virescens*. Boughton *et al.* (2003) speculated that the protease expressed by AcMLF9.ScathL enhanced the efficiency of feeding of the lacewing. No significant differences were found in developmental time, time to onset of oviposition or mean daily egg production of lacewings that fed on virus infected or uninfected larvae. Most of the twelve-spotted lady beetles that fed exclusively on *H. virescens* (either uninfected or virus infected) died. There was no evidence that feeding on AcMLF9.ScathL infected larvae induced any adverse effects in comparison to AcMNPV infected larvae. Furthermore, neither the lacewing nor lady beetle exhibited any feeding preferences for AcMLF9.ScathL-, AcMNPV-, and mock-infected insects. On the basis of these findings, Boughton *et al.* (2003) concluded that there was no greater risk to insect predators by the use of AcMLF9.ScathL in comparison to the use of AcMNPV as a biological pesticide. Smith *et al.* (2000a) have investigated the density and diversity of nontarget predators under field conditions following the application of recombinant baculoviruses (AcMNPV or HzSNPV expressing scorpion toxin LqhIT2 under the *ie1* gene promoter) on cotton. They found that predator densities and diversity were similar between recombinant and wild-type baculovirus treated plots. In contrast, the chemical pesticide (esfenvalerate) treated plots had consistently lower predator populations.

Taken together, these studies indicate that: (1) the amount of baculovirus expressed toxin or protease that accumulates in the larvae is not sufficient to induce any adverse effects on the predator; (2) the baculovirus is not infectious towards the predatory insects; and (3) the toxin or protease encoding gene is not expressed in the predatory insect. In some cases, there appear to be some costs associated with the beneficial insects that prey upon virus infected larvae. However, these costs are significantly lower in comparison to the costs associated with the application of synthetic chemical pesticides. The development of selective recombinant insecticides should augment any IPM program by reducing the impact on nontarget species, including beneficial insects. Consequently, the resurgence of primary pests and outbreaks of secondary pests should be minimized.

10.5.2. Fitness of GM Baculoviruses

Fitness is a term that describes the ability of an organism to produce progeny that survive to contribute to the following generation (Cory, 2000). In order to estimate the relative fitness of a recombinant baculovirus in comparison to the wild-type baculovirus, five key parameters should be assessed: speed of kill, yield, transmission, dispersal, and persistence. Speed of kill (or time to death), virus yield, and transmission rate can be easily determined by laboratory bioassays (as discussed above) and in some cases under field conditions. Although the relationship between speed of kill (generally quantified in terms of LT_{50}) and virus yield is complex, faster speed of kill generally results in dramatically lower virus yields. This correlation between improved speed of kill and reduced virus yield is found regardless of the parental virus that is genetically modified, and for modifications in which an effector gene is inserted or when an endogenous gene is deleted from the genome. For example, the speeds of kill of third, fourth, or fifth instar larvae of *T. ni* that are infected with AcAaIT or AcJHE.KK

are reduced by approximately 30% and 8%, respectively, in comparison to control larvae infected with AcMNPV. These faster speeds of kill result in reductions of approximately 80% and 40% in the yields (polyhedra per milligram of cadaver) of AcAaIT and AcJHE.KK, respectively, in comparison to the yield of AcMNPV (Kunimi *et al.*, 1996). Dramatic reductions of up to 95% in virus yield (polyhedra per microgram of cadaver) are found in second and fourth instar larvae of AcTOX34.4 infected *T. ni* in comparison to control AcMNPV infected larvae (Burden *et al.*, 2000). The corresponding reductions in the mean times to death are 50–60%. A reduction in virus yield is also found by deletion of the *egt* gene of AgMNPV (Pinedo *et al.*, 2003). The mean lethal times of third instar larvae of *A. gemmatalis* infected with various doses of vAgEGTΔ-*lacZ* is reduced by 10–26% in comparison to control larvae infected with the same dose. The yield (polyhedra per gram of cadaver) of vAgEGTΔ-*lacZ* is reduced by approximately 50% in comparison to control larvae infected with the wild-type AgMNPV. Similar results are found in fifth instar larvae of *S. frugiperda* infected with vEGTDEL, which produce 23% fewer polyhedra per insect (the yield of virus per milligram of cadaver is not, however, significantly different) in comparison to control larvae infected with AcMNPV (O'Reilly and Miller, 1991). O'Reilly *et al.* (1991) and Ignoffo *et al.* (2000) speculated that this correlation between improved speed of kill and reduced virus yield results from the considerably reduced size of recombinant baculovirus infected larvae at the time of death.

Milks *et al.* (2001) have focused on intrahost competition between AcAaIT and AcMNPV or AcAaIT and TnSNPV in larvae of *T. ni* that were synchronously or asynchronously infected. They found no differences in the fitness of the genetically modified or wild-type viruses in terms of virus yield. The most important factors in these mixed infections were dose and timing. The virus that was inoculated at the highest dose or the virus that was first inoculated was the one that had the competitive advantage. These findings are not unreasonable when one considers that there are no significant differences in the replication rates of toxin gene carrying and wild-type baculoviruses in cell culture.

A key component of the transmission rate of a virus is its pathogenicity or potency (often quantified in terms of the LD_{50} or LC_{50}). In general, the pathogenicity of a baculovirus is not significantly changed following the insertion of a neurotoxin gene into its genome (McCutchen *et al.*, 1991; Tomalski and Miller, 1992; Prikhod'ko *et al.*,

1996; Harrison and Bonning, 2000b), although there are a few exceptions in which relatively small (threefold or less) increases or decreases have been observed (Chen *et al.*, 2000; Harrison and Bonning, 2000b). Deletion of an endogenous baculovirus gene such as *egt* also results in no significant (O'Reilly and Miller, 1991; Slavicek *et al.*, 1999) or relatively small (Pinedo *et al.*, 2003) differences in pathogenicity. However, the pathogenicity of the virus can be dramatically increased (i.e., lower LD_{50} or LC_{50}) by 4- to 100-fold by genetic modifications that improve the ability of the virus to penetrate the midgut, i.e., by the expression of enhancins (Hayakawa *et al.*, 2000b; Popham *et al.*, 2001) or incorporation of *Bt* toxin into the polyhedra (Chang *et al.*, 2003). In these cases, it is believed that the enhancin or *Bt* toxin helps the virus to pass more easily through the midgut and establish a more rapid systemic infection. In contrast, however, there were no significant differences in the LC_{50} values of recombinant AcMNPVs expressing basement membrane degrading proteases (stromelysin-1, gelatinase A, of cathepsin L) under early or late promoters (Harrison and Bonning, 2001).

Another component of the transmission of a virus is the rate at which the virus and host come into contact. This component is obviously more difficult to quantify in the field. As described above, neurotoxin-induced paralysis will cause the insect to fall off the plant (Cory *et al.*, 1994; Hoover *et al.*, 1995; Sun *et al.*, 2004). This knock-off behavior results in lower levels of foliage contamination, thus the likelihood that a second host insect will come into contact with the recombinant virus (at least during feeding on the plant) will be reduced. In contrast, the wild-type virus infected larvae will most likely die on the plant, thereby increasing the potential for contact between the virus and host. Lee *et al.* (2001) have examined this behavioral effect in a greenhouse microcosm. They found that the wild-type AcMNPV out-competed recombinant viruses (AcAaIT or AcJHE.SG) for a niche in the greenhouse microcosm. AcMNPV and AcJHE epizootics lasted for 8 weeks after the initial release of the virus, whereas the AcAaIT epizootic ended by the fourth week after release. Additionally, AcMNPV polyhedra also increased to greater numbers in the soil in comparison to AcAaIT or AcJHE.SG after 8 weeks. A reduction in wandering behavior resulting from the inactivation of the *ptp* gene may also reduce foliage contamination as discussed above. In both of these examples, the transmission rate of the recombinant baculovirus should be reduced because the likelihood of a host insect and virus coming into contact is reduced.

Dispersal is another parameter that should be assessed in order to determine the fitness of a virus. Abiotic and biotic agents that are known to disperse baculoviruses include rainfall, air currents, predators, parasites, scavengers, and grazing mammals (Fuxa, 1991; Fuxa and Richter, 1994). Predatory and scavenging insects have been found to carry (within their digestive tracts) and disperse recombinant virus (in excrement) at rates of up to 125 cm per day over a period of up to 10 days (Lee and Fuxa, 2000). Thus, recombinant virus induced behavioral changes of the host larvae such as knock-off effects that reduce predation and scavenging will also reduce the dispersal of the virus. Additionally, the ability of the biotic agent itself to disperse the recombinant baculovirus may be reduced. For example, parasitic wasps that develop within larvae of *H. virescens* that are infected with recombinant baculoviruses are significantly smaller than wasps that develop within wild-type virus infected larvae (McCutchen *et al.*, 1996). Thus, the ability of the smaller wasps to travel, and subsequently disperse the recombinant virus, may be reduced because it has less energy reserves and reduced flight capability in comparison to wasps that developed on wild-type virus infected larvae.

Persistence is the final parameter that should be considered when determining the relative fitness of a genetically modified baculovirus. In general, inactivation by sunlight (ultraviolet radiation) is the primary route of inactivation of polyhedra in the environment (Ignoffo and Garcia, 1992; Black *et al.*, 1997; Ignoffo *et al.*, 1997). In contrast, soil and foliage are factors that can protect the virus from sunlight and subsequently increase its persistence (Peng *et al.*, 1999). Knock-off or other behavioral changes induced by the recombinant virus may increase their relative concentrations in the soil such that the recombinant's persistence is increased. In contrast, however, the transmission rate of this recombinant should be reduced because the potential for contact between the virus and host is reduced. The cuticle is another factor that can protect the virus from sunlight such that viral persistence is increased. Fuxa *et al.* (1998) found that the cadavers of *T. ni* that are killed by recombinant viruses (AcAaIT, AcJHE.KK, or AcJHE.SG) take much longer to disintegrate (4–7 days versus 1 day) in comparison to AcMNPV killed *T. ni*. Another report by Ignoffo and Garcia (1996) found, however, that there is no significant difference in the time to cell lysis following death by AcAaIT or AcMNPV (lysis took 1.7 versus 1.5 days, respectively). Clearly, viral fitness is dependent upon a large number of intricately integrated factors. Finally, it is also important to consider as suggested by Hammock (1992) that genetically modified baculoviruses are designed to be biological insecticides and not biological control agents that will become permanently established. Thus, a recombinant baculovirus should by its very design show reduced fitness in comparison to the wild-type.

10.5.3. Movement of the Introduced Gene to Another Organism

The insecticidal efficacy of natural baculoviruses is dramatically improved by insertion of a foreign gene into the genome or inactivation (deletion) of an endogenous gene from the genome or a combination of both. The insertion strategy generally involves the insertion of an effector gene that encodes a protein that is detrimental to the target insect, alters its life cycle or stops it from feeding. The deletion strategy generally involves the inactivation of an endogenous gene (e.g., *egt* or *orf603*) by inserting another gene into its coding sequence. This other gene can be a marker gene such as *lacZ* or an effector gene as described above. In both cases, the genes are placed under a baculoviral or insect promoter. Several critical points should be kept in mind with these strategies. First, the genes are placed under promoters that are active only in insect cells (and in the case of late/very late baculoviral promoters, these promoters also require the products of baculoviral early genes for activity). Thus, should the effector gene and its promoter somehow jump to the genome of a noninsect cell, the gene will not be expressed. Second, the proteins encoded by the effector genes are chosen because they target some critical aspect of the pest insect life cycle or body. The proteins are not biologically active in the noninsects (although it is possible that they may induce an immunological response). Thus, if the effector gene somehow jumps to the genome of a noninsect cell, and if this gene is somehow expressed, detrimental effects will not result.

Genomic variants of baculoviruses are often found in individual field collected insects (Cherry and Summers, 1985; Maeda *et al.*, 1990; Shapiro *et al.*, 1991; Hodgson *et al.*, 2001); this suggests that recombination and/or transposition events commonly occur between baculovirus genomes. Extensive homology between the donor and recipient DNA molecules and replication of both DNA molecules is required for high-frequency recombination to occur (Kamita *et al.*, 2003b). Such conditions may occur when two heterologous viruses (that share some genomic homology) infect the same cell within the same insect. This scenario is the most

likely one in which an effector gene of a GM baculovirus will jump to another organism (i.e., another insect virus). Should the effector gene jump to another virus under these conditions, the fitness of the new recombinant virus will be reduced in comparison to the original GM baculovirus and it too should be rapidly eliminated from the environment. In a second scenario in which the effector gene jumps from the GM baculovirus to the genome of the insect host, the effector gene could cause an adverse effect. However, these effects should be limited to a single individual because once this individual dies the effector gene will also "die." It is also possible that heterologous or random recombination events may lead to the movement of an effector gene to another organism. The same arguments that were made in regard to the homologous recombination based movement can be made in this case. However, the likelihood of heterologous recombination is much lower than the likelihood of homologous recombination.

10.6. Field Testing and Practical Considerations

Laboratory and greenhouse testing has generated a great deal of knowledge about the efficacy, safety, and environmental fate of GM baculovirus pesticides as described above. Mathematical models have also been generated to evaluate the effectiveness and ecological consequences of the release of GM baculoviruses (Dwyer and Elkinton, 1993; Dwyer *et al.*, 1997; Dushoff and Dwyer, 2001). Field testing, however, over both the short-term (e.g., single growing season) and long-term (e.g., multiple seasons and years) is still necessary to confirm the findings of laboratory and greenhouse tests and the accuracy of mathematical models. The commercial potential of GM baculoviruses and practical considerations regarding their use can also be determined by field testing. Issues regarding the commercialization of GM baculovirus insecticides including marketing, *in vivo* and *in vitro* production, formulation, storage, and public acceptance are discussed in detail by Black *et al.* (1997).

Some of the earliest field trials of GM baculoviruses (occlusion-negative AcMNPVs carrying junk DNA or *lacZ* marker gene) were performed in England during the mid to late 1980s (Levidow, 1995; Black *et al.*, 1997). In the USA, the first field trial (a 3-year study) of a GM baculovirus (a *polh* gene deleted AcMNPV that was co-occluded with wild-type AcMNPV) was begun in 1989 (Wood *et al.*, 1994). These early field trials were performed primarily to analyze the environmental persistence of the virus. These trials showed that the persistence of occlusion-negative constructs is exceptionally low. The first field trial to test the efficacy of an occlusion-positive, AaIT expressing AcMNPV (AcST-3) was performed in 1993 (Cory *et al.*, 1994). Cory *et al.* (1994) found that cabbage plants treated with AcST-3 showed 23–29% lower feeding damage (by third instar larvae of *T. ni*) in comparison to wild-type AcMNPV treated cabbage plants. This reduction in feeding damage (approximately 50% reduction) was not as large as that found in the laboratory trials (Stewart *et al.*, 1991). The reduction in feeding damage during the field trial was due to the earlier death of AcST-3 infected larvae in comparison to AcMNPV infected larvae. Cory *et al.* (1994) also found that the yield of AcST-3 was tenfold lower (9.9×10^7 versus 9.6×10^8 polyhedra per larva) in comparison to AcMNPV. Furthermore, they found that under the high (10^8 polyhedra per ml) dose treatments the majority (62%) of the larvae of AcST-3 treated cabbage were knocked off the plant whereas none were knocked off the AcMNPV treated cabbage. They speculated that both the behavioral (knock-off) and pathological (reduced yield per larva) differences between the recombinant and wild-type viruses would have important implications for risk assessment as discussed above.

The American Cyanamid Company performed the first field trials in the USA (one in Georgia and the other in Texas) to test the efficacy of an occlusion-positive, *egt* gene deleted, AaIT expressing AcMNPV in 1995 and 1996 (Black *et al.*, 1997). They found that this virus killed target insects faster than wild-type AcMNPV or an *egt* gene deleted AcMNPV, which resulted in increased control of the target insect. In 1997 and 1998, field trials were conducted by academic scientists and scientists at DuPont Agricultural Products (Smith *et al.*, 2000a) to assess (1) the efficacy of recombinant baculoviruses in protecting cotton, (2) the impact of recombinant virus introduction on predators, and (3) the ability of predators to disperse recombinant baculoviruses. Two occlusion-positive constructs (one based on AcMNPV and the other on HzSNPV) expressing LqhIT2 were tested. Depending upon the timing of the application, they found that the LqhIT2 expressing viruses were able to protect cotton from damage better than the wild-type virus (AcMNPV or HzSNPV) and as well as a synthetic chemical pesticide (esfenvalerate). Predator densities and diversity were similar between the recombinant and wild-type virus treated plots, whereas plots treated with the chemical pesticide had consistently smaller predator populations. Finally, they found that at

2–5 days after the initial application of virus, a very small percentage (0.2%) of predators that were caught and evaluated carried DNA from a recombinant virus. In 1998, Treacy *et al.* (2000) (American Cyanamid Company) conducted field trials in the USA (two in Georgia and one in North Carolina) to test the efficacy of two occlusion-positive constructs (AcMNPV and HzSNPV based) expressing AaIT in protecting cotton. They found that the recombinant HzSNPV (at 5 or 12×10^{11} polyhedra per hectare) protected cotton against heliothine complex larvae at slightly better levels than either the recombinant AcMNPV or *Bt* toxin (Dipel 2X, Abbott Laboratories, at 1121 g (WP) per hectare).

In 2000, Sun *et al.* (2004) conducted field trials in People's Republic of China (in Henan and Hubei provinces) to test the efficacy of HaSNPV-AaIT (an occlusion-positive recombinant HaSNPV expressing AaIT at the *egt* gene locus) in protecting cotton. Cotton plant plots treated with HaSNPV-AaIT showed significantly less damage to squares, flowers, and bolls in comparison to plots treated with the wild-type HaSNPV or HaSNPV-EGTD (HaSNPV with a deletion in its *egt* gene). In 2001 and 2002 (Hubei province), treatment of cotton plants during the entire growing seasons with HaSNPV-AaIT increased the yield of cotton lint by approximately 18% (1250 versus 1023 and 1800 versus 1474 kg per hectare in 2001 and 2002, respectively) in comparison to wild-type HaSNPV treated plots. The yield of cotton lint in plots treated with HaSNPV-AaIT was not significantly different from the yield (1083 and 1994 kg per hectare in 2001 and 2002, respectively) in plots treated with chemical pesticides (λ-Cyhalothrin EC, endosulfan, β-Cypermethrin on different dates).

Numerous field trials have been conducted over the past 10 years at geographically distant sites using several types of GM baculovirus constructs and under different cropping situations. These trials show that GM baculoviruses are effective and safe pesticides with efficacy that can be at levels similar to synthetic chemical pesticides or *Bt* toxin. The field trials also indicate that the recombinant baculovirus will not persist in the environment and has very little, if any, adverse activity against beneficial insects.

10.7. Concluding Remarks

The majority of people in both developed and developing countries are dependent upon others to produce or otherwise provide the minimum requirements of food, water, and/or shelter necessary for survival. A society must carefully manage and allocate its limited resources in order to provide these minimum requirements for all of its members. Over the past five decades, the use of synthetic chemical pesticides has significantly increased the yields of food, fiber, and feed that producers are able to generate from a given amount of land. Although chemical pesticides have without a doubt improved the efficiency of agricultural output and reduced the incidence of disease by killing disease vectors, chemical pesticides are also a source of environmental pollution and both acute and chronic problems for human health. The inappropriate use of chemical pesticides has also helped to generate pesticide resistant insects. These problems are often intensified in developing countries because of poor governmental regulation and training in the appropriate use of these agents. Furthermore, the high costs of the newest generation of synthetic chemical pesticides are precluding their widespread use in many developing countries. These pressures in the agricultural industry are making biological control agents highly attractive as supplements or replacements for synthetic chemical pesticides in integrated pest management schemes.

While biological pesticides avoid chemical residue problems, many such agents lack sufficient potency and speed of action in the field (at least until recently) to be attractive alternatives. As discussed in this chapter, a fresh approach to biological insect control involves using the best genes of nature to enhance insecticidal properties of naturally occurring baculoviruses. Several laboratories, including the authors', have developed NPVs that have the knock-down speed of classical chemical insecticides on the noctuiid complex of insect pests while not harming nontarget species. The viruses are applied like classical insecticides, but they present no residue problems, and they are active on insects resistant to classical insecticides. Industry and academic scientists have overcome most of the limitations associated with natural baculovirus pesticides, especially in terms of effective crop protection, through the use of recombinant DNA techniques and other technologies. Furthermore, should baculoviruses appear to offer sufficient commercial potential for investment, many of the remaining limitations (e.g., those that are specific to a particular cropping situation) can be quickly addressed and overcome. The safety of baculoviruses has been thoroughly investigated and there is no evidence that natural or recombinant baculoviruses provide an increased threat to human or environmental health.

A large number of genetically modified baculovirus pesticides have been described in the scientific literature and, undoubtedly, many more have been tested in academic, governmental, and commercial laboratories. With the currently available genes and parental viruses, it is believed that the fastest speed of kill (median lethal time) that can be achieved is around 48 h because the virus will most likely have to undergo two replication cycles before the host insect is killed. Feeding cessation (median time to feeding cessation) will occur earlier, perhaps as early as 24–36 h postinfection. Feeding cessation at 24 h postinfection should be sufficient to make GM baculovirus pesticides competitive with synthetic chemical insecticides in terms of the protection of many types of crops. In terms of a "best performing" GM baculovirus pesticide, AcMNPV offers the widest host range in terms of infectivity against pest insect species. To this parental virus we would insert at the egt gene locus a gene encoding AaIT fused to a bombyxin signal sequence that is expressed under a chimeric late–very late promoter (e.g., P_{SynXIV}). We prefer the AaIT encoding gene instead of the mite toxin encoding gene (although the mite toxin gene appears to be more insecticidal) because its mode of action is known and the action of AaIT is synergized with pyrethroid insecticides. LqhIT2 is a viable alternative to AaIT. Also, in comparison to the other insect selective toxins such as those from spiders and sea anemones, AaIT and LqhIT2 have undergone more testing. We prefer expression under a baculovirus late-very promoter because of the slightly improved selectivity that such a promoter has in comparison to baculovirus early or constitutive promoters. We prefer the bombyxin signal sequence because it has been shown to efficiently secrete a wide variety of proteins in a wide variety of insect cells. Deletion of the egt gene may reduce the fitness of the virus. Other parental viruses such as HzSNPV or RoMNPV are also viable options for pest species that are not highly susceptible to AcMNPV. Additionally, the coexpression of a protease or polyhedrin-Bt toxin fusion protein may help the virus to penetrate the midgut more quickly resulting in a faster systemic infection.

The safety and effectiveness of GM baculoviruses as pest insect controlling agents has been studied for more than 25 years. During this time, thousands of peer-reviewed studies on the basic biology, host range, ecology, efficacy, safety, and applications of natural and recombinant baculoviruses have been published. Baculoviruses are very commonly used in hundreds of academic and commercial laboratories as protein expression vectors, as gene transfer vehicles, for surface display, and as a model system to study large DNA viruses. Natural baculoviruses are used as effective pest insect control agents in several regions of the world. This chapter primarily reviews (1) major milestones in the development of recombinant baculoviruses for pest insect control, (2) current strategies to generate recombinant baculoviruses with improved feeding prevention and speed of kill, and (3) potential risks associated with the use of recombinant baculoviruses for pest insect control. A tremendous knowledge base is now available regarding the efficacy and safety of genetically modified baculoviruses for pest insect control. Despite this, there are some that would argue that there are unseen and unpredictable risks associated with the release of any genetically modified organism into the environment. The authors agree with this, but would also like to argue that it is impossible to completely avoid all risk associated with the use of any pesticide. We believe that the benefits afforded by GM baculoviruses far outweigh any unseen or unpredicted risks that they pose, especially in comparison to the potential risks associated with the use of chemical pesticides or genetically modified plants. The current research indicates that GM baculovirus pesticides meet or exceed the high standard of expectation set by synthetic chemical insecticides. GM baculovirus pesticides are safe, effective, and ready for immediate use.

This chapter shows that the generation of exceptionally safe and efficient biological pesticides, which readily compete with chemical pesticides, has been achieved. The technology has been tested for many years and the results are not completely unsatisfactory, and indeed are favorable in many cases. However, GM baculoviruses still have limited use in the field. Therefore, it is concluded that these are other reasons why the implementation of an available, safe, and effective pest control technology is not moving at the pace it should have done. Several world-class pesticide manufacturing companies have shown interest in the technology throughout the years. They obtained and licensed critical patents and started their own research programs that positively synergized the efforts of others. On the one hand, this commercial interest has brought excitement and acceleration to the field, on the other hand, the cancellation of their biologicals programs after several years of research have taken a high toll on the commercialization of GM baculoviruses. Were the cancellations based on scientific knowledge or the ineffectiveness or safety of the viruses? Presented here are the studies that indicate otherwise. Was profitability an issue? GM baculoviruses now compete with chemical pesticides and examples of profitable industrial manufacturing schemes of non-GM baculoviruses exist. Was public perception an issue? Recombinant organisms have been safely used for many years now and chemical pesticides have more proven side effects than highly specific baculoviruses. Thus, we believe that the lack of implementation of GM baculoviruses is primarily due to a psychological effect.

Academic laboratories have already done their part by introducing GM baculovirus technology for the benefit of mankind and by presenting proof on its profitability. However, in the process, we may also have been shown that it is quite easy to take the technology and use it without regard to the intellectual property rights. This by itself should not be an excuse because legal ways to protect international patent rights exist. To negate the negative psychological effects of seemingly unsuccessful commercialization is, therefore, the greatest problem in this field of research as we see it. There is also more to do in terms of introducing the end-users to the merits of GM baculoviruses. One thing academic researchers might have done more efficiently is better outreach. This might have induced the agricultural cooperatives to look at this new alternative earlier on and might have helped them to start production on the farm or on a large scale. The intellectual property rights as they are currently held probably exclude this possibility. So why not publicize all relevant patents to encourage the technology rather than burying it? This is a possibility that all private and public patent holders should consider at this point. The Brazilian experience is exemplary and could be a model for other parts of the world. In that instance, even though studies were initiated by the government, cooperatives and private companies eventually showed interest in commercial production of non-GM baculoviruses.

Throughout the years, attempts have been made on many occasions to liaise with the leading scientists in the field in order to achieve a level of organization and to initiate private commercial interest. But the time has now come to put all the work done in perspective by either increasing the cooperation between everyone in the field to obtain a positive result or perhaps accepting the fact that this group of biopesticides will never reach the shelves. We believe that efforts should be directed towards more practical aspects of the field including increased outreach activities and free access to the technology. A discussion involving all interested parties of the problems associated with immediate commercialization will be critical. Countries where GM baculoviruses are actively used could of course provide insight into how to successfully commercialize GM baculoviruses. The studies to date on GM baculoviruses presented in this chapter clearly show that it would be a great waste to abandon this technology.

Acknowledgments

This work was funded in part by grants from the USDA (2003–35302-13499), NIEHS (P30 ES05707), and NIH (AI58267). We thank Dr. B.C. Bonning for critical review of the manuscript and N.M. Wolf for access to data prior to publication.

References

Adachi, T., Takiya, S., Suzuki, Y., Iwami, M., Kawakami, A., *et al.*, 1989. cDNA structure and expression of bombyxin, an insulin-like brain secretory peptide, of the silkmoth, *Bombyxi mori. J. Biol. Chem.* 264, 7681–7685.

Ailor, E., Betenbaugh, M.J., 1999. Modifying secretion and post-translational processing in insect cells. *Curr. Opin. Biotech.* 10, 142–145.

Aronson, A.I., Shai, Y., 2001. Why *Bacillus thuringiensis* insecticidal toxins are so effective: unique features of their mode of action. *FEMS Microbiol. Lett.* 195, 1–8.

Bai, C., Degheele, D., Jansens, S., Lambert, B., 1993. Activity of insecticidal crystal proteins and strains of *Bacillus thuringiensis* against *Spodoptera exempta* (Walker). *J. Invertebr. Pathol.* 62, 211–215.

Barrett, J.W., Krell, P.J., Arif, B.M., 1995. Characterization, sequencing and phylogeny of the ecdysteroid UDP-glucosyltransferase gene from 2 distinct nuclear polyhedrosis viruses isolated from *Choristoneura fumferana. J. Gen. Virol.* 76, 2447–2456.

Barsoum, J., Brown, R., McKee, M., Boyce, F.M., 1997. Efficient transduction of mammalian cells by a recombinant baculovirus having the vesicular stomatitis virus G glycoprotein. *Hum. Gene Ther.* 8, 2011–2018.

Bettencourt, R., Terenius, O., Faye, I., 2002. Hemolin gene silencing by ds-RNA injected into *Cecropia* pupae is lethal to next generation embryos. *Insect Mol. Biol.* 11, 267–271.

Bianchi, F.J.J.A., Snoeijing, I., van der Werf, W., Mans, R.M.W., Smits, P.H., *et al.*, 2000. Biological activity of SeMNPV, AcMNPV, and three AcMNPV deletion mutants against *Spodoptera exigua* larvae (Lepidoptera: Noctuidae). *J. Invertebr. Pathol.* 75, 28–35.

Bischoff, D.S., Slavicek, J.M., 1997. Molecular analysis of an enhancin gene in the *Lymantria dispar* nuclear polyhedrosis virus. *J. Virol.* 71, 8133–8140.

Black, B.C., Brennan, L.A., Dierks, P.M., Gard, I.E., 1997. Commercialization of baculoviral insecticides. In: Miller, L.K. (Ed.), The Baculoviruses. Plenum, New York, pp. 341–387.

Blissard, G., Black, B., Crook, N., Keddie, B.A., Possee, R., *et al.*, 2000. Family Baculoviridae. In: van Regenmortel, M.H.V., Fauquet, C.M., Bishop, D.H.L., Carstens, E.B., Estes, M.K., Lemon, S.M., McGeoch, D.J., Maniloff, J., Mayo, M.A., Pringle, C.R., Wickner, R.B. (Eds.), Virus Toxonomy: Classification and Nomenclature of Viruses. Academic Press, San Diego, pp. 195–202.

Bole, D.G., Hendershot, L.M., Kearney, J.F., 1986. Post-translational association of immunoglobulin heavy chain binding protein with nascent heavy chains in nonsecreting and secreting hybridomas. *J. Cell Biol.* 102, 1558–1566.

Bonning, B.C., Booth, T.F., Hammock, B.D., 1997a. Mechanistic studies of the degradation of juvenile hormone esterase in *Manduca sexta. Arch. Insect Biochem. Physiol.* 34, 275–286.

Bonning, B.C., Boughton, A.J., Jin, H., Harrison, R.L., 2002. Genetic Enhancement of Baculovirus Insecticides. Kluwer Academic/Plenum Publishers, New York.

Bonning, B.C., Hammock, B.D., 1996. Development of recombinant baculoviruses for insect control. *Annu. Rev. Entomol.* 41, 191–210.

Bonning, B.C., Hirst, M., Possee, R.D., Hammock, B.D., 1992. Further development of a recombinant baculovirus insecticide expressing the enzyme juvenile hormone esterase from *Heliothis virescens. Insect Biochem. Mol. Biol.* 22, 453–458.

Bonning, B.C., Possee, R.D., Hammock, B.D., 1999. Insecticidal efficacy of a recombinant baculovirus expressing JHE-KK, a modified juvenile hormone esterase. *J. Invertebr. Pathol.* 73, 234–236.

Bonning, B.C., Roelvink, P.W., Vlak, J.M., Possee, R.D., Hammock, B.D., 1994. Superior expression of juvenile hormone esterase and β-galactosidase from the basic protein promoter of *Autographa californica* nuclear polyhedrosis virus compared to the p10 protein and polyhedrin promoters. *J. Gen. Virol.* 75, 1551–1556.

Bonning, B.C., Ward, V.K., VanMeer, M.M.M., Booth, T.F., Hammock, B.D., 1997b. Disruption of lysosomal targeting is associated with insecticidal potency of juvenile hormone esterase. *Proc. Natl Acad. Sci. USA* 94, 6007–6012.

Booth, T.F., Bonning, B.C., Hammock, B.D., 1992. Localization of juvenile hormone esterase during development in normal and in recombinant baculovirus-infected larvae of the moth *Trichoplusia ni. Tissue Cell* 24, 267–282.

Bortner, C.D., Cidlowski, J.A., 2002. Cellular mechanisms for the repression of apoptosis. *Annu. Rev. Pharmacol. Toxicol.* 42, 259–281.

Boughton, A.J., Obrycki, J.J., Bonning, B.C., 2003. Effects of a protease-expressing recombinant baculovirus on nontarget insect predators of *Heliothis virescens. Biol. Control* 28, 101–110.

Bougis, P.E., Rochat, H., Smith, L.A., 1989. Precursors of *Androctonus australis* scorpion neurotoxins: structures of precursors, processing outcomes, and expression of a functional recombinant toxin II. *J. Biol. Chem.* 264, 19259–19265.

Boyce, F.M., Bucher, N.L.R., 1996. Baculovirus-mediated gene transfer into mammalian cells. *Proc. Natl Acad. Sci. USA* 93, 2348–2352.

Brandt, C.R., Adang, M.J., Spence, K.D., 1978. The peritrophic membrane: ultrastructural analysis and function as a mechanical barrier to microbial infection in *Orgyia pseudotsugata*. *J. Invertebr. Pathol. 32*, 12–24.

Braun, C.J., Siedow, J.N., Williams, M.E., Levings, C.S., **1989**. Mutations in the maize mitochondrial T-urf13 gene eliminate sensitivity to a fungal pathotoxin. *Proc. Natl Acad. Sci. USA 86*, 4435–4439.

Bump, N.J., Hackett, M., Hugunin, M., Seshagiri, S., Brady, K., *et al.*, **1995**. Inhibition of ice family proteases by baculovirus antiapoptotic protein P35. *Science 269*, 1885–1888.

Burden, J.P., Hails, R.S., Windass, J.D., Suner, M.M., Cory, J.S., **2000**. Infectivity, speed of kill, and productivity of a baculovirus expressing the itch mite toxin txp-1 in second and fourth instar larvae of *Trichoplusia ni*. *J. Invertebr. Pathol. 75*, 226–236.

Burges, H.D., **1981**. Safety, safety testing and quality control of microbial pesticides. In: Burges, H.D. (Ed.), Microbiol Control of Pests and Plant Diseases 1970–1980. Academic Press, New York, pp. 737–767.

Cao, J., Ibrahim, H., Garcia, J.J., Mason, H., Granados, R.R., *et al.*, **2002**. Transgenic tobacco plants carrying a baculovirus enhancin gene slow the development and increase the mortality of *Trichoplusia ni* larvae. *Plant Cell Rep. 21*, 244–250.

Caradoc-Davies, K.M.B., Graves, S., O'Reilly, D.R., Evans, O.P., Ward, V.K., **2001**. Identification and *in vivo* characterization of the *Epiphyas postvittana* nucleopolyhedrovirus ecdysteroid UDP-glucosyltransferase. *Virus Genes 22*, 255–264.

Carbonell, L.F., Hodge, M.R., Tomalski, M.D., Miller, L.K., **1988**. Synthesis of a gene coding for an insect specific scorpion neurotoxin and attempts to express it using baculovirus vectors. *Gene 73*, 409–418.

Casu, R.E., Pearson, R.D., Jarmey, J.M., Cadogan, L.C., Riding, G.A., *et al.*, **1994**. Excretory/secretory chymotrypsin from *Lucilia cuprina*: purification, enzymatic specificity and amino acid sequence deduced from mRNA. *Insect Mol. Biol. 3*, 201–211.

Cestele, S., Catterall, W.A., **2000**. Molecular mechanisms of neurotoxin action on voltage-gated sodium channels. *Biochimie 82*, 883–892.

Chang, J.H., Choi, J.Y., Jin, B.R., Roh, J.Y., Olszewski, J.A., *et al.*, **2003**. An improved baculovirus insecticide producing occlusion bodies that contain *Bacillus thuringiensis* insect toxin. *J. Invertebr. Pathol. 84*, 30–37.

Chao, Y.C., Chen, S.L., Li, C.F., **1996**. Pest control by fluorescence. *Nature 380*, 396–397.

Chejanovsky, N., Gershburg, E., **1995**. The wild-type *Autographa californica* nuclear polyhedrosis virus induces apoptosis of *Spodoptera littoralis* cells. *Virology 209*, 519–525.

Chejanovsky, N., Zilberberg, N., Rivkin, H., Zlotkin, E., Gurevitz, M., **1995**. Functional expression of an alpha anti-insect scorpion neurotoxin in insect cells and lepidopterous larvae. *FEBS Lett. 376*, 181–184.

Chen, C.J., Quentin, M.E., Brennan, L.A., Kukel, C., Thiem, S.M., **1998**. *Lymantria dispar* nucleopolyhedrovirus hrf-1 expands the larval host range of *Autographa*

californica nucleopolyhedrovirus. *J. Virol. 72*, 2526–2531.

Chen, X.W., Hu, Z.H., Jehle, J.A., Zhang, Y.Q., Vlak, J.M., **1997**. Analysis of the ecdysteroid UDP-glucosyltransferase gene of *Heliothis armigera* single-nucleocapsid baculovirus. *Virus Genes 15*, 219–225.

Chen, X.W., Sun, X.L., Hu, Z.H., Li, M., O'Reilly, D.R., *et al.*, **2000**. Genetic engineering of *Helicoverpa armigera* single-nucleocapsid nucleopolyhedrovirus as an improved pesticide. *J. Invertebr. Pathol. 76*, 140–146.

Cherry, C.L., Summers, M.D., **1985**. Genotypic variation among wild isolates of two nuclear polyhedrosis viruses isolated from *Spodoptera littoralis*. *J. Invertebr. Pathol. 46*, 289–295.

Clarke, E.E., Tristem, M., Cory, J.S., O'Reilly, D.R., **1996**. Characterization of the ecdysteroid UDP-glucosyltransferase gene from *Mamestra brassicae* nucleopolyhedrovirus. *J. Gen. Virol. 77*, 2865–2871.

Clem, R.J., **2001**. Baculoviruses and apoptosis: the good, the bad, and the ugly. *Cell Death Differ. 8*, 137–143.

Clem, R.J., Fechheimer, M., Miller, L.K., **1991**. Prevention of apoptosis by a baculovirus gene during infection of insect cells. *Science 254*, 1388–1390.

Clem, R.J., Hardwick, J.M., Miller, L.K., **1996**. Antiapoptotic genes of baculoviruses. *Cell Death Differ. 3*, 9–16.

Clem, R.J., Miller, L.K., **1993**. Apoptosis reduces both the *in vitro* replication and the *in vivo* infectivity of a baculovirus. *J. Virol. 67*, 3730–3738.

Coast, G.M., Orchard, I., Phillips, J.E., Schooley, D.A., **2002**. Insect diuretic and antidiuretic hormones. *Adv. Insect Physiol. 29*, 279–400.

Cohen, E., **1987**. Chitin biochemistry: synthesis and inhibition. *Annu. Rev. Entomol. 32*, 71–93.

Condreay, J.P., Witherspoon, S.M., Clay, W.C., Kost, T.A., **1999**. Transient and stable gene expression in mammalian cells transduced with a recombinant baculovirus vector. *Proc. Natl Acad. Sci. USA 96*, 127–132.

Copping, L.G., Menn, J.J., **2000**. Biopesticides: a review of their action, applications and efficacy. *Pest Mgt Sci. 56*, 651–676.

Cory, J.S., **2000**. Assessing the risks of releasing genetically modified virus insecticides: progress to date. *Crop Protect. 19*, 779–785.

Cory, J.S., Hirst, M.L., Williams, T., Halls, R.S., Goulson, D., *et al.*, **1994**. Field trial of a genetically improved baculovirus insecticide. *Nature (Lond.) 370*, 138–140.

Cory, J.S., Myers, H., **2003**. The ecology and evolution of insect baculovirus. *Annu. Rev. Ecol. Evol. Syst. 34*, 239–272.

Croizier, G., Croizier, L., Argaud, O., Poudevigne, D., **1994**. Extension of *Autographa californica* nuclear polyhedrosis virus host range by interspecific replacement of a short DNA sequence in the p143 helicase gene. *Proc. Natl Acad. Sci. USA 91*, 48–52.

Crook, N.E., Clem, R.J., Miller, L.K., **1993**. An apoptosis-inhibiting baculovirus gene with a zinc finger-like motif. *J. Virol. 67*, 2168–2174.

Cusson, M., Palli, S.R., 2000. Can juvenile hormone research help rejuvenate integrated pest management? *Can. Entomol. 132*, 263–280.

Cymborowski, B., Zimowska, G., 1984. Switchover in the sensitivity of the prothoracic glands to juvenile hormone in the cotton leafworm, *Spodoptera littoralis*. *J. Insect Physiol. 30*, 911–918.

Dai, X., Hajos, J.P., Joosten, N.N., vanOers, M.M., Ijkel, W.F.J., et al., 2000. Isolation of a *Spodoptera exigua* baculovirus recombinant with a 10.6 kbp genome deletion that retains biological activity. *J. Gen. Virol. 81*, 2545–2554.

Daimon, T., Hamada, K., Mita, K., Okano, K., Suzuki, M.G., et al., 2003. A *Bombyx mori* gene, BmChi-h, encodes a protein homologous to bacterial and baculovirus chitinases. *Insect Biochem. Mol. Biol. 33*, 749–759.

Darbon, H., Zlotkin, E., Kopeyan, C., Vanrietschoten, J., Rochat, H., 1982. Covalent structure of the insect toxin of the North African scorpion *Androctonus australis* Hector. *Int. J. Pep. Prot. Res. 20*, 320–330.

de Kort, C.A.D., Granger, N.A., 1996. Regulation of JH titers: the relevance of degradative enzymes and binding proteins. *Arch. Insect Biochem. Physiol. 33*, 1–26.

Denli, A.M., Hannon, G.J., 2003. RNAi: an ever-growing puzzle. *Trends Biochem. Sci. 28*, 196–201.

Derksen, A.C.G., Granados, R.R., 1988. Alteration of a lepidopteran peritrophic membrane by baculoviruses and enhancement of viral infectivity. *Virology 167*, 242–250.

Dinan, L., 2001. Phytoecdysteroids: biological aspects. *Phytochemistry 57*, 325–339.

Dougherty, E.M., Gurthrie, K.P., Shapiro, M., 1996. Optical brighteners provide baculovirus activity enhancement and UV radiation protection. *Biol. Control 7*, 71–74.

Du, Q.S., Lehavi, D., Faktor, O., Qi, Y.P., Chejanovsky, N., 1999. Isolation of an apoptosis suppressor gene of the *Spodoptera littoralis* nucleopolyhedrovirus. *J. Virol. 73*, 1278–1285.

Du, X.L., Thiem, S.M., 1997a. Characterization of host range factor 1 (hrf-1) expression in *Lymantria dispar* M nucleopolyhedrovirus- and recombinant *Autographa californica* M nucleopolyhedrovirus-infected IPLB-Ld652Y cells. *Virology 227*, 420–430.

Du, X.L., Thiem, S.M., 1997b. Responses of insect cells to baculovirus infection: protein synthesis shutdown and apoptosis. *J. Virol. 71*, 7866–7872.

Dushoff, J., Dwyer, G., 2001. Evaluating the risks of engineered viruses: modeling pathogen competition. *Ecol. Appl. 11*, 1602–1609.

Dwarakanath, R.S., Clark, C.L., McElroy, A.K., Spector, D.H., 2001. The use of recombinant baculoviruses for sustained expression of human cytomegalovirus immediate early proteins in fibroblasts. *Virology 284*, 297–307.

Dwyer, G., Elkinton, J.S., 1993. Using simple models to predict virus epizootics in gypsy-moth populations. *J. Animal Ecol. 62*, 1–11.

Dwyer, G., Elkinton, J.S., Buonaccorsi, J.P., 1997. Host heterogeneity in susceptibility and disease dynamics: tests of a mathematical model. *Am. Nat. 150*, 685–707.

Eitan, M., Fowler, E., Herrmann, R., Duval, A., Pelhate, M., et al., 1990. A scorpion venom neurotoxin paralytic to insects that affects sodium current inactivation: purification, primary structure, and mode of action. *Biochemistry 29*, 5941–5947.

Elazar, M., Levi, R., Zlotkin, E., 2001. Targeting of an expressed neurotoxin by its recombinant baculovirus. *J. Exp. Biol. 204*, 2637–2645.

Eldridge, R., Horodyski, F.M., Morton, D.B., O'Reilly, D.R., Truman, J.W., et al., 1991. Expression of an eclosion hormone gene in insect cells using baculovirus vectors. *Insect Biochem. 21*, 341–351.

Eldridge, R., O'Reilly, D.R., Hammock, B.D., Miller, L.K., 1992a. Insecticidal properties of genetically engineered baculoviruses expressing an insect juvenile hormone esterase gene. *Appl. Environ. Microbiol. 58*, 1583–1591.

Eldridge, R., O'Reilly, D.R., Miller, L.K., 1992b. Efficacy of a baculovirus pesticide expressing an eclosion hormone gene. *Biol. Control 2*, 104–110.

Evans, H.F., 1986. Ecology and epizootiology of baculoviruses. In: Granados, R.R., Federici, B.A. (Eds.), The Biology of Baculoviruses, vol. I. CRC Press, Boca Raton, pp. 89–132.

Fajloun, Z., Kharrat, R., Chen, L., Lecomte, C., Di Luccio, E., et al., 2000. Chemical synthesis and characterization of maurocalcine, a scorpion toxin that activates Ca^{2+} release channel/ryanodine receptors. *FEBS Lett. 469*, 179–185.

Federici, B.A., 1997. Baculovirus pathogenesis. In: Miller, L.K. (Ed.), The Baculoviruses. Plenum, New York, pp. 33–59.

Feng, Q.L., Ladd, T.R., Tomkins, B.L., Sundaram, M., Sohi, S.S., et al., 1999. Spruce budworm (*Choristoneura fumiferana*) juvenile hormone esterase: hormonal regulation, developmental expression and cDNA cloning. *Mol. Cell. Endocrinol. 148*, 95–108.

Flipsen, J.T.M., Mans, R.M.W., Kleefsman, A.W.F., Knebelmorsdorf, D., Vlak, J.M., 1995. Deletion of the baculovirus ecdysteroid UDP-glucosyltransferase gene induces early degeneration of malpighian tubules in infected insects. *J. Virol. 69*, 4529–4532.

Friedman, S., 1978. Treholose regulation, one aspect of metabolic homeostasis. *Annu. Rev. Entomol. 23*, 389–407.

Friesen, P.D., 1997. Regulation of baculovirus early gene expression. In: Miller, L.K. (Ed.), The Baculoviruses. Plenum, New York, pp. 141–170.

Froy, O., Zilberberg, N., Chejanovsky, N., Anglister, J., Loret, E., et al., 2000. Scorpion neurotoxins: structure/function relationships and application in agriculture. *Pest Mgt Sci. 56*, 472–474.

Furuta, M., Schrader, J.O., Schrader, H.S., Kokjohn, T.A., Nyaga, S., et al., 1997. Chlorella virus PBCV-1 encodes a homolog of the bacteriophage T4 UV damage repair gene *denV*. *Appl. Environ. Microbiol. 63*, 1551–1556.

Fuxa, J.A., Fuxa, J.R., Richter, A.R., **1998**. Host-insect survival time and disintegration in relation to population density and dispersion of recombinant and wild-type nucleopolyhedroviruses. *Biol. Control 12*, 143–150.

Fuxa, J.R., **1991**. Release and transport of entomopathogenic microorganisms. In: Levin, M., Strauss, H. (Eds.), Risk Assessment in Genetic Engineering. McGraw-Hill, New York, pp. 83–113.

Fuxa, J.R., Richter, A.R., **1994**. Distance and rate of spread of *Anticarsia gemmatalis* (Lepidoptera: Noctuidae) nuclear polyhedrosis virus released into soybean. *Environ. Entomol. 23*, 1308–1316.

Gade, G., **2004**. Regulation of intermediary metabolism and water balance of insects by neuropeptides. *Annu. Rev. Entomol. 49*, 93–113.

Gallo, L.G., Corsaro, B.G., Hughes, P.R., Granados, R.R., **1991**. *In vivo* enhancement of baculovirus infection by the viral enhancing factor of a granulosis virus of the cabbage looper, *Trichoplusia ni* (Lepidoptera: Noctuidae). *J. Invertebr. Pathol. 58*, 203–210.

Gershburg, E., Rivkin, H., Chejanovsky, N., **1997**. Expression of the *Autographa californica* nuclear polyhedrosis virus apoptotic suppressor gene *p35* in nonpermissive *Spodoptera littoralis* cells. *J. Virol. 71*, 7593–7599.

Gershburg, E., Stockholm, D., Froy, O., Rashi, S., Gurevitz, M., *et al.*, **1998**. Baculovirus-mediated expression of a scorpion depressant toxin improves the insecticidal efficacy achieved with excitatory toxins. *FEBS Lett. 422*, 132–136.

Ghiasuddin, S.M., Soderlund, D.M., **1985**. Pyrethroid insecticides – potent, stereospecific enhancers of mouse brain sodium channel activation. *Pestic. Biochem. Physiol. 24*, 200–206.

Gilbert, L.I., Granger, N.A., Roe, R.M., **2000**. The juvenile hormones: historical facts and speculations on future research directions. *Insect Biochem. Mol. Biol. 30*, 617–644.

Gill, S.S., Cowles, E.A., Pietrantonio, P.V., **1992**. The mode of action of *Bacillus thuringiensis* endotoxins. *Annu. Rev. Entomol. 37*, 615–636.

Girardie, J., Chaabihi, H., Fournier, B., Lagueux, M., Girardie, A., **2001**. Expression of neuroparsin cDNA in insect cells using baculovirus vectors. *Arch. Insect Biochem. Physiol. 46*, 26–35.

Gomi, S., Majima, K., Maeda, S., **1999**. Sequence analysis of the genome of *Bombyx mori* nucleopolyhedrovirus. *J. Gen. Virol. 80*, 1323–1337.

Gong, C.L., Kobayashi, J., Jin, W., Wu, X.F., **2000**. Inactivation analysis of HcNPV cysteine protease gene and chitinase gene. *Acta Biochim. Biophys. Sinica 32*, 187–191.

Gopalakrishnan, B., Kramer, K.J., Muthukrishnan, S., **1993**. Properties of an insect chitinase produced in a baculovirus gene expression system. *Abstr. Papers Am. Chem. Soc. 205*, 79-Agro.

Gopalakrishnan, K., Muthukrishnan, S., Kramer, K.J., **1995**. Baculovirus-mediated expression of a *Manduca*

sexta chitinase gene: properties of the recombinant protein. *Insect Biochem. Mol. Biol. 25*, 255–265.

Gordon, D., Moskowitz, H., Eitan, M., Warner, C., Catterall, W.A., *et al.*, **1992**. Localization of receptor sites for insect-selective toxins on sodium channels by site-directed antibodies. *Biochemistry 31*, 7622–7628.

Gordon, D., Savarin, P., Gurevitz, M., Zinn-Justin, S., **1998**. Functional anatomy of scorpion toxins affecting sodium channels. *J. Toxicol.-Toxin Rev. 17*, 131–159.

Gould, W.P., Jeanne, R.L., **1984**. *Polistes* wasps (Hymenoptera: Vespidae) as control agents for lepidopterous cabbage pests. *Environ. Entomol. 13*, 150–156.

Goulson, D., **1997**. *Wipfelkrankheit*: modification of host behaviour during baculoviral infection. *Oecologia 109*, 219–228.

Granados, R.R., Federici, B.A., **1986**. The Biology of Baculoviruses I and II. CRC Press, Boca Raton.

Granados, R.R., Fu, Y., Corsaro, B., Hughes, P.R., **2001**. Enhancement of *Bacillus thuringiensis* toxicity to lepidopterous species with the enhancin from *Trichoplusia ni* granulovirus. *Biol. Control 20*, 153–159.

Griffiths, C.M., Barnett, A.L., Ayres, M.D., Windass, J., King, L.A., *et al.*, **1999**. *In vitro* host range of *Autographa californica* nucleopolyhedrovirus recombinants lacking functional *p35*, *iap1* or *iap2*. *J. Gen. Virol. 80*, 1055–1066.

Gröner, A., **1986**. Specificity and safety of baculoviruses. In: Granados, R.R., Federici, B.A. (Eds.), The Biology of Baculoviruses, vol. 1. CRC Press, Boca Raton, pp. 178–196.

Hails, R., **2001**. Nature and genetically modified baculoviruses: environmentally friendly pest control or an ecological threat? *Outlook Agricult. 30*, 171–178.

Hajos, J.P., Vermunt, A.M.W., Zuidema, D., Kulcsar, P., Varjas, L., *et al.*, **1999**. Dissecting insect development: baculovirus-mediated gene silencing in insects. *Insect Mol. Biol. 8*, 539–544.

Hammock, B.D., **1985**. Regulation of juvenile hormone titer: degradation. In: Kerkut, G.A., Gilbert, L.I. (Eds.), Comprehensive Insect Physiology, Biochemistry, and Pharmacology, vol. 7. Pergamon, New York, pp. 431–472.

Hammock, B.D., **1992**. Virus release evaluation. *Nature 355*, 119.

Hammock, B.D., Abdel-Aal, Y.A.I., Mullin, C.A., Hanzlik, T.N., Roe, R.M., **1984**. Substituted thiotrifluoropropanones as potent selective inhibitors of juvenile hormone esterase. *Pestic. Biochem. Physiol. 22*, 209–223.

Hammock, B.D., Bonning, B.C., Possee, R.D., Hanzlik, T.N., Maeda, S., **1990a**. Expression and effects of the juvenile hormone esterase in a baculovirus vector. *Nature 344*, 458–461.

Hammock, B.D., McCutchen, B.F., Beetham, J., Choudary, P.V., Fowler, E., *et al.*, **1993**. Development of recombinant viral insecticides by expression of an insect-specific toxin and insect-specific enzyme in nuclear polyhedrosis viruses. *Arch. Insect Biochem. Physiol.* 22, 315–344.

Hammock, B.D., Wrobleski, V., Harshman, L., Hanzlik, T., Maeda, S., *et al.*, **1990b**. Cloning, expression and biological activity of the juvenile hormone esterase from *Heliothis virescens*. In: Hagedorn, H., Hildebrand, J., Kidwell, M., Law, J. (Eds.), Molecular Insect Science. Plenum, New York, pp. 49–56.

Hanzlik, T.N., Abdel-Aal, Y.A.I., Harshman, L.G., Hammock, B.D., **1989**. Isolation and sequencing of cDNA clones coding for juvenile hormone esterase from *Heliothis virescens*: evidence for a catalytic mechanism of the serine carboxylesterases different from that of the serine proteases. *J. Biol. Chem.* 264, 12419–12425.

Harrison, R.L., Bonning, B.C., **1999**. The nucleopolyhedroviruses of *Rachiplusia ou* and *Anagrapha falcifera* are isolates of the same virus. *J. Gen. Virol.* 80, 2793–2798.

Harrison, R.L., Bonning, B.C., **2000a**. Genetic engineering of biocontrol agents of insects. In: Rechcigl, J.E., Rechcigl, N.A. (Eds.), Biological and Biotechnological Control of Insect Pests. Lewis Publishers, Boca Raton, FL, pp. 243–280.

Harrison, R.L., Bonning, B.C., **2000b**. Use of scorpion neurotoxins to improve the insecticidal activity of *Rachiplusia ou multicapsid nucleopolyhedrovirus*. *Biol. Control* 17, 191–201.

Harrison, R.L., Bonning, B.C., **2001**. Use of proteases to improve the insecticidal activity of baculoviruses. *Biol. Control* 20, 199–209.

Harrison, R.L., Bonning, B.C., **2003**. Comparative analysis of the genomes of *Rachiplusia ou* and *Autographa californica* multiple nucleopolyhedroviruses. *J. Gen. Virol.* 84, 1827–1842.

Hashimoto, Y., Corsaro, B.G., Granados, R.R., **1991**. Location and nucleotide sequence of the gene encoding the viral enhancing factor of the *Trichoplusia ni* granulosis virus. *J. Gen. Virol.* 72, 2645–2651.

Hashimoto, Y., Hayakawa, T., Ueno, Y., Fujita, T., Sano, Y., *et al.*, **2000**. Sequence analysis of the *Plutella xylostella* granulovirus genome. *Virology* 275, 358–372.

Hawtin, R.E., Arnold, K., Ayres, M.D., Zanotto, P.M.D., Howard, S.C., *et al.*, **1995**. Identification and preliminary characterization of a chitinase gene in the *Autographa californica* nuclear polyhedrosis virus genome. *Virology* 212, 673–685.

Hayakawa, T., Ko, R., Okano, K., Seong, S.-I., Goto, C., *et al.*, **1999**. Sequence analysis of the *Xestia c-nigrum* granulovirus genome. *Virology* 262, 277–297.

Hayakawa, T., Rohrmann, G.F., Hashimoto, Y., **2000a**. Patterns of genome organization and content in lepidopteran baculoviruses. *Virology* 278, 1–12.

Hayakawa, T., Shimojo, E., Mori, M., Kaido, M., Furusawa, I., *et al.*, **2000b**. Enhancement of baculovirus infection in *Spodoptera exigua* (Lepidoptera: Noctuidae) larvae with *Autographa californica* nucleopolyhedrovirus or *Nicotiana tabacum* engineered with a granulovirus enhancin gene. *Appl. Entomol. Zool.* 35, 163–170.

Heimpel, A.M., Buchanan, L.K., **1967**. Human feeding tests using a nuclear polyhedrosis virus of *Heliothis zea*. *J. Invertebr. Pathol.* 9, 55–57.

Heimpel, A.M., Thomas, E.D., Adams, J.R., Smith, L.J., **1973**. The presence of nuclear polyhedrosis virus of *Trichoplusia ni* on cabbage from the market shelf. *Environ. Entomol.* 2, 72–75.

Heinz, K.M., McCutchen, B.F., Herrmann, R., Parrella, M.P., Hammock, B.D., **1995**. Direct effects of recombinant nuclear polyhedrosis viruses on selected non-target organisms. *J. Econ. Entomol.* 88, 259–264.

Henrich, V.C., Rybczynski, R., Gilbert, L.I., **1999**. Peptide hormones, steroid hormones, and puffs: mechanisms and models in insect development. Vitamins and hormones. *Adv. Res. Appl.* 55, 73–125.

Herniou, E.A., Luque, T., Chen, X., Vlak, J.M., Winstanley, D., *et al.*, **2001**. Use of whole genome sequence data to infer baculovirus phylogeny. *J. Virol.* 75, 8117–8126.

Herniou, E.A., Olszewski, J.A., Cory, J.S., O'Reilly, D.R., **2003**. The genome sequence and evolution of baculoviruses. *Annu. Rev. Entomol.* 48, 211–234.

Herrmann, R., Moskowitz, H., Zlotkin, E., Hammock, B.D., **1995**. Positive cooperativity among insecticidal scorpion neurotoxins. *Toxicon* 33, 1099–1102.

Hershberger, P.A., Dickson, J.A., Friesen, P.D., **1992**. Site-specific mutagenesis of the 35-kilodalton protein gene encoded by *Autographa californica* nuclear polyhedrosis virus: cell line-specific effects on virus replication. *J. Virol.* 66, 5525–5533.

Hill-Perkins, M.S., Possee, R.D., **1990**. A baculovirus expression vector derived from the basic protein promoter of *Autographa californica* nuclear polyhedrosis virus. *J. Gen. Virol.* 71, 971–976.

Hinton, A.C., Hammock, B.D., **2003a**. *In vitro* expression and biochemical characterization of juvenile hormone esterase from *Manduca sexta*. *Insect Biochem. Mol. Biol.* 33, 317–329.

Hinton, A.C., Hammock, B.D., **2003b**. Juvenile hormone esterase (JHE) from *Tenebrio molitor*: full-length cDNA sequence, *in vitro* expression, and characterization of the recombinant protein. *Insect Biochem. Mol. Biol.* 33, 477–487.

Hirai, M., Kamimura, M., Kikuchi, K., Yasukochi, Y., Kiuchi, M., *et al.*, **2002**. cDNA cloning and characterization of *Bombyx mori* juvenile hormone esterase: an inducible gene by the imidazole insect growth regulator KK-42. *Insect Biochem. Mol. Biol.* 32, 627–635.

Hodgson, D.J., Vanbergen, A.J., Watt, A.D., Hails, R.S., Cory, J.S., **2001**. Phenotypic variation between naturally co-existing genotypes of a lepidopteran baculovirus. *Evol. Ecol. Res.* 3, 687–701.

Hofmann, C., Sandig, V., Jennings, G., Rudolph, M., Schlag, P., *et al.*, **1995**. Efficient gene transfer into human hepatocytes by baculovirus vectors. *Proc. Natl Acad. Sci. USA* 92, 10099–10103.

Hofmann, C., Strauss, M., **1998**. Baculovirus-mediated gene transfer in the presence of human serum or blood facilitated by inhibition of the complement system. *Gene Ther.* 5, 531–536.

Holman, G.M., Nachman, R.J., Wright, M.S., 1990. Insect neuropeptides. *Annu. Rev. Entomol.* 35, 201–217.

Hom, L.G., Volkman, L.E., 2000. *Autographa californica* M nucleopolyhedrovirus *chiA* is required for processing of V-CATH. *Virology 277*, 178–183.

Hoover, K., Schultz, C.M., Lane, S.S., Bonning, B.C., Duffey, S.S., et al., 1995. Reduction in damage to cotton plants by a recombinant baculovirus that knocks moribund larvae of *Heliothis virescens* off the plant. *Biol. Control 5*, 419–426.

Hsu, T.-A., Eiden, J.J., Bourgarel, P., Meo, T., Betenbaugh, M.J., 1994. Effects of co-expressing chaperone BiP on functional antibody production in the baculovirus system. *Prot. Exp. Purif.* 5, 595–603.

Hsu, T.-A., Watson, S., Eiden, J.J., Betenbaugh, M.J., 1996. Rescue of immunoglobulins from insolubility is facilitated by PDI in the baculovirus expression system. *Prot. Exp. Purif.* 7, 281–288.

Hu, Y.C., Tsai, C.T., Chang, Y.J., Huang, J.H., 2003. Enhancement and prolongation of baculovirus-mediated expression in mammalian cells: focuses on strategic infection and feeding. *Biotech. Prog. 19*, 373–379.

Hu, Z.H., Broer, R., Westerlaken, J., Martens, J.W.M., Jin, F., et al., 1997. Characterization of the ecdysteroid UDP-glucosyltransferase gene of a single nucleocapsid nucleopolyhedrovirus of *Buzura suppressaria*. *Virus Res. 47*, 91–97.

Huang, X.P., Davis, T.R., Hughes, P., Wood, A., 1997. Potential replication of recombinant baculoviruses in nontarget insect species: reporter gene products as indicators of infection. *J. Invertebr. Pathol. 69*, 234–245.

Hughes, P.R., Wood, H.A., Breen, J.P., Simpson, S.F., Duggan, A.J., et al., 1997. Enhanced bioactivity of recombinant baculoviruses expressing insect-specific spider toxins in lepidopteran crop pests. *J. Invertebr. Pathol. 69*, 112–118.

Hunter-Fujita, F.R., Entwistle, P.F., Evans, H.F., Crook, N.E., 1998. Insect Viruses and Pest Management. Wiley, Chichester.

Hutvagner, G., Zamore, P.D., 2002. RNAi: nature abhors a double-strand. *Curr. Opin. Genet. Develop. 12*, 225–232.

Ichinose, R., Kamita, S.G., Maeda, S., Hammock, B.D., 1992a. Pharmacokinetic studies of the recombinant juvenile hormone esterase in *Manduca sexta. Pestic. Biochem. Physiol. 42*, 13–23.

Ichinose, R., Nakamura, A., Yamoto, T., Booth, T.F., Maeda, S., et al., 1992b. Uptake of juvenile hormone esterase by pericardial cells of *Manduca sexta. Insect Biochem. Mol. Biol. 22*, 893–904.

Ignoffo, C.M., Garcia, C., 1992. Combinations of environmental factors and simulated sunlight affecting activity of inclusion bodies of the *Heliothis* (Lepidoptera, Noctuidae) nucleopolyhedrosis virus. *Environ. Entomol. 21*, 210–213.

Ignoffo, C.M., Garcia, C., 1996. Rate of larval lysis and yield and activity of inclusion bodies harvested from *Trichoplusia ni* larvae fed a wild or recombinant strain of the nuclear polyhedrosis virus of *Autographa californica. J. Invertebr. Pathol. 68*, 196–198.

Ignoffo, C.M., Garcia, C., Saathoff, S.G., 1997. Sunlight stability and rain-fastness of formulations of baculovirus *Heliothis. Environ. Entomol. 26*, 1470–1474.

Ignoffo, C.M., Wong, J.F.H., McCutchen, W.F., Saathoff, S.G., Garcia, C., 2000. Yields of occlusion bodies from *Heliothis virescens* (Lepidoptera: Noctuidae) and *Helicoverpa (Heliothis) zea* (Lepidoptera: Noctuidae) larvae fed wild or recombinant strains of baculoviruses. *Appl. Entomol. Zool. 35*, 389–392.

Ijkel, W.F.J., van Strien, E.A., Heldens, J.G.M., Broer, R., Zuidema, D., et al., 1999. Sequence and organization of the *Spodoptera exigua* multicapsid nucleopolyhedrovirus genome. *J. Gen. Virol. 80*, 3289–3304.

Ijkel, W.F.J., Westenberg, M., Goldbach, R.W., Blissard, G.W., Vlak, J.M., et al., 2000. A novel baculovirus envelope fusion protein with a proprotein convertase cleavage site. *Virology 275*, 30–41.

Imai, N., Ali, S.E.-S., El-Singabi, N.R., Iwanaga, M., Matsumoto, S., et al., 2000. Insecticidal effects of a recombinant baculovirus expressing scorpion toxin LqhIT2. *J. Seric. Sci. Jpn 69*, 197–205.

Inceoglu, A.B., Kamita, S.G., Hinton, A.C., Huang, Q.H., Severson, T.F., et al., 2001a. Recombinant baculoviruses for insect control. *Pest Mgt Sci. 57*, 981–987.

Inceoglu, B., Lango, J., Wu, J., Hawkins, P., Southern, J., et al., 2001b. Isolation and characterization of a novel type of neurotoxic peptide from the venom of the South African scorpion *Parabuthus transvaalicus* (Buthidae). *Eur. J. Biochem. 268*, 5407–5413.

Jaffe, H., Raina, A.K., Riley, C.T., Fraser, B.A., Holman, G.M., et al., 1986. Isolation and primary structure of a peptide from the corpora cardiaca of *Heliothis zea* with adipokinetic activity. *Biochem. Biophys. Res. Comm. 135*, 622–628.

Jarvis, D.L., Fleming, J.A., Kovacs, G.R., Summers, M.D., Guarino, L.A., 1990. Use of early baculovirus promoters for continuous expression and efficient processing of foreign gene products in stably transformed lepidopteran cells. *Bio/Technology 8*, 950–955.

Jarvis, D.L., Reilly, L.M., Hoover, K., Schultz, C.M., Hammock, B.D., et al., 1996. Construction and characterization of immediate early baculovirus pesticides. *Biol. Control 7*, 228–235.

Je, Y.H., Jin, B.R., Park, H.W., Roh, J.Y., Chang, J.H., et al., 2003. Baculovirus expression vectors that incorporate the foreign protein into viral occlusion bodies. *Biotechniques 34*, 81–87.

Kamita, S.G., Hinton, A.C., Wheelock, C.E., Wogulis, M.D., Wilson, D.K., et al., 2003a. Juvenile hormone (JH) esterase: Why are you so JH specific? *Insect Biochem. Mol. Biol. 33*, 1261–1273.

Kamita, S.G., Maeda, S., Hammock, B.D., 2003b. High-frequency homologous recombination between baculoviruses involves DNA replication. *J. Virol. 77*, 13053–13061.

Kang, K.-D., Lee, E.-J., Kamita, S.G., Maeda, S., Seong, S.-I., **2000**. Ecdysteroid stimulates virus transmission in larvae infected with *Bombyx mori* nucleopolyhedrovirus. *J. Biochem. Mol. Biol. 33*, 63–68.

Kang, W., Tristem, M., Maeda, S., Crook, N.E., O'Reilly, D.R., **1998**. Identification and characterization of the *Cydia pomonella* granulovirus cathepsin and chitinase genes. *J. Gen. Virol. 79*, 2283–2292.

Kataoka, H., Troetschler, R.G., Li, J.P., Kramer, S.J., Carney, R.L., *et al.*, **1989**. Isolation and identification of a diuretic hormone from the tobacco hornworm, *Manduca sexta. Proc. Natl Acad. Sci. USA 86*, 2976–2980.

Keeley, L.L., Hayes, T.K., **1987**. Speculations on biotechnology applications for insect neuroendocrine research. *Insect Biochem. 17*, 639–651.

Khambay, B.P.S., **2002**. Pyrethroid insecticides. *Pestic. Outlook 2*, 49–54.

Kikhno, I., Gutierrez, S., Croizier, L., Croizier, G., Ferber, M.L., **2002**. Characterization of *pif*, a gene required for the *per os* infectivity of *Spodoptera littoralis* nucleopolyhedrovirus. *J. Gen. Virol. 83*, 3013–3022.

Kim, S.H., Park, B.S., Yun, F.Y., Je, Y.H., Woo, S.D., *et al.*, **1998**. Cloning and expression of a novel gene encoding a new antibacterial peptide from silkworm, *Bombyx mori. Biochem. Biophys. Res. Comm. 246*, 388–392.

Kopeyan, C., Mansuelle, P., Sampieri, F., Brando, T., Barhraoui, E.M., *et al.*, **1990**. Primary structure of scorpion anti-insect toxins isolated from the venom of *Leiurus quinquestriatus quinquestriatus. FEBS Lett. 261*, 424–426.

Korth, K.L., Kaspi, C.I., Siedow, J.N., Levings, C.S., **1991**. Urf13, a maize mitochondrial pore-forming protein, is oligomeric and has a mixed orientation in *Escherichia coli* plasma membranes. *Proc. Natl Acad. Sci. USA 88*, 10865–10869.

Korth, K.L., Levings, C.S., **1993**. Baculovirus expression of the maize mitochondrial protein Urf13 confers insecticidal activity in cell cultures and larvae. *Proc. Natl Acad. Sci. USA 90*, 3388–3392.

Kost, T.A., Condreay, J.P., **2002**. Recombinant baculoviruses as mammalian cell gene delivery vectors. *Trends Biotech. 20*, 173–180.

Kozuma, K., Hukuhara, T., **1994**. Fusion characteristics of a nuclear polyhedrosis virus in cultured cells: time course and effect of a synergistic factor and pH. *J. Invertebr. Pathol. 63*, 63–67.

Kramer, K.J., Muthukrishnan, S., **1997**. Insect chitinases: molecular biology and potential use as biopesticides. *Insect Biochem. Mol. Biol. 27*, 887–900.

Kramer, S.F., Bentley, W.E., **2003**. RNA interference as a metabolic engineering tool: potential for *in vivo* control of protein expression in an insect larval model. *Metab. Eng. 5*, 183–190.

Kunimi, Y., Fuxa, J.R., Hammock, B.D., **1996**. Comparison of wild type and genetically engineered nuclear polyhedrosis viruses of *Autographa californica* for

mortality, virus replication and polyhedra production in *Trichoplusia ni* larvae. *Entomol. Exp. Appl. 81*, 251–257.

Kunimi, Y., Fuxa, J.R., Richter, A.R., **1997**. Survival times and lethal doses for wild and recombinant *Autographa californica* nuclear polyhedrosis viruses in different instars of *Pseudoplusia includens. Biol. Control 9*, 129–135.

Lacey, L.A., Frutos, R., Kaya, H.K., Vail, P., **2001**. Insect pathogens as biological control agents: Do they have a future? *Biol. Control 21*, 230–248.

Lafont, R., **2000**. Understanding insect endocrine systems: molecular approaches. *Entomol. Exp. Appl. 97*, 123–136.

Lee, S.Y., Qu, X.Y., Chen, W.B., Poloumienko, A., MacAfee, N., *et al.*, **1997**. Insecticidal activity of a recombinant baculovirus containing an antisense c-myc fragment. *J. Gen. Virol. 78*, 273–281.

Lee, Y., Fuxa, J.R., **2000**. Transport of wild-type and recombinant nucleopolyhedroviruses by scavenging and predatory arthropods. *Microb. Ecol. 39*, 301–313.

Lee, Y., Fuxa, J.R., Inceoglu, A.B., Alaniz, S.A., Richter, A.R., *et al.*, **2001**. Competition between wild-type and recombinant nucleopolyhedroviruses in a greenhouse microcosm. *Biol. Control 20*, 84–93.

Leisy, D.J., Lewis, T.D., Leong, J.A.C., Rohrmann, G.F., **2003**. Transduction of cultured fish cells with recombinant baculoviruses. *J. Gen. Virol. 84*, 1173–1178.

Lepore, L.S., Roelvink, P.R., Granados, R.R., **1996**. Enhancin, the granulosis virus protein that facilitates nucleopolyhedrovirus (NPV) infections, is a metalloprotease. *J. Invertebr. Pathol. 68*, 131–140.

Levidow, L., **1995**. The Oxford baculovirus controversy: Safely testing safety? *Bioscience 45*, 545–551.

Li, J.B., Heinz, K.M., Flexner, J.L., McCutchen, B.F., **1999**. Effects of recombinant baculoviruses on three nontarget heliothine predators. *Biol. Control 15*, 293–302.

Li, Q.J., Donly, C., Li, L.L., Willis, L.G., Theilmann, D.A., *et al.*, **2002**. Sequence and organization of the *Mamestra configurata* nucleopolyhedrovirus genome. *Virology 294*, 106–121.

Li, Q.J., Li, L.L., Moore, K., Donly, C., Theilmann, D.A., *et al.*, **2003**. Characterization of *Mamestra configurata* nucleopolyhedrovirus enhancin and its functional analysis via expression in an *Autographa californica* M nucleopolyhedrovirus recombinant. *J. Gen. Virol. 84*, 123–132.

Liszewski, K., Atkinson, J.P., **1993**. The complement system. In: William, P.E. (Ed.), Fundamental Immunology. Raven Press, New York, pp. 917–939.

Loret, E.P., Hammock, B.D., **1993**. Structure and neurotoxicity of scorpion venom. In: Brownell, P.H., Polis, G. (Eds.), Scorpion Biology and Research. Oxford University Press, Oxford, pp. 204–233.

Lotze, M.T., Kost, T.A., **2002**. Viruses as gene delivery vectors: application to gene function, target validation, and assay development. *Cancer Gene Ther. 9*, 692–699.

Lu, A., Miller, L.K., **1995**. Differential requirements for baculovirus late expression factor genes in 2 cell lines. *J. Virol. 69*, 6265–6272.

Lu, A., Milller, L.K., **1997**. Regulation of baculovirus late and very late gene expression. In: Miller, L.K. (Ed.), The Baculoviruses. Plenum, New York, pp. 193–216.

Lu, A., Seshagiri, S., Miller, L.K., **1996**. Signal sequence and promoter effects on the efficacy of toxin-expressing baculoviruses as biopesticides. *Biol. Control 7*, 320–332.

Lu, L.Q., Du, Q.S., Chejanovsky, N., **2003**. Reduced expression of the immediate-early protein IE0 enables efficient replication of *Autographa californica* multiple nucleopolyhedrovirus in poorly permissive *Spodoptera littoralis* cells. *J. Virol. 77*, 535–545.

Lung, O., Westenberg, M., Vlak, J.M., Zuidema, D., Blissard, G.W., **2002**. Pseudotyping *Autographa californica* multicapsid nucleopolyhedrovirus (AcMNPV): F proteins from group II NPVs are functionally analogous to AcMNPV GP64. *J. Virol. 76*, 5729–5736.

Lung, O.Y., Cruz-Alvarez, M., Blissard, G.W., **2003**. Ac23, an envelope fusion protein homolog in the baculovirus *Autographa californica* multicapsid nucleopolyhedrovirus, is a viral pathogenicity factor. *J. Virol. 77*, 328–339.

Luque, T., Finch, R., Crook, N., O'Reilly, D.R., Winstanley, D., **2001**. The complete sequence of the *Cydia pomonella* granulovirus genome. *J. Gen. Virol. 82*, 2531–2547.

Ma, P.W.K., Davis, T.R., Wood, H.A., Knipple, D.C., Roelofs, W.L., **1998**. Baculovirus expression of an insect gene that encodes multiple neuropeptides. *Insect Biochem. Mol. Biol. 28*, 239–249.

Maeda, S., **1989a**. Expression of foreign genes in insects using baculovirus vectors. *Annu. Rev. Entomol. 34*, 351–372.

Maeda, S., **1989b**. Increased insecticidal effect by a recombinant baculovirus carrying a synthetic diuretic hormone gene. *Biochem. Biophys. Res. Comm. 165*, 1177–1183.

Maeda, S., Kamita, S.G., Kondo, A., **1993**. Host-range expansion of *Autographa californica* nuclear polyhedrosis virus (NPV) following recombination of a 0.6-kilobase pair DNA fragment originating from *Bombyx mori* NPV. *J. Virol. 67*, 6234–6238.

Maeda, S., Mukohara, Y., Kondo, A., **1990**. Characteristically distinct isolates of the nuclear polyhedrosis virus from *Spodoptera litura*. *J. Gen. Virol. 71*, 2631–2640.

Maeda, S., Volrath, S.L., Hanzlik, T.N., Harper, S.A., Majima, K., *et al.*, **1991**. Insecticidal effects of an insect-specific neurotoxin expressed by a recombinant baculovirus. *Virology 184*, 777–780.

Manzan, M.A., Lozano, M.E., Sciocco-Cap, A., Ghiringhelli, P.D., Romanowski, V., **2002**. Identification and characterization of the ecdysteroid UDP-glycosyltransferase gene of *Epinotia aporema* granulovirus. *Virus Genes 24*, 119–130.

Martens, J.W.M., Honee, G., Zuidema, D., Lent, J.W.M.V., Visser, B., *et al.*, **1990**. Insecticidal activity of a bacterial crystal protein expressed by a recombinant baculovirus in insect cells. *Appl. Environ. Microbiol. 56*, 2764–2770.

Martens, J.W.M., Knoester, M., Weijts, F., Groffen, S.J.A., Hu, Z.H., *et al.*, **1995**. Characterization of baculovirus insecticides expressing tailored *Bacillus thuringiensis* Cry1a(B) crystal proteins. *J. Invertebr. Pathol. 66*, 249–257.

Matsumoto, N., Okada, M., Takahashi, H., Ming, Q.X., Nakajima, Y., *et al.*, **1986**. Molecular cloning of a cDNA and assignment of the C-terminal of sarcotoxin IA, a potent antibacterial protein of *Sarcophaga peregrina*. *Biochem. J. 239*, 717–722.

Mazzacano, C.A., Du, X.L., Thiem, S.M., **1999**. Global protein synthesis shutdown in *Autographa californica* nucleopolyhedrovirus-infected Ld652Y cells is rescued by tRNA from uninfected cells. *Virology 260*, 222–231.

McCutchen, B.F., Choundary, P.V., Crenshaw, R., Maddox, D., Kamita, S.G., *et al.*, **1991**. Development of a recombinant baculovirus expressing an insect-selective neurotoxin: potential for pest control. *Bio/Technology 9*, 848–852.

McCutchen, B.F., Hammock, B.D., **1994**. Recombinant baculovirus expressing an insect-selective neurotoxin: characterization, strategies for improvement and risk assessment. In: Hedin, P.A., Menn, J.J., Hollingworth, R.M. (Eds.), Natural and Engineered Pest Management Agents. American Chemical Society, Washington, DC, pp. 348–367.

McCutchen, B.F., Herrmann, R., Heinz, K.M., Parrella, M.P., Hammock, B.D., **1996**. Effects of recombinant baculoviruses on a nontarget endoparasitoid of *Heliothis virescens*. *Biol. Control 6*, 45–50.

McCutchen, B.F., Hoover, K., Preisler, H.K., Betana, M.D., Herrmann, R., *et al.*, **1997**. Interactions of recombinant and wild-type baculoviruses with classical insecticides and pyrethroid-resistant tobacco budworm (Lepidoptera: Noctuidae). *J. Econ. Entomol. 90*, 1170–1180.

McNitt, L., Espelie, K.E., Miller, L.K., **1995**. Assessing the safety of toxin-producing baculovirus biopesticides to a nontarget predator, the social wasp *Polistes metricus* Say. *Biol. Control 5*, 267–278.

Means, J.C., Muro, I., Clem, R.J., **2003**. Silencing of the baculovirus Op-iap3 gene by RNA interference reveals that it is required for prevention of a apoptosis during *Orgyia pseudotsugata* M nucleopolyhedrovirus infection of Ld652Y cells. *J. Virol. 77*, 4481–4488.

Menn, J.J., Borkovec, A.B., **1989**. Insect neuropeptides – potential new insect control agents. *J. Agric. Food Chem. 37*, 271–278.

Merrington, C.L., King, L.A., Posse, R.D., **1999**. Baculovirus expression systems. In: Higgins, S.J., Hames, B.D. (Eds.), Protein Expression: A Practical Approach. Oxford University Press, Oxford, pp. 101–127.

Merryweather, A.T., Weyer, U., Harris, M.P.G., Hirst, M., Booth, T., *et al.*, **1990**. Construction of genetically engineered baculovirus insecticides containing the

Bacillus thuringiensis subsp. *kurstaki* HD-73 delta endotoxin. *J. Gen. Virol. 71*, 1535–1544.

Meyer, A., Hanzmannn, E., Schneiderman, H.A., Gilbert, L.I., Boyette, M., **1970**. The isolation and identification of the two juvenile hormones from the *Cecropia* silk moth. *Arch. Biochem. Biophys. 137*, 190–213.

Milks, M.L., Leptich, M.K., Theilmann, D.A., **2001**. Recombinant and wild-type nucleopolyhedroviruses are equally fit in mixed infections. *Environ. Entomol. 30*, 972–981.

Miller, L.K., **1988**. Baculoviruses as gene expression vectors. *Annu. Rev. Microbiol. 42*, 177–199.

Miller, L.K., **1995**. Genetically engineered insect virus pesticides: present and future. *J. Invertebr. Pathol. 65*, 211–216.

Miller, L.K., **1997**. The Baculoviruses. Plenum, New York.

Miller, L.K., Kaiser, W.J., Seshagiri, S., **1998**. Baculovirus regulation of apoptosis. *Semin. Virol. 8*, 445–452.

Miller, L.K., Lu, A., **1997**. The molecular basis of baculovirus host range. In: Miller, L.K. (Ed.), Plenum, New York, pp. 217–235.

Morris, T.D., Miller, L.K., **1992**. Promoter influence on baculovirus-mediated gene expression in permissive and nonpermissive insect cell lines. *J. Virol. 66*, 7397–7405.

Morris, T.D., Miller, L.K., **1993**. Characterization of productive and non-productive AcMNPV infection in selected insect cell lines. *Virology 197*, 339–348.

Moscardi, F., **1999**. Assessment of the application of baculoviruses for control of Lepidoptera. *Annu. Rev. Entomol. 44*, 257–289.

Moskowitz, H., Herrmann, R., Jones, A.D., Hammock, B.D., **1998**. A depressant insect-selective toxin analog from the venom of the scorpion *Leiurus quinquestriatus hebraeus*, purification and structure/function characterization. *Eur. J. Biochem. 254*, 44–49.

Nakagawa, Y., Lee, Y.M., Lehmberg, E., Herrmann, R., Herrmann, R., *et al.*, **1997**. Anti-insect toxin 5 (AaIT5) from *Androctonus australis*. *Eur. J. Biochem. 246*, 496–501.

Nakagawa, Y., Sadilek, M., Lehmberg, E., Herrmann, R., Herrmann, R., *et al.*, **1998**. Rapid purification and molecular modeling of AaIT peptides from venom of *Androctonus australis*. *Arch. Insect Biochem. Physiol. 38*, 53–55.

Nakai, M., Goto, C., Shiotsuki, T., Kunimi, Y., **2002**. Granulovirus prevents pupation and retards development of *Adoxophyes honmai* larvae. *Physiol. Entomol. 27*, 157–164.

Nijhout, H.F., **1994**. Insect Hormones. Princeton University Press, Princeton, NJ.

Ohba, M., Tanada, Y., **1984**. *In vitro* enhancement of nuclear polyhedrosis virus infection by the synergistic factor of a granulosis virus of the armyworm, *Pseudaletia unipuncta* (Lepidoptera, Noctuidae). *Ann. Virol. 135*, 167–176.

Ohkawa, T., Majima, K., Maeda, S., **1994**. A cysteine protease encoded by the baculovirus *Bombyx mori* nuclear polyhedrosis virus. *J. Virol. 68*, 6619–6625.

Ohno, S., Taniguchi, T., **1981**. Structure of a chromosomal gene for human interferon β. *Proc. Natl Acad. Sci. USA 78*, 5305–5309.

Ooi, B.G., Rankin, C., Miller, L.K., **1989**. Downstream sequences augment transcription from the essential initiation site of a baculovirus polyhedrin gene. *J. Mol. Biol. 210*, 721–736.

O'Reilly, D.R., **1995**. Baculovirus-encoded ecdysteroid UDP-glucosyltransferases. *Insect Biochem. Mol. Biol. 25*, 541–550.

O'Reilly, D.R., **1997**. Auxilary genes of baculoviruses. In: Miller, L.K. (Ed.), The Baculoviruses. Plenum, New York, pp. 267–300.

O'Reilly, D.R., Kelly, T.J., Masler, E.P., Thyagaraja, B.S., Robson, R.M., *et al.*, **1995**. Overexpression of *Bombyx mori* prothoracicotropic hormone using baculovirus vectors. *Insect Biochem. Mol. Biol. 25*, 475–485.

O'Reilly, D.R., Miller, L.K., **1989**. A baculovirus blocks insect molting by producing ecdysteroid UDP-glucosyltransferase. *Science 245*, 1110–1112.

O'Reilly, D.R., Miller, L.K., **1991**. Improvement of a baculovirus pesticide by deletion of the *egt* gene. *Bio/Technology 9*, 1086–1089.

O'Reilly, D.R., Miller, L.K., Luckow, V.A., **1992**. Baculovirus Expression Vectors: A Laboratory Manual. W.H. Freeman, New York.

Pang, Y., Yu, J.X., Wang, L.H., Hu, X.H., Bao, W.D., *et al.*, **2001**. Sequence analysis of the *Spodoptera litura* multicapsid nucleopolyhedrovirus genome. *Virology 287*, 391–404.

Pearson, M.N., Bjornson, R.M., Ahrens, C., Rohrmann, G.F., **1993**. Identification and characterization of a putative origin of DNA-replication in the genome of a baculovirus pathogenic for *Orgyia pseudotsugata*. *Virology 197*, 715–725.

Pearson, M.N., Groten, C., Rohrmann, G.F., **2000**. Identification of the *Lymantria dispar* nucleopolyhedrovirus envelope fusion protein provides evidence for a phylogenetic division of the Baculoviridae. *J. Virol. 74*, 6126–6131.

Pearson, M.N., Russell, R.L., Rohrmann, G.F., **2001**. Characterization of a baculovirus-encoded protein that is associated with infected-cell membranes and budded virions. *Virology 291*, 22–31.

Pelengaris, S., Khan, M., **2003**. The many faces of c-MYC. *Arch. Biochem. Biophys. 416*, 129–136.

Peng, F., Fuxa, J.R., Richter, A.R., Johnson, S.J., **1999**. Effects of heat-sensitive agents, soil type, moisture, and leaf surface on persistence of *Anticarsia gemmatalis* (Lepidoptera: Noctuidae) nucleopolyhedrovirus. *Environ. Entomol. 28*, 330–338.

Petrik, D.T., Iseli, A., Montelone, B.A., Van Etten, J.L., Clem, R.J., **2003**. Improving baculovirus resistance to UV inactivation: increased virulence resulting from expression of a DNA repair enzyme. *J. Invertebr. Pathol. 82*, 50–56.

Pinedo, F.J.R., Moscardi, F., Luque, T., Olszewski, J.A., Ribeiro, B.M., **2003**. Inactivation of the ecdysteroid UDP-glucosyltransferase (*egt*) gene of *Anticarsia* *gemmatalis* nucleopolyhedrovirus (AgMNPV) improves its virulence towards its insect host. *Biol. Control 27*, 336–344.

Ponnuvel, K.M., Nakazawa, H., Furukawa, S., Asaoka, A., Ishibashi, J., et al., 2003. A lipase isolated from the silkworm *Bombyx mori* shows antiviral activity against nucleopolyhedrovirus. *J. Virol. 77*, 10725–10729.

Popham, H.J.R., Bischoff, D.S., Slavicek, J.M., 2001. Both *Lymantria dispar* nucleopolyhedrovirus enhancin genes contribute to viral potency. *J. Virol. 75*, 8639–8648.

Popham, H.J.R., Li, Y.H., Miller, L.K., 1997. Genetic improvement of *Helicoverpa zea* nuclear polyhedrosis virus as a biopesticide. *Biol. Control 10*, 83–91.

Popham, H.J.R., Pellock, B.J., Robson, M., Dierks, P.M., Miller, L.K., 1998a. Characterization of a variant of *Autographa californica* nuclear polyhedrosis virus with a nonfunctional ORF 603. *Biol. Control 12*, 223–230.

Popham, H.J.R., Prikhod'ko, G.G., Felcetto, T.J., Ostlind, D.A., Warmke, J.W., et al., 1998b. Effect of delta-methrin treatment on lepidopteran larvae infected with baculoviruses expressing insect-selective toxins mu-Aga-IV, As II, or Sh 1. *Biol. Control 12*, 79–87.

Possani, L.D., Becerril, B., Delepierre, M., Tytgat, J., 1999. Scorpion toxins specific for Na$^+$-channels. *Eur. J. Biochem. 264*, 287–300.

Prikhod'ko, E.A., Lu, A., Wilson, J.A., Miller, L.K., 1999. *In vivo* and *in vitro* analysis of baculovirus ie-2 mutants. *J. Virol. 73*, 2460–2468.

Prikhod'ko, G.G., Popham, H.J.R., Felcetto, T.J., Ostlind, D.A., Warren, V.A., et al., 1998. Effects of simultaneous expression of two sodium channel toxin genes on the properties of baculoviruses as biopesticides. *Biol. Control 12*, 66–78.

Prikhod'ko, G.G., Robson, M., Warmke, J.W., Cohen, C.J., Smith, M.M., et al., 1996. Properties of three baculoviruses expressing genes that encode insect-selective toxins: mu-Aga-IV, As II, and SH I. *Biol. Control 7*, 236–244.

Regev, A., Rivkin, H., Inceoglu, B., Gershburg, E., Hammock, B.D., et al., 2003. Further enhancement of baculovirus insecticidal efficacy with scorpion toxins that interact cooperatively. *FEBS Lett. 537*, 106–110.

Ribeiro, B.M., Crook, N.E., 1993. Expression of full-length and truncated forms of crystal protein genes from *Bacillus thuringiensis* subsp. *kurstaki* in a baculovirus and pathogenicity of the recombinant viruses. *J. Invertebr. Pathol. 62*, 121–130.

Ribeiro, B.M., Crook, N.E., 1998. Construction of occluded recombinant baculoviruses containing the full-length *cry1Ab* and *cry1Ac* genes from *Bacillus thuringiensis*. *Brazilian J. Med. Biol. Res. 31*, 763–769.

Richards, A.G., Richards, P.A., 1977. The peritrophic membrane of insects. *Annu. Rev. Entomol. 22*, 219–240.

Richardson, C.D., 1995. Baculovirus Expression Protocols. Humana Press, Totowa.

Riddiford, L.M., 1994. Cellular and molecular actions of juvenile hormone I. General considerations and pre-metamorphic actions. In: Evans, P.D. (Ed.), Advances in Insect Physiology. Academic Press, San Diego, pp. 213–274.

Riegel, C.I., Lannerherrera, C., Slavicek, J.M., 1994. Identification and characterization of the ecdysteroid UDP-glucosyltransferase gene of the *Lymantria dispar* multinucleocapsid nuclear polyhedrosis virus. *J. Gen. Virol. 75*, 829–838.

Rodrigues, J.C.M., De Souza, M.L., O'Reilly, D., Velloso, L.M., Pinedo, F.J.R., et al., 2001. Characterization of the ecdysteroid UDP-glucosyltransferase (*egt*) gene of *Anticarsia gemmatalis* nucleopolyhedrovirus. *Virus Genes 22*, 103–112.

Roe, R.M., Venkatesh, K., 1990. Metabolism of juvenile hormones: degradation and titer regulation. In: Gupta, A.P. (Ed.), Morphogenetic Hormones of Arthropods, vol. I. Rutgers University Press, New Brunswick, pp. 126–179.

Roelvink, P.W., Corsaro, B.G., Granados, R.R., 1995. Characterization of the *Helicoverpa armigera* and *Pseudaletia unipuncta* granulovirus enhancin genes. *J. Gen. Virol. 76*, 2693–2705.

Roelvink, P.W., VanMeer, M.M.M., deKort, C.A.D., Possee, R.D., Hammock, B.D., et al., 1992. Dissimilar expression of *Autographa californica* multiple nucleo-capsid nuclear polyhedrosis virus polyhedrin and p10 genes. *J. Gen. Virol. 73*, 1481–1489.

Roller, H., Dahm, K.H., Sweeley, C.C., Trost, B.M., 1967. Die Struktur des Juvenilhormon. *Angew. Chem. 79*, 190–191.

Ruddon, R.W., Bedows, E., 1997. Assisted protein folding. *J. Biol. Chem. 272*, 3125–3128.

Ryerse, J.S., 1998. Basal laminae. In: Harrison, F.W., Locke, M. (Eds.), Insecta, vol. 11A. Wiley-Liss, New York, pp. 3–16.

Sandig, V., Hofmann, C., Steinert, S., Jennings, G., Schlag, P., et al., 1996. Gene transfer into hepatocytes and human liver tissue by baculovirus vectors. *Hum. Gene Ther. 7*, 1937–1945.

Sarkis, C., Serguera, C., Petres, S., Buchet, D., Ridet, J.L., et al., 2000. Efficient transduction of neural cells *in vitro* and *in vivo* by a baculovirus-derived vector. *Proc. Natl Acad. Sci. USA 97*, 14638–14643.

Sato, K., Komoto, M., Sato, T., Enei, H., Kobayashi, M.M., et al., 1997. Baculovirus-mediated expression of a gene for trehalase of the mealworm beetle, *Tenebrio molitor*, in insect cells, SF-9, and larvae of the cabbage armyworm, *Mamestra brassicae*. *Insect Biochem. Mol. Biol. 27*, 1007–1016.

Schnepf, E., Crickmore, N., Van Rie, J., Lereclus, D., Baum, J., et al., 1998. *Bacillus thuringiensis* and its pesticidal crystal proteins. *Microbiol. Mol. Biol. Rev. 62*, 775.

Shanmugavelu, M., Baytan, A.R., Chesnut, J.D., Bonning, B.C., 2000. A novel protein that binds juvenile hormone esterase in fat body tissue and pericardial cells of the tobacco hornworm *Manduca sexta* L. *J. Biol. Chem. 275*, 1802–1806.

Shanmugavelu, M., Porubleva, L., Chitnis, P., Bonning, B.C., 2001. Ligand blot analysis of juvenile hormone esterase binding proteins in *Manduca sexta* L. *Insect Biochem. Mol. Biol. 31*, 51–56.

Shapiro, D.I., Fuxa, J.R., Braymer, H.D., Pashley, D.P., **1991**. DNA restriction polymorphism in wild isolates of *Spodoptera frugiperda* nuclear polyhedrosis virus. *J. Invert. Pathol. 58*, 96–105.

Shinoda, T., Kobayashi, J., Matsui, M., Chinzei, Y., **2001**. Cloning and functional expression of a chitinase cDNA from the common cutworm, *Spodoptera litura*, using a recombinant baculovirus lacking the virus-encoded chitinase gene. *Insect Biochem. Mol. Biol. 31*, 521–532.

Shoji, I., Aizaki, H., Tani, H., Ishii, K., Chiba, T., et al., **1997**. Efficient gene transfer into various mammalian cells, including non-hepatic cells, by baculovirus vectors. *J. Gen. Virol. 78*, 2657–2664.

Slack, J.M., Kuzio, J., Faulkner, P., **1995**. Characterization of V-Cath, a cathepsin L-like proteinase expressed by the baculovirus *Autographa californica* multiple nuclear polyhedrosis virus. *J. Gen. Virol. 76*, 1091–1098.

Slavicek, J.M., Popham, H.J.R., Riegel, C.I., **1999**. Deletion of the *Lymantria dispar* multicapsid nucleopolyhedrovirus ecdysteroid UDP-glucosyl transferase gene enhances viral killing speed in the last instar of the gypsy moth. *Biol. Control 16*, 91–103.

Smith, C.R., Heinz, K.M., Sansone, C.G., Flexner, J.L., **2000a**. Impact of recombinant baculovirus applications on target heliothines and nontarget predators in cotton. *Biol. Control 19*, 201–214.

Smith, C.R., Heinz, K.M., Sansone, C.G., Flexner, J.L., **2000b**. Impact of recombinant baculovirus field applications on a nontarget heliothine parasitoid, *Microplitis croceipes* (Hymenoptera: Braconidae). *J. Econ. Entomol. 93*, 1109–1117.

Smith, I., Goodale, C., **1998**. Sequence and *in vivo* transcription of *Lacanobia oleracea* granulovirus *egt*. *J. Gen. Virol. 79*, 405–413.

Snyder, M., Hunkapiller, M., Yuen, D., Silvert, D., Fristrom, J., et al., **1982**. Cuticle protein genes of *Drosophila* – structure, organization and evolution of 4 clustered genes. *Cell 29*, 1027–1040.

Soderlund, D.M., Clark, J.M., Sheets, L.P., Mullin, L.S., Piccirillo, V.J., et al., **2002**. Mechanisms of pyrethroid neurotoxicity: implications for cumulative risk assessment. *Toxicology 171*, 3–59.

Stewart, L.M., Hirst, M., Ferber, M.L., Merryweather, A.T., Cayley, P.J., et al., **1991**. Construction of an improved baculovirus insecticide containing an insect-specific toxin gene. *Nature 352*, 85–88.

Summers, M.D., Smith, G.E., **1987**. A Manual of Methods for Baculovirus Vectors and Insect Cell Culture Procedures. Texas Agricultural Experiment Station Bulletin No. 1555.

Sun, X., Wang, H., Sun, X., Chen, X., Peng, C., et al., **2004**. Biological activity and field efficacy of a genetically modified *Helicoverpa armigera* single-nucleocapsid nucleopolyhedrovirus expressing an insect-selective toxin from a chimeric promoter. *Biol. Control 29*, 124–137.

Sun, X.L., Chen, X.W., Zhang, Z.X., Wang, H.L., Bianchi, J.J.A., et al., **2002**. Bollworm responses to release of genetically modified *Helicoverpa armigera* nucleopolyhedroviruses in cotton. *J. Invertebr. Pathol. 81*, 63–69.

Taha, A., Nour-el-Din, A., Croizier, L., Ferber, M.C., Croizier, G., **2000**. Comparative analysis of the granulin regions of the *Phthorimaea operculella* and *Spodoptera littoralis* granuloviruses. *Virus Genes 21*, 147–155.

Tanada, Y., **1959**. Synergism between two viruses of the armyworm *Pseudaletia unipuncta* (Haworth) (Lepidoptera, Noctuidae). *J. Insect Pathol. 6*, 378–380.

Tanada, Y., Hukuhara, T., **1968**. A nonsynergistic strain of a granulosis virus of the armyworm, *Pseudaletia unipuncta*. *J. Invertebr. Pathol. 12*, 263–268.

Tanada, Y., Kaya, H.K., **1993**. Insect Pathology. Academic Press, San Diego.

Taniai, K., Inceoglu, A.B., Hammock, B.D., **2002**. Expression efficiency of a scorpion neurotoxin, AaHIT, using baculovirus in insect cells. *Appl. Entomol. Zool. 37*, 225–232.

Tellam, R.L., **1996**. The peritrophic membrane. In: Lehane, M.J., Billingsley, B.F. (Eds.), Biology of the Insect Midgut. Chapman and Hall, New York, pp. 86–114.

Tessier, D.C., Thomas, D.Y., Khouri, H.E., Laliberte, F., Vernet, T., **1991**. Enhanced secretion from insect cells of a foreign protein fused to the honeybee melittin signal peptide. *Gene 98* (2), 177–183.

Thiem, S.M., **1997**. Prospects for altering host range for baculovirus bioinsecticides. *Curr. Opin. Biotech. 8*, 317–322.

Thiem, S.M., Du, X.L., Quentin, M.E., Berner, M.M., **1996**. Identification of a baculovirus gene that promotes *Autographa californica* nuclear polyhedrosis virus replication in a nonpermissive insect cell line. *J. Virol. 70*, 2221–2229.

Thomas, B.A., Church, W.B., Lane, T.R., Hammock, B.D., **1999**. Homology model of juvenile hormone esterase from the crop pest, *Heliothis virescens*. *Proteins: Struct. Funct. Genet. 34*, 184–196.

Thomas, E.D., Heimpel, A.M., Adams, J.R., **1974**. Determination of the active nuclear polyhedrosis virus content of untreated cabbages. *Environ. Entomol. 3*, 908–910.

Tomalski, M.D., Bruce, W.A., Travis, J., Blum, M.S., **1988**. Preliminary characterization of toxins from the straw itch mite, *Pyemotes tritici*, which induce paralysis in the larvae of a moth. *Toxicon 26*, 127–132.

Tomalski, M.D., Hutchinson, K., Todd, J., Miller, L.K., **1993**. Identification and characterization of Tox21a – a mite cDNA encoding a paralytic neurotoxin related to Txp-I. *Toxicon 31*, 319–326.

Tomalski, M.D., Kutney, R., Bruce, W.A., Brown, M.R., Blum, M.S., et al., **1989**. Purification and characterization of insect toxins derived from the mite, *Pyemotes tritici*. *Toxicon 27*, 1151–1167.

Tomalski, M.D., Miller, L.D., **1992**. Expression of a paralytic neurotoxin gene to improve insect baculoviruses as biopesticides. *Bio/Technology 10*, 545–549.

Tomalski, M.D., Miller, L.K., **1991**. Insect paralysis by baculovirus-mediated expression of a mite neurotoxin gene. *Nature 352*, 82–85.

Treacy, M.F., All, J.N., Ghidiu, G.M., **1997**. Effect of ecdysteroid UDP-glucosyltransferase gene deletion on efficacy of a baculovirus against *Heliothis virescens* and *Trichoplusia ni* (Lepidoptera: Noctuidae). *J. Econ. Entomol. 90*, 1207–1214.

Treacy, M.F., Rensner, P.E., All, J.N., **2000**. Comparative insecticidal properties of two nucleopolyhedrovirus vectors encoding a similar toxin gene chimer. *J. Econ. Entomol. 93*, 1096–1104.

Trombetta, E.S., Parodi, A.J., **2003**. Quality control and protein folding in the secretory pathway. *Annu. Rev. Cell Devel. Biol. 19*, 649–676.

Truman, J.W., Riddiford, L.M., **2002**. Endocrine insights into the evolution of metamorphosis in insects. *Annu. Rev. Entomol. 47*, 467–500.

Tumilasci, V.F., Leal, E., Marinho, P., Zanotto, A., Luque, T., *et al.*, **2003**. Sequence analysis of a 5.1 kbp region of the *Spodoptera frugiperda* multicapsid nucleopolyhedrovirus genome that comprises a functional ecdysteroid UDP-glucosyltransferase (*egt*) gene. *Virus Genes 27*, 137–144.

Vail, P.V., Hostertter, D.L., Hoffmann, D.F., **1999**. Development of the multi-nucleocapsid nucleopolyhedroviruses (MNPVs) infectious to loopers (Lepidoptera: Noctuidae: Plusiinae) as microbial control agents. *Integr. Pest Mgt Rev. 4*, 231–257.

Vail, P.V., Jay, D.L., Hunter, D.K., **1973**. Infectivity of a nuclear polyhedrosis virus from the alfalfa looper, *Autographa californica*, after passage through alternate hosts. *J. Invertebr. Pathol. 21*, 16–20.

Vakharia, V.N., Raina, A.K., Kingan, T.G., Kempe, T.G., **1995**. Synthetic pheromone biosynthesis activating neuropeptide gene expressed in a baculovirus expression system. *Insect Biochem. Mol. Biol. 25*, 583–589.

Valdes, V.J., Sampieri, A., Sepulveda, J., Vaca, L., **2003**. Using double-stranded RNA to prevent *in vitro* and *in vivo* viral infections by recombinant baculovirus. *J. Biol. Chem. 278*, 19317–19324.

van Beek, N., Lu, A., Presnail, J., Davis, D., Greenamoyer, C., *et al.*, **2003**. Effect of signal sequence and promoter on the speed of action of a genetically modified *Autographa californica* nucleopolyhedrovirus expressing the scorpion toxin LqhIT2. *Biol. Control 27*, 53–64.

van Meer, M.M.M., Bonning, B.C., Ward, V.K., Vlak, J.M., Hammock, B.D., **2000**. Recombinant, catalytically inactive juvenile hormone esterase enhances efficacy of baculovirus insecticides. *Biol. Control 19*, 191–199.

Vasconcelos, S.D., Cory, J.S., Wilson, K.R., Sait, S.M., Hails, R.S., **1996**. Modified behaviour in baculovirus-infected lepidopteran larvae and its impact on the spatial distribution of inoculum. *Biol. Control 7*, 299–306.

Vlak, J.M., Schouten, A., Usmany, M., Belsham, G.J., Klingeroode, E.C., *et al.*, **1990**. Expression of cauliflower mosaic virus gene I using a baculovirus vector based upon the p10 gene and a novel selection method. *Virology 179*, 312–320.

Wagle, M., Jesuthasan, S., **2003**. Baculovirus-mediated gene expression in zebrafish. *Mar. Biotech. 5*, 58–63.

Wang, P., Granados, R.R., **1997**. An intestinal mucin is the target substrate for a baculovirus enhancin. *Proc. Natl Acad. Sci. USA 94*, 6977–6982.

Wang, P., Hammer, D.A., Granados, R.R., **1994**. Interaction of *Trichoplusia ni* granulosis virus-encoded enhancin with the midgut epithelium and peritrophic membrane of 4 lepidopteran insects. *J. Gen. Virol. 75*, 1961–1967.

Wang, X.Z., Ooi, B.G., Miller, L.K., **1991**. Baculovirus vectors for multiple gene expression and for occluded virus production. *Gene 100*, 131–137.

Washburn, J.O., Chan, E.Y., Volkman, L.E., Aumiller, J.J., Jarvis, D.L., **2003**. Early synthesis of budded virus envelope fusion protein GP64 enhances *Autographa californica* multicapsid nucleopolyhedrovirus virulence in orally infected *Heliothis virescens*. *J. Virol. 77*, 280–290.

Westenberg, M., Wang, H.L., Ijkel, W.F.J., Goldbach, R.W., Vlak, J.M., *et al.*, **2002**. Furin is involved in baculovirus envelope fusion protein activation. *J. Virol. 76*, 178–184.

Whitford, M., Stewart, S., Kuzio, J., Faulkner, P., **1989**. Identification and sequence analysis of a gene encoding gp67, an abundant envelope glycoprotein of the baculovirus *Autographa californica* nuclear polyhedrosis virus. *J. Virol. 63*, 1393–1399.

Wigglesworth, V.B., **1935**. Functions of the *corpus allatum* of insects. *Nature 136*, 338.

Wigglesworth, V.B., **1936**. The function of the *corpus allatum* in the growth and reproduction of *Rhodnius prolixus* (Hemiptera). *Q. J. Microsc. Sci. 79*, 91–121.

Wilson, K.R., O'Reilly, D.R., Hails, R.S., Cory, J.S., **2000**. Age-related effects of the *Autographa californica* multiple nucleopolyhedrovirus *egt* gene in the cabbage looper (*Trichoplusia ni*). *Biol. Control 19*, 57–63.

Woo, S.D., Kim, W.J., Kim, H.S., Jin, B.R., Lee, Y.H., *et al.*, **1998**. The morphology of the polyhedra of a host range-expanded recombinant baculovirus and its parents. *Arch. Virol. 143*, 1209–1214.

Wood, H.A., **1996**. Genetically enhanced baculovirus insecticides. In: Gunasekaran, M., Weber, D.J. (Eds.), Molecular Biology of the Biological Control of Pests and Diseases of Plants. CRC Press, Boca Raton, pp. 91–104.

Wood, H.A., Hughes, P.R., Shelton, A., **1994**. Field studies of the coocclusion strategy with a genetically altered isolate of the *Autographa californica* nuclear polyhedrosis virus. *Environ. Entomol. 23*, 211–219.

Wormleaton, S.L., Winstanley, D., **2001**. Phylogenetic analysis of conserved genes within the ecdysteroid UDP-glucosyltransferase gene region of the slow-killing *Adoxophyes orana* granulovirus. *J. Gen. Virol. 82*, 2295–2305.

Wudayagiri, R., Inceoglu, B., Herrmann, R., Derbel, M., Choudary, P.V., *et al.*, **2001**. Isolation and characterization of a novel lepidopteran-selective toxin from the venom of South Indian red scorpion, *Mesobuthus tamulus*. *BMC Biochem. 2*, 16–23.

Xue, D., Horvitz, H.R., **1995**. Inhibition of the *Caenorhabditis elegans* cell-death protease CED-3 by a CED-3 cleavage site in baculovirus p35 protein. *Nature 377*, 248–251.

Yamamoto, T., Tanada, Y., **1978a**. Phospholipid, an enhancing component in synergistic factor of a granulosis virus of armyworm, *Pseudaletia unipuncta. J. Invertebr. Pathol. 31*, 48–56.

Yamamoto, T., Tanada, Y., **1978b**. Protein components of 2 strains of granulosis virus of armyworm, *Pseudaletia unipuncta* (Lepidoptera, Noctuidae). *J. Invertebr. Pathol. 32*, 158–170.

Yurchenco, P.D., O'Rear, J., **1993**. Supramolecular organization of basement membranes. In: Rohrback, D.H., Timpl, R. (Eds.), Molecular and Cellular Aspects of Basement Membranes. Academic Press, New York, pp. 19–47.

Zheng, Y., Zheng, S., Cheng, X., Ladd, T., Lingohr, E.J., *et al.*, **2002**. A molt-associated chitinase cDNA from the spruce budworm, *Choristoneura fumiferana. Insect Biochem. Mol. Biol. 32*, 1813–1823.

Zlotkin, E., **1991**. Venom neurotoxins – models for selective insecticides. *Phytoparasitica 19*, 177–182.

Zlotkin, E., Fishman, Y., Elazar, M., **2000**. AaIT: from neurotoxin to insecticide. *Biochimie 82*, 869–881.

Zlotkin, E., Fraenkel, G., Miranda, F., Lissitzky, S., **1971**. The effect of scorpion venom on blow fly larvae; a new method for the evaluation of scorpion venom potency. *Toxicon 8*, 1–8.

Zlotkin, E., Gurevitz, M., Fowler, E., Adams, M.E., **1993**. Depressant insect selective neurotoxins from scorpion venom – chemistry, action, and gene cloning. *Arch. Insect Biochem. Physiol. 22*, 55–73.

Zlotkin, E., Kadouri, D., Gordon, D., Pelhate, M., Martin, M.F., *et al.*, **1985**. An excitatory and a depressant insect toxin from scorpion venom both affect sodium conductance and possess a common binding site. *Arch. Biochem. Biophys. 240*, 877–887.

Zlotkin, E., Miranda, F., Rochat, H., **1978**. Venoms of buthinae. In: Bettini, S. (Ed.), Arthropod Venoms. Springer, New York, pp. 317–369.

Zlotkin, E., Moskowitz, H., Herrmann, R., Pelhate, M., Gordon, D., **1995**. Insect sodium channel as the target for insect-selective neurotoxins from scorpion venom. *ACS Symp. Ser. 591*, 56–85.

Zuidema, D., Schouten, A., Usmany, M., Maule, A.J., Belsham, G.J., *et al.*, **1990**. Expression of cauliflower mosaic virus gene I in insect cells using a novel polyhedrin-based baculovirus expression vector. *J. Gen. Virol. 71*, 2201–2210.

A10 Addendum: Genetically Modified Baculoviruses for Pest Insect Control

**S G Kamita, K-D Kang, A B Inceoglu, and
B D Hammock**, University of California, Davis, USA

© 2010 Elsevier B.V. All Rights Reserved

In our original comprehensive review 5 years ago, we extolled the virtues and cautioned against the limitations of genetically modified (GM) baculoviruses for use in pest insect control. At that time we concluded that GM baculoviruses (1) show potency that is comparable to traditional chemical insecticides, (2) pose little or no risk to humans, other nontarget species, and the environment, and (3) can make a near-immediate positive impact on sustainable pest insect control programs. Scientific literature during the past 5 years continues to support these conclusions. Our enthusiastic support of GM baculoviruses as potent and safe biopesticides that can complement or even synergize traditional chemical insecticides and biocontrol strategies has not changed. The field, however, has been slow to embrace our enthusiasm. The high cost of regulatory barriers for GM products coupled with the efficacy and safety of major GM plants as well as new generation insecticides for the control of key noctuid pests makes near-term commercial development of GM baculoviruses unlikely in the United States. In addition, in some other countries, including ones where there is a clear economic advantage of GM baculoviruses, there is a general unease with the use of GM products. A positive trend has been the increased use of natural (i.e., wild type) baculoviruses for augmented biocontrol. This expanded use of natural baculoviruses may serve as a positive step in the acceptance of GM baculovirus as green or biopesticides.

During the past 5 years, several interesting and insightful reviews covering historical and applied aspects of biological control with insect viruses and other microbials were published. Arif (2005), Inceoglu et al. (2006), and Szewczyk et al. (2006) review historical aspects and current developments of natural insect viruses and GM baculovirus for pest insect control. Summers (2006), a pioneer in the development of GM baculoviruses for protein expression, reviews how advances in the development of baculoviruses as protein expression vectors were key in the development of GM baculovirus

biopesticides. Hynes and Boyetchko (2006), Lord (2005), Hajek et al. (2007), Whetstone and Hammock (2007), and Rosell et al. (2008) review formulation strategies and the relative importance of baculoviruses (both natural and GM) with respect to other microbials and the role of baculoviruses as delivery systems of insecticidal agents. Gelernter (2007) and Kunimi (2007) review the current status of the use of baculoviruses and other microbials in Asia.

In our original review, the genetic modifications of the baculovirus genome were categorized into three major approaches (1) insertion of hormone and enzyme genes, (2) insertion of insect-selective toxin genes, and (3) genome modifications. When multiple approaches were combined into a single construct, the modifications resulted in at best a roughly 60% improvement in the speed of kill relative to the wild-type virus with reductions in feeding damage of roughly 70% (comparison of GM baculovirus- and mock-infected insects). During the past 5 years, various groups (see the following paragraphs) continued with single and technology-stacked approaches with similar improvements in efficacy. Although the efficacy of current GM baculoviruses is sufficient under many crop protection scenarios, we believe that further improvements (e.g., by improving toxin folding and/or expression, identifying alternative insect-selective toxins, modifying the virus backbone, improving formulation) can be generated, if necessary, by sustained efforts by private and public sectors so that this technology is more attractive for commercial agriculture.

During the past 5 years, Rajendra et al. (2006), Jinn et al., (2006), and Choi et al. (2008) continued to study improved insecticidal efficacy by expressing lepidopteran-selective toxin genes under various promoters. Shim et al. (2009) developed a "stacked" construct that targets the host at three levels: the gut, nervous system, and systemic baculovirus infection. Constructs designed to express cathepsin B-like (Hong-Lian et al., 2008) and L-like (Li et al., 2008; Sun et al., 2009) proteases have also been tested in the laboratory and field. The cathepsin L-like cysteine protease expressing construct (HearNPV-cathL) of Sun *et al.* (2009) was designed to protect cotton against larval cotton bollworm, a species that has developed resistance to chemical and/or *Bacillus thuringiensis* insecticides. Sun *et al.* conducted field trials with this construct in 2004 and 2005 in China and found that the efficacy of HearNPV-cathL is similar to that of pyrethroid insecticide (λ-Cyhalothrin) treatment in terms of cotton boll protection against the bollworm. In addition, they found that the density of beneficial insects, such as lady beetles, was similar between baculovirus-treated and untreated plots and higher than that found in pyrethroid insecticide-treated plots in the 2005 field trial. Applications for commercial sale of GM baculoviruses for the protection of cotton against *Helicoverpa* were submitted to the government in China but the government has yet to respond (Prof. Bryony Bonning, personal communications).

During the past 5 years, several primary studies investigated optimal and innovative production methodologies of natural and GM baculoviruses in both cultured insect cells (Salem and Maruniak, 2007; Micheloud *et al.*, 2009) and insect larvae (Lasa *et al.*, 2007a; van Beek and Davis, 2007). The importance of optical brighteners (Lasa *et al.*, 2007b; Ibargutxi *et al.*, 2008) and feeding stimulants (Lasa *et al.*, 2009) on baculovirus formulations were also investigated. Other studies evaluated use of baculoviruses for the protection of nonmajor food crops and in situations where traditional chemical insecticides are not acceptable (Prater *et al.*, 2006; Kunimi, 2007; Grzywacz *et al.*, 2008; Sciocco *et al.*, 2009). These studies continue to emphasize that natural and potentially GM baculoviruses are particularly valuable under conditions where traditional chemical insecticides have become ineffective, economically prohibitive, or have lost favor with the general public. Use of the baculovirus delayed early *39K* or very late *p10* promoters to drive foreign gene expression in mammalian CHO cells, an indicator of potential safety, has been further investigated by Regev *et al.* (2006). Studies have also further addressed the competitive fitness and within-host fitness of GM baculoviruses in which the endogenous *egt* gene has been deleted (Zwart *et al.*, 2009).

Use of GM baculoviruses for evaluating genes that are useful for insect control will certainly continue. For commercial development, though, sustained and coordinated public and/or private efforts such as those of the Sun and Hu groups at the Wuhan Institute of Virology in China are needed. The GM baculovirus construct developed by DuPont is another example of efficacy that can be obtained with a coordinated effort to optimize toxin, expression, virus, production, formulation, and other strategies (Dr. Nikolai van Beek, personal communications). Unfortunately, the DuPont virus and data on its evaluation are not in the public domain. Work of the past 5 years continues to show that both natural baculoviruses and GM baculoviruses hold clear and substantial benefits for crop protection. Advantages of green pesticides, in particular biopesticides that can be produced locally, are most obvious in developing countries where older and more dangerous pest control materials are often used and where high costs of both modern pesticides and GM plants are often prohibitive. Since many technologies for production and use of natural and GM baculoviruses are common, increased use of natural baculoviruses may open doors to the use of GM baculoviruses. GM baculoviruses remain a viable alternative to classical chemical pesticides and GM crops. These viruses should be utilized if alternative methods of pest insect control are required because of resistance, high cost, and changes in the regulatory environment or public opinion.

Acknowledgments

This work was funded in part by a grant #2007–35607–17830 from the USDA. We thank Drs. Bryony Bonning and Nikolai van Beek for insightful discussions.

References

Arif, B.M., **2005**. A brief journey with insect viruses with emphasis on baculoviruses. *J. Invertebr. Pathol 89*, 39–45.

Choi, J.Y., Wang, Y., Kim, Y.-S., Kang, J.N., Roh, J.Y., Woo, S.-D., Jin, B.R., Je, Y.H., **2008**. Insecticidal activities of recombinant *Autographa californica* nucleopolyhedrovirus containing a scorpion neurotoxin gene using promoters from *Cotesia plutellae* bracovirus. *J. Asia Pac. Entomol 11*, 155–159.

Gelernter, W.D., **2007**. Microbial control in Asia: a bellwether for the future? *J. Invertebr. Pathol 95*, 161–167.

Grzywacz, D., Mushobozi, W.L., Parnell, M., Jolliffe, F., Wilson, K., **2008**. Evaluation of *Spodoptera exempta* nucleopolyhedrovirus (SpexNPV) for the field control of African armyworm (*Spodoptera exempta*) in Tanzania. *Crop Prot 27*, 17–24.

Hajek, A.E., McManus, M.L., Delalibera, I., **2007**. A review of introductions of pathogens and nematodes for classical biological control of insects and mites. *Biol. Control 41*, 1–13.

Hong-Lian, S., Du-Juan, D., Jin-Dong, H., Jin-Xin, W., Xiao-Fan, Z., **2008**. Construction of the recombinant baculovirus AcMNPV with cathepsin B-like proteinase and its insecticidal activity against *Helicoverpa armigera*. *Pestic. Biochem. Physiol 91*, 141–146.

Hynes, R.K., Boyetchko, S.M., **2006**. Research initiatives in the art and science of biopesticide formulations. *Soil Biol. Biochem 38*, 845–849.

Ibargutxi, M.A., Munoz, D., Bernal, A., de Escudero, I.R., Caballero, P., **2008**. Effects of stilbene optical brighteners on the insecticidal activity of *Bacillus thuringiensis* and a single nucleopolyhedrovirus on *Helicoverpa armigera*. *Biol. Control 47*, 322–327.

Inceoglu, A.B., Kamita, S.G., Hammock, B.D., **2006**. Genetically modified baculoviruses: a historical overview and future outlook. *Adv. Virus Res 68*, 323–360.

Jinn, T.R., Tu, W.C., Lu, C.I., Tzen, J.T.C., **2006**. Enhancing insecticidal efficacy of baculovirus by early expressing an insect neurotoxin, LqhIT2, in infected *Trichoplusia ni* larvae. *Appl. Microbiol. Biotechnol 72*, 1247–1253.

Kunimi, Y., **2007**. Current status and prospects on microbial control in Japan. *J. Invertebr. Pathol 95*, 181–186.

Lasa, R., Caballero, P., Williams, T., **2007a**. Juvenile hormone analogs greatly increase the production of a nucleopolyhedrovirus. *Biol. Control 41*, 389–396.

Lasa, R., Ruiz-Portero, C., Alcazar, M.D., Belda, J.E., Caballero, P., Williams, T., **2007b**. Efficacy of optical brightener formulations of *Spodoptera exigua* multiple nucleopolyhedrovirus (SeMNPV) as a biological insecticide in greenhouses in southern Spain. *Biol. Control 40*, 89–96.

Lasa, R., Williams, T., Caballero, P., **2009**. The attractiveness of phagostimulant formulations of a nucleopolyhedrovirus-based insecticide depends on prior insect diet. *J. Pestic. Sci 82*, 247–250.

Li, H.R., Tang, H.L., Sivakumar, S., Philip, J., Harrison, R.L., Gatehouse, J.A., Bonning, B.C., **2008**. Insecticidal activity of a basement membrane-degrading protease against *Heliothis virescens* (Fabricius) and *Acyrthosiphon pisum* (Harris). *J. Insect Physiol 54*, 777–789.

Lord, J.C., **2005**. From Metchnikoff to Monsanto and beyond: the path of microbial control. *J. Invertebr. Pathol 89*, 19–29.

Micheloud, G.A., Gioria, V.V., Perez, G., Claus, J.D., **2009**. Production of occlusion bodies of *Anticarsia gemmatalis multiple nucleopolyhedrovirus* in serum-free suspension cultures of the saUFL-AG-286 cell line: influence of infection conditions and statistical optimization. *J. Virol. Methods 162*, 258–266.

Prater, C.A., Redmond, C.T., Barney, W., Bonning, B.C., Potter, D.A., **2006**. Microbial control of black cutworm (Lepidoptera: Noctuidae) in turfgrass using *Agrotis ipsilon* multiple nucleopolyhedrovirus. *J. Econ. Entomol 99*, 1129–1137.

Rajendra, W., Hackett, K.J., Buckley, E., Hammock, B.D., **2006**. Functional expression of lepidopteran-selective neurotoxin in baculovirus: potential for effective pest management. *Biochim. Biophys. Acta 1760*, 158–163.

Regev, A., Rivkin, H., Gurevitz, M., Chejanovsky, N., **2006**. New measures of insecticidal efficacy and safety obtained with the 39K promoter of a recombinant baculovirus. *FEBS Lett 580*, 6777–6782.

Rosell, G., Quero, C., Coll, J., Guerrero, A., **2008**. Biorational insecticides in pest management. *J. Pestic. Sci 33*, 103–121.

Salem, T.Z., Maruniak, J.E., **2007**. A universal transgene silencing approach in baculovirus-insect cell system. *J. Virol. Methods 145*, 1–8.

Sciocco, A., Bideshi, D.K., Johnson, J.J., Federici, B.A., **2009**. Nucleopolyhedrovirus from the western avocado leafroller, *Amorbia cuneana*: isolation and characterization of a potential viral control agent. *Biol. Control 49*, 154–159.

Shim, H.J., Choi, J.Y., Li, M.S., Wang, Y., Roh, J.Y., Woo, S.-D., Jin, B.R., Je, Y.H., **2009**. A novel recombinant baculovirus expressing insect neurotoxin and producing occlusion bodies that contain *Bacillus thuringiensis* Cry toxin. *J. Asia Pac. Entomol 12*, 217–220.

Summers, M.D., **2006**. Milestones leading to the genetic engineering of baculoviruses as expression vector systems and viral pesticides. *Adv. Virus Res 68*, 3–73.

Sun, X.L., Wu, D., Sun, X.C., Jin, L., Ma, Y., Bonning, B.C., Peng, H.Y., Hu, Z.H., **2009**. Impact of *Helicoverpa armigera* nucleopolyhedroviruses expressing a cathepsin L-like protease on target and nontarget insect species on cotton. *Biol. Control 49*, 77–83.

Szewczyk, B., Hoyos-Carvajal, L., Paluszek, M., Skrzecz, W., de Souza, M.L., **2006**. Baculoviruses – re-emerging biopesticides. *Biotechnol. Adv 24*, 143–160.

van Beek, N., Davis, D.C., **2007**. Baculovirus insecticide production in insect larvae. In: Murhammer, D.W. (Ed.), Methods in Molecular Biology: Baculovirus and Insect Cell Expression Protocols, vol. 338. Humana Press Inc, Totowa, JH, pp. 367–378.

Whetstone, P.A., Hammock, B.D., **2007**. Delivery methods for peptide and protein toxins in insect control. *Toxicon 49*, 576–596.

Zwart, M.P., van der Werf, W., van Oers, M.M., Hemerik, L., van Lent, J.M.V., de Visser, J.A.G.M., Vlak, J.M., Cory, J.S., **2009**. Mixed infections and the competitive fitness of faster-acting genetically modified viruses. *Evol. Appl 2*, 209–221.

11 Entomopathogenic Fungi and their Role in Regulation of Insect Populations

M S Goettel, Agriculture and Agri-Food Canada, Lethbridge, AB, Canada
J Eilenberg, The Royal Veterinary and Agricultural University, Frederiksberg C, Denmark
T Glare, AgResearch, Canterbury, New Zealand

© 2010, 2005 Elsevier B.V. All Rights Reserved

11.1. Introduction

Entomopathogenic fungi play an important role in the regulation of insect populations. There is a diverse array of fungal insect pathogenic species from within four different classes. Adaptations range from obligate pathogens of specific insect species to generalists capable of infecting many host species to species that are facultative pathogens. Fungal epizootics are common in some insect species, while others are rarely affected. The earliest studies with entomopathogenic fungi were carried out in the 1800s, and these concentrated on developing ways of managing diseases that were devastating the silkworm industry (Steinhaus, 1975). In fact, the germ theory was first demonstrated by Bassi (1835 as cited by Steinhaus, 1975) using silkworms and the muscardine fungus, which was later named *Beauveria bassiana* in his honor. Bassi's studies on the disease in silkworms enabled him to extrapolate his findings into the field of human diseases. The stimulus for the idea of using entomopathogens to manage pest insects came largely from the silkworm disease studies. However, early attempts in using fungi as microbial control agents of pest insects were soon overshadowed by the development of chemical insecticides. It was not until the late 1950s that attempts to use entomopathogenic fungi for insect pest management resurfaced. To date, there are many commercial products available worldwide, based on less than 10 fungal species

(Shah and Goettel, 1999; Copping, 2001), but the potential for the development of many more remains high.

In this article, an overview of entomopathogenic fungi and their current and potential role in the regulation of pest insect populations is provided, emphasizing as much as possible the developments over the last 20 years. Reviews dealing with entomopathogenic fungi and their development as microbial control agents include those of Ferron (1985), McCoy *et al.* (1988), Evans (1989), Ferron *et al.* (1991), Glare and Milner (1991), Roberts and Hajek (1992), Tanada and Kaya (1993), Hajek and St. Leger (1994), Boucias and Pendland (1998), Wraight and Carruthers (1999), and several chapters in Butt *et al.* (2001) and Upadhyay (2003). For methods and techniques used to study these pathogens, readers are referred to Butt and Goettel (2000), and the chapters in Lacey (1997) and Lacey and Kaya (2000).

11.2. Entomopathogenic Fungi

Insect pathogens have a long history of recognition despite the relatively recent understanding of microbial infections. Descriptions of entomopathogenic fungi can be found in drawings from several centuries ago. Japanese descriptions of silkworm infections by muscardine fungi (probably *Beauveria bassiana*) and *Cordyceps* spp. infections of a number of insects date from around the nineth century (Samson *et al.*, 1988). Perhaps because of their prominent and often brightly colored fruiting bodies, *Cordyceps* species feature in many of the earliest references and drawings of entomopathogenic fungi. Species of *Cordyceps* have also been of practical value in several countries. Maori, the native people of New Zealand, were known to use extracts of *Cordyceps robertsii* as a pigment for traditional face tattoos (moko), while the Chinese have used *Cordyceps* infected caterpillars as traditional medicines for many years. Other entomopathogenic fungi had been recorded by the late eighteenth and through the nineteenth century. Flies infected by entomophthoralean fungi, where the distinct halo formed by forcibly discharged primary conidia could clearly be seen with the naked eye, were, for example, observed by Goethe (1817–1822).

Classification of these interesting fungi began with *Cordyceps* and other obvious insect-infecting fungi. While Cooke (1892) listed four groups parasitic to insects (*Cordyceps*, Entomophthorales, Laboulbeniales, and opportunitistic fungi such as *Cladosporium* and *Penicillium*), interest in the diversity of these fungi did not really blossom until the 1970s. Currently, at least 90 genera and more than 700 species, representing all the major classifications of fungi, have been implicated in diseases of insects.

With the advent of molecular comparisons as a tool in classification, the distinction between protozoa and fungi has been revisited and revised. Fungal taxonomy is still in a state of flux. As presently delimited, the kingdom Fungi may constitute a monophyletic group that shares some characters with animals, such as chitinous structures, storage of glycogen, and mitochondrial UGA coding for tryptophan. The Chytridiomycetes possess the primitive character of a single, smooth, posteriorly inserted flagellum and are basal to the other fungi (Cavalier-Smith, 1987; Barr, 1992). It does not include the Oomycetes, which have been moved to the kingdom Chromista with some algae (Leipe *et al.*, 1994).

While some phylogenists have moved the classes between kingdoms, the entomopathogens remain within four major classes: the Oomycetes, Chytridiomycetes, Ascomycetes, and Zygomycetes. Some unusual insect pathogenic species are found in other fungal classes, but most can be found in the Ascomycetes and Zygomycetes. A distinction must be made between disease causing species that in many cases result in mortality of the insect and those that live on insects, usually without causing mortality. The latter group includes species of Laboulbeniomycetes, which can be obligate parasites on insects, but have little effect on the health of their hosts (Tanada and Kaya, 1993) and Basidiomycota such as *Septobasidium* and *Uredinella*.

One of the most confusing aspects of fungal taxonomy is the treatment of species only known by an asexual stage. Many such species have been classified in the Deuteromycota within the class Hyphomycetes, which is essentially a form-class. This has led to many species having both a name for the sexual (teleomorph) stage and the asexual (anamorph) stage with no link between the names. With the advent of molecular identification tools, it has been possible to link many anamorph and teleomorph stages, or to substantially predict the teleomorph genera for a species described only through an anamorph stage. Many common insect pathogenic species are currently classified in the Hyphomycetes, such as *B. bassiana* and *Metarhizium anisopliae*, but with modern technology, it can be demonstrated that their affinities are with the Ascomycete genus *Cordyceps*. Whether separate pleomorphic and anamorphic names are still required is a current topic of debate around the International Code of Botanical Nomenclature.

As molecular analysis has the ability to link anamorphic stages with appropriate teleomorphic genera, the use of a dual nomenclature system is probably superfluous, but may be hard to lose in the short term.

The main classes that contain entomopathogenic species are reviewed in the following sections. Examples of insects infected with entomopathogenic fungi are presented in **Figures 1–15**.

11.2.1. Phylum Oomycota

Oomycetes are characterized by cellulose containing coenocytic hyphae, biflagellate zoospores, and usually contain no chitin. Sexual reproduction can occur between gametangia (antheridia and oogonia) on the same or different hyphae. While a number of species are saprophytes and parasites of animals and plants, two genera contain species pathogenic to mosquito larvae. The best studied species is *Lagenidium giganteum* (Lagenidiales), a pathogen of mosquito larvae (Glare and Milner, 1991; Kerwin and Petersen, 1997). Other species of *Lagenidium* can cause infections in aquatic crustaceans, such as crabs (e.g., Hatai *et al.*, 2000).

Leptolegnia spp. (Saprolegniales) have also been reported to be pathogenic to mosquitoes, chironomids, and several other Diptera. These species are thought to infect via a secondary zoospore formed after encystment of the primary zoospore derived from sporangia (Zattau and McInnis, 1987).

11.2.2. Phylum Chytridiomycota

Chrytridiomycetes are characterized by cell walls containing chitin and no cellulose. Posteriorly uniflagellate zoospores and gametes settle and grow into a thallus, which becomes either a resting spore or coenocytic hyphae. This group is considered basal to the fungal branch under SSU rRNA phylogenetic comparisons. The Blastocladiales genus *Coelomomyces* contains most of the common entomopathogenic Chytridiomycetes. There are over 70 entomopathogenic species described in the *Coelomomyces* (Lucarotti *et al.*, 1985). These species have been described from Diptera (mainly mosquitoes) and Heteroptera. Several species have an obligate intermediate host, such as copepods. They are characterized by formation of both thick walled resistant sporangia and flagellate zoospores. Other entomopathogenic species are known from *Coelomycidium* (Blastocladiales) and *Myriophagus* (Chytridiales); the former is found on blackflies and mosquitoes, and the latter has been reported as a pathogen on dipterous pupae by Sparrow, 1939 and Karling, 1948 (in Samson *et al.*, 1988).

11.2.3. Phylum Zygomycota

Traditionally, the Zygomycota are separated on the basis of often nonseptate, multinuclear hyphae, and production of zygospores by copulation between gametangia. However, molecular analyses have not found the Zygomycota to be monophyletic

Figure 1 Larvae of the European wireworm, *Agriotus obscurus*, killed by *Metarhizium anisopliae*. The progression of the emergence of the fungus from the cadavers and subsequent development of fungal structures is seen from left to right, with the larva on the right completely covered by the green spores of the fungus. (Photo courtesy of T. Kabaluk.)

(Jensen *et al.*, 1998), and some groups, such as the Glomales, may be placed elsewhere eventually (e.g., Bruns *et al.*, 1993). *Basidiobolus ranarum* and other *Basidiobolus* species, traditionally considered as Zygomycetes, have now been placed in the Chytridiomycetes, despite being nonflagellate (Nagahama *et al.*, 1995).

Within the true Zygomycota, the class Trichomycetes contains species often associated with insects. While Sweeney (1981) described the species *Smittium morbosum* (Trichomycetes) as a pathogen of mosquitoes, most associations of the Trichomycetes are symbiotic or weakly parasitic rather than true pathogens (Beard and Adler, 2002; Cafaro, 2002). Some species of *Mucor* (Mucorales) are occasionally associated with insect mortality.

The majority of entomopathogenic species within Zygomycota are contained in one order, the Entomophthorales. More than 200 entomopathogenic Entomophthorales species have been recognized. They commonly cause spectacular epizootics and are characterized in all genera but one (*Massospora*), by the production of forcibly discharged primary conidia. Many species are capable of producing various types of secondary conidia from the primary conidia and, in some cases, infection is obligatorily through a secondary conidium. Many species are also capable of producing resting spores, long-lived zygospores or azygospores. These fungi are generally obligate pathogens in nature, and many species are presently difficult or impossible to culture on artificial media.

Until around the 1960s, most entomopathogenic Entomophthorales were contained within a single genus, *Entomophthora*, but several years of taxonomic revision have placed the species into several genera. The main genera containing entomopathogenic species are *Conidiobolus*, *Entomophaga*, *Entomophthora*, *Erynia*, *Furia*, *Massospora*, *Neozygites*, *Pandora*, *Strongwellsea*, and *Zoophthora*.

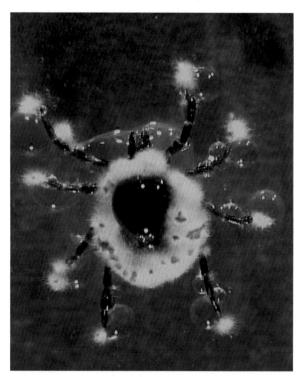

Figure 2 Tick (*Ixodes ricinus*) killed by *Metarhizium anisopliae*. (Photo courtesy of C. Nielsen.)

Figure 3 Adult cabbage fly (*Delia radicum*) killed by *Beauveria bassiana*. This fungus produces clusters of white conidia. (Photo courtesy of J. Eilenberg.)

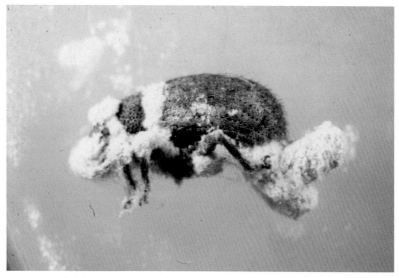

Figure 4 Adult weevil killed by *Paecilomyces farinosus*. The fungus produces extensive external growth and spores. (Photo courtesy of J. Martin.)

Figure 5 Pupae of the lepidopteran species *Dasyneura pudibunda* killed by *Cordyceps militaris*. The ascospores are produced in the orange asci. (Photo courtesy of J. Eilenberg.)

Figure 6 Alate termite (*Macrotermes* sp.) killed by *Cordycepioideus* sp. The fungus has only been found on alate termites. (Photo courtesy of J. Eilenberg.)

11.2.4. Ascomycota (and Deuteromycota)

Fungi from Ascomycota have septate and haploid mycelia and the sexual spores, ascospores, are produced in an ascus on a fruiting body, the ascomata. Typically, eight ascospores are produced in each ascus. No motile zoospores are produced. There has been continuing taxonomic confusion over classification of the teleomorphic (sexual) stages and the anamorphic (asexual) conidial stage when they are not produced by a single occurrence. *Cordyceps*, a genus containing many (>300) entomopathogenic species is probably the best known Ascomycete because the sexual and sometimes asexual stages are produced on highly visible stroma. The genus

Ascosphaera, which is heterothallic and sexually dimorphic, contains species responsible for chalkbrood disease in bees. Unusually for entomopathogenic fungi, *Ascosphaera* spores are ingested by bee larvae and germinate in the gut (Gilliam and Vandenberg, 1990).

Many strains within Ascomycetes have apparently lost the ability to form the sexual stages, the teleomorph forms, and were classified in the form-class Hyphomycetes. Entomopathogenic species are known from over 40 genera, but the most important entomopathogenic species occur in *Aschersonia*, *Aspergillus*, *Beauveria*, *Culicinomyces*, *Gibellula*, *Hirsutella*, *Hymenostilbe*, *Lecanicillium* (formerly *Verticillium*), *Metarhizium*, *Nomuraea*, *Paecilomyces*, *Sorosporella*, and *Tolypocladium*.

Figure 7 Small Diptera (Chironomidae) killed by *Erynia conica*. The dead insect is covered by conidiophores from which conidia are discharged. (Photo courtesy of J. Eilenberg.)

Figure 9 Adult cantharid beetle (*Rhagonycha fulva*) infected with *Entomophthora muscae*. Conidia have been discharged from the abdomen and clusters of conidia are seen on the wings. (Photo courtesy of J. Eilenberg.)

Figure 8 Adult dipteran (*Ptychoptera contaminata*) killed by *Entomophaga* (= *Eryniopsis*) *ptychopterae*. No fungus growth is visible before incubation in a moist chamber. (Photo courtesy of J. Eilenberg.)

Figure 10 Small muscoid fly (*Coenosia testaceae*) infected with *Strongwellsea* sp. This fungus causes an abdominal hole in the host, while still alive. Conidia are projected from the hole and the infectious fly can disperse the disease. (Photo courtesy of J. Eilenberg.)

Most of these genera have been linked to one or several ascomycete genera. Such linkages can be demonstrated either biologically (an insect infected with an anamorph dies and a teleomorph is produced) or by using molecular tools showing the genetic relationship between anamorphs and teleomorphs (Huang *et al.*, 2002; Liu *et al.*, 2002). With the advent of molecular identification, many more links between anamorph and teleomorph

stages will be made and, as discussed above, the need for the form-class Hyphomycetes will be unnecessary.

11.2.5. Basidiomycota

There are very few Basidiomycetes that have been implicated in insect pathogenesis. Some publications list the Septobasidiales genera *Septobasidium* and *Uredinella* as pathogenic to insects, but most

Figure 11 Colony of cereal aphids (*Sitobion avenae*). One individual (swollen, whitish-orange) is infected by *Pandora neoaphidis* and the discharged primary conidia may result in infection of the rest of the colony. (Photo courtesy of T. Steenberg.)

Figure 13 Adult *Scatophaga stercoraria* killed by a member of the *Entomophthora muscae* complex. White bands of conidiophores emerge from the abdomen, which then produce forcibly discharged conidia. (Photo courtesy of A. Bruun Jensen.)

Figure 12 Ant infected with *Pandora formicae*. Before death, the ant climbs to the top of the canopy, turns the head downwards, and grasps the plant with its legs. The conidiophores then emerge and conidia are discharged, showering the understory, where presumably more hosts are present. (Photo courtesy of J. Eilenberg.)

Figure 14 Adult *Delia radicum* filled with bright orange resting spores of *Strongwellsea castrans*. (Photo courtesy of J. Eilenberg.)

are considered symbiotic to insects, such as scales (Samson *et al.*, 1988).

11.2.6. Approaches to Classification

Traditionally, fungal taxonomy has been based on morphological, developmental, and physiological characteristics, from which the current structure of species, genera, and classes has emerged. While these

characteristics worked very well for many years, the complexity of fungal life has led to use of some characteristics in classification that are not the result of coevolution, but parallel evolution of a common function. This has led to the incorrect placement of some species and genera in the phylogenetic scheme. In addition, the use of indirect methods, which evaluate phenotypic traits, have not been useful at

Figure 15 Earwig (*Forficula auricolaria*) infected with *Zoophthora forficulae*. The insect is completely covered with the fungus and is firmly attached to the vegetation by sticky rhizoids produced by the fungus. (Photo courtesy of J. Eilenberg.)

the subspecies level. Identification of specific isolates is necessary in biocontrol, both as a regulatory requirement and to advance understanding of ecology and epizootiology (see Section 11.4).

There have been many different approaches to classifying insect pathogenic fungi. In some cases, fungi have been defined functionally or ecologically, rather than phylogenetically, especially when techniques to directly determine phylogeny did not exist. The entomogenous fungi provide an example where the entomogenous habitat was seen as a defining criterion in many cases, whereas morphology and physiology formed the basis of many of the earlier efforts with other fungi. Early descriptions of *M. anisopliae* refer to *Entomophthora anisopliae* (Zimmermann *et al.*, 1995), recognizing the relationship between two common entomogenous groups over dissimilarities in morphology and physiology. The mode and morphology of conidiation has been a main character used for classification.

In recent times, classification systems have attempted to take a more phylogenetic approach to systematics (e.g., Humber, 1984), but a true phylogenetic classification had not emerged for most organisms until the advent of molecular biology. Other approaches are still being explored. Chemotaxonomy has been attempted with some entomopathogenic groups such as *Beauveria*, utilizing the commercially available carbohydrate utilization strips system of API (i.e., API50CH) (Todorova *et al.*, 1994; Rath *et al.*, 1995). Isoenzymes have been investigated for their ability to discriminate between fungal strains (e.g., May *et al.*, 1979; St. Leger *et al.*, 1992b), but their power in phylogenetic classification is limited.

Molecular techniques that target RNA and DNA have been applied at every level of the taxonomic hierarchy. Many techniques have been applied, some of which directly sequence genes, and some of which represent differences in DNA sequence, such as random amplified polymorphic DNA (RAPD) and restriction fragment length polymorphism (RFLP) approaches. Pulse field gel electrophoresis (PFGE) allows the separation of DNA fragments as large as chromosomes and can be used for whole genome restriction digest comparisons. Chromosome length polymorphism was used by Viaud *et al.* (1996) to study nine isolates of *B. bassiana*. Mitochondrial DNA has been used to estimate intraspecies variation in *Lecanicillium* (= *Verticillium*) *lecanii* and *M. anisopliae* isolates (Typas *et al.*, 1998). There are a number of reviews on the use of molecular characterization for entomogenous fungi (e.g., Clarkson, 1992; Driver and Milner, 1998).

With the development of molecular biology as a mainstream tool in classification, it became possible to test assumptions formed by studies based on morphology and other characteristics reliant on phylogenetic relationships. In macroevolution, the seven kingdoms are based largely on sequencing of small subunit ribosomal genes. While classifying fungi solely on the sequence of any gene is not wise, a general consensus is being reached over the evolutionary position of the fungi in respect to other groups based on rRNA gene sequences (e.g., Berbee and Taylor, 2001). Molecular analysis has shown that a number of species and genera classified as fungi, or in the kingdom Eumycota, are actually grouped with the Chromista. Many species have been placed in the fungi simply on the basis of hyphal growth, but are now being reconsidered.

The genera of entomopathogenic fungi are being reconsidered on the basis of sequencing of conserved genes and introns. Recent efforts to clarify classification of entomopathogenic fungi demonstrate the utility of molecular characterization at several levels. Zare *et al.* (2000), Gams and Zare (2001), Zare and Gams (2001) have redefined the genus *Verticillium* using rDNA sequencing, placing all insect pathogens in the genus *Lecanicillium*. Similarly, sequencing of large subunit rRNA gene revealed *Paecilomyces* may not be monophyletic (Obornik *et al.*, 2001). Sequencing of the small subunit rDNA has also been used to examine phylogenetic relationships among the order Entomophthorales (Jensen *et al.*, 1998). These molecular studies supported the use of spore discharge characteristics as an identifying characteristic for Entomophthorales.

Molecular studies have also been useful in positioning individual species, such as *Basidiobolus ranarum*, previously placed in the Zygomycota, as a nonflagellate species of chytrid (Nagahama *et al.*, 1995). The use of sequence level analysis also suggested reassigning of species such as *Eryniopsis ptychopterae* to *Entomophaga* (Hajek *et al.*, 2003).

As noted above, the use of DNA and RNA comparisons has never been more useful than when considering the fungal species that never reproduce sexually, because the characters of sexual reproduction are the basis for much of the traditional alignment of family, genera, and species. While the higher classification of fungi is in a state of flux, the form-class Hyphomycetes is likely to disappear in the near future, with DNA analysis making the need for a group for which a sexual stage is not known redundant. Linking of anamorphs and teleomorphs is now relatively straightforward. For example, sequencing of the ITS-5.8s region of nuclear RNA has recently been used to confirm a teleomorph, *Cordyceps bassiana*, of *Beauveria bassiana* (Huang *et al.*, 2002).

Strain characterization is a useful attribute, particularly for biocontrol, and in some cases, molecular techniques have assisted the placement of an important strain into species. The taxonomic position of *M. anisopliae* strain IMI 330189 used in the LUBILOSA program for locust control was confusing, as it was originally classified as *M. flavoviride* based on morphology of the conidia and some other features. Driver *et al.* (2000) used molecular techniques to revise the genus *Metarhizium* and demonstrate subspecific relationships among *M. anisopliae* and *M. flavoviride* strains, based largely on the sequence of the ITS-5.8s regions of rRNA. Sequencing of the ITS-5.8s region of *M. anisopliae* strain IMI 330189 revealed it to be a *M. anisopliae* subspecies, which they named *M. anisopliae* var. *acridum*. However, species cannot be erected on sequence data alone, and several of the subspecies "clades" identified by Driver *et al.* (2000) can only be delimited by sequence data at present.

The main drawback with molecular approaches to classification is the lack of agreement on standard techniques. This is not generally a problem for use in strain identification, but the choice of method does influence the relationships among species and genera. The use of rRNA sequences has become widespread, but as the studies of Rehner (2003) show, reliance on single region sequences may be misleading. Rehner (2003) used a multigene sequencing approach to clarify the taxonomic position of *Beauveria* species strains. This approach is a step beyond what is currently known about the molecular taxonomy of most entomopathogenic fungi and is likely to be a common approach in the future.

The contribution of molecular techniques to the development of entomopathogenic fungi has been enormous. The techniques have been used to clarify evolutionary relationships (e.g., Jensen *et al.*, 1998; Driver *et al.*, 2000) as well as allow development of theories of evolution around these, often obligate, pathogens. However, there are a number of difficulties with some current studies, such as incorrect typification and lack of holotypes (Samson, 1995). Few of the species have yet been studied in detail.

11.3. Biology and Pathogenesis

Survival of entomopathogenic fungi requires a delicate balance of interaction between the fungus, host, and the environment. In general, the life cycle of entomopathogenic fungi involves an infective spore stage, which germinates on the cuticle of the host, forming a germ tube that penetrates the cuticle and invades the hemocoel of the insect host (Hajek and St. Leger, 1994). The fungus then multiplies within the insect and kills it; death is due to toxin production by the fungus or multiplication to inhabit the entire insect. Under favorable environmental conditions, the fungus grows out of the cadaver, and forms conidiophores or analogous structures and sporulates. Alternatively, many species form some type of resting stage capable of surviving periods of adverse conditions before forming or releasing a type of spore. Spores need new hosts, so the fungus needs a strategy for dissemination. There are many variations and even exceptions to this generalized life cycle. However, the important point is that the environment and host are crucial to the survival and reproduction of the fungus.

11.3.1. Encounter with Host

The role of the host is both passive, in that it is the main nourishment for the fungus to produce reproductive structures in most cases, and active, in that the fungus needs to contact the host under suitable conditions. The role of environmental variables for terrestrial fungi is hard to overstate. For some fungi, such as conidia producing Hyphomycetes, germination, penetration, and sporulation can all require high humidity. Temperature is usually limiting, with infection occurring within a certain range. Hyphomycetes generally have temperature optima between 20 and 30 °C, while Entomophthorales, require 15–25 °C, suggesting that Entomophthorales are the more temperate fungi. Individual isolates within a single species can vary in temperature optima. Similar temperatures are usually required for germination, sporulation, and discharge.

Some species are more dependent on high humidity than others. *Lecanicillium lecanii*, a hyphomycete, can require up to 16 h of 100% relative humidity (RH) at the leaf level in order to cause high mortality rates in whiteflies (Milner and Lutton, 1986). In other cases, humidity requirements appear lower, such as *Entomophthora muscae* infecting Diptera (Mullens and Rodriguez, 1985). However, it is often difficult to measure the actual humidity that the fungus is exposed to, due to microclimatic variability. Leaf transpiration may create adequate humidity at the leaf surface, which is difficult to measure (Vidal *et al.*, 2003), allowing infection to occur even at low ambient humidity conditions (Fargues *et al.*, 2003). Dew period can also be more important than ongoing high humidity (e.g., Milner and Lutton, 1986). Entomophthorales are often very dependent on >95% RH for sporulation and conidial germination, despite the occasional exception such as *E. muscae* (see Section 11.3.4).

Soil presents a range of abiotic and biotic factors that influences the life cycle of fungi. Moisture can be more constant and higher in the soil environment than on plants, but spore movement is more restricted than in the epigeal environments. The level of organic matter and pH can be important in the eventual infections levels, as well as in survival of propagules. In this way, soil is more similar to aquatic environments. In soil, other microbes and small invertebrates (e.g., springtails; Broza *et al.*, 2001) can consume fungi directly, compete saprophytically, or be directly or indirectly antagonistic to the entomopathogens. It is not surprising that fungi are more effective in sterilized than unsterilized soils in the laboratory (e.g., Inglis *et al.*, 1998). Microbes such as bacteria resident within insect hosts can compete within infected cadavers and, if the fungus does not have mechanisms to exclude these microbes, the fungus will be unable to colonize the cadaver sufficiently to allow sporulation.

Other abiotic conditions that influence encounters between the host and pathogens are wind and sunlight. Wind can assist spore dispersal but will also decrease humidity and remove free water in some cases. Sunlight and UV can be detrimental to the persistence of infectivity of most fungi. For example, Inglis *et al.* (1997a) found that the poor efficacy of *B. bassiana* against rangeland grasshoppers in a Canadian field trial was a result of conditions of temperature and light exposure, as the fungus was more virulent in greenhouses and shaded cages than in the open rangeland. However, there is significant variation in susceptibility to UV among different species and strains within species (Fargues *et al.*, 1996).

Aquatic species of fungi are not subject to the influence of humidity, as they are generally infective only in water. Periodic drying of habitats is a constant problem, but most aquatic entomopathogens have strategies to survive drying. Similar to terrestrial environments, temperature can be a limiting factor. Chemical factors, such as salinity or organic pollutants in the water are also important.

Meteorological data do not always correlate with the occurrence of epizootics, partly because microclimatic conditions are not the same as macroclimatic conditions. There are still many gaps in our knowledge of the interaction between the pathogen, host, and environment, as demonstrated by the lack of prediction of epizootic and disease occurrence in most fungi.

11.3.2. Specificity and Host Range

Specificity of entomopathogenic fungi varies widely between genera, within genera, and even among strains of a species. The so-called muscardine fungi, *B. bassiana* and *M. anisopliae*, have well-documented broad host ranges, which include hundreds of insect species from several classes. Veen (1968) listed 204 insects naturally infected by *M. anisopliae*. In the intervening years, many more hosts have been reported and include species of Lepidoptera, Diptera, Coleoptera, Hymenoptera, and Homoptera. Coleoptera are particularly prevalent hosts for *M. anisopliae* with over 70 scarab species included in the list of Veen (1968). Broad host lists, such as for *M. anisopliae*, are often misleading, as individual strains of entomopathogenic fungi can vary in their specificity. In addition, definition and delimitation of species can be arbitrary in these often asexual fungi, as demonstrated by recent molecular analyses of *Metarhizium* (i.e., Driver *et al.*, 2000) and *Beauveria* (Rehner, 2003; see Section 11.2.6). For biocontrol applications, the isolate is more important than the species as a unit, partly due to variation in virulence to hosts among strains. There are implications in host range for determining the nontarget safety of fungal based bioinsecticides (Goettel *et al.*, 1990a; Vestergaard *et al.*, 2003). Based on published host ranges of the species *B. bassiana* or *M. anisopliae* it would be very difficult to convince regulators of their environmental safety, but it is commonly recognized that strains of *B. bassiana* and *M. anisopliae* are more restricted in their hosts than the species. A single strain of any entomopathogenic fungus rarely attacks both beneficial and pest species.

Some entomopathogenic fungi have been recorded from as limited a host range as a single species of insect. In the Entomophthorales, several species are known from only one host, but it is difficult to know if this is a true reflection of host range or the result of limited study. Few species have been cultured and even fewer have been bioassayed against a number of insects to determine potential host range. *Zoophthora radicans* (Entomophthorales) has been described from over 80 insect species from Diptera, Coleoptera, Lepidoptera, and Homoptera, but there is clear indication that strains are restricted, generally, to some species in a single insect class (Milner and Mahon, 1985). The individual strains are generally more pathogenic to the insects more closely related to the original host, but there are many exceptions.

Other groups of entomopathogenic fungi are limited in host range. *Coelomomyces opifexi*, for example, is known from only two mosquito species and a copepod intermediator (Glare and Milner, 1991). The many species of *Cordyceps* represent both broad and narrow host range species (e.g., Kobayasi, 1941), but here, again, information is limited.

In many cases, entomopathogenic fungi will kill certain species only in special situations. *M. anisopliae* is rarely recorded naturally as a pathogen of mosquitoes, but in the laboratory, many strains are highly pathogenic to mosquito larvae (Daoust and Roberts, 1982), indicating that other factors than susceptibility are important in occurrence of disease in nature. In some cases, the fungus can only kill weakened or stressed hosts. Behavioral avoidance can also lead to nonsusceptibility in the field, when laboratory bioassays indicate susceptibility. For example, *Aspergillus flavus* can be isolated from many wasp (*Vespula* spp.) nests in New Zealand and is highly pathogenic in the laboratory to *Vespula* (Glare *et al.*, 1996), but the hygienic behavior of wasps is such that infections in healthy nests are not seen. *Zoophthora phalloides* is a pathogen of aphids, but shows distinct preferences in the field for some species. Remaudiére *et al.* (1981) recorded 72% of *Myzus ascalonicus* infected with *Z. phalloides* on a single bush, but only 6% of *Myzus ornatus* on the same bush. Conversely, *Pandora neoaphidis* on the same bush infected 71% of *M. ornatus* and only 1% of *M. ascalonicus*. The difference in susceptibility is likely due in part to the mode of infection by the two fungal species, as much as to possible differences in resistance among the aphid species. *Zoophthora phalloides* infects via the sessile capilliconidia, making it more effective against mobile hosts, while *P. neoaphidis* uses

forcibly discharged primary and secondary conidia, which are more effective when they land directly on a host. Unsurprisingly, *M. ascalonicus* is a more mobile aphid than *M. ornatus*.

These results highlight the difference between laboratory and field susceptibility. If results from laboratory assays are to be used to predict activity in the field, pertinent environmental and exposure parameters must be incorporated as much as possible in the bioassay design (Butt and Goettel, 2000). For instance, in the laboratory, most bioassays do not allow for avoidance of the pathogen through biological, ecological, or behavioral methods. In the field, an insect may never come in contact with sufficient inoculum to succumb to infection. An example is caterpillars and *E. maimaiga*, a pathogen of gypsy moth. Hajek *et al.* (1996) examined the field incidence of caterpillars infected with *E. maimaiga* where high gypsy moth infections were occurring. They found only two individual caterpillars (1 of 318 *Malacosoma disstria* and 1 of 96 *Catocala ilia*) of a total of 1511 larvae from 52 species belonging to seven lepidopteran families infected with *Entomophaga maimaiga*. In the laboratory, more species were found to be susceptible and high percentages of the few species found infected in the field were infected in the laboratory. Despite *E. maimaiga* being a pathogen of the Lymantriidae in the laboratory, species other than gypsy moth are unlikely to be infected in the field unless they spend periods of time near the leaf litter where the fungus is sporulating (Hajek *et al.*, 2000). Similarly, differential infections of mosquito hosts can be linked to position in the water profile, as bottom feeding species are more likely to come in contact with settling inoculum than surface feeders (Sweeney, 1981).

In most entomopathogenic fungi, there is differential virulence towards life stages of insects that are susceptible. Unlike bacterial, viral, and protozoan pathogens, most fungi directly penetrate the cuticle and do not need to be ingested. Therefore, entomopathogenic fungi have the potential to be active against nonfeeding stages such as pupae. Aquatic species, such as *Lagenidium* and *Coelomomyces*, rarely attack adult stages, although they can persist in them (see Section 11.3.4). Hyphomycetes such as *Metarhizium* and *Beauveria* often infect both adult and larval stages. Keller and others have used the ability of *Beauveria brongniartii* to infect both larval and adult stages of the European cockchafer, *Melolontha melolontha*, to develop biopesticide strategies based on spraying adults in order to transfer inoculum to the larvae in the soil (Keller *et al.*, 1989). Most fungi, even if they attack all life stages, show variability in virulence between the stages. Even within the larval stage, there is usually variability in susceptibility. For example, mosquito larvae were more susceptible to *Culicinomyces clavisporus* at earlier instars than later ones (Panter and Russell, 1984), while the scarab *Costelytra zealandica* is more resistant to *M. anisopliae* at the second compared to the third instar (Glare, 1994), indicating there is no general rule regarding which developmental stage will be more susceptible.

The development of resistance to entomopathogenic fungi in insect populations has rarely been documented. When the pea aphid, *Acyrthosiphon pisum*, was introduced to Australia, it became resistant to the pathogen, *P. neoaphidis*. Soon after the development of resistance, the fungus developed strains that were fully pathogenic for pea aphid, indicating a rapid genetic capacity to overcome resistance (Milner, 1982, 1985).

Genotyping using various molecular methods is becoming increasingly common in the identification of entomopathogenic fungi (see Section 11.2.6). However, the methods used to date rarely demonstrate a link between genotype and host preference where sufficient strains have been examined to give a meaningful result. There are exceptions, such as *B. bassiana* pathogenic to *Ostrinia nubilalis* and *Sitona* weevils, where some homogeneity among isolates from these hosts was found (Viaud *et al.*, 1996; Maurer *et al.*, 1997). In most cases, the molecular methods do not target genetic regions involved in specificity, but are aimed at random, nonspecific regions, or regions involved in basic cellular processes, such as ribosomal and mitochondrial regions. This suggests host specificity evolves relatively rapidly. Targeting of genes involved directly in virulence related events may lead to better matching between molecular identification and host range.

11.3.2.1. Vertebrate safety

Fungi in general are well known allergens that produce toxins and are capable of infecting vertebrates. Fungal epizootics may be responsible for drastic reductions in invertebrate populations, populations that may be an important food source for some vertebrates. Consequently, the actions of entomopathogenic fungi may compromise vertebrate health and well-being in several ways.

Fungi are an important source of human allergens. However, entomopathogenic fungi are not typical human allergens (Gumowski *et al.*, 1991), indicating that exposure to fungi generated from natural epizootics does not generally induce allergenic reactions. Allergy problems have been documented in workers frequently exposed to certain entomopathogenic fungi such as *B. bassiana* (Saik *et al.*, 1990) and

crude extracts from *M. anisopliae* contain potent allergens (Ward *et al.*, 1998). However, there were no reports of human hypersensitivity during mass production of *B. bassiana* by the Mycotech Corporation involving more than 30 000 person-hours of exposure (Goettel and Jaronski, 1997). Screening for allergies in 145 Tate and Lyle (Reading, UK) employees exposed to *L. lecanii* during its manufacture and of 31 employees involved in its field testing in greenhouses, concluded that the fungus is not significantly allergenic, even to a worker population that was more highly allergic to other allergens than average (Eaton *et al.*, 1986). Since all fungi are potentially allergenic, it is unreasonable and impractical to expect a total absence of allergic potential in microbial insecticides based on fungal pathogens. However, certainly every effort should be made to avoid human exposure during manufacture and application.

For the most part, entomopathogenic fungi are not virulent towards vertebrates (Saik *et al.*, 1990; Siegel and Shadduck, 1990). However, there are some exceptions. Certain species of entomopathogens such as *Conidiobolus coronatus* or *A. flavus* are well-known pathogens of vertebrates and, because of this, they are generally not considered for development as commercial biopesticides. An exception may be *Paecilomyces lilacinus*. Although as a species, it is known to cause human and other vertebrate infections and is an important emerging nosocomial fungal pathogen (Goettel *et al.*, 2001), the strain P 251, isolated in the Philippines, has been successfully developed as a safe microbial nematicide by the Australian Technology Innovation Corporation (Copping, 2001). This is another example of how fungal pathogens must be judged at the strain level, and not at the species level.

Although there have been no reports of infections in vertebrates directly resulting from the use of fungal pathogens as microbial control agents, there are several documented cases of "chance" occurrences. For instance, several entomopathogenic Hyphomycetes have been recovered from captive cold-blooded vertebrates (Austwick, 1983; Gonzalez Cabo *et al.*, 1995), although it is generally believed that these animals have been severely stressed by chilling. Laboratory challenge assays demonstrated pathogenicity of *B. bassiana* to silverside fish embryos and fry (Genthner and Middaugh, 1992), and of *M. anisopliae* to grass shrimp embryos (Genthner *et al.*, 1997) and silverside fish embryos and fry (Genthner and Middaugh, 1995). There have been several cases of human infection by entomopathogenic fungi reported for species presently in use as microbial control agents, with one fatality in an immunoincompetent child, but none of these have been implicated from the commercial use of specific strains (Goettel *et al.*, 2001; Vestergaard *et al.*, 2003). The most recently reported case concerns the deep tissue infection of an immunosuppressed female by a *Beauveria* sp. (Henke *et al.*, 2002). The isolate in question was unable to grow *in vitro* at 37 °C, which indicates that the inability to grow at this temperature *in vitro* is insufficient to rule out possible infection and growth in mammalian tissues. Interestingly, this isolate was more virulent against a test insect, the Colorado potato beetle, as compared to a well-known entomopathogenic strain of *B. bassiana*.

Entomopathogenic fungi produce an array of metabolites (see Section 11.3.3.1 below), many of which are toxic or carcinogenic to vertebrates (Strasser *et al.*, 2000; Vey *et al.*, 2001). However, any hazard to vertebrates would require exposure to significant levels of these metabolites. Current evidence suggests that, unless infected insects are actively pursued and ingested (e.g., insect cadavers infected with *B. bassiana* are widely consumed in China for medicinal purposes), the risk to vertebrates should be minimal. To date, no detrimental effects have been noted in vertebrates fed insect cadavers. For instance, ring-necked pheasant chicks given feed coated with *M. anisopliae* var. *acridum* (as *M. flavoviride*) spores or infected grasshoppers for 5 days showed no significant changes in weight, growth rate, behavior, or mortality rate as compared to controls fed noninoculated food or noninfected grasshoppers (Smits *et al.*, 1999). Strasser *et al.* (2000) speculate that entomopathogenic fungi should pose no obvious risk to humans because toxin quantities produced *in vivo* are usually far lower than those produced *in vitro* and, therefore, toxin levels should never rise to harmful levels in the environment.

Fungal epizootics may deplete an important food source for certain vertebrates. However, at the same time, depletion of a pest is the desired effect in the use of an entomopathogenic fungus in microbial control. The possible effects on vertebrates due to the depletion of a food resource as a result of fungal epizootics, either natural or induced, are poorly documented. However, for the most part, depletion of the target host is the desired effect and can be more or less regulated in an integrated pest management program where fungi are used in large amounts (also called inundation). However, special attention must be paid if exotic fungi are to be used in the classical sense of biocontrol (Goettel and Hajek, 2001).

Stringent regulatory requirements address vertebrate safety issues. Before any entomopathogenic fungus can be registered as a microbial control agent, it must first undergo stringent testing for potential harmful effects on vertebrates, including mammals (see Laird *et al.*, 1990; Siegel, 1997). In principle, no harmful agent should ever make it onto the market. However, with the increasing development of immunosuppressive therapy in medicine, the increase in iatrogenic factors and nosocomial diseases, and the advent of new infectious diseases such as AIDS, the list of opportunistic fungi causing deep mycoses is increasing (Chabasse, 1994), and this includes some entomopathogens. Entomopathogens must be carefully scrutinized, but at the same time, regulations must not become unjustly stringent so as to significantly deter registration of safe and useful products.

11.3.3. Mode of Action and Host Reactions

Entomopathogenic fungi kill the host by a variety of means, from starvation through multiplication in the host, to production of toxins. Given the pathogenic habit, it is hardly surprising that entomopathogenic fungi produce extracellular enzymes and toxins. Entomopathogenic fungi produce a variety of chitinases and proteases, which aid penetration of the host physical defenses.

The main barrier to infection of insects by entomopathogenic fungi is the cuticle. For the majority of fungi, the route of infection is directly through the cuticle, not after ingestion. The fungus, therefore, needs enzymatic and/or physical means to penetrate this thick, multilayered shell. The process of infection starts with the spore contacting the cuticle of a host. In many cases, the conidia are adhesive to the cuticle, or secrete adhesive mucus as the conidium swells during pregermination (Hajek and St. Leger, 1994). Capilliconidia, the infective sessile spores of some *Zoophthora*, have a drop of adhesive on the end of the spore to assist attachment to passing insects (Glare *et al.*, 1985). Zoospores of *Lagenidium* encyst on the host surface, ensuring attachment (Kerwin and Petersen, 1997).

The exact mechanisms used by each fungus for penetration of the host cuticle may differ, but there are general processes and structures involved. *Metarhizium anisopliae*, for example, develops a number of specific structures to assist the anchoring and penetration of germ tubes, which arise from conidia (St. Leger, 1993). Appressoria and infection pegs assist the penetration of germ tubes. Once the penetrative tube has passed through the layers of the cuticle and epidermis, the fungus can proliferate in the hemocoel. For some hyphomycete fungi, such as *M. anisopliae*, this is initially performed by blastospores, while for some Entomophthorales, proliferation can be by wall-less protoplasts (e.g., Butt *et al.*, 1981; Bidochka and Hajek, 1998).

Injection of entomopathogenic fungi directly into the hemolymph can demonstrate that the cuticle is a major barrier to infection. However, the insect also has cellular and humoral methods of defense against fungal invasion. For example, when Bidochka and Hajek (1998) injected *Entomophaga aulicae* into gypsy moth larvae, a nonpermissive host, there was no evidence of restricted fungal growth, fungus specific induction of plasma proteins, or hemoctye encapsulation, but higher levels of phenoloxidase and prophenoloxidase-activating trypsin activity were found up to 96 h postchallenge. Bidochka and Hajek (1998) suggested that surface components of protoplasts, such as glycoproteins, may be implicated in activation of zymogenic trypsins in the insect, which in turn activate the prophenoloxidase cascade as a nonpermissive response.

11.3.3.1. Toxin production

Some entomopathogenic fungi produce toxins, some of which are insecticidal and may assist in pathogenesis. Others produce metabolites that are antimicrobial, either to restrict saprophytic competition or protect against antagonistic microorganisms. Toxic metabolites from entomopathogenic fungi have been reviewed extensively by Roberts (1981) and, more recently, by Vey *et al.* (2001). The safety aspects of fungal metabolites have become an important issue, with implications for risk assessments, and these were reviewed by Strasser *et al.* (2000; see also Section 11.3.2.1).

Among the fungal metabolites that assist pathogenicity are the destruxins of *Metarhizium* spp. First described in 1961 (Kodaira, 1961), these cyclodepsipeptides are toxic to a number of insects. Susceptibility varies considerably, with a 30-fold difference between silkworm larvae and *Galleria* (Roberts, 1981). While the basic structure of destruxins is based on five amino acids and α-hydroxy acid, a number of isomers and congeners have been described (Vey *et al.*, 2001). Destruxins have also been described from *Aschersonia* sp. (Krassnoff and Gibson, 1996) and from some plant pathogenic fungi. As well as toxicity at high doses, some destruxins reduce growth and reproduction (e.g., Brousseau *et al.*, 1996). There is possibly a relationship between the chemical structure of the various destruxins and activity, with some forms more active against certain insect species (Vey *et al.*, 2001). Destruxins can also be repellent or act as an antifeedant (Robert and Riba, 1989; Amiri *et al.*, 1999; Thomsen and Eilenberg, 2000). Production of destruxins by some species and strains has been

linked to increased host range, suggesting a role in pathogenic determination (Amiri-Besheli *et al.*, 2000). Synthetic analogs to destruxins have been tested against lepidopteran insects and also proved toxic, and thus are potentially useful as novel pesticides (Thomsen and Eilenberg, 2000).

Some strains of *Beauveria* spp. produce beauvericin, a depsipeptide metabolite, which has shown toxicity to a number of invertebrates (Hamill *et al.*, 1969; Roberts, 1981). Beauvericin has also been isolated from *Paecilomyces fumosoroseus* mycelium and the plant pathogenic fungus *Fusarium* spp. Activity has been classified as moderate against insects (Vey *et al.*, 2001) and is not effective against all insects (Champlin and Grula, 1979). *Beauveria bassiana* also produces beauverolides, isarolides, and bassianolides, all cyclotetradepsipeptides, the former of which has been shown to be toxic to cockroaches (Frappier *et al.*, 1975; Suzuki *et al.*, 1977). Bassianolides have been shown to have activity against silkworms (Kanaoka *et al.*, 1978). A further two nonpeptide toxins, bassianin and tenellin, have been isolated from *Beauveria* spp. and shown to inhibit the erythrocyte membrane ATPases (Jeffs and Khachatourians, 1997), but little else is known.

Linear peptidic efrapeptins are produced by *Tolypocladium* species and have some insecticidal and miticidal effects (Matha *et al.*, 1988; Krasnoff *et al.*, 1991). The species of *Tolypocladium* produce efrapeptins in varying amounts (Bandani *et al.*, 2000). Bandani and Butt (1999) reported antifeedant and growth inhibitory properties.

Hirsutellin A is produced by *Hirsutella thompsonii* and is not proteolytic, but is toxic to a range of insects including *Galleria mellonella*, the mosquito, *Aedes aegypti* (Mazet and Vey, 1995), and mites (Omoto and McCoy, 1998). It is produced during liquid fermentation, and the sequence and amino acid composition is unlike any other known protein (Mazet and Vey, 1995). Other fungal toxins known to have some insecticidal properties include *Aspergillus* spp. Several toxins have been isolated from cultures of *Aspergillus*, such as kojic acid (Beard and Walton, 1969) and some mosquitocidal toxins (Toscano and Reeves, 1973). The production of aflatoxins by some strains of *Aspergillus* restricts interest in the group as insect pathogens, as these compounds have carcinogenic and tumor producing properties as well as some insecticidal properties.

It has not been demonstrated whether all entomopathogenic fungi produce toxins in the disease process and, indeed, if toxins are required for virulence. Quesada-Moraga and Vey (2003) demonstrated that *B. bassiana* can be pathogenic to the locust regardless of toxin production. Furthermore, *in vivo* passage through a host or repeated subculture on artificial medium had a variable effect on toxin production; virulence and toxicogenic activity of one isolate was dependent on the mycological media that the inoculum was produced on, whereas virulence and toxicogenic activity of another isolate was greatly increased after two passages through the host.

In some cases, toxins are suspected, but not conclusively demonstrated. Some of the lower fungi, such as *Coelomycidium, Coelomomyces,* and the Entomophthorales, may possess only weak toxins, if any at all. It is more likely that they overcome their hosts by utilizing the nutrients and invading vital tissues (Roberts, 1981). Culture filtrates from entomophthoralean fungi injected into greater wax moth larvae (*Galleria*) resulted in blackening similar to that found in fully infected larvae, suggesting that toxins were active, but none were identified (Roberts, 1981). A short-lived cell lytic factor is thought to be responsible for death in lepidopteran hosts infected with *E. aulicae* (Milne *et al.*, 1994).

Antibiotics are produced by entomopathogenic fungi in order to exclude saprophytes and resident microbes that compete for nutrients in the cadavers. Oosporein, a red-colored dibenzoquinone, is produced by strains of *Beauveria* spp. and has antiviral and antibacterial properties. Oosporein was found to inhibit the herpes simplex virus-1 DNA polymerase (Terry *et al.*, 1992) and is active against Gram-positive, but not Gram-negative, bacteria (references in Vey *et al.*, 2001). Beauvericin has shown antibiotic activity against bacteria (Ovchinnikov *et al.*, 1971), and destruxin E has antivirus activity against nucleopolyhedrovirus (Quiot *et al.*, 1980). The antibiotic phomalactone has been described from *H. thompsonii* var. *synnematos* (Krasnoff and Gupta, 1994) and was inhibitory to the entomopathogenic fungi *Beauveria, Tolypocladium,* and *Metarhizium.* Members within the *Cordyceps* also produce a number of metabolites that may be weak toxins or antibiotics. *Cordyceps* infected caterpillars are used as a traditional medicine in parts of Asia, and this may be partly based on the production of cordycepin, a weak antibiotic, by *Cordyceps* Zabra *et al.* (1996) reported that metabolites from *Pandora neoaphidis* had antibacterial activity.

11.3.3.2. Behavioral responses There have been relatively few detailed studies on the behavioral responses by hosts as a result of infection by entomopathogenic fungi. In the early stages of infection, in many cases, there are no noticeable symptoms; however, several days prior to death, symptoms become evident and include reduced feeding, activity, and coordination. Other responses include increased feeding, behavioral fever, altered mating or oviposition preferences, and positive or negative photo- or geotropism. Most responses are seemingly adaptations that favor either the host or the pathogen, while others may be the result of depletion of the host's nutritional reserves and the process of dying.

Reduced feeding has been reported in several insects. Examples include grasshoppers and locusts infected with *M. anisopliae* var. *acridum* (Prior *et al.*, 1992; Thomas *et al.*, 1997; Arthurs and Thomas, 2000), gypsy moth larvae infected with *E. maimaiga* (Hajek, 1989), *Plutella xylostella* larvae infected with *Z. radicans* (Furlong *et al.*, 1997), and larvae of the Colorado potato beetle infected with *B. bassiana* (Fargues *et al.*, 1994). In contrast, no change in consumption of food has been reported in some insects, such as *Plathypena scabra* (Lepidoptera, Noctuidae) infected with *Nomuraea rileyi* (Thorvilson *et al.*, 1985) and *Cerotoma arcuata* (Coleoptera; Chrysomelidae) infected with *B. bassiana* (Lord *et al.*, 1987). Increased feeding has been reported in *Lygus hesperus* (Noma and Strickler, 2000) and Colorado potato beetles infected with *B. bassiana* (Fargues *et al.*, 1994). With the Colorado potato beetle, however, phagostimulation occurred only within the first 24 h after inoculation. Thereafter, there was no significant effect on consumption until day 2, and after that, consumption decreased significantly, resulting in an overall reduction in food consumption.

Intuitively, one could surmise that increased feeding during early stages of infection might favor either the host or pathogen. The pathogen might be favored if such increased feeding increases resources provided to it, although spore production of *E. maimaiga* in starved postinoculated gypsy moth larvae equaled production on fed larvae, suggesting, at least for this system, that larval feeding during the period of pathogen incubation within the host is not necessary for fungal development (Hajek, 1989). However, it may favor the host if such feeding provides more resources to fight off the pathogen. Most probably, decreased feeding just prior to death is a result of an overall shutting down of metabolic functions as the host nears death.

Some insects respond to fungal infection by altering their thermoregulatory behavior and achieving a higher than normal body temperature by basking in the sun or orienting on warmer surfaces. This is called "behavioral fever" and has been shown to either significantly slow the progress of infection or at times even eliminate the disease. This phemonenon has been observed in fungal infections of *E. muscae* in flies (Watson *et al.*, 1993; Kalsbeek *et al.*, 2001a) and of *B. bassiana* and *M. anisopliae* var. *acridum* in grasshoppers and locusts (Inglis *et al.*, 1996b, 1997a; Blanford *et al.*, 1998; Blanford and Thomas, 2001). Clearly, the higher temperatures achieved during thermoregulation and/or behavioral fever exceed the pathogen's thermal threshold and are consequently detrimental to its growth and proliferation (Carruthers *et al.*,1992; Inglis *et al.*, 1996a, 1997b; Ouedraogo *et al.*, 2003). However, the ability to thermoregulate is possible only when environmental conditions are favorable and, consequently, behavioral fever is often not possible or is interrupted. For instance, a higher death rate of caged *B. bassiana* inoculated grasshoppers occurred in shaded than in nonshaded areas (Inglis *et al.*, 1997a). Even though some insects such as acridids infected by *M. anisopliae* var. *acridum* cannot escape infection, behavioral fever acts to prolong survival; only infected desert locusts that were allowed to "fever" were capable of surviving to adulthood, mating, and producing viable offspring (Elliot *et al.*, 2002). In addition to slowing the growth of fungal propagules within the host, elevated temperatures may increase the host immunity by maintaining high hemocyte population levels (Ouedraogo *et al.*, 2003) and inducing hemolymphal proteins (Ouedraogo *et al.*, 2002).

Insects infected with an entomopathogenic fungus may also alter their behavior during mating and oviposition. Housefly males preferably seek out *E. muscae*-killed females in their attempts to copulate (Møller, 1993). These males may themselves become infected or they may transmit conidia to uninfected females during copulation (Watson *et al.*, 1993). This behavior favors the dissemination of the fungus. The response to and production of sex pheromones by *Z. radicans* infected male and female *P. xylostella* moths is inhibited, thereby disrupting mating (Reddy *et al.*, 1998). Female carrot flies infected with *E. schizophorae* (as *E. muscae*) were unable to recognize their host plants and oviposited their eggs haphazardly instead of their normal deposition near the food plants (Eilenberg, 1987). It was concluded that, even though infection did not alter fecundity or fertility, *E. muscae* infected flies do not contribute to the development of the carrot fly population, as their eggs have no chance of survival on a nonhost plant. It is unclear whether this behavior provides any benefit to either the host or the pathogen.

Fungal infections are often responsible for inducing aberrant behaviors at or near host death. The result of such behaviors often are of benefit to the pathogen. The most common of these is to cause the infected host to move to elevated positions just prior to death. This is most common in insects succumbing to entomophthoralean infections. For instance, just prior to death, grasshoppers infected with *Entomophaga grylli* climb to the top of the canopy where they die firmly grasping onto the plant substrate by their first two pairs of legs (Carruthers *et al.*, 1997). Flies infected with *E. muscae* exhibit highly stereotyped behavior, which is characterized by four events: last locomotory movement to an elevated position, the last extension of the proboscis to the substrate, the start of upward wing movement, and the end of upward wing movement (Krasnoff *et al.*, 1995). Consequently, the flies die in elevated positions, with the proboscis extended and attached to the substrate, the abdomen angled away from the substrate, and the wings raised above the thorax. Since conidia are forcibly discharged by entomophthoralean fungi, this behavior acts to help distribute the spores over a wider area, allowing the spores to shower on susceptible hosts below. **Figure 16** shows an example of disease

Figure 16 Field situation showing the transmission of *Entomophthora* sp. (a) An adult cadaver of *Rhagonycha fulva* (Coleoptera) killed by *Entomophthora* sp. hangs from a leaf, attached by its mandibles. (b) An uninfected individual is attracted and is likely to get inoculated by physical contact with the conidia present on the cadaver and the leaf. (Photo courtesy of J. Eilenberg.)

transmission between cantharid hosts due to the exposed position of the fungus killed specimen.

11.3.4. Spore Production on Host

Spores are the main source of reproduction in entomopathogenic fungi. Among the groups, a number of strategies have been adopted to maximize the chances of finding new hosts to infect, and many different types of spores and sporulation structures have evolved to adapt to different habitats. Host death is a necessary prerequisite for sporulation in most, but not all, entomopathogenic fungi. The entomophthoralean fungus *Strongwellsea castrans*, for example, sporulates from one or two abdominal holes in still living flies (Eilenberg *et al.*, 2000a). However, in the majority of cases, spores are produced after the host has died and the fungus has fully colonized the cadaver.

In the asexual Ascomycetes (largely still classified as Hyphomycetes), production of robust primary conidia in large quantities is the main strategy for reproduction. Conidial production is dependent upon the size of the cadaver and nutritional factors, as well as species and strain of the fungus. In general, infections of a preferred host will generate 10^7–10^9 conidia/cadaver for terrestrial hosts (e.g., Glare, 1987; Luz and Fargues, 1998). For the fungus *N. rileyi* infecting *Helicoverpa armigera* and *H. punctigera* caterpillars, there was an approximately five- to tenfold increase in the number of spores produced in cadavers between 2nd, 3rd, and 5–6th instars (Glare, 1987). Under similar environmental conditions, conidial numbers appear to be mainly dependent upon the surface area of the host (Kish and Allen, 1976).

Most species from the Entomophthorales can forcibly discharge the primary spores, ejecting them away from the cadaver, and hopefully towards new hosts. Furthermore, many species then produce secondary conidia from the primary conidia, the spores of which are also forcibly discharged, and in some cases, capilliconidia are produced, which are detached from the capilli. Capilliconidia are sessile spores formed on top of a long capillary tube and represent a different strategy for infecting hosts, where new hosts encounter the spore when moving about. In some species, capilliconidia are the primary infective spore, e.g., *Z. phalloides* (Glare *et al.*, 1985) and *Neozygites floridana* (Nemoto and Aoki, 1975). Capilliconidia of some species can survive for days at reduced humidities (Uziel and Kenneth, 1991). **Figure 17** demonstrates how conidia may germinate and produce conidia of higher orders in the species *Entomophaga ptychopterae*.

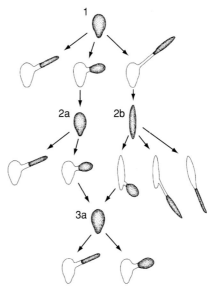

Figure 17 Conidia formation and germination in *Entomophaga* (formerly *Eryniopsis*) *ptychopterae*. A discharged primary conidium (1) can germinate with a germ tube or produce a secondary conidium of two different types: type 2a, which is similar in shape to the primary conidium or type 2b, which is narrow, ellipsoid, elongated, and produced on long conidiophores. Each type of secondary conidia can germinate with a germ tube or produce a tertiary conidium similar to the primary one (3a). This conidium can germinate with a germ tube or produce a quarternary conidium similar to the primary conidium. In addition, a type 2b secondary conidium can produce a tertiary conidium similar to a secondary conidium type 2b.

Production of primary and secondary spores by entomophthoralean fungi is generally very dependent on environmental conditions, as high humidity (even free water in some cases) is required for spore production (e.g., Steinkraus and Slaymaker, 1994). *Zoophthora phalloides* will only sporulate on aphid hosts at 100% RH and peak spore discharge takes place between 10 and 20°C (Glare *et al.*, 1986). This species produces up to 30 000 conidia per aphid cadaver. In the gypsy moth pathogen *E. maimaiga*, primary conidial production can be as high as 2.6×10^6 conidia per 5th instar cadaver (Shimazu and Soper, 1986). Not all entomophthoralean fungi depend on conditions of high humidity; for example, *Entomophthora muscae* can sporulate at just 50% RH (Kramer, 1980).

Inoculum density can influence eventual spore production on the cadavers. The aquatic *Culicinomyces clavisporus* produced less spores when the host was inoculated with higher doses of the pathogen (Cooper and Sweeney, 1986), and this is often the case for entomophthoralean fungi. This may be due to lack of complete colonization of the host by the fungus, or death due to the release of toxins rather than infection and multiplication.

Entomopathogenic fungi active in aquatic environments usually produce fewer spores than those infecting terrestrial insects. The mosquito pathogen *C. clavisporus* produces around 10^4 conidia/cadaver (Cooper and Sweeney, 1986). *Culicinomyces* produces nonmotile, true conidia, as do other Hyphomycetes. Other aquatic insect pathogenic fungi belong to fungal groups that can produce both motile and nonmotile spores. The oomycete *L. giganteum* produces both motile, biflagellate zoospores and sexually produced oospores (Kerwin and Petersen, 1997). The zoospores are released from both zoosporangial vesicles and oospores. Up to 20 000 sporangia can develop in a mosquito larval host, each releasing zoospores (Couch and Romney, 1973). Oospore production has been demonstrated to require sterols (Kerwin and Washino, 1983). These resistant and long-lived spores are produced within host cadavers. Another group of mosquito pathogens, *Coelomomyces* spp. (Chytridiomycetes) also has two infective stages. Resistant sporangia from mosquitoes release motile haploid zoospores, while the biflagellate diploid zygotes form within or after release from the obligate secondary host, usually a copepod (Lucarotti *et al.*, 1985). The zygotes and zoospores are similar. Sporangia in infected mosquitoes can number from 100 to 15 000, releasing up to 5000 motile zoospores from each sporangium (Pillai and O'Loughlin, 1972).

Fluctuating conditions of temperature or humidity interrupt maximum spore production for most fungi. In addition, light and dark cycle influence the sporulation of some entomopathogens, but not others. At least nine entomophthoralean fungi cause host mortality in diurnal rhythms (Milner *et al.*, 1984; Mullens, 1985; Tyrrell, 1987; Eilenberg, unpublished data), triggered by the onset of light (dawn) (Milner *et al.*, 1984). This timing of host death seems important to allow the fungus to maximize the early morning dew period for sporulation. Conidial discharge of *Z. radicans* demonstrates circadian rhythms under alternating light and dark regimes, with more conidia discharged in the dark (Yamamoto and Aoki, 1983). *Erynia/Pandora neoaphidis* produced around 500 000 primary conidia from infected aphids under ideal conditions, but changing from high to lower humidities and back often stopped, or at least severely reduced, the number of conidia produced (Glare and Milner, 1991).

Resting stages represent a different strategy for entomopathogenic fungi. While primary spores are produced in large quantities by members of the Entomophthorales, they last for only a few days; however, resting spores can survive for months to even years. This allows them to survive periods of unfavorable conditions and cause infections the next season through production of germ conidia. Resting spores are produced within infected cadavers. Factors that influence the production of resting spores include developmental stage of the host, temperature, humidity, and inoculum density (e.g., Shimazu, 1979; Glare *et al.*, 1989). With *Z. radicans*, which infects aphids, there is evidence of cytoplasmic based genetic elements involved in resting spore production, with more resting spores produced when aphids are inoculated with dual strains (Glare *et al.*, 1989). The ability to produce resting spores varies with individual isolates. With *E. maimaiga* infecting gypsy moth, the primary determinant of resting spore production was larval instar, with infections of later instars producing the most resting spores (Hajek and Shimazu, 1996).

11.3.5. Transmission and Dispersal

Dispersal of infective propagules is a crucial component of survival, and entomopathogenic fungi use different strategies to maximize the chances of encountering new hosts. For Hyphomycetes and some other groups that produce copious numbers of spores, wind, rain, and invertebrates play a role in transmission. Generally, wind is a major aid for the distribution of conidia. The minimum airspeed required for dislodgement of *N. rileyi* was calculated to be 2.7 km h^{-1} (Garcia and Ignoffo, 1977).

Entomophthorales generally forcibly discharge the primary conidia, increasing the distribution of these short-lived conidia. Distribution can also be aided by the type of spore formed, with both the forcibly discharged and sessile capilliconidia produced by some Entomophthorales. The aphid pathogen *Neozygites fresenii* discharges about 3000 primary spores per cadaver, and Steinkraus *et al.* (1999) detected up to 90 000 primary conidia per cubic meter of air during the nighttime in a Louisiana cotton crop, indicating how successful forcible dispersal aided by air movement can be. However, for *N. fresenii*, most infections come from the capilliconidia, which are more resistant to the environment than primary conidia. With *Z. phalloides*, it has been demonstrated how aphids collect capilliconidia on their legs when walking across leaf surfaces, making this spore type ideal for infecting low-density mobile hosts (Glare *et al.*, 1985).

Another mechanism to assist dispersal is through growth of hyphae out of a cadaver. The *Cordyceps* spp. and some of their anamorphs produce extensive stroma, which can grow more than 30 cm (e.g., Evans, 1982) and produce sexual and/or asexual spores at the end. Even common asexual species such as *B. bassiana* sometimes grow several centimeters from the infected cadaver in soil. Movement of spores in the soil environment is much more limited than distribution in water or air. However, there is some movement of spores in soil, whether down the soil profile by water movement, or through distribution by invertebrates.

Insect behavior can assist dispersal. Some fungus mediated behavioral changes have been recorded, such as "summit disease" where infected grasshoppers climb to the top of plants to die (see Section 11.3.3.2). This may aid spore dispersal, but it has also been suggested that it is a mechanism by which the host reduces infection through increased exposure to UV. Movement of insects such as scavengers (e.g., collembolans; Dromph, 2001), natural enemies (e.g., coccinellid predators or parasitoids; Roy and Pell, 2000; see also Section 11.3.6), and pollinators (e.g., honeybees; Butt *et al.*, 1998) can act to disperse fungal inoculum. In some situations, insects may actually be attracted to diseased cadavers. For instance, adult *Rhagonycha fulva* have been observed to visit cohort cadavers (**Figure 16**). Use of insects to disseminate beneficial microbial control agents is further discussed by Vega *et al.* (2000).

Insect behavior can also lead to avoidance of infection. There is evidence that some insects actively avoid spores of entomopathogenic fungi. Scarab larvae were found to move away from soil containing *M. anisopliae* mycelia, but not from conidia (Villani *et al.*, 1999). The same species laid eggs preferentially in areas where mycelia were present, presumably because the mycelia respiration mimics plant root growth. Social insects are particularly adept at behavioral methods to avoid disease. For instance, termites have been shown to avoid conidia of *M. anisopliae* (Milner and Staples, 1996), and the common vespid wasps and honeybees use hygienic behavior to eject diseased individuals from the nest to reduce disease transmission (Gilliam *et al.*, 1983; Harcourt, unpublished data).

Some fungi use the host to aid dispersal directly. The fungus *S. castrans* sporulates from live hosts (**Figure 18**), thereby increasing distribution. Transmission in this species is also assisted by collembola, which can colonize the abdominal holes where the conidia are produced (Griffiths, 1985).

The production of resting spores is a major factor in transmission of many fungi, allowing the fungus to avoid unfavorable environmental conditions or lack of host. Resting spores do not synchronically germinate for most species of the Entomophthorales, increasing the temporal distribution of inoculum and the chances the fungus will eventually encounter new hosts. These spores can be long lived.

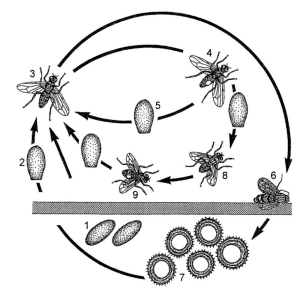

Figure 18 Life cycle of *Strongwelsea castrans*. (1) Pupae of *D. radicum* overwinter in the soil. (2) During spring, it is assumed that infective conidia of *S. castrans* are produced and discharged from resting spores in the soil. (3) Adult *D. radicum* emerge from pupae during spring and become infected by the conidia; one to two abdominal holes appear in the insect as a result of infection. (4) Conidia are forcibly discharged from the abdominal holes up to approximately 30 µm while the fly is still alive. (5) Discharged conidia, eventually after replicative conidiation, infect other adult *D. radicum*. Several successive infection cycles can take place during one season in the host population. (6) After midsummer, some *S. castrans* infected *D. radicum* develop resting spores instead of conidia. Following an incubation period, flies filled with resting spores die and drop to the soil surface. (7) Thick-walled resting spores survive in the soil layers during winter. (8) An alternative dipteran host gets infected. (9) A hole develops and primary conidia are discharged.

Persistence of *E. maimaiga* resting spores was found to be up to 6 years in forest soils (Weseloh and Andreadis, 1997), while the infection of the 17-year cicada by *Massospora* apparently depends upon resting spore germination over the full time frame (Soper *et al.*, 1976).

Transmission in many terrestrial fungi is dependent upon high humidity, with moisture required for sporulation and germination of conidia. It is not surprising that fungi have evolved to take advantage of the early morning dew through timing of host death. Timing death to occur just prior to dusk allows many entomophthoraleans to sporulate during the humidity of early morning. Despite the active discharge, conidia of Entomophthorales rarely reach above crop canopies and for *Pandora neoaphidis*, did not travel more than 1 m, even with wind assistance (Glare and Milner, 1991).

The ability to attack multiple life stages is important in disease transmission. Winged insects can be responsible for the spread of disease, especially between populations and regions. Movement between sites for aquatic fungi, such as mosquito pathogens, is likely to be through adults, whether the adults are infected or just carrying spores. For instance, *Coelomomyces* infected adult female mosquitoes mate and take blood normally, but oviposit sporangia instead of eggs (Lucarotti, 1987). Small mammals and birds may also spread infective spores after ingesting diseased cadavers (Christensen *et al.*, 1977). Transmission between adults and larvae has been demonstrated for some Hyphomycetes, such as the beech-boring beetles of *Platypus* (Coleoptera) inoculated with *B. bassiana* (Glare *et al.*, 2002) and diamondback moth with *Z. radicans* (Pell *et al.*, 2001).

In aquatic environments, motile spores, such as the zoospores of *L. giganteum* and *Coelomomyces* spp., are able to move through the water in search of hosts. Even the resting stage, oospores and sporangia, produce zoospores, combining motility with long-term survival. Zoospores are the main dispersal stage for *L. giganteum* and are actively attracted to a chemotactic agent emitted by mosquito larvae (Domnas, 1981). *Coelomomyces* spp. zoospores may also be attracted to larvae in water (Kerwin, 1983; Federici and Lucarotti, 1986). Zoospores can travel up to 10 cm in still water (Jaronski, 1982) and many meters in natural systems, but cannot overcome physical barriers to move between discontinuous systems. This is not the only method entomopathogenic species employ to infect hosts in aquatic environments. The aquatic *Culicinomyces* infect hosts using sessile spores, which can persist in water for up to 8 days before settling (Sweeney, 1985).

Lastly, the intentional introduction of fungi as biocontrol agents can be a significant means of dispersal in some fungi, and is discussed in more detail in Section 11.6.

11.3.6. Interactions between Pathogens and Other Natural Enemies

Most insects have a variety of diverse natural enemies, which include predators, parasitoids, and numerous pathogens. Many of these compete for the same resource and, inevitably, with each other. Insects can become coinfected by two or more pathogens or by a combination of parasitoid and pathogen(s). Predators or parasitoids can prey upon pathogen infected hosts. Fungi may infect nontarget insects, including natural enemies, and natural enemies can also act to disperse fungal inoculum. The outcome of all of these interactions can be either nil, antagonistic, additive, or synergistic. For the most part, interactions between fungal entomopathogens and other natural enemies is positive (Roy and Pell, 2000).

The same insect population can be attacked by several species of fungi. Studies on aphids, like cereal aphids (*Diuraphis noxia, Rhopalosiphon padi, Sitobion avenae*, and others) and their fungal pathogens from Entomophthorales (*Conidiobolus obscurus, Entomophthora planchoniana, P. neoaphidis*, and others), have often documented this situation at the population level (Keller and Suter, 1980; Dedryver, 1981; Feng *et al.*, 1990). Each fungus species will contribute to the overall mortality in the aphid population, but may differ in their prevalence. For example, some studies documented that in cereal systems, *P. neoaphidis* was the most common species in 2 out of 4 years, while *E. planchoniana* was the most common in the other 2 years (Eilenberg *et al.*, unpublished data). In carrot flies, *Entomophthora schizophorae* is normally the only common fungal species, but one year in August, *Conidiobolus pseudapiculatus* was the most common species (Eilenberg, 1988; Eilenberg and Philipsen, 1988).

Different fungal species can affect different host stages in the same host population. For instance, the chrysomelid beetle, *Coleomera lanio*, is susceptible to *B. bassiana* and *M. anisopliae*, but the former infects larvae, pupae, and adults while the latter infects only pupae (Silveira *et al.*, 2002).

The same host specimen can be naturally infected by two fungi simultaneously, but it seems that this normally occurs at lower rates than would be expected from a statistical point of view. In laboratory studies, Inglis *et al.* (1999a) studied the effects of temperature on the competitive infection and colonization of grasshoppers coinoculated with *B. bassiana* and *M. anisopliae* var. *acridum*. They found that coapplication of the two fungi did not significantly affect the prevalence of mortality, but that in some instances, total fungal populations in the hemocoel of coinoculated grasshoppers were smaller than those in grasshoppers treated with each fungus alone, indicating a degree of interspecies antagonism. More *B. bassiana* than *M. anisopliae* was recovered from coinoculated grasshoppers incubated at a constant temperature of 25 °C, but as the amplitude of temperature differences increased, more *M. anisopliae* than *B. bassiana* was recovered, indicating that temperature influenced their competitiveness within the host cadaver. In contrast, using an avirulent strain of *B. bassiana* and a virulent strain of *M. anisopliae* var. *acridum*, and the desert locust as a host, Thomas *et al.* (2003) were able to demonstrate that even an avirulent pathogen can alter the virulence and reproduction of a second, highly virulent pathogen. Locusts preinoculated with the avirulent fungus prior to inoculation with the virulent fungus exhibited a reduction in survival time and a significant decrease in sporulation of the virulent fungus compared to individuals inoculated with the virulent fungus alone.

Insect pathogenic fungi may also interact with other types of insect pathogens. Eilenberg *et al.* (2000a) documented that the fungus *S. castrans* and the microsporidium *Cystosporogenes deliaradicae* occurred commonly together in adult *Delia radicum*. In the same study, the bacterium *Bacillus thuringiensis* was also found sporulating in the abdomen in two fungus infected hosts, but serotyping and toxicity tests proved that the bacterial isolates belonged to serotypes that were avirulent to the host. As mentioned above, avirulent insect pathogens may play an important role in host–pathogen dynamics for the virulent insect pathogens. Another example of coinfection is found in the gypsy moth, *L. dispar*, which is naturally infected by the fungus *E. maimaiga* and the nuclear polyhedrosis virus LdNPV. Both pathogens are highly virulent and laboratory studies demonstrated that, depending on dosage and timing, there was sometimes an apparent synergistic interaction between the two pathogens, thereby significantly increasing overall mortality (Malakar *et al.*, 1999).

Insect pathogenic fungi can interact with parasitoids (Goettel *et al.*, 1990a). The two types of natural enemies may compete for the host's tissues and can, therefore, be regarded as antagonistic to one another. For instance, none of the parasitoids *Cotesia plutellae* and *Diadegma semiclausum* discriminated between healthy and *Zoophthora radicans* infected host *P. xylostella* larvae in oviposition experiments, although such infection always resulted in the death of the immature parasitoids (Furlong and Pell, 2000). The fungus also infects *D. semiclausum*, and can thus in several ways be regarded as potentially antagonistic in field situations. In contrast, some parasitoids are able to discriminate between healthy and infected hosts, avoiding oviposition in infected cadavers (Goettel *et al.*, 1990a). The parasitoid *Aphidius ervi* initiated fewer attacks on colonies of the aphid *Acyrtosiphon padi* that were subjected to the fungus *P. neoaphidis* on the previous day than on uninoculated colonies (Pope *et al.*, 2002). Several aspects in the interactions between the parasitoid *Aphelinus asychis*, the pathogen *P. farinosus*, and the host *D. noxia* demonstrate the potential of synergistic or additive interactions: the parasitoid discriminates fungal infected hosts thereby avoiding ovipositing in them; parasitoids emerging from fungal inoculated host populations are healthy; and there is a possibility that the fungus

may be transmitted by the parasitoids (Mesquita and Lacey, 2001).

Insect predators can, of course, be infected by insect pathogenic fungi. Some insect pathogenic fungi occur in both predator and prey insects in specific ecosystems, but infection levels in the predators are usually low (Vestergaard *et al.*, 2003). It is still unknown whether the genotypes of the fungal isolates occurring in the different insect populations are identical and, furthermore, whether transmission of infection between predator and prey population is of significant importance. Predators can be antagonistic to fungal pathogens simply by eating fungus killed cadavers or parts thereof, thereby reducing subsequent fungal sporulation and spread. Roy *et al.* (1998) documented this effect in their studies of the predator *Coccinella septempunctata*, the fungus *P. neoaphidis*, and the prey and host aphid *Acyrthosiphon pisum*. The same predator could, however, vector the fungal disease to uninfected hosts and thereby enhance fungal dissemination and epizootic potential; this effect was found to be enhanced due to the increased activity of the pea aphid, *A. pisum*, to avoid the predators, at the same time resulting in a higher uptake of inoculum of *P. neoaphidis* and consequently higher infection levels (Roy *et al.*, 1998, 2001).

11.4. Evaluation of and Monitoring Fate

Evaluation of use of entomopathogenic fungi in biological control is an extremely important, yet complex, task. The most obvious evaluation criterion is control of the pest population. In situations where measurement of pest population changes is not difficult, the success or failure of biocontrol can be seemingly straightforward. However, even in such situations, the sublethal effects, such as reduced feeding, fecundity, or crop damage, should not be overlooked. In many situations, the actual number of pests is unimportant compared to the damage they cause or disease they vector. In these cases, evaluation depends more on reducing damage or disease than the pest populations themselves, the ultimate evaluation criterion being crop quality or yield.

With changes in regulations and concerns over safety of introduced agents, more emphasis is being placed on monitoring the fate of fungi (and other biological agents) after release. Monitoring the fate of entomopathogenic fungi in the environment can be important in ecological studies, for commercial evaluations, for measuring establishment, and for regulation and safety. In particular, there is concern and interest in the release and monitoring of genetically modified organisms in the environment, including entomopathogenic fungi. Bidochka (2001) identified three main benefits from monitoring the fate of fungi: (1) identification of the introduced agent in a biological impact assessment, (2) economic cost benefit assessment, and (3) valuable information on fungal epidemiology.

Despite many years of research into entomopathogenic fungi, there remains an inadequate understanding of epizootiology, due in part to lack of specific methods for tracking individual isolates in the environment. The fate of released fungi, the genetic structure of fungal populations, and how to measure these components has not been well explored.

There are several possible ways in which to study persistence of artificially augmented fungal inoculum levels, including leaf and soil washes, leaf impressions, differential staining, bioassay of field collected substrates, and monitoring of sentinel insects (Goettel *et al.*, 2000). Comparisons can be made to untreated areas. When there are no longer differences in population levels between treated and untreated areas, it can be concluded that the augmented fungus has returned to background levels. The number of fungal propagules in the environment has often been used as a measure of the potential success of biocontrol applications. Propagule loadings have been measured by colony counts on semiselective media, where the fungus can be cultured, or through spore traps for fungi that discharge their spores into the air, such as the Entomophthorales. However, these methods do not distinguish between applied and background strains.

Methods that have been used to study the fate of entomopathogenic fungi at the specific strain level include phenotypic markers, vegetative compatibility, mating types and, more recently, a plethora of molecular approaches, which allow specific tracking of individual strains in most environments. When strains can be cultured, studies have used markers such as colony appearance or antibiotic resistance, but these techniques suffer from genetic instability and alterations due to environmental factors. Other nonmolecular based markers have been investigated, such as commercially available carbohydrate utilization or enzyme production strips system of API (i.e., API50CH; APIZyms), but with limited success (St. Leger *et al.*, 1986; Todorova *et al.*, 1994; Rath *et al.*, 1995). These systems can be used to distinguish groups, but are unlikely to distinguish individual isolates in most cases. Therefore their usefulness in tracking released strains is usually minimal.

Vegetative compatibility groups (VCGs) are subspecific groups within the same fungal species from which isolates can fuse hyphae to form heterokaryons and, thus, exchange genetic information (the parasexual cycle). Couteaudier and Viaud (1997) demonstrated that compatibility groups, within *Beauveria* spp. in Europe, could be used to delimit genetically incompatible groups and aid monitoring.

Many entomopathogenic fungi cannot be cultured and the lack of simple methods for isolation and specific strain characterization has made it difficult to conduct ecological studies on fungal persistence. The most important advances in monitoring the fate of entomopathogenic fungi in the environment have come from the application of molecular biology techniques. Precursors to DNA techniques were the use of allozymes to identify strains, and use of enzyme-linked immunosorbent assay (ELISA) for tracking in the environment. Application of allozyme analysis to strain detection in the environment has been attempted, although single isolate identification amongst strains from a single host has not proved possible. For example, the allozymes of *Beauveria* and *Metarhizium* spp. have a high degree of variability, indicative that the species had maintained a large effective population size for a long time (St. Leger *et al.*, 1992b). In some cases, isoenzymes have not been useful in separating applied and background strains (e.g., Trzebitzky and Lochelt, 1994) or showed little correlation with virulence (Reineke and Zebitz, 1996), but Lin and Lin (1998) found esterase patterns of *Beauveria* isolates correlated with geographical distribution and host. Allozymes have also been used with the Entomophthorales. Silvie *et al.* (1990) used allozymes to study the fate of *P. neoaphidis* released for greenhouse aphid control. Milner and Mahon (1985) used similar techniques to distinguish between an introduced Israeli strain of *Z. radicans* for aphid control and endemic strains pathogenic to flies and caterpillars. Hajek *et al.* (1990) used allozymes to confirm that the gypsy moth pathogen causing epizootics in North America was *E. maimaiga*. In further studies using allozymes and RFLPs, they demonstrated that *E. maimaiga* and *E. aulicae*, two very similar species, were attacking different caterpillars in the same area (Hajek *et al.*, 1991b). ELISA was also developed for *E. maimaiga* based on allozymes, but was cross-reactive for *E. aulicae* (Hajek *et al.*, 1991a).

Application of DNA techniques is making major advances in monitoring the fate of entomopathogenic fungi in the environment. There are numerous studies that have demonstrated the power of molecular biology in strain, species, and genera separation (see Section 11.2.5). These same techniques are useful when applied to monitoring fungal isolates in the environment. For example, Enkerli *et al.* (2001) used 10 microsatellite markers to characterize *B. brongniartii* and could use these markers to demonstrate that strains applied in Switzerland in 1983 were still active over 8 years later. Specific identification of *B. brongniartii*, based on polymerase chain reaction (PCR) primers designed to group 1 intron insertions in the 28s rRNA gene, was used to identify and monitor an introduced strain for control of the scarab pest, *Hoplochelus marginalis*, in the Reunion Islands (Neuvéglise *et al.*, 1997). Other techniques, such as chromosomal length polymorphisms, RAPDs, and RFLP have all been used to assist monitoring strains in the environment. Molecular approaches to phylogeography of insect pathogens are also developing. Using a molecular genetics approach, Bidochka *et al.* (2001) found that genotypes of *M. anisopliae* in Canada were more influenced by habitat than host.

An extension of the use of molecular techniques for distinguishing strains important in biological control is the insertion of sequences (tags) or genes which can be used to specifically assist monitoring. For example, insertion of a green fluorescent protein gene was used to allow detection and tracking of a genetically modified *M. anisopliae* (Hu and St. Leger, 2002).

Media based detection methods, such as selective agars, can determine the number of viable propagules in the environment, but generally cannot allow separation between applied and background strains. Molecular approaches can allow separation of strains, but while these molecular approaches are extremely powerful, most techniques do not provide information on the activity (viability) of the propagules detected. Even though fungi can be detected, it is not possible to say whether the detected fungus is able to infect hosts. Use of *Galleria* or other sentinel insects, such as in the soil baiting method (Zimmermann, 1986), can demonstrate virulence, but gives no quantification of propagules. A combination of methods is required to monitor applied fungi in the environment, to provide information on the number of viable propagules of the applied fungus present over time.

11.5. Epizootiology and Its Role in Suppressing Pest Populations

Entomopathogenic fungi are well known for their ability to rapidly decimate insect outbreaks through spectacular epizootics, and it is principally this property that has spurred interest in the use of fungi for pest management. The term "epizootic"

is used to describe a situation where the proportion of infected individuals is high or at least higher than the enzootic condition, i.e., the disease condition that is usually of low prevalence and constantly present in a host population (Tanada and Kaya, 1993). Prevalence is defined as the number or proportion of infected individuals at a given point in time (Fuxa and Tanada, 1987). Natural epizootics caused by fungi are an important factor in regulating pest insect populations, often alleviating the need for other interventions, such as application of chemical pesticides. Here, some examples of fungi with a role in suppressing pest populations in epigeal, soil, and aquatic environments are provided.

11.5.1. Epigeal Environment

Epizootics often occur within the epigeal environment. Infection may spread rapidly directly from infected insects to uninfected insects, from spores on leaves to insects feeding on the leaves, and also via spores present in the air.

High prevalence is often found among aphids (e.g., Feng *et al.*, 1991; Hollingsworth *et al.*, 1995; Hatting *et al.*, 2000). Hollingsworth *et al.* (1995) monitored the fungus *N. fresenii* in populations of *Aphis gossypii* in cotton in Arkansas, USA. Epizootics were common and prevalences could reach 90%. Such prevalences were detrimental to the aphid populations, which actually began to decline when the prevalence reached 15%. Also, aphids living in cold environments suffer from fungus infections. Nielsen *et al.* (2001) sampled *Elatobium abietinum* on sitka spruce in Iceland. Both *Entomophthora planchoniana* and *N. fresenii* were present in the aphids and were responsible for epizootics during the autumn, when daily mean temperatures were below 15 °C. However, the effect on the host population was not documented during this study.

Another group of insects often found infected with fungal pathogens are dipterans. Steinkraus *et al.* (1993) studied the prevalence of *Entomophthora muscae* in populations of *Musca domestica* and found prevalences of up to ~75% towards the end of the season. In this case, as in other cases, the epizootic builds up during the season. In studies of *M. domestica* and *Entomophthora schizophorae*, Six and Mullens (1996) documented that maximum daily temperatures higher than 26–28 °C correlated with low prevalences.

Entomopthoralean fungi from the *Entomophaga grylli* species complex are important pathogens of acridids worldwide (Carruthers *et al.*, 1997). They commonly cause epizootics, particularly following periods of high humidity or rain. Such epizootics have frequently led to dramatic reductions in acridid populations, saving farmers and ranchers millions of dollars. The epizootics are very noticeable, as thousands of cadavers can be seen adhering to the tops of vegetation (see Section 11.3.3.2).

The host density-dependent development of epizootics was documented in the *L. dispar* populations infected with *E. maimaiga*. This fungus has shown remarkable dispersal over a period of several years (or maybe even decades) in north-eastern USA (Hajek, 1997, 1998, 1999). An interesting adaptation to the host is seen, since first instar larvae produce infective conidia of the fungus while larger larvae produce resting spores for winter survival. The effect on the host population is significant and the fungus is responsible for decimating large populations of this destructive, introduced pest.

Recent examples of epizootics in other insect groups from the epigeal environment include *N. parvispora* on thrips (*Frankliniella occidentalis*) in Spain (Montserrat *et al.*, 1998), *Zoophthora phytonomi* on weevils in alfalfa in the USA (Morris *et al.*, 1996), and *Entomophthora leyteensis* on whiteflies in the Philippines (Villacarlos *et al.*, 2003). It appears that epizootics occur worldwide and in all major insect orders.

11.5.2. Soil Environment

Due to the cryptic nature of insects inhabiting the soil environment, natural epizootics most probably occur much more often than reported. It appears that the main source of inoculum of several important Hyphomycetes is the soil and epizootics have often been observed among soil living stages of insects (Steenberg *et al.*, 1995; Fuhrer *et al.*, 2001). Using insects as baits, a number of studies have demonstrated the presence of spores from the genera *Beauveria*, *Metarhizium*, *Paecilomyces*, and *Tolypocladium* in soil (Klingen *et al.*, 2002; Keller *et al.*, 2003). Among Ascomycetes, *Cordyceps* spp. have often been observed on insect cadavers on the soil surface. A study in Colombia documented, for example, that species of *Cordyceps* were common in populations of ants (Sanjuan *et al.*, 2001). The effect of *Cordyceps* infections on insect populations is in need of further investigation.

The biological role of the soil ecosystem for the insect pathogenic fungi is, however, not clarified. The soil may simply act as a reservoir for a population of spores, which then infect host insects, or the soil ecosystems can be involved in a more complex interaction with insect pathogenic fungi. For instance, *B. brongniartii* was found in localities where the host, *Melolontha melolontha*, was absent for approximately 40 years (Keller *et al.*, 2003).

Either the fungus can survive by using other hosts or by saprophytic growth, or the spores can survive for many years. Hu and St. Leger (2002) suggest that the rhizosphere is an important interphase between insects, insect pathogenic fungi, and plants. Insects not susceptible to certain fungi may act to disperse fungal inoculum. Broza *et al.* (2001) and Dromph (2001) both documented that collembolans can be involved in transmission of insect pathogenic fungi without being infected themselves.

The presence of inoculum in the upper soil surface layers is important for the initiation of epizootics in the epigeal environment during spring. Nielsen *et al.* (2003) demonstrated that inoculum of epigeal, entomophthoralean pathogens (*C. obscurus*, *P. neoaphidis*) of cereal aphids (*Sitobion avenae* and *R. padi*) was present in the soil surface in early spring. Aphids became infected by these fungi when crawling on the soil surface. The epizootic would then develop in the epigeal environment.

11.5.3. Aquatic Environment

The aquatic environment is very different from terrestrial environments. Most insect pathogenic fungi infecting aquatic species have motile spores, such as zoospores, which can be actively attracted to the host (see Section 11.3.4). In aquatic environments, disease can be influenced by a range of abiotic and biotic factors, some of which are not experienced in other environments. For example, salinity, pH, organic and inorganic pollutants, and water movement can alter or stop infection. Salinity affects both *Coelomomyces* and *Lagenidium* infections. No *C. opifexi* infections were detected in salt water pools with over 20% salt, although the mosquito larval hosts could be detected in pools with twice that level of salt (Pillai, 1971). *Lagenidium giganteum* is even more susceptible to salinity with less than 0.05% salt levels in water required for high infection levels (Merriam and Axtell, 1982a). *Culicinomyces clavisporus* was more resistant to salt, infecting at 1.3% salt (Merriam and Axtell, 1982b). Temperature is also a major influence on disease and epizootiology, as with terrestrial fungi.

There has been substantial interest in some aquatic-active entomopathogenic fungi, mainly due to the pest status of mosquitoes and blackflies. However, many of these fungal species have complex life cycles unsuited to biocontrol, or have limited prevalence in most situations. *Coelomomyces* have an obligate intermediatory host (also referred to as an "alternate host"), often a copepod, between infecting mosquito larvae. This has both limited the occurrence of the pathogen, as well as its use in intentional introductions. Copepod abundance was found to be the critical factor in epizootics caused by *C. punctatus* (Apperson *et al.*, 1992). Pillai (1971) sampled over 400 pools in New Zealand and detected *C. opifexi* in only 21. When pools that had both mosquito and copepod hosts were intentionally inoculated, disease levels persisted for over 2 years, but disease incidence remained low (mean <16%). It seems that the pathogen remained in equilibrium with its hosts, even in a closed system of saltwater pools. There is evidence that *L. giganteum* persists better in high density than in low density populations, where the fungus may struggle to recycle.

Unlike the zoospore-infective species, the hyphomycete mosquito pathogen *C. clavisporus* is infective after ingestion. This reduces the need for a motile stage, as conidia present in the water are ingested, rather than needing to find and attach to the host. However, settling of conidia becomes a major factor in epizootiology. Applying high rates of conidia to ponds can kill up to 100% of mosquito larvae but there is no residual prevalence after 2 days (Sweeney, 1981). However, more recently, Seif and Shaarawi (2003) found that *C. clavisporus* was active for 5 days against mosquito larvae in brackish ponds in Egypt.

The ability to persist through times of habitat drying is a useful attribute for aquatic pathogens. *Lagenidium* and *Coelomomyces* both produce thick-walled resting stages (oospores and sporangia). *Lagenidium* can be unaffected by periodic drying (e.g., Fetter-Lasko and Washino, 1983).

Lagenidium giganteum and *C. clavisporus* have been the subject of development as biopesticides, which has resulted in epizootiological studies and inundative release. *Lagenidium giganteum* is now the basis of a commercially available mosquitocide. However, research on *C. clavisporus* peaked in the 1970–80s in Australia, and did not lead to a commercial product. Attempts at introducing *Coelomomyces* as a biocontrol agent in the classical sense into the Tokelau Islands provided inconclusive results (Laird, 1985).

11.6. Development as Inundative Microbial Control Agents

Inundation biological control is defined as

> The use of living organisms to control pests when control is achieved exclusively by the released organisms themselves
>
> Eilenberg *et al.* (2001)

It is attractive from the point of view that the effect on the target host is obtained with the released agent itself and is often regarded as *the* strategy in biological control with microorganisms. Terms like "biopesticide," "biological pesticide," or "mycopesticide" are frequently used to describe inundation biological control with fungi. These terms also reflect that this strategy is in many ways an "insecticidal approach" in that the inoculum has to be applied directly to the crop or host, often several times per season, which is attractive from a commercial point of view.

Inundation biological control is, however, not necessarily the most sustainable way to control pests, especially if the biological control agent is used simply as a replacement for chemical insecticides. Replacement of a chemical with a microbial control agent could result in the same type of problems associated with chemicals, namely development of populations resistant to the control agent, be it chemical or biological, as has occurred with the bacterial microbial agent *B. thuringiensis* (Tabashnik, 1994). Although no instances of the development of resistance to an entomopathogenic fungus have been documented to date, it must be realized that inundative use of entomopathogenic fungi has been limited and, therefore, the prerequisites for selection of a resistant population would, in most instances, not have been met as yet.

Many Hyphomycetes are well suited for the inundative strategy based on their relatively wider host ranges, ease of production, formulation and application, and a relatively long shelf life. Several species of fungi from Hyphomycetes have been developed commercially and are currently for sale worldwide (see Section 11.6.2 below).

Entomophthoralean fungi possess some characteristics that allow them to fit into the inundation biological control paradigm, while other characteristics suggest that these fungi would fail under these conditions (Pell *et al.*, 2001). The generally high virulence of many species is a useful characteristic for this strategy, while the presence of replicative, infective but short-lived conidia hampers the development of "easy to use products." A further drawback is the fact that isolation *in vitro* is difficult and has been achieved only for very few species. Efficient mass production methods are mostly lacking as is knowledge on formulation.

In greenhouses, entomophthoralean fungi may have a better chance. The most thorough recent attempts were made by Shah *et al.* (2000). Experimental preparations based on alginate granules and mycelial mats of *P. neoaphidis* caused up to 36% infections in the potato aphid *Macrosiphon euphorbiae* after 14 days, although better results were obtained with *B. bassiana*.

In most temperate countries, greenhouses are the only place where inundative biological control of crops with entomophthoralean fungi could be recommended at present. Preparations of *P. neoaphidis* could be tested against aphids in greenhouses and may, due to some horizontal transmission effects afterwards, prove to have some advantages over other biological control agents used, including *Lecanicillium lecanii*. For some of the greenhouse pest–pathogen systems mentioned in Section 11.7, there could also be some prospects in inundation. Whatever the environment, a most promising release strategy could be a combination: an inundation with some immediate effect but including a significant inoculative effect.

11.6.1. Mass Production

The success of any commercial microbial control agent depends largely on the ability to produce adequate supplies of the agent to economically manage the target pest or pests. Entomopathogenic fungi produce different types of spores, depending on the growth environment. Aerial conidia are produced on the surface of cadavers, while blastospores, sometimes referred to as "hyphal bodies" or protoplasts, are produced in the host's hemocoel. Consequently, the fermentation medium and method will also dictate the type of spore that will be obtained. In submerged culture, primarily mycelia and blastospores are produced while aerial conidia are produced on the surface of liquid media or on solid substrates.

Efficient mass production methods are mostly lacking for entomophthoralean fungi as is knowledge on their formulation. Despite the development of methods for mass producing several species, to date, there are no commercialized products based on entomophthoralean fungi. Due to the short-lived nature of entomophthoralean conidia, attempts at mass production have targeted primarily mycelia or resting spores (Pell *et al.*, 2001).

Many insect cadavers killed by fungi can be found in the environment in a dehydrated state; often, transfer of these cadavers to conditions of high humidity results in massive conidiogenesis and spore release (e.g., Tyrrell, 1988). This phenomenon has spurred the development of "dried mycelia" as a potential way of mass producing and using entomopathogenic fungi (McCabe and Soper, 1985). Mycelia are mass produced in deep fermentation, air dried in the presence of a sugar desiccant protectant, granulated, and stored dehydrated under low temperature. Upon rehydration, the fungus sporulates on the granule surface. This method is suitable both for entomophthoralean and hyphomycetous fungi. Despite some successful field demonstrations using dried mycelia against several insect pests (e.g., Pereira and Roberts, 1991; Krueger *et al.*, 1992; Krueger and Roberts, 1997; Booth *et al.*, 2000; Shah *et al.*, 2000), this method has not yet been adopted commercially.

Some entomopathogenic fungi produce conidia or resting spores in submerged culture. Examples are resting spores of the entomophthoraleans, *Conidiobolus obscurus* (Latgé, 1980) and *E. maimaiga* (Kogan and Hajek, 2000), and conidia of the hyphomycetous fungi *Culicinomyces clavisporus* (Cooper and Sweeney, 1986) and *M. anisopliae* var. *acridum* (Jenkins and Prior, 1993). Submerged conidia of some species perform as well as aerially produced ones (e.g., *M. anisopliae* var. *acridum*; Jenkins and Thomas, 1996). For other species this is not so, e.g., submerged conidia of *C. clavisporus* performed worse than aerial conidia (Cooper and Sweeney, 1986). To date, there are no commercialized products based on conidia or resting spores produced in submerged culture.

Most Hyphomycetes produce blastospores in submerged culture, but for the most part, these are thin walled and short lived, and with few exceptions, are used primarily to produce starter batches for use in inoculation of aerial culture (see Section 11.6.2). In some cases, blastospores are more virulent than aerial conidia (e.g., Jackson *et al.*, 1997; Lacey *et al.*, 1999), and several commercial products consist of blastospores produced in submerged culture (e.g., *L. lecanii* and *P. fumosoroseus*; see below).

Solid state fermentation is used for production of aerial conidia of a number of commercially produced Hyphomycetes (see below). The solid substrates can be either nutritive, such as bran, cereals, or rice, or nonnutritive, such as vermiculite, clay, or sponge. Production of *B. bassiana* and *M. anisopliae* on rice is widely practiced in South America (Alves *et al.*, 2003), Africa (Cherry *et al.*, 1999), and China (Feng *et al.*, 1994).

More details on the production of entomopathogenic fungi can be found in Bartlett and Jaronski (1988), Jenkins and Goettel (1997), Jenkins *et al.* (1998), and Wraight *et al.* (2001).

11.6.2. Commercialized Products

There are many commercial products available worldwide, most based on fungi from six to seven species from Hyphomycetes (Shah and Goettel, 1999; Copping, 2001; Wraight *et al.*, 2001; Alves *et al.*, 2003). Hyphomycetes are very well suited for commercialization due to their relatively wider host ranges, ease of production, shelf life, persistence, and ease of application.

11.6.2.1. *Beauveria bassiana* The most widely used species available commercially is *B. bassiana*. Products based on this species are available for use against a very wide variety of insect pests, from banana weevils (*Cosmopolites sordidus*) in Brazil (Alves *et al.*, 2003), pine caterpillars (*Dendrolimus* spp.) in China (Feng *et al.*, 1994), to the European corn borer, and greenhouse aphids in the Western world (Shah and Goettel, 1999; Copping, 2001). Formulations consist of aerially produced conidia produced as wettable powders or in emulsifiable oil.

The largest applications of *B. bassiana* have occurred in China, where an estimated 10 000 tons of conidia were produced annually for decades to control a wide range of forest and crop pests (Feng *et al.*, 1994). Most of these conidia were produced in state-owned, mechanized Chinese production facilities. However, with the recent transition from a state-planned and directed system to a market driven one, *B. bassiana* products have come to be considered as marginal and have not attracted the necessary attention from investors (Feng, 2003).

A solid culture process based on a packed bed of a proprietary starch granule has been developed to produce conidia that are marketed as Botanigard and Mycotrol in the USA and elsewhere (Bradley *et al.*, 1992). The products are used primarily to control aphids and whiteflies in commercial greenhouses. With host ranges that include over 700 species of arthropods (see Goettel *et al.*, 1990a), the market potential for more commercial products based on isolates of *B. bassiana* is large.

11.6.2.2. *Beauveria brongniartii* Products based on *B. brongniartii* are available against a wide variety of coleopteran, homopteran, lepidopteran, and dipteran pests in flowers, vegetables, oil palms, and other crops in Colombia (Alves *et al.*, 2003), against cerambicid beetles in Japan (Wraight *et al.*, 2001), and against the European cockchafer and other white grubs in Europe (Copping, 2001). Conidia are produced in solid-state fermentation either on barley kernels or clay granules. In Peru, the fungus has been widely used to control the Andean potato weevil (Alves *et al.*, 2003). The soil beneath potato storage facilities is treated with spores of the fungus prior to introduction of the potato tubers. Weevils become infected when the larvae drop to the soil to pupate.

11.6.2.3. *Lecanicillium lecanii* *Lecanicillium* (= *Verticillium*) *lecanii* is primarily marketed for control of greenhouse aphids, whiteflies, and thrips (Copping, 2001; Wraight *et al.*, 2001); however, products are also available in Colombia and Peru

for control of lepidopteran, homopteran, and dipteran pests of flowers, vegetables, and other crops (Shah and Goettel, 1999; Alves *et al.*, 2003). Blastospores produced in submerged fermentation or conidia produced in solid-state fermentation are used in the various products. Previously, the requirement for very high humidity was a serious detriment in adopting this fungus for use in greenhouses, where high humidities also favor development of plant diseases. However, development of an improved formulation that decreases the humidity requirements of the fungus has alleviated much of this problem (Ravensberg, personal communication).

11.6.2.4. *Metarhizium anisopliae* (including var. *acridum*) Products based on *M. anisopliae* are available for a wide range of pests, including sugarcane spittle bugs (*Mahanarva* spp.) in Brazil (Alves *et al.*, 2003), red-headed cockchafer in Australia (Shah and Goettel, 1999), and termites in the USA (Copping, 2001). Products based on *M. anisopliae* var. *acridum* are available for control of grasshoppers and locusts in Africa (Lomer *et al.*, 2001) and Australia (Copping, 2001). Production systems vary, but most depend on culture in small bags or containers using rice as the substrate (Mendonca, 1992; Jenkins *et al.*, 1998). Such simple technologies may be more appropriate and economically feasible for smaller scale technologies, especially in developing countries, than larger, more sophisticated and capital intensive industrial scale production facilities (Swanson, 1997; Cherry *et al.*, 1999).

11.6.2.5. *Paecilomyces fumosoroseus* Formulations of *P. fumosoroseus* are marketed against whiteflies, thrips, spider mites, and aphids in Latin America, Europe, and North America (Copping, 2001; Alves *et al.*, 2003). Blastospores, produced in submerged culture, are dehydrated and formulated in a water dispersible bran granule or a water dispersible powder. The products are used primarily in greenhouse applications.

11.6.2.6. Others Numerous other products are currently available in many countries. Some examples include "*Entomophthora virulenta*" (= *Conidiobolus thromboides*) for control of whiteflies in Colombia (Alves *et al.*, 2003), *N. rileyi* for control of lepidopterans in Colombia (Shah and Goettel, 1999; Alves *et al.*, 2003), *P. lilacinus* for control of plant parasitic nematodes in Australia (Copping, 2001), and the Oomycete, *L. giganteum*, for control of mosquito larvae (Shah and Goettel, 1999; Copping, 2001).

Several products are based on microbial "cocktails." For instance, in Colombia, a product called

"Microbiol Completo" consists of a mixture of the fungi B. bassiana, M. anisopliae, N. rileyi, P. fumosoroseus, and the bacterium Bacillus thuringiensis. This mixture is marketed for use against a wide variety of pest lepidopterans, coleopterans, hemipterans, dipterans, and acari (Alves et al., 2003). Another cocktail, "Microbiol H.E.," contains a mixture of B. bassiana, M. anisopliae, M. flavoviride, N. rileyi, P. fumosoroseus, P. lilacinus, V. lecanii, and Hirsutella thompsonii, which includes activity against nematodes in addition to the pests listed under the Microbio Completo. Microbial cocktails not only increase host spectrum, but also may ensure activity under different environmental conditions. For instance, results from a field simulation study against grasshoppers suggested that B. bassiana itself would be more efficacious under cooler, cloudy conditions, although M. anisopliae var. acridum (as flavoviride) would be more efficacious under warmer, sunny conditions, while a cocktail of both fungi would be efficacious under both conditions (Inglis et al., 1997b).

11.6.3. Improvement of Efficacy

There are many factors that can affect the efficacy of biopesticides, such as slow speed of kill, moderate efficacy, poor storage characteristics, lack of field persistence after application, and expense (see Section 11.3; see also Inglis et al., 2001; Pell et al., 2001). Many of these factors can be overcome by strain selection, genetic improvement (St. Leger and Screen, 2001), formulation (Wraight et al., 2001), and application (Bateman and Chapple, 2001).

11.6.3.1. Strain selection
Entomopathogenic fungal species are composed of often genetically heterogeneous collections of strains, which can have different properties on key characteristics, such as virulence to pest species. Natural variation between strains of any given species has long been used in the search for biocontrol fungi. The selection among strains for the more virulent, better environmental persistence, or some other important variable, is a constant theme in the literature. A common first step in the search for fungal control agents for any pest insect is to assay strains from a number of locations to select the most pathogenic (e.g., Vicentini et al., 2001; Shaw et al., 2002).

In addition to utilizing natural variation, there have been efforts to improve biocontrol fungi through selection. Improving virulence by passage through hosts has been attempted, sometimes with success. Ogarkov and Ogarkova (1999) reported an up to twofold increase in virulence of B. bassiana to Trialeurodes vaporariorum after passage through the

host. Lagenidium giganteum increased virulence to two mosquito hosts by bi-weekly passage through another mosquito (Washino, 1981). Fargues and Robert (1983) restored lost virulence of M. anisopliae through serial passage through a scarab. These are all indicative of the potential for strain improvement through selection.

Improvement of other characteristics of entomopathogenic fungi have been investigated through selection. Increasing the endogenous reserves of conidia to enhance viability and desiccation tolerance was proposed by Hallsworth and Magan (1994). By growing cultures of B. bassiana, M. anisopliae, and P. farinosus under different conditions to obtain conidia with a modified polyol and trehalose content, conidia with increased intracellular levels of glycerol and erythritol were isolated. These conidia germinated more quickly and at lower water activity than unselected conidia (Hallsworth and Magan, 1995). Conidia with increased trehalose germinated more slowly but had better shelf life than unselected conidia.

More recently, there have been attempts to use biotechnological techniques, not just in direct genetic improvement of fungi, but to assist strain selection. As described in Sections 11.2.6 and 11.4, biochemical and molecular techniques have been used to identify specific strains of entomopathogenic fungi, allowing for more specific selection of strains following exposure to discriminatory conditions. For example, protoplast fusion has been used to combine characteristics of strains of Beauveria to improve the biocontrol potential. Fusion of a strain of B. bassiana from Leptinotarsa decemlineata with an insecticidal toxin-producing strain of B. sulfurescens, resulted in recovery of some di-auxotrophic mutants with enhanced activity against L. decemlineata and the caterpillar O. nubilalis (Couteaudier et al., 1996; Viaud et al., 1998). Molecular analysis demonstrated that the resulting mutants were mixtures of the two genotypes. Some of the resulting strains were classified as hypervirulent, as they killed the hosts more quickly than the parental strains. Virulence was stable after passage through the host, suggesting protoplast fusion may be an alternative method to direct genetic engineering for improving the biocontrol efficiency of entomopathogenic fungi (Viaud et al., 1998).

11.6.3.2. Genetic modification
The use of genetic manipulation to overcome perceived limitations of entomopathogenic fungi for use as biocontrol agents is attractive. The research so far is limited when compared to other groups of organisms, but some interesting research has been done. The most

advanced research has been on the hyphomycete fungi *M. anisopliae* and *B. bassiana*.

Transformation systems have been developed for some entomopathogenic fungi. The first transformations of an entomopathogenic fungus were by Bernier *et al.* (1989) and Goettel *et al.* (1990b), who expressed a benomyl resistance gene in *M. anisopliae*. Electroporation and biolistics have also been used to transform *M. anisopliae* (St. Leger *et al.*, 1995). Polyethylene glycol (PEG) mediated transformation of protoplasts is another method for transformation of entomopathogenic fungi, as done with *P. fumosoroseus* and *P. lilacinus* (Inglis *et al.*, 1999b) using benomyl as the selective agent. Sandhu *et al.* (2001) developed a heterologous transformation system for *B. bassiana* and *M. anisopliae* based on the use of the *Aspergillus nidulans* nitrate reductase gene (niaD), after EMS treatment of protoplasts and regeneration on chlorate medium.

Genetic modification of entomopathogenic fungi by direct manipulation has been led by St. Leger, and the area was recently reviewed by St. Leger and Screen (2001), covering both insect and plant pathogenic fungi. In the numerous publications, these researchers described the identification and cloning of protease genes, particularly the *pr1* cuticle degrading protease gene from *M. anisopliae* (St. Leger *et al.*, 1992a). Several other entomopathogenic Hyphomycetes (*A. flavus*, *B. bassiana*, *Paecilomyces farinosus*, *Tolypocladium niveum*, and *L. lecanii*) also have *pr1*-type enzymes (St. Leger *et al.*, 1991, 1992a). Efforts to produce more effective strains of fungi have included the overexpression of *pr1* by insertion of multiple copies, resulting in increased speed of kill of a host insect, although transformants had very poor sporulation ability (St. Leger, 2001). Modification of *pr1* gene expression in *M. anisopliae* resulted in melanization and cessation of feeding 25–30 h earlier than the wild-type disease in caterpillars (St. Leger *et al.*, 1996). The group has gone on to place the *pr1* gene under control of a constitutive promoter and included a green fluorescent protein encoding gene for tracking the fungi after field release (Hu and St. Leger, 2002). Genes involved in regulation of PR1 expression have also been identified from *M. anisopliae* (Screen *et al.*, 1997, 1998).

Chitinase, produced by many insect pathogenic fungi and implicated in pathogenesis, has also been the target of molecular studies. Several chitinase gene sequences from *Beauveria* and *Metarhizium* have been submitted to GenBank and described in publications (e.g., Bogo *et al.*, 1998). Overexpression of extracellular chitinase has also been demonstrated for *M. anisopliae* var. *acridum*, but did not alter virulence to the caterpillar, *M. sexta*, compared to the wild-type fungus (Screen *et al.*, 2001).

11.6.3.3. Formulation and application strategies

There are a number of limitations that still affect the use of fungi. Formulation and application of fungal propagules has the potential to overcome some of these limitations. Jones and Burges (1998) defined four basic functions of formulation: (1) stabilization of the agent during production, distribution, and storage; (2) aiding the handling and application of the product to the target in the most appropriate manner and form; (3) protection of the agent from harmful environmental factors at the target site, thereby increasing persistence; and (4) enhancing efficacy of the agent at the target site by increasing activity, reproduction, contact, and interaction with the target pest.

Progress has been made in formulation science, and formulations for entomopathogens have been comprehensively reviewed by Burges (1998). Early formulations of entomopathogenic fungi often used just a weak detergent to get the hydrophobic spores into solution, without any efforts to use the formulations to improve efficacy. More recently, the use of oil as a formulation ingredient has been successfully used to overcome several problems. In the LUBILOSA program developing *M. anisopliae* var. *acridum* for control of locusts in Africa, the combination of oil formulations with ultra low volume (ULV) spraying has led to successful development of a biopesticide (Lomer *et al.*, 2001). Although it would seem impossible to use an entomopathogenic fungus requiring high humidity against locusts in the hot, dry conditions prevalent in much of their environment, studies have shown that ambient humidity may not be as critical as once thought (Fargues *et al.*, 1997; see also Section 11.3.1). Formulating *Metarhizium* conidia in nonevaporative diluents such as oils allowed the conidia to attach, spread (Inglis *et al.*, 1996a), and germinate on susceptible locusts, even at low ambient humidities (Bateman, 1997).

Formulation can also assist in extending shelf stability of fungal propagules generally desired by commercial producers. Spore survival after mass production varies greatly, depending on the species, but few entomopathogenic species of fungi can achieve the 1–2 years of shelf stability at formulation. Formulation in clay material is known to improve the shelf life of many conidial preparations (e.g., Shi, 1998). This appears to be due to the ability of clay to reduce the moisture content of the spores, as spores survive longer with less than 5% moisture (Wraight *et al.*, 2001). Moore *et al.* (1996) have shown that survival of conidia of *M. anisopliae* was highest at low (<5%) relative humidity, and similar results have been found for other Hyphomycetes. Some oils can also maintain a shelf life of over 1 year for some Hyphomycetes, possibly due to maintenance of low spore moisture content.

Several groups have examined formulation of hyphal material from members of the Entomophthorales, which are more fragile than Hyphomycete conidia. McCabe and Soper (1985) patented a process of drying the mycelium of *Z. radicans* and coating it with sugar as a method for long-term storage and, more recently, Shah *et al.* (1999) demonstrated algination as a method for formulating *Pandora (Erynia) neoaphidis* mycelium. Formulations of dried mycelia of *M. anisopliae* and *B. bassiana* were developed by coating with a sugar solution before drying (Pereira and Roberts, 1990). Entomopathogenic fungi have also been formulated as granules (e.g., Reinecke *et al.*, 1990) or, in some cases, applied without removal from the grains used as a production medium.

Persistence is a major issue with fungal biopesticides. One of the major factors affecting persistence is ultraviolet (UV) radiation, which rapidly degrades fungal spores. Use of various UV protectants in formulations has been tested, but with only limited success (Burges, 1998; Wraight *et al.*, 2001). The most effective method to reduce UV degradation of spores has been to use application techniques that place the spores in the appropriate position to rapidly infect the host or to avoid UV, such as by applying subsurface. Subsurface application of entomopathogenic fungi has been used for *M. anisopliae* against scarabs (e.g., Logan *et al.*, 2000) and other insect pests, but introducing large amounts of fungal inoculum into the soil and securing an even spread remains problematic. Methods that have been used include using seed drills for subsurface application of granules and hand application. The problems of spread of conidia after application to soil has led to the *Melolontha* researchers developing an area-wide approach based on augmentative applications of *B. brongniartii* for long-term suppression of pest populations (Hajek *et al.*, 2001).

Compatibility between production, formulation, and application techniques is vital to successful use of microbial biopesticides. Application is a very specialized area of agricultural research, and has both its own technological and specific issues related to achieving adequate control, as well as coverage. Application technology can ensure infective propagules contact the target pest. Much of this technology is derived from our experience with chemical pesticides and may not be the most

appropriate for use with fungal pathogens. Bateman and Chapple (2001) have recently reviewed spray application of mycopesticides in detail. Advances have been made with fungi, such as the use of ULV application technology. ULV has been used with the oil formulated *M. anisopliae* for locust control, as the droplets are resistant to evaporation (Lomer *et al.*, 2001). Electrostatically charged ULV sprayers have been investigated for better coverage on leaf undersides (Sopp *et al.*, 1989). Hydraulic spray systems have been used to apply water based formulations on crops, with air-blast and air-assist technologies primarily used for low-volume applications in fields and orchards. There are many variables to be considered in spray application, whether with ULV or with more conventional equipment. Droplet size is an important consideration, as there is a trade-off between small, more aerial droplets for better spread and large droplets, which contain more propagules and settle more rapidly. The size of the area to which fungal propagules are to be applied can alter the type of application equipment used. For large-scale applications, such as aerial applications, it is possible to use ULV applications of undiluted materials in some cases, whereas it can be difficult to get sufficient coverage on small areas using the undiluted sprays.

A number of novel application methods have been developed that do not rely on methods developed for chemical pesticide application technology. These methods, which rely on the behavior of the pest, may more closely mimic natural infection routes and have more chance of success in situations where conventional spray application is impossible, such as dispersed and highly mobile pests. "Lure and infect" uses an attractant to bring a susceptible insect to a trap containing a pathogen. After contamination in the trap, the insect leaves the trap and infects others, as demonstrated by the use of *Z. radicans* against diamondback moth. Furlong *et al.* (1995) showed that using pheromone lures to attract moths to traps containing sporulating *Z. radicans* can result in contamination and spread of the fungus through the target population. Auto-dissemination similarly uses trapping of adults in pathogen laced attractant traps then using those adults to disseminate the fungus into the population. This technique was successful in the Azores for introducing *M. anisopliae* to *Popillia japonica* populations (Klein and Lacey, 1999), but whether population suppression would occur depends on the control ability of the fungus. Pheromone impregnated pellets containing *B. bassiana* spores have also been tested against grain borers (Smith *et al.*, 1999).

11.7. Development as Inoculative Microbial Control Agents

Inoculation biological control is defined as

> The intentional release of a living organism as a biological control agent with the expectation that it will multiply and control the pest for an extended period, but not that it will do so permanently.
> Eilenberg *et al.* (2001)

This is desirable because only a limited amount of inoculum (*in vivo* or *in vitro* acquired material) is released at each treatment. The inoculum is expected to proliferate and control the target pest over time, but additional releases would normally be needed in the future (e.g., next cropping season). Entomophthoralean fungi may fit very well in inoculation biological control because they have the ability to establish epizootics quickly in target pest populations, have a narrow host range, and are able to persist in the target insect population.

Inoculation biological control is being used in northeastern USA to initiate epizootics of *E. maimaiga* in populations of *L. dispar* (Hajek and Webb, 1999; Pell *et al.*, 2001). Resting spores are collected on sites with high populations of the target insect and high prevalence of the fungus, and then spread on soil in areas without the fungus, but with host infestation. Although the fungus is well established throughout the host's distribution area, the host continues to spread, and cadavers are distributed along the leading edge of this spread in an effort to halt the destruction caused by this introduced pest.

There are other possibilities for inoculative biological control using fungi, such as inoculation of a small amount of *B. brongniartii* to control soil dwelling beetles from Scarabaeidae (Eilenberg *et al.*, 2000b). Another possibility would be use of *E. muscae* and *S. castrans* against cabbage root flies (*D. radicum* and *D. floralis*). Moreover, indoor, inoculative releases to control *M. domestica* could be based on *in vivo* cultures of *E. muscae* or *E. schizophorae* (e.g., Kuramoto and Shimazu, 1997). Conidia are sufficiently persistent in the environment for this release (Kalsbeek *et al.*, 2001b). Indoor stables are favorable for studies on fungus dispersal in a confined environment.

Dissemination is essential for the success of inoculation biological control. Pell *et al.* (1993) and Furlong *et al.* (1995) demonstrated that autodissemination of *Z. radicans* was possible in field populations of the diamondback moth *P. xylostella* in Malaysia by using a pheromone trap to attract adult insects. After the adults left the trap, they dispersed the fungus to both larvae and adults, and larvae became infected. Further experiments of inoculative releases in England demonstrate that it is possible to induce epizootics in *P. xylostella* using *Z. radicans* (Pell and Wilding, 1994). Other examples are provided by Vega *et al.* (2000).

11.8. Use in Classical Biocontrol

Classical biological control is defined as

> The intentional introduction of an exotic biological control agent for permanent establishment and long-term pest control
> Eilenberg *et al.* (2001)

It is desirable from the point of view that limited inoculative releases of an organism will result in a long-lasting control of a pest (often an introduced pest species). Because the introduction of an exotic species has the potential of irreversible direct effects on nontarget species, or indirect effects through host depletion, it is important to gain as much information on the potential candidate's host range and epizootiology as possible prior to its introduction (Hajek and Butler, 2000; Hajek and Goettel, 2000; Goettel and Hajek, 2001; Goettel *et al.*, 2001).

There is great potential in the use of entomopathogenic fungi in classical biological control, especially against introduced pests. Fungi from Entomophthorales possess several characteristics favoring their use in classical biological control; they can establish epizootics quickly, they have mostly a narrow host range, and they persist in pest populations and in the environment. Unfortunately, in most cases, it is not known whether or not a given species is already present in a locality, and the tools needed for detailed studies of their spread after release (e.g., isolate specific characterization) are usually not fully developed (see Section 11.4).

As early as 1909, *E. maimaiga* infected gypsy moths, *L. dispar*, were collected in Japan and released in Boston, USA, in an attempt to control introduced *L. dispar*, but the fungus was not observed in local *L. dispar* populations for many years. It was, however, discovered in northeastern USA in 1989 and has now spread into many other areas within North America where *L. dispar* infestations were prevalent. It has not been ascertained if the fungus now commonly present is a result of the early release or if it was later accidentally imported (Hajek *et al.*, 1995; Pell *et al.*, 2001). A recent introduction of *E. maimaiga* into Bulgaria was successful (Pilarska *et al.*, 2000).

Another example of an establishment in a completely new environment was the release of an Israeli isolate of *Z. radicans* to control the clover aphid *Therioaphis trifolii* in Australia. The experiments were limited in time and the area studied was small, yet it was shown that dispersal of the fungus up to at least 65 m took place within 2 months of release (Milner *et al.*, 1982). Another isolate of this same species was released in Illinois in 1984 to control the potato leafhopper, *Empoasca fabae* (McGuire *et al.*, 1987a). Epizootics of *Z. radicans* then occurred in *E. fabae* and contributed to regulation of the population, and there was evidence to suggest that one of the isolates might have become established (McGuire *et al.*, 1987b).

Neozygites floridana from South America has been considered for classical biological control of the cassava green mite (*Mononychellus tanajore*) in Africa (Elliot *et al.*, 2000). The data on dispersal in plots in Brazil indicated that the fungus would be capable of spread under African conditions and contribute to long-term control of the pest. However, it was concluded that *N. floridana* would not be capable of providing adequate control of the mite if used as the sole control agent. Releases of the Brazilian isolate were made in Benin in 1999 (Hountondji *et al.*, 2002). Postrelease monitoring revealed that the introduced isolate had become established, with infection prevalences of up to 36%, 48 weeks after initial introduction.

The use of fungi for classical biological control is at the present time a rather difficult issue. First, one must find a potential agent that does not already occur in the chosen environment. Then, laboratory studies are warranted to document that the proposed agent has potential for classical biological control. Finally, the authorities must be convinced that the agent for release and establishment poses no harm to the native fauna, and this is difficult, but not impossible, to accomplish (Hajek and Goettel, 2000; Goettel and Hajek, 2001). Today, the latter is difficult to achieve, due to the present attitude in society that more or less demands zero risk to biodiversity.

11.9. Role in Conservation Biocontrol

Conservation biological control (sometimes also called habitat manipulation) is defined as

> Modification of the environment or existing practices to protect and enhance specific natural enemies or other organisms to reduce the effect of pests
> Eilenberg *et al.* (2001)

It is probably the most attractive among all strategies for biological control, yet the least studied.

No agents are released, but the farming systems and practice are modified in order to stimulate the naturally occurring predators, parasitoids, and insect pathogens, to regulate the pest insect populations below the damage threshold (Barbosa, 1998). Sufficient regulation by entomophthoralean fungi and other natural enemies may actually already exist for the vast number of crop herbivorous insect and mite species, which never do any damage because of low densities. The conservation strategy also fits well into modern sustainable agriculture, where actions to make direct use of the naturally occurring enemies of pests are taken into consideration, including increasing biodiversity and supporting a positive interaction between the crop, the flora, and fauna within the field and field margins, and landscape elements (Gurr and Wratten, 2000; Landis *et al.*, 2000). Fungi from the order Entomophthorales fit very well into this strategy (Pell *et al.*, 2001). They have the ability to establish natural epizootics depending on the host insect and a number of biotic and abiotic factors. A number of entomophthoralean species exist in each field, forest, private garden, greenhouse, stable and nature area. Certainly, this diversity of species (and isolates from each species) could be explored and exploited in conservation biological control.

Each system should be analyzed adequately. Studies on the epizootiology of *N. fresenii* in the cotton aphid, *Aphis gossypii*, in Arkansas have demonstrated that it is possible to monitor and to some extent predict the epizootic development of the fungus in the target insect population (Hollingsworth *et al.*, 1995; Steinkraus *et al.*, 1996; Pell *et al.*, 2001). The monitoring of the pest insect can therefore be supplemented with knowledge on the symptoms of fungal disease in the host, which would allow extension officers to modify their recommendations of intervention; e.g., if the pest population is high, but fungus infected, no pesticide treatment would be required as the pest population would soon succumb to the disease.

Habitat manipulation and modification of existing practices are often proposed as means to promote infections caused by entomopathogenic fungi (Pell *et al.*, 2001): more hedges, less spraying with chemicals (e.g., fungicides), and irrigation are all, at least theoretically, supportive to development of natural fungal epizootics among pest insects. It was recently shown that soil-occurring insect pathogenic fungi from Hyphomycetes are found in higher densities in organic grown fields compared to conventionally treated fields (Klingen *et al.*, 2002). To date, however, it has not been possible to document that a specified practice will result in a sufficiently high infection level *every time* in field situations. Indoors it has been shown that intensive spraying with water supports the development of natural epizootics of *Erynia ithacensis* in fungus sciarid gnats (Huang *et al.*, 1992), and such manipulation should be better studied and eventually adopted.

Among biotic factors influencing the success of entomopathogenic fungi are their interactions with other natural enemies. Such interactions may be positive, since predators and parasitoids may assist in the transmission of these fungi as passive vectors (see Sections 11.3.5 and 11.3.6). A better understanding of the interactions between entomopathogenic fungi and other natural enemies would allow us to use different biological control agents in different strategies. There is much potential for conservation biological control using entomopathogenic fungi, however, more studies are needed before we can take full advantage of this naturally occurring resource. Especially warranted are studies to clarify the importance of soil as a reservoir of inoculum and to determine whether crop rotation is supportive to infection of entomophthoralean fungi and other insect pathogenic fungi.

More information is needed to determine how landscape elements affect early initiation of epizootics. In both cases, the inclusion of studies of insect pathogens should be promoted strongly to complement the more detailed existing knowledge of interactions among predators, parasitoids, host and nonhost insects, cropping systems, and landscape elements. Despite the wish to make entomopathogenic fungi operational for conservation biological control, much work still remains. Compared with knowledge on predators and parasitoids, the knowledge on insect pathogens in natural systems is limited, and much observational and descriptive work is certainly still needed.

11.10. What Does the Future Hold?

Compared to other technologies, development of entomopathogenic fungi as inundative, inoculative, and classical microbial control agents is still in its infancy. Inundative application of mass produced spores has long been seen as an attractive alternative to the classical use of fungi for control. A number of fungal based biopesticides are now being produced. But, despite the number of "products" based on entomopathogenic fungi currently commercially produced for pest control throughout the world, the market share is still well below the potential suggested by their role in natural insect control. Although great strides have been made in improving methods of production, formulation, and application, a great many improvements are still possible. Production of species "cocktails," especially combining microbials from different classes or even with chemical or botanical antagonists is only just beginning. To date, all commercial products based on entomopathogenic fungi have been developed from wild-type field strains. In many cases, these products are developed with the insecticidal paradigm in mind, that is, strains with the widest host range are formulated to be applied by conventional equipment. These products are often difficult to market to clientele that are used to conventional chemical insecticides and expect tank mixing possibilities with either herbicides or fungicides (and this is often not compatible), a rapid kill (fungi are usually slower acting), and a broad spectrum of activity (even fungi with the broadest host spectrum have reduced spectra as compared to most chemicals in present use).

With only a few studies attempting genetic engineering on entomopathogenic fungi, it is hard to determine when or not modification of fungi will increase their efficacy in biocontrol. The current regulatory environment in many countries makes such research expensive, and public acceptability without demonstrated consumer (rather than farmer) gains may be some time off. Strain improvement through guided selection strategies is more likely to find acceptance in the short term.

In addition to development of entomopathogenic fungi for use in inundative applications, we see great potential in their use in conservation and classical biological control. Whatever strategies are contemplated, they should be considered with the intention not just of replacing chemical pesticides with entomopathogenic fungi, but contributing to the development of sustainable agriculture, horticulture, and forestry as well as the preservation of biodiversity and nature. With increasing public and political pressure to implement integrated pest management strategies, it is inevitable that use of entomopathogenic fungi will play an increasingly important role (Lacey and Goettel, 1995).

References

Alves, S.B., Pereira, R.M., Lopes, R.B., Tamai, M.A., 2003. Use of entomopathogenic fungi in Latin America. In: Upadhyay, R.K. (Ed.), Advances in Microbial Control of Insect Pests. Kluwer, New York, pp. 193–211.

Amiri, B., Ibrahim, L., Butt, T.M., 1999. Antifeedant properties of destruxins and their use with the entomogenous fungus *Metarhizium anisopliae* for improved control of crucifer pests. *Biocon. Sci. Technol.* 9, 487–498.

Amiri-Besheli, B., Khambay, B., Cameron, S., Deadman, M., Butt, T.M., 2000. Inter- and intra-specific variation in destruxin production by the insect pathogenic *Metarhizium*, and its significance to pathogenesis. *Mycol. Res. 104*, 447–452.

Apperson, C.S., Federici, B.A., Tarver, F.R., Stewart, W., 1992. Biotic and abiotic parameters associated with an epizootic of *Coelomomyces punctatus* in a larval population of the mosquito *Anopheles quadrimaculatus*. *J. Invertebr. Pathol. 60*, 219–228.

Arthurs, S., Thomas, M.B., 2000. Effects of a mycoinsecticide on feeding and fecundity of the brown locust *Locustana pardalina*. *Biocont. Sci. Technol. 10*, 321–329.

Austwick, P.K.C., 1983. Some mycoses of reptiles. *RMVN 18*, 779.

Barr, D.J.S., 1992. Evolution and kingdoms of organisms from the perspective of a mycologist. *Mycologia 84*, 1–11.

Bandani, A.R., Butt, T.M., 1999. Insecticidal, antifeedant and growth inhibitory activities of efrapeptins, metabolites of the entomogenous fungus *Tolypocladium*. *Biocon. Sci. Technol. 9*, 499–506.

Bandani, A.R., Khambay, B.P.S., Faull, J., Newton, R., Deadman, M., et al., 2000. Production of efrapeptins by *Tolypocladium* species and evaluation of their insecticidal and antimicrobial properties. *Mycol. Res. 104*, 537–544.

Barbosa, P. (Ed.), 1998. Conservation Biological Control. Academic Press, San Diego.

Bartlett, M.C., Jaronski, S.T., 1988. Mass production of entomogenous fungi for biological control of insects. In: Burge, R.N. (Ed.), Fungi in Biological Control Systems. Manchester University Press, Manchester, pp. 61–85.

Bateman, R., 1997. The development of a mycoinsecticide for the control of locusts and grasshoppers. *Outlook Agric. 26*, 13–18.

Bateman, R., Chapple, A., 2001. The spray application of mycopesticide formulations. In: Butt, T., Jackson, C., Magan, N. (Eds.), Fungal Biocontrol Agents. CABI Press, Wallingford, pp. 289–309.

Beard, C.E., Adler, P.H., 2002. Seasonality of Trichomycetes in larval black flies from South Carolina, USA. *Mycologia 94*, 200–209.

Beard, R.L., Walton, G.S., 1969. Kojic acid as an insecticidal mycotoxin. *J. Invertebr. Pathol. 14*, 53–59.

Berbee, M.L., Taylor, J.W., 2001. Fungal molecular evolution: gene trees and geologic time. In: McLaughlin, D.J., McLaughlin, E.G., Lenke, P.A. (Eds.), The Mycota: a Comprehensive Treatise on Fungi as Experimental Systems for Basic and Applied Research, Systematics and Evolution, Part B. Springer GmbH, Berlin. pp. 229–245.

Bernier, L., Cooper, R.M., Charnley, A.K., Clarkson, J.M., 1989. Transformation of the entomopathogenic fungus *Metarhizium anisopliae* to benomyl resistance. *FEMS Microbiol. Lett. 60*, 261–266.

Bidochka, M.J., 2001. Monitoring the fate of biocontrol fungi. In: Butt, T.M., Jackson, C., Magan, N. (Eds.), Fungi as Biocontrol Agents. CAB International, Wallingford, pp. 193–218.

Bidochka, M.J., Hajek, A.E., 1998. A nonpermissive entomophthoralean fungal infection increases activation of insect prophenoloxidase. *J. Invertebr. Pathol. 72*, 231–238.

Bidochka, M.J., Kamp, A.M., Lavender, T.M., Dekoning, J., De Croos, J.N., 2001. Habitat association in two genetic groups of the insect-pathogenic fungus *Metarhizium anisopliae*: uncovering cryptic species? *Appl. Environ. Microbiol. 67*, 1335–1342.

Blanford, S., Thomas, M.B., 2001. Adult survival, maturation, and reproduction of the desert locust *Schistocerca gregaria* infected with the fungus *Metarhizium anisopliae* var *acridum*. *J. Invertebr. Pathol. 78*, 1–8.

Blanford, S., Thomas, M.B., Langewald, J., 1998. Behavioural fever in the Senegalese grasshopper, *Oedaleus senegalensis*, and its implications for biological control using pathogens. *Ecol. Entomol. 23*, 9–14.

Bogo, M.R., Rota, C.A., Pinto, H., Ocampos, M., Correa, C.T., et al., 1998. A chitinase encoding gene (chit1 gene) from the entomopathogen *Metarhizium anisopliae*: isolation and characterization of genomic and full-length cDNA. *Curr. Microbiol. 37*, 221–225.

Booth, S.R., Tanigoshi, L., Dewes, I., 2000. Potential of a dried mycelium formulation of an indigenous strain of *Metarhizium anisopliae* against subterranean pests of cranberry. *Biocont. Sci. Technol. 10*, 659–668.

Boucias, D.G., Pendland, J.C., 1998. Principles of Insect Pathology. Kluwer Academic Publishers, Boston.

Bradley, C.A., Black, W.E., Kearns, R., Wood, P., 1992. Role of production technology in mycoinsecticide development. In: Leatham, G.F. (Ed.), Frontiers in Industrial Mycology. Chapman and Hall, New York, pp. 160–179.

Brousseau, C., Charpentier, G., Bellonick, S., 1996. Susceptibility of spruce budworm, *Choristoneura fumiferana* Clemens, to destruxins, cyclodepsipeptidic mycotoxins of *Metarhizium anisopliae*. *J. Invertebr. Pathol. 68*, 180–182.

Broza, M., Pereira, R.M., Stimac, J.L., 2001. The nonsusceptibility of soil Collembolla to insect pathogens and their potential as scavengers of microbial pesticides. *Pedobiology 45*, 523–534.

Bruns, T.D., Vilgalys, R., Barns, S.M., Gonzalez, D., Hibbett, D.S., et al., 1993. Evolutionary relationships within the fungi: analysis of nuclear small subunit rRNA sequences. *Molec. Phylogen. Evol. 1*, 231–241.

Burges, H.D., 1998. Formulation of mycoinsecticides. In: Burges, H.D. (Ed.), Formulation of Microbial Biopesticides. Kluwer, Dordrecht, pp. 131–185.

Butt, T.M., Goettel, M.S., 2000. Bioassays of Entomogenous fungi. In: Navon, A., Ascher, K.R.S. (Eds.), Bioassays of Entomopathogenic Microbes and Nematodes. CAB International Press, Wallingford, pp. 141–195.

Butt, T.M., Beckett, A., Wilding, N., 1981. Protoplasts in the *in vivo* life cycle of *Erynia neoaphidis*. *J. Gen. Microbiol. 127*, 417–421.

Butt, T.M., Jackson, C.W., Magan, N., 2001. Fungi as Biocontrol Agents. CABI Publishing, Wallingford.

Butt, T.M., Carreck, N.L., Ibrahim, L., Williams, I.H., 1998. Honey-bee-mediated infection of pollen beetles (*Meligethes aeneus* Fab.) by the insect-pathogenic fungus, *Metarhizium anisopliae*. *Biocont. Sci. Technol.* 8, 533–538.

Cafaro, M.J., 2002. Species richness pattern in symbiotic gut fungi (Trichomycetes). *Fungal Divers.* 9, 47–56.

Carruthers, R.I., Larkin, T.S., Firstencel, H., Feng, Z., 1992. Influence of thermal ecology on the mycosis of a rangeland grasshopper. *Ecology 73*, 190–204.

Carruthers, R.I., Ramos, M.E., Larkin, T.S., Hostetter, D.L., Soper, R.S., 1997. The *Entomophaga grylli* (Fresenius) Batko species complex: its biology, ecology, and use for biological control of pest grasshoppers. *Mem. Entomol. Soc. Canad.* 171, 329–353.

Cavalier-Smith, T., 1987. The origin of fungi and pseudofungi. In: Rayner, A.D.M., Brasier, C.M., Moore, D. (Eds.), Evolutionary Biology of the Fungi. Cambridge University Press, Cambridge, pp. 339–353.

Chabasse, D., 1994. New opportunistic fungi appearing in medical pathology. A general review. *J. Mycol. Med.* 4, 9–28 (in French).

Champlin, F.R., Grula, E.A., 1979. Noninvolvement of beauvericin in the entomopathogenicity of *Beauveria bassiana*. *Appl. Environ. Microbiol.* 37, 1122–1125.

Cherry, A.J., Jenkins, N.E., Heviefo, G., Bateman, R., Lomer, C.J., 1999. Operational and economic analysis of a West African pilot-scale production plant for aerial conidia of *Metarhizium* spp. for use as a mycoinsecticide against locusts and grasshoppers. *Biocont. Sci. Technol. 9*, 35–51.

Christensen, J.B., Fetter-Lasko, J.L., Washino, R.K., Husbands, R.C., Kaffman, E.E., 1977. A preliminary field study employing *Lagenidium giganteum*, a fungus, as a possible biological control agent against the pasture mosquito *Aedes nigromaculis*. *Proc. Calif. Vector Mosq. Control Assoc. 45*, 105.

Clarkson, J.M., 1992. Molecular approaches to the study of entomopathogenic fungi. In: Lomer, C.J., Prior, C. (Eds.), Biological Control of Locusts and Grasshoppers. CAB International Press, Wallingford, pp. 191–199.

Cooke, M.C., 1892. Handbook of Australian Fungi. Williams and Norgate, London.

Cooper, R.D., Sweeney, A.W., 1986. Laboratory studies on the recycling potential of the mosquito pathogenic fungus *Culicinomyces clavisporus*. *J. Invertebr. Pathol.* 48, 152–158.

Copping, L.G. (Ed.), 2001. The Biopesticide Manual, 2nd edn. British Crop Protection Council, Farnham.

Couch, J.N., Romney, S.V., 1973. Sexual reproduction in *Lagenidium giganteum*. *Mycologia 65*, 250–252.

Couteaudier, Y., Viaud, M., 1997. New insights into population structure of *Beauveria bassiana* with regard to vegetative compatibility groups and telomeric restriction fragment length polymorphisms. *FEMS Microbiol. Ecol. 22*, 175–182.

Couteaudier, Y., Viaud, M., Riba, G., 1996. Genetic nature, stability, and improved virulence of hybrids from protoplast fusion in *Beauveria*. *Microbial Ecol. 32*, 1–10.

Daoust, R.A., Roberts, D.W., 1982. Virulence of natural and insect passaged strains of *Metarhizium anisopliae* to mosquito larvae. *J. Invertebr. Pathol. 40*, 107–117.

Dedryver, C.A., 1981. Biology of the cereal aphids in the west of France 2. Spatial and temporal distribution and field pathogenicity of 3 species of Entomophthoraceae. *Entomophaga 26*, 381–394.

Domnas, A.J., 1981. Biochemistry of *Lagenidium giganteum* infection of mosquito larvae. In: Davidson, E.W. (Ed.), Pathogenesis of Invertebrate Microbial Diseases. Allanheld, Totowa, pp. 425–449.

Driver, F., Milner, R.J., 1998. PCR applications to the taxonomy of entomopathogenic fungi. In: Bridge, P.D., Arora, D.K., Reddy, C.A., Elander, R.P. (Eds.), Applications of PCR in Mycology. CABI Press, Wallingford, pp. 153–186.

Driver, F., Milner, R.J., Trueman, J.W.H., 2000. A taxonomic revision of *Metarhizium* based on a phylogenetic analysis of rDNA sequence data. *Mycol. Res. 104*, 134–150.

Dromph, K., 2001. Dispersal of entomopathogenic fungi by collembolans. *Soil Biol. Biochem. 33*, 2047–2051.

Eaton, K.K., Hennessy, T.J., Snodin, D.J., McNulty, D.W., 1986. *Verticillium lecanii* allergological and toxicological studies on work exposed personnel. *Ann. Occup. Hyg. 30*, 209–217.

Eilenberg, J., 1987. Abnormal egg-laying behaviour of female carrot flies (*Psila rosae*) induced by the fungus *Entomophthora muscae*. *Entomol. Expl. Appl. 43*, 61–65.

Eilenberg, J., 1988. Occurrence of fungi from Entomophthorales in a population of carrot flies (*Psila rosae* F.). Results 1985 and 1986. *IOBC/WPRS Bull. 11*, 53–59.

Eilenberg, J., Philipsen, H., 1988. The occurrence of Entomophthorales on the carrot fly (*Psila rosae* F.) in the field during two successive seasons. *Entomophaga 33*, 135–144.

Eilenberg, J., Hajek, A.E., Lomer, C., 2001. Suggestions for unifying the terminology in biological control. *BioControl 46*, 387–400.

Eilenberg, J., Damgaard, P.H., Hansen, B.M., Pedersen, J.C., Bresciani, J., et al., 2000a. Natural co-prevalence of *Strongwellsea castrans*, *Cystosporogenes deliaradicae* and *Bacillus thuringiensis* in the host, *Delia radicum*. *J. Invertebr. Pathol. 75*, 69–75.

Eilenberg, J., Vestergaard, S., Harding, S., Nielsen, C., 2000b. Potential for microbial control of Scarbaeidae and Curculionidae in Christmas tree and greenery production. In: Christensen, C.J. (Ed.), Improvements in Christmas Tree and Greenery Quality. Ministry of Environment, Danish Forest and Landscape Research Institute, Report No. 7, Copenhagen, pp. 22–23.

Elliot, S.L., Blanford, S., Thomas, M.B., 2002. Host-pathogen interactions in a varying environment: temperature, behavioural fever and fitness. *Proc. R. Soc. Lond. 269*, 1599–1607.

Elliot, S.L., De Moraes, G.J., Delalibera, I., Da Silva, C.A.D., Tamai, M.A., et al., 2000. Potential of the mite-pathogenic fungus *Neozygites floridana* (Entomophthorales: Neozygitaceae) for control of the cassava green mite *Mononychellus tanajore* (Acari: Tetranychidae). *Bull. Entomol. Res. 90*, 191–200.

Enkerli, J., Widmer, F., Gessler, C., Keller, S., 2001. Strain-specific microsatellite markers in the entomopathogenic fungus *Beauveria brongniartii*. *Mycol. Res. 105*, 1079–1087.

Evans, H.C., 1982. Entomogenous fungi in tropical forest ecosystems: an appraisal. *Ecol. Entomol. 6*, 47–60.

Evans, H.C., 1989. Mycopathogens of insects of epigeal and aerial habitats. In: Wilding, N., Collins, N.M., Hammond, P.M., Webber, J.F. (Eds.), Insect–Fungus Interactions. Academic Press, London, pp. 205–238.

Fargues, J.F., Robert, P.H., 1983. Effects of passage through scarabeid hosts on virulence and host specificity of two strains of the entomopathogenic Hyphomycete *Metarhizium anisopliae*. *Can. J. Microbiol. 29*, 576–583.

Fargues, J., Delmas, J.C., Lebrun, R.A., 1994. Leaf consumption by larvae of the Colorado Potato Beetle (Coleoptera: Chrysomelidae) infected with the entomopathogen, *Beauveria bassiana*. *J. Econ. Entomol. 87*, 67–71.

Fargues, J., Ouedraogo, A., Goettel, M.S., Lomer, C.J., 1997. Effects of temperature, humidity and inoculation method on susceptibility of *Schistocerca gregaria* to *Metarhizium flavoviride*. *Biocont. Sci. Technol. 7*, 345–356.

Fargues, J., Goettel, M.S., Smits, N., Ouedraogo, A., Vidal, C., et al., 1996. Variability in susceptibility to simulated sunlight of conidia among isolates of entomopathogenic Hyphomycetes. *Mycopathologia 135*, 171–181.

Fargues, J., Vidal, C., Smits, N., Rougier, M., Boulard, T., et al., 2003. Climatic factors on entomopathogenic Hyphomycetes infection of *Trialeurodes vaporariorum* (Homoptera: Aleyrodidae) in Mediterranean glasshouse tomato. *Biol. Control 28*, 320–331.

Federici, B.A., Lucarotti, C.J., 1986. Structure and behaviour of the meiospore of *Coelomomyces dodgei* encystment in the copepod host *Acanthocyclops vernalis*. *J. Invertebr. Pathol. 48*, 259–268.

Feng, M.G., 2003. Microbial control of insect pests with entomopathogenic fungi in China: a decade's progress in research and utilization. In: Upadhyay, R.K. (Ed.), Advances in Microbial Control of Insect Pests. Kluwer, New York, pp. 213–234.

Feng, M.G., Johnson, J.B., Halbert, S.E., 1991. Natural control of cereal aphids (Homoptera, Aphididae) by entomopathogenic fungi (Zygomycetes, Entomophthorales) and parasitoids (Hymenoptera, Braconidae and Encyrtidae) on irrigated spring wheat in Southwestern Idaho. *Environ. Entomol. 20*, 1699–1710.

Feng, M.G., Johnson, J.B., Kish, L.P., 1990. Survey of entomopathogenic fungi naturally infecting cereal aphids (Homoptera: Aphididae) of irrigated grain crops in Southwestern Idaho. *Environ. Entomol. 19*, 1534–1542.

Feng, M.G., Poprawski, T.J., Khachatourians, G.G., 1994. Production, formulation and application of the entomopathogenic fungus *Beauveria bassiana* for insect control: current status. *Biocontr. Sci. Technol. 4*, 3–34.

Ferron, P., 1985. Fungal control. In: Kerkut, G.A., Gilbert, L.I. (Eds.), Comprehensive Insect Physiology, Biochemistry and Pharmacology. Pergamon, Oxford, pp. 313–346.

Ferron, P., Fargues, J., Riba, G., 1991. Fungi as microbial insecticides against pests. In: Arora, D.K., Ajello, L., Mukerji, K.G. (Eds.), Handbook of Applied Mycology, Vol. 2: Humans, Animals, and Insects. Dekker, New York, pp. 665–706.

Fetter-Lasko, J.L., Washino, R.K., 1983. In-situ studies on seasonality and recycling pattern in California USA of *Lagenidium giganteum* an aquatic fungal pathogen of mosquitoes. *Environ. Entomol. 12*, 635–640.

Frappier, F., Ferron, P., Pais, M., 1975. Chimie des champignons entomopathogenes – le beauvellide, nouveau cyclodepsipeptide isole d'un *Beauveria tenella*. *Phytochemistry 14*, 2703–2705.

Fuhrer, E., Rosner, S., Schmied, A., Wegensteiner, R., 2001. Studies on the significance of pathogenic fungi in the population dynamics of the lesser spruce sawfly, *Pritiphora abietina* Christ. (Tenthredinidae). *J. Appl. Entomol. 125*, 235–242.

Furlong, M.J., Pell, J.K., 2000. Conflicts between a fungal entomopathogen, *Zoophthora radicans*, and two larval parasitoids of the diamond back moth. *J. Invertebr. Pathol. 76*, 85–94.

Furlong, M.J., Pell, J.K., Reddy, G.V.P., 1997. Premortality effects of *Zoophthora radicans* infection in *Plutella xylostella*. *J. Invertebr. Pathol. 70*, 214–220.

Furlong, M.J., Pell, J.K., Choo, O.P., Rahman, S.A., 1995. Field and laboratory evaluation of a sex pheromone trap for the autodissemination of the fungal entomopathogen *Zoophthora radicans* (Entomophthorales) by the diamondback moth, *Plutella xylostella* (Lepidoptera: Yponomeutidae). *Bull. Entomol. Res. 85*, 331–337.

Fuxa, J.R., Tanada, Y., 1987. Epizootiology of Insect Diseases. Wiley, New York.

Gams, W., Zare, R., 2001. A revision of *Verticillium* Sect. *Prosteata*. III. Generic classification. *Nova Hedwigia 72*, 329–337.

Garcia, C., Ignoffo, C.M., 1977. Dislodgement of conidia of *Nomuraea rileyi* from cadavers of cabbage looper, *Trichoplusia ni*. *J. Invertebr. Pathol. 30*, 114–116.

Genthner, F.J., Middaugh, D.P., 1992. Effects of *Beauveria bassiana* on the embryos of the Inland Silverside fish (*Menidia beryllina*). *Appl. Environ. Microbiol. 58*, 2840–2845.

Genthner, F.J., Middaugh, D.P., 1995. Nontarget testing of an insect control fungus: effects of *Metarhizium anisopliae* on developing embryos of the inland silverside fish *Menida beryllina*. *Dis. Aquat. Org. 22*, 163–171.

Genthner, F.J., Foss, S.S., Gals, P.S., 1997. Virulence of *Metarhizium anisopliae* to embryos of the grass shrimp *Palaemonetes pugio*. *J. Invertebr. Pathol. 69*, 157–164.

Gilliam, M., Vandenberg, J.D., **1990**. Fungi. In: Morse, R.A., Nowogrodzki, R. (Eds.), Honey Bee Pests, Predators, and Diseases. Cornell University Press, Ithaca, pp. 64–90.

Gilliam, M., Taber, S., Richardson, G.V., **1983**. Hygienic behavior of honey bees in relation to chalkbrood disease. *Apidologie 14*, 29–39.

Glare, T.R., **1987**. Effect of host species and light conditions on production of conidia by *Nomuraea rileyi*. *J. Invertebr. Pathol. 50*, 67–69.

Glare, T.R., **1994**. Stage-dependant synergism using *Metarhizium anisopliae* and *Serratia entomophila* against *Costelytra zealandica*. *Biocontr. Sci. Technol. 4*, 321–329.

Glare, T.R., Milner, R.J., **1991**. Ecology of entomopathogenic fungi. In: Arora, D.K., Mukeriji, K.G., Pugh, J.G.F. (Eds.), Handbook on Applied Mycology, Vol. 2: Humans, Animals, and Insects. Dekker, New York, pp. 547–612.

Glare, T.R., Chilvers, G.A., Milner, R.J., **1985**. Capilliconidia as infective spores in *Zoophthora phalloides* (Entomophthorales). *Trans. Brit. Mycol. Soc. 85*, 463–470.

Glare, T.R., Harris, R.J., Donovan, B.J., **1996**. *Aspergillus flavus* as a pathogen of wasps, *Vespula* spp., in New Zealand. *N. Z. J. Zool. 23*, 339–344.

Glare, T.R., Milner, R.J., Chilvers, G.A., **1986**. The effect of environmental factors on the production, discharge and germination of primary conidia of *Zoophthora phalloides* Batko. *J. Invertebr. Pathol. 48*, 275–283.

Glare, T.R., Milner, R.J., Chilvers, G.A., **1989**. Factors affecting the production of resting spores by *Zoophthora radicans* in the spotted alfalfa aphid, *Therioaphis trifolii* f. *maculata*. *Canad. J. Bot. 67*, 848–855.

Glare, T.R., Placet, C., Nelson, T.L., Reay, S.D., **2002**. Potential of *Beauveria* and *Metarhizium* as control agents of pinhole borers (*Platypus* spp.). *N. Z. Plant Protect. J. 55*, 73–79.

Goethe, J.W. von, **1817–1822**. Zur morphologie. Erfahrung, betrachtung, folgerung durch lebens-ereignisse verbunden erster band. Edition Kuhn, D., (1954). Morphologisher Hefte, Neunter Band, Hermann Böhlaus Nachfolger, Weimar.

Goettel, M.S., Hajek, A.E., **2001**. Evaluation of nontarget effects of pathogens used for management of arthropods. In: Wajnberg, E., Scott, J.K., Quimby, P.C. (Eds.), Evaluating Indirect Ecological Effects of Biological Control. CABI Press, Wallingford, pp. 81–97.

Goettel, M.S., Jaronski, S.T., **1997**. Safety and registration of microbial agents for control of grasshoppers and locusts. *Mem. Entomol. Soc. Canad. 171*, 83–99.

Goettel, M.S., Inglis, G.D., Wraight, S.P., **2000**. Fungi. In: Lacey, L.A., Kaya, H.K. (Eds.), Field Manual of Techniques for the Application and Evaluation of Entomopathogens. Kluwer Academic Press, Dordrecht, pp. 255–282.

Goettel, M.S., Hajek, A.E., Siegel, J.P., Evans, H.C., **2001**. Safety of fungal biocontrol agents. In: Butt, T.,

Jackson, C., Magan, N. (Eds.), Fungal Biocontrol Agents – Progress, Problems and Potential. CABI Press, Wallingford, pp. 347–375.

Goettel, M.S., Poprawski, T.J., Vandenberg, J.D., Li, Z., Roberts, D.W., **1990a**. Safety to nontarget invertebrates of fungal biocontrol agents. In: Laird, M., Lacey, L.A., Davidson, E.W. (Eds.), Safety of Microbial Insecticides. CRC Press, Boca Raton, pp. 209–231.

Goettel, M.S., St. Leger, R.J., Bhairi, S., Jung, M.K., Oakley, B.R., et al., **1990b**. Pathogenicity and growth of *Metarhizium anisopliae* stably transformed to benomyl resistance. *Curr. Genetics 17*, 129–132.

Gonzalez Cabo, J.F., Espejo Serrano, J., Barcena Asensio, M.C., **1995**. Mycotic pulmonary disease by *Beauveria bassiana* in a captive tortoise. *Mycoses 38*, 167–169.

Griffiths, G.C.D., **1985**. *Hypogastrura succinea* (Collembola Hypogastruridae) dispersed by adults of the cabbage maggot *Delia radicum* (Diptera Anthomyiidae) infected with the parasitic fungus *Strongwellsea castrans* (Zygomycetes Entomophthoraceae). *Canad. Entomol. 117*, 1063–1064.

Gumowski, P.I., Latgé, J.-P., Paris, S., **1991**. Fungal allergy. In: Arora, D.K., Mukeriji, K.G., Rugh, J.G.F. (Eds.), Handbook of Applied Mycology, Vol. 2: Humans, Animals, and Insects. Dekker, New York, pp. 163–204.

Gurr, G., Wratten, S., **2000**. Biological Control: Measures of Success. Kluwer, Dordrecht.

Hajek, A., **1989**. Food consumption by *Lymantria dispar* (Lepidoptera: Lymantriidae) larvae infected with *Entomophaga maimaiga* (Zygomycetes: Entomophthorales). *Environ. Entomol. 18*, 723–727.

Hajek, A.E., **1997**. Fungal and viral epizootics in gypsy moth (Lepidoptera: Lymantriidae) populations in Central New York. *Biol. Cont. 10*, 58–68.

Hajek, A.E., **1998**. Pathology and epizootiology of *Entomophaga maimaiga* infections in forest Lepidoptera. *Microbiol. Molec. Biol. Rev. 63*, 814–835.

Hajek, A., **1999**. Pathology and epizootiology of *Entomophaga maimaiga* infections in forest Lepidoptera. *Microbiol. Molec. Biol. Rev. 63*, 814–835.

Hajek, A.E., Butler, L., **2000**. Predicting the host range of entomopathogenic fungi. In: Follett, P.A., Dunn, J.J. (Eds.), Nontarget Effects of Biological Control. Kluwer, Dordrecht, pp. 263–276.

Hajek, A.E., Goettel, M.S., **2000**. Guidelines for evaluating effects of entomopathogens on nontarget organisms. In: Lacey, L.A., Kaya, H.K. (Eds.), Field Manual of Techniques for the Application and Evaluation of Entomopathogens. Kluwer, Dordrecht, pp. 847–868.

Hajek, A.E., St. Leger, R.J., **1994**. Interactions between fungal pathogens and insect host. *Ann. Rev. Entomol. 39*, 293–322.

Hajek, A.E., Shimazu, M., **1996**. Types of spores produced by *Entomophaga maimaiga* infecting the gypsy moth *Lymantria dispar*. *Canad. J. Bot. 74*, 708–715.

Hajek, A.E., Webb, R.E., **1999**. Inoculative augmentation of the fungal entomopathogen *Entomophaga maimaiga* as a homeowner tactic to control gypsy moth (Lepidoptera: Lymantriidae). *Biol. Cont. 14*, 11–18.

Hajek, A.E., Humber, R.A., Elkington, J.S., **1995**. Mysterious origin of *Entomophaga maimaiga* in North America. *Am. Entomol. 41*, 31–42.

Hajek, A.E., Wraight, S.P., Vandenberg, J.D., **2001**. Control of Arthropods Using Pathogenic Fungi. *Fungal Divers. Res. Ser. 6*, 309–347.

Hajek, A.E., Butler, L., Liebherr, J.K., Wheeler, M.M., **2000**. Risk of infection by the fungal pathogen *Entomophaga maimaiga* among Lepidoptera on the forest floor. *Environ. Entomol. 29*, 645–650.

Hajek, A.E., Butt, T.M., Strelow, L.I., Gray, S.M., **1991a**. Detection of *Entomophaga maimaiga* (Zygomycetes Entomophthorales) using ELISA. *J. Invertebr. Pathol. 58*, 1–9.

Hajek, A.E., Humber, R.A., Walsh, S.R.A., Silver, J.C., **1991b**. Sympatric occurrence of two *Entomophaga aulicae* (Zygomycetes Entomophthorales) complex species attacking forest Lepidoptera. *J. Invertebr. Pathol. 58*, 373–380.

Hajek, A.E., Jensen, A.B., Thomsen, L., Hodge, K., Eilenberg, J., **2003**. PCR-RFLP is used to investigate relations among species in the entomopathogenic genera *Eryniopsis* and *Entomophaga*. *Mycologia 95*, 262–268.

Hajek, A.E., Humber, R.A., Elkinton, J.S., May, B., Walsh, S.R.A., *et al.*, **1990**. Allozyme and restriction fragment length polymorphism analyses confirm *Entomophaga maimaiga* responsible for 1989 epizootics in North American gypsy moth populations. *Proc. Natl Acad. Sci. USA 87*, 6979–6982.

Hajek, A.E., Butler, L., Walsh, S.R.A., Silver, J.C., Hain, F.P., *et al.*, **1996**. Host range of the gypsy moth (Lepidoptera: Lymantriidae) pathogen *Entomophaga maimaiga* (Zygomycetes: Entomophthorales) in the field versus laboratory. *Environ. Entomol. 25*, 709–721.

Hallsworth, J.E., Magan, N., **1994**. Effects of KCl concentration on accumulation of acyclic sugar alcohols and trehalose in conidia of three entomopathogenic fungi. *Letters Appl. Microbiol. 18*, 8–11.

Hallsworth, J.E., Magan, N., **1995**. Manipulation of intracellular glycerol and erythritol enhances germination of conidia at low water availability. *Microbiol. Reading 141*, 1109–1115.

Hamill, R.L., Higgens, C.E., Boaz, H.E., Gorman, M., **1969**. The structure of beauvericin, a new depsipeptide antibiotic toxic to *Artemia salina*. *Tetrahedron Lett. 49*, 4255–4258.

Hatai, K., Roza, D., Nakayama, T., **2000**. Identification of lower fungi isolated from larvae of mangrove crab, *Scylla serrata*, in Indonesia. *Mycoscience 41*, 565–572.

Hatting, J.L., Poprawski, T.J., Miller, R.M, **2000**. Prevalences of fungal pathogens and other natural enemies of cereal aphids (Homoptera: Aphididae) in wheat under dryland and irrigated conditions. *BioControl 45*, 179–199.

Henke, M.O., de Hoog, G.S., Gross, U., Zimmermann, G., Kraemer, D., *et al.*, **2002**. Human deep tissue infection with an entomopathogenic *Beauveria* species. *J. Clin. Microbiol. 40*, 2698–2702.

Hollingsworth, R.G., Steinkraus, D.C., McNew, R.W., **1995**. Sampling to predict fungal epizootics in cotton aphids. *Environ. Entomol. 24*, 1414–1421.

Hountondji, F.C.C., Lomer, C.J., Hanna, R., Cherry, A.J., Dara, S.K., **2002**. Field evaluation of Brazilian isolates of *Neozygites floridana* (Entomophthorales: Neozygitaceae) for the microbial control of cassava green mite in Benin, West Africa. *Biocontr. Sci. Technol. 12*, 361–370.

Hu, G., St. Leger, R., **2002**. Field studies using a recombinant mycoinsecticide (*Metarhizium anisopliae*) reveal that it is rhizosphere competent. *Appl. Environ. Microbiol. 68*, 6383–6387.

Huang, B., Li, C.R., Li, Z.G., Fan, M.Z., Li, Z.Z., **2002**. Molecular identification of the teleomorph of *Beauveria bassiana*. *Mycotaxon 81*, 229–236.

Huang, Y., Zhen, B., Li, Z., **1992**. Natural and induced epizootics of *Erynia ithacensis* in mushroom hothouse populations of yellow-legged fungus gnats. *J. Invertebr. Pathol. 60*, 254–258.

Humber, R.A., **1984**. Foundations for an evolutionary classification of the Entomophthorales (Zygomycetes). In: Wheeler, Q., Blackwell, M. (Eds.), Fungus–Insect Relationships: Perspectives in Ecology and Evolution. Columbia University Press, New York, pp. 166–183.

Inglis, G.D., Johnson, D.L., Goettel, M.S., **1996a**. Effect of bait substrate and formulation on infection of grasshopper nymphs by *Beauveria bassiana*. *Biocontr. Sci. Technol. 6*, 35–50.

Inglis, G.D., Johnson, D.L., Goettel, M.S., **1996b**. Effects of temperature and thermoregulation on mycosis by *Beauveria bassiana* in grasshoppers. *Biol. Contr. 7*, 131–139.

Inglis, G.D., Johnson, D.L., Goettel, M.S., **1997a**. Effects of temperature and sunlight on mycosis (*Beauveria bassiana*) (Hyphomycetes: Sympodulosporae) of grasshoppers under field conditions. *Environ. Entomol. 26*, 400–409.

Inglis, G.D., Duke, G.M., Kawchuk, L.M., Goettel, M.S., **1999a**. Influence of oscillating temperatures on the competitive infection and colonization of the migratory grasshopper by *Beauveria bassiana* and *Metarhizium flavoviride*. *Biol. Control 14*, 111–120.

Inglis, G.D., Goettel, M.S., Butt, T.M., Strasser, H., **2001**. Use of hyphomycetous fungi for managing insect pests. In: Butt, T., Jackson, C., Magan, N. (Eds.), Fungal Biocontrol Agents. CABI Press, Wallingford, pp. 23–69.

Inglis, G.D., Johnson, D.L., Cheng, K.-J., Goettel, M.S., **1997b**. Use of pathogen combinations to overcome the constraints of temperature on entomopathogenic Hyphomycetes against grasshoppers. *Biol. Contr. 8*, 143–152.

Inglis, G.D., Johnson, D.L., Kawchuk, L.M., Goettel, M.S., **1998**. Effect of soil texture and soil sterilization on susceptibility of ovipositing grasshoppers to *Beauveria bassiana*. *J. Invertebr. Pathol. 71*, 73–81.

Inglis, P.W., Tigano, M.S., Valadares-Inglis, M.C., **1999b**. Transformation of the entomopathogenic fungi,

Paecilomyces fumosoroseus and *Paecilomyces lilacinus* (Deuteromycotina: Hyphomycetes) to benomyl resistance. *Gen. Mol. Biol. 22*, 119–123.

Jackson, M.A., McGuire, M.R., Lacey, L.A., Wraight, S.P., **1997**. Liquid culture production of desiccation tolerant blastospores of the bioinsecticidal fungus *Paecilomyces fumosoroseus*. *Mycol. Res. 10*, 35–41.

Jaronski, S., **1982**. Oomycetcs in mosquito control. In: Proc. 3rd Int. Colloq. Invertebr. Pathol., September 6–10, 1982. Brighton, England, pp. 420–424.

Jeffs, L.B., Khachatourians, G.G., **1997**. Toxic properties of *Beauveria* pigments on erythrocyte membranes. *Toxicon 35*, 1351–1356.

Jenkins, N.E., Goettel, M.S., **1997**. Methods for mass-production of microbial control agents of grasshoppers and locusts. *Mem. Entomol. Soc. Canad. 171*, 37–48.

Jenkins, N.E., Prior, C., **1993**. Growth and formation of true conidia by *M. flavoviride* in a simple liquid medium. *Mycol. Res. 97*, 1489–1494.

Jenkins, N.E., Thomas, M.B., **1996**. Effect of formulation and application method on the efficacy of aerial and submerged conidia of *M. flavoviride* for locust and grasshopper control. *Pesticide Sci. 46*, 299–306.

Jenkins, N.E., Heviefo, G., Langewald, J., Cherry, A.J., Lomer, C.J., **1998**. Development of mass production technology for aerial conidia for use as mycopesticides. *Biocontr. News Infor. 19* (1), 21N–31N.

Jensen, A.B., Gargas, A., Eilenberg, J., Rosendahl, S., **1998**. Relationships of the insect-pathogenic order Entomophthorales (Zygomycota, Fungi) based on phylogenetic analyses of nuclear small subunit ribosomal DNA sequences (SSU rDNA). *Fungal Gen. Biol. 24*, 325–334.

Jones, K.A., Burges, H.D., **1998**. Technology of formulation and application. In: Burges, H.D. (Ed.), Formulation of Microbial Biopesticides. Kluwer, Dordrecht, pp. 7–30.

Kalsbeek, V., Mullens, B.A., Jespersen, J.B., **2001a**. Field studies of *Entomophthora* (Zygomycetes: Entomophthorales) – induced behaviour fever in *Musca domestica* (Diptera: Muscidae) in Denmark. *Biol. Contr. 21*, 264–271.

Kalsbeek, V., Pell, J.K., Steenberg, T., **2001b**. Sporulation by *Entomophthora schizophorae* (Zygomycetes: Entomophthorales) from housefly cadavers and the persistence of primary conidia at constant temperatures and relative humidities. *J. Invertebr. Pathol. 77*, 149–157.

Kanaoka, M., Isogai, A., Murakoshi, S., Ichione, M., Suzuki, A., *et al.*, **1978**. Bassianoilide, a new insecticidal cyclodepsipeptide from *Beauveria bassiana* and *Verticillium lecanii*. *Agric. Biol. Chem. 42*, 629–635.

Keller, S., Suter, H., **1980**. Epizootiologische Untersuchungen über das *Entomophthora*-Auftre-ten bei feldbaulich wichtigen Blattlausarten. *Acta Oecologica 1*, 63–81.

Keller, S., Kessler, P., Schweizer, C., **2003**. Distribution of insect pathogenic fungi in Switzerland with special reference to *Beauveria brongniartii* and *Metarhizium anisopliae*. *BioControl 48*, 307–319.

Keller, S., Keller, E., Schweizer, C., Auden, J.A.L., Smith, A., **1989**. Two large field trials to control the cockchafer *Melolontha melolontha* L. with the fungus *Beauveria brongniartii* (Sacc). Petch. In: BCPC Mono. No. 43 "Progress and Prospects in Insect Control." pp. 183–190.

Kerwin, J.L., **1983**. Biological aspects of the interactions between *Coelomomyces psoropborae* zygotes and the larvae of *Culiseta inornata*: host-mediated factors. *J. Invertebr. Pathol. 41*, 224–232.

Kerwin, J.L., Petersen, E.E., **1997**. Fungi: Oomycetes and Chytridiomycetes. In: Lacey, L.A. (Ed.), Manual of Techniques in Insect Pathology. Academic Press, San Diego, pp. 251–268.

Kerwin, J.L., Washino, R.K., **1983**. Sterol induction of sexual reproduction in *Lagenidium giganteum*. *Exp. Mycol. 7*, 109–115.

Kish, L.P., Allen, G.E., **1976**. Conidial production of *Nomuraea rileyi* on *Pseudoplusia includens*. *Mycologia 68*, 436–439.

Klein, M.G., Lacey, L.A., **1999**. An attractant trap for autodissemination of entomopathogenic fungi into populations of the Japanese Beetle *Popillia japonica* (Coleoptera: Scarabaeidae). *Biocontr. Sci. Technol. 9*, 151–158.

Klingen, I., Eilenberg, J., Meadow, R., **2002**. Impact of farming systems, field margins and bait insect on the findings of insect pathogenic fungi in soil. *Agric. Ecosys. Environ. 91*, 191–198.

Kobayasi, Y., **1941**. The genus *Cordyceps* and its allies. *Sci. Rep. Tokyo Bunrika Daigaku Sect. B 5*, 53–260.

Kodaira, Y., **1961**. Biochemical studies on the muscardine fungi in the silkworms, *Bombyx mori*. *J. Fac. Text. Sci. Technol. Sinshu Uni. Sericult. 5*, 1–68.

Kogan, P.H., Hajek, A.E., **2000**. Formation of azygospores by the insect pathogenic fungus *Entomophthora maimaiga* in cell culture. *J. Invertebr. Pathol. 75*, 193–201.

Kramer, J.P., **1980**. The housefly mycosis caused by *Entomophthora muscae*: influences of relative humidity on infectivity and conidial germination. *J. N.Y. Ent. Soc. 88*, 236–240.

Krassnoff, S., Gibson, D.M., **1996**. New destruxins from the entomopathogenic fungus *Aschersonia* sp. *J. Nat. Prod. 59*, 485–489.

Krasnoff, S.B., Gupta, S., **1994**. Identification of the antibiotic phomalactone from the entomopathogenic fungus *Hirsutella thompsonii* var. *synnematosa*. *J. Chem. Ecol. 20*, 293–302.

Krasnoff, S.B., Watson, D.W., Gibson, D.M., Kwan, E.C., **1995**. Behaviorial effects of the entomopathogenic fungus, *Entomophthora muscae* on its host *Musca domestica*: postural changes in dying hosts and gated pattern of mortality. *J. Insect Physiol. 41*, 895–903.

Krasnoff, S.B., Gupta, S., St. Leger, R.J., Renwick, J.A., Roberts, D.W., **1991**. Antifungal and insecticidal properties of efrapeptins: metabolites of the fungus *Tolypocladium niveum*. *J. Invertebr. Pathol. 58*, 180–188.

Krueger, S.R., Roberts, D.W., **1997**. Soil treatment with entomopathogenic fungi for corn rootworm (*Diabrotica* spp.) larval control. *Biol. Contr. 9*, 67–74.

Krueger, S.R., Villani, M.G., Martins, A.S., Roberts, D.W., 1992. Efficacy of soil applications of *Metarhizium anisopliae* (Metsch.) Sorokin conidia, and standard and lyophilized mycelial particles against scarab grubs. *J. Invertebr. Pathol.* 59, 54–60.

Kuramoto, H., Shimazu, M., 1997. Control of house fly populations by *Entomophthora muscae* (Zygomycotina: Entomophthorales) in a poultry house. *Appl. Entomol. Zool.* 32, 325–331.

Lacey, L.A., 1997. Manual of Techniques in Insect Pathology. Academic Press, San Diego.

Lacey, L.A., Goettel, M.S., 1995. Current developments in microbial control of insect pests and prospects for the early 21st century. *Entomophaga* 40, 1–25.

Lacey, L.A., Kaya, H.K. (Eds.), 2000. Field Manual of Techniques in Invertebrate Pathology. Kluwer, Dordrecht.

Lacey, L.A., Kirk, A.A., Millar, L., Mercadier, G., Vidal, C., 1999. Ovicidal and larvicidal activity of conidia and blastospores of *Paecilomyces fumosoroseus* (Deuteromycotina: Hyphomycetes) against *Bemisia argentifolii* (Homoptera: Aleyrodidae) with a description of a bioassay system allowing prolonged survival of control insects. *Biocontr. Sci. Technol.* 9, 9–18.

Laird, M., 1985. Use of *Coelomomyces* in biological control: introduction of *Coelomomyces stegomyiae* into Nukunono, Tokelau Islands. In: Couch, J.N., Bland, C.E. (Eds.), The Genus Coelomomyces. Academic Press, New York, pp. 369–390.

Laird, M., Lacey, L.A., Davidson, E.W., 1990. Safety of Microbial Insecticides. CRC Press, Boca Raton.

Landis, D.A., Wratten, S.D., Gurr, G.M., 2000. Habitat management to conserve natural enemies of arthropod pests in agriculture. *Annu. Rev. Entomol.* 45, 175–201.

Latgé, J.P., 1980. Sporulation de *Entomophthora obscura* Hall & Dunn en culture liquide. *Canad. J. Microbiol.* 26, 1038–1048.

Leipe, D.D., Wainright, P.O., Gunderson, J.H., Porter, D., Patterson, D.J., et al., 1994. The stramenopiles from a molecular perspective: 16S-like rRNA sequences from *Labyrinthuloides minuta* and *Cafeteria roenbergensis*. *Phycologia.* 33, 369–377.

Lin, H., Lin, H.F., 1998. Testing strain types of *Beauveria bassiana* by means of electrophoretic patterns of esterase isoenzyme. *Acta Phytophylacica Sinica 25*, 315–320.

Liu, Z.Y., Liang, Z.Q., Liu, A.Y., Yao, Y.J., Hyde, K.D., et al., 2002. Molecular evidence for teleomorph-anamorph connections in *Cordyceps* based on ITS-5.8S rDNA sequences. *Mycol. Res.* 106, 1100–1108.

Logan, D.P., Robertson, L.N., Milner, R.J., 2000. Review of the development of *Metarhizium anisopliae* as a microbial insecticide, BioCane™, for the control of greyback canegrub *Dermolepida albohirtum* (Waterhouse) (Coleoptera: Scarabaeidae) in Queensland sugarcane. *Bulletin-OILB-SROP 23*, 131–137.

Lomer, C.J., Bateman, R.P., Johnson, D.L., Langewald, J., Thomas, M., 2001. Biological Control of Locusts and Grasshoppers. *Annu. Rev. Entomol.* 46, 667–702.

Lord, J.C., Magalhaes, B.P., Roberts, D.W., 1987. Effects of the fungus *Beauveria bassiana* (Bal.) Vuill behavior, oviposition, and susceptibility to secondary infections of adult *Ceroptoma arcuata* (Olivier, 1791) (Coleoptera: Chrysomelidae). *Ann. Soc. Entomol. Brasil 16*, 187–197.

Lucarotti, C.J., 1987. *Coelomomyces stegomyiae* infection in adult *Aedes aegypti*. *Mycologia 79*, 362–369.

Lucarotti, C.J., Federici, B.A., Chapman, H.C., 1985. Progress in the development of *Coelomomyces* fungi for use in integrated mosquito control programs. In: Laird, M., Miles, J.W. (Eds.), Integrated Mosquito Control Methodologies. Academic Press, New York, pp. 251–268.

Luz, C., Fargues, J., 1998. Factors affecting conidial production of *Beauveria bassiana* from fungus-killed cadavers of *Rhodnius prolixus*. *J. Invertebr. Pathol.* 72, 97–103.

Malakar, R., Elkington, J.S., Hajek, A.E., Burand, J.P., 1999. Within-host interactions of *Lymantria dispar* (Lepidoptera: Lymantridae) nucleopolyhedrosis virus and *Entomophaga maimaiga* (Zygomycetes: Entomophthorales). *J. Invertebr. Pathol.* 73, 91–100.

Matha, V., Weiser, J., Olejnicek, J., 1988. The effect of tolypin in *Tolypocladium niveum* crude extract against mosquito and blackfly larvae in the laboratory. *Folia Parasitol. 35*, 381–383.

Maurer, P., Couteaudier, Y., Girard, P.A., Bridge, P.D., Riba, G., 1997. Genetic diversity of *Beauveria bassiana* and relatedness to host insect range. *Mycol. Res. 101*, 159–164.

May, B., Roberts, D.W., Soper, R.S., 1979. Intraspecific genetic variability in laboratory strains of *Entomophthora* as determined by enzyme electrophoresis. *Exp. Mycol. 3*, 289–297.

Mazet, I., Vey, A., 1995. Hirsutellin A, a toxic protein produced in vitro by *Hirsutella thompsonii*. *Microbiol. Reading 141*, 1343–1348.

McCabe, D., Soper, R.S., 1985. Preparation of an entomopathogenic fungal insect control agent. US Patent 4,530,834.

McCoy, C.W., Samson, R.A., Boucias, D.G., 1988. Entomogenous fungi. In: Ignoffo, C., Mandava, N.B. (Eds.), CRC Handbook of Natural Pesticides, Vol. 5: Microbial Insecticides, Part A: Entomogenous Protozoa and Fungi. CRC Press, Boca Raton, pp. 151–236.

McGuire, M.R., Maddox, J.V., Ambrust, E.J., 1987a. Effect of temperature on distribution and success of introduction of an *Empoasca fabae* (Homoptera: Cicadellidae) isolate of *Erynia radicans* (Zygomycetes: Entomophthoraceae). *J. Invertebr. Pathol. 50*, 291–301.

McGuire, M.R., Morris, M.J., Ambrust, E.J., Maddox, J.V., 1987b. An epizootic caused by *Erynia radicans* (Zygomycetes: Entomophthoraceae) isolated from *Empoasca fabae* (Homoptera: Cicadellidae). *J. Invertebr. Pathol. 50*, 78–80.

Mendonca, A.F., 1992. Mass production, application and formulation of *Metarhizium anisopliae* for control of sugarcane froghopper, *Mahanarva posticata*, in Brazil. In: Lomer, C.J., Prior, C. (Eds.), Biological Control of Locusts and Grasshoppers. CABI Press, Wallingford, pp. 239–244.

Merriam, T.L., Axtell, R.C., **1982a**. Salinity tolerance of 2 isolates of *Lagenidium giganteum* (Oomycetes: Lagenidiales) a fungal pathogen of mosquito larvae. *J. Med. Entomol. 19*, 388–393.

Merriam, T.L., Axtell, R.C., **1982b**. Evaluation of the entomogenous fungi *Culicinomyces clavosporus* and *Lagenidium giganteum* for control of the salt marsh mosquito *Aedes taeniorhynchus. Mosq. News 42*, 594–602.

Mesquita, A.L.M., Lacey, L.A., **2001**. Interactions among the entomopathogenic fungus, *Paecilomyces fumosoroseus* (Deuteromycotina: Hyphomycetes), the parasitoid *Aphelinus asychis* (Hymenoptera: Aphelinidae), and their aphid host. *Biol. Cont. 22*, 51–59.

Milne, R., Wright, T., Welton, M., Budau, C., Gringorten, L., *et al.*, **1994**. Identification and partial purification of a cell-lytic factor from *Entomophaga aulicae*. *J. Invertebr. Pathol. 64*, 253–259.

Milner, R.J., **1982**. On the occurrence of pea aphids, *Acyrthosiphon pisum*, resistant to isolates of the fungal pathogen, *Erynia neoaphidis. Entomol. Exp. Appl. 32*, 23–27.

Milner, R.J., **1985**. Distribution in time and space of resistance to the pathogenic fungus, *Erynia neoaphidis* in the pea phid, *Acyrthosiphon pisum. Entomol. Exp. Appl. 37*, 235–240.

Milner, R.J., Lutton, G.G., **1986**. Dependence of *Verticillium lecanii* (Fungi: Hyphomycetes) on high humidities for infection and sporulation using *Myzus persicae* (Homoptera: Aphididae) as host. *Environ. Entomol. 15*, 380–382.

Milner, R.J., Mahon, R.J., **1985**. Strain variation in *Zoophthora radicans:* a pathogen on a variety of insect hosts in Australia. *J. Austral. Entomol. Soc. 24*, 195–198.

Milner, R.J., Staples, J.A., **1996**. Biological control of termites: results and experiences within a CSIRO project in Australia. *Biocontr. Sci. Technol. 6*, 3–9.

Milner, R.J., Holdom, D.G., Glare, T.R., **1984**. Diurnal patterns of mortality in aphids infected by entomophthoran fungi. *Entomol. Exper. Appl. 36*, 37–42.

Milner, R.J., Soper, R.S., Lutton, G.G., **1982**. Field release of an Israeli strain of the fungus *Zoophthora radicans* (Brefeld) Batko for biological control of *Therioaphis trifolii* (Monell) *f. maculata. J. Austral. Entomol. Soc. 21*, 113–118.

Møller, A.P., **1993**. A fungus infecting domestic flies manipulates sexual behaviour of its hosts. *Behav. Ecol. Sociobiol. 33*, 403–407.

Montserrat, M., CastaZé, C., Santamaria, S., **1998**. *Neozygites parvispora* (Zygomycota: Entomophthorales) causing an epizootic in *Frankliniella occidentalis* (Thysanoptera: Thripidae) in Spain. *J. Invertebr. Pathol. 71*, 165–168.

Moore, D., Douro-Kpindou, O.K., Jenkins, N.E., Lomer, C.J., **1996**. Effects of moisture content and temperature on storage of *Metarhizium flavoviride* conidia. *Biocontr. Sci. Technol. 6*, 51–61.

Morris, M.J., Roberts, S.J., Maddox, J.V., Armbrust, E.J., **1996**. Epizootiology of the fungal pathogen *Zoophthora phytonomi* (Zygomycetes: Entomophthorales) in field populations of alfalfa weevil (Coleoptera: Curculionidae) larvae in Illinois. *Great Lakes Entomol. 29*, 129–140.

Mullens, B.A., **1985**. Host age, sex, and pathogen exposure level as factors in the susceptibility of *Musca domestica* to *Entomophthora muscae. Entomol. Exp. Appl. 37*, 33–39.

Mullens, B.A., Rodriguez, J.L., **1985**. Dynamics of *Entomophthora muscae* (Entomophthorales: Entomopthoraceae) conidial discharge from *Musca domestica* (Diptera: Muscidae) cadavers. *Environ. Entomol. 14*, 317–322.

Nagahama, T., Sato, H., Shimazu, M., Sugiyama, J., **1995**. Phylogenetic divergence of the entomophthoralean fungi: evidence from nuclear 18S ribosomal RNA gene sequences. *Mycologia 87*, 203–209.

fNemoto, H., Aoki, J., **1975**. *Entomophthora floridana* (Entomophthorales: Entomophthoraceae) attacking the sugi spider mite, *Oligonychus honoensis* (Acarina: Tetranychidae) in Japan. *Appl. Entomol. Zool. 10*, 90–95.

Neuvéglise, C., Brygoo, Y., Riba, G., **1997**. 28s rDNA group-I introns: a powerful tool for identifying strains of *Beauveria brongniartii. Mol. Ecol. 6*, 373–381.

Nielsen, C., Eilenberg, J., Harding, S., Oddsdottir, E., Haldorsson, G., **2001**. Geographical distribution and host range of Entomophthorales infecting the green spruce aphid *Elatobium abietinum* Walker in Iceland. *J. Invertebr. Pathol. 78*, 72–80.

Nielsen, C., Hajek, A.E., Humber, R.A., Bresciani, J., Eilenberg, J., **2003**. Soil as an environment for winter survival of aphid-pathogenic Entomopththorales. *Biol. Contr. 28*, 92–100.

Noma, T., Strickler, K., **2000**. Effects of *Beauveria bassiana* on *Lygus hesperus* (Hemiptera: Miridae) feeding and oviposition. *Environ. Entomol. 29*, 394–402.

Obornik, M., Jirku, M., Dolezel, D., **2001**. Phylogeny of mitosporic entomopathogenic fungi: is the genus *Paecilomyces* polyphyletic? *Canad. J. Microbiol. 47*, 813–819.

Ogarkov, B.N., Ogarkova, G.R., **1999**. Influence of strain passages on virulence of entomopathogenic fungi from Moniliales. *Mikolog. Fitopatolog. 33*, 432–436.

Omoto, C., McCoy, C.W., **1998**. Toxicity of purified fungal toxin hirsutellin A to the citrus rust mite *Phyllocoprura oleivora. J. Invertebr. Pathol. 72*, 319–322.

Ouedraogo, R.M., Cusson, M., Goettel, M.S., Brodeur, J., **2003**. Inhibition of fungal growth in thermoregulating locusts, *Locusta migratoria*, infected by the fungus *Metarhizium anisopliae* var *acridum. J. Invertebr. Pathol. 82*, 103–109.

Ouedraogo, R.M., Kamp, A., Goettel, M.S., Brodeur, J., Bidochka, M.J., **2002**. Attenuation of fungal infection in thermoregulating *Locusta migratoria* is accompanied by changes in hemolymph proteins. *J. Invertebr. Pathol.* 81, 19–24.

Ovchinnikov, Y.A., Ivanov, V.T., Mikhaleva, I.I., **1971**. The synthesis and some properties of beauvericin. *Tetrahed. Lett.* 2, 159–162.

Panter, C., Russell, R.C., **1984**. Rapid kill of mosquito larvae by high concentrations of *Culicinomyces clavisporus* conidia. *Mosq. News* 44, 242–244.

Pell, J.K., Wilding, N., **1994**. Preliminary caged field trials using the pathogen *Zoophthora radicans* Brefeld (Zygomycetes: Entomophthorales) against the diamond-back moth *Plutella xylostella* L. (Lepidoptera: Yponomeutidae) in the U.K. *Biocontr. Sci. Technol.* 4, 71–75.

Pell, J.K., Macaulay, E.D.M., Wilding, N., **1993**. A pheromone trap for dispersal of the pathogen *Zoophthora radicans* Brefeld. (Zygomycetes: Entomophthorales) amongst populations of the diamond-back moth, *Plutella xylostella* L. (Lepidoptera: Yponomeutidae). *Biocontr. Sci. Technol.* 3, 315–320.

Pell, J.K., Eilenberg, J., Hajek, A.E., Steinkraus, D.C., **2001**. Biology, ecology and pest management of Entomophthorales. In: Butt, T.M., Jackson, C.W., Magan, N. (Eds.), Fungi as Biocontrol Agents. CAB International Press, Wallingford, pp. 71–153.

Pereira, R.M., Roberts, D.W., **1990**. Dry mycelium preparations of entomopathogenic fungi, *Metarhizium anisopliae* and *Beauveria bassiana*. *J. Invertebr. Pathol.* 56, 39–46.

Pereira, R.M., Roberts, D.W., **1991**. Alginate and cornstarch mycelial formulations of entomopathogenic fungi, *Beauveria bassiana* and *Metarhizium anisopliae*. *J. Econ. Entomol.* 84, 1657–1661.

Pilarska, D., McManus, M., Hajek, A.E., Herard, F., Vega, F.E., *et al.*, **2000**. Introduction of the entomopathogenic fungus *Entomophaga maimaiga* Hum., Shim. & Sop. (Zygomycetes: Entomophthorales) to a *Lymantria dispar* (L.) (Lepidoptera: Lymantridae) population in Bulgaria. *J. Pest Sci.* 73, 125–126.

Pillai, J.S., **1971**. *Coelomomyces opifexi* (Coelomomycetaceae: Blastocladiales). Part 1. Its distribution and the ecology of infection pools in New Zealand. *Hydrobiologia* 38, 425–436.

Pillai, J.S., O'Loughlin, I.H., **1972**. *Coelomomyces opifexi* (Pillai and Smith) (Coeolomomycetaceae: Blastocladiales). 2. Experiments in sporangial germination. *Hydrobiologia* 40, 77–86.

Pope, T., Croxson, E., Pell, J.K., Godfray, H.C.J., Muller, C.B., **2002**. Apparent competition between two species of aphids via the fungal pathogen *Erynia neoaphidis* and its interaction with the aphid parasitoid *Aphidius ervi*. *Ecol. Entomol.* 27, 196–203.

Prior, C., Moore, D., Reed, M., LePatourel, G., Abraham, Y.J., **1992**. Reduction of feeding by the dessert locust, *Schistocerca gregaria*, after infection with *Metarhizium flavoviride*. *J. Invertebr. Pathol.* 60, 304–307.

Quesada-Moraga, E., Vey, A., **2003**. Intra-specific variation in virulence and *in vitro* production of macromolecular toxins active against locust among *Beauveria bassiana* strains and effects of *in vitro* and *in vivo* passage on these factors. *Biocontr. Sci. Technol.* 13, 323–340.

Quiot, J.-M., Vey, A., Vago, C., Pais, M., **1980**. Action antivirale d'une mycotoxine. Etude d'une toxine de l'hyphomycete *Metarhizium anisopliae* (Metsch.) Sorok. en culture cellulaire. *CR Acad. Sci. Ser. D (Paris)* 291, 763–766.

Rath, A.C., Carr, C.J., Graham, B.R., **1995**. Characterization of *Metarhizium anisopliae* strains by carbohydrate utilization (API50CH). *J. Invertebr. Pathol.* 65, 152–161.

Reddy, G.V.P., Furlong, M.J., Pell, J.K., Poppy, G.M., **1998**. *Zoophthora radicans* infection inhibits the response to and production of sex pheromone in the diamondback moth. *J. Invertebr. Pathol.* 72, 167–169.

Rehner, S.A., **2003**. A multigene phylogeny of *Beauveria*: new insights into species diversity, biogeography, host affiliation and life history. In: Proc. 36th Ann. Meet. Soc. Invertebr. Pathol., July 26–30, 2003. Vermont, USA, p. 102 (abstract only).

Reineke, A., Zebitz, C.P.W., **1996**. Protein and isoenzyme patterns among isolates of *Beauveria brongniartii* with different virulence to European cockchafer larvae (*Melolontha melolontha* L.). *J. Appl. Entomol.* 120, 307–315.

Reinecke, P., Andersch, W., Stenzel, K., Hartwig, J., **1990**. BIO 1020, a new microbial insecticide for use in horticultural crops. *Brighton Crop Prot. Conf.: Pests Disease* 1, 49–84.

Remaudiére, G., Latgé, J.-P., Michel, M.F., **1981**. Ecologie comparee des entomophthoracees pathogenes de pucerons en France littorale et continentale. *Entomophaga* 26, 157–178.

Robert, P.H., Riba, G., **1989**. Toxic and repulsive effect of spray *per os* and systemic application of destruxin E to aphids. *Mycopathologia* 108, 170–183.

Roberts, D.W., Hajek, A.E., **1992**. Entomopathogenic fungi as bioinsecticides. In: Leatham, G.F. (Ed.), Frontiers in Industrial Mycology. Chapman and Hall, New York, pp. 144–159.

Roy, H.E., Pell, J.K., **2000**. Interactions between entomopathogenic fungi and other natural enemies: implications for biological control. *Biocontr. Sci. Technol.* 10, 737–752.

Roy, H.E., Pell, J.K., Alderson, P.G., **2001**. Targeted dispersal of the aphid pathogenic fungus *Erynia neoaphidis* by the aphid predator *Coccinella septempunctata*. *Biocontr. Sci. Technol.* 11, 99–110.

Roy, H.E., Pell, J.K., Clark, S.J., Alderson, P.G., **1998**. Implications of predator foraging on aphid pathogen dynamics. *J. Invertebr. Pathol.* 71, 236–247.

Saik, J.E., Lacey, L.A., Lacey, C.M., **1990**. Safety of microbial insecticides to vertebrates-domestic animals and wildlife. In: Laird, M., Lacey, L.A., Davidson, E.W.

(Eds.), Safety of Microbial Insecticides. CRC Press, Boca Raton, pp. 115–132.

Samson, R.A., 1995. Constraints associated with taxonomy of biocontrol fungi. *Canad. J. Bot.* 73 (Suppl. 1), S83–S88.

Samson, R.A., Evans, H.C., Latgé, J.-P., 1988. Atlas of Entomopathogenic Fungi. Springer, Berlin.

Sandhu, S.S., Kinghorn, J.R., Rajak, R.C., Unkles, S.E., 2001. Transformation system of *Beauveria bassiana* and *Metarhizium anisopliae* using nitrate reductase gene of *Aspergillus nidulans*. *Indian J. Exp. Biol.* 39, 650–653.

Sanjuan, T., Henao, L.G., Amat, G., 2001. Spatial distribution of *Cordyceps* spp. (Ascomycotina: Clavicipitaceae) and its impact on the ants in forests of the Amazonia Colombian foothill. *Rev. Biol. Tropic.* 49, 945–955.

Screen, S.E., Hu, G., St. Leger, R.J., 2001. Transformants of *Metarhizium anisopliae* sf. *anisopliae* overexpressing chitinase from *Metarhizium anisopliae* sf. *acridum* show early induction of native chitinase but are not altered in pathogenicity to *Manduca sexta*. *J. Invertebr. Pathol.* 78, 260–266.

Screen, S., Bailey, A., Charnley, K., Cooper, R., Clarkson, J., 1997. Carbon regulation of the cuticle-degrading enzyme PR1 from *Metarhizium anisopliae* may involve a trans-acting DNA-binding protein CRR1, a functional equivalent of the *Aspergillus nidulans* CREA protein. *Curr. Gen.* 31, 511–518.

Screen, S., Bailey, A., Charnley, K., Cooper, R., Clarkson, J., 1998. Isolation of a nitrogen response regulator gene (nrr1) from *Metarhizium anisopliae*. *Gene* 221, 17–24.

Seif, A.I., Shaarawi, F.A., 2003. Preliminary field trials with *Culicinomyces clavosporus* against some Egyptian mosquitoes in selected habitats. *J. Egyptian Soc. Parasitol.* 33, 291–304.

Shah, P.A., Goettel, M.S., 1999. Directory of Microbial Control Products and Services. Society for Invertebrate Pathology, Gainesville.

Shah, P.A., Aebi, M., Tuor, U., 2000. Infection of *Macrosiphum euphorbiae* with mycelial preparations of *Erynia neoaphidis* in a greenhouse trial. *Mycol. Res.* 104, 645–652.

Shaw, K.E., Davidson, G., Clark, S.J., Ball, B.V., Pell, J.K., et al., 2002. Laboratory bioassays to assess the pathogenicity of mitosporic fungi to *Varroa destructor* (Acari: Mesostigmata), an ectoparasitic mite of the honeybee, *Apis mellifera*. *Biol. Control* 24, 266–276.

Shi, Z.M., 1998. Technology for conidial preparation of *Beauveria bassiana*. In: Li, Y.W., Li, Z.Z., Liang, Z.Q., Wu, J.W., Wu, Z.K., Xu, Q.F. (Eds.), Study and Application of Entomogenous Fungi in China. Academic Periodical Press, Beijing, pp. 114–115.

Shimazu, M., 1979. Resting spore formation of *Entomophthora sphaerosperma* Fresenius (Entomophthorales: Entomophthoraceae) in the brown plant hopper, *Nilaparvata lugens* (Stal) (Hemiptera: Delphacidae). *Appl. Entomol. Zool.* 14, 383–388.

Shimazu, M., Soper, R.S., 1986. Pathogenicity and sporulation of *Entomophaga maimaiga* Humber, Shimazu,

Soper and Hajek (Entomophthorales Entomophthoraceae) on larvae of the gypsy moth *Lymantria dispar* L. (Lepidoptera Lymantriidae). *Appl. Entomol. Zool.* 21, 589–596.

Siegel, J.P., 1997. Testing the pathogenicity and infectivity of entomopathogens to mammals. In: Lacey, L.A. (Ed.), Manual of Techniques in Insect Pathology. Academic Press, San Diego, pp. 325–336.

Siegel, J.P., Shadduck, J.A., 1990. Safety of microbial insecticides to vertebrates – humans. In: Laird, M., Lacey, L.A., Davidson, E.W. (Eds.), Safety of Microbial Insecticides. CRC Press, Boca Raton, pp. 101–114.

Silveira, R.D., dos Anjos, N., Zanuncio, J.C., 2002. Natural enemies of *Coelomera lanio* (Coleoptera: Chysomelidae) in the region of Vicosa, Minas Gerais, Brasil. *Revista Biol. Trop.* 50, 117–120.

Silvie, P., Dedryver, C.A., Tanguy, S., 1990. Experimental application of *Erynia neoaphidis* (Zygomycetes Entomophthorales) mycelium on aphid populations of lettuces in a commercial glasshouse study of the inoculum spread by isoenzyme profiles. *Entomophaga* 35, 375–384.

Six, D., Mullens, B., 1996. Seasonal prevalence of *Entomophthora muscae* and introduction of *Entomophthora schizophorae* (Zygomycotina: Entomophthorales) in *Musca domestica* (Diptera: Muscidae) populations in California dairies. *Biol. Cont.* 6, 315–323.

Smith, S.M., Moore, D., Karanja, L.W., Chandi, E.A., 1999. Formulation of vegetable fat pellets with pheromone and *Beauveria bassiana* to control the larger grain borer, *Prostephanus truncatus* (Horn). *Pest. Sci.* 55, 711–718.

Smits, J.E., Johnson, D.L., Lomer, C.J., 1999. Pathological and physiological response of ring-necked pheasant chicks following dietary exposure to the fungus *Metarhizium flavoviride*, a biocontrol agent for locusts in Africa. *J. Wildlife Dis.* 35, 194–203.

Soper, R.S., Delyzer, A.J., Smith, L.F.R., 1976. The genus *Massospora* entomopathogenic for cicadas. Part 2. Biology of *Massospora levispora* and its host *Okanagana rimosa* with notes on *Massospora cicadina* on the periodical cicadas. *Ann. Entomol. Soc. Am.* 69, 89–95.

Sopp, P.I., Gillespie, A.T., Palmer, A., 1989. Application of *Verticillium lecanii* for the control of *Aphis gossypii* by a low-volume electrostatic rotary atomiser and a high-volume hydraulic sprayer. *Entomophaga* 34, 417–428.

St. Leger, R.J., 1993. Biology and mechanisms of insect-cuticle invasion by deuteromycete fungal pathogens. In: Beckage, N.C., Thomson, S.N., Federici, B.A. (Eds.), Parasites and Pathogens of Insects Vol. 2. Pathogens. Academic Press, New York, pp. 211–229.

St. Leger, R.J., 2001. Notification of intent to release a transgenic strain of *Metarhizium anisopliae* (document submitted by the University of Maryland as required by FIFRA) www.epa.gov/pesticides/biopesticides/otherdocs/release_notification.htm (16/08/01).

St. Leger, R., Screen, S., 2001. Prospects for strain improvement of fungal pathogens of insects and weeds.

In: Butt, T.M., Jackson, C.W., Magan, N. (Eds.), Fungi as Biocontrol Agents. CABI Publishing, Wallingford, pp. 219–237.

St. Leger, R.J., Staples, R.C., Roberts, D.W., 1991. Changes in translatable mRNA species associated with nutrient deprivation and protease synthesis in *Metarhizium anisopliae*. *J. Gen. Microbiol.* 137, 807–815.

St. Leger, R.J., Charnley, A.K., Cooper, R.M., Slots, J., 1986. Enzymatic characterization of entomopathogens with the API ZYM system. *J. Invertebr. Pathol.* 48, 375–376.

St. Leger, R.J., Frank, D.C., Roberts, D.W., Staples, R.C., 1992a. Molecular cloning and regulatory analysis of the cuticle-degrading-protease structural gene from the entomopathogenic fungus *Metarhizium anisopliae*. *Eur. J. Biochem.* 204, 991–1001.

St. Leger, R.J., Joshi, L., Bidochka, M.J., Roberts, D.W., 1996. Construction of an improved mycoinsecticide overexpressing a toxic protease. *Proc. Natl Acad. Sci. USA* 93, 6349–6354.

St. Leger, R.J., Allee, L.L., May, R., Staples, R.C., Roberts, D.W., 1992b. World-wide distribution of genetic variation among isolates of *Beauveria* spp. *Mycol. Res.* 96, 1007–1015.

St. Leger, R.J., Shimizu, S., Joshi, L., Bidochka, M.J., Roberts, D.W., 1995. Co-transformation of *Metarhizium anisopliae* by electroporation or using the gene gun to produce stable GUS transformants. *FEMS Microbiol. Lett.* 131, 289–294.

Steenberg, T., Langer, V., Esbjerg, P., 1995. Entomopathogenic fungi in predatory beetles (Col.: Carabidae and Staphylinidae) from agricultural fields. *Entomophaga* 40, 77–85.

Steinhaus, E.A., 1975. Disease in a Minor Chord. Ohio State University Press, Columbus.

Steinkraus, D.C., Slaymaker, P.H., 1994. Effect of temperature and humidity on formation, germination and infectivity of *Neozygites fresenii* (Zygomycetes: Neozygitaceae) from *Aphis gossypii* (Homoptera: Aphididae). *J. Invertebr. Pathol.* 64, 130–137.

Steinkraus, D.C., Geden, C.J., Rutz, D.A., 1993. Prevalence of *Entomophthora muscae* (Cohn) Fresenius (Zygomycetes: Entomophthoraceae) in house flies (Diptera: Muscidae) on dairy farms in New York and induction of epizootics. *Biol. Contr.* 3, 93–100.

Steinkraus, D.C., Hollingworth, R.G., Boys, G.O., 1996. Aerial spores of *Neozygites fresenii* (Entomophthorales: Neozygitaceae): density, periodicity, and potential role in cotton aphid (Homoptera: Aphididae) epizootics. *Environ. Entomol.* 25, 48–57.

Steinkraus, D.C., Howard, M.N., Hollingsworth, R.G., Boys, G.O., 1999. Infection of sentinel cotton aphids (Homoptera: Aphididae) by aerial conidia of *Neozygites fresenii* (Entomophthorales: Neozygitaceae). *Biol. Cont.* 14, 181–185.

Strasser, H., Vey, A., Butt, T., 2000. Are there any risks in using entomopathogenic fungi for pest control, with particular reference to the bioactive metabolites of *Metarhizium*, *Tolypocladium* and *Beauveria* species? *Biocont. Sci. Technol.* 10, 717–735.

Suzuki, A., Kanaoka, M., Isogai, A., Murakoshi, S., Ichinoe, M., et al., 1977. Bassianolide, a new insecticidal cyclodepsipeptide from *Beauveria bassiana* and *Verticillium lecanii*. *Tetrahedron Lett.* 25, 2167–2170.

Swanson, D., 1997. Economic feasibility of two techniques for production of a mycopesticide in Madagascar. *Mem. Entomol. Soc. Canad.* 171, 101–113.

Sweeney, A.W., 1981. Preliminary field tests of the fungus *Culicinomyces* against mosquito larvae in Australia. *Mosq. News* 43, 290–297.

Sweeney, A.W., 1985. The potential of the fungus *Culicinomyces clavisporus* as a biocontrol agent for medially important Diptera. In: Laird, M., Miles, J.W. (Eds.), Integrated Mosquito Control Methodologies. Academic Press, New York, pp. 269–284.

Tabashnik, B.E., 1994. Evolution of resistance to *Bacillus thuringensis*. *Annu. Rev. Entomol.* 39, 47–79.

Tanada, Y., Kaya, H.K., 1993. Insect Pathology. Academic Press, San Diego.

Terry, B.J., Liu, W.C., Cianci, C.W., Proszynski, E., Fernandes, P., et al., 1992. Inhibition of herpes simplex virus type 1 DNA polymerase by the natural product oosporein. *J. Antibiotics* 45, 286–288.

Thomas, M.B., Blanford, S., Lomer, C.J., 1997. Reduction of feeding by the variegated grasshopper, *Zonocerus variegatus*, following infection by the fungal pathogen, *Metarhizium flavoviride*. *Biocontr. Sci. Technol.* 7, 327–334.

Thomas, M.B., Watson, E.L., Valverde-Garcia, P., 2003. Mixed infections and insect-pathogen interactions. *Ecol. Lett.* 6, 183–188.

Thomsen, L., Eilenberg, J., 2000. Time-concentration mortality of *Pieris brassicae* (Lepidoptera: Pieridae) and *Agrotis segetum* (Lepidoptera: Noctuidae) larvae from different destruxins. *Environ. Entomol.* 29, 1041–1047.

Thorvilson, H.G., Pedigo, L.P., Lewis, L.C., 1985. Soybean leaf consumption by *Nomuraea rileyi* (Fungi: Deuteromycotina)-infected *Plathypena scabra* (Lepidoptera: Noctuidae) larvae. *J. Invertebr. Pathol.* 46, 265–271.

Todorova, S.I., Cote, J.C., Martel, P., Coderre, D., 1994. Heterogeneity of two *Beauveria bassiana* strains revealed by biochemical tests, protein profiles and bioassays on *Leptinotarsa decemlineata* (Col.: Chrysomelidae) and *Coleomegilla maculata lengi* (Col.: Coccinellidae) larvae. *Entomophaga* 39, 159–169.

Toscano, N.C., Reeves, E.L., 1973. Effect of *Aspergillus flavus* mycotoxin on *Culex* mosquito larvae. *J. Invertebr. Pathol.* 22, 55–59.

Trzebitzky, C., Lochelt, S., 1994. Wiedererkennung rückisolierter Stämme von *Beauveria brongniartii* (Sacc.) Petch aus freilandversuchen durch isoenzymanalyse. *Z. Pflanzenkr. Pflanzensch.* 101, 519–526.

Typas, M.A., Mavridou, A., Kouvelis, V.N., 1998. Mitochondrial DNA differences provide maximum intraspecific polymorphism in the entomopathogenic fungi

Verticillium lecanii and *Metarhizium anisopliae*, and allow isolate detection/identification. In: Bridge, P., Couteaudier, Y., Clarkson, J. (Eds.), Molecular Variability of Fungal Pathogens. CAB International Press, Wallingford, pp. 227–237.

Tyrrell, D., **1987**. Induction of periodic mortality in insect larvae infected with *Entomophaga aulicae*. In: Ann. Meeting Soc. Invertebr. Pathol. Gainsville, Fla., p. 53.

Tyrrell, D., **1988**. Survival of *Entomophaga aulicae* in dried insect larvae. *J. Invertebr. Pathol. 52*, 187–188.

Upadhyay, R.K. (Ed.), **2003**. Advances in Microbial Control of Insect Pests. Kluwer Academic/Plenum, New York.

Uziel, A., Kenneth, R.G., **1991**. Survivial of primary conidia and capilliconidia at different humidities in *Erynia* (subgen. *Zoophthora*) spp. and *Neozygites fresenii* (Zgyomycotina: Entomophthorales), with special emphasis on *Erynia radicans*. *J. Invertebr. Pathol. 58*, 118–126.

Veen, K.H., **1968**. Reserches sur la maladie, due a *Metarhizium anisopliae* chez le criquet perlerin. *Med. Landbouwbogeschool Wageningen 68*, 1–77.

Vega, F.E., Dowd, P.F., Lacey, L.A., Pell, J.K., Jackson, D.M., et al., **2000**. Dissemination of beneficial microbial agents by insects. In: Lacey, L.A., Kaya, H.K. (Eds.), Field Manual of Techniques in Invertebrate Pathology. Kluwer, Dordrecht, pp. 153–177.

Vestergaard, S., Cherry, A., Keller, S., Goettel, M., **2003**. Hyphomycete fungi as microbial control agents. In: Hokkanen, H.M.T., Hajek, A.E., (Eds.), Environmental Impacts of Microbial Insecticides. Kluwer, Dordrecht, pp. 35–62.

Vey, A., Hoagland, R.E., Butt, T.M., **2001**. Toxic metabolites of fungal biocontrol agents. In: Butt, T., Jackson, C., Magan, N. (Eds.), Fungal Biocontrol Agents. CABI Publishing, Wallingford, pp. 311–346.

Viaud, M., Couteaudier, Y., Levis, C., Riba, G., **1996**. Genome organization in *Beauveria bassiana*: electrophoretic karyotype, gene mapping, and telomeric fingerprint. *Fungal Genet. Biol. 20*, 175–183.

Viaud, M., Couteaudier, Y., Riba, G., **1998**. Molecular analysis of hypervirulent somatic hybrids of the entomopathogenic fungi *Beauveria bassiana* and *Beauveria sulfurescens*. *Appl. Environ. Microbiol. 64*, 88–93.

Vicentini, S., Faria, M., de Oliveira, M.R.V., **2001**. Screening of *Beauveria bassiana* (Deuteromycotina: Hyphomycetes) isolates against nymphs of *Bemisia tabaci* (Genn.) biotype B (Hemiptera: Aleyrodidae) with description of a new bioassay method. *Neotrop. Entomol. 30*, 97–103.

Vidal, C., Fargues, J., Rougier, M., Smits, N., **2003**. Effect of air humidity on the infection potential of hyphomycetous fungi as mycoinsecticides for *Trialeurodes vaporariorum*. *Biocontr. Sci. Technol. 13*, 183–198.

Villacarlos, L.T., Meija, B.S., Keller, S., **2003**. *Entomophthora leyteensis* Villacarlos & Keller sp.nov. (Entomophthorales: Zygomycetes) infecting *Tetraleurodes acaciae* (Quaintance) (Insecta, Hemiptera: Aleyrodidae)

a recently introduced whitefly on *Gliricidia sepium* (Jaq.) Walp. (Fabaceae) in the Philippines. *J. Invertebr. Pathol. 83*, 16–22.

Villani, M.G., Allee, L.L., Diaz, A., Robbins, P.S., **1999**. Adaptive strategies of edaphic arthropods. *Annu. Rev. Entomol. 44*, 233–256.

Ward, M.D., Sailstad, D.M., Selgrade, M.K., **1998**. Allergic responses to the biopesticide *Metarhizium anisopliae* in Balb/c mice. *Toxicol. Sci. 45*, 195–203.

Washino, R.K., **1981**. Biocontrol of mosquitoes associated with California rice fields with special reference to the recycling of *Lagenidium giganteum* Couch and other microbial agents. In: Laird, M. (Ed.), Biocontrol of Medical and Veterinary Pests. Praeger, New York, pp. 122–139.

Watson, D.W., Mullens, B.A., Petersen, J.J., **1993**. Behavioral fever response of *Musca domestica* (Diptera: Muscidae) to infection by *Entomophthora muscae* (Zygomycetes: Entomophthorales). *J. Invertebr. Pathol. 61*, 10–16.

Weseloh, R.M., Andreadis, T.G., **1997**. Persistence of resting spores of *Entomophaga maimaiga*, a fungal pathogen of the gypsy moth, *Lymantria dispar*. *J. Invertebr. Pathol. 69*, 195–196.

Wraight, S.P., Carruthers, R.I., **1999**. Production, delivery and use of mycoinsecticides for control of insect pests of field crops. In: Hall, F.R., Menn, J.J. (Eds.), Methods in Biotechnology, Vol. 5: Biopesticides: Use and Delivery. Humana Press, Totowa, pp. 233–269.

Wraight, S.P., Jackson, M.A., de Lock, S.L., **2001**. Production, stabilization and formulation of fungal biocontrol agents. In: Butt, T.M., Jackson, C.W., Magan, N. (Eds.), Fungi as Biocontrol Agents. CABI Publishing, Wallingford, pp. 253–287.

Yamamoto, M., Aoki, J., **1983**. Periodicity of conidial discharge of *Erynia radicans*. *Trans. Mycol. Soc. Japan 24*, 487–496.

Zabra, A., Piatkowski, J., Greb-Markiewicz, B., Bujak, J., **1996**. Secondary metabolites produced by entomopathogenic fungi of the genera *Zoophthora* and *Paecilomyces*. *Bull-OILB-SROP 19*, 196–199.

Zare, R., Gams, W., **2001**. A revision of *Verticillium* section *Prostrata*. IV. The genera *Lecanicillium* and *Simplicillium*. *Nova Hedwigia 73*, 1–50.

Zare, R., Gams, W., Culham, A., **2000**. A revision of *Verticillium* sect. *Prostrata*. I. Phylogenetic studies using ITS sequences. *Nova Hedwigia 71*, 465–480.

Zattau, W. C., McInnis, T., Jr., **1987**. Life cycle and mode of infection of *Leptolegnia chapmanii* (Oomycetes) parasitizing *Aedes aegypti*. *J. Invertebr. Pathol. 50*, 134–145.

Zimmermann, G., **1986**. The "*Galleria* bait method" for detection of entomopathogenic fungi in soil. *Z. Ang. Entomol. 102*, 213–215.

Zimmermann, G., Papierok, B., Glare, T., **1995**. Elias Metschnikoff, Elie Metchnikoff or Ilya Ilich Mechnikov (1845–1916): a pioneer in insect pathology, the first describer of the entomopathogenic fungus *Metarhizium anisopliae* and how to translate a Russian name. *Biocont. Sci. Technol. 5*, 527–530.

A11 Addendum: Entomopathogenic Fungi and Their Role in Regulation of Insect Populations, 2004–2009

T R Glare, Lincoln University, Lincoln, Christchurch, New Zealand
M S Goettel, Agriculture & Agri-Food Canada, Lethbridge, Canada
J Eilenberg, University of Copenhagen, Frederiksberg, Denmark

© 2010 Elsevier B.V. All Rights Reserved

A11.1. Introduction

Our review article (Goettel *et al.*, 2004) covered aspects of the biology, ecology, and pathogenesis of these fascinating and useful insect-killing microbes. Herein we briefly update that review with key findings for the period 2004–2009.

A11.2. Systematics

Seminal studies utilizing multigene phylogenies have reclassified relationships between higher level phyla and provided greater understanding of the evolutionary stability of morphological traits once used for high level divisions. Hibbett *et al.* (2007) have comprehensively revised the fungal kingdom down to the level of order. Of importance to the entomopathogenic fungi, the taxonomy of the Ascomycota and Basidiomycota has been revised and combined in the subkingdom Dikarya and the Zygomycota and Chytridiomycota, recognized as polyphyletic have been reclassified. The Entomophthorales, containing many insect pathogenic species, is now in the subphylum Entomophthoromycotina. In the Ascomycota, Sung *et al.* (2007) revised the class Sordariomycetes (=Pyrenomycetes), which contained important entomopathogenic genera such as *Cordyceps*, *Torrubiella*, and *Hypocrella* in the family Clavicipitaceae. *Cordyceps* spp. (Clavicipitaceae) were documented to be polyphyletic and have been reassigned into the three families: the Cordycipitaceae, Ophiocordycipitaceae, and Clavicipitaceae.

Apart from the reclassification of families and genera, several new genera have been proposed, based on both morphological and genetic considerations. For example, a new genus, *Metacordyceps*, is proposed for some *Cordyceps* species in Clavicipitaceae s. s. and includes teleomorphs linked to *Metarhizium* and related anamorphs (Sung *et al.*, 2007).

Several multigene studies were conducted on significant entomopathogenic genera including *Beauveria*, *Metarhizium*, and *Paecilomyces*. In *Metarhizium*, a multigene phylogenetic approach combined with morphological considerations of the *M. anisopliae* complex recognized eight species, including *M. anisopliae and M. acridum* stat. nov (Bischoff *et al.*, 2009). The authors also reinstated *M. brunneum*, which is the correct identity of the commonly used F52/Ma43/BIPESCO 5 biocontrol strain.

Rehner and Buckley (2005) found six well supported clades within *Beauveria* using ITS and EF-1α. One group is a large globally distributed *B. bassiana* (including teleomorph *Cordyceps staphylinidaecola*). Interestingly, a large group of "*B. cf bassiana*" type clustered out relatively distant from the global origin *B. bassiana* clade.

Paecilomyces has been found to be phylogenetically diverse (Luangsa-ard *et al.*, 2004, 2005), although a number of common insect pathogenic species were monophyletic in the "*Isaria*" clade, with teleomorphic relationship in *Cordyceps* (Luangsa-ard *et al.*, 2005). Species in *Lecanicillium* (formerly *Verticillium*) have been largely resolved by multigene analysis (Kouvelis *et al.*, 2008).

The new genetic comparisons have allowed consideration of the implications of habitat, such as host spectrum, on fungal evolution. Pathogenicity for invertebrates has arisen and been lost multiple times over several branches of the fungal evolutionary tree (Humber, 2008; Vega *et al.*, 2009). Humber (2008) considered proximity to new hosts as encouraging gain or loss of entomopathogenicity over the fungi, and interkingdom host jumping between plants and animals especially within the Clavicipitaceae was found.

A11.3. Approaches to Classification and Development of PCR Tools for Ecological Studies

The maturation of DNA-based phylogenetics has been the hallmark of the last years, with multigene analyses now as the standard method for classification. Studies have indicated DNA/RNA regions which give the best differentiation. The use of 16s rDNA has been largely superseded by more specific gene regions. Bischoff *et al.* (2009) found that of all gene regions used to identify *Metarhizium* spp., the 5′ region of EF-1α was the most informative.

In addition, strain and species identification has been the subject of several studies using other techniques than sequence comparison. Oulevey *et al.* (2009) used microsatellite markers to separate genotypes of *M. anisopliae* collected in Switzerland, with a set of 16 common markers able to distinguish around 92% of genotypes identified with a broader set of markers. Chemotaxonomic analyses of hopane triterpenes were also used as markers for *Hypocrella* and *Moelleriella* species and their *Aschersonia* anamorphs (Isaka *et al.*, 2009).

Within the Entomophthorales, species-specific primers were developed for *Zoophthora radicans*, *Pandora blunckii* (Guzman-Franco *et al.*, 2008), and *Pandora neoaphidi*s (Fournier *et al.*, 2008). Such tools may be very useful for studying host–pathogen relationships. To detect resting spores of *Entomophaga maimaiga* in soil, real time PCR was employed by Castrillo *et al.* (2007). In light of these new studies, to fully characterize fungal strains would require (1) morphological data on fungus, (2) data on host range, and (3) molecular

characterization, as shown for *Entomophthora muscae sensu lato* (Jensen *et al.*, 2006).

A11.4. New Insights into an Ecological Role of Entomopathogenic Fungi

Insect pathogenic fungi have recently been shown to provide protection against insects, plant parasitic nematodes, and plant pathogens (Ownley *et al.*, 2010; Vega *et al.*, 2009). *B. bassiana* and other entomopathogenic species have been reported endophytic in corn, tomato, cocoa, pine, opium poppy, date palm, bananas, and coffee. The mode of action largely remains unknown although the entomopathogens are known to produce fungal metabolites that cause feeding deterrence or antibiosis. Some endophytic strains have been shown to infect insects in bioassay but field reports of infection after contact with endophytic plants are currently lacking.

In addition to their endophytic abilities, there have been significant and far reaching advances made in the soil- and plant-associated ecology of entomopathogenic fungi. Not only have recent phylogenies stressed the host jumping of plant and animal/invertebrate abilities among fungi (e.g., Humber, 2008), but new studies have demonstrated that entomopathogenic fungi colonize the rhizosphere and act as plant pathogen antagonists (see reviews by Jaronski, 2007; Ownley *et al.*, 2010; Bruck, 2009)

Building on the work of Hu and St Leger (2002) where higher survival of a green-fluorescent protein expressing strain of *M. anisopliae* in the cabbage rhizosphere was shown, an expanded field trial is underway with *M. anisopliae* strains expressing two fluorescent protein genes, a deletion mutant of a gene highly expressed in the haemolymph of the host and required for immune evasion (Wang and St. Leger, 2006) and a mutant of MAD2, a plant adhesive protein (Wang and St. Leger, 2007a). These field experiments are designed to investigate the adaptations of *M. anisopliae* to live in the soil and rhizosphere (St. Leger, 2008).

A11.5. Potential in Conservation Biological Control

Recently, more attention has been given to insect pathogenic fungi for their potential in conservation biological control although a lack of knowledge of fungal ecology, epizootiology, life cycles, and trophic interactions is still hampering their exploitation in conservation biological control (Pell *et al.*, 2010). Recent studies on *Pandora neoaphidis* infecting cereal aphids have revealed the importance of interaction between different host aphid species, other natural enemies (predators, parasitoids), and fungal pathogens (Baverstock *et al.*, 2008; Roy *et al.*, 2008).

A11.6. Production

The production of microsclerotia by *M. anisopliae in vitro* has been found through varying the concentration of carbon and carbon/nitrogen concentrations (Jaronski and Jackson, 2008). These tight hyphal bundles are more resistant to desiccation and could be rehydrated to produce hyphae, sporulate, and infect the sugar beet maggot (*Tetanops myopaeformis*). The production of microslerotia in liquid culture could provide a novel method for biopesticide production against soil dwelling pests, as well as a possible increased persistence for *Metarhizium* in soil.

In a very different approach to application, the entomophthoralean pathogen *Neozygites fresenii* (Entomophthorales: Neozygitaceae) has been collected as cadavers of the host *Aphis gossypii* (Homoptera: Aphididae), dried using salt or silica gel and stored frozen (Steinkraus and Boys, 2005) to be used to inoculate cotton fields for aphid control in subsequent seasons.

A11.7. Role of Metabolites

A number of studies have advanced knowledge on the genetics and function of secondary metabolites and toxins from entomopathogens, especially *Beauveria* and *Metarhizium*, which can be useful in understanding infection processes and developing biocontrol. Large EST or genome studies have demonstrated regulation of known enzyme or toxin genes during exposure to the cuticle or other conditions (see next section) and several studies demonstrated the involvement of well known metabolites in virulence.

Bassianolide, a cyclooligomer depsipeptide secondary metabolite from *B. bassiana,* was shown to be a highly significant virulence factor through targeted inactivation studies. Disruption of bassianolide did not affect another metabolite, beauvericin (Xu *et al.*, 2009), another cyclodepsipeptide, which was identified as a nonessential virulence factor during infection of *Galleria mellonella, Spodoptera exigua,* and *Helicoverpa zea* (Xu *et al.*, 2008). Beauvericin was also highly toxic *in vitro* to cells of the fall armyworm, *S. exigua* (Fornelli *et al.*, 2004). However, Eley *et al.* (2007) showed that another metabolite of *B. bassiana*, tenellin, had no role in virulence.

The cyclic depsipeptides destruxins, produced by *M. anisopliae,* have insecticidal, antiviral, and phytotoxic abilities and are also studied for their toxicity to cancer cells. Gene expression studies on *Drosophila melanogaster* following injection of destruxin showed a novel role for destruxin A in specific suppression of the humoral immune response in insects (Pal *et al.*, 2007).

Subtilisins (Pr1) are known to be involved in virulence of some entomopathogenic fungi. *Metarhizium* strains with broad host ranges expressed up to 11 subtilisins during growth on insect cuticle (Bagga *et al.*, 2004) and up to 8 in *Beauveria* (Cho *et al.*, 2006a). Pr1 was also shown to be upregulated during mycelial emergence in the host (Small and Bidochka, 2005), suggesting that, as the nutrition within the host is depleted, Pr1 is upregulated to assist breaching the host cuticle again.

A zinc-dependent metalloprotease, ZrMEP1, was isolated from *Zoophthora radicans*, the first report of this type of metalloprotease from an entomopathogenic fungus. It appears to have a role in the infection process (Xu *et al.*, 2006).

A11.8. Advances in Molecular Genetics

Recent studies identified many genes involved in the infection process of fungi. For *Metarhizium* and *Beauveria*, they show different gene expression depending on growth form, host, and environment. Cho *et al.* (2006a, b), conducted extensive expression sequence tag (EST) analysis of *B. bassiana* cDNA-libraries from conidia, blastospores, and under different growth conditions, with around 4000 sequences isolated. The evidence demonstrates highly plastic gene expression depending on cDNA library. Pathan *et al.* (2007) used analyses of gene expression patterns through cDNA-AFLPs of a *B. bassiana* isolate grown on cuticular extracts of various insects and synthetic medium. In general, they found the activity on cuticular extracts from diverse insects was similar, suggesting a relatively generic response to the penetration of cuticle may be indicative of a broad host range. In contrast, genes expressed on synthetic medium were quite different from those on cuticle.

Freimoser *et al.* (2005) examined the response of *M. anisopliae* to different insect cuticles using cDNA microarrays constructed from ESTs. They found unique expression responses for different insect cuticles, indicating the fungus could react specifically to species of insects. *M. anisopliae* had several forms of catabolic enzymes which were regulated by different sugar levels. This provided more evidence that the fungus could respond to nutrition in different environments.

Examination of gene expression patterns to understand pathogenic and saprophytic adaptations of various strains of entomopathogenic fungi was further advanced by Wang *et al* (2009). They used heterologous hybridization of genomic DNA from specialist strains of *Metarhizium* to genes from the generalist strain Ma2575. Approximately 7% of Ma2575 genes were highly divergent or absent in specialist strains. The absence of genes from the specialist strains was taken as an indication of loss of ability to utilize some substrates/environments. These included genes involved in toxin biosynthesis and sugar metabolism in root exudates, so the specialists are losing genes required to live in alternative hosts or as saprophytes.

Wang *et al.* (2005) identified *M. anisopliae* genes which are upregulated in the presence of cuticle, haemolymph, and root extract, providing more insights into the variable habitats able to be colonized by this entomopathogen. The demonstration of specific genes which enable *M. anisopliae* to adhere to insects (Mad1) and roots (Mad2) clearly documented that the fungus has abilities as a rhizosphere colonizer (Wang and St. Leger, 2007a).

Other studies have identified genes in addition to those already mentioned, for example genes encoding cuticle degrading enzymes (Fang *et al.*, 2005), a perilipin-like protein that regulates appressorium turgor pressure and differentiation (Wang and St. Leger, 2007b), sporulation-related genes (Wu *et al.*, 2008), and a G protein (Fang *et al.*, 2007, 2008).

Modification of entomopathogenic fungi to express exotic proteins to improve performance has been used successfully in *Metarhizium* and *Beauveria*. The expression of scorpion neurotoxin AAIT by transgenic strains of *Metarhizium* and *Beauveria* led to a decrease in mean lethal concentration (LC$_{50}$) required to kill hosts. In *M. anisopliae*, Wang and St Leger (2007c) placed AAIT under control of a promoter which only expressed in the insect haemolymph and found a 22-fold reduction of the LC$_{50}$ of the transgenic strain against *Manducta sexta* and a 9-fold reduction against the mosquito, *Aedes aegypti*. This strain was also effective against the coleopteran coffee pest, *Hypothenemus hampei* with the LC$_{50}$ reduced 16-fold and survival time of the pest reduced by 20% (Pava-Ripoll *et al.*, 2008). *B. bassiana* expressing AAIT and the cuticle degrading protease Pr1A (from *M. anisopliae*) resulted in transgenic strains that required less conidia to kill host insects and a 40% reduction in the median lethal time (Lu *et al.* 2008). A double transformant, expressing both genes, was not more effective, as it appears that the protease can degrade AAIT.

Similarly, Fan *et al.* (2007) overexpressed *B. bassiana* chitinase and showed increased virulence against the aphid, *Myzus persicae*. Similarly, Fang *et al.* (2009) found that a mixture of a *B. bassiana* protease Pr1A homolog (CDEP1) and a chitinase Bbchit1 degraded insect cuticle *in vitro* more efficiently than either CDEP1 or Bbchit1 alone. The double transformant resulted in 60.5% reduction in the LC$_{50}$.

A11.9. Conclusion

It is clear that understanding the genetics, biology, and ecology of entomopathogenic fungi is entering a new era. New insights into the ecological roles these fungi occupy have been strengthened by advances in the "omics," bringing perception on the genetics underpinning substrate utilization and pathogenicity determinants. The phylogenetic relationships between the fungi are also becoming clearer, showing new and interesting links to other fungal groups. We may yet see entomopathogenic fungi fully live up to their potential as widespread and versatile control agents of invertebrate pests.

References

Bagga, S., Hu, G., Screen, S.E., St. Leger, R.J., **2004**. Reconstructing the diversification of subtilisins in the pathogenic fungus *Metarhizium anisopliae*. *Gene* 324, 159–169.

Baverstock, J., Baverstock, K.E., Clark, S.J., Pell, J.K., **2008**. Transmission of *Pandora neoaphidis* in the presence of co-occurring arthropods. *J. Invertebr. Pathol.* 98, 356–359.

Bischoff, J.F., Rehner, S.A., Humber, R.A., **2009**. A multilocus phylogeny of the *Metarhizium anisopliae* lineage. *Mycologia* 101, 512–530.

Bruck, D.J., **2009**. Fungal entomopathogens in the rhizosphere. *BioControl* 55, 103–112.

Castrillo, L.A., Thomsen, L., Juneja, P., Hajek, A.E., **2007**. Detection and quantification of *Entomophaga maimaiga* resting spores in forest soil using real-time PCR. *Mycol. Res.* 111, 324–331.

Cho, E.M., Boucias, D., Keyhani, N.O., **2006a**. EST analysis of cDNA libraries from the entomopathogenic fungus *Beauveria* (*Cordyceps*) *bassiana*. II. Fungal cells sporulating on chitin and producing oosporein. *Microbiology* 152, 2855–2864.

Cho, E.M., Liu, L., Farmerie, W., Keyhani, N.O., **2006b**. EST analysis of cDNA libraries from the entomopathogenic fungus *Beauveria* (*Cordyceps*) *bassiana*. I. Evidence for stage-specific gene expression in aerial conidia, in vitro blastospores and submerged conidia. *Microbiology* 152, 2843–2854.

Eley, K.L., Halo, L.M., Song, Z., Powles, H., Cox, R.J., Bailey, A.M., Lazarus, C.M., Simpson, T.J., **2007**.

Biosynthesis of the 2-pyridone tenellin in the insect pathogenic fungus *Beauveria bassiana*. *ChemBioChem* 8, 289–297.

Fan, Y., Fang, W., Guo, S., Pei, X., Zhang, Y., Xiao, Y., Li, D., Jin, K., Bidochka, M.J., Pei, Y., 2007. Increased insect virulence in *Beauveria bassiana* strains overexpressing an engineered chitinase. *Appl. Environ. Microbiol.* 73, 295–302.

Fang, W., Leng, B., Xiao, Y., Jin, K., Ma, J., Fan, Y., Feng, J., Yang, X., Zhang, Y., Pei, Y., 2005. Cloning of *Beauveria bassiana* chitinase gene Bbchit1 and its application to improve fungal strain virulence. *Appl. Environ. Microbiol.* 71, 363–370.

Fang, W., Pei, Y., Bidochka, M.J., 2007. A regulator of a G protein signaling (RGS) gene, cag8, from the insect–pathogenic fungus *Metarhizium anisopliae* is involved in conidiation, virulence and hydrophobin synthesis. *Microbiology* 153, 1017–1025.

Fang, W., Scully, L.R., Zhang, L., Pei, Y., Bidochka, M.J., 2008. Implication of a regulator of G protein signalling (BbRGS1) in conidiation and conidial thermotolerance of the insect pathogenic fungus *Beauveria bassiana*. *FEMS Microbiol. Lett.* 279, 146–156.

Fang, W., Feng, J., Fan, Y., Zhang, Y., Bidochka, M.J., St. Leger, R.J., Pei, Y., 2009. Expressing a fusion protein with protease and chitinase activities increases the virulence of the insect pathogen *Beauveria bassiana*. *J. Invertebr. Pathol.* 102, 155–159.

Fornelli, F., Minervini, F., Logrieco, A., 2004. Cytotoxicity of fungal metabolites to lepidopteran (*Spodoptera frugiperda*) cell line (SF-9). *J. Invertebr. Pathol.* 85, 74–79.

Fournier, A., Enkerli, J., Keller, S., Widmer, F., 2008. A PCR-based tool for the cultivation-independent monitoring of *Pandora neophidis*. *J. Invertebr. Pathol.* 99, 49–56.

Freimoser, F.M., Hu, G., St. Leger, R.J., 2005. Variation in gene expression patterns as the insect pathogen *Metarhizium anisopliae* adapts to different host cuticle or nutrient deprivation in vitro. *Microbiol* 151, 361–371.

Goettel, M.S., Eilenberg, J., Glare, T.R., 2004. Entomopathogenic fungi and their role in regulation of insect populations. In: Gilbert, L.I., Iatrou, K., Gill, S. (Eds.), Comprehensive Molecular Insect Science, Vol. 6. Elsevier, London, pp. 361–406.

Guzman-Franco, A.W., Atkins, S.D., Alderson, P.G., Pell, J.K., 2008. Development of species-specific diagnostic primers for *Zoophthora radicans* and *Pandora blunckii*: two co-occurring fungal pathogens of the diamondback moth, *Plutella xylostella*. *Mycol. Res.* 112, 1227–1240.

Hibbett, D.S., et al., 2007. A higher-level phylogenetic classification of the fungi. *Mycol. Res.* 111, 509–547.

Hu, G., St Leger, R.J., 2002. Field studies using a recombinant mycoinsecticide (*Metarhizium anisopliae*) reveal that it is rhizosphere competent. *Appl. Environ. Microbiol.* 68, 6383–6387.

Humber, R.A., 2008. Evolution of entomopathogenicity in fungi. *J. Invertebr. Pathol.* 98, 262–266.

Isaka, M., Hywel-Jones, N.L., Sappan, M., Mongkolsamrit, S., Saidaengkham, S., 2009. Hopane

triterpenes as chemotaxonomic markers for the scale insect pathogens *Hypocrella* s. lat. and *Aschersonia*. *Mycol. Res.* 113, 491–497.

Jaronski, S.T., 2007. Soil ecology of the entomopathogenic ascomycetes: a critical examination of what we (think) we know. In: Maniana, K., Ekesi, S. (Eds.), Use of Entomopathogenic Fungi in Biological Pest Management. Research SignPosts, Trivandrum, India, pp. 91–143.

Jaronski, S.T., Jackson, M.A., 2008. Efficacy of *Metarhizium anisopliae* microsclerotial granules. *Biocontrol Sci. Technol.* 18, 849–863.

Jensen, A.B., Thomsen, L., Eilenberg, J., 2006. Value of host range, morphological, and genetic characteristics within the *Entomophthora muscae* species complex. *Mycol. Res.* 110, 941–950.

Kouvelis, V.N., Sialakouma, A., Typas, M.A., 2008. Mitochondrial gene sequences alone or combined with ITS region sequences provide firm molecular criteria for the classification of *Lecanicillium* species. *Mycol. Res.* 112, 829–844.

Lu, D., Pava-Ripoll, M., Li, Z., Wang, C., 2008. Insecticidal evaluation of *Beauveria bassiana* engineered to express a scorpion neurotoxin and a cuticle degrading protease. *Appl. Microbiol. Biotechnol.* 81, 515–522.

Luangsa-ard, J.J., Hywel-Jones, N.L., Samson, R.A., 2004. The polyphyletic nature of *Paecilomyces sensu lato* based on 18S generated rDNA phylogeny. *Mycol. Res.* 109, 581–589.

Luangsa-ard, J.J., Hywel-Jones, N.L., Manoch, L., Samson, R.A., 2005. On the relationships of *Paecilomyces* sect. *Isarioidea* species. *Mycologia* 96, 773–780.

Oulevey, C., Widmer, F., Kolliker, R., Enkerli, J., 2009. An optimized microsatellite marker set for detection of *Metarhizium anisopliae* genotype diversity on field and regional scales. *Mycol. Res.* doi:10.1016/j.mycres.2009.06.005.

Ownley, B.H., Gwinn, K.D., Vega, F.E., 2010. Endophytic fungal entomopathogens with activity against plant pathogens: ecology and evolution. *BioControl* 55, 113–128.

Pal, S., St Leger, R.J., Wu, L.P., 2007. Fungal peptide Destruxin A plays a specific role in suppressing the innate immune response in *Drosophila melanogaster*. *J. Biol. Chem.* 282, 8969–8977.

Pathan, A.A.K., Devi, K.U., Vogel, H., Reineke, A., 2007. Analysis of differential gene expression in the generalist entomopathogenic fungus *Beauveria bassiana* (Bals.) Vuillemin grown on different insect cuticular extracts and synthetic medium through cDNA-AFLPs. *Fungal Genet. Biol* 44, 1231–1241.

Pava-Ripoll, M., Posada, F.J., Momen, B., Wang, C., St. Leger, R.J., 2008. Increased pathogenicity against coffee berry borer, *Hypothenemus hampei* (Coleoptera: Curculionidae) by *Metarhizium anisopliae* expressing the scorpion toxin (AaIT) gene. *J. Invertebr. Pathol.* 99, 220–226.

Pell, J.K., Hannam, J.J., Steinkraus, J.J., 2010. Conservation biological control using fungal entomopathogens. *BioControl* 55, 187–198.

Rehner, S.A., Buckley, E., 2005. A *Beauveria* phylogeny inferred from nuclear ITS and EF1-α sequences: evidence for cryptic diversification and links to *Cordyceps* teleomorphs. *Mycologia 97*, 84–98.

Roy, H.E., Baverstock, J., Ware, R.L., Clark, S.J., Majerus, M.E.N., Baverstock, K.E., Pell, J.K., 2008. Intraguild predation of the aphid pathogenic fungus *Pandora neoaphidis* by the invasive coccinellid *Harmonia axyridis*. *Ecol. Entomol. 33*, 175–182.

Small, C.L.N., Bidochka, M.J., 2005. Up-regulation of Pr1, a subtilisin-like protease, during conidiation in the insect pathogen *Metarhizium anisopliae*. *Mycol. Res. 109*, 307–313.

St. Leger, R.J., 2008. Studies on adaptations of *Metarhizium anisopliae* to life in the soil. *J. Invertebr. Pathol. 98*, 271–276.

Steinkraus, D.C., Boys, G.O., 2005. Mass harvesting of the entomopathogenic fungus, *Neozygites fresenii*, from natural field epizootics in the cotton aphid, *Aphis gossypii*. *J. Invertebr. Pathol. 88*, 212–217.

Sung, G.H., Hywel-Jones, N.L., Sung, J.M., Luangsa-ard, J.J., Shrestha, B., Spatafora, J.W., 2007. Phylogenetic classification of *Cordyceps* and the clavicipitaceous fungi. *Stud. Mycol. 57*, 5–59.

Vega, F.E., Goettel, M.S., Blackwell, M., Chandler, D., Jackson, M.A., Keller, S., Koike, M., Maniania, N.K., Monzon, A., Ownley, B.H., Pell, J.K., Rangel, D.E.N., Roy, H.E., 2009. Fungal entomopathogens: new insights on their ecology. *Fungal Ecol 2*, 1–11.

Wang, C., St. Leger, R.J., 2006. A collagenous protective coat enables *Metarhizium anisopliae* to evade insect immune responses. *PNAS 103*, 6647–6652.

Wang, C., St. Leger, R.J., 2007a. The MAD1 adhesin of *Metarhizium anisopliae* links adhesion with blastospore production and virulence to insects: the MAD2 adhesin enables attachment to plants. *Eukaryotic Cell 6*, 808–816.

Wang, C., St. Leger, R.J., 2007b. The *Metarhizium anisopliae* perilipin homolog MPL1 regulates lipid metabolism, appressorial turgor pressure, and virulence. *J. Biol. Chem. 282*, 21110–21115.

Wang, C., St. Leger, R.J., 2007c. A scorpion neurotoxin increases the potency of a fungal insecticide. *Nat. Biotechnol 25*, 1455–1456.

Wang, C., Hu, G., St. Leger, R.J., 2005. Differential gene expression by *Metarhizium anisopliae* growing in root exudate and host (*Manduca sexta*) cuticle or hemolymph reveals mechanisms of physiological adaptation. *Fungal Genet. Biol. 42*, 704–718.

Wang, S., Leclerque, A., Pava-Ripoll, M., Fang, W., St. Leger, R.J., 2009. Comparative genomics using microarrays reveals divergence and loss of virulence-associated genes in host-specific strains of the insect pathogen *Metarhizium anisopliae*. *Eukaryotic Cell 8*, 888–898.

Wu, J., Ridgway, H.J., Carpenter, M.A., Glare, T.R., 2008. Identification of novel genes associated with sporulation in *Beauveria bassiana* using suppression subtractive hybridisation. *Mycologia 100*, 20–30.

Xu, J., Baldwin, D., Kindrachuk, C., Hegedus, D.D., 2006. Serine proteases and metalloproteases associated with pathogenesis but not host specificity in the Entomophthoralean fungus *Zoophthora radicans*. *Can. J. Microbiol. 52*, 550–559.

Xu, Y., Orozco, R., Wijeratne, E.M.K., Gunatilaka, A.A.L., Stock, S.P., Molnár, I., 2008. Biosynthesis of the cyclooligomer depsipeptide beauvericin, a virulence factor of the entomopathogenic fungus *Beauveria bassiana*. *Chem. Biol. 15*, 898–907.

Xu, Y., Wijeratne, E.M.K., Espinosa-Artiles, P., Gunatilaka, A.A.L., Molnár, I., 2009. Combinatorial mutasynthesis of scrambled beauvericins, cyclooligomer depsipeptide cell migration inhibitors from *Beauveria bassiana*. *ChemBioChem 10*, 345–354.

12 Insect Transformation for Use in Control

P W Atkinson, University of California, Riverside, CA, USA
D A O'Brochta, University of Maryland Biotechnology Institute, College Park, MD, USA
A S Robinson, FAO/IAEA Agriculture and Biotechnology Laboratory, Seibersdorf, Austria

© 2010, 2005 Elsevier B.V. All Rights Reserved

12.1. Introduction

In the previous series of *Comprehensive Insect Physiology, Biochemistry, and Pharmacology*, the corresponding article by Whitten (1985) dealing, briefly, with insect transformation technology raised two points about the problems that the extension of this technology into the field would most likely face. The prescience of these forecasts is even more remarkable given that it was written shortly after the development of a genetic transformation technology for *Drosophila melanogaster* and so was authored some 10 years before the repeatable transposable element-mediated transformation of nondrosophilid pest insect species.

In writing this review we have chosen not to regurgitate information already presented in recent reviews on the subject (Robinson and Franz, 2000; Atkinson, 2002; Handler, 2002; Robinson, 2002; Wimmer, 2003; Robinson *et al.*, 2004), and we will not discuss the issue of the development of a regulatory framework for the use of this technology. Two recent reports on the regulation of the release of transgenic arthropods have recently been published and the reader is referred to these for illumination as to the state of this issue (National Research Council, 2002; Pew Initiative on Food and Biotechnology, 2004). It would also be difficult to improve on Whitten's (1985) review, which comprehensively examined the history, conceptual basis, and application of genetic control in insects. Rather, under examination are two points raised by Whitten as a means to assess critically the current status of the application of genetically engineered insects to help solve problems in medicine and agriculture.

It must first be said that, some 20 years after the publication of Whitten's review, there is not one single example of the use of genetically engineered pest insects in the field. This is despite the first report of the transformation of an anopheline mosquito by Miller *et al.* (1987), and then, 8 years later, the first of several published reports ultimately describing five new transposable elements used to transform a range of nondrosophilid insect species, e.g., the Mediterranean fruit fly, *Ceratitis capitata*, and the yellow fever mosquito, *Aedes aegypti* (review: Robinson *et al.*, 2004).

In his 1985 review, Whitten stated:

> Clearly it is unlikely that our technical ability to transform a species will prove lacking: rather, it is our uncertainty concerning what genetic modifications are desirable that will usually prove to be the major impediment.

The current state of the field supports this prediction. Repeatable genetic transformation technologies for nondrosophilid insects have now existed for 9 years. While the development of these did prove to be more problematic than expected and awaited the discovery of new transposable elements such as *Minos* (Franz *et al.*, 1994), *Mos1* (Medhora *et al.*, 1991), *piggyBac* (Cary *et al.*, 1989), and *Hermes* (Warren *et al.*, 1994), the application of these technologies to practical problems in medical, veterinary, and economic entomology has not occurred. Several reasons exist for this, the most important one being the difficulty in discovering genetic and biochemical systems that might prove to be effective genetic control strategies in the field. Even as recently as 2001, the most elegant examples of using genetic-based strategies to selectively eliminate females (e.g., for the augmentation of a sterile insect technology) occurred in *D. melanogaster* and with experimental population sizes several orders of magnitude smaller than those required for the mass-rearing of insects in sterile insect technique (SIT) programs (Heinrich and Scott, 2000; Thomas *et al.*, 2000). Both experiments utilized conventional *P* element-mediated genetic transformation of *Drosophila*. As discussed, research into the manipulation of genetic and biochemical systems has produced interesting advances to prevent the transmission of pathogens through mosquitoes and, using model systems, proof of principle tha transmission of pathogens can be diminished in genetically modified mosquitoes has been demonstrated (Ito *et al.*, 2002; Moreira *et al.*, 2002). While significant for the control of mosquitoes and the human diseases they vector, these experiments have little bearing on the development of corresponding technology in agriculturally important insect species, such as *C. capitata* and the many lepidopteran species that feed on crops. Strategies for introducing and testing genetic markers, such as the green fluorescent protein (GFP), into an SIT strain for the purpose of allowing easy identification of this strain from the field strain, have been discussed and such strains have been generated in the Caribbean fruit fly, *Anastrapha suspensa*, in which the green fluorescence has been found to be detectable up to 4 weeks postmortem (Handler, 2002). Strains having similar properties have also been developed in *C. capitata* (Franz, personal communication). The stability of this protein, although very important for its use as a marker, has raised concerns about the transfer of the protein to nontarget organisms through predation. Significant effort has been expended towards unraveling the genetic basis of sexual development in *C. capitata* with the goal being to use these genes and their promoters as the basis of new genetic sexing strategies. Important genes such as *double-sex* and *transformer* have been isolated and characterized, allowing comparisons to be made of their function with their structural homologs in *D. melanogaster* (Saccone *et al.*, 2002). Sex-specific promoters from *C. capitata* have also been isolated (Christophides *et al.*, 2000). To date, however, no genetically engineered strains of *C. capitata* designed for a genetic sexing strategy have been generated or tested. In lepidopterans, the only progress in pest species has been the genetic transformation of *Pectinophora glossypiella* using the *piggyBac* element and the consequent application for permits to examine the robustness of this genetically engineered strain in outdoor cage experiments (Peloquin *et al.*, 2000).

Another factor contributing to the slow development of the technology is that genetic transformation of nondrosophilid insects is still not a routine robust technology. In this regard Ashburner *et al.* (1998) defined robustness:

> by the property that anyone skilled in the art can carry out the procedure, as is the case with transformation of *Drosophila*, a technique that has been performed in hundreds of laboratories world-wide.

Indeed, one of the reasons that *C. capitata* and *A. aegypti* are most frequently the subject of transformation experiments is the relative ease with which their embryos can be handled and injected. Combined with this is the very small number of laboratories in which nondrosophilid transformation is practiced. This is compounded by, in the case of tephritid fruit flies, quarantine laws that prevent many pest species from being reared in regions where they are classified as being exotic, thereby ruling out many research groups from developing genetically engineered strains of tephritids. Some 9 years after the successful demonstration of *Minos* transposable element-mediated transformation of *C. capitata* by Loukeris *et al.* (1995), the comparison with the rapid spread of *D. melanogaster* transformation protocols through the *Drosophila* community following the publication of this technique in 1982 by Rubin and Spradling (1982) could not be more stark. By 1991, *Drosophila* transformation was a routine technique that permeated *Drosophila* genetics. This has not been achieved as yet with mosquitoes, let alone many tephritids, lepidopterans, or coleopterans.

Another reason concerns the nature of the transposable elements used to transform nondrosophilid species. As mentioned above, the identification of new transposable elements such as *Mos1*, *Minos*, *Hermes*, *piggyBac*, and *Tn5* (Rowan *et al.*, 2004), was a major step in achieving genetic transformation of these species. With the exception of *Tn5*, which has been developed as a transformation agent of bacteria, we know little about the genetic and biochemical behavior of these elements, particularly with respect to their behavior in nonhost insect species. Possible problems about deploying transposable elements, and the transgenes contained within them, was forecast by Whitten (1985):

> Thus the selfishness of DNA is clearly subject to rather strict rules of etiquette, presumably imposed by evolutionary constraints on its host. These constraints may force those workers who see practical benefits flowing from the availability of transposable element vectors for genomic modification of particular plants or animals, to tailor a system suited to the special genetic constraints of their particular candidate species.

Recently, the behavior of *Hermes, piggyBac*, and *Mos1* elements in *A. aegypti* was described by O'Brochta *et al.* (2003) while the immobility of *Mos1* in transgenic lines of *A. aegypti* was reported by Wilson *et al.* (2003). These studies reveal that, while these transposable elements can be used to genetically transform this mosquito species, they are then rendered relatively immobile in following generations. Furthermore, the mechanism of *Hermes* element transposition appears to differ between germline nuclei and somatic nuclei of *A. aegypti*. Both mechanisms are transposase dependent, yet cut-and-paste transposition occurs in somatic nuclei while a mechanism that results in the transposition of both the transposable element and flanking sequences occur in the germline. This difference in mechanisms of *Hermes* element transposition has not been seen in any of the higher dipteran species – *D. melanogaster* (O'Brochta *et al.*, 1996; Guimond *et al.*, 2003), *C. capitata* (Michel *et al.*, 2001), and *Stomoxys calcitrans* (Lehane *et al.*, 2000) – in which *Hermes* has been used as a gene vector. It is clear we have a very limited knowledge of how any of these transposable elements function either in their original host species, or in species into which they are introduced. This lack of knowledge, combined with the problem of identifying genetic systems they would carry into a population so as to be used in genetic control programs, has contributed to the inability to bring these transgenic-based control programs to fruition. In particular, those who propose transposable element-based strategies for spreading beneficial transgenes through field populations of insects need to take into account that the ability of a given element to transform a species does not necessarily mean that it can also subsequently spread through this same species. That the *P* element of *D. melanogaster* can do this in this species is no guarantee that other elements can also do the same. Kidwell and Ribeiro (1992) specify three critical parameters needed for a transposable element to spread through a population. These are:

- the basic reproductive rate of the insects possessing the transposable element;
- the infectivity of the element (which is a function of its transposition frequency, its ability to undergo replicative transposition, and its propensity to transpose to loci unlinked to the original donor site); and
- the size of the target insect population.

In a subsequent article (Ribeiro and Kidwell, 1994), they proposed seven questions that needed to be resolved if one is to believe that a transposable element can spread genes through an insect population. These are: (1) the instability of chimeric transposable elements containing transgenes, (2) stabilization of these transgenes following their fixation in the target population (if, indeed, this is possible), (3) the biochemical properties of the transposase, (4) the possibility of repression systems becoming active as the copy number of the transposable element increases in the genome, (5) position effects, (6) the promoters chosen to control expression of the transgene, and (7) the length of the chimeric transposable element. It must be emphasized that none of these determines how well a given transposable element will function as a gene vector in achieving transformation in the first place. It is also poignant that we know next to nothing about the answers to any of these questions for the four "new" transposable elements used to transform pest insect species.

12.2. The Status of Transgenic-Based Insect Control Programs

What then, is the current status of programs in which transgenic insects have been proposed to utilize transgenic insects? They can be categorized into three broad categories:

1. Those that are designed to impose a genetic load on the target population in order to reduce population size.
2. Those that are designed to only eliminate the pest phenotype of the insects in a receiving population without altering population size.
3. Those that are designed to increase the fitness of biological control agents or increase the production of products excreted by insects, such as honey or silk.

12.2.1. Load Imposition on the Target Population

The sterile insect technique is the classic example of controlling insects through the imposition of a genetic load. The method was developed during the 1940s and 1950s and is now a mature and widely accepted method for insect population suppression and, in some cases, eradication. The pest species is mass-reared and irradiated to induce dominant lethal mutations and rearrangements at high frequencies in the germlines of these insects. The mass-reared insects, preferably only males, are released into the receiving, field population where they mate with wild fertile insects. Fertile females mating with released sterile males are nonproductive with a net reproductive rate (R_o) of zero. In many species a small proportion of the females mate more than once so the net effect on R_o is a function of sperm competitiveness in multiple-mated females. Successive releases in multiple generations continues to drive R_o down until either the population size is below its economic threshold (at which point it does not impact upon the economy of the relevant commodity) or until it is eradicated. The advantages of using this method of control have been discussed elsewhere (Hendrichs, 2000) but clearly rest on the species specificity of the method and its low environmental impact. Nonetheless implementing this "clean" control method faces some constraints including fitness costs associated with the development of specific strains and the effects of mass rearing, radiation, handling, and release on many aspects of field behavior (Robinson and Franz, 2000). Transgenic technology has the potential to positively impact existing SIT programs by improving aspects of insect production and monitoring and also has the potential to increase the use of SIT to include species that have otherwise been intractable to this technology.

Perhaps the biggest impact transgenic insect technology could have on any SIT program is to create strains of insects that permit the removal or elimination of females. The most significant effect that this would have on an SIT program is the reduction of the costs of mass-rearing and release because fewer insects need to be reared and released. In other programs, for example those that might involve vectors of human disease, removal of females prior to release may be essential since females would still be capable of biting and transmitting disease. Transgenic technology offers numerous options for creating strains that permit the selective removal of females prior to release. Conceptually the problem is one of expressing genes in one sex or the other that can be easily selected for or against. Using transgenic technology specificity of expression can be achieved by a variety of ways. First, expression can be controlled through the use of sex-specific promoters in combination with genes resulting in lethality. Sex-specific promoters from the vitellogenin and yolk protein genes have been characterized, and are examples of female-specific promoters that have been tested in transgenic insects with respect to possible applications in insect control programs (Heinrich and Scott, 2000; Kokoza et al., 2000; Thomas et al., 2000). These promoters are active in adult stages and would have limited application in SIT programs. They would not be useful for eliminating females at the larval stage and so any cost savings resulting from rearing both sexes would still be accrued. For eliminating females prior to mass-rearing, the challenge

remains to identify promoters or other regulatory sequences that selectively eliminate females during, or immediately following, embryogenesis.

12.2.2. Challenges of Long-Term Gene Introduction into Natural Populations

Transgenic strategies relying on load imposition to reduce or eliminate a population involve the short-term introduction of "effector" genes (usually dominant or conditionally dominant lethals) into the gene pool of native populations. As described above for the SIT, approaches relying on inundative releases of nongenetically engineered mass-reared insects carrying dominant lethal mutations have been successful. More subtle approaches have been proposed whereby deleterious genes can be transmitted over the course of a few generations before the load effects are encountered (Schliekelman and Gould, 2000a, 2000b). The most ambitious plans involve the permanent and stable alteration of the genotypes of wild insects. These ideas present enormous and novel challenges to insect geneticists. Introducing new laboratory-produced genotypes into populations is relatively simple (e.g., inundative releases of radiation-sterilized flies), but maintaining those genotypes in the population and, in fact, having them increase in frequency is an unprecedented undertaking in applied insect genetics.

An allele can increase in frequency in populations for a variety of reasons. For example, if there is a fitness advantage associated with a genotype then over time this allele is expected to become more abundant. If selection pressures are sufficiently high then the forces of natural selection can result in relatively rapid changes in allele frequencies. The global spread of insecticide resistance in *Culex* mosquitoes is one of many such examples of selection driven increase in gene frequencies (Raymond, 1991). The fitness costs associated with transgenic mosquitoes are largely unknown and is an area of research in need of attention. A few reports on the ability of transgenics to compete with wild conspecifics or on fitness estimates based on life table analyses, consistently indicate that the process of transgenesis decreases the fitness of the host insect (Catteruccia et al., 2003; Irvin et al., 2004). The sources of these fitness costs have not been precisely determined but are expected to be partitioned between costs associated with inbreeding during the process of creating a transgenic line of insects, transgene expression, and mutagenesis associated with transgene integration. It should be remembered, however, that genetic sexing strains generated through standard Mendelian genetics are less fit than wild-type strains but are still successfully used in SIT programs. In *Drosophila* it has been estimated that transposing *P* elements results in a 1% reduction in viability per integration event, on average (Eanes *et al.*, 1988). Thus, transgene integration is likely to result in fitness costs to the host organism. Expression of transgenes might also have effects on the fitness of the host organism. General costs associated with the metabolic load imposed by transgene expression are expected but have not been specifically measured (Billingsley, 2003). Individual transgenes, including marker genes, which have been generally considered to be "neutral," are likely to have costs associated with their expression. It is reasonable to assume, like others (Tiedje *et al.*, 1989) that transgenic insects will be less fit relative to the nontransgenic host insect. Understanding the magnitude of fitness costs associated with transgenesis will permit insect geneticists to allow for any fitness costs in a given experiment. However, given that natural selection is unlikely to favor the increase in frequency of transgenes in populations, insect geneticists are still faced with the problem of driving a deleterious gene into populations quickly and effectively enough to alleviate the problem, economic or medical, caused by the target pest species.

Deleterious genes can be spread through populations by a number of mechanisms. For example, hitchhiking is a genetic phenomenon whereby genes closely linked to a particular gene under positive selection are also selected despite potentially negative fitness costs associated with them. Hitchhiking is a well-established phenomenon and insect geneticists could conceivably exploit it if a gene under strong positive selection could be identified and isolated. An alternative to hitchhiking is to link the deleterious transgene to a genetic element with transmission-skewing properties. Genetic segregation distorter systems and transposable elements have enhanced transmission potential under certain conditions. That is, organisms heterozygous for such elements give rise to a disproportionate number of gametes with the element or gene as compared to what would be expected based on simple Mendelian inheritance. A notable example of a transposable element with the ability to rapidly infiltrate populations of naive insects is the *P* element of *D. melanogaster*. Indeed, the natural history of *P* has been promoted to the status of a paradigm, and optimistic views by some vector biologists about the prospects of driving anti-*Plasmodium* genes into native populations of *Anopheles gambiae*, thereby eliminating malaria transmission, are based on exploiting the anticipated mobility

properties of a *P*-like transposable element. It has been implicitly assumed that solution of the gene transformation problem in nondrosophilid insects through the development of transposable element based gene vectors would simultaneously solve the problem of gene drivers. This is an erroneous assumption.

Transposable elements differ in their abilities to spread through populations as illustrated by the differences observed between actively transposing *hobo* and *P* elements in *D. melanogaster*. The dynamics of movement and the final state achieved was element specific. A given element, such as the *P* element, may have drastically different drive potential in one species as compared to another. For example, while *P* elements effectively invade and spread in small laboratory cages of *D. melanogaster* they are relatively ineffective in *D. simulans* (Kimura and Kidwell, 1994).

There are a number of properties of a transposable element that are likely to influence its ability to spread through wild populations and these have already been listed based on previous studies (Kidwell and Ribeiro, 1992; Ribeiro and Kidwell, 1994). If we focus on the mechanism of transposition, then, clearly, activity of an element alone is not sufficient for an element to spread. Spread will require that transposition is associated with replication of the element to insure a net gain in allele frequency. Elements that transpose by a cut-and-paste mechanism are not expected to spread unless they are replicated in association with transposition. Class II transposable elements employ a number of mechanisms by which they can replicate. Element excision leaves a double-stranded break in the host DNA that is highly reactive and can initiate strand invasion and recombination using the sister chromatid as a template. Template-directed gap repair results in a copy of the moving element replacing the copy that has excised as a result of transposition. The frequency with which this type of repair takes place varies from element to element. The timing of transposition during the cell cycle is also a factor. Class II elements can also be replicated during the transposition process by timing their jumps to correspond to S phase of the cell cycle. If an element transposes after it has been replicated and targets an unreplicated region of a chromosome there will be a net gain of one element in the genome of one of the resulting mitotic products. The *Mos1* element of *D. mauritiana* appears to transpose during S phase in transgenic lines of *A. aegypti*; however, the stability of this element in these insects limits its use as a genetic drive mechanism in at least this species (O'Brochta *et al.*, 2003). The proximity of the new insertion sites relative to the donor site is also an important factor in determining the ability of the transposable element to spread. A highly active element that moves only, or even predominately, to tightly linked sites would not be a suitable agent to spread genes through an insect population, while an element with a reduced transposition rate that moved to unlinked loci either on the same or different chromosomes would be a viable spreading agent.

As mentioned above (see Section 12.1), at present little or no information exists as to the mobility properties of the four transposable elements used to genetically transform nondrosophilid insect species. Indeed, even for the relatively well-characterized P element of D. melanogaster, little is known about its mode of movement within and between chromosomes. This element does show a tendency to insert in or near the 5′ ends of genes, perhaps due to a relaxation of the DNA double helix during gene transcription. It has also been suggested that this element recognizes a structural feature at the insertion site rather than a strict canonical motif (Liao et al., 2000). This may well be true of other transposable elements and may well be an important factor in determining transposable element spread, but this remains an underexplored area of research.

12.2.3. Engineering of Beneficial Insects

Progress in this area has been limited to the stable introduction of genes, using the *piggyBac* transposable element, into the silkworm, *Bombyx mori* (Tamurua et al., 2000). Initially these experiments have been proof of principle experiments in which enchanced green fluorescent protein (EGFP) was used as a genetic marker to demonstrate that transformation could be achieved. Recently, Imamura et al. (2003) demonstrated that the GAL4/UAS system functions sufficiently in transgenic B. mori to enable tissue-specific expression of a reporter gene to occur. Transformation frequencies using the *piggyBac* transposable element as the gene vector were in the order of several percent. These experiments pave the way for gene identification using enhancer trapping in *Bombyx* which will further elevate the use of this species as a model system for other lepidopteran species. The extension of these techniques into practical benefits of silk production remains a challenge. While there have been reports of sperm-mediated transformation of the honeybee, *Apis mellifera* (Robinson et al., 2000), this technology has not yet been exploited by the honey industry and, since honey is a food, genetic engineering of its source may encounter public

resistance. Similarly, the initial report of genetic transformation of the predatory mite *Metaseiulus occidentalis*, which is used as a biological control agent, has not been pursued in field applications (Presnail and Hoy, 1992).

12.3. Conclusion

Genetic transformation technologies have been successfully extended into selected nondrosophilid species using transposable elements. This significant and exciting success has, however, being confined to the laboratory where it has enabled novel genotypes to be constructed and tested. These technologies have yet to be extended to the field, despite many years elapsing since genetic transformation protocols were first established for key pest species such as C. capitata and A. aegypti. If the full potential and benefits of genetic modification of medically and agriculturally significant insect species is to be realized then there must be a stronger linkage between the formulation of ideas and the subsequent timely and safe testing of these in transgenic insect strains in the laboratory, and in the field. Key to this is improving the robustness of transgenic technology in these insect species. Alternatively, the wisdom of establishing a handful of insect transformation centers that would provide this service to the community needs to be explored. This may be particularly attractive for species, such as A. gambiae, that remain difficult to transform. Providing a central transformation center may encourage researchers to develop and test new concepts, confident that at least the transgenic insects containing the desired transgenes will be routinely produced in a timely manner. It is critical to demonstrate in the laboratory and then in field cage experiments clear and concrete examples of how transgenic insect technology is beneficial to the general public so that arguments about the benefits of these new approaches can be clearly made to this interested and undoubtedly concerned audience.

References

Ashburner, M., Hoy, M.A., Peloquin, J.J., 1998. Prospects for the genetic transformation of arthropods. *Insect Mol. Biol. 7*, 201–213.

Atkinson, P.W., 2002. Genetic engineering in insects of agricultural importance. *Insect Biochem. Mol. Biol. 32*, 1237–1242.

Billingsley, P.F., 2003. Environmental constraints on the physiology of transgenic mosquitoes. In: Takken, W., Scott, T.W. (Eds.), Ecological Aspects for Application of

Genetically Modified Mosquitoes. Kluwer Academic Publishers, Dordrecht, pp. 149–161.

Cary, L.C., Goebel, M., Corsaro, B.G., Wang, H.G., Rosen, E., et al., 1989. Transposon mutagenesis of baculoviruses: analysis of Trichoplusia ni transposon IFP2 insertions within the FP-locus of nuclear polyhedrosis viruses. Virology 172, 156–169.

Catteruccia, F., Godfray, H.C., Crisanti, A., 2003. Impact of genetic manipulation on the fitness of Anopheles stephensi mosquitoes. Science 299, 1225–1227.

Christophides, G.K., Livadaras, I., Savakis, C., Komitopoulou, K., 2000. Two medfly promoters that have originated by recent gene duplication drive distinct sex, tissue and temporal expression patterns. Genetics 156, 173–182.

Eanes, W.F., Wesley, C., Hey, J., Houle, D., Ajioka, J.W., 1988. The fitness consequences of P element insertion in Drosophila melanogaster. Genet. Res. 52, 17–26.

Franz, G., Loukeris, T.G., Dialektaki, G., Thompson, C.R., Savakis, C., 1994. Mobile Minos elements from Drosophila hydei encode a two-exon transposase with similarity to the paired DNA-binding domain. Proc. Natl Acad. Sci. USA 91, 4746–4750.

Guimond, N.D., Bideshi, D.K., Pinkerton, A.C., Atkinson, P.W., O'Brochta, D.A., 2003. Patterns of Hermes element transposition in Drosophila melanogaster. Mol. Gen. Genom. 268, 779–790.

Handler, A.M., 2002. Prospects for using genetic transformation for improved SIT and new biocontrol methods. Genetica 116, 137–149.

Heinrich, J.C., Scott, M.J., 2000. A repressible female-specific lethal genetic system for making transgenic strains suitable for a sterile-release program. Proc. Natl Acad. Sci. USA 97, 8229–8232.

Hendrichs, J., 2000. Use of the Sterile Insect Technique against key insect pests. Sust. Devel. Int. 2, 75–79.

Imamura, M., Nakai, J., Inoue, S., Quan, G.X., Kanda, T., et al., 2003. Targeted gene expression using the GAL4/UAS system in the silkworm Bombyx mori. Genetics 165, 1329–1340.

Irvin, N., Hoddle, M.S., O'Brochta, D.A., Carey, B., Atkinson, P.W., 2004. Assessing fitness costs for transgenic Aedes aegytpi expressing the green fluorescent protein marker and transposase genes. Proc. Natl Acad. Sci. USA 101, 891–896.

Ito, J., Ghosh, A., Moreira, L.A., Wimmer, E.A., Jacobs-Lorena, M., 2002. Transgenic anopheline mosquitoes impaired in transmission of a malaria parasite. Nature 417, 387–388.

Kidwell, M.G., Ribeiro, J.M.C., 1992. Can transposable elements be used to drive refractoriness genes into vector populations? Parasitol. Today 8, 325–329.

Kimura, K., Kidwell, M.G., 1994. Differences in P element population dynamics between the sibling species Drosophila melanogaster and Drosophila simulans. Genet. Res. 63, 27–38.

Kokoza, V., Ahmed, A., Cho, W.L., Jasinskiene, N., James, A.A., et al., 2000. Engineering blood-meal activated systemic immunity in the yellow fever mosquito,

Aedes aegypti. Proc. Natl Acad. Sci. USA 97, 9144–9149.

Lehane, M.J., Atkinson, P.W., O'Brochta, D.A., 2000. Hermes-mediated genetic transformation of the stable fly, Stomoxys calcitrans. Insect Mol. Biol. 9, 531–538.

Liao, G.C., Rehm, E.J., Rubin, G.M., 2000. Insertion site preferences of the P transposable element in Drosophila melanogaster. Proc. Natl Acad. Sci. USA 97, 3347–3451.

Loukeris, T.G., Livadras, I., Arca, B., Zabalou, S., Savakis, C., 1995. Gene transfer into the Medfly, Ceratitis capitata, using a Drosophila hydei transposable element. Science 270, 2002–2005.

Medhora, M., Maruyama, K., Hartl, D.L., 1991. Molecular and functional analysis of the mariner mutator element Mos1 in Drosophila. Genetics 128, 311–318.

Michel, K., Stamenova, A., Pinkerton, A.C., Franz, G., Robinson, A.S., et al., 2001. Hermes-mediated germline transformation of the Mediterranean fruit fly, Ceratitis capitata. Insect Mol. Biol. 10, 155–162.

Miller, L.H., Sakai, R.K., Romans, P., Gwadz, R.W., Kantoff, P., et al., 1987. Stable integration and expression of a bacterial gene in the mosquito, Anopheles gambiae. Science 237, 779–781.

Moreira, L.A., Ito, J., Ghosh, A., Devenport, M., Zieler, H., et al., 2002. Bee venom phospholipase inhibits malaria parasite development in transgenic mosquitoes. J. Biol. Chem. 277, 40839–40843.

National Research Council, 2002. Animal Biotechnology: Science-Based Concerns. The National Academies Press, Washington, DC.

O'Brochta, D.A., Sethuraman, N., Wilson, R., Hice, R.H., Pinkerton, A.C., et al., 2003. Gene vector and transposable element behavior in mosquitoes. J. Exp. Biol. 206, 3823–3834.

O'Brochta, D.A., Warren, W.D., Saville, K.J., Atkinson, P.W., 1996. Hermes, a functional non-drosophilid gene vector from Musca domestica. Genetics 142, 907–914.

Peloquin, J.J., Thibault, S.T., Miller, T.A., 2000. Genetic transformation of the pink bollworm Pectinophora gossypiella with the piggyBac element. Insect Mol. Biol. 9, 323–333.

Pew Initiative on Food and Biotechnology, 2004. "Bugs in the System? Issues in the Science and Regulation of Genetically Modified Insects." Washington, DC, pp. 109.

Presnail, J.K., Hoy, M.A., 1992. Stable genetic transformation of a beneficial arthropod, Metaseiulus occidentalis (Acari: Phytoseiidae), by a microinjection technique. Proc. Natl Acad. Sci. USA 89, 7732–7736.

Raymond, M., 1991. Worldwide migration of amplified insecticide resistance genes in mosquitoes. Nature 350, 151–153.

Ribeiro, J.M., Kidwell, M.G., 1994. Transposable elements as population drive mechanisms: specification of critical parameter values. J. Med. Entomol. 31, 10–16.

Robinson, A.S., 2002. Mutations and their use in insect control. Mutat. Res. 511, 113–132.

Robinson, A.S., Franz, G., Atkinson, P.W., 2004. Insect transgenesis and its potential role in agriculture and human health. Insect Biochem. Mol. Biol. 34, 113–120.

Robinson, A.S., Franz, G., 2000. The application of transgenic insect technology in the sterile insect technique. In: Handler, A.M., James, A.A. (Eds.), Insect Transgenesis: Methods and Applications. CRC Press, Boca Raton, FL, pp. 307–318.

Robinson, K.O., Ferguson, H.J., Cobey, S., Vassein, H., Smith, B.H., 2000. Sperm-mediated transformation of the honey bee, *Apis mellifera. Insect Mol. Biol. 9*, 625–634.

Rowan, K.H., Orsetti, J., Atkinson, P.W., O'Brochta, D.A., 2004. *Tn5* as an insect gene vector. *Insect Biochem. Mol. Biol. 34*, 695–705.

Rubin, G.M., Spradling, A.C., 1982. Genetic transformation of *Drosophila* with transposable element vectors. *Science 218*, 348–353.

Saccone, G., Pane, A., Polito, L.C., 2002. Sex determination in flies, fruitflies and butterflies. *Genetica 116*, 15–23.

Schliekelman, P., Gould, F., 2000a. Pest control by the introduction of a conditonal lethal trait on multiple loci: potential, limitations, and optimal strategies. *J. Econ. Entomol. 93*, 1543–1565.

Schliekelman, P., Gould, F., 2000b. Pest control by the release of insects carrying a female-killing allele on multiple loci. *J. Econ. Entomol. 93*, 1566–1579.

Tamura, T., Thibert, C., Royer, C., Kanda, T., Abraham, E., et al., 2000. Germline transformation of the silkworm *Bombyx mori* L. using a *piggyBac* transposon-derived vector. *Nature Biotechnol. 18*, 81–84.

Thomas, D.D., Donnelly, C.A., Wood, R.J., Alphey, L.S., 2000. Insect population control using a dominant, repressible, lethal genetic system. *Science 287*, 2474–2476.

Tiedje, J.M., Colwell, R.K., Grossman, Y.L., Hodson, R.E., Lenski, R.E., et al., 1989. The planned introduction of genetically engineered organisms: ecological considerations and recommendations. *Ecology 70*, 298–315.

Warren, W.D., Atkinson, P.W., O'Brochta, D.A., 1994. The *Hermes* transposable element from the housefly, *Musca domestica*, is a short inverted repeat-type element of the *hobo, Ac*, and *Tam3 (hAT)* element family. *Genet. Res. 64*, 87–97.

Whitten, M.J., 1985. The conceptual basis for genetic control. In: Kerkut, G.A., Gilbert, L.I. (Eds.), Comprehensive Insect Physiology, Biochemistry and Pharmacology, vol. 12. Pergamon, Oxford, pp. 465–528.

Wilson, R., Orsetti, J., Klocko, A.K., Aluvihare, C., Peckham, E., et al., 2003. Post-integration behavior of a *Mos1* mariner gene vector in *Aedes aegypti. Insect Biochem. Mol. Biol. 33*, 853–863.

Wimmer, E.A., 2003. Innovations: applications of insect transgenesis. *Nature Rev. Genet. 4*, 225–232.

A12 Addendum: Insect Transformation for Use in Control

P W Atkinson, University of California, CA, USA

© 2010 Elsevier B.V. All Rights Reserved

A12.1. Introduction

The past 5 years witnessed some extremely promising developments in the laboratory studies of the genetic manipulation of pest insects, yet demonstration of any of these in a field environment remains elusive, partly because of both the underlying complexity of generating the necessary insect strains and public misgivings concerning the release of genetically engineered pest insects. Most promisingly, these developments have occurred in alternative strategies in genetic control with the logical expectation that at least one of these should emerge as a true candidate for large field cage studies followed by, pending success and regulatory approval, limited field trials. This addendum briefly summarizes these developments and also outlines areas in which technological progress still needs to be made.

A12.2. Progress in strategies Dependent on the Release of Sterile Insects

The sterile insect technique (SIT) has proved to be a cost-effective method of pest insect control. Genetic innovations in improving strains available for SIT have been hampered by a lack of genetic tools available for use in these species; however, some recent outcomes illustrate that use of site-specific recombination systems can lead to rather elegant genetic manipulations in transgenic strains. The ability of serine site-specific ΦC31 integrase from a broad host range bacteriophage of *Streptomyces* to function correctly in a wide range of organisms has now been successfully extended to insect pests such as the Mediterranean fruit fly, *Ceratitis capitata*, and the mosquito *Aedes aegypti* (Nimmo *et al.*, 2006; Schetelig *et al.*, 2009b). An advantage of this integration system is that it enables site-specific insertion of a transgene, provided that the target site is present in the genome. This integration system is introduced into the genome of pest species by transposon-mediated transformation, which can remain a rate-limiting step, especially in *Ae. aegypti*, for implementation of these strategies. The ΦC31 site-specific recombination system is mechanistically different from tyrosine site-specific recombinases such as the Flp recombinase from yeast and the Cre recombinase from bacteriophage P1, which enjoy use in the construction of novel genetic strains in *D. melanogaster* but are yet ineffectively employed in other insects (Horn and Handler, 2005). A particularly useful feature of the ΦC31 system is that it relies on two different integration sites, *attP* and *attB*, which, once combined, form two different sites, *attR* and *attL*, that are not recognized by the integrase (Thorpe and Smith, 1998). As a consequence, integrants are stable, even in the presence of integrase. A particularly elegant use of this system is demonstrated by its use to remove one of the terminal inverted repeats (TIRs) of a *piggyBac* transposon following its integration into the *C. capitata* genome rendering the remainder of the *piggyBac* transposon, together with the transgene it contains, stable in the genome (Schetelig *et al.*, 2009b). This system relied on the prior introduction of the genetically tagged *piggyBac* transposon containing an *attB* integration site; however, transformation efficiencies in this pest species using the *piggyBac*, *Minos*, or *Hermes* transposons can be reasonably high.

Use of an efficient transposon-mediated transformation technology with multiple fluorescent protein genes when combined with subsequent deployment of site-specific recombination does therefore permit sophisticated genetic manipulations common in *D. melanogaster* genetics to be deployed in *C. capitata* and in principle, any insect in which an efficient genetic transformation technology exists and genetic crosses can be performed. Indeed the strategy to remove a single transposon TIR to generate stability was first demonstrated in *D. melanogaster*, using a tyrosine site-specific recombination system on transgenic insects, with the same outcome as subsequently reported by Schetelig *et al.* (2009a, b). It is also sobering to note that the demonstration of ΦC31 integrase activity in *Ae. aegypti* was made some three years before the use of this in the elegant experiments of Scheteligh *et al.* (2009a, b) and that its use in *Ae. aegypti* is still confined to the generation of transgenic lines, rather than their subsequent modification (Nimmo *et al.*, 2006).

Progress has also been made in the identification of tissue- and stage-specific promoters that can direct the expression of desired effector genes. Once again *C. capitata*, because of its relative ease of genetic manipulation, provides a poignant example of what can be achieved (despite the absence of a genome project) and yet challenges remain. For example, Schetelig *et al.* (2009a, b) identified and cloned five native promoters active during blastoderm cellularization in embryonic development of *C. capitata* and placed them upstream of the tetracycline-induced transactivator gene that activates expression of a second gene, leading to embryonic lethality through programmed cell death. An advantage of this strategy is that death occurs before larval development and so before damage to the fruit takes place. Multiple transgenic lines containing each transgenic construct were generated, crossed, and 60 combinations were examined for embryonic lethality (Schetelig *et al.*, 2009a). Only one exhibited complete lethality, the authors finding that position effects played a significant role in determining the level of expression from each of the transgenes. These data show that engineered embryonic lethality systems can be extended from *D. melanogaster* to pest insects such as *C. capitata* and in doing so provide significant augmentations to the SIT. That a large number of lines needed to be analyzed to obtain a single combination that exhibited the desired properties is a concern and illustrates the problems that position effects can generate in these insects.

These developments built on earlier attempts to extend a dominant lethal-based autocidal technology in *C. capitata* using a one-component autoregulatory systems in which the tetracycline transactivator regulates its own expression (Gong *et al.*, 2005). It is reasonable to predict that completion of genome sequencing of *C. capitata* and other pests will lead to the identification of more promoter sequences that can be used in this next generation

of strategies for augmentation of the SIT. β2-tubulin germ-line specific promoters for limiting expression of transgenes to testes in *C. capitata*, *Anastrepha suspensa*, *Anopheles stephensi*, and *Ae. aegypti* have been identified as the *vasa* promoter from *An. gambiae* that directs gene expression to the female germ-line (Catteruccia *et al.*, 2005; Smith *et al.*, 2007; Scolari *et al.*, 2008; Papathanos *et al.*, 2009; Zimowska *et al.*, 2009) has been. Use of these promoters will enable germ-line specific gene expression that can further enhance genetic control strategies in pest insects.

A12.3. Progress in Strategies of Population Replacement

Two significant recent developments have been the demonstration of a maternal-effect selfish genetic drive system in *D. melanogaster* and the demonstration that homing endonucleases can function in Anopheles mosquitoes (Chen *et al.*, 2007; Windbichler *et al.*, 2007). Both are important initial steps in the development of potential genetic drive systems in pest insects. The maternal-effect system developed for *D. melanogaster* is an ingenious adaptation of a naturally occurring gene drive phenomenon (called Medea) described in the red flour beetle, *Tribolium castaneum* (Thomson and Beeman, 1999). The strategy is based on the ability of a zygotically expressed antidote to rescue a maternally expressed toxin. Individuals in which rescue cannot be achieved die during embryonic development, resulting in a driving force that establishes the antidote in the population. Linkage of a beneficial gene to the antidote should ensure that this gene also becomes established in the population at high frequencies. In the *D. melanogaster* Medea-like system, a maternal-effect promoter drives the expression of two miRNAs that silence the expression of an endogenous gene required for normal dorsal ventral development (Chen *et al.*, 2007). The antidote consists of an early zygotic promoter that drives expression of this developmental gene; however, it lacks the target-binding sites of the two miRNAs. Zygotes lacking the antidote die and so the antidote gene can become fixed in the population. The success of this synthetic system is, in part, dependent on the vast knowledge about genetic control of development in *D. melanogaster*. Clearly, it is a challenge identifying analogous promoters in pest insects, in which there may not be a strict conservation of the roles of specific genes in development. Moreover, the ability of each miRNA to function has to be empirically determined to avoid nonspecific effects which could compromise the efficacy of the strategy. Nonetheless, the fact that Medea systems exist naturally in insects indicates that maternal-effect drive systems can be developed and applied to pests in the field.

Homing endonuclease genes are found in bacteria and fungi and act by integrating themselves into specific target sites that are recognized and cut by the endonuclease. Once excised, the rate of integration into these sites is extremely high meaning that if the rate of excision can be elevated, then any additional beneficial gene tightly linked to the homing endonuclease should also be able to spread though a population. The challenges are to increase the low rates of excision of these homing endonuclease genes and to determine whether these systems can function in pest insect species. Progress in this area was made by the demonstration that two homing endonucleases, *I-SceI* and *I-PpoI*, can function in *An. gambiae* cell culture and that the former can also function in plasmid assays performed in embryos (Windbichler *et al.*, 2007). The same authors also determined that the *I-PpoI* endonuclease may be able to specifically target a naturally occurring target site within rDNA residing on the X chromosome which, if validated, could be the basis of a plausible population strategy.

A12.4. Challenges that Remain

Most, if not all, of the strategies described earlier rely on the use of transposons for the initial introduction of new DNA sequences to the pest insect. The *piggyBac* transposon is proving to be a widely used gene vector in both insects and higher vertebrates and its recent use as a mutagen and enhancer trap in *T. castaneum* illustrates what can be achieved with this transposon (Trauner *et al.*, 2009). However, transformation rates in mosquitoes remain low and remobilization of transposons following transformation of *Ae. aegypti* occurs at a near-negligible frequency. Whether this is due to the presence of an efficient piRNA or endo-siRNA system or due to more localized effects of silencing within chromatin remains unknown (Brennecke *et al.*, 2007; Ghildiyal *et al.*, 2008). This deficiency combined with the absence of a gene replacement technology based on homologous recombination is perhaps the remaining technological bottleneck preventing the full exploitation of elegant genetic control strategies in many insect pest species. The outcomes of the past five years have shown that, pending successful introduction to the genome, many of the strategies shown to work in *D. melanogaster* should function in other insects.

References

Brennecke, J.B., Aravin, A.A., et al., 2007. Discrete small RNA-generating loci as master regulators of transposon activity in Drosophila. *Cell 128*, 1089–1103.

Catteruccia, F., Benton, J.P., et al., 2005. An Anopheles transgenic sexing strain for vector control. *Nat. Biotechnol. 23*, 1414–1417.

Chen, C.-H., Huang, H., et al., 2007. A synthetic maternal-effect selfish genetic element drives population replacement in Drosophila. *Science 316*, 597–600.

Ghildiyal, M., Seitz, H., et al., 2008. Endogenous siRNAs derived from transposons and mRNAs in Drosophila somatic cells. *Science 320*, 1077–1081.

Gong, P., Epton, M.J., et al., 2005. A dominant lethal genetic syste for autocidal control of the Mediterranean fruitfly. *Nat. Biotechnol. 23*, 453–456.

Horn, C., Handler, A.M., 2005. Site-specific genomic targeting in Drosophila. *Proc. Natl. Acad. Sci. USA 102*, 12483–12488.

Nimmo, D.D., Alphey, L., et al., 2006. High efficiency site-specific engineering of the mosquito genome. *Insect Mol. Biol. 15*, 129–136.

Papathanos, P.A., Windbichler, N., et al., 2009. The *vasa* regulatory region mediates germline expression and maternal transmission of proteins in the malaria mosquito Anopheles gambiae: a versatile tool for genetic control strategies. *BMC Mol. Biol. 10*, 65.

Schetelig, M.F., Caceres, C., et al., 2009a. Conditional embryonic lethality to improve the sterile insect technique in Ceratitis capitata (Diptera: Tephritidae). *BMC Biol. 7*, 4.

Schetelig, M.F., Scolari, F., et al., 2009b. Site-specific recombination for the modfication of the transgenic strains of the Mediterranean fruit fly *Cerattis capitata*. *Proc. Natl. Acad. Sci. USA 106*(43), 18171–18176.

Scolari, F., Schetelig, M.F., et al., 2008. Fluorescent sperm marking to improve the fight against the pest insect *Ceratitis capitata* (Wiedemann: Diptera: Tephritidae). *Nat. Biotechnol. 25*(1), 76–84.

Smith, R.C., Walter, M.F., et al., 2007. Testis-specific expression of the ß2 tubulin promoter of Aedes aegypti and its application as a genetic sex-separation marker. *Insect Mol. Biol. 16*, 61–71.

Thomson, M.S., Beeman, R.W., 1999. Assisted suicide of a selfish gene. *J. Hered. 90*(1), 191–194.

Thorpe, H.M., Smith, M.C.M., 1998. In vitro site-specific integration of bacteriophage DNA catalyzed by a recombinase of the resolvase/integrase family. *Proc. Natl. Acad. Sci. USA 95*, 5505–5510.

Trauner, J., Schinko, J., et al., 2009. Large-scale insertional mutagenesis of a coleopteran stored grain pest, the red flour beetle *Tribolium castaneum*, identifies embryonic lethal mutations and enhancer traps. *BMC Biol. 7*, 73.

Windbichler, N., Papathanos, P.A., et al., 2007. Homing endonuclease mediated gene targeting in Anopheles gambiae cells and embryos. *Nucl. Acids Res. 35*(17), 5922–5933.

Zimowska, G.J., Nirmala, X., et al., 2009. The beta2-tubulin gene from three tephritid fruit fly species and use of its promoter for sperm marking. *Insect Biochem. Mol. Biol. 39*, 508–515.

Subject Index

Cross-reference terms in italics are general cross-references, or refer to subentry terms within the main entry (the main entry is not repeated to save space). Readers are also advised to refer to the end of each article for additional cross-references – not all of these cross-references have been included in the index cross-references.

The index is arranged in set-out style with a maximum of four levels of heading. Major discussion of a subject is indicated by bold page numbers. Page numbers suffixed by *T* and *F* refer to Tables and Figures respectively. *vs.* indicates a comparison.

To save space in the index the following abbreviations have been used:
ETH – ecdysis triggering hormone
GPCRs – G protein-coupled receptors
PBAN – pheromone biosynthesis activating neuropeptide
PDV – polydnaviruses
PTTH – prothoracicotropic hormone
QSAR – qualitative structure-activity relation
RDL – resistance to dieldrin

This index is in letter-by-letter order, whereby hyphens and spaces within index headings are ignored in the alphabetization. Prefixes and terms in parentheses are excluded from the initial alphabetization.

pyrethroid efficacy group (PEG), 21
selection pressure, 21
knockdown resistance *(kdr)*, 14
metabolic resistance *see* Pyrethroids,
metabolic resistance
target-site resistance, 19
kdr-type mechanism, 20
knockdown resistance, 19–20,
20–21
para-type sodium channel, 20
site of action, 10
knockdown action, 10
sodium channels
interactions, 31, 33*fF*
investigations, 12
structure-activity relationship, 6
actin site binding, 9
chirality, 8
piperonyl butoxide (PBO), 8
quantitative structure-activity
relationship, 9
segment A, 6
segment A+B, 6
segment C, 8
segment G, 9
segments D, 8
segments E, 8
selectivity, 9
seven-segment model, 6, 8*fF*
stereochemistry, 6–8
synergism, 9
synthetic pyrethroids, 1
activity spectrum, 1–2
application, 1
sales, 2
toxicity values, 230*tT*
type I, 11
type II, 11
Pyrethrum cinerariaefolium, 1
Pyrethrum extract, 1
Pyriproxifen
juvenile hormone analog, 151
Pyriproxyfen, juvenile hormone analog,
149*fF*
Pyrroles, toxicity values, 230*tT*

Q

Quasi cyclic conformation, 72

R

Rachiplusia ou
insect selective toxin genes, 343–344
Recombinant HaSNPV, 350
Reporter transactivation response,
144–145
Reproduction
azadirachtin effects, 194
Responsive genes, juvenile hormone, 148
Resting stage, entomopathogenic fungi,
404
Resting state, sodium channel blockers,
43, 43*fF*
Reticulitermes flavipes, 83

Reticulitermes santonensis, juvenile
hormone analog insecticides, 152*tT*
Reticulitermes speratus
juvenile hormone analog insecticides,
152*tT*
Retinoid X receptor (RXR)
molting hormone ecdysone receptors,
124
RH-5849 (non-steroidal ecdysone agonist)
ecdysone agonist, 124–125
structure, 123*fF*
RH-131039, 130
structure, 130*fF*
Rhagoletis pomonella, 152*tT*
Rhagonycha fulva
dispersal, 405
Entomophthora muscae infection,
392*fF*
Rhipicephalus sanguineus, 103
Rhithropanopeus harrisii, 156*tT*
Rhizoglyphus robini, 16
Rhodnius prolixus
azadirachtin, reproduction, 194–195
juvenile hormone
molecular mode, putative, 148
Rhopalosiphon padi
entomopathogenic fungi interactions,
406
neonicotinoid insecticides, 82–83
Rhopalosiphum padi, 80
Rhyaciona buoliana, 139
Rhyzopertha dominica, juvenile hormone
analog insecticides, 152*tT*
Rice, neonicotinoid insecticides, 81
Ring systems, neonicotinoid insecticides,
65
Ring system *vs.* noncyclic structure,
neonicotinoids, 72
RNA interference (RNAi)
juvenile hormone esterase, baculovirus,
337

S

S4 segments, 46–47
S6 transmembrane segments
blocker binding face, 48*fF*
sequence alignment, 47*fF*
site 10, 46–47
Saccharomyces cereviseae
bisacylhydrazine, selective toxicity, 137
Saccharopolysopra pogona
21-butenyl spinosyns, 211–212
spinosyns structure–activity
relationships, 235–236
Saccharopolysopra spinosa, spinosyns,
208
structure–activity relationships, 235–236
Saccharopolyspora erythaea, spinosyns
biosynthesis, 213
Salannin, 187*fF*
Salmonella enterica (typhi), pathogenicity
island, 317*fF*
Sarcophaga falculata
toxins, 341–342

Sarcoptes scabiei, imidacloprid, 103
Scathophaga stercoraria
killed by *Entomophthora muscae*,
393*fF*
Scenedesmus subspicatus, neonicotinoid
insecticides, safety profile, 94
Schistocerca americana
toxins, multiple synergistic, 353–354
Schistocerca gregaria
azadirachtin
antifeedant effects, 192
mode of action studies, 197
neem tree, 186
nicotinic acetylcholine receptor
(nAChR) interactions, selectivity,
insect over vertebrate, 84
Scorpion
toxins
selective genes, 341
Scott's Py III (pyrethroids), 23*fF*
Secondary messenger systems
Secreted alkaline phosphatase (SEAP),
baculovirus insect control safety,
362
Segment A+B, pyrethroids, 6
Segment A, pyrethroids, 6
Segregation distorter systems, 443
Septobasidium, entomopathogenic fungi,
392–393
Serratia entomophila
pADAP conjugative virulence plasmid,
tc-like loci, 316*fF*
toxin complexe (tc) genes, 317
gene organization, 318
Serratia entomophila pathogenicity (sep)
genes, 317
Signal recognition particle (SRP), insect
selective toxin genes, 351–352
Silafluofen, 5*fF*
Single-electrode voltage clamp method
(SEVC), 223
Site 10 (insects), 45
molecular nature, 46
BTX action, 46–47
S4 segments, 46–47
blocker binding, 46–47
S6 segments, 46–47
sodium channel blocker insecticides
(SCBIs), 45
Sitobion avenae, 80
entomopathogenic fungi interactions,
406
neonicotinoid insecticides, 82–83
Pandora infection, 393*fF*
Slowly adapting stretchreceptor (SASR),
sensory nerve activity, 41–42
Smittium morbosum, entomopathogenic
fungi, 389–390
Sodium channel
blocker insecticide *see* Sodium channel
blocker insecticide (SCBI)
pyrethroids, 12, 31, 33*fF*
Sodium channel blocker insecticides
(SCBIs), 36, 58–59
action at site 10, 45

Printed in the United States
By Bookmasters